This is the first book to deal comprehensively with Spain's tectonic and sedimentary history over the past sixty or so million years. During these Tertiary times, Spain has suffered compressional collision between France and Africa, and its Atlantic and Mediterranean coasts have been further modified by extensional rifting. This study will therefore be of interest to earth scientists generally because of the insights it provides into continental crustal deformation.

Spain contains some of the best exposed outcrop geology in Europe. Because it includes sectors of two separate foreland basins, and an intervening craton with basins that have been influenced by extensional and strike–slip deformation, it provides excellent material for the development and testing of theories on the study of sedimentary basin formation and filling.

This volume contains a collection of specially written articles dealing with all aspects of these studies. Because of the nature of the subject, much of the data is presented in diagrammatic form.

This book will be of value to earth scientists in research positions, whether academic or industrial.

Tertiary basins of Spain: the stratigraphic record of crustal kinematics

World and regional geology series
Series Editors: M.R.A. Thomson and J.A. Reinemund

This series comprises monographs and reference books on studies of world and regional geology. Topics will include comprehensive studies of key geological regions and the results of some recent IGCP projects.

Geological Evolution of Antarctica M.R.A. Thomson, J.A. Crame & J.W. Thomson (eds.)
Permo-Triassic Events in the Eastern Tethys W.C. Sweet, Yang Zunyi, J.M. Dickins & Yin Hongfu (eds.)
The Jurassic of the Circum-Pacific G.E.G. Westermann (ed.)
Global Geological Record of Lake Basins vol. 1 E. Gierlowski-Kordesch & K. Kelts (eds.)
Earth's Glacial Record M. Deynoux, J.M.G. Miller, E.W. Domack, N. Eyles, I.J. Fairchild & G.M. Young (eds.)
The Quaternary History of Scandinavia J.J. Donner
Tertiary Basins of Spain: The Stratigraphic Record of Crustal Kinematics P.F. Friend & C.J. Dabrio (eds.)

Tertiary basins of Spain

the stratigraphic record of crustal kinematics

EDITED BY

PETER F. FRIEND

Department of Earth Sciences, University of Cambridge

AND

CRISTINO J. DABRIO

Departamento de Estratigrafía, Facultad de Ciencias Geológicas and Instituto de Geología Económica, CSIC, Universidad Complutense, Madrid, Spain

CAMBRIDGE
UNIVERSITY PRESS

CAMBRIDGE UNIVERSITY PRESS
Cambridge, New York, Melbourne, Madrid, Cape Town, Singapore, São Paulo

Cambridge University Press
The Edinburgh Building, Cambridge CB2 2RU, UK

Published in the United States of America by Cambridge University Press, New York

www.cambridge.org
Information on this title: www.cambridge.org/9780521461719

First published 1996
This digitally printed first paperback version 2005

A catalogue record for this publication is available from the British Library

Library of Congress Cataloguing in Publication data

Tertiary basins of Spain : the stratigraphic record of crustal
kinematics / edited by Peter F. Friend and Cristino J. Dabrio.
p. cm. – (World and regional geology series)
Includes bibliographical references.
ISBN 0 521 46171 5
1. Geology, Stratigraphic – Tertiary. 2. Geology, Structural – Spain. 3. Basins (Geology) – Spain. I. Friend, P.F. II. Dabrio, Cristino J. III. Series.
QE691.T465 1995
551.7′8′0946 – dc20 94-21724 CIP

ISBN-13 978-0-521-46171-9 hardback
ISBN-10 0-521-46171-5 hardback

ISBN-13 978-0-521-02198-2 paperback
ISBN-10 0-521-02198-7 paperback

Contents

Contributors

ALONSO, J.L., Departamento de Geología, Universidad de Oviedo, Arias de Velasco s/n, 33005 Oviedo, Spain.

ALONSO ZARZA, A.M., Departamento de Petrología y Geoquímica, Facultad de Ciencias Geológicas, and Instituto de Geología Económica, CSIC, Universidad Complutense, 28040 Madrid, Spain.

ALVAREZ DE BUERGO, E., Lucaz Oil Company of Spain, Alfonso XII 15, 28040 Madrid, Spain.

ÁLVAREZ SIERRA, M.A., Departamento de Paleontología, Facultad de Ciencias Geológicas, and Instituto de Geología Económica, CSIC, Universidad Complutense, 28040 Madrid, Spain.

ANADÓN, P., Institut de Ciències de la Terra 'Jaume Almera', CSIC, Martí i Franquès s/n, 08028 Barcelona, Spain.

ARRIBAS, J., Departamento de Petrología y Geoquímica, Facultad de Ciencias Geológicas, and Instituto de Geología Económica, CSIC, Universidad Complutense, 28040 Madrid, Spain.

ARRIBAS, M.E., Departamento de Petrologia y Geoquímica, Facultad de Ciencias Geológicas, and Instituto de Geología Económica, CSIC, Universidad Complutense, 28040 Madrid, Spain.

BACELAR, J., ENDESA, Mina de Puentes, Oficina del Tresuro, As Pontes de García Rodríguez, 15320 La Coruña, Spain.

BANDA, E., Institut de Ciències de la Terra 'Jaume Almera', CSIC, Martí i Franquès s/n, 08028 Barcelona, Spain.

BARBA, P., Departamento de Geología (Estratigrafía), Universidad de Salamanca, Plaza de la Merced s/n, 37008 Salamanca, Spain.

BENTHAM, P., ALAME Area, AMOCO Production Company, PO Box 3092, Houston, TX 77253–3092, USA.

BETZLER, Ch., Geologisch. Paläontologisches Institut, Senckenberganlage 32–34, 60054 Frankfurt am Main, Germany.

BOND, J., Mobil North Sea Limited, Mobil Court, 3 Clements Inn, London, WC2A 2EB, United Kingdom.

BRAGA, J.C., Departamento de Estratigrafía y Paleontología, and Instituto Andaluz de Geología Mediterránea, CSIC, Universidad de Granada, Campus de Fuentenueva, 18071 Granada, Spain.

BURBANK, D.W., Department of Geological Sciences, University of Southern California, Los Angeles, CA 90089–0740, USA.

CABRERA, L., Departamento de Geologia Dinàmica, Geofísica i Paleontologia, Universitat de Barcelona, Zona Universitària de Pedralbes, Martí i Franquès s/n, 08028 Barcelona, Spain.

CALVET, F., Departamento de Geoquímica, Petrologia i Prospecció Geològica, Universidad de Barcelona, Zona Universitària de Pedralbes, Martí i Franquès s/n, 08028 Barcelona, Spain.

CALVO, J.P., Departamento de Petrología y Geoquímica, Facultad de Ciencias Geológicas, and Instituto de Geología Económica, CSIC, Universidad Complutense, 28040 Madrid, Spain.

CAÑAVERAS, J.C., Departamento de Petrología y Geoquímica, Facultad de Ciencias Geológicas, and Instituto de Geología Económica, CSIC, Universidad Complutense, 28040 Madrid, Spain.

CAPDEVILA, J., Departamento de Geologia (Estratigrafia), Universitat Autònoma de Barcelona, 08193 Bellaterra, Spain.

CIVIS, J., Departamento de Geología (Paleontología), Universidad de Salamanca, Plaza de la Merced s/n, 37008 Salamanca, Spain.

COSTA, J.M., Bona-sort 36, 1–2, 08193 Bellaterra, Spain.

DAAMS, R., Departamento y UEI de Paleontología, Facultad de Ciencias Geológicas (UCM), and Instituto de Geología Económica (CSIC-UCM), 28040 Madrid, Spain.

DABRIO, C.J., Departamento de Estratigrafía, Facultad de Ciencias Geológicas, and Instituto de Geología Económica, CSIC, Universidad Complutense, 28040 Madrid, Spain.

DE VICENTE, G., Departamento de Geodinámica, Facultad de Ciencias Geológicas, Universidad Complutense, 28040 Madrid, Spain.

DÍAZ-MOLINA, M., Departamento de Estratigrafía, Facultad de Ciencias Geológicas, and Instituto de Geología Económica, CSIC, Universidad Complutense, 28040 Madrid, Spain.

DOCHERTY, C., Institut de Ciències de la Terra 'Jaume Almera', CSIC, Martí i Franquès s/n, 08028 Barcelona, Spain.

ELLIOTT, T., Department of Earth Sciences, University of Liverpool, PO Box 147, Liverpool L69 3BX, United Kingdom.

FERNÁNDEZ, J., Departamento de Estratigrafía y Paleontología, and Instituto Andaluz de Geología Mediterránea, CSIC, Universidad de Granada, Campus de Fuentenueva, 18071 Granada, Spain.

FERRÚS, B., Departamento de Geologia Dinàmica, Geofísica i Paleontologia, Universitat de Barcelona, Zona Universitària de Pedralbes, Martí i Franquès s/n, 08028 Barcelona, Spain.

FLORES, J.A., Departamento de Geología (Paleontología), Universidad de Salamanca, Plaza de la Merced s/n, 37008 Salamanca, Spain.

FRIEND, P.F., Department of Earth Sciences, University of Cambridge, Downing Street, Cambridge CB2 3EQ, United Kingdom.

GARCÍA DEL CURA, M.A., Laboratorio de Petrología Aplicada, Instituto de Geología Económica, CSIC, and División de Geología, Universidad de Alicante, Apartado 99, 03690 San Vicente del Raspeig, Spain.

GARCÍA-RAMOS, J.C., Departamento de Geología, Universidad de Oviedo, Arias de Velasco s/n, 33005–Oviedo, Spain.

GINER, J., Departamento de Geodinámica, Facultad de Ciencias Geológicas, Universidad Complutense, 28040 Madrid, Spain.

GÓMEZ FERNÁNDEZ, J.J., Departamento de Estratigrafía, Facultad de Ciencias Geológicas, and Instituto de Geología Económica, CSIC, Universidad Complutense, 28040 Madrid, Spain.

GONZÁLEZ, A., Departamento de Geología (Estratigrafía), Universidad de Zaragoza, 50009 Zaragoza, Spain.

GONZÁLEZ-CASADO, J.M., Departamento de Geología y Geoquímica, Universidad Autónoma de Madrid, 28049 Madrid, Spain.

GONZÁLEZ DELGADO, J.A., Departamento de Geología (Paleontología), Universidad de Salamanca, Plaza de la Merced s/n, 37008 Salamanca, Spain.

HOGAN, P.J. Department of Geological Sciences, University of Southern California, Los Angeles, CA 90089–0740, USA.

HOYOS, M., Museo Nacional de Ciencias Naturales, CSIC, José Gutiérrez Abascal, 2, 28006 Madrid, Spain.

JURADO, M.J., Institut de Ciències de la Terra 'Jaume Almera', CSIC, Martí i Franquès s/n, 08028 Barcelona, Spain.

LENDÍNEZ, A., ENADIMSA, DR. Esquerdo 138, 28007 Madrid, Spain.

LLOYD, M.J., Department of Earth Sciences, University of Cambridge, Downing Street, Cambridge CB2 3EQ, United Kingdom.

MAESTRO-MAIDEU, E., Departamento de Geologia (Estratigrafia), Universitat Autònoma de Barcelona, 08193 Bellaterra, Spain.

MARTÍN, J.M., Departamento de Estratigrafía y Paleontología, and Instituto Andaluz de Geología Mediterránea, CSIC, Universidad de Granada, Campus de Fuentenueva, 18071 Granada, Spain.

MARTÍN-SERRANO, A., Instituto Tecnológico y Geominero de España (ITGE), Ríos Rosas 23, 28003 Madrid, Spain.

MARTÍNEZ DEL OLMO, W., Repsol Exploración SA, Paseo de la Castellana 278–280, 28046 Madrid, Spain.

MAS, R., Departamento de Estratigrafía, Facultad de Ciencias Geológicas, and Instituto de Geología Económica, CSIC, Universidad Complutense, 28040 Madrid, Spain.

MCELROY, R., Department of Earth Sciences, University of Cambridge, Downing Street, Cambridge CB2 3EQ, United Kingdom.

MEDIAVILLA, R., Instituto Tecnológico y Geominero de España (ITGE), Ríos Rosas 23, 28003 Madrid, Spain.

MELÉNDEZ-HEVIA, F., San Marcos 39, 3° C, 28004 Madrid, Spain.

MOISSENET, E., UER de Geographie, Université de Paris I, 191 rue Sant Jacques, 75005 Paris, France. (deceased, see appreciation)

MONTENAT, CH., IGAL, Centre Polytechnique Saint-Louis, 13 Boulevard de l'Hautil, 95092 Cergy-Pontoise Cedex, France.

MUÑOZ, A., Departamento de Geología (Estratigrafía), Universidad de Zaragoza, 50009 Zaragoza, Spain.

MUÑOZ-MARTÍN, A., Departamento de Geodinámica, Facultad de Ciencias Geológicas, Universidad Complutense, 28040 Madrid, Spain.

ORDÓÑEZ, S., Laboratorio de Petrología Aplicada, División de Geología, Universidad de Alicante, Apartado 99, 03690 San Vicente del Raspeig, Spain.

OTT D'ESTEVOU, P., IGAL, Centre Polytechnique Saint-Louis,13 Boulevard de l'Hautil, 95092 Cergy-Pontoise Cedex, France.

PARDO, G., Departamento de Geología (Estratigrafía), Universidad de Zaragoza, 50009 Zaragoza, Spain.

PELÁEZ-CAMPOMANES, P., Departamento de Paleontología, Facultad de Ciencias Geológicas, and Instituto de Geología Económica, CSIC, Universidad Complutense, 28040 Madrid, Spain.

PÉREZ, A., Departamento de Geología (Estratigrafía), Universidad de Zaragoza, 50009 Zaragoza, Spain.

PIERSON D'AUTREY, L., SIMECSOL, 8 avenue Newton, 92350 Le Plessis-Robinson, France.

PUIGDEFÁBREGAS, C., Servei Geològic de Catalunya, Avinguda del Parallel 71, 08004 Barcelona, Spain.

PULGAR, J.A., Departamento de Geología, Universidad de Oviedo, Arias de Velasco s/n, 33005–Oviedo, Spain.

REMACHA, E., Departamento de Geologia (Estratigrafía), Universitat Autònoma de Barcelona, 08193 Bellaterra, Spain.

RIAZA, C., Repsol Exploración SA, Paseo de la Castellana 278–280, 28046 Madrid, Spain.

RIBA, O., Departamento de Geologia Dinàmica, Geofísica i Paleontologia, Universitat de Barcelona, Zona Universitària de Pedralbes, Martí i Franquès s/n, 08028 Barcelona, Spain.

ROCA, E., Departamento de Geologia Dinàmica, Geofísica i Paleontologia, Universitat de Barcelona, Zona Universitària de Pedralbes, Martí i Franquès s/n, 08028 Barcelona, Spain.

RODRÍGUEZ-ARANDA, J.P., Departamento de Petrología y Geoquímica, Facultad de Ciencias Geológicas, and Instituto de Geología Económica, CSIC, Universidad Complutense, 28040 Madrid, Spain.

RODRÍGUEZ-FERNÁNDEZ, J., Instituto Andaluz de Geología Mediterránea, CSIC, Universidad de Granada, Campus de Fuentenueva, 18071 Granada, Spain.

RODRÍGUEZ-PASCUA, M.A., Departamento de Geodinámica, Facultad de Ciencias Geológicas, Universidad Complutense, 28040 Madrid, Spain.

SÁEZ, A., Departamento de Geologia Dinàmica, Geofísica i Paleontologia, Universitat de Barcelona, Zona Universitària de Pedralbes, Martí i Franquès s/n, 08028 Barcelona, Spain.

SANTANACH, P.F., Departamento de Geologia Dinàmica, Geofísica i Paleontologia, Universitat de Barcelona, Zona Universitària de Pedralbes, Martí i Franquès s/n, 08028 Barcelona, Spain.

SANTISTEBAN, J.I., Instituto Tecnológico y Geominero de España (ITGE), Ríos Rosas 23, 28003 Madrid, Spain.

SANZ DE GALDEANO, C.M., Instituto Andaluz de Geología Mediterránea, CSIC, Universidad de Granada, Campus de Fuentenueva, 18071 Granada, Spain.

SANZ-MONTERO, M.E., Departamento de Petrología y Geoquímica, Facultad de Ciencias Geológicas, and Instituto de Geología Económica, CSIC, Universidad Complutense, 28040 Madrid, Spain.

SERRA ROIG, J., Departamento de Geologia (Estratigrafia), Universitat Autònoma de Barcelona, 08193 Bellaterra, Spain.

SIERRO, F.J., Departamento de Geología (Paleontología), Universidad de Salamanca, Plaza de la Merced s/n, 37008 Salamanca, Spain.

SMITH, A.G., Department of Earth Sciences, University of Cambridge, Downing Street, Cambridge CB2 3EQ, United Kingdom.

SORIA, J., División de Geología, Universidad de Alicante, Apartado 99, 03690 San Vicente del Raspeig, Spain.

TORNE, M., Institut de Ciències de la Terra 'Jaume Almera', CSIC, Martí i Franquès s/n, 08028 Barcelona, Spain.

TORTOSA, A., Departamento de Petrología y Geoquímica, Facultad de Ciencias Geológicas, and Instituto de Geología Económica, CSIC, Universidad Complutense, 28040 Madrid, Spain.

TURNER, J., School of Earth Sciences, University of Birmingham, PO Box 363, Birmingham, West Midlands B15 2TT, United Kingdom.

VAN DER MEULEN, A.J., Institute of Earth Sciences, University of Utrecht, Budapestlaan 4, 3508 TA Utrecht, The Netherlands.

VAN GELDER, A., Facultiet de Runitelijke Wetenschappen, Postbus 80.115, 3508 TA Utrecht, The Netherlands.

VERGÉS, J., Departamento de Geologia Dinàmica, Geofisica i Paleontologia, Universitat de Barcelona, Zona Universitària de Pedralbes, Martí i Franquès s/n, 08028 Barcelona, Spain.

VILLENA, J., Departamento de Geología (Estratigrafía), Universidad de Zaragoza, 50009 Zaragoza, Spain.

VINCENT, S.J., Department of Earth Sciences, University of Liverpool, PO Box 147, Liverpool L69 3BX, United Kingdom.

VISERAS, C., Departamento de Estratigrafía y Paleontología, and Instituto Andaluz de Geología Mediterránea, CSIC, Universidad de Granada, Campus de Fuentenueva, 18071 Granada, Spain.

Preface

Reasons for this book

In recent years, much high quality geological work of international interest has been carried out in Spain. This book has been prepared in order to bring together one aspect of this work for the first time, and make it available in the English language.

Studies of outcrop geology depend critically on the quality of the outcrops available, and Spain is very fortunate in this respect. Limited vegetation and weathering reflecting the relative aridity of much of the climate, are combined with topographic incision reflecting the relatively high elevation of much of the country, to produce exposure quality which is some of the best in Europe.

The chapters that make up this book will fully demonstrate the exceptional interest of the Tertiary sedimentary record of Spain. Here we shall simply point out that for a 'middle-sized' country (about 800 km square), Spain has a very wide range of Tertiary geological features, covering two distinct orogenic belts, with the intervening craton, and bounding continental margin structures. The sedimentary basins provide much of the best evidence for the evolution of these major features. The basins range widely in size, and have formed under orogenic loading, compression, extension, transpression and transtension. Clear examples of all these basins will be found in this book.

Organisation of this book

The book opens with a General section, which has been designed to provide necessary background, and review material to introduce the relevant geology for Spain as a whole. It also includes two chapters reviewing economic aspects of the Tertiary basins.

The rest of the book consists of short articles grouped in four regions. Although the boundaries of these regions have had to be arbitarily defined, they do broadly correspond to the obvious major features of the Tertiary geology of Spain. Within each of the four regions (Fig. 1), we have tried to arrange the chapters so that neighbouring areas are generally close to each other, and that more detailed studies are preceded by chapters that appear to provide helpful preliminary material. In a very general way, the sequence of regions and their chapters may be visualised in terms of a transect that starts with the Mediterranean basin off north-east Spain, and procedes anticlockwise, via the East, West, Centre and South, to finish in the Mediterannean basins off south-west Spain.

Acknowledgements

This book is not the proceedings of a meeting. It has come together as a result of the action of ourselves, the editors, making contact and having discussions with scientists we knew to be working on these basins. We apologise to those who escaped our net; we are all the poorer for their escape. Our problem has been that responses have been so enthusiastic that we have been discouraged from pursuing further contacts for fear of running out of space.

As editors we are extremely grateful to many people who have informally read and reviewed individual articles. At a relatively late stage, the following also provided overall reviews of sections of the book: Drs Penny Barton, Tony Dickson, Alan Smith, Dave Snyder and Nigel Woodcock (all from the Department of Earth Sciences, University of Cambridge, UK) and Dr Gary Nichols (Royal Holloway College, University of London, UK)

PETER F. FRIEND
CRISTINO J. DABRIO

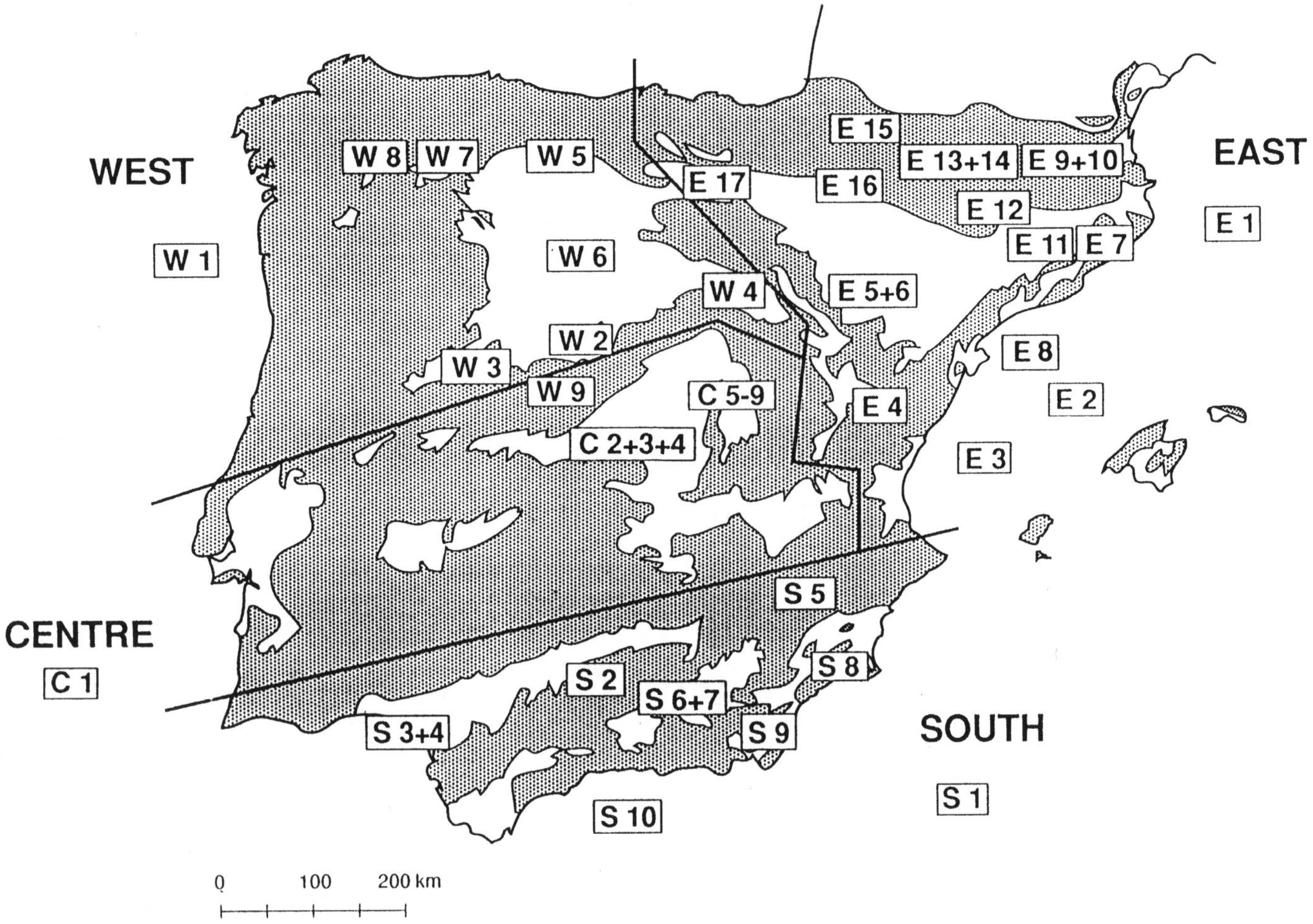

Fig. 1. Index map for locations of chapters.

Dedication to Professor Oriol Riba I Arderiu

C. PUIGDEFÁBREGAS

It was with great pleasure that I accepted the invitation to write an appreciation of Professor Oriol Riba, to whom this book is dedicated.

Most of the contributors to this book know Riba and his work well, but, for the benefit of other readers, I offer a brief chronology of some of the events in his life so far.

Oriol Riba was born in 1923 in Barcelona. He got his degree in Natural Sciencies in 1949 in Barcelona and presented his doctoral thesis *Estudio Geológico de la Sierra de Albarracin* in 1952 in Madrid. He was research scientist of the Consejo Superior de Investigaciones Cientificas between 1953 and 1959 and worked in oil exploration during the same period. He became Professor at the University of Zaragoza (1960–1969) and, later on, at the University of Barcelona (1969–1988), being now Professor Emeritus since 1988.

My first strong memories of Riba date from 1966, at the University of Zaragoza, where I had my first job, and Riba his first University chair. In the evening, on my way home, I often, or was often, stopped at the end of the corridor, where Professor Riba's door was invitingly open. The result was generally a chat covering not only geology, but often cultural or political topics. Good humour and personal anecdotes were typical. I particularly remember his account of a field visit to the wet Dutch tidal flats (Fig. 1), when he, a rather short man, dressed in the spare trousers and boots of Van Straaten, a rather tall man, followed in the tall man's long steps until finally overcome and collapsing in a mud pool!

By the mid sixties, Riba had accumulated a profound knowledge of the geology of the Ebro basin, particularly from his years in hydrocarbon exploration with CIEPSA. Riba is basically a field geologist, and his work during that period was in many ways the work of an exploration pioneer. It must have been an unforgettable experience to trace beds across the country riding his two-stroke Guzzi 65 that periodically *caught a pearl*, because of the fatal mixture of bad petrol and worse oil.

He was able to advance from this level of rather primitive logistic conditions when, a few years later, the first coverage of aerial photographs became available. He quickly understood the power of the new tool, and applied to the photo-interpretation of surface bedding traces the same geometrical thinking now commonly used in seismic interpretations. His remarkable maps of bedding traces in the area of La Rioja and Solsona, were constructed on the basis of the same criteria of geometric relationship used today in seismic stratigraphy. In my opinion, the development of this approach was Riba's first major contribution to our science.

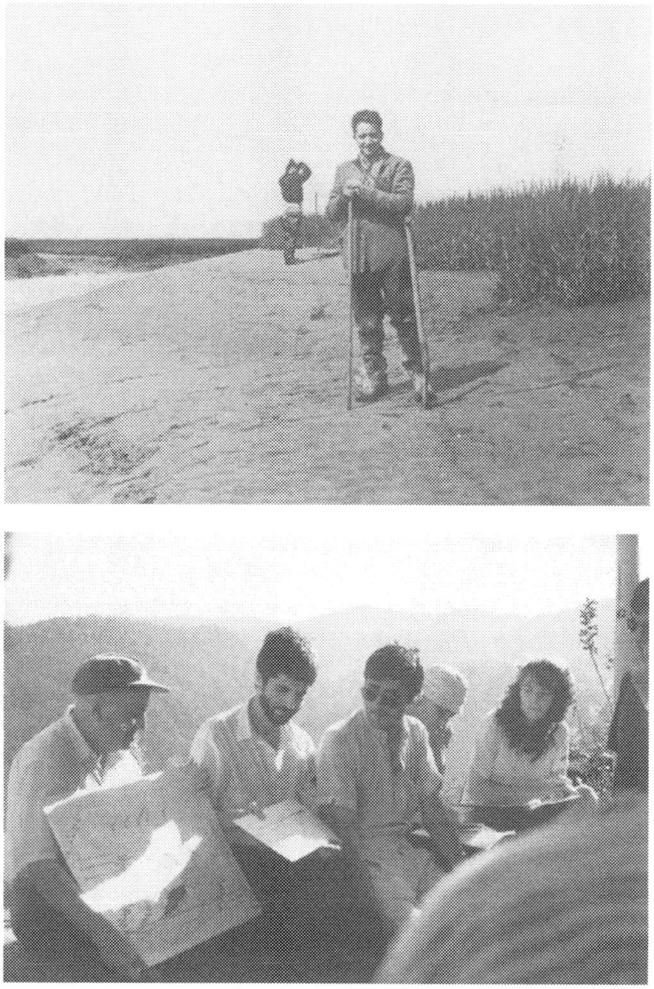

Fig. 1 (above) Professor Oriol Riba on the Dutch tidal flats; (below) Professor Oriol Riba with young colleagues.

Riba was making other contributions during that period of the

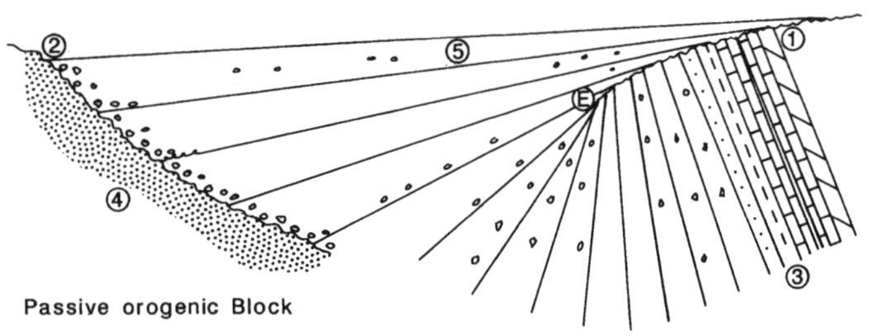

Fig. 2. 'A genetic model of syntectonic unconformity', redrawn after Riba (1976). (1), syntectonic angular unconformity of active orogenic front. (2), syntectonic climbing angular unconformity (non-marine progressive overlap of passive orogenic block). (3), preorogenic sequence. (4), passive orogenic block with buried topography. (5) syntectonic sequence. (E), articulation axis of envelope surfaces.

late fifties and sixties. Unhappy with the prevailing autocthonous interpretation of the Pyrenean structures, he was the first to recognise features such as the outwards migration of depocentres in the Ebro foreland basin of the Pyrenees. At a time when sedimentology was just developing, as a distinct speciality (Riba is a founding member of the International Association of Sedimentologists), he recognised and correctly interpreted the Caspe paleochannels in the southern Ebro basin. In spite of attention by subsequent workers, no fundamental improvement has been made to his original interpretation of these paleochannels.

The work by Riba on the syntectonic progressive unconformities of Sant Llorenç de Morunys (Fig. 2) also dates from the sixties, although it was not until the eighties that it became known more widely, when the English-speaking community could read the paper in *Sedimentary Geology*. Although few people outside the immediate Pyrenean or Spanish community have realised the general importance of these structures, Riba's careful description and theoretical analysis is now perhaps to be seen as his most significant contribution to the wider scientific community. The evidence these unconformities provide of the interaction of crustal movements with the simultaneous and continuing sedimentation cannot be denied, and once seen and accepted, opens the door to new and exciting ideas in crustal tectonics and sedimentation that have inspired many of us today.

Since 1969, when Riba accepted the Professorship of the Department of Stratigraphy at the University of Barcelona, a younger generation of sedimentary geologist has been growing up around him (Fig. 1). Many Spanish and foreign geologists have, in different ways, benefitted from the extraordinary ability to combine acute field observation with ambitious and energetic synthesis. Riba has been involved not only as author, but as supervisor and coordinator of large mapping projects and geological syntheses. He has supervised PhD theses, and worked on the urban geology of Barcelona and geological topics related to historical research.

The breadth of his cultural interests is hardly surprising, both inside and outside the sciences, because his father and mother are regarded as some of the foremost contributors to Catalan literature.

Selected References from the work of Oriol Riba

Oriol Riba is author of more than two hundred papers and about a hundred unpublished technical reports. He is also author or co-author of eleven sheets of the Geological Map of Spain at 1:200 000, fourteen sheets of the Geological Map of Spain at 1:50 000 and the lithological Map of Spain at 1:500 000.

Now, Professor Emeritus at the University of Barcelona, he keeps on surprising us: in the light of recent ideas on thrust tectonics and new seismic and well data, he goes back to his thirty-year-old but perfectly valid field data and surprises us with the admirable paper on La Rioja published in this very volume.

The following are a few references I have selected from amongst his published material:

1955 'Sédimentation du Tertiaire continental de la partie Ouest du Bassin de l'Ebre'. *Verh. 4th Int. Sediment. Kongr. Braunschweig-Göttingen, 1954. Geol. Rundschau*, 43, 2.

1959 'Estudio geológico de la Sierra de Albarracín'. *Inst.* 'L. Mallada'. *CSIC*. 16.

1962 'Los torrenos yesíferos en España'. *1er. Col. Int. Obras Públicas en los Terrenos Yesíferos. Sevilla-Madrid-Zaragoza, 1962. Serv. Geol. obr. Públ.* 1 Vol. Inst. de Edafología, CSIC (col. J. Benayas).

1966 'Estratigrafía del Terciario continental de Navarra y Rioja'. *Not. y Com. IGME*, 90.

1967 'Terciario continental de la parte Este de la Depresión Central Catalana'. *Act. Geol. Hisp.*, nº 1.

1967 'Sedimentación de paleocanales terciarios en la zona de Caspe-Chiprana (Prov. de Zaragoza)'. *An. Edaf.* 26.

1970 'Mapa Litológico de España' Esc. 1:500 000. *Inst. Edafología del CSIC, Serv. Geol. O.P. & IGME.*

1976 'Syntectonic unconformities of the Alt Cardener. Spanish Pyrenees: a genetic interpretation'. *Sedim. Geol.* 15. pp. 213–233.

1981 'Geologia Marina de la Conca Mediterrània Balear durant el Neogen'. *Mem. R. Ac. Cien. Art. Barcelona*, nº 805.

1989 'Unidades tectosedimentarias y Secuencias deposicionales'. *Rev. Soc. Geol. Esp.* 2. Granada.

1992 'La Conca de l'Ebre'. *In*: 'Historia natural dels Països Catalans' Vol. 2. Enciclopedia Catalana SA Barcelona.

1993 'Morfologia de la Rambla barcelonina', treballs de la *Societat Catalana de Geografia*, vol. 7, nº 33–34.

Memorial, Etienne Moissenet 1941–1994

Etienne Moissenet died suddenly of a brain haemorrhage, in Paris, on Wednesday, 23 February 1994. He was 52 years old. At the time of his death, he was teaching a course on Stratigraphy and Geomorphology for students of Archaeology at the Geographical Institute of the Université de Paris-I, where he was Assistant Professor of Geography.

Etienne was born on June 3, 1941, in Dijon, France, where he grew up. In 1961, he moved to Saint Cloud to study at the Ecole Normale Supérieure. There he met Professor Pierre Birot, one of the most influential geomorphologists of the time. Under the guidance of Professor Birot, he completed his *Diplôme* on 'Les glacis d'érosion de la Combe de Die'. In 1966 he took the opportunity to start an academic teaching career, joining the Sorbonne University as Assistant Professor under Birot. Etienne then began his research work on the Iberian Ranges. He was initially supported, from 1969 to 1972 by a grant from the 'Casa de Velasquez', an organisation devoted to cultural cooperation between Spain and France, and he worked in Madrid for some three months before starting detailed field work. In 1972, Etienne joined the Université de Paris-I, and became involved in teaching course-work for much of every year. Over the full period of some 23 years of his field studies in Spain, he was accompanied on most of his summer (and other) field campaigns by his wife, Nicole.

Etienne rapidly realised the importance of fossil mammals in dating both the nonmarine Tertiary of the Teruel Trough, and the later morphogenetic features of the Iberian Range. From L. Thaler of Montpellier, he learnt the washing methods developed in searching for micromammal materials, and the search for these materials in the Tertiary deposits became an obsession with him. To establish the age of the basin fills of the Iberian Range, Etienne contacted two mammal palaeontologists who collaborated with him throughout the rest of his life: Rafael Adrover from Teruel, and Pierre Mein from Lyon. The first collaborative work by these friends was published in 1976. Starting with detailed field work in the Jucar and Cabriel basins in 1973, Etienne went on to develop a new stratigraphic framework and dating for all the Tertiary basins of the Iberian Range. He was then able to turn to geomorphological and tectonic problems as well. In the early 1980s he collaborated in work for the 1/50000 map of the Instituto Geológico y Minero de España. taking a very active part based on his wide knowledge of the geology, stratigraphy and dating of the Teruel Graben and other smaller basins. This knowledge was incorporated in the 1/200000 map synthesis of Teruel. His energetic and detailed fieldwork (more than 200 fossil localities were found in the Teruel graben) resulted in the recognition of Pliocene deposits in the upper parts of the Teruel Graben fill.

After some preliminary contacts with E.H. Lindsay in 1982, Etienne started magnetostratigraphic work in the Neogene Cabriel Basin, in collaboration with Alfredo Pérez González. Similar studies combining magnetostratigraphy and biostratigraphy were extended to the Teruel basin in order to provide a chronostratigraphic framework for the potential mammal stage stratotype for the Early and Middle Pliocene: the Alfambrian.

Over the years, Etienne was author of more than 30 papers on biostratigraphy, magnetostratigraphy, geomorphology, tectonics and the role of diapirism in the origin of some of the Iberian Range.

Etienne made many friends in Spain, not only Earth Scientists, but farmers, innkeepers and other people from the Sierra. He spoke Spanish fluently with a distinct French Aragonese accent! He will be remembered for his kindness, determination and patience, and as one of the giants of the investigation of the Iberian Range. Etienne Moissenet will be sorely missed.

Pere Anadón
Nicole Moissenet
Oriol Riba

Part G
General

G1 Tertiary stages and ages, and some distinctive stratigraphic approaches

P.F. FRIEND

Abstract

A standard scheme of stages for the Tertiary is presented, and their ages discussed. Sequence stratigraphy was becoming established as an approach to the analysis of the Spanish Tertiary before it became widely used internationally. Geomagnetic polarity reversal stratigraphy has been shown to be an important approach to the dating of some non-marine successions.

Introduction

Basin stratigraphy generally provides the best key to the geometry and timing of development of the basin. Because much of this book reports recent work on the stratigraphic analysis of the various local Tertiary successions, this chapter has been written to introduce Tertiary stratigraphic terminology for those unfamiliar with it, and to draw attention to some of the special techniques used on the Spanish material.

The Tertiary stages

A geologic time scale 1989 or *GTS 89* (Harland *et al.*, 1990) is the starting point of this introduction. It has been widely used because of its broad and balanced coverage.

Fig. 1 presents the scheme of eras, sub-eras, periods and stages proposed for standard use in *GTS 89*. The stages are, to a great extent, the basic units for indicating the ages of rocks in the Tertiary successions, and are now based on global correlations using planktonic oceanic faunas and floras. Fig. 1 also shows correlations between the various standard stages and certain stages and divisions used more locally in Spain. These local schemes have been developed where correlations with the standard stages are particularly uncertain. For example, the stages of column (5) (Fig. 1) have been developed using faunas of benthonic foraminifera, and are therefore applied where deposits include these fossils and lack planktonic forms (e.g. Luterbacher *et al.*, 1991). The stages and Mammal Neogene zones (Mein, 1975, 1990) of columns (6) to (8) have been developed where marine faunas are completely lacking and the biostratigraphy depends on land faunas (consisting mainly of mammals) (see Chapters C3 and C5).

Ages of the Tertiary stages

Fig. 2 presents some recent time-scale age estimates for the boundaries between the standard Tertiary stages. These time scales are those of Harland *et al.* (1990), Cande and Kent (1992), and Odin and Odin (1990). The greatest difference in the three estimates for any particular stage boundary is 4.0 Ma in the middle Eocene, but is only 1.5 Ma in the Neogene, and the differences are less for stage boundaries before and after these two maxima.

Sequence stratigraphy

A *depositional sequence* has been defined (Mitchum *et al.*, 1977) as:

> a stratigraphic unit of a relatively conformable succession of genetically related strata and bounded at its top and base by unconformities or their correlative conformities

Fig. 3 illustrates some typical sequence relationships.

Sequences matching the above definition form the major subdivisions in most of the Tertiary basins of Spain, and the recognition and study of depositional sequences have contributed greatly to recent advances. Indeed, it was in the continental (non-marine) basin fills of Spain that A.G. Megias developed his pioneering use of 'Tectosedimentary units' or 'TSUs' (1973, doctoral thesis, University of Granada, and subsequent publications), which appear to be identical to the depositional sequences defined above. His TSUs are bounded by 'rupturas sedimentarias' (Castillian), sometimes translated as 'ruptures' (English) but better translated as breaks or boundaries.

The belief that many depositional sequences are a response to global sea-level change (Vail *et al.*, 1977) has such exciting implications in terms of dating events and correlating changes, that it has had a major impact on stratigraphic work over the last 15 years. But it is now clear (Galloway, 1989; Miall, 1986, 1991) that careful analysis must be carried out before a particular depositional sequence can be interpreted as caused by global sea-level change. In fact, most non-marine, inland sequences, particularly, seem likely to have been formed primarily by tectonic or climatic factors.

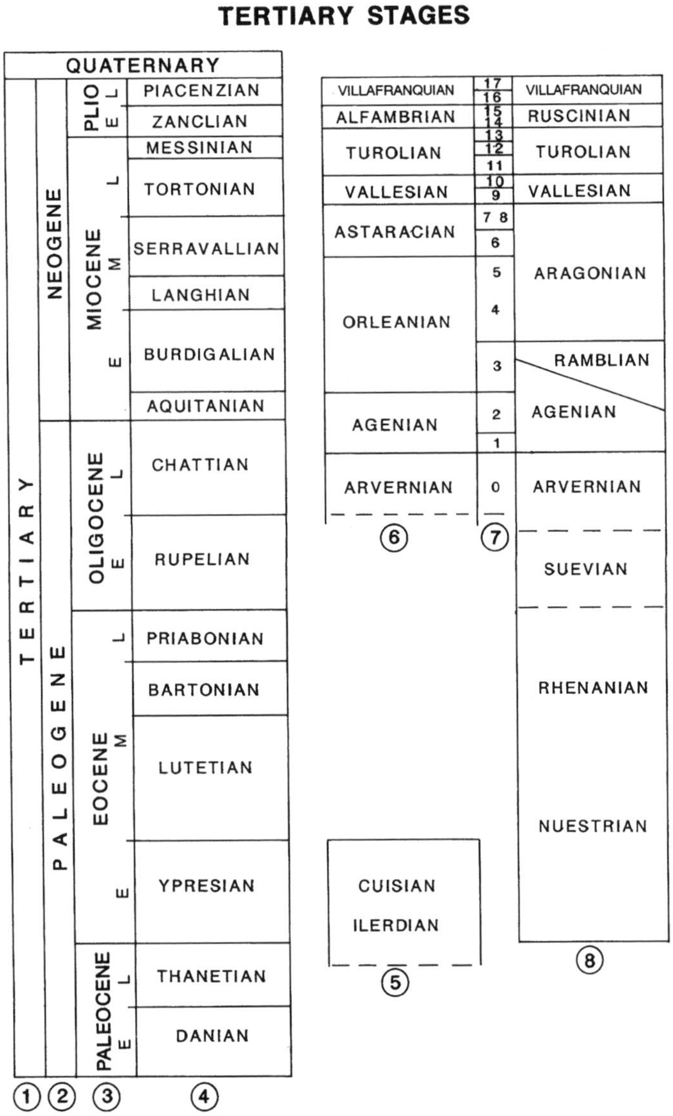

TERTIARY STAGES
QUATERNARY
TERTIARY
NEOGENE
PALEOGENE
PLIO
MIOCENE
OLIGOCENE
EOCENE
PALEOCENE
PIACENZIAN
ZANCLIAN
MESSINIAN
TORTONIAN
SERRAVALLIAN
LANGHIAN
BURDIGALIAN
AQUITANIAN
CHATTIAN
RUPELIAN
PRIABONIAN
BARTONIAN
LUTETIAN
YPRESIAN
THANETIAN
DANIAN
VILLAFRANQUIAN
ALFAMBRIAN
TUROLIAN
VALLESIAN
ASTARACIAN
ORLEANIAN
AGENIAN
ARVERNIAN
RUSCINIAN
ARAGONIAN
RAMBLIAN
SUEVIAN
RHENANIAN
NUESTRIAN
CUISIAN
ILERDIAN
(1) (2) (3) (4) (5) (6) (7) (8)

Fig. 1. Tertiary stratigraphic names used in this book. Columns (1) to (4) show the eras (1), sub-eras (2), periods (3) and stages (4), according to the *GTS 89* time scale (Harland *et al.*, 1990). Conventional divisions of the periods into early (E), middle (M) and late (L) are shown. The boundaries between the units in columns (1) to (4) are spaced vertically according to the ages assigned in the *GTS 89* (see Fig. 2). Column (5) shows stages based on benthonic foraminifera (Luterbacher *et al.*, 1991). Columns (6) to (8) are Neogene mammal ages (6), the MN zones (7) of European mammal biostratigraphy and European continental stages (8) from charts by Calvo *et al.* (this volume) and Gomez *et al.* (this volume).

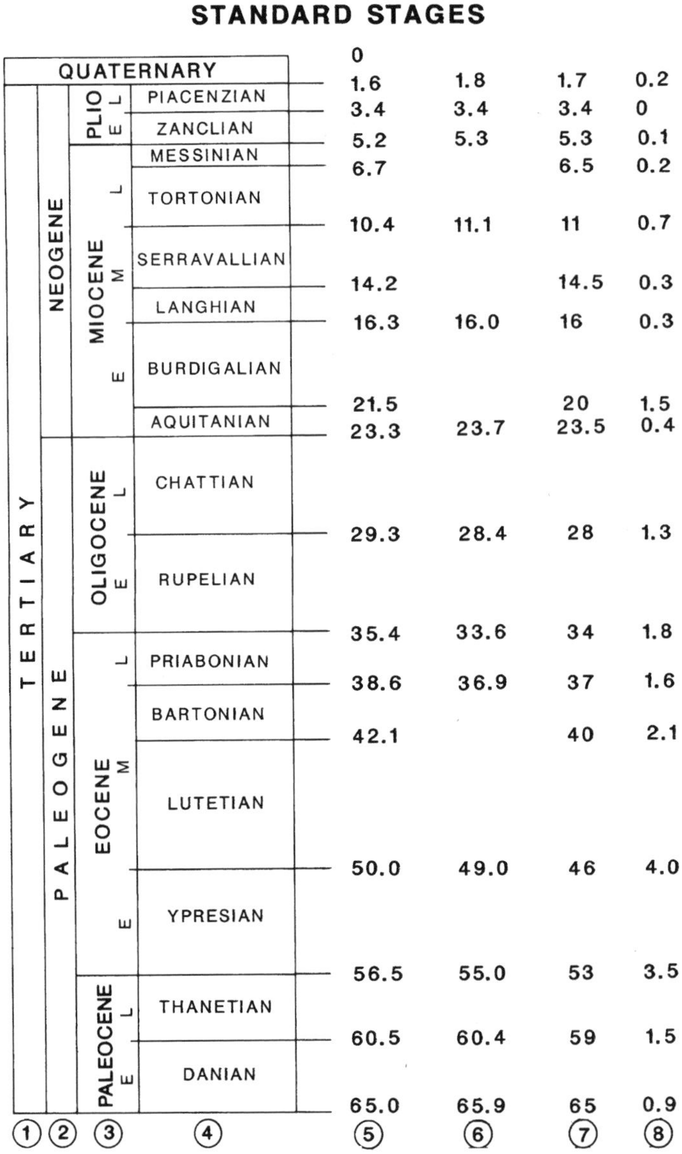

STANDARD STAGES
QUATERNARY
TERTIARY
NEOGENE
PALEOGENE
PIACENZIAN
ZANCLIAN
MESSINIAN
TORTONIAN
SERRAVALLIAN
LANGHIAN
BURDIGALIAN
AQUITANIAN
CHATTIAN
RUPELIAN
PRIABONIAN
BARTONIAN
LUTETIAN
YPRESIAN
THANETIAN
DANIAN
0
1.6 1.8 1.7 0.2
3.4 3.4 3.4 0
5.2 5.3 5.3 0.1
6.7 6.5 0.2
10.4 11.1 11 0.7
14.2 14.5 0.3
16.3 16.0 16 0.3
21.5 20 1.5
23.3 23.7 23.5 0.4
29.3 28.4 28 1.3
35.4 33.6 34 1.8
38.6 36.9 37 1.6
42.1 40 2.1
50.0 49.0 46 4.0
56.5 55.0 53 3.5
60.5 60.4 59 1.5
65.0 65.9 65 0.9
(1) (2) (3) (4) (5) (6) (7) (8)

Fig. 2. Recently quoted ages (in Ma) for the standard Tertiary stages. Columns (1) to (4) as explained in the caption for Fig. 1. Column (5) contains the ages quoted in *GTS 89* (Harland *et al.*, 1990). Column (6) contains ages indicated by Cande and Kent (1992). Column (7) contains ages quoted by Odin and Odin (1990). Column (8) shows the maximum difference in the ages quoted in columns (5) to (7).

Geomagnetic polarity reversal stratigraphy

One significant approach to the dating of long successions in the largely unfossiliferous continental basins of northern Spain has been the work on paleomagnetic reversal stratigraphy by Burbank and his group (see Chapters E11, E13 and E14). Whereas the first of these chapters uses the *GTS 89* reversal time scale, the second and third have used the polarity reversal scale developed by Cande and Kent (1992). Fig. 4 compares the ages attributed by Cande and Kent for the top of the highest normal polarity interval for each of the standard ocean-floor anomaly intervals, with the equivalent ages quoted in *GTS 89* (Harland *et al.*, 1990). The Cande and Kent ages are generally younger than the *GTS 89* ages, by a maximum of 1.5 Ma at ages of about 40 Ma, and the discrepancies are less for both older and younger Tertiary anomalies.

References

Cande, S.C. and Kent, D.V. (1992). A new geomagnetic polarity time-scale for the late Cretaceous and Cenozoic. *Journal of Geophysical Research*, vol. 97, no. B10; pp. 13917–13951.

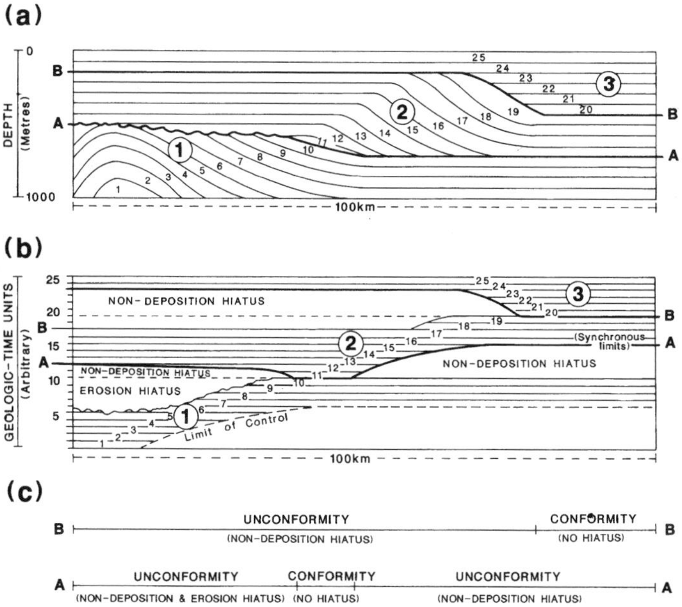

Fig. 3. Possible scale and relationships between three depositional sequences. (a) Depth–location relationships along a profile, with time-units (1) to (25), and (b), time–location relationships along the same profile, to illustrate the pattern of deposition, non-deposition hiatus and erosion hiatus, and (c) the changes in the sequence boundaries (A) and (B) across the profile. (Redrawn from Mitchum *et al.*, 1977.)

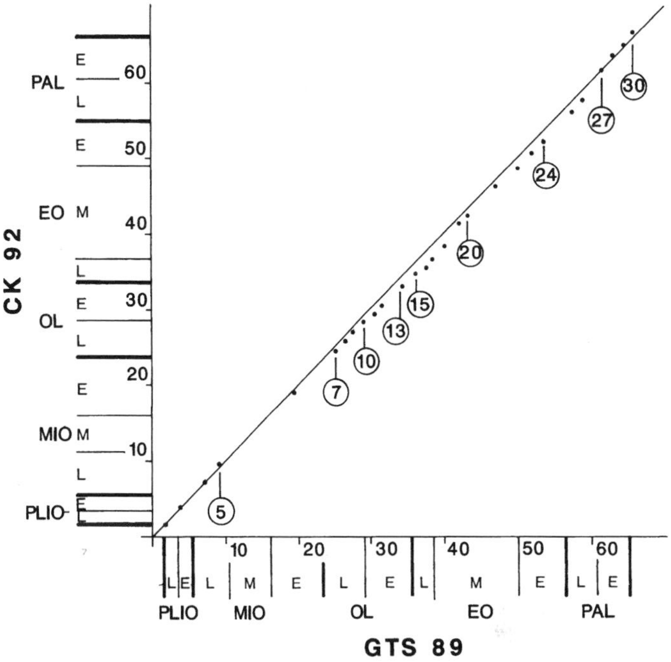

Fig. 4. The relationship between ages estimated in the *GTS 89* time scale (Harland *et al.*, 1990) and the *CK 92* time scale (Cande and Kent, 1992). Each point on the graph represents the ages quoted for the top of the youngest normal polarity interval in each of the major numbered magnetic anomalies conventionally recognised in ocean-floor anomaly mapping. The straight line on the graph will coincide with points of identical age on the two scales.

Galloway, W.E. (1989). Genetic stratigraphic sequences in basin analysis. *Bulletin of the American Association of Petroleum Geologists*, vol. 73, pp. 125–154.

Harland, W.B., Armstrong, R.L., Cox, A.V., Craig, L.E., Smith, A.G. and Smith, D.G. (1990). *A geologic time scale 1989*, Cambridge University Press, Cambridge, 263 pp.

Luterbacher, H.P., Eichenseer, H., Betzler, C. and Hurk, A. van den (1991). Carbonate–siliciclastic depositional systems in the Paleogene of the south-Pyrenean foreland basin: a sequence stratigraphic approach. In McDonald, D. (ed.), *Sea level changes at active plate margins. Special Publications of the International Association of Sedimentologists*, Vol. 12, pp. 391–407.

Mein, P. (1975). *Resultats du Groupe de Travail des vertebrès*. Report on activity of the RCMNS Working Groups, Bratislava, pp. 77–81.

Mein, P. (1990). Updating of MN zones. In Lindsay, Fahlbusch and Mein (eds.) *European Neogene Mammal Chronology*, NATO ASI Series, A, 180, pp. 73–90.

Miall, A.D. (1986). Eustatic sea-level changes interpreted from seismic stratigraphy: a critique of the methodology with particular reference to the North Sea Jurassic record. *Bulletin of the American Association of Petroleum Geologists*, vol. 70, pp. 131–137.

Miall, A.D. (1991). Stratigraphic sequences and their chronostratigraphic correlation. *Journal of Sedimentary Petrology*, vol. 61, pp. 497–505.

Mitchum, R.M., Vail, P.R. and Thompson, S. (1977). The depositional sequence as a basic unit for stratigraphic analysis. *Memoir of the American Association of Petroleum Geologists*, vol. 26, pp. 53–62.

Odin, G.S. and Odin, C. (1990). Sommaire geochronique. *Geochronique* No. 35.

Vail, P.R., Mitchum, R.M., Todd, R.G., Widmier, J.M., Thompson, S., Sangree, J.B., Bubb, J.N. and Hatlelid, W.G. (1977). Seismic stratigraphy and global changes of sea-level. *Memoir of the American Association of Petroleum Geologists*, vol. 26 pp. 49–212.

G2 Cenozoic latitudes, positions and topography of the Iberian Peninsula

A.G. SMITH

Abstract

Four maps of the Iberian peninsula at 65, 35, 16 and 0 Ma show the positions of Iberia relative to Europe and Africa as estimated from ocean-floor spreading data. The maps themselves are in a global paleomagnetic reference frame. The Cenozoic changes in latitude of Seville, Madrid and Barcelona are also shown. In early Paleocene time the latitude of Seville was as low as 30°. The origins of the present high topography of the Spanish meseta are unclear.

Introduction

The chapters in this volume discuss the Tertiary basins in Spain. The nature of their sedimentary fill is controlled partly by tectonics, paleoclimate and paleotopography. This short chapter comments on all of these features from a broader viewpoint than that which has necessarily been adopted by most of the accompanying chapters and also raises some general questions. In particular, this chapter highlights three topics: 1. Cenozoic positions of the Iberian peninsula relative to Eurasia and Africa as estimated from published Atlantic ocean-floor spreading data; 2. Cenozoic paleolatitudinal changes of the peninsula; and 3. the peninsula's anomalously high topography.

Cenozoic position and orientation

In Cenozoic time the Iberian peninsula has behaved as part of a microplate directly linked to Africa and Eurasia by orogenic belts. In principle, the position of the peninsula relative to the stable parts of the Eurasian and African plates can be determined by fixing Iberia and restoring the orogenic belts to their late Cretaceous shapes and then adjusting the positions of Eurasia and Africa accordingly. However, the uncertainties in this approach are considerable because it involves palinspastically restoring fold and thrust belts, unstraining the more deformed zones as well as recognizing and evaluating all major strike–slip deformation.

A more general approach is to use the Atlantic ocean-floor spreading data to reposition Iberia, Eurasia and Africa. Africa may be repositioned relative to Europe by using the spreading history of the central Atlantic between Africa and North America (Klitgord & Schouten, 1990), and of the North Atlantic between North America and Europe (Srivastava & Tapscott, 1986). Iberia can be repositioned relative to Europe from the data of Olivet *et al.* (1984) and Sibuet *et al.* (1980). The magnetic anomaly scale and numerical time scale is that of Harland *et al.* (1990). Fig. 1 shows four maps: for 65 Ma (the Cretaceous–Tertiary boundary); 35 Ma (approximately the Eocene–Oligocene boundary); 16 Ma (approximately the Early–Middle Miocene boundary); and 0 Ma (present-day) based on these data. All four maps show an arbitrary northern boundary

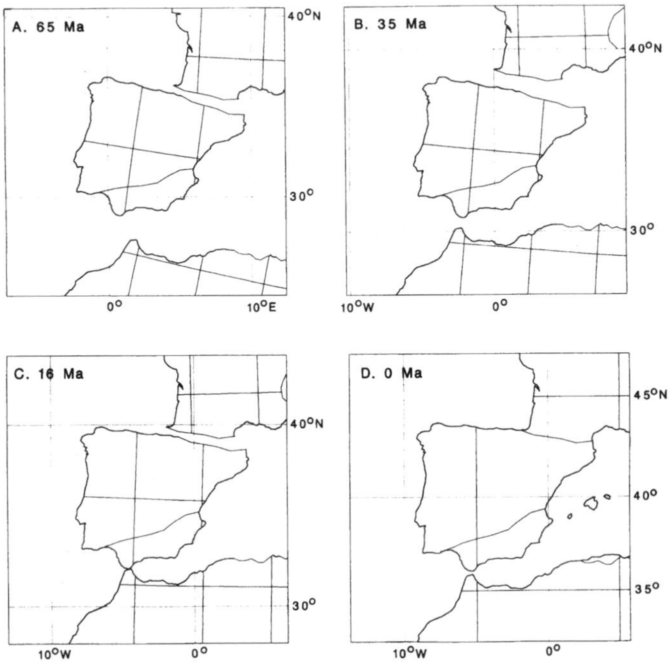

Fig. 1. Maps of Iberia, Europe and Africa in a paleomagnetic latitude frame of reference. Present-day latitude and longitude grids are shown for convenience. Iberia has been repositioned relative to Europe using data from the Bay of Biscay. Europe and Africa have been repositioned relative to one another from ocean-floor spreading data. A. 65 Ma (Cretaceous–Tertiary); B. 35 Ma (Eocene–Oligocene); C. 16 Ma (Early–Middle Miocene) and D. 0 Ma (present-day).

to Iberia in the Pyrenees; the Betics have been retained in their present-day position relative to Iberia as a whole; fragments of a formerly continuous continental area to the E of Iberia (the Balearic Islands, the Grande and Petit Khabylie of N Africa, Corsica, Sardinia and small fragments now forming part of southern Italy and northern Sicily) have been omitted. The *relative positions* of Eurasia, Africa and Iberia are those implied by the spreading data, but the reassembly has been rotated into a global paleomagnetic reference frame (see below).

Because of the uncertainties in the spreading data these maps should not be used as evidence of movements on a scale of 100–200 km. However, the maps are reliable on a smaller scale, i.e. for larger effects. The following effects are probably real: 1. there is very little rotation of Spain with respect to Africa during the Cenozoic; 2. there is a small rotation relative to Europe in the same period; 3. Iberia was separated both from southern Europe and the Betics and/or Africa in early Cenozoic time; 4. the removal of these spaces presumably by compression to give rise to orogenic belts; and 5. there have been no large (>200 km) strike–slip motions between Iberia and either Africa or Europe in Cenozoic time. Strike–slip motion of <200 km may have occurred, but is not resolvable from these data. The constraints provided by the ocean-floor data must be satisfied by all Cenozoic tectonic syntheses.

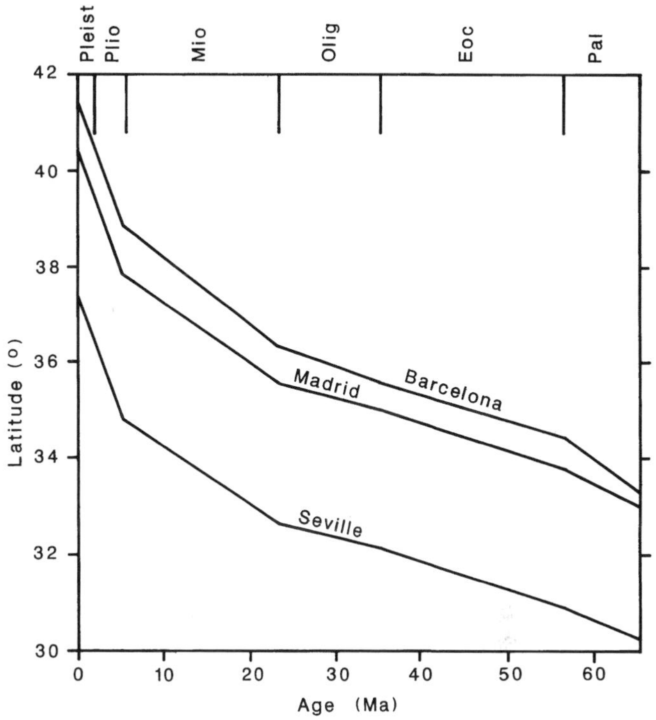

Fig. 2. The Cenozoic changes in latitude of Seville, Madrid and Barcelona, estimated from reconstructions placed in a global paleomagnetic frame of reference.

Paleolatitudes

Paleomagnetic pole positions give directly the former latitude of a point on the Earth's surface. Though there are some reliable Cenozoic pole positions from the Iberian peninsula, they are relatively few in number and not well distributed in time. The most reliable method for determining the paleolatitude of a point in the Iberian peninsula is to link the peninsula to a larger reference continent, such as Eurasia, by using ocean-floor speading data. The global paleomagnetic pole can be determined relative to this reference continent (e.g. Besse & Courtillot, 1991) and hence the paleolatitude of any point on Eurasia and any continental fragments that can be linked to it can be found. The poles used are those in the ORACLE database version 1 of Lock & McElhinny (1991). Fig. 2 shows the estimated changes in latitude of the sites of three Spanish cities – Barcelona, Madrid and Seville – throughout the Cenozoic Era. The numerical ages are taken from Harland *et al.* (1990). The uncertainties are difficult to estimate but are probably less than 2°.

The overall changes are relatively small (8°). However, the southern part of the Iberian peninsula was close to 30° in the Paleocene – at the same latitude as parts of the northern Sahara today. Other things being equal, one might therefore expect the climate to have been more arid in such areas than it was in later Cenozoic time.

Paleotopography

The peninsula is remarkable for its high topography, averaging several hundred metres. If isostatically balanced, an increase in thickness over that of a normal crust at sea level would be in the range 3–6 km. High topography is often associated with orogenic belts due to crustal thickening by collision and underplating. Though the peninsula is bordered by the Pyrenees to the N and the Betics to the S, the lateral extent of the high topography in the Iberian peninsula is much greater than is found adjacent to other orogenic belts of a similar age, such as the Alps and the Carpathians. Seismic studies (see Chapter G6) show that the crust under the Spanish meseta has an average seismic structure, though it is not clear whether the anticipated increase in thickness would be detectable. The high topography appears unlikely to be an orogenic effect.

High topography in non-orogenic areas is often associated with 'hot-spots', but large-scale Cenozoic hot-spot activity is absent. Even central Spain seems to have been near sea level during Late Cretaceous time (C.J. Dabrio, pers. commun.), but there is no obvious Cenozoic cause for the high topography. Thus, the ultimate origin of the topography and what sustains its present height is a key problem in regional Iberian studies.

References

Besse, J. and Courtillot, V. (1991). Revised and synthetic apparent polar wander paths of the Africa, Eurasian North American and Indian plates, and true polar wander since 200 Ma. *Journal of Geophysical Research*, **96**, 4029–4050.

Harland, W.B., Armstrong, R.L., Cox, A.V., Craig, L.E., Smith, A.G. and

Smith, D.G. (1990). *A geologic time scale 1989*. Cambridge University Press, Cambridge.

Klitgord, K.D. and Schouten, H. (1990). Plate kinematics of the central Atlantic. In Vogt, P.R. and Tucholke, B.E. (eds.) *The western North Atlantic region*. Vol. **M** (Geology of North America). Geological Society of America, Boulder, Colorado, pp. 351–378.

Lock, J. and McElhinny, M.W. (1991). The global paleomagnetic database: design, installation and use with ORACLE. *Surveys in Geophysics*, **12**, 317–491.

Olivet, J-L., Bonnin, J., Beuzart, P. and Auzende, J-M. (1984). *Cinematique de l'Atlantique nord et central, scientific report 54*. Centre national pour l'exploitation des oceans. 108 pp.

Sibuet, J-C., Ryan, W.B.F., Arthur, M., Barnes, R., Blechsmidt, G., De Charpel, O., De Graciansky, P.L., Habib, D., Iaclarino, S., Johnson, D., Lopatin, B.G., Maldonado, A., Montadert, L., Moore, D.G., Morgan, G.E., Mountain, G., Rehault, J.P., Sigal, J. and Williams, C.A. (1980). Deep drilling results of Leg 47b (Galicia Bank area) in the framework of the early evolution of the North Atlantic Ocean. *Philosophical Transactions of the Royal Society of London*, **A294**, 51–61.

Srivastava, S.P. and Tapscott, C.R. (1986). Plate kinematics of the North Atlantic. In Vogt, P.R. and Tucholke, B.E. (eds.) *The western North Atlantic region*. Vol. **M** (Geology of North America). Geological Society of America, Boulder, Colorado, pp. 379–404.

G3 Tertiary tectonic framework of the Iberian Peninsula

C.M. SANZ DE GALDEANO

Abstract

During the Tertiary, Iberia was strongly deformed. It was located between the large EuroAsiatic and African plates, and was displaced east relative to the opening Atlantic, along with N–S convergence. During the Eocene, major compressional deformation occurred in the Pyrenees and Bay of Biscay, and this continued in the Oligocene. The opening of new basins in the Mediterranean led to the compression and formation of the Iberian Cordillera. The Betic–Riffian Internal zones, strongly deformed in the Palaeocene, were expelled westwards towards the end of the Oligocene and in the Neogene. In the Betic and Iberian Cordilleras, regional uplift with radial extension coexisted with N–S compression, starting in the Late Miocene. The interior of the Iberian massif, particularly the Spanish Central System, was also deformed under these regimes, and the inner basins were formed. The Ebro and Guadalquivir Basins were clearly foreland basins.

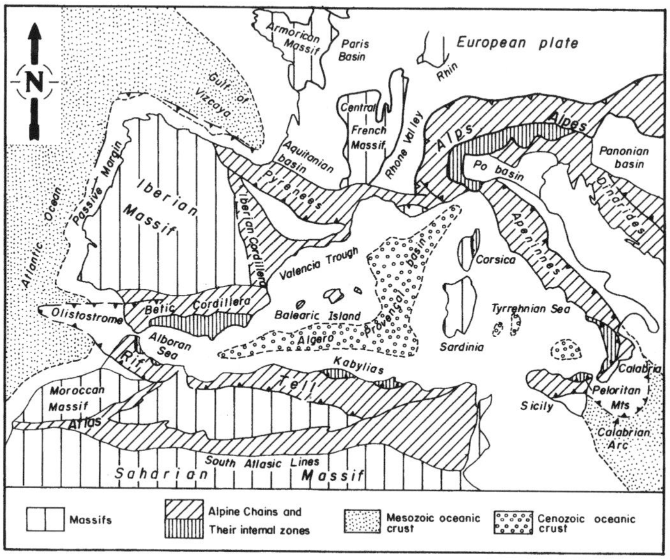

Fig. 1. General situation of Iberia.

Introduction

The Iberian Peninsula has existed as a lithospheric plate located between the African and European plates (Fig. 1) throughout the Mesozoic and Tertiary. From the beginning of the Mesozoic until late in the Early Cretaceous its movements mainly coincided with those of Europe, whereas from that time to the Oligocene it moved together with Africa. Since the Oligocene or Early Miocene, the Iberian Peninsula has been joined to Europe (Malod, 1989).

Uchupi (1988) and to a lesser extent Boillot *et al.* (1984) provide syntheses of the general tectonic situation of the Iberian Peninsula throughout the Mesozoic and Cenozoic. This chapter will concentrate on its evolution during the Cenozoic.

The nucleus of the Iberian Peninsula is the Iberian (or Hesperian) Massif (Fig. 2), which was deformed in the Hercynian. Several Alpine domains and cordilleras are located around this mainly Palaeozoic massif. Thus, to the north is the Bay of Biscay, with oceanic crust formed during the Cretaceous. To the northeast we find the Pyrenean Cordillera, consisting of a Palaeozoic nucleus and a cover of Mesozoic and Tertiary sediments. The Iberian Cordillera is located to the east of the Iberian Massif, and the Catalan Coastal Range to the southeast of the Pyrenees. The Betic Cordillera is situated to the south and southeast. To the west, the peninsula is bounded by, and forms a passive margin to, the Atlantic Ocean.

Two large Tertiary basins (Duero and Tagus) are found in the Iberian Massif, separated from each other by the Spanish Central System (a cordillera mainly controlled by the Alpine movement of late Hercynian faults). The Guadalquivir Basin is located south of the Massif, while the Ebro Basin is somewhat further away between the Iberian Cordillera and the Pyrenees. Numerous smaller Tertiary basins are found in all the domains mentioned above.

Principal evolutionary features of the Iberian Peninsula during the Mesozoic

In order to understand the evolution of the Iberian Peninsula during the Tertiary, we must place it in its geological context, particularly with regard to the processes occurring in the Western Mediterranean. Many of these processes were caused by

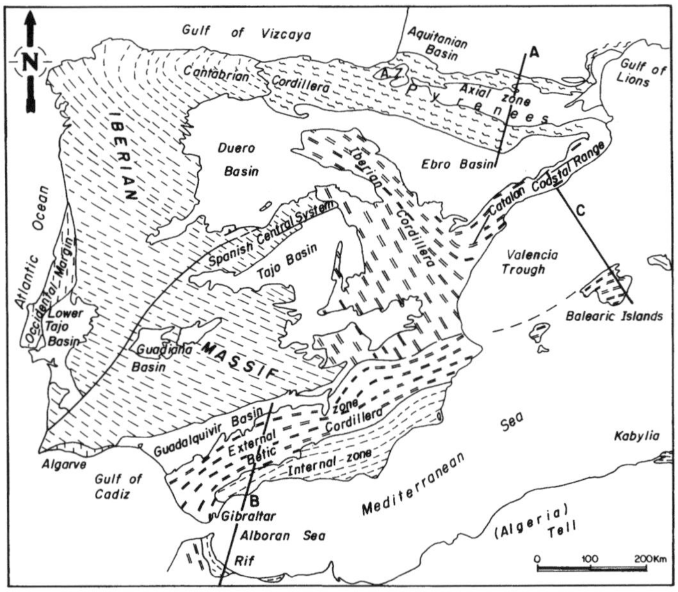

Fig. 2. Highly simplified geological scheme of the principal domains in the Iberian Peninsula. Lines, A, B and C mark locations of sections shown in Fig. 5. (Simplified and modified from Julivert *et al.*, 1972).

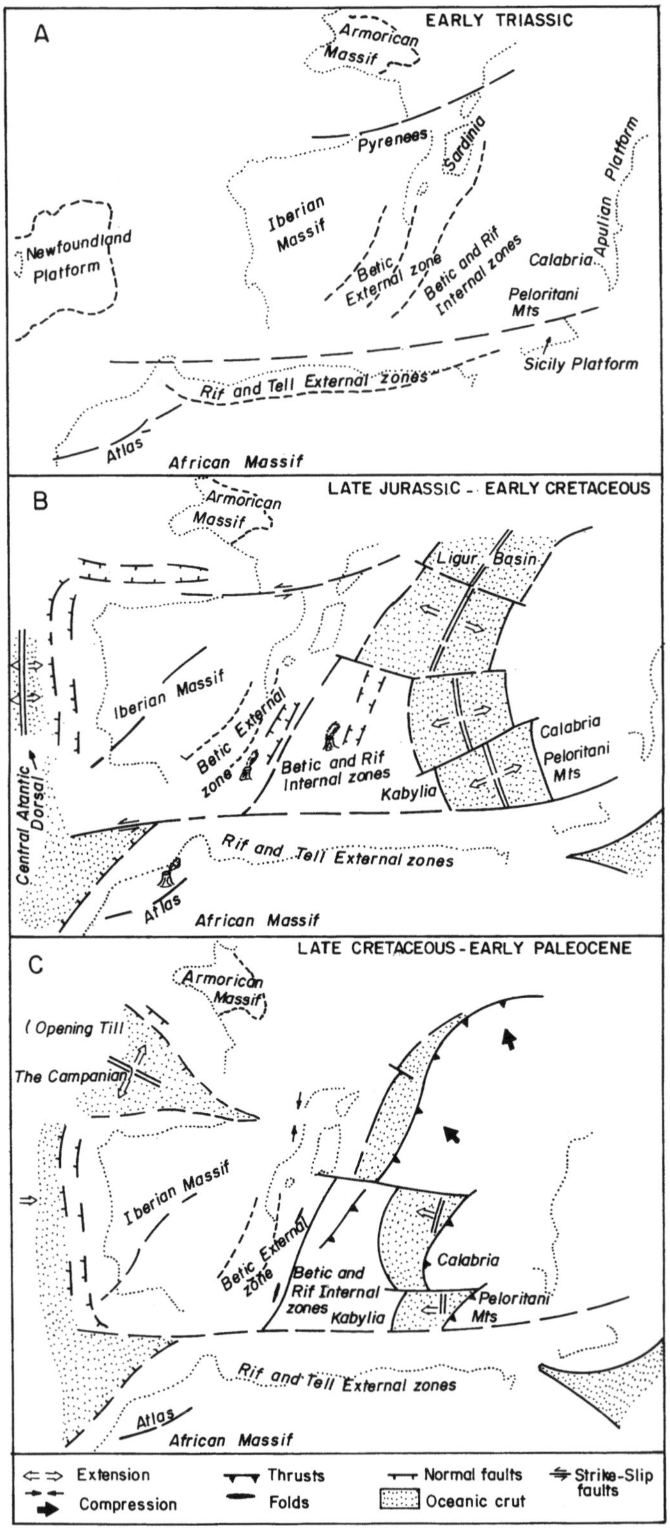

Fig. 3. Highly simplified interpretative schemes of the evolution of Iberia. A: Early Triassic. B: Late Jurassic–Early Cretaceous. C: Late Cretaceous–Early Palaeocene.

interactions between the African and European plates, which in turn were related to processes of oceanic growth, especially to the opening of the Atlantic. We must therefore give an albeit schematic outline of the evolution of Iberia during the Mesozoic.

Most palaeogeographic reconstructions locate the Iberian Massif opposite the coasts of Tunisia and Algeria during the Triassic (Fig. 3A). Throughout this period, the Jurassic and part of the Cretaceous (Fig. 3B), the Western Mediterranean underwent an important process of extension, crustal thinning and even creation of oceanic crust. The latter occurred in the possibly highly complex Ligurian Basin (Biju-Duval *et al.*, 1977; Dercourt *et al.*, 1986), which probably extended from the Betic–Riffian Internal Zone to beyond the basin of the Internal Zone of the Alps. During this process the opening of the South Atlantic gradually displaced Africa to the east, so that Iberia moved relatively over 1200 km westwards.

The Central and North Atlantic began to open during the Late Jurassic. Later, during the Aptian–Albian, this gradual opening caused the Iberian Massif to rotate counter-clockwise (some 30 degrees in all), thus inducing the opening of the Bay of Biscay that separated western France from northwest Spain. This process continued at least until the Campanian (Olivet *et al.*, 1982; Boillot *et al.*, 1984) and may have occurred together with some displacement of Iberia towards the south, southwest or southeast, although this question is still the subject of debate. At the same time the first convergent movement took place between Africa and Europe and the first compressive deformations occurred in the internal zones of the Alps.

During the rest of the Late Cretaceous, Iberia did not as a rule undergo important deformation. To the west the Atlantic contin-

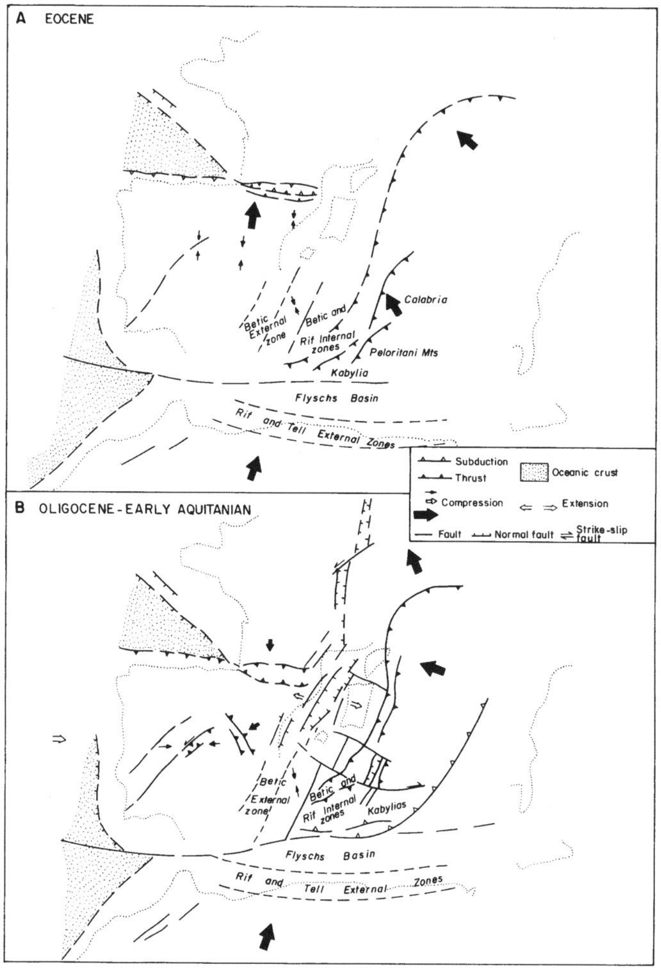

Fig. 4. Interpretative schemes of the evolution of Iberia. A: Eocene. B: Oligocene–Early Aquitanian.

ued to open, while, to the east, the Betic–Riffian Internal Zone seems to have been affected by important tectonic instability and even subduction (De Jong, 1991). The first compressive movements and nappe formation took place in the Eastern Pyrenees (Muñoz *et al.*, 1988), while extension continued in the rest of this cordillera (Ducasse *et al.*, 1986). (Fig. 3C).

Tectonic evolution during the Palaeocene and Eocene (Fig. 4A)

Displacement of the African plate relative to the European plate was clearly convergent at the beginning of the Tertiary and throughout the Palaeocene (Tapponier, 1977). In the Western Mediterranean this brought about subduction of the oceanic crust previously formed in the Ligurian Basin, which clearly affected the internal zones of the Alps and Betic Cordilleras, which, in the case of the latter, were located some 500 km east of their present position. Although weak, approximately N–S, compressive movements affected Iberia, and there was no very important deformation, not even in the Betic external zones. To the west the Atlantic

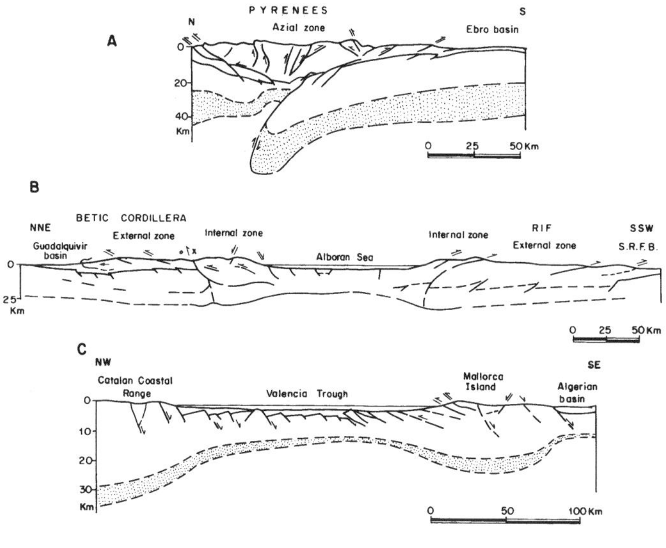

Fig. 5. Schematic geological cross-sections; their positions are marked in Fig. 2. A: Interpretation of the geological structure of the Pyrenees. Simplified from Muñoz (1992) and modifications suggested by Velasque *et al.* (1989). B: Interpretation of the cross-section from the Betic Cordillera to the Rif. C: Structure of the Valencia trough. Simplified from Martí *et al.* (1992) and Fontboté *et al.* (1990). Lower crust stippled.

continued to open as a passive margin and, to the north, there was no significant deformation, and deposition continued in most of the Pyrenees.

A new stage began towards the end of the Palaeocene or in the Early Eocene, in which collision took place after subduction. The Betic–Riffian internal zones, which did not as yet form part of Iberia, underwent an important stage of tectonic structuring and metamorphism.

This Eocene compression was transmitted to the entire Iberian plate with multiple consequences. The most significant of these occurred in the Pyrenees, which were heavily deformed forming important nappes (Megías, 1988) (Pyrenean Eocene phase). Modern deep seismic profiles show that the Iberian plate is subducted beneath the European plate (Roure *et al.*, 1989; Suriñach *et al.*, 1993) (Fig. 5A), even though there may not previously have been a large volume of oceanic crust.

The foreland basin (Ebro Basin) in the Southern Pyrenees (Anadón *et al.*, 1985) migrated southward (simultaneously) with the progressive deformation of the cordillera, while the northern margin was partially incorporated into the younger thrust sheets (Muñoz *et al.*, 1988).

The compressive movements were transmitted further west during this Eocene phase, so that the northwestern margin of Iberia tended to become superposed on the Bay of Biscay; i.e., part of the oceanic crust in the Bay of Biscay was subducted beneath Iberia (Boillot *et al.*, 1984). Part of the Cantabrian coast (N and NW Iberia) was also deformed.

There was some deformation in the Catalan Coastal Range, with approximately N–S compression (continuing into part of the Oligocene) (Guimerà, 1983, 1988) and important development of

NE–SW sinistral strike–slip faults (Anadón *et al.*, 1985). The crust that was to become the Iberian Cordillera from the Oligocene onwards (Viallard, 1989) was also affected.

These compressional episodes also had an effect on the Spanish Central System during the Eocene, causing crustal thickening (Vegas *et al.*, 1990) and reactivation of Late Hercynian faults.

Tectonic evolution during the Oligocene (Fig. 4B)

Some of the processes active in the Eocene continued in the Oligocene. Thus, most of the nappe structure visible at present in the Betic Internal Zone formed at this time. Tectonic instability caused deformation and unconformities in the Betic External Zone.

Basically N–S or NW–SE compressions continued in the Pyrenees (Megías, 1988) (Oligocene phase), causing more displacements of nappes and southward migration of the foreland (Ebro Basin).

The deformation occurring in the Cantabrian zone (NW Iberia) and the Bay of Biscay was much weaker during the Oligocene (and later in the Neogene) than in the Eocene (Boillot *et al.*, 1984).

However, a new phenomenon developed during the Oligocene, adding to the convergence of Africa and Europe. This was the approximately E–W extension affecting Central Europe, which reached the Western Mediterranean through the Rhône Valley (Boillot *et al.*, 1984). The present Gulf of Lion opened in NE Iberia and the eastward displacement and gradual counter-clockwise rotation of Corsica and Sardinia began. Further south the Valencia Trough opened, which in turn caused eastward displacement and gradual rotation of the Balearic Isles.

This extensional process in the Western Mediterranean (with the initial opening of the Valencia Trough, Fontboté *et al.*, 1990; Maillard *et al.*, 1992), together with the constant eastward drift of Iberia due to Atlantic opening, compressed the eastern sector of Iberia (with an approximately E–W compression direction). Particularly during the Late Oligocene (Viallard, 1985; Simón, 1990) and until the earliest Miocene, this gave rise to the structuring of the Iberian Cordillera, which is a clearly intracontinental cordillera and presents a generally moderate degree of deformation.

Within the Iberian Massif the reactivation of orginally late Hercynian faults, which determined most of the Alpine structure of the Spanish Central System, also to a large extent controlled formation of the Duero and Tagus Basins, as well as many other smaller ones (Vegas, 1975). The age of these structures is not accurately known. Although in general terms we can accept that some deformation began to occur at the end of the Cretaceous (Vegas *et al.*, 1986), the first well-developed episode happened during the Oligocene–Early Miocene (Capote *et al.*, 1990), which coincides with the deformation of the Iberian Cordillera.

Tectonic evolution during the Neogene (Fig. 6)

The extension that began in the Oligocene in the Western Mediterranean continued to develop throughout the Neogene. The eastward migration and counter-clockwise rotation of Corsica and Sardinia took place as the Provençal Basin opened, with formation of oceanic crust (Rehault *et al.*, 1984). However, opening of the Valencia Trough did not significantly continue beyond the Early Miocene.

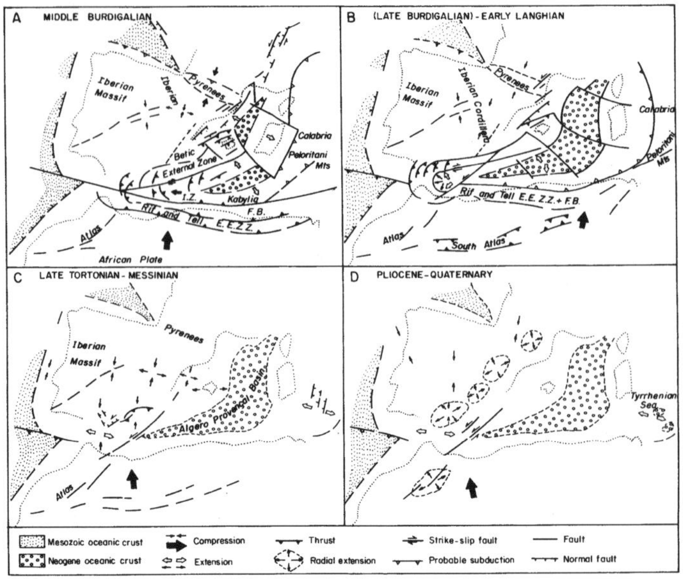

Fig. 6. Reconstruction of the main features of the tectonic evolution of Iberia during the Neogene. A: Middle Burdigalian. B: (Late Burdigalian) – Langhian. C: Late Tortonian–Messinian. D: Pliocene–Quaternary.

The Provençal Basin extended southwards, developing a new approximately E–W direction and thus forming the Algerian–Provençal Basin, which also contained oceanic crust. The westward continuation of the Algerian Basin caused opening of the Alboran Sea or Basin (Sanz de Galdeano, 1990a), in which oceanic crust was not clearly created, although the continental crust is noticeably thinned (Fig. 5B). Boillot *et al.* (1984) consider the Algerian Basin to be a back-basin connected to possible subduction of Africa.

This extensional process and the formation of the Algerian Basin were linked with compression on the margins. Towards the south it caused the formation of the Kabylias nappes in the Tell Cordillera in Algeria (not dealt with here). To the north and west it led to the mainly westward expulsion of the Betic–Riffian internal zones, which were originally some 500 km east of their present position according to the interpretations of Andrieux *et al.* (1971), Durand Delga (1980), Durand Delga and Fontboté (1980), Sanz de Galdeano (1983 and 1990a), Wildi (1983), Maldonado (1985), Martín Algarra (1987), and Sanz de Galdeano and Rodríguez-Fernández (Chapter S2), etc. The external zones of both the Betic Cordillera and the Rif were deformed by the advance of these internal zones, beginning towards the end of the Early Burdigalian. The Betic foreland basin (Guadalquivir Basin, formerly North Betic Strait) was then formed and the numerous basins in the Betic Cordillera also began to form at this time (Sanz de Galdeano and Vera, 1992).

Further east, compression also took place during part of the Burdigalian and Langhian in the Balearic Isles. This was transmitted to the Valencia Trough, where some tectonic inversion occurred and overthrusts formed (Fontboté *et al.*, 1990; Banda and Santanach, 1992) (Fig. 5C).

The compression also affected the Pyrenees and some overthrusts formed. However, from the Middle Miocene onwards, NE–SW sinistral strike–slip faults (Cabrera *et al.*, 1988) led to the formation of basins such as the Cerdaña Basin, cutting the Pyrenean structural trend.

Some approximately N–S compression also occurred in the Iberian Cordillera during part of the Miocene (Simón Gómez and Paricio Cardona, 1988). From the Middle–Late Miocene onwards, there was simultaneous WNW–ESE extension (Moissenet, 1989) or radial extension, particularly from the Pliocene onwards (Simón Gómez, 1990), and approximately N–S compression.

NW–SE to NNW–SSE compression occurred throughout the Early and Middle Miocene in the Spanish Central System. Between the Late Miocene and the Quaternary its direction changed to N–S (Capote *et al.*, 1990), causing reverse faults with imbricate thrusts (de Vicente *et al.*, 1992) and strike–slip faults, which contributed to the uplift of the system and clearer separation of the Duero and Tagus Basins, as well as forming many small basins.

Compressions of NW–SE to almost N–S in direction are also detected in Portugal, at least from the Tortonian on.

Finally, mention must be made of the process of regional uplift with radial extension described by Simón Gómez in the Iberian Cordillera (and which may also have affected the Valencia Trough). This process also affected most of the Betic Cordillera from the Tortonian and more especially the Pliocene onwards (Sanz de Galdeano and López Garrido, 1991). Some of this was associated with a N–S compression, although the mechanism causing it is not clear.

Summary

Trapped between the EuroAsiatic and African plates, Iberia underwent very important deformation throughout the Tertiary. The opening of the Atlantic caused general displacement towards the east, but also led to N–S convergence. During the Eocene, compression produced important deformation in the Pyrenees and the Bay of Biscay. This deformation continued in the Oligocene, but the opening of new basins in the Mediterranean led to compression and formation of the Iberian Cordillera. The Betic–Riffian Internal Zones, which had been deformed throughout the Palaeocene, were expelled westward towards the end of the Oligocene and particularly in the Neogene, when they reached their present position. From the Late Miocene on, regional uplift with radial extension coexisted with approximately N–S compressions in at least the Iberian and Betic Cordilleras.

These Tertiary deformations also affected the interior of the Iberian Massif, especially the Spanish Central System, and controlled the evolution of the inner basins. The Ebro and Guadalquivir Basins were clearly characterised as foreland basins.

Acknowledgments

This paper is a contribution by Project PB91–0079 (DIGCYT) and Working Group 4085 Análisis y dinámica de cuencas (Junta de Andalucía).

References

Anadón, P., Cabrera, L., Guimerà, J. and Santanach, P. (1985). Paleogene strike–slip deformation and sedimentation along the southeastern margin of the Ebro Basin. In *Society of Economic Paleontologists and Mineralogists*. Special Publ. no. 37, *Strike–slip deformation, basin formation and sedimentations*. K.T. Biddle and N. Christie-Blick (eds.), pp. 303–318.

Andrieux, J., Fontboté, J.M. and Mattauer, M. (1971). Sur un modèle explicatif de l'Arc de Gibraltar. *Earth Planet. Sci. Lett.*, **12**, 191–198.

Banda, E. and Santanach, P. (1992). The Valencia trough (western Mediterranean): an overview. *Tectonophysics*, **208**, 183–202.

Biju-Duval, B., Dercourt, J. and Le Pichon, X. (1977). From the Tethys to the Mediterranean seas: a plate tectonic model of the evolution of the Western Alpine System. In B. Biju-Duval and L. Montadert (eds.), *Structural history of the Mediterranean Basin*. Technip, Paris, pp. 143–164.

Boillot, G., Montadert, L., Lemoine, M. and Biju-Duval, B. 1984. *Les margins continentales actuelles et fossiles autour de la France*. Masson, Paris, 342 pp.

Cabrera, L., Roca, E. and Santanach, P. (1988). Basin formation at the end of a strike–slip fault: the Cerdanya Basin (eastern Pyrenees). *J. Geol. Soc.*, London, **145**, 261–268.

Capote, R., de Vicente, G. and González Casado, J.M. (1990). Evolución de las deformaciones alpinas en el Sistema Central Español (S.C.E.). *Geogaceta*, **7**, 20–22.

De Jong K. (1991). Tectono-metamorphic studies and Radiometric dating in the Betic Cordilleras (SE Spain), with implications for the dynamics of extension and compression in the western Mediterranean area. Thesis, University of Amsterdam, 204 pp.

De Vicente, G., González Casado, J.M., Bergamín, J.F., Tejero, R., Babín, R., Rivas, A., Enrile, H.J.L., Giner, J., Sánchez Serrano, F., Muñoz, A. and Villamor, P. (1992). Alpine structure of the Spanish Central System. III *congreso Geológico de España.*, **1**, 284–288.

Dercourt, J., Zonenshain, L.P., Ricou, L.E., Kazim, V.G., Le Pichon, X., Knipper, A.L., Grandjaquet, C., Sbortshikow, I.M., Geyssant, J., Lepvrier, C., Pechersky, D.H., Boullin, J., Sibuet, J.C., Savostin, L.A., Sorokhtin, O., Westphal, M., Bazhenov, M.L., Lauer, J.P. and Biju-Duval, B. (1986). Geological evolution of the Tethys belt from the Atlantic to the Pamirs since the Lias. *Tectonophysics*, **123**, 241–315.

Ducasse, L., Muller, J. and Velasque, P.C. (1986). La chaîne pyrénéo-cantabrique: subduction hercynienne, rotation crétacée de l'Ibérie et subductions alpines différentielles. *C.R. Acad. Sci. Paris*, **303**, (5), 419–423.

Durand-Delga, M. (1980). La Méditerranée occidentale: étape de sa genèse et problèmes structuraux liés à celle-ci. *Livre Jubilaire de la Soc. Géol. de France*, 1830–1980.

Durand-Delga, M. and Fontboté, J.M. (1980). Le cadre structural de la Méditerranée occidentale. 26 Congrès. Géol. Internatl., Paris. Les Chaînes alpines issues de la Téthys. *Mém. B.R.G.M.*, **115**, 67–85.

Fontboté, J.M., Guimerà, J., Roca, E., Sabat, F., Santanach, P. and Fernández-Ortigosa, F. (1990). The Cenozoic geodynamic evolution of the Valencia trough (Western Mediterrean). *Rev. Soc. Geol. España*, **3** (3–4), 249–259.

Guimerà, J. (1983). Evolution de la déformation alpine dans le NE de la Chaîne Ibérique et dans la Chaîne Côtière Catalane. *C.R. Acad. Sci. Paris*, **297**, 425–430.

Guimerà, J. (1988). Estudi estructural de l'enllaç entre la Serralada Iberica i la Serralada Costanera Catalana. Thesis, University of Barcelona. 2 vols., 600 pp. (unpublished).

Julivert, M., Fontboté, J.M., Ribeiro, A. and Conde, I. (1972). *Mapa tectónico de la Península Ibérica y Baleares a escala 1:1000.000*. IGME.

Maillard, A., Mauffret, A., Watts, A.B., Torné, M., Pascal, G., Buhl, P. and Pinet, B. (1992). Tertiary sedimentary history and structure of the Valencia trough (western Mediterranean). *Tectonophysics*, **203**, 57–75.

Maldonado, A. (1985). Evolution of the Mediterranean Basins and a detailed reconstruction of the Cenozoic paleogeography. In Margalet, R. (ed.), *Western Mediterranean: key environments*. Pergamon Press, pp. 17–59.

Malod, J.A. (1989). Ibérides et plaque ibérique. *Bull. Soc. Géol. France*, **8** (5), 927–934.

Martí, J., Mitjavila, J., Roca, E. and Aparicio, A. (1992). Cenozoic magmatism of the Valencia trough (western Mediterranean): relationship between structural evolution and volcanism. *Tectonophysics*, **203**, 145–165.

Martín Algarra, A. (1987). Evolución geológica alpina del contacto entre las zonas Internas y las zonas Externas de la Cordillera Bética (Sector Occidental). Thesis, University of Granada, 1308 pp.

Megías, A.G. (988). La tectónica pirenaica en relación con la evolución alpina del margen noribérico. *Rev. Soc. Geol. España.*, **1** (3–4), 365–372.

Moissenet, E. (1989). Les fossés néogènes de la Chaîne ibérique: leur évolution dans le temps. *Bull. Soc. Géol. France*, **8**, (5), 919–926.

Muñoz, J.A. (1992). Evolution of a continental collision belt: ECORS Pyrenees crustal balanced cross-sections. In K.R. McClay (ed.), *Thrust Tectonics*, Chapman and Hall, London, pp. 235–246.

Muñoz, J.A., Casas, J.M., Martínez, A. and Vergés, J. (1988). An introduction to the structure of the Southeastern Pyrenees, the Ter-Freser cross-section. *Symposium on the Geology of the Pyrenees and Betics* (Barcelona), Excursion guide. 86 pp.

Olivet, J.L., Bonnin, J., Beuzart, P. and Auzende, J.M. (1982). Cinématique des plaques et paléogéographie: une revue. *Bull. Soc. Géol. France*, **7** (24), 875–892.

Rehault, J.P., Boillot, G. and Mauffret, A. (1984). The Western Mediterranean Basin geological evolution. *Marine Geology*, **55**, 447–477.

Roure, F., Choukroune, P., Berastegui, X., Muñoz, J.A., Villien, A., Matheron, P., Bareyt, M., Séguret, M., Camara, P. and Déramond, J. (1989). ECORS deep seismic data and balanced cross sections: geometric constraints on the evolution of the Pyrenees. *Tectonics*, **8**, 41–50.

Sanz de Galdeano, C. (1983). Los accidentes y fracturas principales de las Cordilleras Béticas. *Estud. Geol.*, **39**, 157–165.

Sanz de Galdeano, C. (1990a). Geologic evolution of the Betic Cordilleras in the Western Mediterranea, Miocene to the present. *Tectonophysics*, **172**, 107–119.

Sanz de Galdeano, C. (1990b). Le prolongación hacia el sur de las fosas y desgarres del Norte y Centro de Europa: Una propuesta de interpretación. *Rev. Soc. Geol. España*, **3** (1–2), 231–241.

Sanz de Galdeano, C. and López Garrido, A.C. (1991). Tectonic evolution of the Malaga basin (Betic Cordillera). Regional implications. *Geodinam. Acta*, **5**, 173–186.

Sanz de Galdeano, C. and Vera, J.A. (1992). Stratigraphic record and palaeogeographical context of the Neogene basins in the Betic Cordillera, Spain. *Basin Res.* **4**, 21–36.

Simón Gómez, J.L. (1989). Late Cenozoic stress field and fracturing in the Iberian Chain and Ebro Basin (Spain). *J. Struct. Geol.*, **11** (3), 285–294.

Simón Gómez, J.L. (1990). Algunas reflexiones sobre los modelos tectónicos aplicados a la Cordillera Ibérica. *Geogaceta*, **8**, 123–129.

Simón Gómez, J.L. and Paricio Cardona, J. (1988). Sobre la compresión neógena en la Cordillera Ibérica. *Estud. Geol.*, **44** (3–4), 271–283.

Suriñach, E., Marthelot, J.M., Gallart, J., Daignières, M. and Hirn, A. (1993). Seismic images and evolution of the Iberian crust in the Pyrenees. *Tectonophysics*, **221**, 67–80.

Tapponier, P. (1977). Evolution tectonique du système alpin en Méditerranée: poinçonnement et écrasement rigide-plastique. *Bull. Soc. Géol. France*, **7** (19), 437–460.

Uchupi, E. (1988). The Mesozoic–Cenozoic geologic evolution of Iberia. A tectonic link between Africa and Europe. *Rev. Soc. Geol. España*, **1** (3–4), 257–294.

Vegas, R. (1975). Wrench (transcurrent) fault system of the southwestern Iberian Peninsula: paleogeographic and morphostructural implications. *Geol. Rundschau*, **64**, 266–278.

Vegas, R., Vázquez, J.T. and Marcos, A. (1986). Tectónica alpina y morfogénesis en el Sistema Central español: Modelo de deformación intracontinental distribuida. *Geogaceta*, **1**, 24–25.

Vegas, R., Vázquez, J.T., Suriñach, E. and Marcos, A. (1990). Model of distributed deformation, block rotations and crustal thickening for the formation of the Spanish Central system. *Tectonophysics*, **184**, 367–378.

Velasque, P.C., Ducasse, L., Muller, J. and Scholten, R. (1992). The influence of inherited extensional structures on the tectonic evolution of an intracratonic chain: the example of the Western Pyrenees. *Tectonophysics*, **162**, 243–264.

Viallard, P. (1985). Ibérides et Ibérie: un exemple de relations entre tectogenèse intracontinentale et tectonique des plaques. *C.R. Acad. Sci. Paris*, **300** (6), 217–222.

Viallard, P. (1989). Décollement de couverture et décollement médio-crustal dans une chaîne intraplaque: variations verticales du style tectonique des Ibérides (Espagne). *Bull. Soc. Géol. France*, **8** (5), 913–918.

Wildi, W. (1983). La chaîne tello-rifaine (Algérie, Maroc, Tunisie): structure, stratigraphie et évolution du Trias au Miocène. *Rev. Géol. Dyn. Geogr. Phys.*, **24**, 201–297.

G4 Deep crustal expression of Tertiary basins in Spain

E. BANDA

Abstract

The Variscan crust of the Iberian massif is 32–35 km thick, although this thins towards the Atlantic margin. To the north and south in the Pyrenees and Betics, the crustal structure is much more complex, but a crustal thickness of up to 50 km is well established in the Pyrenees. In the related foreland basins (Ebro and Guadalquivir) lithospheric flexing appears to have taken place without significant modification of the crust. The Neogene Valencia trough and Alboran basins formed by thinning of the crust, and minimum thicknesses of 15 km or less have been measured in each case.

Introduction

The shallow structure and stratigraphy of the Tertiary basins of Spain are relatively well known from industrial exploration and field geology. However, because sedimentary basins reflect deformation processes on a lithospheric scale, knowledge of the structure and mechanical properties of the lithosphere in which basins develop is essential for a complete understanding of basin evolution. Therefore, a brief introduction to lithospheric structure, which will be limited to the crust and uppermost mantle, in relation to Tertiary basins in Spain is presented.

The Tertiary basins of Spain (Fig. 1a) have generally formed by deformation of the Variscan (Hercynian) crust of the Iberian Massif. The crustal structure of Iberia has been extensively explored by seismic methods including refraction/wide-angle reflection and near-vertical deep reflection (Fig. 1b). Consequently, the basement of the Tertiary basins in Spain, i.e., the crystalline crust of pre-basin formation, can be considered to be reasonably well known. In some instances Mesozoic modification of the crust linked with Mesozoic basin development may have been significant and needs to be taken into account when studying Tertiary basins.

Foreland basins are essentially mechanical basins formed by the loading and downsagging of the lithosphere in front of orogenic belts, their width and depth being controlled by the flexural rigidity of the lithosphere and, to a lesser extent, by the load configuration. In Spain major foreland basins are the Ebro and Guadalquivir basins, associated with the Pyrenees and Betic Alpine chains respectively (Fig. 1a). The Ebro foreland basin seems to have been influenced also by deformation and loading of the Iberian Chain and Catalan Coastal Ranges (Desegaulx and Moretti, 1988; Zoetemeijer *et al.*, 1990). The Betic foreland extends to the north of the Betic metamorphic complexes, and possibly also offshore to the west of Gibraltar linked with the east–west emplacement of the Gibraltar thrust (Balanyá and García-Dueñas, 1987).

Offshore basins in the Atlantic developed essentially during Mesozoic times as passive margins and will not be discussed here. The same applies to the Cantabrian (Bay of Biscay) margin where limited subduction of oceanic lithosphere may have taken place in the late Mesozoic and Early Tertiary (Boillot *et al.*, 1979).

Offshore basins in the Mediterranean, developed in the Tertiary, namely the Valencia trough and the Alboran sea (Fig. 1a), have also been extensively studied (see Chapters E2, E3 and S10). It has been shown that these basins correspond to rift-type basins, although they carry a particular imprint related to the recent tectonic history of the Mediterranean region.

Major interior basins, such as the Duero and Tagus basins (Fig. 1a), correspond to intraplate settings that show moderate Alpine deformation. They have little deep expression, as expected from their moderately compressional origin.

The Iberian crust

The crust of the Iberian Massif (Variscan) is easily characterized, from the available deep seismic studies, by its upper, middle and lower crust (Fig. 2). The upper crust has an average *P*-wave velocity of 6.0 km/s with very little variation around this value. Available near-vertical deep reflection surveys show a fairly transparent upper crust in the Variscan external zones of Cantabria. In this area upper crustal reflections are correlated with thin-skinned tectonics related to the building of the Variscan chain (Perez-Estaún *et al.*, 1994). The base of the upper crust is distinguished in refraction and wide-angle reflection surveys at about 12–15 km depth, where the velocity increases to 6.3–6.5 km/s. The lower crust is well recognized with both methods. The top is found around 20–22 km and the bottom (Moho) at 32–35 km depth, shallowing towards the Atlantic margin where minimum values of

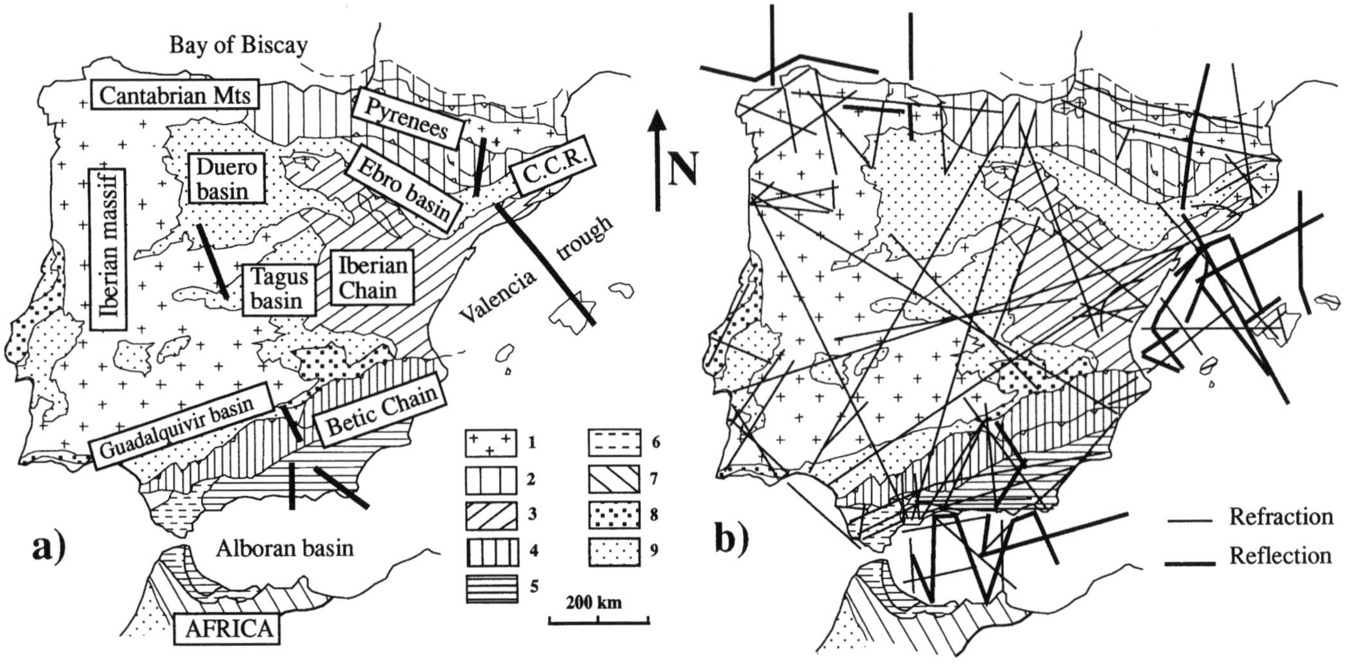

Fig. 1. (a) Main Tertiary basins and geological units of the Iberian Peninsula. 1: Hercynian basement; 2: Deformed Mesozoic cover of the Pyrenean chain; 3: Mesozoic of the Iberian Chain; 4: Mesozoic external units of the Betics; 5: Internal units of the Betic-Rif realm; 6: Flysch units of the Gibraltar arc; 7: Mesozoic of the African margin; 8: Undeformed Mesozoic cover; 9: Tertiary onshore basins. Thick lines indicate the location of crustal sections shown in Figs. 3 and 4. C.C.R. indicates the Catalan Coastal Ranges (after Vegas and Banda, 1982). (b) Approximate location of most refraction and wide-angle reflection (thin lines), and deep near-vertical reflection (thick lines) surveys in Iberia.

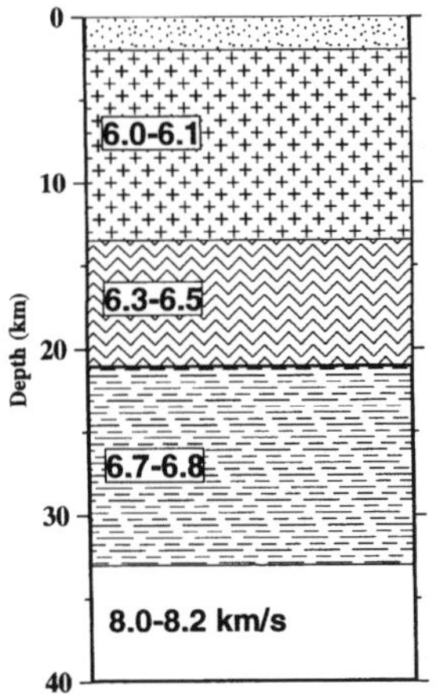

Fig. 2. Crustal column of the Hercynian crust in the Iberian Peninsula. The column cannot be taken as a geological image but rather as an average of velocity (V_p in km/s) distribution and thicknesses of the upper, middle and lower crust as inferred from refraction and wide-angle reflection techniques.

28 km have been recorded (Córdoba *et al.*, 1988; ILIHA DSS Group, 1993). The average velocity of the lower crust is around 6.8 km/s although local deviations have been detected. The upper mantle in the whole of the Iberian Peninsula displays normal *P*-wave velocities around 8.1 km/s, except in the Mediterranean borders and basins where it can reach values as low as 7.7 km/s.

In contrast to a well-characterized Variscan crust in the Iberian Massif, the crust of the Alpine chains is more complex and therefore difficult to explore by seismic methods. Thickening of the crust in the Pyrenees up to 50 km is well established, while the internal structure, which is highly heterogeneous, is only known in detail along a few sections. The southern foreland basin, the Ebro basin, however, has been well imaged by deep near-vertical reflection surveys (ECORS-Pyrenees Team, 1988; Gallart *et al.*, 1994). Likewise, the metamorphic complexes of the Betic chain show a complex internal structure. Crustal thickening is here more difficult to assess from available wide-angle and near vertical reflection studies (Banda *et al.*, 1993; García-Dueñas *et al.*, 1994). Moreover, the structure of the external Betics, the Mesozoic paleomargin and Guadalquivir foreland basin is not well known at depth due to scarcity of data. Abrupt thinning towards the Mediterranean has been observed from refraction data and in a recent study that combines seismic and gravity data (Torné and Banda, 1992) (Fig. 4).

Foreland basins

The Ebro foreland basin was formed during the Paleogene as a response to the formation of the Pyrenees. The basin is located between the South Pyrenean Zone and the northern border of the Iberian Chain and Catalan Coastal Ranges (Fig. 1a). Seismic reflection and well data show that the basin deepens towards the Pyrenees, with maximum sediment thicknesses close to the South

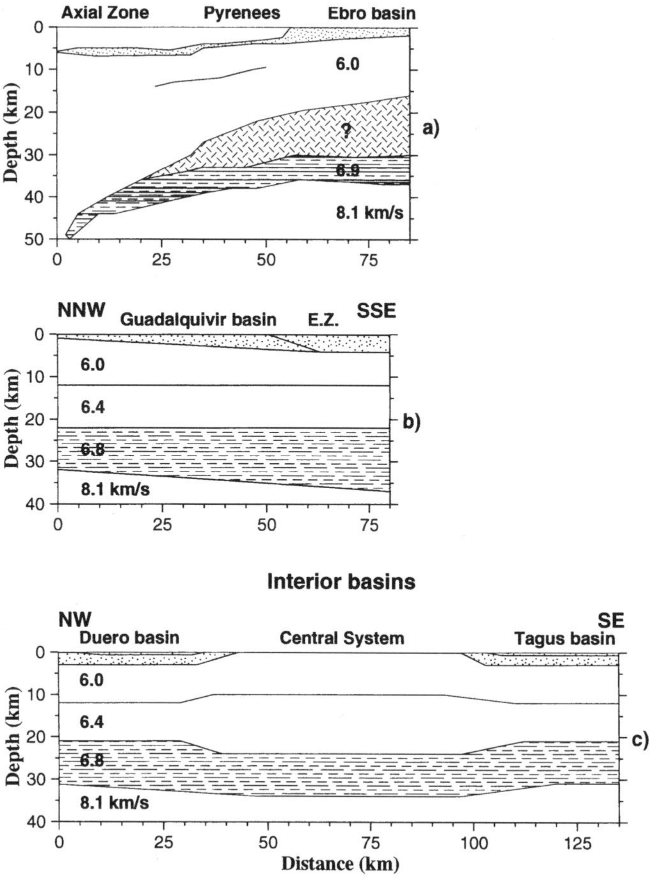

Fig. 3. Crustal and velocity structure (V_p in km/s) of the Spanish foreland basins and interior basins. (a) Ebro basin. Combined refraction and near-vertical reflection image of the crust. Note thickening of the crust to about 50 km under the Axial Zone of the Pyrenees. The lower crust is tentatively divided into two main units. Both of them appear as reflective in reflection surveys. (Mainly from ECORS Pyrenees Team, 1988 and Banda and Berástegui, 1990.) (b) Guadalquivir basin. Seismic section from refraction and reflection surveys (Banda *et al.*, 1993; García-Dueñas *et al.*, 1994). The lower crust is moderately reflective. E.Z.: External zones of the Betics. (c) Duero and Tagus basins north and south of the Central System respectively (modified from Suriñach and Vegas, 1988).

Pyrenean Frontal Thrusts (Fig. 3a). The basin's deep structure does not differ essentially from the description of the characteristic Variscan crust of the Iberian Massif discussed above. In contrast, its shallow structure is complex and controlled by the southward emplacement of thrust sheets. The deep crust shows a reflective character with its top dipping to the north at a rather shallow depth, about 7 s TWT (two-way time) on average (21 km), and is about 5 s thick (16 km), drastically thinning to the north. Banda and Berástegui (1990) interpreted, on the basis of seismic fabrics, that the lower crust is formed by different units (Fig. 3a). The upper part of the lower crust is thought to display detachments that may have been active during Mesozoic extension and subsequent Alpine compression. The ascription of these units to the lower crust is, however, at variance with other interpretations.

The Guadalquivir basin is a young, asymmetrical basin striking ENE–WSW that developed in front of the Betic chain. Commercial seismic profiles reveal NW–SE dipping events within the sedimentary sequence, which attains maximum thicknesses of about 4000–4500 m (Fig. 3b). To the west maximum sedimentary thickness is located under the Betic front while eastward it is progressively overthrust by the Betic nappes. Middle Miocene to Recent sediments constitute the autochtonous sedimentary infill of the basin.

The northernmost part of the only available deep seismic reflection profile (close to the Guadalquivir river) displays a typical Variscan crust with its reflective lower crust dipping south and a Moho depth of about 35 km (Fig. 3b). Similar to the Ebro basin, the formation of the Guadalquivir basin seems to have been caused by lithospheric flexing without significant modification of the crust.

Interior basins

The Duero and Tagus basins are Tertiary basins separated by the Central System, which is an outcrop of the Hercynian basement, slightly uplifted and overthrust during moderate Alpine shortening deformation (Fig. 3c). It is likely that both basins were created during this compressional event.

Investigation of crustal structure in the interior basins of Duero and Tagus has only been attempted by refraction and wide-angle reflection techniques. The results correspond to the description of the Variscan crust given above and show a fairly constant crustal thickness, indicating little or no effect of basin formation at crustal levels. This may be explained by the relatively thin sedimentary cover as well as by the intraplate compressional character. However, the southern border of the Duero basin seems to be affected by a moderate crustal thickening of about 2 km (Suriñach and Vegas, 1988) (Fig. 3c). According to these authors, thickening of the crust under the Central System would favour a localized Alpine reorganization affecting the whole crust.

The intracrustal discontinuity at about 12–15 km depth, detected as the transition from the upper crust to the middle crust in wide-angle surveys, has been interpreted in some places as a detachment surface. Unfortunately, deep near-vertical seismic reflection results are not available yet in the interior basins to confirm or rule out that interpretation.

Rift-type basins

The Valencia trough and Alboran basins are Neogene basins formed by thinning of the continental crust because of the Neogene to Recent extensional tectonics that affected, in different ways, the Western Mediterranean. They differ from other Mediterranean basins (e.g., the Tyrrhenian and Provençal basins) in that extension was not sufficient to have generated oceanic crust.

The crustal structure of the Valencia trough is well known from a number of geological and geophysical studies that have been carried out during the last decade (Chapters E2 and E3). The Neogene–Quaternary sedimentary cover is well developed along its western border (e.g., the Ebro Platform) and the centre of the trough, whereas on its eastern border, the Balearic promontory,

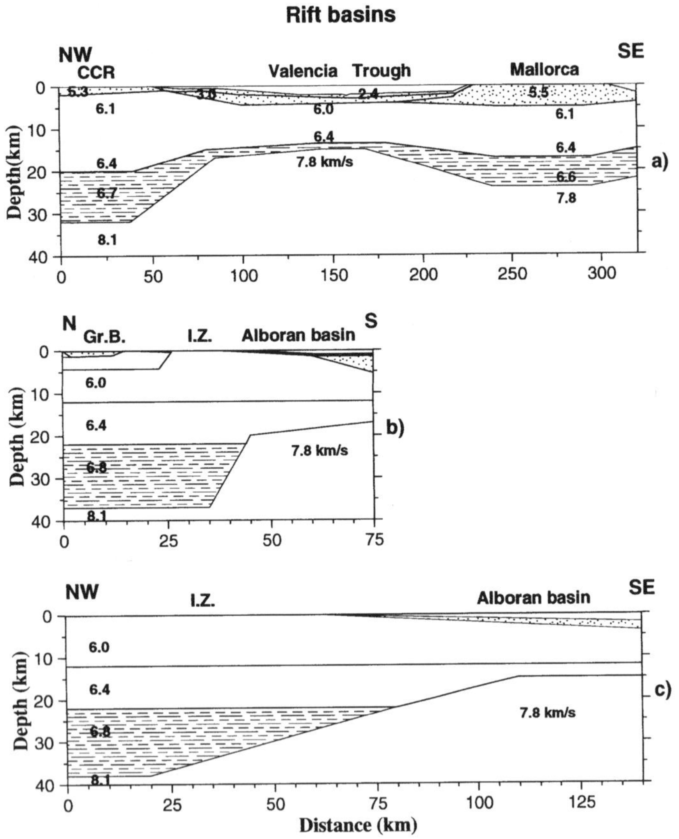

Fig. 4. Crustal and velocity structure (V_p in km/s) of the Valencia trough and Alboran basin. (a) Valencia trough. Note the differential crustal thinning between the upper–middle crust and the lower crust. CCR: Catalan Coastal Ranges. (b) Abrupt crustal thinning from the central Betics to the Alboran basin. Gr.B.: Granada Basin; I.Z.: Internal zones of the Betics. (c) Gentle crustal thinning from the eastern Betics to the Alboran Sea. (Mainly from Torné *et al.*, 1992 and Torné and Banda, 1992).

these deposits are dramatically reduced or are absent. The upper-middle crust is of variable thickness (6–9 km) with an average *P*-wave velocity of 6.2 km/s. The lower crust is highly reflective along the Ebro platform, thinning dramatically towards the axis of the trough yielding a total crustal thickness of less than 14 km (Fig. 4a). A low upper-mantle velocity of 7.7–7.8 km/s is widespread beneath the basin indicating the presence of an anomalous uppermost mantle or a transitional crust-mantle zone.

The crustal structure asymmetry has also been observed at lithospheric scales (Watts and Torné, 1992a). These authors proposed that the eastern margin is a constructional margin that is underlain by a broad region of lithospheric thinning. They also conclude that simple stretching models can explain the thermal and subsidence history of the northern part of the trough, although the same model fails to explain the observed data in the southern end of the area. This is basically due to the problem posed by the lack of knowledge of the pre-rift lithospheric structure (Banda and Santanach, 1992). From three-dimensional backstripping studies, Watts and Torné (1992b) found that the lithosphere has acquired little or no strength since the time of rifting.

The development of the Alboran basin since the Early Miocene appears to have involved major lithospheric thinning. Here, the lithosphere corresponds to that of the Alboran crustal domain. Compared with the Valencia trough, its deeper structure is not well known. The basin appears as a segmented basin with structural highs and local (small) sub-basins where the sediment thickness can be as much as 7 km.

Available seismic data in the Alboran sea indicate an average crustal thickness of 15 km in the central part of the basin based on a few studies (Boloix and Hatzfeld 1977; Hatzfeld, 1976; among others). As in the Valencia trough area, the uppermost mantle velocities are in the range of 7.6–7.8 km/s, which again points to the presence of an anomalous uppermost mantle structure (Fig. 4b,c). Torné and Banda (1992), using gravity and seismic data, have studied the style of crustal thinning from the thickened crust of the Betics to the attenuated crust beneath the basin. These authors conclude that thinning in the western and central part of the basin occurs in an area 15 to 30 km wide, while to the east the thinning occurs in a wider area of about 100 km (Fig. 4b,c). While the latter crustal attitude resembles that of rifted passive margins, the steep rise of the Moho in the western and central part of the basin could correspond to a model of extensional collapse as proposed by Platt and Vissers (1989). Alternatively, it can also be compared with margins originated by strike–slip mechanisms (Scrutton, 1982), although in the absence of oceanic crust, there would have to be some crustal thinning due to detachment extensional tectonics (Balanyá and García-Dueñas, 1987). The difference in styles of thinning agrees with subsidence studies that show a different pattern of tectonic subsidence from east to west (Docherty and Banda, 1992). However, with the available data we cannot reach a conclusion about the origin of crustal thinning.

Acknowledgements

This manuscript has been improved through critical reading by Penny Barton, Craig Docherty, Manel Fernández, Maria J. Jurado and Montse Torné, to whom I am very grateful. This work has been partly funded by NATO grant GRG/890570.

References

Balanyá, J.C. and García-Dueñas, V. (1987). Les directions structurales dans le Domain d'Alboran de part et d'autre du Detroit de Gibraltar. *Comptes Rendues Academie de Sciences de Paris*, **304**: 929–933.

Banda, E. and Berastegui, X. (1990). Seismic fabrics in deep reflection profiles. In B. Pinet and C. Bois (eds.), *The potential of hydrocarbon exploration*, Editions Technip, Paris, pp. 73–75.

Banda, E. and Santanach, P. (1992). The Valencia trough (western Mediterranean): an overview. *Tectonophysics*, **208**: 183–202.

Banda, E., Gallart, J., Dañobeitia, J.J. and Makris, J. (1993). Lateral variations of the crust in the Iberian Peninsula. New evidence from the Betic Cordillera. *Tectonophysics*, **221**: 53–66.

Boillot, G., Auxietre, J.L., Dunand, J.P., Dupeuble, P.A. and Mauffret, A. (1979). The northwestern Iberian margin: a Cretaceous passive margin deformed during Eocene. In M. Talwani, W. Hayet and

W.B.F. Ryan (eds.), *Deep drilling results in the Atlantic Ocean: continental margin and paleoenvironmental*, AGU, Washington, DC, pp. 138–153.

Boloix, M. and Hatzfeld, D. (1977). Preliminary results of the 1974 seismic refraction experiment in the Alboran Sea. *Publication Institute of Geophysics Polish Academy of Sciences*, **A-4** (115): 365–368.

Córdoba, D., Banda, E. and Ansorge, J. (1988). *P*-wave velocity-depth distribution in the Hercynian crust of northwest Spain. *Physics of the Earth and Planetary Interiors*, **51**: 235–248.

Desegaulx, P. and Moretti, I. (1988). Subsidence history of the Ebro basin. *Journal of Geodynamics*, **10**: 9–24.

Docherty, J.I.C. and Banda, E. (1992). A note on the subsidence history of the northern margin of the Alboran basin. *Geo-Marine Letters*, **12**: 82–87.

ECORS Pyrenees Team (1988). The ECORS deep reflection seismic survey across the Pyrenees. *Nature*, **331**: 508–511.

Gallart, J., Vidal, N. and Dañobeitia J.J. (1994). Lateral variations in the deep crustal structure at the northeastern Iberian margin imaged from reflection seismic methods. *Tectonophysics*, **232**: 59–75.

García-Dueñas, V., Banda, E., Torné, M., Córdoba, D. and ESCI-Béticas Working Group (1994). A deep seismic reflection survey across the Betic Chain (Southern Spain): first results. *Tectonophysics*, **232**: 77–89.

Hatzfeld, D. (1976). Etude sismologique et gravimetrique de la structure profonde de la mer d'Alboran: mise en evidence d'un manteau anormal. *Comptes Rendues Academie de Sciences de Paris*, **283**: 1021–1024.

ILIHA DSS Group (1993). Deep seismic sounding investigation of lithospheric heterogeneity and anisotropy beneath the Iberian Peninsula. *Tectonophysics*, **221**: 35–51.

Perez-Estaún, A., Pulgar, J.A., Banda, E., Alvarez-Marrón, J. and ESCI-N Research Group (1994). Crustal structure of the external Variscides in NW Spain from deep seismic reflection profiling. *Tectonophysics*, **232**: 91–118.

Platt, J.P and Vissers, R.L.M. (1989). Extensional collapse of thickened continental lithosphere: a working hypothesis for the Alboran sea and Gibraltar arc. *Geology*, **17**: 540–543.

Scrutton, R.A. (1982). Crustal structure and development of sheared passive margins. In Scrutton, R.A. (ed.), *Dynamics of passive margins. Geodynamic Series*, **6**: 133–140.

Suriñach, E. and Vegas, R. (1988). Lateral inhomogeneities of the Hercynian crust in central Spain. *Physics of the Earth and Planetary Interiors*, **51**: 226–234.

Torné, M. and Banda, E. (1992). Crustal thinning from the Betic Cordillera to the Alboran Sea. *Geo-Marine Letters*, **12**: 76–81.

Torné, M., Pascal, P., Buhl, P., Watts, A.B. and Mauffret, A. (1992). Crustal and velocity structure of the Valencia trough (Western Mediterranean). Part I: A combined refraction/wide-angle reflection and near vertical reflection study. *Tectonophysics*, **203**: 325–344.

Vegas, R. and Banda, E. (1982). Tectonic framework and Alpine evolution of the Iberian Peninsula, *Earth Evolution Sciences*, **4**: 320–343.

Watts, A.B. and Torné (1992a). Subsidence history, crustal structure, and thermal evolution of the Valencia trough: a young extensional basin in the Western Mediterranean. *Journal of Geophysical Research*, **97**: 20021–20042.

Watts, A.B. and Torné, M. (1992b). Crustal structure and the mechanical properties of extended continental lithosphere in the Valencia trough. *Journal of the Royal Society of London*. **149**: 813–827.

Zoetemeijer, R., Desegaulx, P., Cloetingh, S., Roure, F. and Moretti, I. (1990). Lithospheric dynamics and tectonic-stratigraphic evolution of the Ebro basin. *Journal of Geophysical Research*, **95**: 2701–2711.

G5 Oil and gas resources of the Tertiary basins of Spain

F. MELÉNDEZ-HEVIA AND E. ALVAREZ DE BUERGO

Abstract

Gas has been produced from the Sub-Pyrenean and Guadalquivir basins, the two Tertiary foreland basins in north and south Spain. Oil and some gas are found in the post-orogenic offshore Gulf of Valencia extensional basin.

General overview

Oil and/or gas have been found in Spain in two different kinds of sedimentary basin: syn-orogenic basins, namely the Sub-Pyrenean and Guadalquivir basins, and in the post-orogenic offshore Gulf of Valencia rift basin. The first two only produce gas, although there are indications that oil-prone source rocks are present, while the last one is an oil province which also contains gas.

The Sub-Pyrenean basin

The Sub-Pyrenean basin is located in northern Spain, along the southern flank of the Pyrenees Mountains, an Alpine thrust belt, formed between late Cretaceous and early Miocene times as a direct consequence of the collision between the Iberian and European plates. Simultaneously, a foreland basin developed south of the new mountain belt and was folded and thrust by the advancing Pyrenean nappes. The main deposits are marine and Eocene in age, and these are covered by continental Oligocene and lower Miocene 'molasse-type' clastics (Fig. 1).

Three sub-basins developed along this margin of the Pyrenees; from west to east these are the Jaca, Tremp and Ampurdán sub-basins. In each of these sub-basins the stratigraphic section and the tectonic style are different. During Eocene times, when the hydrocarbon-bearing deposits were being formed, the basin was open to the west, where the Bay of Biscay had just formed during Cretaceous times. The general structure of the basin consisted of a northern platform attached to the Pyrenees and mainly carbonate, a central trough with deep-water sediments, and a southern edge against the highlands of the stable 'Meseta'. The syn-sedimentary tectonic activity of the Pyrenees produced frequent slides of platform materials into the deep turbiditic central trough, and these formed giant slumps or olistostromes (commonly called 'megabeds').

The *Ampurdán sub-basin* has a very distinct restricted facies, specially within the Armancies Formation (Cuisian, Early Eocene), which is a strongly oil-prone source. This Formation commonly produces oil seeps at the surface, and a few tar mines were very actively mined during the first half of this century, although they have now been abandoned. The few wells drilled in this basin have failed to establish commercial production, although there have been oil shows.

Despite the relatively large number of wells drilled in the *Tremp sub-basin*, only minor shows of gas have been encountered, and commercial activity there has ceased.

In contrast, the *Jaca sub-basin* has attracted a lot of exploration, and this culminated in the discovery of the '*Serrablo*' gas field in 1979 (Fig. 2). The reservoir is a series of submarine shallow-water carbonate slumps that slid from the carbonate platform to the north into the central, deep-water turbiditic trough. In this particular field, two of these 'megabeds' have been mapped and named 'Aurin', at a depth of some 1600–1700 m, and 'Jaca', at some 2500 m. The reservoirs are shallow-water carbonates, fractured and with chaotic structure produced by the slumping. The reserves of the field have been estimated to be 90 BCF (billions of cubic feet) of dry gas. The field has not been completely depleted because it has been transformed into an underground gas storage facility to adjust, seasonally, the gas transfer from Europe through the Pyrenees. The source rocks are dark basinal Eocene shales, several thousand metres thick, which, although not very rich, have enough generating potential to have sourced the field. The structure of the field is fairly complicated, with at least two imbricate thrusts and a broad intervening anticline in the Eocene section (Fig. 3), where the reservoirs are located.

The Guadalquivir basin

The Guadalquivir basin is located in southern Spain, along the northern flank of the Betic Thrust Belt, which formed during Miocene times as a consequence of the collision between the Iberian and African plates. This foreland basin developed north of,

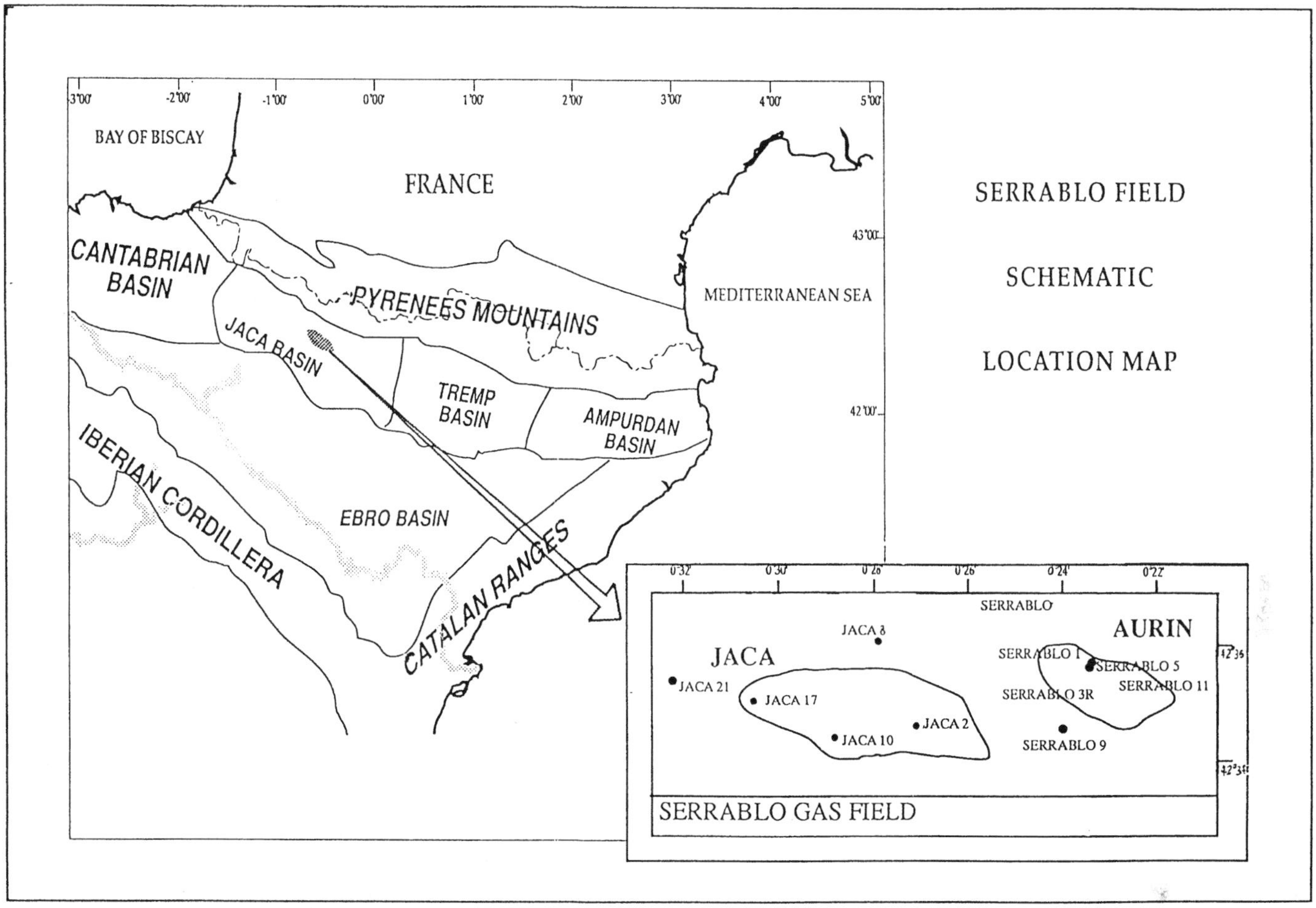

Fig. 1. Sub-Pyrenean basin in northern Spain showing location and details of the 'Serrablo' gas field (inset).

and simultaneously with, the new mountain chain. It acted as the marine connection between the Atlantic and Tethys Oceans during most of the Neogene, and finally became closed during the late Miocene, probably reflecting the so-called 'Messinian event' (Fig. 4).

This basin is a relatively shallow feature, with only some 500 m depth to the basement at its eastern end, increasing westwards to some 1200 m at the present-day shoreline, and extending offshore to slightly greater depths. Its fill is mainly clastics, sourced from the stable 'Meseta' to the north, but also from the advancing Betic nappes to the south, particularly in the upper part of the section that was deposited as the new mountains were rising most actively. The synorogenic character of the basin is demonstrated by the presence of huge masses of olistostrome material derived from the Betic nappes along the southern edge of the basin.

The sands are located mainly at four stratigraphic levels, at the base ('basal transgressive calcarenite'), in the Middle and Late Miocene (Guadalquivir Formation), and in the Pliocene which unconformably overlies the Miocene. The most prolific level is within the Guadalquivir Formation, where a series of shallow marine sand units (formed in beach, shoal, tidal channel and tidal bar environments) passes into deep-water turbidites (Fig. 5).

All discoveries so far are of dry gas only, biogenic in origin. The source is the Miocene shales, several hundreds of metres thick, not very rich, but probably generating gas from the whole section. This basin has not been deformed by strong tectonics and the trapping mechanism is basically stratigraphic. In any locality, it tends to be the uppermost sand body in the section that contains the gas. The limited extent of these sand bodies means that seals are present laterally as well as vertically, and this results in small accumulations and consequently small gas fields. There is no evidence of a possible oil source.

Several discoveries have been made to date. The first ones were offshore in the Gulf of Cádiz, all grouped together as the '*Atlántida*' fields in the early 1970s, and with total combined reserves of some 250 BCF. However, their dispersed character has made them subcommercial and consequently they have not been put into production yet.

These early discoveries led the oil companies to extend the exploration onshore, where despite the shallowness of the basin, several commercial gas fields have been found: '*Marismas*', with some 30 BCF of reserves, is presently producing some 14MUCFD; and '*Las Barreras*', with some 4.5 BCF, '*El Romeral*', with some 4.5 BCF, and '*El Ruedo*', with some 5 BCF, all of them still under

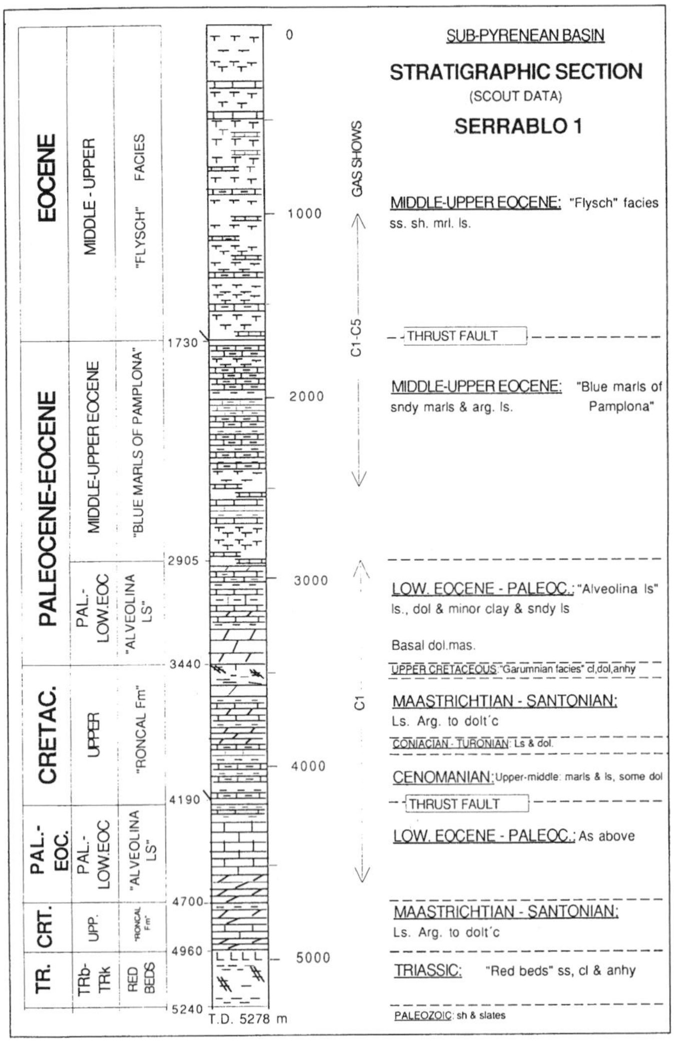

Fig. 2. Stratigraphic section of 'Serrablo #1' well (discovery well, formerly 'Jaca #1', schematic).

evaluation. In these discoveries the high quality of the seismic work has been very important in delineating the geometry of the reservoirs, and in particular the use of the direct detection technique called AVO (amplitude versus offset) has dramatically increased the rate of success (Fig. 5).

Gulf of Valencia

The Gulf of Valencia basin is located off the shore of eastern Spain, in the western Mediterranean. More than 100 hydrocarbon exploration wells have been drilled in this NE–SW elongated, Tertiary post-orogenic basin that is bounded by the Spanish coast to the west and the Balearic islands to the east (Fig. 6). The Spanish mainland coast dips gently below the Mediterranean to a maximum water depth of more than 2000 m in the centre of the basin, where an aborted rift is recognised; in contrast, the Balearic coastal side is abrupt. The Gulf has been influenced by Eocene and Oligocene compressional phases, which were followed by extensional episodes that created several rift sub-basins of decreasing age from north to south. This was part of a European rift system that extended from the North Sea to northern Africa.

The Rosas basin is the northernmost sub-basin and closest to the Pyrenees; it is the only one filled with non-marine Paleocene sediments overlain by marine Oligocene and Miocene. This indicates that while the rest of the Gulf of Valencia was still uplifted and subject to karstification, a graben formed in this area and became rapidly filled with fluvial, lacustrine and evaporitic deposits in response to its fast subsidence. Most of the highs surrounding the graben are cored by Palaeozoic rocks, and most of the Mesozoic carbonates have been eroded. Well information shows that oil has been generated in this sub-basin, although it has not accumulated in commercial quantities. The main source-rock appears to be in the older Tertiary sediments.

Closer to the Ebro delta, the Tarragona trough has been filled by more than 3000 m of Tertiary sediments, dated as Miocene and Plio-Quaternary. This thick section is subdivided into three groups: the Alcanar Group (Early to Middle Miocene); the Castellón Group (Middle and Late Miocene); and the Ebro Group (Plio-Quaternary). The Messinian event occurred between the deposition of the last two Groups, and is marked by a strongly erosional unconformity that quite often cuts down into the Alcanar Group or even deeper (Fig. 7). The Castellón Group consists of a basal shaly formation that grades upwards and laterally into a sandy formation with interbedded sands and shales with planktonic and benthonic foraminifera. The Ebro Group consists of shelf edge to slope deposits which show strong progradation in the seismic profiles, and grade laterally and upwards into shallow marine facies.

The Alcanar Group (late Early Miocene to early Middle Miocene) is the first transgressive unit deposited on top of the paleorelief formed after the folding and subsequent erosion of the Mesozoic during Paleogene times. The Group reaches a maximum thickness of some 1000 m in the centre of the trough, thinning to almost disappear on top of the surrounding highs (Fig. 8). The Group consists of a basal member of breccias, thin shallow marine carbonates on top of structural highs, and low-energy inner-shelf marls and deeper marine massive clays mixed with volcanic components, deposited in restricted-shelf to open marine environments. Along the margins of this trough there are belts of shallow marine clastics (for example, calcarenites and sand bar deposits) and carbonate build-ups. The anoxic black shales are good source-rocks, strongly oil prone, and sourcing at least three of the four commercial discoveries: 'Casablanca' (126×10^6 bbls produced, and still on production); 'Tarraco' (14.4×10^6 bbls, depleted); and 'Dorada' (16.6×10^6 bbls, depleted). The source of the fourth field, 'Amposta' (56×10^6 bbls, depleted) has been thought to be different, perhaps the basal Cretaceous Salsadella Formation, because of the much higher sulphur content and lower ^{0}API, although other factors, such as an intense sulphurous influx due to volcanic activity, immaturity of the oil and/or biodegradation due to meteoric water influx from the nearby coast could explain the differences.

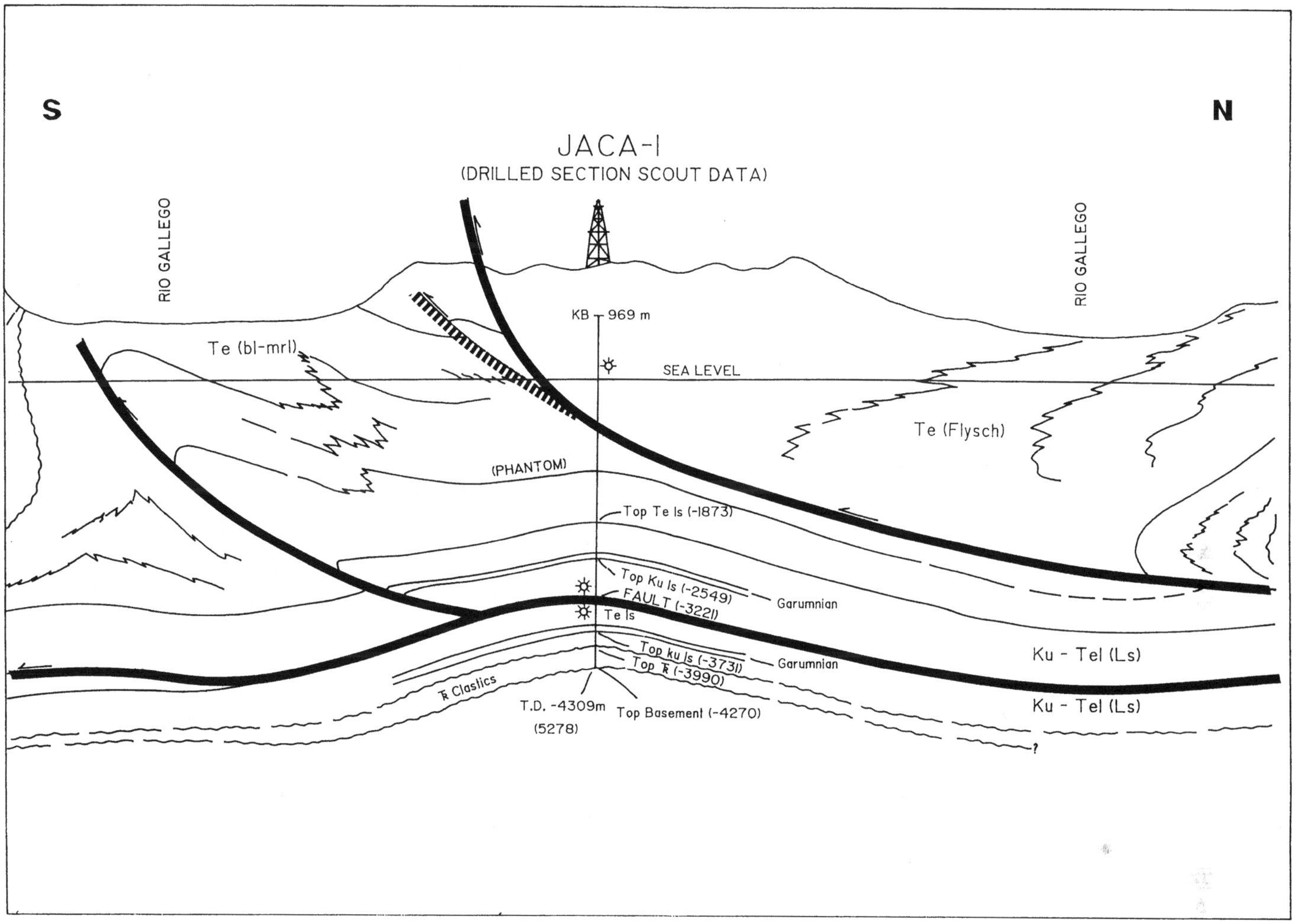

Fig. 3. Schematic structural cross-section across the 'Serrablo' gas field.

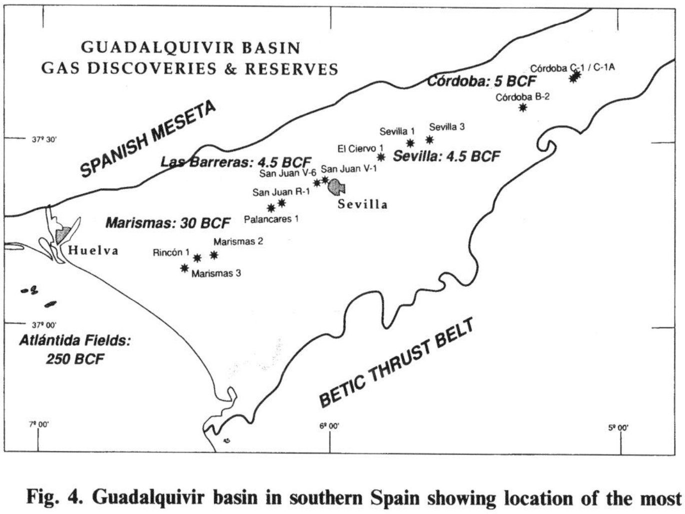

Fig. 4. Guadalquivir basin in southern Spain showing location of the most important discoveries and gas reserves.

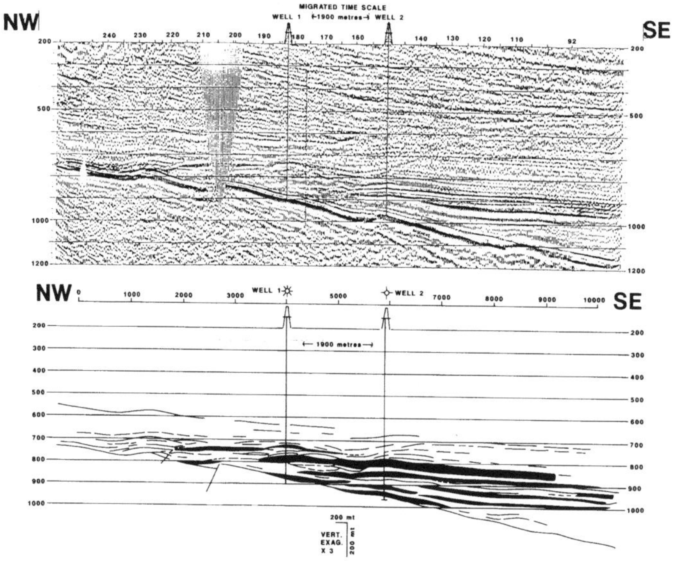

Fig. 5. Seismic dip section in the 'Marismas' area, showing the location of the reservoirs and their facies interpretation, as well as two characteristic wells.

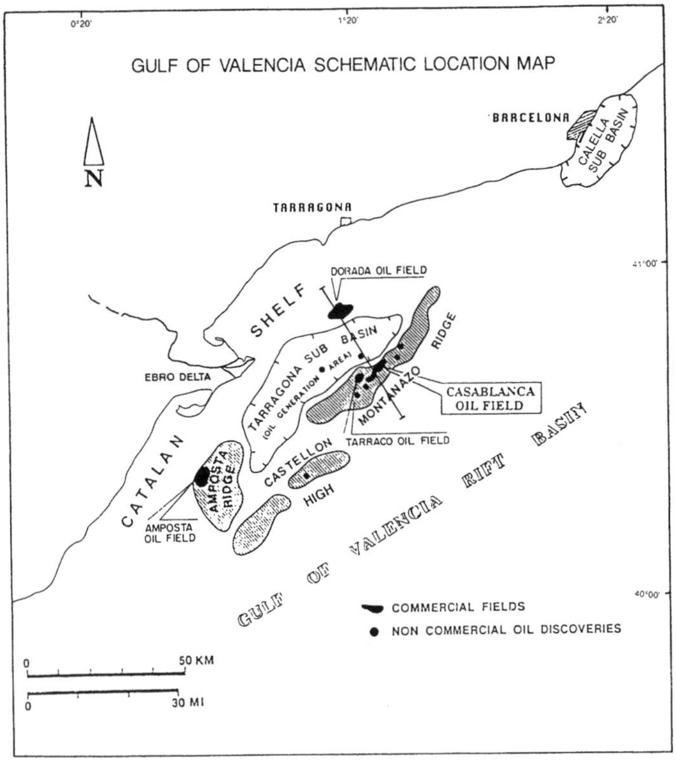

Fig. 6. Gulf of Valencia rift basin off the Spanish Mediterranean coast showing location of the Calella and Tarragona sub-basins and the major oil fields and discoveries.

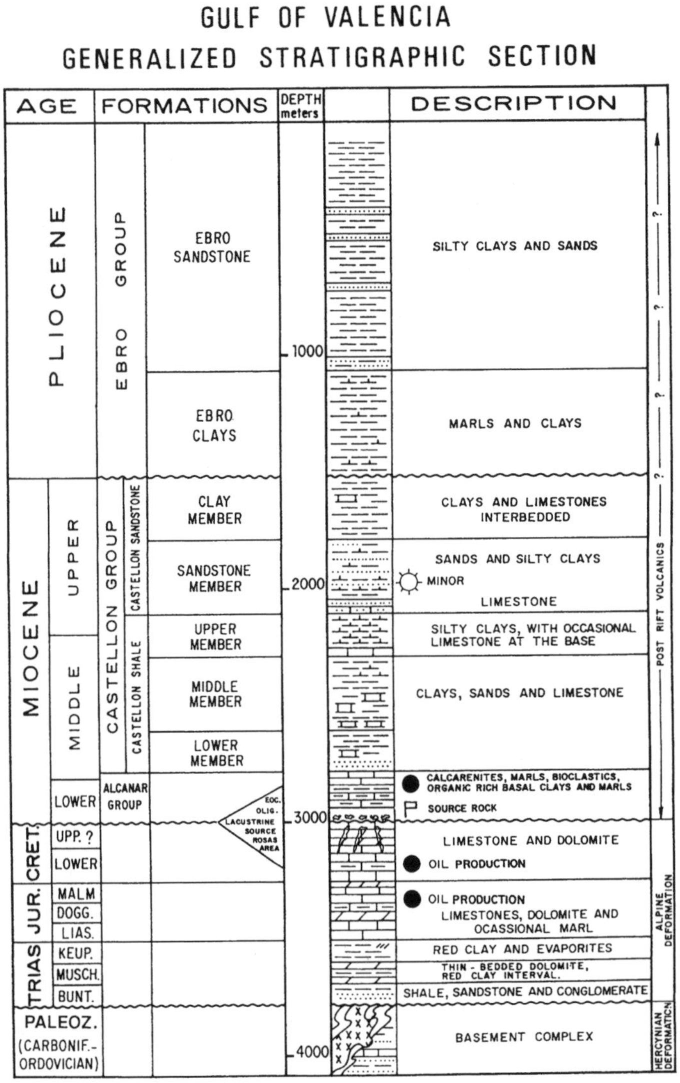

Fig. 7. Gulf of Valencia generalized stratigraphic section showing the oil production in karsted mesozoic carbonates and 'Alcanar' group sediments, the presence of gas in the 'Castellón' sands and the oil source in the basal 'Alcanar'.

The reservoirs of these fields are the karst-weathered carbonates of the Jurassic and Cretaceous, and the seal is either the Alcanar Group or the shales of the Castellón Group. The clastics and carbonates of the Alcanar Group, and some of the Castellón sands have test-yielded oil and gas in a few wells, although not in commercial quantities, so far, due to low permeability: Amposta West, Casablanca-5, Castellón-B-2, Castellón-B-4, Castellón-B-9, Castellón-C-1, Castellón-C-3, Castellón-D-1, Delta-C-2, Delta-D-2, Delta-E-1, Montanazo-C-1, Pulpo-1, Salmonete-1, Angula-1, Sardina-1, Tarragona-F-1 and at Amposta South field (with 120 BCF gas estimated reserves, although 85% is CO_2).

Other sub-basins are distinguished in the southern Gulf of Valencia, but none seems to have any generative potential.

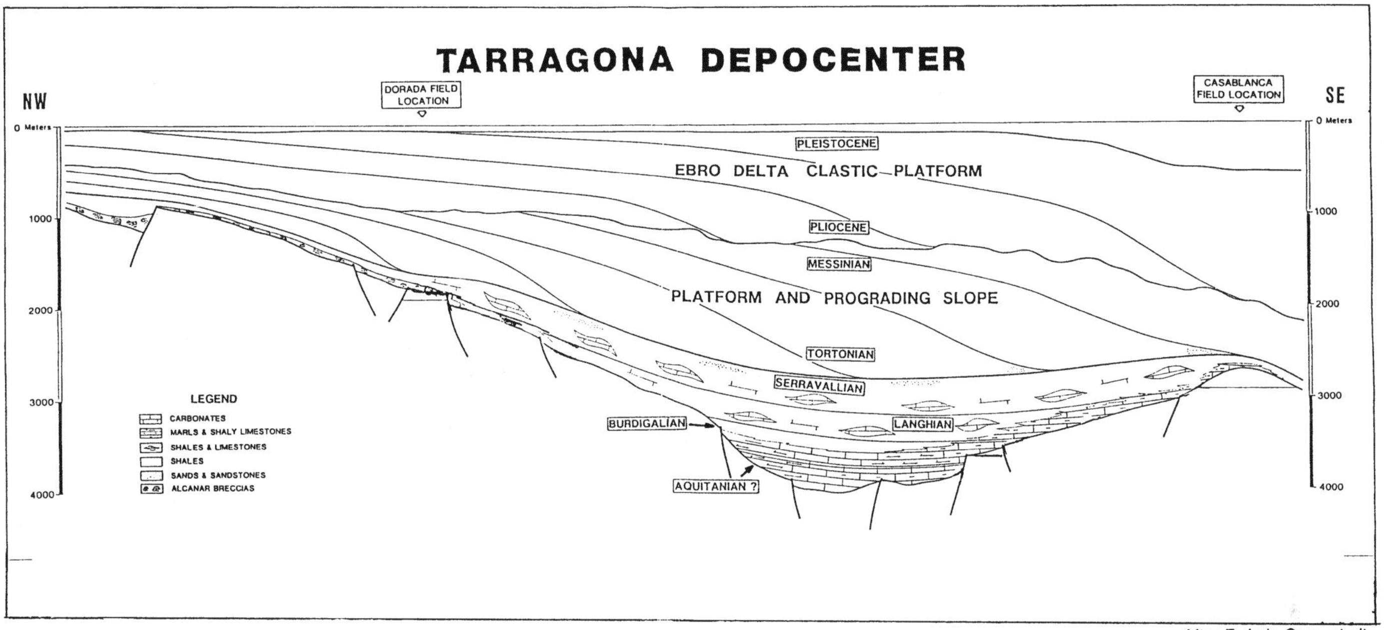

After F. de la Cruz et alt 1987

Fig. 8. Structural cross-section across the 'Tarragona trough' between the 'Dorada' and 'Casablanca' fields, schematic (see Fig. 6 for location).

G6 Mineral resources of the Tertiary deposits of Spain

M.A. GARCÍA DEL CURA, C.J. DABRIO AND S. ORDÓÑEZ

Abstract

Spain is the most self-sufficient country for minerals in the EU. A major proportion of these Spanish mineral resources has been obtained from Tertiary materials. The main materials exploited in Tertiary basins have been: brown coal and lignites, potassium salts, sodium salts (sulphates and chlorides), diatomite, sepiolite and other absorbent clays, bentonites, celestine, pumice, and also dimension (building-) stones and ceramics, portland cement and plaster raw materials.

Pb–Zn–Ag and gold alunite volcanogenic ores, related to Neogene volcanism, besides *Au*-placers have been mined from Roman times. Minor *Cu* and *Mn* occurrences are also reviewed.

The *brown coal* mines of Galician basins have provided all the significant production of Spain: more than 17 Mt. Low-quality Oligocene lignites in the eastern part of the Ebro Basin and Balearic Islands are less important from the economic point of view. Other occurrences are in the Guinzo de Limia, Guadix Baza, Granada (Arenas del Rey) and Alcoy basins.

The Spanish Tertiary basins (continental and marine) (Oligocene–Miocene) are filled by thick evaporites in which are obtained *potassium salts and sodium salts* (*sulphates and chlorides*). The Montevives *celestine* mine is located in the evaporitic unit of the Granada basin (Miocene), and provides all of the Spanish celestine production. Spain is the world's third largest producer of celestine.

The Madrid basin and the minor Calatayud basin supply the whole of Spanish *sepiolite* production. The most important Spanish *attapulgite* production is obtained from the El Cuervo mine (Sevilla and Cádiz provinces). The genesis of the Cabo de Gata *bentonite* deposits is thought to be by hydrothermal alteration and halmyrolsis of Neogene volcanic rocks. The Madrid basin bentonites and 'pink clays' have been interpreted as an early diagenetic, even edaphic Mg-rich, attapulgitization of illite clays. The most important areas of *ceramic clay* production in Spain are located in the Guadalquivir basin (Bailén area) and the Madrid basin (La Sagra – Alcalá de Henares).

The continental Neogene basin of the Hellín area supplies 90% of Spanish *diatomite* production.

Introduction

The estimated value of mineral production (oil and gas excluded) in 1990 in Spain was about $4325 million, which represented over 1% of the country's Gross National Product. With a level of supply estimated at 35% of the domestic consumption of mineral raw materials, Spain is the most self-sufficient country in the EU (Mañana, 1992).

In 1990 the whole production of brown coal (20.9 Mt), potassium salts (0.78 Mt), sodium sulphate salts (0.71 Mt), bentonites (0.15 Mt), sepiolite-attapulgite (0.56 Mt), diatomite (92 kt), celestine (80 kt), and pumice was extracted from Tertiary rocks (ITGE, 1992). The Tertiary basins also provide a large proportion of lignites, natural stones, including commercial marbles, ceramic raw materials, cement raw materials (limestone, clays and gypsum), aggregates (crushed limestone and basalt) and gypsum for plaster manufacture.

This chapter presents a review of the most important mines, both active and historical, and occurrences located in – or related to – the sedimentary rocks of the Tertiary basins in Spain.

Metallic ores

The most important ore deposits are the *Zn–Pb–Ag* volcanogenic stockworks (network ore-bodies) mined in Neogene volcanic rocks in south-eastern Spain (Fig. 1). The origin and interpretation of these ore deposits are controversial, but in all cases the Neogene vulcanism played an essential role in the ore genesis (Manteca & Ovejero, 1992). Silver has been mined since Roman times in La Unión–Cartagena district. Supergene enriched carbonates and sulphides in gossans were mined in the nineteenth century. The main deposits mined in this district are lead, silver, zinc, and minor baryte and iron ores. After the Spanish civil war, the production of the district rose to 90 Mt, producing lead (1.2 Mt), zinc (1.6 Mt) and silver (1.5 kt). Two Pb–Zn–Ag stockworks have been described recently in the Mazarrón district; one of these deposits contains 2.9% zinc, 1% lead, 28 g/t silver and 9 Mt ore, and the other contains 2.3% zinc, 0.7% lead, 20 g/t silver in a 5 Mt ore (Rodríguez, 1992).

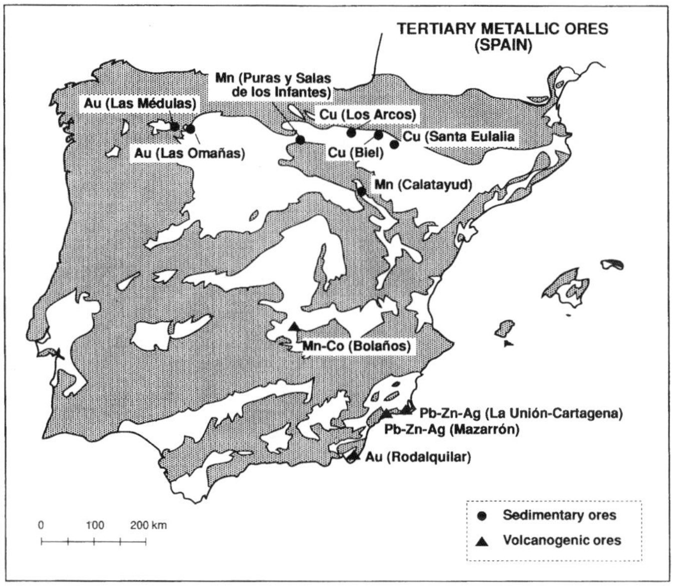

Fig. 1. Schematic location map of Tertiary metallic ores (see text).

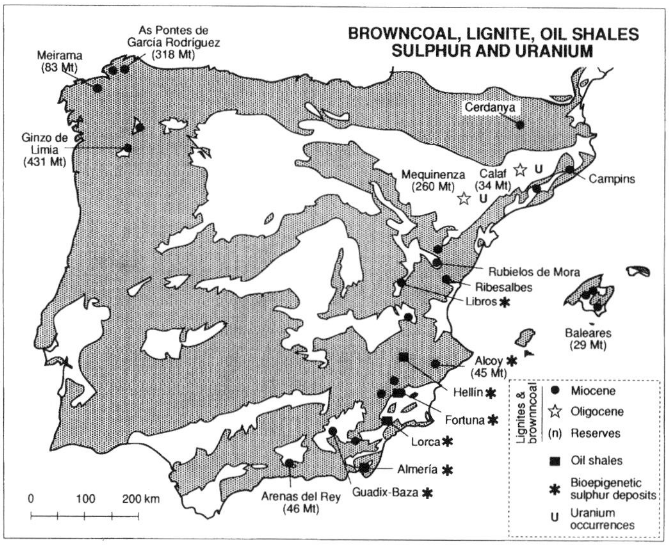

Fig. 2. Schematic location map of brown coal lignites and sulphur deposits, and oil shales and uranium occurrences (see text).

The Rodalquilar *gold–alunite* field is related to calderas in the Cabo de Gata Neogene volcanic area (Fig. 1). Arribas Rosado (1992) described ore grades up to 10 g/t. Mined since 1970s, its production reached 85 kg in 1989.

The NW Iberian Peninsula, NW of the Duero Basin and the El Bierzo Basin, was mined for *gold* in Roman times (Pérez García & Sánchez Palencia, 1985, 1992). More than 7×10^5 m^3 of gold-bearing sediments have been removed in over 600 mines. The total amount of gold exploited by the Romans in the alluvial deposits was about 200 t (Porter & Alvarez Moran, 1992). Gold grades up to 300 mg Au/m^3 have been described in the alluvial fan deposits of the Las Médulas Formation in the Bierzo Basin (see Chapter W9). Channel fill deposits contain gold flakes over 70 microns across (Pérez García & Sánchez Palencia, 1985). There are reports of gold grades up to 50 mg Au/m^3 in the Omañas Formation in the northern Duero Basin.

Some *copper* occurrences have been described in the Oligocene conglomerates of the north Ebro basin. The main ore minerals are cuprite, calcosine, galena and native copper, all them related to granular cementation of conglomerates. Ore deposition is thought to be induced by the oxidising action of infiltrational water on the formation waters of the Oligocene sediments. The thickness of individual conglomerate layers reaches 8 m. The major occurrence is the Biel Mine (Zaragoza), with estimated reserves about 500 kt and copper contents around 1.7–2%.

Small *oxide–hydroxide manganese* ores have been mined in the Campos de Calatrava (Bolaños district) since the nineteenth century, (Crespo Zamorano, 1988 a and b). The ore deposits are related to Pliocene alluvial-fan sediments of the Guadiana Basin. Manganese ores occur as granular cement in conglomerates and also as manganiferous crusts. Both types may be related to low-temperature thermal waters derived from the Campos de Calatrava vulcanism. Grade in these ores reaches 3.59% manganese, and 0.051% cobalt, with inferred resources of 15.6 Mt (Crespo Zamorano, 1988a and b). Occurrences of manganese oxides–hydroxides as granular cement ores have been cited in the western part of the Ebro and Calatayud basins.

Lignites, brown coal, oil shales and related sulphur and uranium

At present, the Tertiary basins of the Iberian Peninsula produce up to 17 Mt per year of brown coal, and some important quantities of lignite (Fig. 2). Spain was the second largest sulphur-producing country in Europe during the first part of the present century and the Tertiary basins were the main suppliers of this production. In the 1970s native sulphur mining stopped as a consequence of sulphur recovery as a bi-product of oil and gas desulphuration.

The sulphur brimstone was obtained from bioepigenetic sulphur deposits. The 1912 Spanish sulphur production was 14 kt, and at the same time Italy, the European leader, reached a production of 60 kt. The most important sulphur mines since Roman times are located in Coto Menor (Hellín basin). Up to eight sulphur seams, less than 1 m thick, have been exploited, with a 16% sulphur content. At the height of its activity this mine produced more than 50 kt per year. The rock age of the Hellin sulphur deposits is Late Miocene to Pliocene. Other bioepigenetic sulphur deposits and occurrences have been described in the Baza, Fortuna, Lorca and Almería basins (Reyes *et al.*, 1992), the Libros basin (Anadón *et al.*, 1989), and Alcoy (Reyes, pers. commun.).

The increase of crude oil prices since 1971 has resulted in the increase of oil shale and coal exploration in the Tertiary basins (Reyes & Feixas, 1984; Reyes & Crespo, 1984). Laminated oil shales in the Neogene basins of SE Spain (Almería, Fortuna, Lorca,

Table 1. *Properties of oil shale deposits in the Cerdanya, Campins, Ribielos and Libros basins of eastern Spain (see Fig. 2 for the location of these basins)*

Basin	Age	Thickness (m)	TOC (%)	Yield (l/t)
Cerdanya	Middle to Late Miocene	250	>5	—
Campins	Upper Oligocene	150	11.5	50
Ribesalbes	Lower Miocene	100	1–15	87
Rubielos	Lower Miocene	250	—	30
Libros	Upper Miocene	10	1–2.6	20–70

Note:
TOC = total organic carbon.

Table 2. *Uranium reserves in lignite seams of the Calaf, Sta. Maria de Queralt (village to the SSW of Calaf), and Mequinenza areas (see Fig. 2 for locations)*

Area	Reserves[1]	Resources[1]
Calaf	15.3	26.3
Sta Maria de Queralt	13.6	26.3
Mequinenza – Fraga – Almaret	51.9	93.4

Note:
[1]Reserves US $15–30/lb U_3O_8

Hellín), located in a marine pre-evaporitic environment of Tortonian–Messianian age, have been described (Reyes *et al.*, 1984). On the other hand, Anadón *et al.* (1989) have described some oil shale occurrences in the Cerdanya, Campins, Ribesalbes, Rubielos and Libros basins (see Fig. 2). Some technical properties – TOC (total organic carbon), oil yield (l/t) and geological data (age, oil bearing shales unit thickness) – have been included in Table 1.

Radioactivity anomalies in the lignite seams of the Calaf and Mequinenza areas (Ebro basin) (Ramírez Ortega, 1966) led to the investigation of the uranium contents and distribution in these Tertiary basins. As a result of this exploration, the uranium resources of the lignite seams of both basins have been estimated as up to 80.8 kt of U (which sells at US$15–30 per pound of U_3O_8) (Martín Delgado, 1975) (Table 2). At present, severe difficulties in extracting uranium from lignites, in addition to the availability of important Spanish high-quality uranium in proven quantity, make these Tertiary deposits uneconomic.

Fossil fuels (oil and gas excluded)

Brown coals of the Galicia basins

Several Tertiary basins occur along two dextral slip fault zones in NW Spain. Brown coals were formed in the terrigenous deposits of these basins, as a result of the evolution of the alluvial and limnic systems. The largest coal deposits recorded so far in the zone occur in the As Pontes de García Rodríguez and Meirama basins (littoral basins) and the Xinzo da Limia basin (SE Galicia) (see Chapter W9). The age of the Tertiary coal-bearing sediments is not clear, because there is a controversial disparity in the dates interpreted from mammal vertebrate biozonation and from palynological biozonation (Chapter W9).

As Pontes de García Rodríguez basin This basin is located in the NW–SE-striking structural corridor that extends for 55 km, via Pedroso – As Pontes – Moinonoro (Santanach *et al.*, 1988). It is an elongate basin with 7 km maximal length and a width varying from 1.5 to 2.5 km (Bacelar *et al.*, 1991). As mentioned earlier, the As Pontes basin is a compressional basin partly controlled by a dextral strike–slip fault. The sedimentary fill was alternately overthrust by, and onlapping, the Hercynian basement during the successive deformation phases. The coal seams appear to have formed as a consequence of relative restriction and expansion between the limnic and alluvial systems in response to the varying subsidence of the deformation phases. Two depocentres developed in the basin: one located in the NW, the West Field, and another located in the SE, the East Field. The fill of the As Pontes basin has been divided into four sedimentary units represented in both coal fields. The thickness of individual seams varies from 0.5 to 25 m, and displays continuity along the basin, although there are lateral changes to terrigenous sediments in the marginal facies. In the basin, 19 coal seams have been identified. Three main lignite types have been distinguished: brown coal, xiloide and pyropisitic lignite, the last mainly formed by resins. The vitrinite reflectance varies from 0.1 to 0.4, the average being 0.3 (Martín Calvo, 1973). The average calorific value varies from 1600 to 2200 kcal/kg. The waste/lignite ratio average in the open pit mines is up to three. The 1979 reserves of the As Pontes basin were 318 Mt, and the 1989 lignite production in the basin was 12.6 Mt.

Meirama basin This basin is located in the NW–SE-striking structural corridor via Lendo–Meirama–Baimil (Santanach *et al.*, 1988). The surface area of the basin is about 1.5 km^2; it has an elongate shape, with a long axis length of about 2.6 km, parallel to the main dextral strike–slip fault. The thickness of the basin fill is up to 250 m and it is longitudinally folded. The production in 1989 was estimated as up to 5 Mt, and the proved resources are nearly 80 Mt.

Xinzo da Limia basin This basin is located in the Villalba–Maceda–Xinzo da Limia structural lineation, which trends NNE–SSW. The sediment fill of the basin is up to 250 m thick, and up to 130 m of this may be lignite seams. The quality of the lignite is brown coal, and the proven reserves may attain 431 Mt. The economic potential of the deposit was investigated recently by ENDESA (Baltuille, pers. commun.).

Lignites of the Ebro Basin

Several coal sequences developed in Oligocene times in the Tertiary Ebro basin. Despite the fact that the individual seams of coal are thin and subeconomic, mining has taken place since the nineteenth century in the Calaf and Mequinenza areas. In both areas, the organic biomarkers identified by Gorch *et al.* (1992) show that the original organic matter derived mainly from higher plants and bacteria.

Calaf area The coal-bearing stratigraphic section formed in the Segarra lacustrine system and consists of limestones, marls, terrigenous and even evaporitic sediments of Early Oligocene age (Gorch *et al.*, 1992). The paludine–lacustrine, coal-bearing intervals consist mainly of limestone and grey mudstone beds that interfinger with lenticular channel-fill sandstones showing cross-bedding and ripple laminations. The grey massive mudstones are interbedded with coal seams ranging from a few centimetres to 0.8 m in thickness. Organo-sulphur compounds (Gorch *et al.*, 1992) are mainly responsible for the high sulphur content of the lignites. The coals consist mainly of lignite to sub-bituminous coal with calorific values that vary from 6.400 to 7.042 kcal/kg (Cabrera & Sáez, 1987), a sulphur content that varies from 2.98 to 8.36% and ash content ranging from 16.46% to 24.18%. The reserves in the Calaf area are up to 34 Mt (IGME, 1985). From the petrological point of view, the coals are vitrinite-rich with minor amounts of exinite and inertite. The vitrinite reflectance ranges from 0.4 to 0.7 (Martín Calvo, 1973). This author pointed out the relationship between the uranium content and the organic matter content, and that uranium is preferentially associated with humines and humic acid fractions. Local values of up to 0.180% U_3O_8 content have been recorded, although average values are nearer to 0.020% U_3O_8.

Mequinenza area The coal-bearing stratigraphic interval formed in the Los Monegros lacustrine system and consists of limestones, marls, and evaporitic and terrigenous sediments of the Late Oligocene. The immediate coal-bearing strata of the Los Monegros lacustrine system consist mainly of pale brown to grey micritic limestones, grey mudstones and minor sandstone lenses and sheets. Thin lenticular coal seams interfinger with the lower and middle part of the unit. Individual seams are generally thin with sharp and well-defined boundaries. The average thickness is less than 0.3 m, and varies from a few centimetres to 0.9 m. The coal deposits consist mainly of an ash-rich lignite, with 30–53% average ash content, a high sulphur content, ranging from 1.54 to 11.88%, and calorific values ranging from 3.500 to 5.500 kcal/kg (Cabrera & Sáez, 1987). The proven reserves are 260 Mt (IGME, 1985).

Table 2 presents data on the uranium resources in both the Calaf and the Mequinenza coal areas.

Other Tertiary coal basins

The only mines active at present and not already discussed are in Balearic Basins. These mines produce up to 14 kt/year of lignite. The identified reserves of lignite in these basins are up to 71.4 Mt, and the proven reserves are 29 Mt (Fig. 2).

In the Balearic Islands, lacustrine deposits have been described, of Early Oligocene, Ludian–Sannoisian age (Colom, 1983) and of Middle Eocene age (Ramos-Guerrero *et al.*, 1989). Only the deepest lacustrine facies of the central part of Mallorca contains lignite seams (Sta María, Binisalem, Alaró, Inca, Lloseta and Selva).

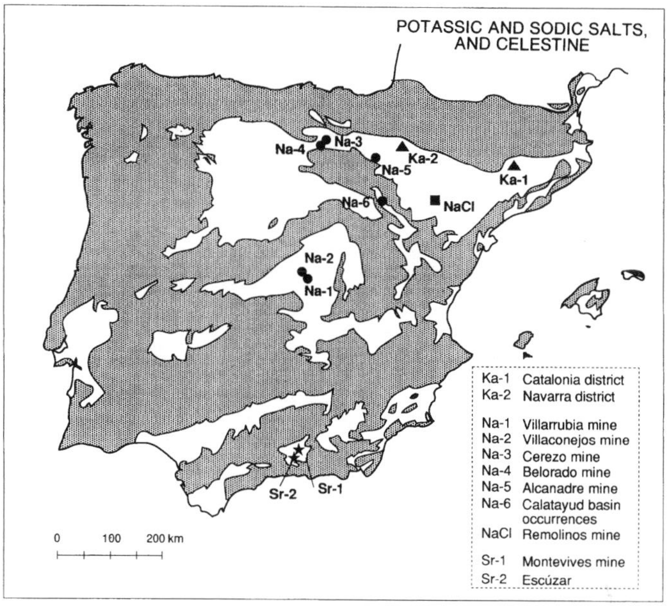

Fig. 3. Schematic location map of potassic salts, sodic salts and celestine mines (see text).

Recently, Ramos-Guerrero *et al.* (1989) have described the Binisalem member, which consists mainly of an alternation of limestone (bioclastic wackestone) and lignites. The coals consist mainly of a humic matrix that contains exinic macerals, such as resinite and cutinite (Ramos Guerrero *et al.*, 1989) with calorific values up to 4.6 kcal/kg, 1.7% sulphur content and 18.5% ash content (Colom, 1983).

Lignite occurrences in Tertiary basins have been cited in the Cerdanya, Alcoy, Guadix–Baza and Granada basins (Arenas del Rey). In the Prats–Alp area of the Neogene Cerdanya basin some lignite beds have been identified in unit C of Anadón *et al.* (1989) (Fig. 2). The lignite-bearing sediments of the Alcoy, Guadix–Baza and Granada basins have been reported as Late Miocene – Pliocene. Lignites from these basins are mainly soft brown coals.

In the Neogene Madrid and Duero basins, in the lower evaporitic units, and also in the limnic shallow marly sediments, minor organic matter occurrences (lignites and/or oil shales) have been cited.

Potassic and sodic salts, and celestine

Spanish Tertiary basins, both continental and marine, contain thick evaporitic deposits from most ages of the Tertiary. Potassium salts, sodium salts (sulphate and minor chloride), celestine and gypsum are obtained from these Tertiary basins. In this section we focus on potassic and sodic salts and celestine. (Fig. 3).

Potassic salts

The Spanish production of potassic salts has decreased from 860 kt of K_2O in 1981 to 585 kt in 1991. All the production is concentrated in two mineral districts: Catalonia and Navarra.

Table 3. *Estimated mineral reserves of potassic salts in the Catalonia and Navarra districts*

District	Amount of resource (kt) Proven	Probable	Possible
Catalonia	10470	9185	16645
Navarra	2000	8000	12000

Source: From ITGE (1992).

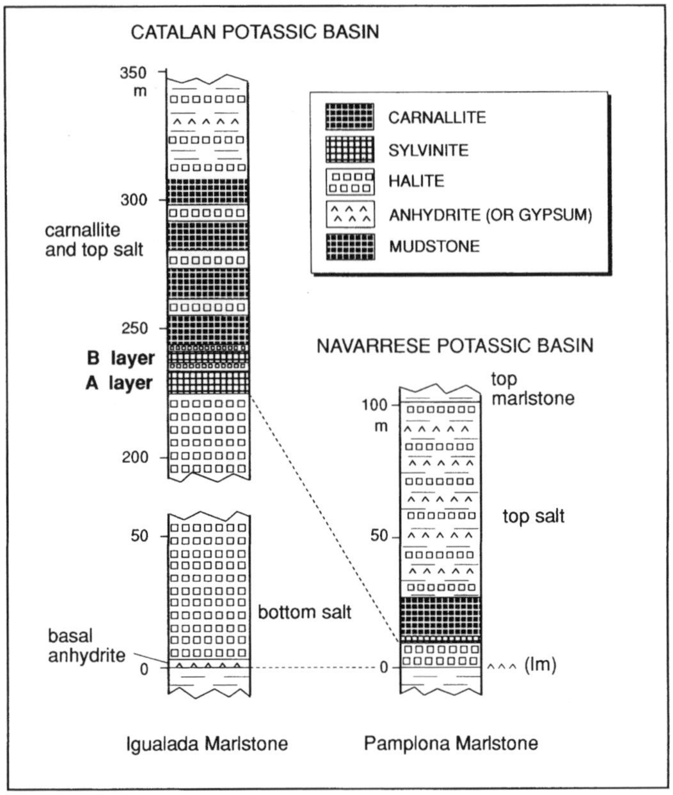

Fig. 4. Stratigraphic section correlation between Catalonian and Navarrese potassic basins. After Rosell (1983).

Estimated mineral resources and reserves in these areas are presented in Table 3.

Salinity increased dramatically in the Sub-Pyrenean basin during the Late Eocene–Early Oligocene (Ludian); most probably it was this salinity crisis that was responsible for the deposition of potassic salts in the basin. Deep drill holes reach potassic salts in the central part of the Ebro basin below a thick cover of Oligocene–Neogene sediments. Pueyo (1975) and Rosell (1983) interpreted the deposition of salt as related to marine brine with a low content of magnesium sulphate. The basin underwent active nappe (overthrust) tectonics after the deposition of the salts and, according to Rosell (1983), the Catalonia district remained in the autochthonous zone, whereas the Navarrese district was transported by thrusting. As a result, erosion removed the seams of potassic salts in the cores of the anticlines of the Navarrese district, while the anticlinal cores were preserved in the Catalonia district under a thick cover of Oligocene sediments.

Catalonia district The structure of the Catalan district is an E–W-trending syncline, with salt thicknesses ranging from 150 to 500 m (Pueyo, 1975) as a consequence of local tectonics. In the Cardona diapir the thickness of potassic salts reaches 2000 m. The main, and best-known mines (Cardona, Balsareny and Suria) are located in this district.

The stratigraphic section includes, from bottom to top (Fig. 4):

1. Marine marls (Igualada marls).
2. 4 to 5 m, laminated anhydrite.
3. 130 to 200 m, massive halite (sal de muro).
4. 5 to 20 m, lower potassic unit, consisting of decimetre sequences of terrigenous clay, carbonates, sulphates, halite and silvite. This unit is divided into two by a thick layer of massive halite.
5. 40 to 80 m, upper potassic unit (carnallitic unit) including several seams (three or four in the Suria mine, and more in other localities). It is interesting to note that the content of halite (low-grade potassic salt) in this unit increases when the seams are thinner and less numerous.
6. 85 to 120 m, grey mudstones, or transition unit. Some intercalations of halite have been described in the lower part of this unit.
7. 500 m (or more), lacustrine reddish deposits (top).

Navarrese district

As in the former case, the potassic salts are associated with marine marls (Pamplona marls). The only mine (Subiza shaft) is located in a synclinal basin (the so-called Pamplona potassic basin or Sierra del Perdón potassic basin), but a research programme has only recently been initiated.

A general stratigraphic succession includes:

1. Deep marine marls (bottom).
2. 0.6 to 1 m, laminated anhydrite unit.
3. Almost 10 m, lower banded, massive halite unit (sal de muro).
4. 2 m, lower potassic unit, formed by 18 silvinite–halite couplets; some of them are used as markers.
5. 12 m, carnallitic unit, made up of eight halite–carnallite couplets.
6. Top halite unit, made up of sequences of decimetre- to metre-thick red mudstones and centimetre-thick halite layers.
7. Up to 50 m, top marls, mudstones with some anhydrite layers.

It should be noted that this stratigraphic section is very similar to the one in the Catalonia district (Fig. 4). The only difference is the total thickness.

Table 4. *Reserves of sodium salts in Spanish mining districts*

Company	Raw material	Plant location	Basin	Extractive method	Proved reserve (Mt)
Crimidesa	Glauberite	Cerezo de Río Tirón	Ebro Trench	a	19
Foret	Thenardite	Villarrubia	Madrid basin	b	15.5
Minera S. Marta	Glauberite	Belorado	Ebro Trench	a	63
Sulquisa	Glauberite	Villaconejos	Madrid basin	a	57

Notes:
(a) Open-pit solution; (b) underground mine.
Source: From ITGE (1991)

Sodium salts

For many years, Spain has been the European leader in natural sodium sulphate production, and all of it comes from Tertiary continental basins. The 1990 Spanish natural sodium sulphate production, in terms of Na_2SO_4 content, amounted to 714 kt, 240 kt from thenardite only in the Madrid basin, and the remainder from glauberite in both the Madrid and Ebro basins, particularly the Tertiary Trench or Bureba Corridor in the western part of the Ebro basin (ITGE, 1992). Table 4 presents some data on the reserves of the mining districts.

There are two active mines in the Madrid basin (Fig. 5), one located near Villaconejos–Colmenar de Oreja, and another located near the eighteenth-century mine in Villarrubia de Santiago. There are many no-longer-active mines in the Madrid basin, but, in the past, the most common method of recovering salts was to use saline springs and wells (e.g. Espartinas, Carabaña, Carcaballana, Loeches), and some of these were even being exploited in Roman times.

The mine located near Villaconejos–Colmenar de Oreja (the Fátima mine), is an open-pit dissolution mine. The mineral glauberite is preferentially dissolved in pools located over a bed of glauberite–anhydrite that is 27 m thick. The brine is recovered by pumping from wells and sent to an evaporation plant where top-quality sodium sulphate precipitates. The source brines of the evaporation plant are introduced into pools and then recirculated (Ordóñez *et al.*, 1982).

The mine located near Villarrubia de Santiago ('El Castellar') is a pillar and room underground mine that works a 5–8 m thick thenardite bed with minor glauberite. The mineral extracted is processed in an evaporation plant to obtain the commercial sodium sulphate (Ordóñez *et al.*, 1982).

The Neogene stratigraphic record of the Madrid basin is summarised by Megias *et al.* (1983) and by Calvo *et al.* (Chapter C2). The sodium sulphate mines are located in the Lower Unit. The sedimentological interpretation of the Lower Unit or Saline Unit has been discussed by Ordóñez *et al.* (1992) and by Ordóñez & García del Cura (1995). The mineralogical and petrological features of the economic saline deposits have been described by García del Cura (1979), García del Cura *et al.* (1979) and Ortí *et al.* (1979). The only paleontological data obtained from the Saline Unit are the flora found in an exploration drill hole near the Villaconejos mine. This flora indicates a Late Oligocene – Early Miocene age for the Saline Unit (Alvarez Ramis *et al.*, 1989).

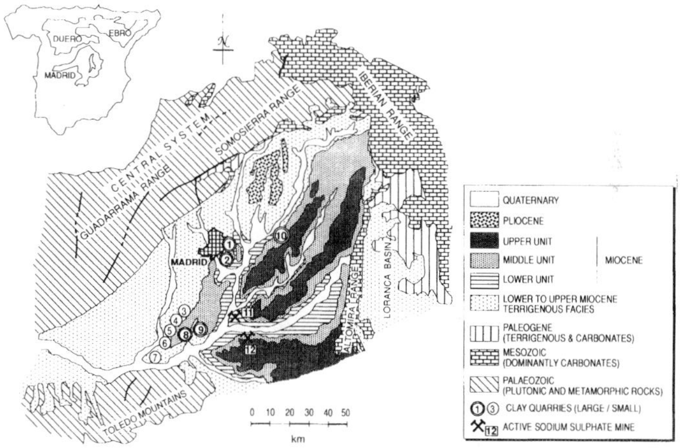

Fig. 5. Location map of main active mines in Madrid basin: 1, Vicálvaro sepiolite mine; 2, Almodóvar sepiolite mine; 3–6, Yepes – Cabañas de La Sagra mines (bentonites and sepiolite); 7, La Sagra ceramic raw clay mines; 8 and 9, Esquivias special clay mines; 10 Alcalá ceramic raw clay mines; 11, Villaconejos ('Fátima mine') sodium sulphate mine (glauberite); and 12, Villarrubia ('El Castellar') sodium sulphate mine (glauberite and thenardite).

The 'El Castellar' mine is placed in the Upper Saline Subunit. The sedimentology and facies distribution of this Upper Saline Subunit are poorly understood, and are, at present, under review by the authors of this chapter. However, it is possible to point out some distinctive features of this Upper Saline Subunit, one being the presence of thenardite as the main mineral, and another being the local and restricted character of this Subunit. In this Upper Subunit it is possible to identify at least six repetitions of the following sedimentary dm to m sequence: 1. reddish mudstone containing interstitial halite; 2. muddy terrigenous sediment ± halite ± glauberite; 3. massive thenardite ± glauberite. The thenardite ± glauberite bed mined in Villarrubia is up to 8 m thick (Fig. 6). The massive thenardite mined there displays a typical blue colour and is thought to be a secondary mineral from mirabilite. This soft

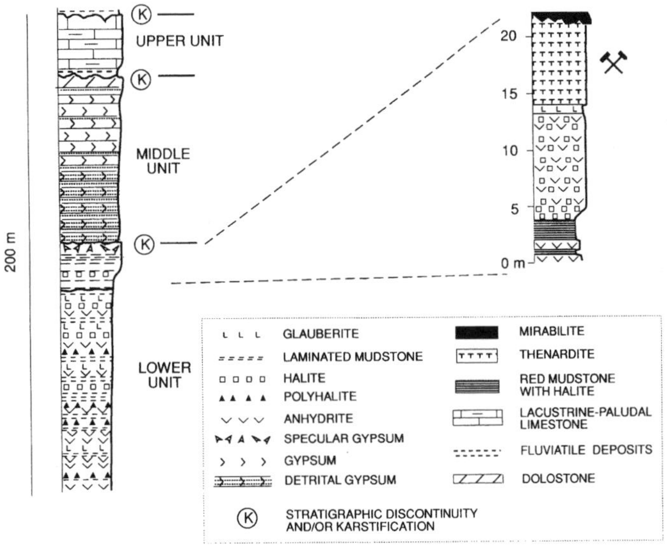

Fig. 6. Location of Villarrubia thenardite ± glauberite mine. Stratigraphic section of central part of Madrid basin and a detailed section of thenardite seam mined there.

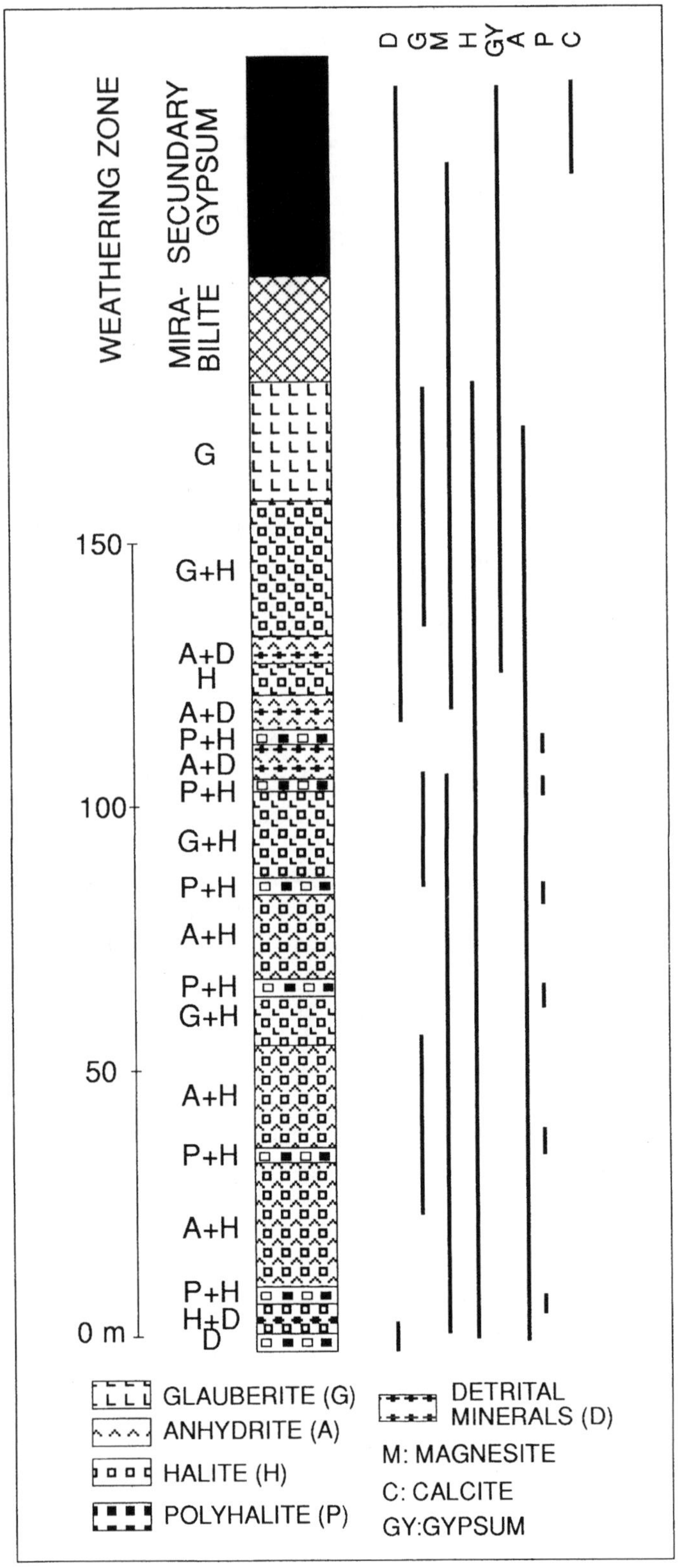

Fig. 7. Schematic mineralogical section of an exploration drill hole located close to the Villaconejos glauberite mine.

and wet crystalline mirabilite precursor of thenardite formerly contained some precipitated idiomorphic crystals of glauberite that are now enclosed in massive thenardite.

The glauberite mine of Villaconejos exploits the lower Saline Subunit (Fig. 7). This Subunit is thought to have formed in a perennial saline lake surrounded by a wide mudflat. As pointed out by Ordóñez *et al.* (1983) and Utrilla *et al.* (1992), brines of this saline lake were derived mainly from the weathering of Upper Cretaceous–Paleocene marine evaporites. Saline mineral associations in this stratigraphic interval are reported in Fig. 7. Glauberite occurs as a massive bed of idiomorphic glauberite crystals, with minor magnesite marls interbedded with anhydrite and micritic magnesite. Glauberite cements and nodular anhydrite have recently been interpreted to result from the early diagenetic glauberitisation of primary anhydrite in the muddy lacustrine belt during episodes of low lake water levels (Ordóñez & García del Cura, 1993).

The Cerezo de Río Tirón, Santa Marta and Belorado glauberite mines are located in the Tertiary Trench, 'La Bureba Corridor', that connects the Ebro and Duero basins. Glauberite mines are located in the Cerezo Evaporite Formation of uncertain Late Miocene age, and are restricted to the central part of a small basin. The Evaporite Formation displays sharp lateral facies changes to non-economic detrital sediments through saline mudstones with nodular anhydrite. A perennial saline model has been proposed to explain the general features of the Evaporite Formation (Menduiña *et al.*, 1984). The detrital sediments laterally associated with the saline deposits have been interpreted as the distal part of a prograding lacustrine deltaic system, and this model has been used in the exploration of the Santa Marta mine, which started production in 1989.

The recovery method used in Cerezo de Río Tirón and Belorado is similar to that described in the Villaconejos mine, that is to say, an

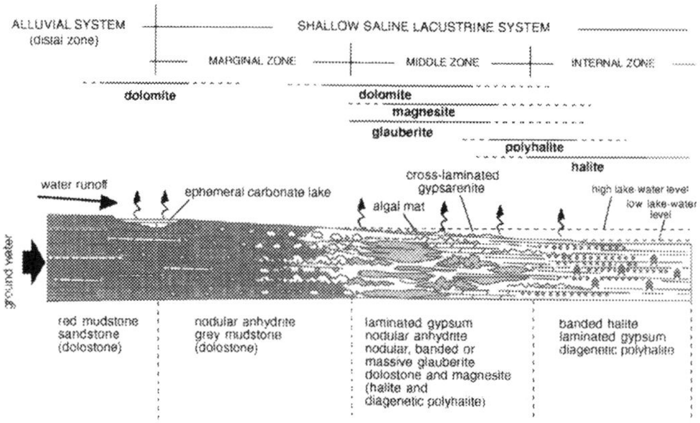

Fig. 8. Paleoenvironmental and sedimentological model of Alcanadre non-active mine. Adapted from Salvany & Ortí (1992).

open pit solution mine cut down to a glauberite seam; the resulting brine is the raw material for top-quality commercial sodium sulphate.

The Cerezo Evaporite Formation stratigraphic interval contains alternations of metre-thick seams of microcrystalline glauberite and non-economic muddy anhydrite. The number of glauberite seams is up to eight, but the recovery methods, and the commercial productivity of the seams, are conditioned by the present-day topography and by the impoverishment of the upper seams by infiltration of meteoric waters.

Some no-longer-active underground mines with glauberite have been described from the western Ebro basin. These glauberite deposits are located in the Lerín Formation of Late Oligocene–Early Miocene age (Salvany, 1984; Salvany & Ortí, 1992, 1994). These authors have recently proposed a paleogeographical and sedimentological model to explain the main features of the Alcanadre mine zone (Fig. 8). The thickness of the overburden and the thinness of the glauberite beds result in high values of the stripping ratio and the submarginal economic character of these occurrences.

Some glauberite occurrences in the central part of the Ebro basin and below the halite beds of La Real mine (Remolinos–Zaragoza) have been described (Fernández Nieto & Galán, 1979; García Veigas *et al.*, 1991). Some outcrops of glauberite beds have also been reported in the Calatayud basin (IGME, 1980; Sánchez Moral *et al.*, 1993).

Halite rock salt is present in both continental and marine Tertiary basins of the Iberian Peninsula. The 1990 Spanish production of sodium chloride was 3.3 Mt, 19.8% of which was obtained as a byproduct of Tertiary potassic salt recovery, and 3.2% was obtained from the Tertiary Ebro basin. A high proportion of the remainder was obtained from the natural evaporation of marine brines, and a smaller proportion from Triassic rock salt deposits. The rock salt deposits of Remolinos and Torres de Berrellen in the Zaragoza Formation (Lower Miocene) of the central part of the Ebro basin have been mined for a long time. Proved reserves amount to more than 16 Mt NaCl. The halite deposit is up to 100 m thick, and consists of decimetre layers of halite with minor nodular anyhdrite (Ortí & Pueyo, 1977). The lateral continuity of the deposit is over 30 km along the Ebro river valley (Ortí, 1990).

Celestine

Spanish celestine production has risen from about 19 kt in 1980 to more than 80 kt in 1990. Spain is the third country in the world ranking of production of this mineral. Celestine is produced from the Montevives mine, about 12 km southwest of Granada (Granada Province). The Montevives ore grade is 80% $SrSO_4$, with proven reserves of 3 Mt. The regularity of the ore body in this case enables the deposit to be competitively concentrated by the flotation method.

The Montevives mine is located in a small hill in the central part of the Tertiary Granada basin. The host sedimentary materials are of Middle to Late Miocene age. Recently, Rubio Navas (1990) has pointed out the near domic structure of the ore body and the tectonic character of the celestine ore body outcrop.

The celestine ore is strata bound, up to 40 m thick, and it dips 20° to 50° towards the northwest. The celestine mineralisation displays a mm-scale, clear–dark banded texture similar to the primary stromatolite textures of the host limestone and marls. The rich ore displays a massive structure and earthy and/or microcrystalline textures with obliteration of primary structures. The ore body is well stratified in layers ranging from 20 to 50 cm thick. In addition to celestine, the mineral paragenesis of the Montevives ore deposits includes calcite, dolomite, quartz, gypsum, strocianite, Fe–Mn oxides and hydroxides, phyllosilicates, etc.

The Escúzar village celestine occurrence is an E–W-striking outcrop with more than 10 km lateral continuity – the Escúzar Celestine Belt of Martín *et al.* (1984). The host rock of the celestine ore is also a stromatolitic limestone, and the thickness of the deposit is roughly 20 m. The average celestine content is lower than at Montevives, and rarely rises to 55%. A significant feature of this occurrence is the presence of a karstic surface at the top of the deposit, with dolines and other karstic depressions filled by low-grade brecciated ore and stromatolite carbonates. Escúzar has estimated reserves of about 1 Mt.

Both the Montevives and Escúzar celestine deposits have been interpreted as the results of early diagenetic cementation of stromatolitic carbonates when dessication processes followed the fall in sea-level (Martín *et al.*, 1984). In this model, the strontium is derived from marine connate waters.

Other celestine occurrences have been cited in the Duero Tertiary basin (Ordóñez *et al.*, 1980), and Ebro basin (Rubio Navas, 1990).

Special clays (sepiolite, attapulgite and bentonites), ceramic clays and diatomites

Sepiolite, attapulgite and bentonite deposits are commonly associated in the Tertiary clayey formations, and in consequence these minerals are mined together or in the same areas. The official statistical data of these special clays show some contradictory interpretations because the uses of the special clays are similar

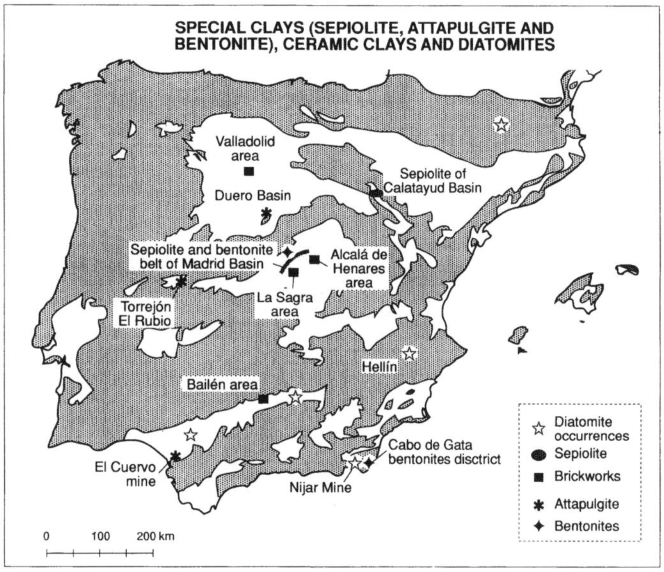

Fig. 9. Schematic location map of special clays (sepiolite, attapulgite and bentonite), ceramic clays and diatomites (see text).

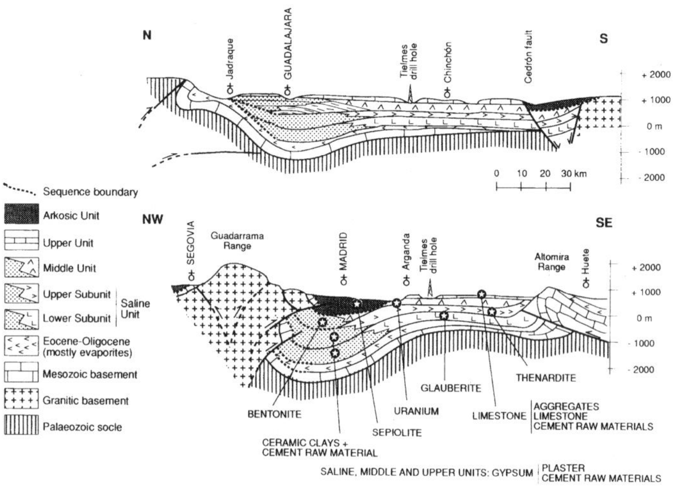

Fig. 10. Schematic cross-section of the Neogene Madrid basin (adapted from Megias *et al.*, 1983) showing stratigraphic location of main mined mineral deposits: glauberite, thenardite, sepiolite, bentonite, ceramic clays, Portland cement raw materials, dimension stone (Colmenar limestone) aggregates (crushed limestones), lime raw materials and uranium occurrences.

and generally connected with the physical properties of absorption, exchange capacity, etc. As a consequence, we also include in this section the diatomite deposits because these physical properties are also their most characteristic feature.

Sepiolite

The 515 kt produced in Spain in 1990 were obtained entirely from the Tertiary basins: 453 kt from the Madrid basin, 61 kt from the Calatayud basin and 1 kt from the Duero basin. The sepiolite resources in the Madrid basin reach 100 Mt, and the estimated production capacity is more than 1 Mt/year. Sepiolite is also present in the attapulgite deposit of Lebrija (El Cuervo mine Guadalquivir basin), with a sepiolite grade of up to 5% (Fig. 9).

In the Madrid basin, sepiolite is quarried at Vicálvaro, Vallecas, Parla and San Blas, to the south of the city of Madrid (Fig. 5), and Yunclillos (prov. Toledo) (Galán & Castillo, 1984; Ordóñez *et al.*, 1992). The sepiolite seams occur at the top of fining-upward sequences of arkosic sandstones. The economic sepiolite seams occur at an intermediate position between the lacustrine deposits of the Middle Unit of the Madrid basin and arkosic sediments derived from granitoid and high-grade metamorphic source areas that outcrop in the northwestern part of the Madrid basin (Fig. 10).

The classic sepiolite deposits of Vallecas–Vicálvaro were exploited three centuries ago to obtain the light rough stone of Madrid buildings, and special refractory clays. The areal extent of these deposits is almost 6.6 km^2, and they consist of two exploitable layers of sepiolite with sharp lateral changes of facies to the neighbouring abundant arkosic sandstones. The thickness of the upper layer, the richest in sepiolite, is up to 10 m, and it is separated from the lower sepiolite layer by more than 15 m of non-economic muddy sandstones. The lower sepiolite layer is 1 to 5 m thick. A classic mammal site (Cerro Almódovar site) of Middle to Late Aragonian age is located in the muddy marls at the top of the lower sepiolite layer. The sepiolite grade varies between 65% and more than 95%, being accompanied by smectites, quartz, illite, feldspars and carbonates. This is the world's most important known deposit of sepiolite (Galán & Castillo, 1984).

The Yunclillos sepiolite deposits extend over 3 km^2 and there are two sepiolite-containing layers. The lower sepiolite layer is richer than the upper layer and its thickness is up to 3 m. The upper layer contains smectite and nodular chert. The location is close to Vallecas and Vicálvaro. In Paracuellos del Jarama (east of the city of Madrid) there are important reserves of sepiolite, probably subeconomic because of the thick cover and its location in an urban area. The seams of the Paracuellos sepiolite deposits are interbedded with burrowed muddy arkosic sandstones and, in places, with reworked sepiolite and paleosols (Alonso *et al.*, 1986). The genesis of the Paracuellos sepiolite deposits has been reported as edaphic to paludal in a distal alluvial fan environment (Calvo *et al.*, 1986). Some uranium vanadates (yuyamunite) have been formed in the edaphic environment, associated with dolocretes and with vertebrate remains in the marls (Arribas, 1963). In our opinion, it is possible to define a *sepiolite belt* along the Madrid arkosic trench, elongated from NW to SE from Yunclillos to Paracuellos del Jarama, with many sepiolite occurences as at north La Sagra, north Esquivias, Parla, North Getafe, etc.

In the Calatayud basin, sepiolite marls were exploited a few years ago in the Isabel mine (Orera). The annual production of this mine, as we have pointed out before, was up to 61 kt in 1990 (ITGE, 1992). The sepiolite deposits of the Calatayud basin are interpreted as lacustrine deposits in a brackish shallow lacustrine environment. Arauzo *et al.* (1989) describe the mineralogical paragenesis: dolomite, illite, smectite and quartz, with minor amounts of calcite,

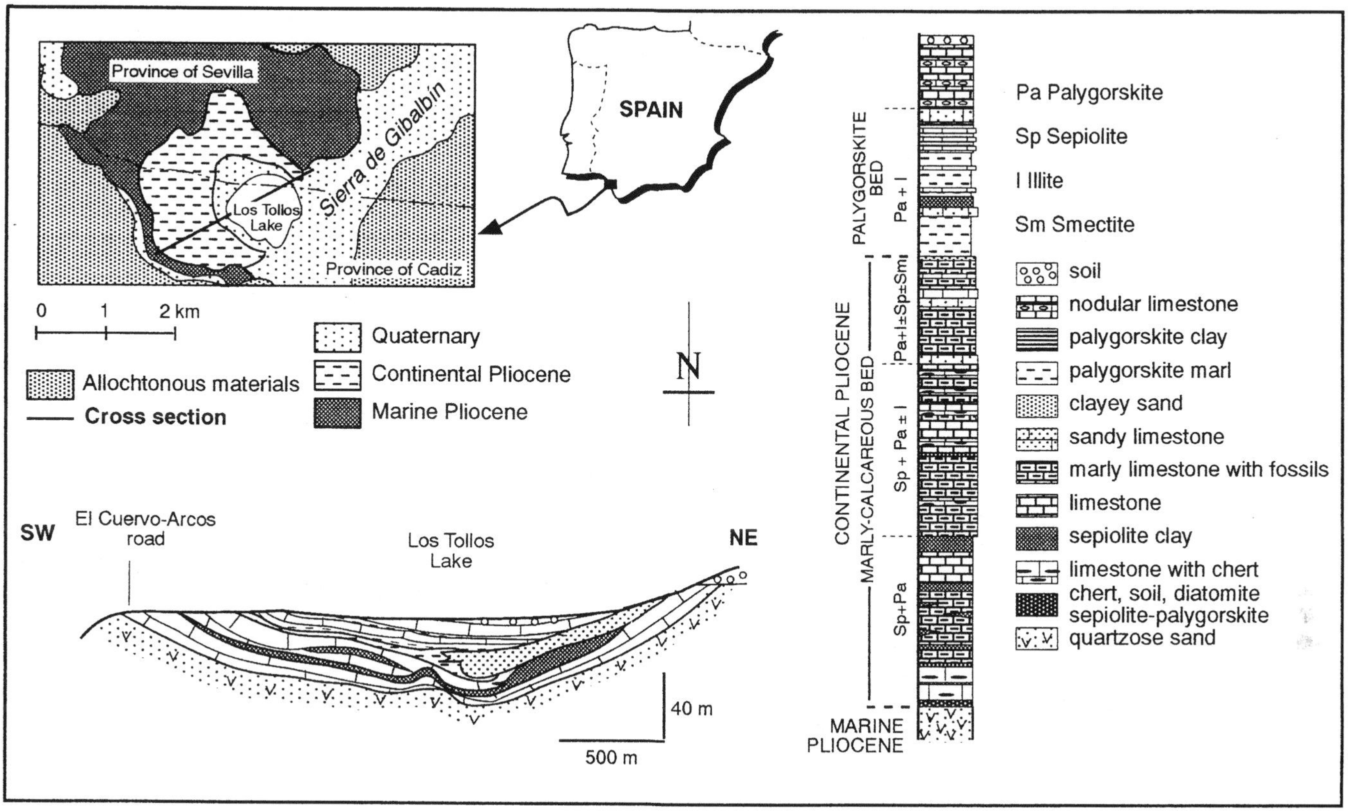

Fig. 11. Geological location of El Cuervo attapulgite mine, cross-section and detailed stratigraphic section. After Galán & Ferrero (1982).

feldspars, chlorite and kaolinite in the Mara area, and sepiolite alternating with smectites in the Orera area. Early diagenetic vadose to edaphic processes of carbonate–siliceous (due to diatoms and/or detrital clays) sediments may explain the genesis of the sepiolite marls. Other occurrences of sepiolite marls south of Guadalajara (Madrid basin) and Cuestas Facies (Duero basin) may be interpreted in a similar way.

Attapulgite

The 1990 Spanish attapulgite production was 54 kt. The most important production comes from the El Cuervo attapulgite mine (Sevilla and Cádiz provinces) and only 4 kt was from the Torrejón el Rubio attapulgite deposit (Fig. 9).

The El Cuervo mine is located in a Pliocene to (?)Quaternary lacustrine deposit (Galán & Ferrero, 1982) (Fig. 11). Two main units are distinguished: a lower marly layer and an upper palygorskitic (attapulgitic) layer. The thickness of the attapulgite layer varies from 3 to 30 m with an attapulgite grade of between 35 and 75% and with minor sepiolite (0–30%) and calcite. It includes some fossiliferous limestone beds interbedded in the attapulgite seam. The estimated resources are up to 30 Mt.

The Torrejón el Rubio deposit is located in the small Tertiary fault basin of Torrejón el Rubio (Galán & Castillo, 1984). The source area and the basement of the basin are deeply weathered slates. The lowest basin fill is composed of weathered slate material and this layer is overlain by a terrigenous–muddy bed with a thickness that varies from 6 to 50 m. On top of this bed (0.5–4 m) an attapulgite grade of up to 70% can be measured. Paragenesis of the attapulgite deposit is, in addition to attapulgite, palygorskite ± illite ± smectite (saponites) ± chlorites ± sepiolite ± quartz ± feldspars and dolomite. The genesis of the attapulgite deposits appears to be a result of a magnesium-rich weathering process that affected clayey, probably illite-rich, sediments (Fig. 12).

Some attapulgite occurrences have been described in the lacustrine sequences of the Tertiary Duero basin: García del Cura & López Aguayo (1974); Pozo & Leguey (1985); in Sacramenia: Martín Pozas *et al.* (1983); and in Bercimuel: Suárez *et al.* (1989) and Suárez *et al.* (1991). Other less-important attapulgite occurrences have been described in which the mineral forms a cement of the proximal detrital facies of the Madrid and Duero basins (Ordóñez *et al.*, 1977; Megías *et al.*, 1982; Leguey *et al.*, 1984a and b). The paragenesis of these attapulgite occurrences includes palygorskite ± cristobalite ± dolomite, interpreted as late diagenetic cement of the arkosic sandstones.

Bentonites

The 1990 Spanish production of Ca-bentonite was 151 kt, and this came from two districts: Cabo de Gata (Almería province) and Villaluenga (Toledo province).

The Cabo de Gata district deposits are associated with the

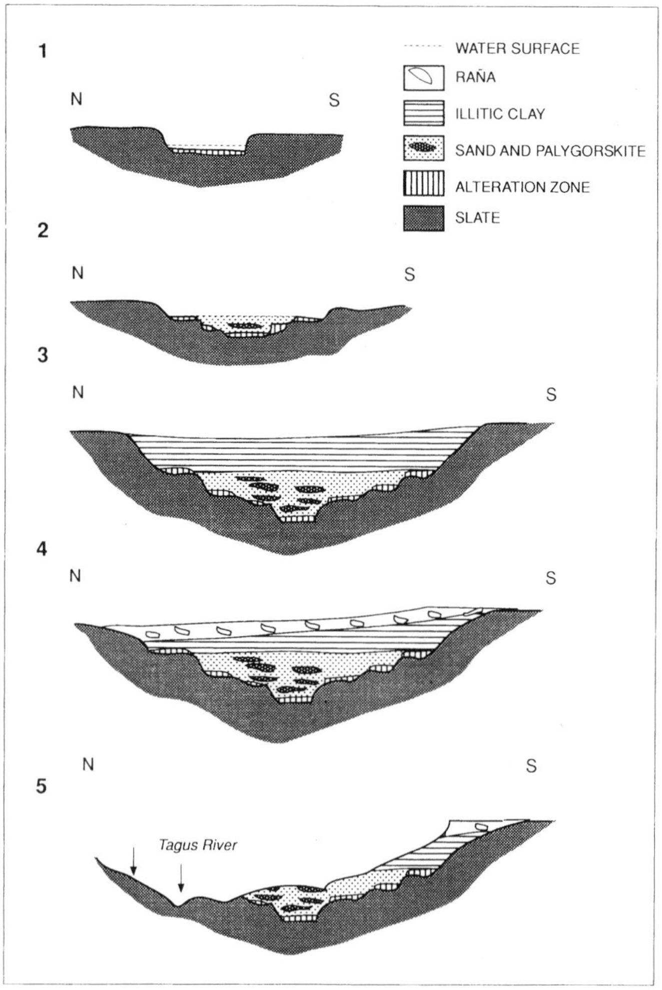

Fig. 12. Schematic interpretation of Torrejón el Rubio attapulgite deposit from the 1, weathering of substrate; 2, clayey sand sedimentation and early diagenetic (vadose) magnesium-rich attapulgite, probably related to low sedimentation; 3, silting up sedimentation; 4, alluvial fan ('raña') deposits; 5, fluvial network trenching. After Galán & Castillo (1984).

Neogene calc-alkaline volcanism of SE Spain, with an age of the volcanism varying between 8 and 17 Ma.

The main bentonite mine is in the La Serrata–Los Trancos deposit, and the production of this mine is up to 100 kt/year. The proven reserves are estimated as 3.5 Mt (J. Teodoro, pers. commun.). According to Doval (1992) the average purity of the smectites is up to 98%. The genesis of these bentonite deposits is low temperature (about 40–70 °C) hydrothermal alteration of volcanic rocks (Leone *et al.*, 1983): about 70 °C in the Sierra de Gata and about 40 °C in the Serréta de Nijar. The mineralogical paragenesis of the Serreta de Nijar deposits has been described by Caballero *et al.* (1983) as formed mainly of smectites, the only phyllosilicate present in the fine fractions belonging to the montmorillonite–beidellite–nontronite series, along with jarosite, pyrolusite and alpha-trydimite (probably neoformed) and quartz, plagioclase and potassium feldspars; inherited minerals from the parent rocks; the parent rocks are rhyolites, dacites and trachytes. The exchange cation capacity is $Na^+ > Mg^{2+} > Ca^{2+} > K^+$, and the chemical composition of the bentonite displays wide variations even within the same deposit. The most important bentonite deposits at Serreta de Nijar are closely related to N5OE striking fissures and commonly display vein-fill features.

The bentonites of the South Cabo de Gata zone consist of smectite, jarosite, zeolite and trydimite, as neoformed minerals, and of plagioclase, potassium feldspar, amphiboles and micas as minerals inherited from the parent materials (Caballero *et al.*, 1985).

In addition to this volcanogenic interpretation of these bentonite deposits, F. Ferrero (1984) has pointed out the probable influence of halmyrolytic processes in the genesis of some of the blanket bentonite deposits. The geochemical data seem to exclude a marine origin for the alteration solution, and favour the idea that the solutions resulted from a system of meteoric waters heated by a geothermal cycle (Gaballero *et al.*, 1985).

The Villaluenga district (Toledo province) is located in the Middle Unit of the Neogene Madrid basin (Middle Aragonian). It occurs in a transitional zone between the outer marginal facies and inner chemical facies of the lacustrine sequences. These bentonite deposits usually occur in laterally continuous levels 0.4–2 m thick. Commonly, the bentonites are intercalated in the lacustrine sequences with distal deltaic micaceous sands and dolomitic carbonates, some chert nodules and other mudstones. The mineralogy of the bentonites is essentially Mg-rich smectites, mostly stevensite and saponite (Galán *et al.*, 1986; Pozo *et al.*, 1991), and their potential as backfilling and sealing material in high-level radioactive waste disposal is very interesting (Cuevas, 1992; Cuevas *et al.*, 1993). The reserves of the Madrid basin are probably up to 10 Mt, located in a closed stratigraphic position from Cerro del Aguila (Villaluenga) to the southern part of the city of Madrid, markedly parallel to the sepiolite belt. The reserves of the mined deposits at Cerro del Aguila (Villaluenga) are up to 0.7 Mt.

Some 'pink clays' formed by interstratified kerolite, most probably stevensite with interlayered kerolite related to the bentonite deposits, are found (Martín de Vidales *et al.*, 1989, 1991). These 'pink clays' have been used as an oil discolourant. Pink clay layers are associated with dessication levels developed over bentonitic clay beds (green colour). The possible resources of pink clays in the Madrid basin are up to 2 Mt.

Ceramic clays

Almost all the marginal and basal formations of the continental Tertiary basins of the Iberian Peninsula contain clay sediments that are being used as ceramic raw materials. In this facies there are large volumes of material that are potential sources of ceramic clays. They have a variable mineralogical composition, but illite is generally the most abundant phyllosilicate mineral. Commercial interest in these clay-rich sediments depends on the physical properties of the crude materials: plasticity, drying capacity without shrinkage, extrusionability, firing temperature, and, after firing, degree of efflorescence, permeability and compressive strength, etc. However, the distance from the location of the clay-rich sediment deposits to the potential consumption centres may be the most important economic factor, because haulage costs may quickly make the price prohibitive.

Recently, González Díaz (1992) has reviewed Spanish ceramic clays. The most important ceramic-clay production areas are located in the Guadalquivir basin (Bailén) and in Madrid basin (La Sagra–Alcalá de Henares) (Fig. 9).

The ceramic raw materials mined in the Bailén area are of Miocene age, and have an average mineralogical composition of 20% quartz, 12% feldspars, 25% calcite and 54% phyllosilicate, mainly illite and smectite (González *et al.*, 1986).

The ceramic raw materials exploited in La Sagra (Toledo province) and Alcalá de Henares (Madrid province) are located in the Lower Unit of the Neogene Madrid basin (Lower Aragonian). The average content of illitic clays is up to 50% (Menduiña, 1988; García *et al.*, 1990). Resources amount to 500 Mt. These clay-rich sediments are used to add a top-quality silicate component to Portland cement (García Calleja *et al.*, 1991). The potteries production at La Sagra reaches 2–5 Mt per year.

Clays from the basal Unit (Tierra de Campos) of the Duero basin in the area near Valladolid are also exploited for use as ceramic raw materials.

Diatomites

Spanish diatomite production during 1990 was 107 kt, lower than the 1986 production. Spanish diatomite deposits occur in Neogene formations in the southern part of Spain in the Guadalquivir Basin (Porcuna and Martos), the Prebetic intramontane basins and the Internal Betics intramontane Late Miocene basins (Sorbas, Vera, and Níjar in Almería and Murcia provinces) (Calvo, 1984; Regueiro *et al.*, 1993).

The Prebetic area supplies more than 90% of the total Spanish diatomite production. The Hellín diatomite district extends over more than 100 km², and the diatomite occurs here in several basins that were either separated or episodically interconnected: the Cenajo Basin, Camarillas Basin, Calderones Basin, Elche de la Sierra Basin and Hijar Basin (see Table 5). These five intramontane basins outcrop along the Mundo and Segura river valleys. A schematic stratigraphic column for these basins has been proposed recently by Elízaga & Calvo (1988) (Fig. 13). Sedimentological and isotope data have been reviewed recently by Bellanca *et al.* (1989). The diatomaceous sediments are close to Facies E just above a 30–50 m thick megaslump. These megaslumps are associated with volcanism and/or a probably tectonic reactivation of the basins. The K/Ar age of the volcanism measured in the Monegrillo is up to 5 Ma (Bellón *et al.*, 1981). The megaslump-level offers the best marker in prospecting for diatomite in the district (Regueiro *et al.*, 1993).

Ten per cent of Spanish diatomite production is from the Níjar Basin (Almería); these diatomaceous sediments are located in mm-thick laminated shales formed by couplets of dark organic-rich oil shale and clear diatomite-rich laminae. The age of the deposits spans the interval from Late Tortonian to Early Messinian (Reyes *et al.*, 1984).

Table 5. *Reserves and resources of diatomite deposits in the Hellín district*

	Cenajo	Camarillas	Calderones	Elche	Hijar
r I	7.5	—	—	—	—
r II	18.7	29	—	—	—
r III	—	—	—	—	5
R I	152	—	—	—	—
R II	133	12	—	337	—
R III	8.8	—	34	—	—

Notes:
Reserves (r) and identified resources (R) in Mt for the different basins. I $SiO_2 < 30\%$; II $30\% < SiO_2 < 70\%$; III $SiO_2 > 70\%$.
Source: Adapted from Regueiro *et al.* (1993).

Diatomaceous sediment occurrences, sometimes mines, have been cited in the Guadalquivir Basin, for example at Sanlúcar de Barrameda, Porcuna, and Martos. Earlier in this chapter we have cited the diatomaceous sediments in the El Cuervo attapulgite deposit. Other occurrences have been cited in the Madrid basin (Calvo *et al.*, 1988), and in the Cerdanya basin. In this last occurrence, dated as Middle–Late Miocene, Anadón *et al.* (1989) have described a diatomaceous mudstone facies with fine lamination.

Dimension (building) stone

Spain plays a very important role in the world stone industry. The natural stone production of Spain accounts for 13.2% of the world total (Lombardero & Regueiro, 1992).

Some Tertiary limestones of the Inner Prebetic Units have great purity and soundness, are easily polished and may be considered as 'commercial marbles' or marbles of type C, according to the Marble Institute of America (MIA). The quarries are grouped mainly in three areas: Coto Pinoso and Peña de Zafra in Alicante province and Sierra de la Puerta in Murcia province. The most famous quarries are located in Coto Pinoso, where the 'Crema Marfil' marble is obtained; the 1989 production of blocks was up to 10^5 m³, and the proven reserves are 7×10^3 m³. From a petrographic point of view the Coto Pinoso limestone may be classified as a biosparite–biomicrite (Llopis & López Jimeno, 1991).

The 'Colmenar de Oreja Limestone' has been obtained from the Upper Unit of the Madrid basin (Fig. 10), and it has been used since the eighteenth century in the construction of buildings and monuments in Madrid (Dapena *et al.*, 1988). According to the ASTM (American Society for Testing of Materials) Standard C–568–79, Colmenar limestone is among the most suitable for use for the outside of buildings (Dapena *et al.*, 1988). Colmenar limestones are paludine–lacustrine biosparites with less than 2.45% porosity. At present, Colmenar limestones and other similar limestones of the Upper Tertiary Unit of Madrid are intensively used as crushed

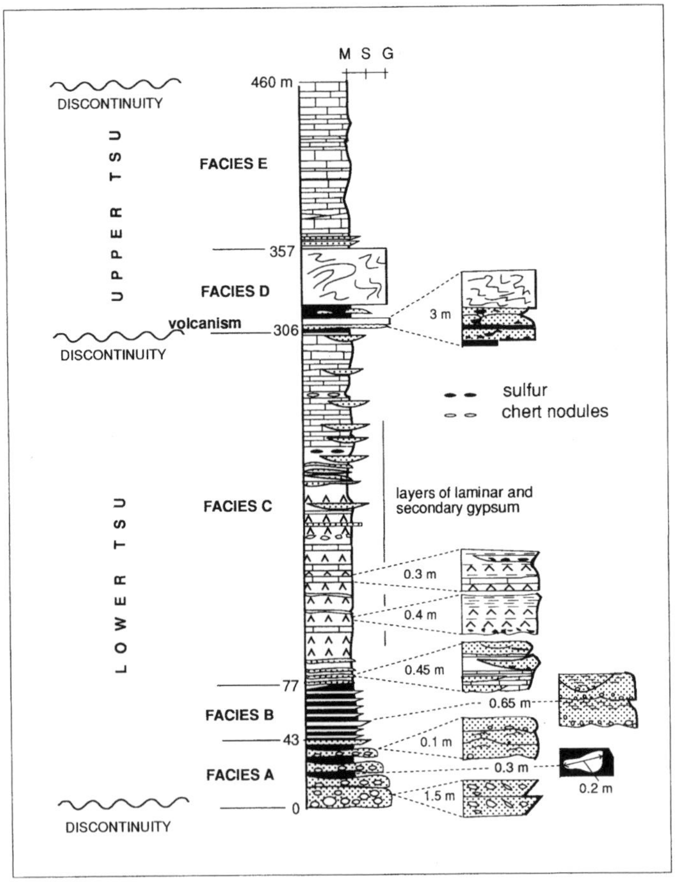

Fig. 13. Schematic stratigraphic section of Cenajo basin: A, conglomeratic basal beds; B, sandy layers in laminated oil shales; C, this subunit consists of alternations of laminated detrital carbonate and gypsum; D, megaslump subunit probably related to eruptive volcanism (5.7 ± 0.3 Ma, after Bellón *et al.*, 1981); E, diatomaceous sediments. After Elízaga & Calvo (1988). 'TSU' refers to Tecto sedimentary unit (chapter G1).

aggregate raw materials, cement raw materials, and lime raw materials (> 98% $CaCO_3$) and fillers (García Calleja, 1991; García Calleja *et al.*, 1991; García del Cura *et al.*, 1993).

References

Alvarez Ramis, C., Fernández Marrón, M.T., Garcia del Cura, M.A. and Ordóñez, S. (1989). Preliminary data and paleoecological characteristics of the Saline Unit flora from the Tertiary Madrid basin. *Abstracts II European Paleobotanical Conf.* Madrid: 19.

Anadón, P., Cabrera, L., Juliá, R., Roca, E. and Rosell, L. (1989). Lacustrine oil–shale basins in Tertiary grabens from NE Spain (Western European rift system). *Palaeogeogr., Palaeoclimatol., Palaeoecol.*, **70**: 7–28.

Arauzo Perez, M., González López, J.M. and López Aguayo, F. (1989). Primeros datos sobre la mineralogía y génesis del yacimiento de sepiolita de Mara (Prov. de Zaragoza). *Bol. Soc. Esp. Mineral.*, **12**: 329–340.

Arribas, A. (1963). Mineralogía y metalogenia de los yacimientos españoles de uranio: Paracuellos del Jarama (Madrid). *Bol. R. Soc. Española Hist. Nat.*, **61**: 63–65.

Arribas Rosado, A. (1992). Los yacimientos de oro del sureste peninsular. In García Guinea, J. and Martínez Frías, J. (eds.), *Recursos Minerales de España*, Textos Universitarios, 15. C.S.I.C.: 875–890.

Bacelar, J., Alonso, M., Kaiser, C., Sánchez, M., Cabrera, L., Ferrus, B., Sáez, A. and Santanach, P. (1991). Relaciones tectónica-sedimentación de la Cuenca Cenozoica de As Pontes (A Coruña). *Congreso del Grupo Español del Terciario*: 25–27.

Bellanca, A., Calvo, J.P., Censi, P., Elízaga, E. & Neri, R. (1989). Evolution of lacustrine diatomite carbonate cycles of Miocene age, southeastern Spain: petrology and isotope geochemistry. *J. Sed. Petrol.*, **59**: 45–52.

Bellon, H., Bizon, G., Calvo, J.P., Elízaga, E., Gaudant, J. and López, N. (1981). Le volcan de Cerro de Monegrillo (province de Murcia): âge radiométrique et corrélations avec les sédiments Neogènes du Bassin de Hellín. *C.R. Acad. Sci. Paris*, **2eme série, II**: 1035–1038.

Caballero, E., Fernández Porto, M.J., Linares, J. and Reyes, E. (1983). Las bentonitas de la Serrata de Níjar (Almería). Mineralogía, geoquímica y mineralogénesis. *Estudios Geol.*, **39**: 121–140.

Caballero, E., Reyes, E., Linares, J. and Huertas, J. (1985). Hydrothermal solutions related to bentonite genesis, Cabo de Gata region, Almeria, SE Spain. *Mineral. Petrograph. Acta*, **29A**: 187–196.

Caballero, E., Reyes, E., Yusta, A., Huertas, F. and Linares, J. (1985). Las bentonitas de la zona sur de Cabo de Gata (Almería). Geoquímica y Mineralogía. *Acta Geológica Hispánica*: **20**: 267–287.

Cabrera, Ll. and Sáez, A. (1987). Coal deposition in carbonate-rich shallow lacustrine systems: the Calaf and Mequinenza sequences (Oligocene, Eastern Ebro basin, NE Spain). *J. Geol. Soc. Lond.*, **144**: 451–461.

Calvo, J.P. (1984). Los yacimientos españoles de diatomitas. In Leguey, S., Menduiña, J. and Ordóñez, S. (eds.) *Segundo Curso de Rocas Industriales*: 330–347.

Calvo, J.P., Pozo, M. and Servant-Vildary, S. (1988). Lacustrine diatomite deposits in the Madrid Basin (Central Spain). *Geogaceta*, **4**: 14–17.

Colom, G. (1983). *Los lagos del Oligoceno de Mallorca*. Caja 'Sa Nostra', Palma de Mallorca, 166 pp.

Crespo Zamorano, A. (1988a). Primeras notas sobre los yacimientos cobaltíferos en el Plioceno del Campo de Calatrava (Ciudad Real). *Bol. Soc. Española Mineral.*, **11**: 149–152.

Crespo Zamorano, A. (1988b). Depósitos de manganesos cobaltíferos en Ciudad Real. *Congreso Internacional de Minería y Metalurgia*. Oviedo.

Cuevas, J. (1992). *Caracterización de esmectitas magnésicas de la Cuenca de Madrid como materiales de sellado*. Publicación Técnica no. 04/92. ENRESA, Madrid, 188 pp.

Cuevas, J., Pelayo, M., Rivas, P. and Leguey, S. (1993). Characterization of Mg-clays from the Neogene of the Madrid Basin and their potential as backfilling and sealing material in high level radioactive waste disposal. *Appl. Clay Sci.*, **7**: 199–214.

Dapena, E., Ordóñez, S. and García del Cura, M.A. (1988). Study of limestone rock used in the construction of palaces in Madrid during the 18th and 19th centuries. In P.G. Marinos and G.C. Koukis (eds.) *The engineering geology of ancient works, monuments and historical sites*. Balkema. Rotterdam: 683–690.

Doval, M. (1992). Bentonitas. In J. García Guinea and J. Martínez Frías (eds.), *Recursos Minerales de España*. Textos Universitarios, 15, C.S.I.C.: 45–69.

Elízaga, E. and Calvo, J.P. (1988). Evolución sedimentaria de las cuencas lacustres neógenas de la zona prebética (Albacete). *Bol. Geol. Mineral*, **99**: 837–846.

Fernández Nieto, C. and Galán, E. (1979). Mineralogía de los depósitos de sal de Remolinos (Zaragoza). *Soc. Esp. Mineral.*, **vol. extra 1**: 51–65.

Ferrero, F. (1984). Los depósitos de sepiolita, bentonita y attapulgita en España. In Leguey, S., Menduiña, J. and Ordóñez, S. (eds.)

Segundo Curso de Rocas Industriales. Fundación Universidad Empresa, Madrid: 126–135.

Galán, E. and Castillo, A. (1984). Sepiolite–palygorskite in Spanish Tertiary basins: genetical patterns in continental environments. In Singer and Galán (eds.), *Palygorskite–Sepiolite: occurrences, genesis and uses.* Elsevier: 87–124.

Galán, E. and Ferrero, A. (1982). Palygorskite–sepiolite clays of Lebrija, Southern Spain. *Clay. Clay Mineral.* **30**: 191–199.

Galán, E., Brell, J.M., La Iglesia, A. and Robertson, R.H.S. (1975). The Cáceres palygorskite deposit. Spain. In Bailey, S. (ed.), *Proc. Int. Clay Conf. México* Applied Publ. Ltd. Illinois: 81–94.

García, E., Brell, J.M., Doval, M. and Navarro, J.V. (1990).Caracterización mineralógica y estratigrafía de las formaciones neógenas del borde Sur de la Cuenca del Tajo (Comarca de La Sagra). *Bol. Geol. Min.*, **101**: 945–956.

García Calleja, M.A. (1991). Estudio petrológico y geoquímico de las materias primas de la Cuenca de Madrid para su uso en la industria cementera. PhD Thesis. U.C.M. Madrid, 463 pp.

García Calleja, M.A., Soriano, J. and Ordóñez, S. (1991). Materias primas para la fabricación de clinker de cemento Portland en el area de Madrid. *IV Congreso de Geoquímica de España*: 493–506.

García del Cura, M.A. (1979). Las sales sódicas, calcosódicas y magnésicas de la Cuenca del Tajo. *Serie Universitaria*, **109**, Fundación Juan March, Madrid, 39 pp.

García del Cura, M.A. and López Aguayo, F. (1974). Estudio mineralógico de las facies detritico – calcáreas del Terciario de la Zona Centro-Oriental de la Cuenca del Duero (Aranda de Duero). *Estud. Geol.*, **30**: 503–513.

García del Cura, M.A., Ordoñez, S. and Lopez Aguayo, F. (1979). Estudio petrológico de la Unidad Salina de la Cuenca del Tajo. *Estud. Geol.*, **35**, 325–339.

García del Cura, M.A., Ordóñez, S. & González, J.A. (1991). Los carbonatos biogénicos de los episodios terminales del relleno Neógeno de la Cuenca de Madrid. *Comunicaciones I Congreso del Grupo Español del Terciario*: 136–139.

García del Cura, M.A., Ordóñez, S. and Sánchez Moral, S. (1992). Un yacimiento de glauberita en facies evaporíticas marginales: Mina Consuelo (San Martín de la Vega – Madrid). *Actas III Congreso Geológico de España y VIII Congreso Latinoamericano de Geología*, **3**: 372–377.

García del Cura, M.A., Ordóñez, S., González, J.A., Dapena, E., Cañaveras, J.C., Díaz Alvarez, M.C., García Calleja, M.A., Galindo, E., Martínez Alfaro, P.E., Calvo, J.P., La Iglesia, A. and Sanz, E. (1993). Las canteras de calizas de los interfluvios de los ríos Jarama–Tajuña–Tajo en la Comunidad de Madrid: valoración de recursos, impacto ambiental y propuestas de restauración. 84 pp. + maps (unpublished).

García Guinea, J. and Martinez Frías, J. (eds.) (1992). *Recursos minerales de España.* Textos Universitarios, 15, C.S.I.C. Madrid, 1448 pp.

García Veigas, J., Fernández Nieto, C. and Orti, F. (1991). Nota sobre la mineralogía y petrología de la Formación Zaragoza en el sondeo Purasal. *Bol. Soc. Española Mineral.*, **14–1**: 82–83.

González Díez, I. (1992). Arcillas comunes. In García Guinea, J. and Martinez Frías, J. (eds.) *Recursos Mineral. España*, C.S.I.C.: 95–112.

González, I., Renedo, E. and Galan, E. (1986). Clay materials for structural clay products from the Bailen area Southern Spain. *Uppsala Symposium Clay Minerals–Modern Society, 1985.* Nordic Society for Clay Minerals: 77–90.

Gorch, R., De Las Heras, F.X., Grimalt, J., Albaiges, J., Sáez, A. and Cabrera, Ll. (1992). Biomarcadores en los lignitos lacustres oligocénicos de la Cuenca del Ebro (Calaf, Mequinenza). *III Congreso Geológico de España, Simp.* **I**: 88–97.

Griffiths, J. (1991). Spain's minerals: mixed fortunes. *Industrial Minerals*, **285**, June: 23–47.

IGME (1985). *Actualización del inventario de recursos nacionales de carbón.* MIE Madrid: 217 pp.

ITGE (1991). *Panorama Minero 1989.* MIE, Madrid, 428 pp.

ITGE (1992). *Panorama Minero 1990.* MIE, Madrid, 719 pp.

Leguey, S., Martín de Vidales, J. and Casas, J. (1984a). Diagenetic palygorskite in marginal continental detrital deposits located in the South of the Tertiary Duero basin (Segovia). In Singer and Galán (eds.) *Palygorskite–Sepiolite: occurrences, genesis and uses. Developments in Sedimentology*, 37. Elsevier: 149–156.

Leguey, S., Menduiña, J. and Ordóñez, S. (1984b). *Segundo Curso de Rocas Industriales.* Apuntes Universidad Empresa, Madrid, 494 pp.

Leone, G., Reyes, E., Cortecci, G., Pochini, A. and Linares, J. (1983). Genesis of bentonites from Cabo de Gata, Almeria, Spain: a stable isotope study. *Clay Mineral.* **18**: 227–238.

Llopis, L. and López-Jimeno, C. (1991). Crema Márfil, un mármol con futuro. *Canteras y Explotaciones*, **March**: 50–58.

Llopis, L., López-Jimeno, C. and Mazadiego, L.F. (1992). Rocas ornamentales de Alicante. *Canteras y Explotaciones*, **January**: 103–113.

Lombardero, M. and Regueiro, M. (1992). Spanish natural stone: cladding the world. *Industrial Minerals*, **300**: 81–97.

Mañana, R. (1992). A synthetic overview of the Spanish mining industry. AITEMIN. 15 pp.

Manteca, J.I. and Ovejero, G. (1992). Los yacimientos Zn, Pb, Ag–Fe del distrito minero de la Unión-Cartagena, Bética Oriental. In J. García Guinea and J. Martínez Frías (eds.) *Recursos Minerales de España.* C.S.I.C.: 1085–1102.

Martín, J.M., Ortega Huertas, M. and Torres Ruiz, J. (1984). Genesis and evolution of strontium deposits of the Granada basin (Southeastern Spain): evidence of diagenetic replacement of a stromatolite belt. *Sediment. Geol.* **39**: 281–298.

Martín Calvo, M. (1973). Sobre la petrogénesis de algunas litofacies españolas con fases urano-orgánicas. PhD Thesis. U.C.M. Madrid: 306 pp.

Martín Delgado, J. (1975). Pasado, presente y futuro de la investigación uranífera en España. *Energía Nuclear*, **19**: 321–329.

Martín Pozas, J.M., Martín Vivaldi, J. and Sanchez Camazano, M. (1983). El yacimiento de sepiolita–palygorskita de Sacramenia, Segovia. *Bol. Geol. Min.*, **94**: 112–120.

Martín de Vidales, J.L., Pozo, M. and Leguey, S. (1989). Kerolite–stevensite mixed layers from Neogene Madrid Basin. Genetic implications (abstract). *International Clay Conference, Strasbourg*: 246 pp.

Martín de Vidales, J.L., Pozo, M. Alia, J.M., Garcia Navarro, F. and Rull, F. (1991). Kerolite–stevensite mixed-layers from the Madrid Basin, Central Spain. *Clay Minerals*, **26**: 329–342.

Megías, A.G., Leguey, S. and Ordóñez, S. (1982). Interpretación tectosedimentaria de la génesis de fibrosos de la arcilla en series detríticas continentales (Cuencas de Madrid y del Duero) España. *5º Congreso Latinoamericano de Geología, Argentina, Actas* **II**: 427–439.

Megías, A.G., Ordóñez, S. and Calvo, J.P. (1983). Nuevas aportaciones al conocimiento geológico de la Cuenca de Madrid. *Rev. Material Procesos Geol.*, **1**: 163–191.

Menduiña, J. (1988). Geología y significado económico de las arcillas cerámicas de la Cuenca de Madrid. PhD Thesis. U.C.M. Madrid. 305 pp.

Menduiña, J., Ordóñez, S. and García del Cura, M.A. (1984). Geología del yacimiento de glauberita de Cerezo del Río Tirón (Provincia de Burgos). *Bol. Geol. Mineral.*, **95**: 33–51.

Meseguer Pardo, J. (1924). Estudio de los yacimientos de azufre en las provincias de Albacete y Murcia. *Bol. Inst. Geol.* **45**: 1–84.

Michell, B. (1979). *Geography and resource analysis.* Longman, London, 399 pp.

Ordóñez, S., Brell, J.M. Calvo, J.P and Lopez Aguayo, F. (1977). Contribución al conocimiento mineralógico del borde SW de la Cuenca del Tajo (Toledo–San Martin de Pusa). *Estud. Geol*, **33**: 467–472.

Ordóñez, S. and García del Cura, M.A. (1992). El sulfato sódico natural en España: Las sales sódicas de la Cuenca de Madrid. In J. García Guinea and J. Martínez Frías (eds.), *Recursos Minerales de España*. Textos Universitarios, 15, C.S.I.C.: 1229–1250.

Ordóñez, S. and García del Cura, M.A. (1994). Sodium–calcium sulfate salt deposition and diagenesis in Tertiary saline lakes: Madrid Basin (Spain). *Spec. Publ. SEPM*, **150**: 229–238.

Ordóñez, S., López Aguayo, F. and García del Cura, M.A. (1980). Contribución al conocimiento sedimentológico del Sector Centro-Oriental de la Cuenca del Duero (Sector Roa-Baltanás). *Estud. Geol.* **36**: 361–369.

Ordóñez, S., Menduiña, J. and García del Cura, M.A. (1982). El sulfato sódico natural en España. *Tecniterrae*, **46**: 16–33

Ordóñez, S., Fontes, J.Ch. and García del Cura, M.A. (1983). Contribución al conocimiento de la sedimentogénesis evaporítica de las cuencas neógenas de Madrid y del Duero, en base a datos de isótopos estables (^{13}C, ^{18}O y ^{34}S). *Com. X Cong. Nac. Sedimentología, Menorca*: 49–52.

Ortega Huertas, M. (1992). Yacimientos de estroncio en España: geología e interés económico. In J. García Guinea and J. Martínez Frías (eds.). *Recursos Minerales de España.* Textos Universitarios, 15, C.S.I.C., 429–438.

Ortí, F. (1990). Observaciones sobre la formación Zaragoza y unidades evaporítcas adyacentes (Mioceno Continental). In F. Ortí and J.H. Salvany (eds.) *Formaciones evaporíticas de la Cuenca del Ebro y cadenas periféricas, y de la zona de Levante.* Universidad de Barcelona: 117–119.

Ortí, F. and Pueyo, J.J. (1977). Asociación halita bandeada-anhidrita nodular del yacimiento de Remolinos, Zaragoza (sector central de la Cuenca del Ebro). Nota petrogenética. *Rev. Inst. Inv. Geol. Dip. Prov. Barcelona*, **32**: 167–202.

Ortí, F. and Salvany, J.M. (eds.) (1990). *Formaciones evaporíticas de la Cuenca del Ebro y cadenas periféricas, y de la zona de Levante.* Universidad de Barcelona. 306 pp.

Ortí, F., Pueyo, J.J. and San Miguel, A. (1979). Petrogénesis del yacimiento de sales sódicas de Villarrubia de Santiago, Toledo (Terciario continental de la Cuenca del Tajo). *Bol. Geol. Min.* **90**: 347–373.

Pérez García, L.C. and Sánchez-Palencia, F.J. (1985). Los sedimentos auríferos en la Antiguedad. *Investigación y Ciencia*, **104**: 64–75.

Pérez García, L.C. and Sánchez-Palencia, F.J. (1992). Los yacimientos de oro de Las Médulas de Carucedo (León). In J. García Guinea and J. Martínez Frías (eds.). *Recursos Minerales de España.* Textos Universitarios, 15, C.S.I.C.: 862–873.

Porter, D.H. and Alvarez Morán, B. (1992). Mineralizaciones de oro del Noroeste de España. In J. García Guinea and J. Martínez Frías (eds.). *Recursos Minerales de España.* Textos Universitarios, 15, C.S.I.C.: 849–860.

Pozo, M. and Leguey, S. (1985). Distribution of clay minerals in central facies, Duero Basin, Valladolid province, Spain. *Mineral. Petrograph. Acta*: **29-A**: 344–345.

Pozo, M., Cuevas, J., Moreno, A., Redondo, R. and Leguey, S. (1991). Caracterización de arcillas magnésicas bentoníticas en la zona de Yuncos (Toledo). *Bol. Geol. Min.* **102**: 893–904.

Pueyo, J.J. (1975). Estudio petrológico y geoquímico de los yacimientos potásicos de Cardona, Suria, Sallent y Balsareny (Barcelona, España). PhD Thesis. Universidad de Barcelona. 351 pp.

Ramírez, A. (1966). Las Cuencas lignito-uraníferas de Calaf y Ebro-Segre. *Energía Nuclear*, **10**: 458–464.

Ramos-Guerrero, E., Cabrera, Ll. and Marzo, M. (1989). Un modelo de sedimentación carbonatada lacustre-palustre en el Eoceno Medio de Mallorca. *XI Congreso Español Sedimentología, Comunicaciones*: 75–78.

Regueiro, M., Calvo, J.P., Elízaga, E. and Calderón, V. (1993). Spanish diatomite: geology and economics. *Industrial Minerals*, **306**, March 1993: 57–67.

Reyes, J.L. and Feixas, J.C. (1984). Las pizarras bituminosas: definición, composición y clasificación. *I Congreso Español de Geología*, **II**: 817–827.

Reyes, J.L., Crespo, V., Feixas, J.C. and Zapata, M.J. (1984). La sedimentación evaporítica en las cuencas neógenas del SE peninsular. *I Congreso Español de Geología*, **II**: 803–815.

Reyes, J.L., Zapatero, M.A., Feixas, J.C. and Avila, J. (1992). El azufre biogénico de las Cuencas Neógenas del Sureste. *III Congreso geológico de España*, **3**: 410–417.

Rodríguez, P. (1992). Distrito de Mazarrón Zn-Pb-Ag. Mineralizaciones, potencial y trabajos de evaluación. *Tierra y Tecnología*, **3**: 28–33.

Rosell, L. (1983). Estudi petrològic, sedimentològic i geoquímic de la formació de sals potassiques de Navarra (Eocè superior). PhD Thesis. Universitat de Barcelona, 321 pp.

Rosell, L. (1990). La Cuenca Potásica Surpirenaica. In F. Ortí and J.M. Salvany (eds.), *Formaciones evaporíticas de la Cuenca del Ebro y cadenas periféricas, y de la zona de Levante.* Universidad de Barcelona: 89–95.

Rubio Navas, J. (1990). *Inventario nacional de recursos de estroncio, 1989.* ITGE. MIE Madrid, 199 pp.

Salvany, J.M. (1989). Las Formaciones evaporíticas del Terciario continental de la Cuenca del Ebro en Navarra y La Rioja. Litoestratigrafía, petrología y sedimentología. PhD Thesis. Universidad de Barcelona, 397 pp.

Salvany, J.M. and Ortí, F. (1992). El yacimiento glauberítico de Alcanadre: Procesos sedimentarios y diagenéticos (Mioceno inferior, Cuenca del Ebro). In García Guinea and Martínez Frías (eds.), *Recursos minerales de España.* Textos Universitarios, 15, C.S.I.C.: 1251–1274.

Salvany, J.M. and Ortí, F. (1994). Miocene glauberite deposits of Alcanadre, Ebro basin: sedimentary and diagenetic processes. *Spec. Publ SEPM*, **50**: 203–215.

Sánchez Moral, S., Hoyos, M., Ordóñez, S., García del Cura, M.A. and Cañaveras, J.C. (1993). Genesis de epsomita infiltracional por dedolomitización en ambiente sulfatado árido: eflorescencias en la Unidad Inferior Evaporíta de la Cuenca de Calatayud. *Congreso de Geoquimica de España.* Soria, 1993: 24–29.

Santanach, P., Baltuille, J.M., Cabrera, Ll., Monge, C., Sáez, A. and Vidal Romaní, J.R. (1988). Cuencas terciarias gallegas relacionadas con corredores de fallas direccionales. *II Congreso Geológico de España, Simposios*: 123–133.

Singer, A. and Galán, E. (eds.) (1984). *Palygorskite–sepiolite:occurrences, genesis and uses. Developments in Sedimentology*, 37. Elsevier. Amsterdam, 352 pp.

Suárez, F., Armenteros, I., Martín Pozas, J.M. and Navarrete, J. (1989). El yacimiento de palygorskita de Bercimuel (Segovia): Génesis y propiedades tecnológicas. *Estud. Geol.* **26**: 27–46.

Suárez, F., Flores, L., Añorbe, M., Díez, J.A., Navarrete, J. and Martín Pozas, J.M. (1991). Mineralogical and textural characterisation of the Bercimuel Paligorskite (Segovia, Spain). *Proc. 7th Euroclay Conf., Dresden, 1991*: 1019–1023.

Utrilla, R., Pierre, C., Ortí, F. and Pueyo, J.J. (1992). Oxygen and sulphur isotope composition as indicators of the origin of Mesozoic and Cenozoic evaporites from Spain. *Chem. Geol.*, **102**: 229–244.

Vázquez Guzman, F. (1983). *Depósitos minerales de España.* IGME, Madrid. 153 pp.

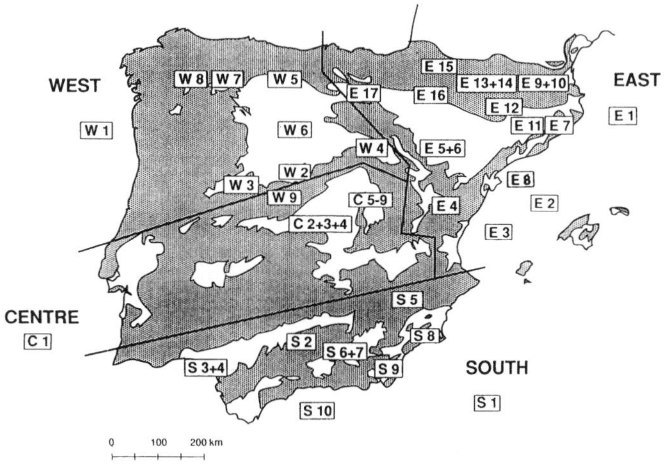

Part E
East

The sharp ridge to the right of centre is La Gronsa (819 m), 6 km south of Horta de Sant Joan, in the Catalan coastal range, between Alcañiz and the Ebro delta. Syndepositional uplift of the southeast margin of the Ebro basin has created a clear example of one of Riba's 'progressive unconformities' (see p. xvi). This is marked by stratal thinning towards the margin to the right, progressively greater dips in the older strata, and overstep of the older strata by some of the younger (photo: Jaume Vergés).

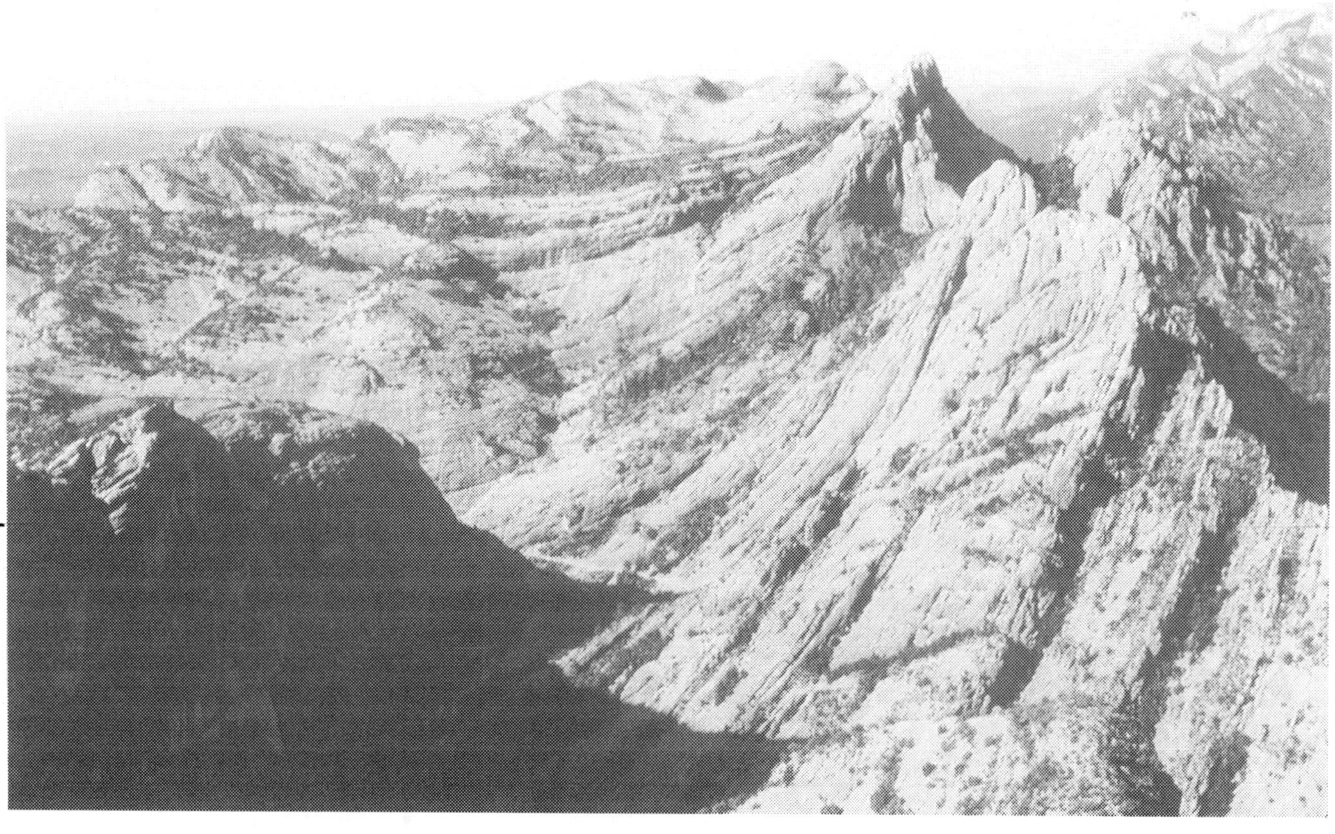

E1 Geological setting of the Tertiary basins of Northeast Spain

P. ANADÓN AND E. ROCA

Abstract

This introductory chapter outlines the geology of the following major features: the Pyrenees, the Catalan coastal range, the Iberian range, the southern Foreland basins of the Pyrenees, including the Ebro basin, and the Mediterranean Valencia trough.

Introduction

The present structure of the NE Iberian Plate (Fig. 1) results mainly from the convergence between the African, Iberian and Eurasian plates. Initiated during the Cenomanian–Turonian (Dewey *et al.*, 1973; Dercourt *et al.*, 1986; Dewey *et al.*, 1989), this convergence occurred in two well-differentiated stages (Srivastava *et al.*, 1990; Roest & Srivastava, 1991): a first Late Cretaceous–Middle Oligocene stage in which the convergence took place between Eurasia and Iberia, and a second Late Oligocene–Quaternary stage in which the convergence moved to the Iberian–Africa boundary.

During the first stage (mainly Paleogene in age), the convergence and later continental collision between the Iberian and Eurasian plates led to the building of the Pyrenean fold-and-thrust belt in the northern margin of the Iberian plate and the development of large compressive Paleogene structures in the northern inner parts of this plate (i.e. the Iberian Range and the Catalan Coastal Range in the NE Iberian Plate; Figs. 2 and 3). Mainly related to the building of the Pyrenees, the Ebro foreland basin also developed during this Paleogene stage.

After the conclusion of the collision between the Iberian and Eurasian plates and related to the beginning of the subduction of the African plate beneath the Iberian plate, during the Neogene, most of NE Iberia was subjected to generalized rifting related to the opening of back-arc basins in the Western Mediterranean. The Valencia Trough is the most prominent of the extensional basins generated during this stage (Fig. 2). Other Neogene grabens, often resulting in the tectonic inversion of Paleogene compressive structures, also developed in the Eastern Pyrenees, Catalan Coastal Range and Iberian Range (Fig. 3). Synchronously with these rifting processes, the Betic-Balearic thrust-and-fold belt developed until the Middle Miocene in the neighbouring areas of the Balearic Islands, recording the collision between the Iberian and African plates.

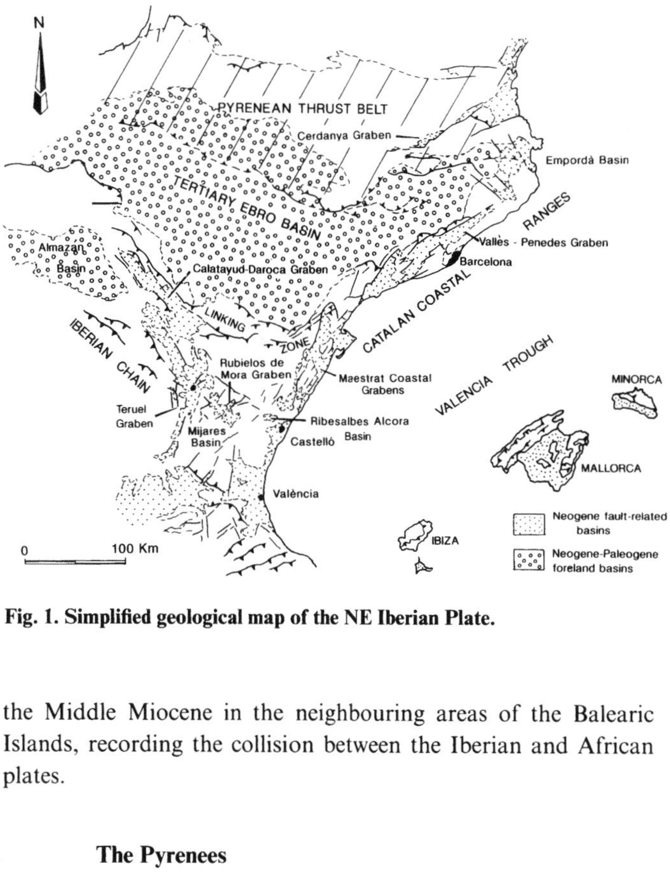

Fig. 1. Simplified geological map of the NE Iberian Plate.

The Pyrenees

The Pyrenees are an E–W-trending mountain belt that extends linearly between the Golf du Lion (Western Mediterranean) and the Bay of Biscay (Atlantic Ocean). Formation and development of the Pyrenean orogen was related to the progessive N–S continental collision between Iberia and Eurasia that led to the subduction of the Iberian plate beneath the European one (ECORS Pyrenees team, 1988; Muñoz, 1992). This subduction, in the upper crustal levels, generated the development of a basement-involved antiformal stack in the middle of the chain (Axial Zone), bounded by imbricate cover thrust systems, mainly south-directed thrusts to the south (Cover Upper thrust sheets of the Southern Pyrenees) and

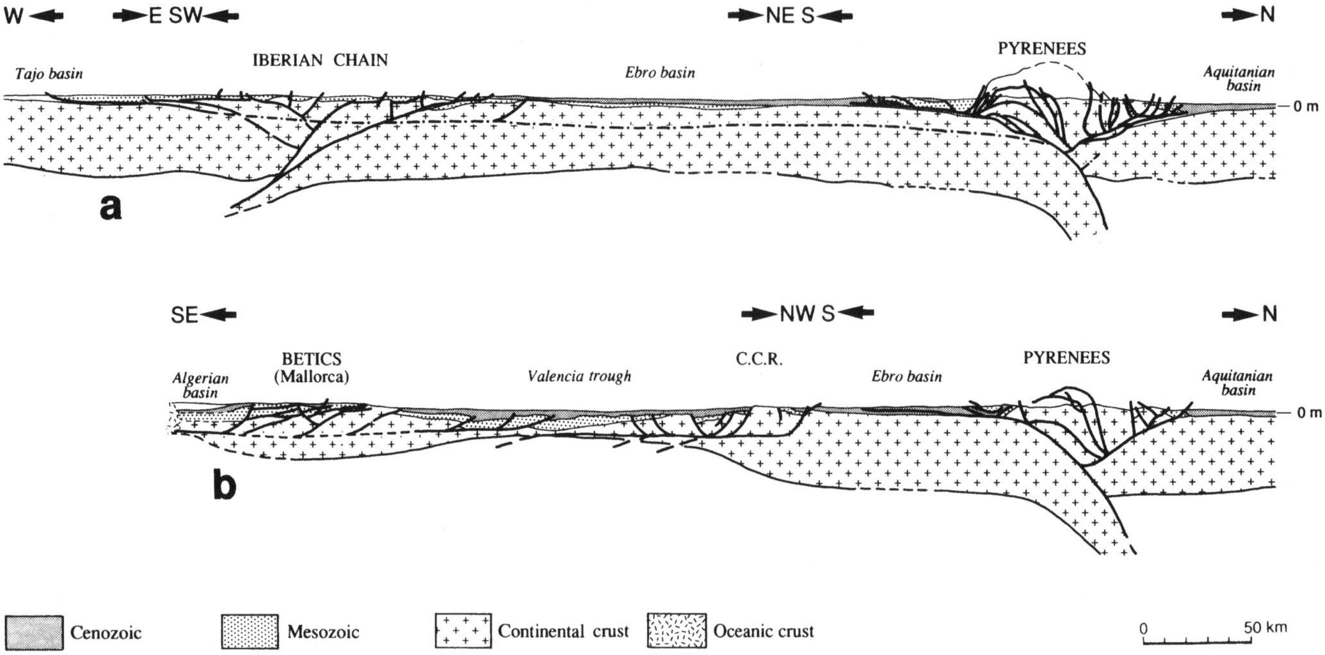

Fig. 2. Geological cross-sections along two lines (a and b) through the main Cenozoic structural units of the NE Iberian Plate. The deep structure of the Iberian Range (cross-section a) has been drawn following two different models. In the first model (continuous lines), the whole crust has been affected by Paleogene compressive tectonics and the presence of an incipient subduction zone is assumed (based on Salas and Casas, 1993). In the second model (discontinuous and dotted line, cross-section 2a), only the upper crust is affected by Paleogene compression, which implies the presence of a detachment in the middle crust under the whole Ebro basin (Guimerà and Alvaro, 1990).

mainly north-directed thrusts to the north (Northern Pyrenees) (Fig. 2). The thrusts of the basement antiformal stack are south-directed and they involve Hercynian rocks and, locally, Triassic cover rocks. Detached within the Middle Triassic evaporites and lutites, the Cover Upper thrust sheets (=South Pyrenean Central Unit of Séguret, 1972) are made of Mesozoic, mainly platform sequences, and Paleogene rocks. These thrust sheets are well developed in the Central Pyrenees where they cover a wide area of the autochthonous non-deformed Paleogene series of the Ebro Basin, which unconformably overlie very reduced Mesozoic series or basement rocks (Cámara & Klimovitz, 1985; Muñoz, 1992).

The emplacement of all these thrust units began in the Maastrichtian–Late Cretaceous (Simó & Puigdefàbregas, 1985; Vergés & Martínez, 1988) and continued until the Oligocene in the Eastern Pyrenees (Sáez, 1987; Vergés & Muñoz, 1990), and until the Early Miocene in the Central and Western Pyrenees (Puigdefàbregas & Soler, 1973).

Since the Oligocene (Arthaud *et al.*, 1981), in the eastern parts of the Pyrenean orogen, this compressive structure was affected by extensional processes that generated the development of diverse horst and graben systems (Fig. 1). In this superimposed extensional structure, two chronologically and geometrically well-differentiated systems have been distinguished in the Eastern Pyrenees. The first one consists of ENE–WSW grabens (i.e. Seu d'Urgell, Cerdanya, Conflent and Roselló basins) and has developed from the Early Miocene to the present in the eastern areas of the Axial Zone (see Chapter E10). The second system is only developed in the southern parts of the easternmost Pyrenees and is characterized by NW–SE horst and grabens (i.e. Empordà basin) that were generated from the Late Miocene (Calvet, 1985; Agustí *et al.*, 1990). This last system also cuts across the northern parts of the Catalan Coastal Range (i.e. Selva graben).

The Catalan Coastal Range

The Catalan Coastal Range is a NE–SW-trending belt which separates the Valencia Trough from the Ebro Basin (Fig. 1). Its structure is controlled by a set of right stepping, en échelon, basement faults whose strike changes from ENE–WSW to NE–SW along the range. Active as extensional faults during the Mesozoic (Anadón *et al.*, 1979; Roca & Guimerà, 1992), the motion of these SE-dipping faults has changed at different times during the Cenozoic (Fig. 3).

First, during the Paleogene, these faults moved as sinistral convergent strike–slip faults giving rise to the building of a mountain chain that bounded the Ebro Basin towards the SE (Fontboté, 1954; Guimerà, 1984; Anadón *et al.*, 1985). In the northern parts of the range, the convergent wrenching on these faults produced a set of NW-directed thrust slices of Hercynian rocks that overthrusted the Triassic rocks and the Paleogene sequences of the Ebro Basin. To the south, where the Mesozoic cover is thicker and detached over a relatively thick Upper Triassic evaporitic interval (Keuper), the Paleogene motion of these faults generated the folding of the thick Mesozoic cover and, locally, the thrusting of the detached cover over the Mesozoic and Paleogene sequences of the Ebro Basin margins. The timing of the beginning of the compressive deforma-

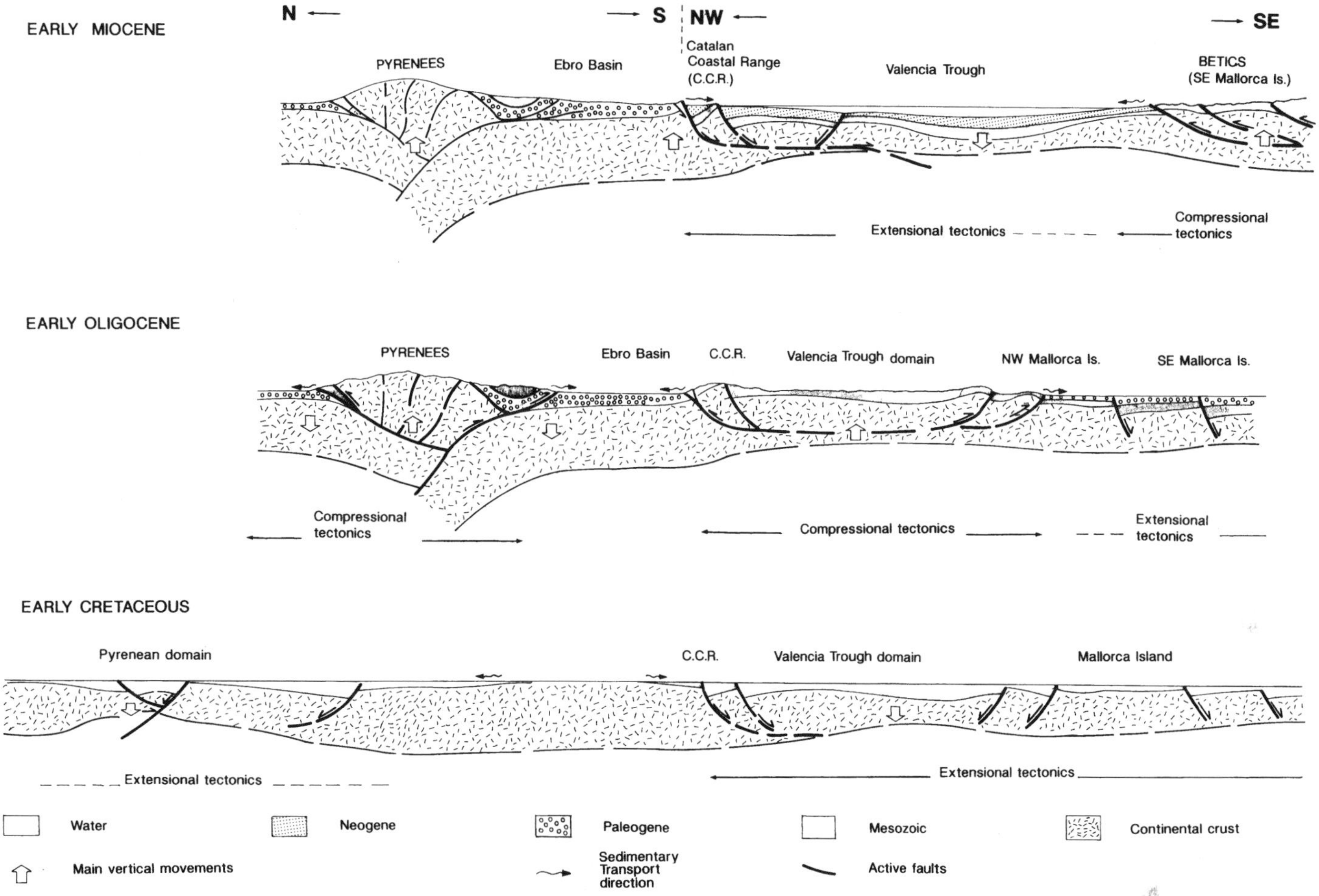

Fig. 3. Series of simplified sketches based on the cross-section shown in Fig. 2b, showing the major trends of the Tertiary geodynamic evolution of the NE Iberian Plate. The vertical scale is slightly exaggerated and north–south directions have been reversed relative to Fig. 2b.

tion was diachronous, starting in the Early Eocene in the northern part of the chain and in Middle to Late Eocene to the south (Guimerà, 1984; Anadón *et al.*, 1985). The youngest compressive deformation occurred in the Late Oligocene (Anadón *et al.*, 1985).

From the Late Oligocene–Early Miocene, the motion of the major basement faults changed as a consequence of the opening of the Valencia Trough (Fontboté *et al.*, 1989, 1990). During this Neogene second period, the Catalan Coastal Range was affected by extensional processes that gave rise to the inversion of most of the major Paleogene compressive faults (Fig. 3). The normal motion of these faults generated the development of a widespread ENE–WSW to NE–SW graben and horst system along the whole range (Bartrina *et al.*, 1992; Roca & Guimerà, 1992). The grabens are bounded towards the NW by normal faults with kilometre-scale slip and are infilled by terrigenous and carbonate sequences, up to 3 km thick, which range in age from the Early Miocene to present (Bartrina *et al.*, 1992).

The Iberian Range

The Iberian Range is a NW–SE intraplate chain generated in an area relatively far from the Iberian Plate margins (Fig. 1). It constitutes the southwestern boundary of the Ebro Basin and its present structure was mainly formed by the Paleogene inversion of an intracontinental NW–SE Mesozoic rift system that included some E–W extensive and thick basins (e.g. Maestrat and Cameros basins; Álvaro *et al.*, 1979; Guimerà & Álvaro, 1990).

The Paleogene structure of the Iberian Range is strongly controlled by the existence of a thick Upper Triassic evaporitic and lutitic interval that separates two differently structured levels: the Hercynian basement – with Permian and Lower Triassic tegument – and the Jurassic to Tertiary cover (Richter & Teichmüller, 1933; Viallard, 1983; Guimerà & Álvaro, 1990). Whereas the basement is mainly controlled by NW–SE convergent strike–slip faults and E–W N-verging basement thrusts, the cover has diverse fold and thrust systems that, locally, have been deformed by the movement of strike–slip faults developed in the basement. All these compressive structures are mainly well developed in the margins of the range and they verge outward from the axis of the belt (Stille, 1931); that is they verge towards the N and NE (Ebro Basin) in the northeastern part of the chain, and towards the W and SW (Tagus–La Mancha Basin) in the western part of the chain. The beginning of the compressive deformation of the Iberian Range is not clear, although it is supposed that it began during the Latest Eocene. In contrast, its end has been well dated as Late Oligocene–Early Miocene in the eastern parts (Viallard, 1973; Guimerà, 1987) and Early to Middle (?) Miocene in the northwestern parts (Riba *et al.*, 1983; Colomer & Santanach, 1988; Muñoz-Jiménez, 1992).

Superimposed on this compressional structure there is a complex system of Neogene grabens (see Chapter E4) which developed synchronously with the opening of the Valencia Trough. Generated from the Latest Oligocene (?)–Earliest Miocene up to the present (Crusafont *et al.*, 1966; Agustí *et al.*, 1988; Moissenet, 1989), these grabens display different orientations and ages along the chain (Fig. 1). Close to the Mediterranean coast and in the SE parts of the chain, the grabens are mainly Early Miocene in age and they are oriented NNE–SSW, oblique or transverse to the chain (e.g. Teruel graben). In the central and northwestern parts (e.g. Calatayud–Daroca graben), they are parallel to the main trend of the chain (NW–SE) and younger.

The Southern Foreland Basin(s) of the Pyrenees: The Ebro Basin

The Ebro Basin is a broadly triangular feature with an E–W trending asymmetric sedimentary trough which comprises an autochthonous, relatively intact part, and some piggy-back basins (Fig. 1). The evolution of this foreland basin complex was mainly linked to the emplacement of the South Pyrenean thrust sheets during the Paleogene (Séguret, 1972; Labaume *et al.*, 1985; Muñoz *et al.*, 1986; Puigdefàbregas *et al.*, 1986; 1992). Some parts of the Ebro basin were detached from the main sedimentary areas and moved in a piggy-back fashion on the emplacing thrust sheets (Ori & Friend, 1984). Other basins (piggy-back basins *sensu stricto*) developed exclusively during the propagation of the underlying thrust sheets (e.g. La Pobla de Segur basin; Mellere, 1992). The southward propagation of the compressive deformation led to migration of the facies depocentres in the foreland from N to S from the Paleocene to the Upper Miocene (Riba *et al.*, 1983). The final deformation of the southern foreland basin took place during the late Oligocene–Early Miocene leading in the Early Miocene to the Ebro Basin present configuration.

Apart from the Pyrenees the evolution of the Ebro Basin is also linked to the compressive building of the Iberian Range and the Catalan Range, which bound the Ebro Basin, respectively to the SW and to the E (Fig. 1). Although the entire southern margins of the Ebro Basin were tectonically active during Paleogene time, their activity was minor in comparison with that of the Pyrenean margin during the same period. Accordingly, the southern margins may be regarded as relatively passive margins of the basin (Anadón *et al.*, 1986). As in the Pyrenees, the development of progressive unconformities in the alluvial fan deposits located in the margins of these ranges denote the synsedimentary building of these edifices, which include thrusts, folds and strike–slip faults (Riba, 1973; Anadón *et al.*, 1986). Major thrusting along the Pyrenean and Iberian margins and basinward migration of the deformation led to the reduction of the extension of the original foreland basin. In the Rioja area (Western Ebro Basin) the width of the basin has been reduced by about 70% (see Chapter E17).

In general, the base of the Tertiary sequences in the Ebro Basin dips towards the Pyrenees (Riba *et al.*, 1983). This surface is very shallow in the southern parts of the basin and over 5000 m deep in the northern part of the foreland (Fig. 2). Two main depocentres, separated by the Monegros high, have been distinguished. In the first (Navarra–La Rioja), the base of the Tertiary sequences is 5000 m deep, whereas in the second, located in the eastern part (Catalonia), it is 3600 m deep.

Related to the beginning of the upbuilding of the Pyrenean chain (Puidefàbregas & Souquet, 1986), the first Ebro Basin fill deposits (Upper Cretaceous–Paleocene in age) include both continental and marine deposits in the western part of the Pyrenean trough. From then until the Late Eocene, the main trough of the southern foreland was connected to the Atlantic Ocean. Marine deposition was widespread in the northern areas of the Ebro foreland basin, whereas in the southern areas non-marine deposition took place. In the northern areas, strongly subsiding troughs infilled by turbiditic sediments developed during Lower and Middle Eocene. These basins were infilled by deltaic systems during the Late Eocene. Two important transgressive maxima affected zones relatively distant from the main sedimentary trough in the Eastern Ebro Basin. The Ilerdian transgression initiated the deposition of shallow marine carbonates over extensive platforms. The Bartonian ('Biarritzian') transgression was simultaneous with coarse alluvial fan deposition in the Catalan Coastal Range leading to formation of fan deltas along the eastern margin of the Ebro Basin (Anadón *et al.*, 1985; Puigdefàbregas *et al.*, 1986). In the Late Eocene, a widespread regression initiated the deposition of marine evaporites in two main zones (the Navarra or Pamplona Basin and the Catalan Evaporite Basin). From the Late Eocene to the Late Miocene the Ebro Basin became closed and non-marine deposition occurred. Alluvial fan and fluvial sedimentation predominated near the basin margins. In the inner parts of the basin a thick succession of fluvial and shallow marine, carbonate and evaporitic deposits was formed.

The Valencia Trough

Located between the Iberian Peninsula and the Balearic Islands, the Valencia Trough is a NE–SW oriented basin, developed upon a very thin continental crust (Gallart *et al.*, 1990; Dañobeitia *et al.*, 1992; Watts & Torné, 1992). It belongs to the complex system of Neogene basins with thinned crust created in the Mediterranean during the Alpine orogeny. The Valencia Trough was formed in the Late Oligocene to Early Miocene, coeval with the opening of the Provençal Basin (Burrus, 1984; Roca & Desegaulx, 1992).

Well known from an extensive program of drilling and seismic reflection profiling by the oil and gas industry, the northwestern and southeastern margins of the Valencia Trough show noticeable differences in their structure and geological evolution. The northwestern margin ('Catalan–Valencian domain' from Fontboté *et al.*, 1990) includes the Neogene rift basins developed in the Catalan Range and in the southeastern part of the Iberian Range (see above). This margin shows the typical features of a rifted margin (Fig. 2). It is characterized by a widespread system of Neogene horst and graben bounded by ENE–WSW to N–S oriented normal faults

(Dañobeitia *et al.*, 1990; Roca & Desegaulx, 1992). In contrast, the southeastern margin ('Betic–Balearic domain') structure is more complex (Fig. 2). Here, the pre-Neogene rocks and the Early–Middle Miocene basin fill deposits are involved in a NE–SW thrust-and-fold belt that has been affected by listric extensional faults, Late Miocene to present-day in age (Fontboté *et al.*, 1990). This thrust-and-fold belt represents the northeastern prolongation of the Betic orogen, which crops out in the Balearic Islands and eastern Prebetic Zone. Hence, the present-day structure of the Valencia Trough appears to be the result of the superposition of Late Oligocene–Early Miocene crustal thinning and subsequent Early–Middle Miocene crustal thickening in the Balearic Promontory (Fontboté *et al.*, 1990; Roca & Desegaulx, 1992).

The Valencia Trough is filled by Late Oligocene to Quaternary sediments up to 6 km thick, which unconformably overlie an eroded and heterogeneous substratum of Paleozoic and Mesozoic rocks (Stoeckinger, 1976; Soler *et al.*, 1983; Clavell & Berástegui, 1991). Strongly influenced by the synsedimentary tectonic activity and the sea-level variations, the features of these sediments are different in the two margins of the Valencia Trough. In the northwestern margin, the basin fill consists of: (a) a set of Early–Middle Miocene synrift deposits formed by terrigenous and carbonate sediments deposited in a wide range of continental and shallow marine environments; and (b) a set of Middle Miocene–Quaternary postrift deposits of terrigenous platform facies. In the southeastern margin, it includes: (a) a set of Late Oligocene–Earliest Miocene carbonate to terrigenous platform deposits; (b) a set of Early–Middle Miocene terrigenous deposits formed in deep marine environments and correlative with the emplacement of the thrust sheets; and (c) a set of Late Miocene–Quaternary synrift deposits formed on terrigenous to carbonate marine platforms.

References

Agustí, J., Anadón, P., Ginsburg, L., Mein, P. and Moissenet, É. (1988). Araya et Mira: nouveaux gisements de mammifères dans le Miocène inférieur-moyen des Chaînes Ibériques orientales et méditerranéennes. Conséquences stratigraphiques et structurales. *Paleontol. Evol.*, **22**, 83–101.

Agustí, J., Domènech, R., Julià, R. and Martinell, J. (1990). Evolution of the Neogene basin of Empordà (NE Spain). In *Iberian Neogene basins: field guidebook* (ed. J. Agustí, R. Domènech, R. Julià and J. Martinell). *Paleontol. Evol. (spec. Publ.)*, **2**, 251–267.

Álvaro, M., Capote, R. and Vegas, R. (1979). Un modelo de evolución geotectónica para la Cadena Celtibérica. *Acta Geol. Hisp.*, **14**, 172–177.

Anadón, P., Colombo, F., Esteban, M., Marzo, M., Robles, S., Santanach, P. and Solé Sugrañes, Ll. (1979). Evolución tectonoestratigráfica de los Catalánides. *Acta Geol. Hisp.*, **14**, 242–270.

Anadón, P., Cabrera, Ll., Guimerà, J. and Santanach, P. (1985). Paleogene strike–slip deformation and sedimentation along the southeastern margin of the Ebro Basin. In *Strike–slip deformation, basin formation and sedimentation* (ed. K. Biddle and N. Christie-Blick). *Spec. Publ. Soc. Econ. Paleontol. Mineral.*, **37**, 303–318.

Anadón, P., Cabrera, L., Colombo, F., Marzo, M. and Riba, O. (1986). Syntectonic intraformational unconformities in alluvial fan deposits, eastern Ebro Basin margins (NE Spain). In *Foreland basins* (ed. P. Allen and P. Homewood). *Int. Assol. Sedimentol., Spec. Publ.*, **8**, 256–271.

Arthaud, F., Ogier, M. and Séguret, M. (1981). Géologie et géophysique du Golfe de Lion et de sa bordure nord. *Bull. Bur. Rech. Géol. Min.*, Série 2, Section I, **3**, 175–193.

Bartrina, M.T., Cabrera, L., Jurado, M.J., Guimerà, J. and Roca, E. (1992). Evolution of the central Catalan margin of the Valencia trough (western Mediterranean). *Tectonophysics*, **203**, 219–247.

Burrus, J. (1984). Contribution to a geodynamic synthesis of the Provençal Basin (North-western Mediterranean). *Mar. Geol.*, **55**, 247–269.

Calvet, M. (1985). Néotectonique et mise en place des reliefs dans l'est des Pyrénées; l'exemple du horst des Albères. *Rev. Géog. Phys. Géol. Dyn.*, **26**, 119–130.

Cámara, P. and Klimowitz, J. (1985). Interpretación geodinámica de la vertiente centro-occidental surpirenaica. *Estud. Geol.*, **41**, 391–404.

Clavell, E. and Berástegui, X. (1991). Petroleum geology of the Gulf of Valencia. In *Generation, accumulation and production of Europe's hydrocarbons* (ed. A.M. Spencer). Oxford University Press, Oxford, 355–368.

Colomer, M. and Santanach, P. (1988). Estructura y evolución del borde sur-occidental de la Fosa Calatayud–Daroca. *Geogaceta*, **4**, 29–31.

Crusafont, M., Gautier, F. and Ginsburg, L. (1966). Mise en évidence du Vindobonien inférieur continental dans l'est de la province de Teruel (Espagne). *C. R. Somm. Soc. Géol. France*. **1**, 30–32.

Dañobeitia, J.J., Alonso, B. and Maldonado, A. (1990). Geological framework of the Ebro continental margin and surrounding areas. *Mar. Geol.*, **95**, 265–287.

Dañobeitia, J.J., Arguedas, M., Gallart, J., Banda, E. and Makris, J. (1992). Deep crustal configuration of the Valencia trough and its Iberian and Balearic borders from extensive refraction and wide-angle reflection seismic profiling. *Tectonophysics*, **303**, 37–55.

Dercourt, J., Zonenshain, L.P., Ricou, L.E., Kazmin, V.G., Le Pichon, X., Knipper, A.L., Grandjacquet, C., Sbortshikov, I.M., Geyssant, J., Lepvrier, C., Pechersky, D.H., Boulin, J., Sibuet, J.C., Savostin, L.A., Sorokhtin, O., Westphal, M., Bazhenov, M.L., Lauer, J.P. and Biju Duval, B. (1986). Geological evolution of the Tethys belt from the Atlantic to the Pamirs since the Lias. *Tectonophysics*, **123**, 241–315.

Dewey, J.F., Pitman III, W.C., Ryan, W.B.F. and Bonnin, J. (1973). Plate tectonics and evolution of the Alpine System. *Geol. Soc. Am. Bull.*, **84**, 3137–3180.

Dewey, J.F., Helman, M.L., Turco, E., Hutton, D.H.W. and Knott, S.D. (1989). Kinematics of the western Mediterranean. In *Alpine tectonics* (ed. M.P. Coward, D. Dietrich and R.G. Park). *Geol. Soc. London, Spec. Publ.*, **45**, 265–283.

ECORS Pyrenees team (1988). The ECORS deep seismic survey across the Pyrenees. *Nature*, **331**, 508–511.

Fontboté, J.M. (1954). Las relaciones tectónicas de la depresión del Vallés-Penedés con la Cordillera Prelitoral Catalana y con la Depresión del Ebro. *Tomo homenaje Prof. E. Hernández-Pacheco. R. Soc. Esp. Hist. Nat.*, 281–310.

Fontboté, J.M., Guimerà, J., Roca, E., Sàbat, F. and Santanach, P. (1989). Para una interpretación cinemática de la génesis de la cuenca catalano-balear: datos estructurales de sus márgenes emergidos. In *Libro homenaje a Rafael Soler*. AGGEP, Madrid, 37–51.

Fontboté, J.M., Guimerà, J., Roca, E., Sàbat, F., Santanach, P. and Fernández-Ortigosa, F. (1990). Cenozoic geodynamic evolution of the Valéncia trough (western Mediterranean). *Rev. Soc. Geol. Esp.*, **3**, 249–259.

Gallart, J., Rojas, H., Díaz, J. and Dañobeitia, J.J. (1990). Features of deep crustal structure and the onshore-offshore transition of the Iberian flank of the Valencia trough (Western Mediterranean). *J. Geodynam.*, **12**, 233–252.

Guimerà, J. (1984). Paleogene evolution of deformation in the northeastern Iberian Peninsula. *Geol Mag.*, **121**, 413–420.

Guimerà, J. (1987). Comentarios sobre 'Aportaciones al conocimiento de la compresión tardía en la cordillera ibérica centro-oriental: la cuenca neógena inferior del Mijares (Teruel–Castellón)', de J. Paricio y J.L. Simón Gómez. *Estud. Geol.*, **43**, 63–69.

Guimerà, J. and Álvaro, M. (1990). Structure et évolution de la compression alpine dans la Chaîne ibérique et la Chaîne côtière catalane (Espagne). *Bull. Soc. Géol. Fr.*, Série 8, **VI**, 339–348.

Labaume, P., Séguret, M. and Seyve, C. (1985). Evolution of a turbiditic foreland basin and analogy with an accretionary prism: example of the Eocene South-Pyrenean basin. *Tectonics*, **4**, 661–685.

Mellere, D. (1992). I conglomerati di Pobla de Segur: stratigrafia fisica e relazioni tettonica-sedimentazione. PhD Thesis. Università degli Studi di Padova. 203 pp.

Moissenet, É. (1989). Les fossés néogènes de la Chaîne Ibérique: leur évolution dans le temps. *Bull. Soc. Géol. Fr.*, Série 8, **V**, 919–926.

Muñoz, J.A. (1992). Evolution of a continental collision belt: ECORS–Pyrenees crustal balanced cross-section. In *Thrust tectonics* (ed. K.R. McClay). Chapman and Hall, London, 235–246.

Muñoz, J.A., Martínez, A. and Vergés, J. (1986). Thrust sequences in the eastern Spanish Pyrenees. *J. Struct. Geol.*, **8**, 399–405.

Muñoz-Jiménez, A. (1992). *Análisis tectonosedimentario del Terciario del sector occidental de la cuenca del Ebro (comunidad de la Rioja).* Ed. Instituto de Estudios Riojanos, Logroño, 347 pp.

Ori, G.G. and Friend, P.F. (1984). Sedimentary basins formed and carried piggyback on active thrust sheets. *Geology*, **12**, 475–478.

Puigdefàbregas, C. and Soler, M. (1973). Estructura de las Sierra Exteriores Pirenaicas en el corte del río Gállego (prov. de Huesca). *Pirineos*, **109**, 5–15.

Puigdefàbregas, C. and Souquet, P. (1986). Tecto-sedimentary cycles and depositional sequences of the Mesozoic and Tertiary from the Pyrenees. *Tectonophysics*, **129**, 173–203.

Puigdefàbregas, C., Muñoz, J.A. and Marzo, M. (1986). Thrust belt development in the eastern Pyrenees and related depositional sequences in the southern foreland basin. In *Foreland basins* (ed. P. Allen and P. Homewood). *Int. Assoc. Sedimentol., Spec. Publ.*, **8**, 229–246.

Puigdefàbregas, C., Muñoz, J.A. and Vergés, J. (1992). Thrusting and foreland basin evolution in the Southern Pyrenees. In *Thrust tectonics* (ed. K.R. McClay). Chapman and Hall, London, 247–254.

Riba, O., Reguant, S. and Villena, J. (1983). Ensayo de síntesis estratigráfica y evolutiva de la cuenca terciaria del Ebro. *Libro jubilar J.M. Ríos. Geologia de España II.* IGME, Madrid, 131–159.

Richter, G. and Teichmüller, R. (1933). Die Entwicklung der Keltiberischen Ketten. *Abh. Ges. Wiss. Göttingen Math. Phys. Kl.*, Ser. 3, **7**, 1067–1186.

Roca, E. and Desegaulx, P. (1992). Analysis of the geological evolution and vertical movements in the Valencia Trough area, Western Mediterranean. *Mar. Petrol. Geol.*, **9**, 167–185.

Roca, E. and Guimerà, J. (1992). The Neogene structure of the eastern Iberian margin: structural constraints on the crustal evolution of the Valencia trough (western Mediterranean). *Tectonophysics*, **203**, 203–218.

Roest, W.R. and Srivastava, S.P. (1991). Kinematics of the plate boundaries between Eurasia, Iberia, and Africa in the North Atlantic from the Late Cretaceous to present. *Geology*, **19**, 613–616.

Sáez, A. (1987) Estratigrafía y sedimentología de las formaciones lacustres del tránsito Eocene–Oligoceno del NE de la cuenca del Ebro. Tesis doctoral, Universitat de Barcelona, 352 pp.

Séguret, M. (1972). *Étude tectonique des nappes et séries décollées de la partie centrale du versant sud des Pyrénées.* Publications de l'Université de Sciences et Technique de Languedoc, série Géologie Structurale, **2**, Montpellier, 155 pp.

Simó, A. and Puigdefàbregas, C. (1985). Transition from shelf to basin on an active slope, Upper Cretaceous, Tremp area, southern Pyrenees. In *6th Europ. Reg. Mtg. Sedimentol. I.A.S., Lleida 1985. Excursion Guidebook* (ed. M.D. Milá and J. Rosell), pp. 63–108. Lérida, Spain.

Soler, R., Martínez del Olmo, W., Megías, A.G. and Abeger, J.A. (1983). Rasgos básicos del Neógeno del Mediterráneo español. *Mediterránea, Ser. Geol.*, **1**, 71–82.

Srivastava, S.P., Roest, W.R., Kovacs, L.C., Oakey, G., Lévesque, S., Verhoef, J. and Macnab, R. (1990). Motion of Iberia since the Late Jurassic: Results from detailed areromagnetic measurements in the Newfoundland Basin. *Tectonophysics*, **184**, 229–260.

Stille, H. (1931). Die Keltiberische Scheitelung. *Nachr. Ges. Wiss. Göttingen Math. Phys. Kl*, **10**, 138–164.

Stoeckinger, W. (1976). Valencia gulf offer deadline nears. *Oil–Gas J.*, **March 29**, 197–204.

Vergés, J. and Martínez, A. (1988). Corte compensado del Pirineo oriental: geometría de las cuencas de antepaís y edades de emplazamiento de los mantos de corrimiento. *Acta Geol. Hisp.*, **23**, 95–106.

Vergés, J. and Muñoz, J.A. (1990). Thrust sequences in the southern central Pyrenees. *Bull. Soc. Géol. Fr.*, Série 8, **VI**, 265–271.

Viallard, P. (1973). Recherches sur le cycle alpin dans la Chaîne Ibérique sud-occidentale. Thèse d'etat. Université Paul Sabatier, Toulouse, 445 pp.

Viallard, P. (1983). Le décollement de couverture dans la Chaîne Ibérique méridionale (Espagne): effet des raccourcissements différentiels entre substratum et couverture. *Bull. Soc. Géol. Fr.*, Série 7, **XXV**, 379–387.

Watts, A.B. and Torné, M. (1992). Subsidence history, crustal structure, and thermal evolution of the Valencia Trough: a young extensional basin in the western Mediterranean. *J. Geophys. Res.*, **97**, 20021–20041.

E2 The lithosphere of the Valencia trough: a brief review

M. TORNÉ

Abstract

Deep seismic profiling and gravity data show that the continental crust under the Valencia trough has been thinned by up to a factor of three in the centre, and that the uppermost mantle there has an unusually low density. Structural mapping in the borders of the Valencia trough shows important asymmetry, and there are asymmetrical features in the geophysical modelling. There are many aspects of the basin formation that are still to be worked out.

Introduction

The past decades have seen a rapid advance in our understanding of the deep structure of extended continental lithosphere. This advance has come primarily because of the integration of geological and geophysical data sets into self-consistent models of basin structure. The investigation of the Valencia Trough is an excellent example of the integrated use of geological and geophysical data to study the structure of a rift-type basin that developed in a region of convergence between the Eurasian and African plates (Fig. 1). The Trough is a small NE–SW-oriented basin in the Western Mediterranean located between the northeastern border of the Iberian Peninsula and the northeastern prolongation of the external zones of the Betic Chain (the Balearic Promontory) (Fig. 1). It is thought to be a basin that formed during a rifting event that began during Late Oligocene–Early Miocene. Its northwestern border (the Catalan–Valencian domain) is characterized by extensional tectonics related to the Neogene extensional event, while its southeastern border (the Balearic Promontory) shows folding and thrusting during the Early–Middle Miocene and subsequent extension (Fontboté *et al.*, 1990). For a more detailed discussion the reader is referred to Banda and Santanach (1992a) where extensive regional geological and geophysical studies of the basin are presented.

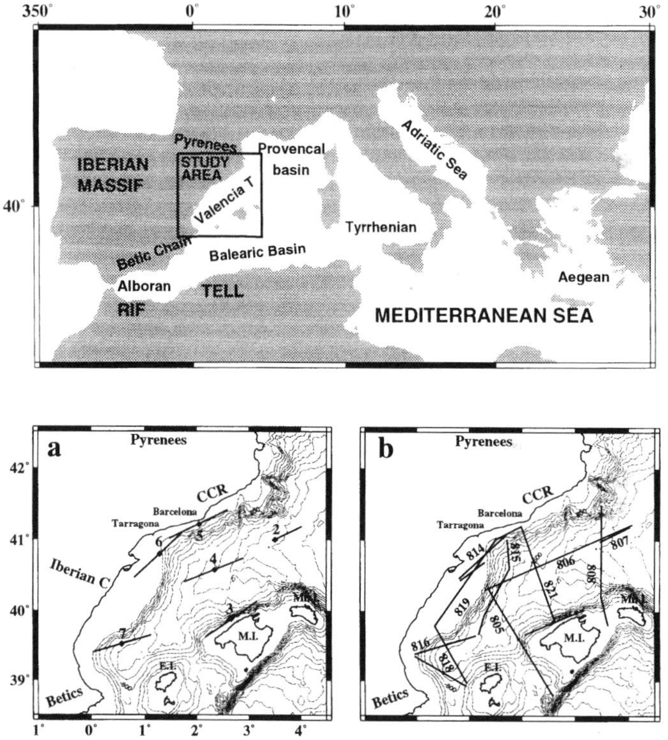

Fig. 1. Upper panel shows a global view of the study area. Lower panels show location maps of the two-ship multichannel seismic profiles collected during the VALSIS-II experiment in the Valencia Trough. Dotted contour lines are bathymetry at 200 m interval. (a) Location of common depth point (DCP) profiles. (b) Location of individual expanding spread profile (ESP) profiles. Diamonds indicate location of midpoint areas. CCR = Catalan Coastal Ranges; E.I. = Eivissa (Ibiza) Island; M.I. = Mallorca Island; Mn. I. = Menorca Island. These islands form the Balearic Promontory. (Fig. 1a and b taken from Torné *et al.*, 1992.)

Although the trough appears symmetric on bathymetric maps, seismic, gravity and geoid data suggest a fundamental asymmetry in its deep crustal and lithospheric structure. The results of seismic and gravity studies (e.g. Dañobeitia *et al.*, 1992; Torné *et al.*, 1992) indicate that the trough is underlain by thinned continental crust – except at its northeastern border where oceanic crust is present (Pascal *et al.*, 1992) – and low-density upper-mantle material, both suggesting an upward perturbation of the lithosphere–asthenosphere boundary.

Studies of the basin using different data sets have resulted in a wide range of structural models. Dewey *et al.* (1989) suggest that the

Western Mediterranean basins developed as Western Pacific-type back-arc basins behind a subduction zone located to the southeast, although there is no clear evidence of such a subduction zone. Doblas and Oyarzun (1990) propose a simple shear model in which extension occurred above a major SE-directed intracrustal detachment surface. More recently, Vegas (1992) has suggested an escape-tectonic model that would account for the co-existence and progressive migration of compressional and extensional strain fields, while Roca and Desegaulx (1992) have postulated a foreland-type basin. A difficulty in applying any extensional model in the area is the possible co-existence of compression and extension at the first stages of the basin formation.

The purpose of this chapter is to summarize the main findings on the deep structure of the trough and to highlight several questions that, despite the large amount of data available, still remain unanswered.

Crustal and lithospheric structure

Crustal structure

Large amounts of geological and geophysical data are available from the trough due to an intense program carried out by industry (Lanaja, 1987) and academic institutions over the past decades (Banda and Santanach, 1992a). Previous seismic, gravity, geoid, heat flow and well data studies (e.g. Watts *et al.*, 1990; Dañobeitia *et al.*, 1992; Foucher *et al.*, 1992; Torné *et al.*, 1992; Watts and Torné 1992a, b) indicate that the trough is underlain by a continental crust that has been thinned by a factor of three in its central part, where the total crustal thickness – including water depth – is less than 14 km (Fig. 2). The crust thickens asymmetrically toward its flanks, reaching values of about 20–23 km offshore from the Iberian Peninsula and 23 km in the Mallorca margin (Fig. 2a, b). Along the axis of the trough, the crust thins from 16 km in its southwestern border to less than 13 km in its northeastern border where oceanic crust is present (e.g., Pascal *et al.*, 1992; Torné *et al.*, 1992) (Fig. 2c).

The observed differential thinning of the upper crust relative to the lower crust – which appears either to be missing or to be reduced to a 1–2 km thick layer in the central part of the trough (Fig. 2) (Banda and Santanach, 1992b; Dañobeitia *et al.*, 1992; Torné *et al.*, 1992) – and the well-constrained low-*P*-wave upper-mantle velocities have been used by Banda *et al.* (1992) to propose a mixing hypothesis. These authors argue that a variable mixture of lower crust and upper mantle would explain both the observed differential crustal stretching and the low-*P*-wave upper-mantle velocities widely observed in continental extended areas.

Lithospheric structure

A first approach to understand better the geodynamic processes governing basin formation involves the analysis of the structure of the lithosphere–asthenosphere boundary. Recent deep reflection and refraction studies combined with gravity, geoid, heat-flow and subsidence data have permitted the development of different approaches to deduce the lithospheric structure and the amount of extension across the trough.

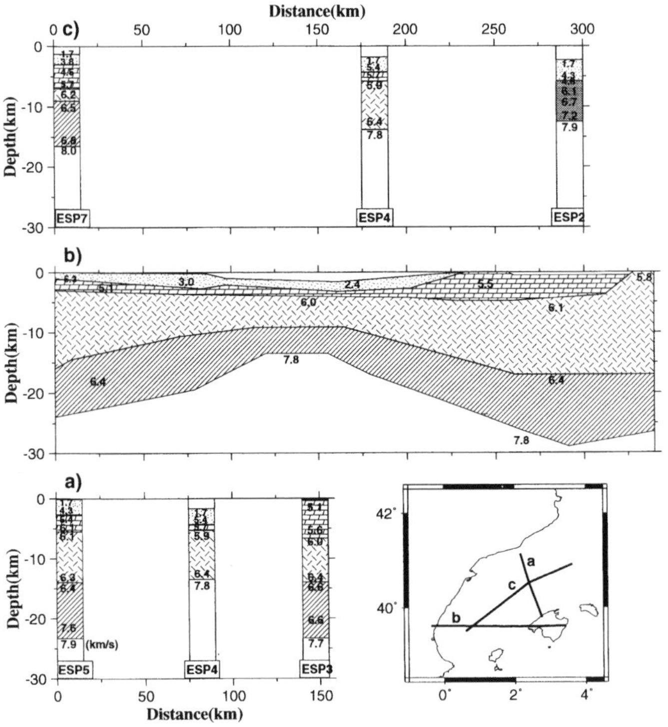

Fig. 2. Crustal section determined from: (a) ESPs 5, 4 and 3 located across the trough north to the left; (b) refraction seismic profiles and (c) ESPs 7, 4 and 2, located along the axis of the trough, west to the left. Lower right inset shows location of the profiles. Profiles a and c modified from Torné *et al.*, 1992. Profile b modified from Dañobeitia *et al.*, 1992.

Fig. 3 (a, b, c, d) summarizes the main results of a finite rifting model, applied along a profile cutting across the basin, by Watts and Torné (1992a) (Line 821, Fig. 1b). As shown the observed tectonic subsidence requires crustal stretching factors increasing from about 1.4 beneath the flanks of the trough to about 3 in its axis, while the observed tectonic uplift requires different amounts of extension in the crust and upper mantle. This model also indicates that the lithosphere thins as much as 65 km at the centre of the basin, which roughly coincides with previous thermal studies carried out by Fernàndez *et al.*, 1990.

As would be expected, temperatures are elevated beneath the axis of the trough compared with its flanks. The gravity effect of the low-density region in the mantle is shown in Fig. 3g. The low-density region causes a long-wavelength large-amplitude (110 mGal) gravity-anomaly low. Fig. 4 shows, however, that the trough is characterized by a Bouguer anomaly high of about 100–130 mGal. The high indicates that the gravity effect of the low-density region is more than compensated by the gravity effects of density differences in the overlying crust.

A parameter that is sensitive to the deep structure of the lithosphere is the geoid. Fig. 4 (inset) shows that the axis of the trough is associated with a geoid anomaly low of about 130–150 cm, while its flanks are associated with geoid anomaly highs of up to 100

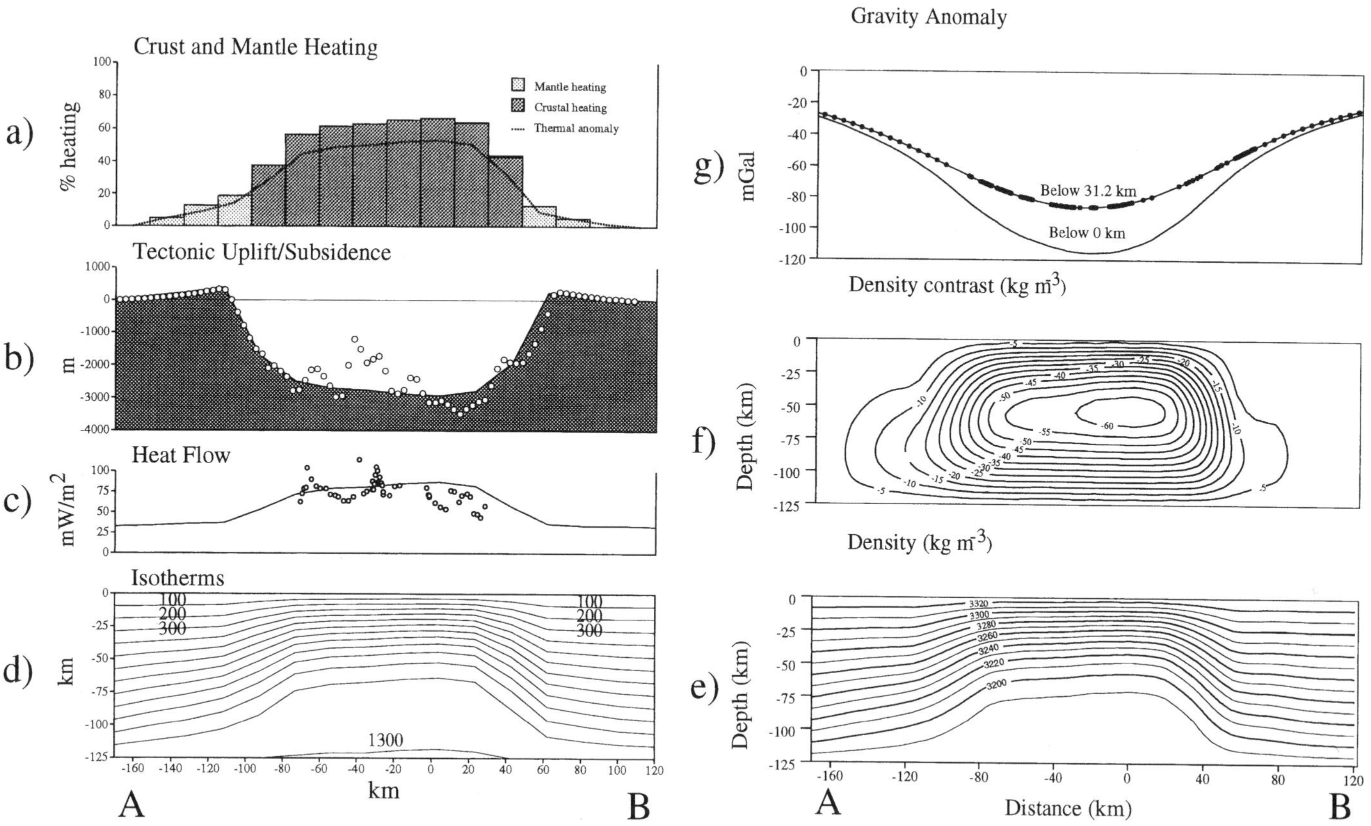

Fig. 3. Thermal model for the Valencia Trough based on a finite rifting time of 16 Ma. (a) Crustal and mantle heating. (b) Comparison of observed tectonic subsidence and uplift (open circles) with a calculated profile based on an elastic thickness (Te) that depends on the depth of the 450 °C oceanic isotherm since rifting. (c) Comparison of observed heat flow values (open circles) based on Foucher *et al.*, 1992 and calculated heat flow values (solid line). (d) Isotherms in 100 °C intervals. (e) Density distribution corresponding to the isotherms shown in (d). (f) Density contrast obtained from the temperature difference between a standard column and a thinned lithospheric column. (g) Calculated gravity effect of the thermal model shown in (a). Taken from Watts and Torné (1992a).

cm. The observed geoid and gravity anomalies were compared by Watts and Torné (1992a) with different density lithospheric models (Fig. 4). From this modelling these authors conclude that the best-fitting model is one in which the region of lithospheric thinning continues uninterrupted from beneath the axis of the trough and beneath the Balearic Promontory further to the SE with a slight thickening beneath the Balearic Promontory. If this is true it could be argued that the trough it is not a basin itself but forms part of a wider basin, perhaps a South Balearic basin.

Although the same thermal model that explains the observed tectonic subsidence/uplift and geoid and gravity anomalies in the central part of the trough also appears to be in accord with the observed heat-flow data (Foucher *et al.*, 1992), there are difficulties in applying it to other parts of the basin. This is particularly true at its southern end where heat flow values are higher and the tectonic subsidence less than expected.

A second approach to studying the evolution of rift-type basins is to compare the present-day crustal and lithospheric structure with that predicted by backstripping, crustal restoration and sediment loading. This technique has been used by several authors (e.g. Barton and Wood, 1984; Watts, 1988; Dyment, 1990) to investigate whether the crustal structure beneath a particular basin has been modified by additional processes (e.g. previous stretching events, underplating, uncompensated loading, etc.). Assuming that sediment loading is one of the main processes that modify the post-rift crustal structure in extensional basins, the backstripped Moho would reflect – in the absence of other modifying processes – the crustal structure due to extension.

Watts and Torné (1992b) have recently studied the Valencia Trough using this approach. These authors have used 3D backstripping techniques to restore the expected crustal structure based on the tectonic/subsidence uplift and sediment-loading history of the trough and compare it with that measured seismically. From their 3D backstripping results (Fig. 5) it is deduced that the trough is associated with a broad region of subsidence flanked by uplift in its margins. Maximum subsidence values of about 4.2–4.4 km are reached in the NE part of the trough where the velocity–depth profile (Pascal *et al.*, 1992; Torné *et al.*, 1992) closely resembles that of oceanic crust. Elsewhere, the tectonic subsidence is in the range of 3.0–4.0 km, which is to be expected assuming that the central part of the trough has been stretched by up to a factor of three.

Watts and Torné (1992b) compared the seismic Moho and the flexed Moho for different values of Te (elastic thickness), where the flexed Moho is the backstripped Moho modified for the effects of

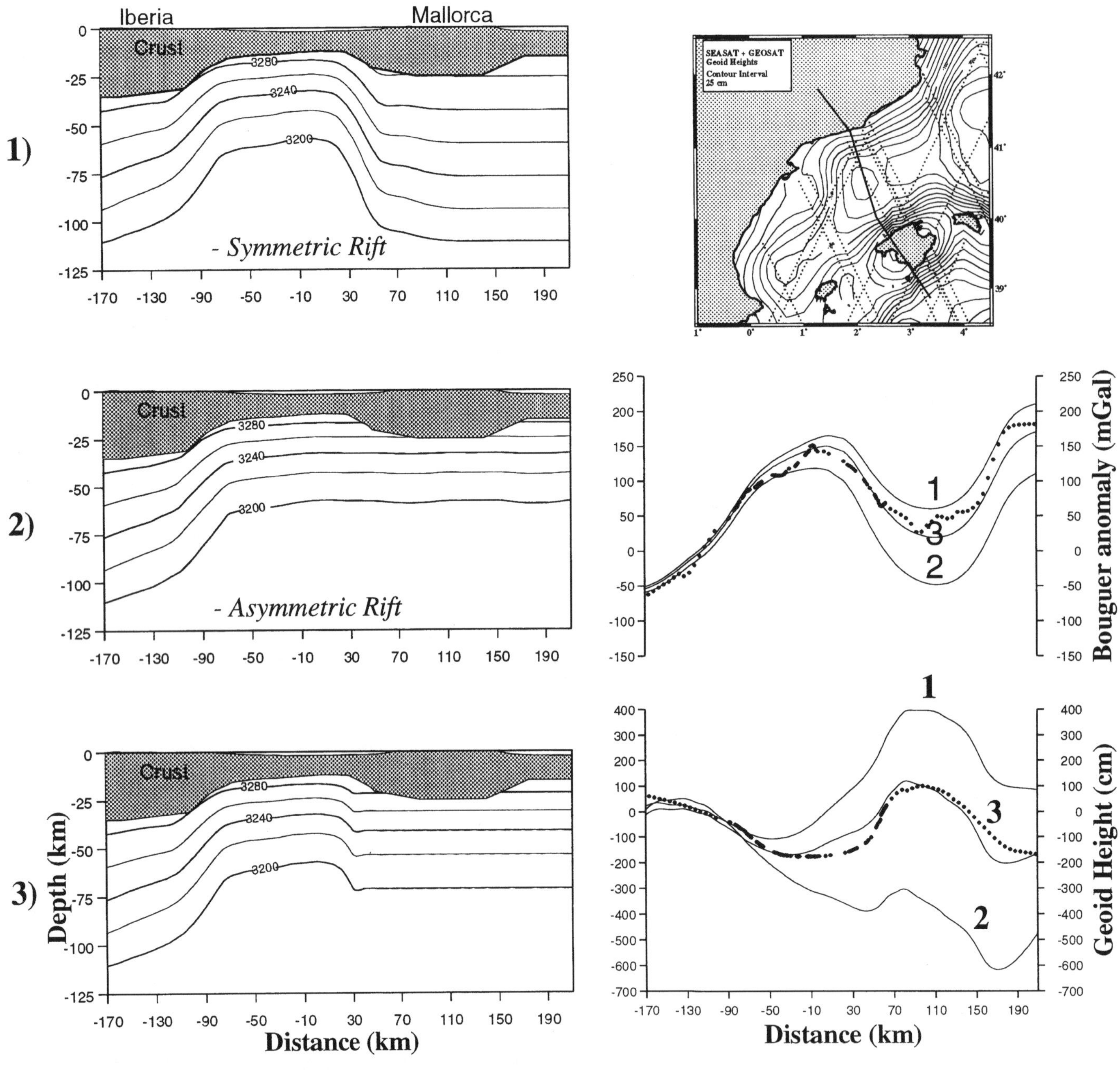

Fig. 4. Comparison (lower right) of observed (solid circles) and calculated (thin lines) geoid and gravity anomalies (profile 821 in Fig. 1) for different density lithospheric models shown in left panel. Model 1) Symmetric rift; model 2) Asymmetric rift; 3) Combination of both models 1 and 2 (modified from Watts and Torné, 1992a). Upper right inset shows geoid anomalies over the Valencia Trough. The geoid anomaly map is based on averages of sea surface heights obtained from SEASAT and GEOSAT. A regional field based on a PGS 110 model to degree and order 16 has been removed from the observed heights prior to smoothing and gridding (taken from Watts and Torné, 1992a). Thick line shows the location of model profile. Dashed lines indicate satellite tracks.

sediment loading. These authors concluded that the best overall fit is achieved for a Te = 0 (local isostasy model), which is also supported by gravity data. Fig. 6 shows a comparison between the seismic Moho and the flexed Moho for a Te = 0. Fig. 7 shows that there is a close agreement between the long-wavelength computed anomaly of this model and the observed sediment-corrected Bouguer anomaly. From these results Watts and Torne (1992b) computed a residual Moho depth anomaly map by substracting the flexed Moho from the seismic one (Fig. 6). This map shows that the residual depths are less than ± 2 km over the central part of the basin, while the largest anomalies occur beneath the Ebro platform and Balearic promontory where the flexed Moho is deeper than the seismic Moho.

From these results the authors conclude that a model where the

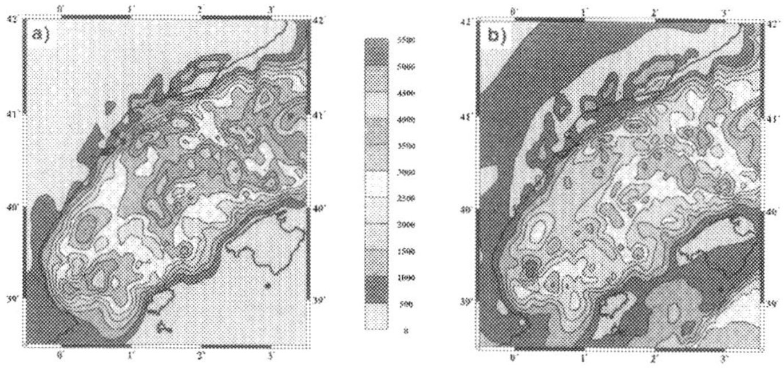

Fig. 5. (a) Depth to the base of the Neogene based on IGME (Lanaja, 1987) and Maillard *et al.* (1992). (b) Tectonic subsidence/uplift obtained by flexurally backstripping the Neogene sediment thickness. The map has been obtained using a Te=5 km and densities of the water, sediments, and mantle of 1030, 2400, and 3300 kg/m^3 respectively (modified from Watts and Torné, 1992b).

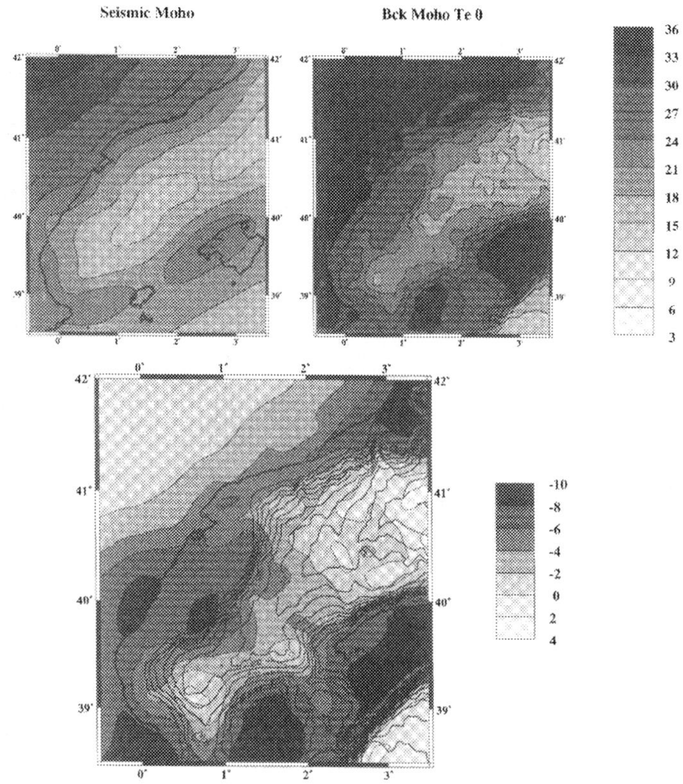

Fig. 6. Comparison of the seismic and flexed Moho for a value of Te=0 (local isostasy model). (top left) Seismic Moho. (top right) Flexed Moho for Te=0. (Taken from Watts and Torné (1992b)). (bottom) Computed residual Moho depth anomaly map by substracting the flexed Moho from the seismic one.

lithosphere has acquired little or no strength satisfactorily explains the overall crustal structure of the trough. They argue that the anomaly observed along the Ebro platform can be explained by the incomplete stress relaxation of the lithosphere during loading, suggesting that from the Pliocene there has been little or no contribution to the tectonic subsidence/uplift in the Ebro delta.

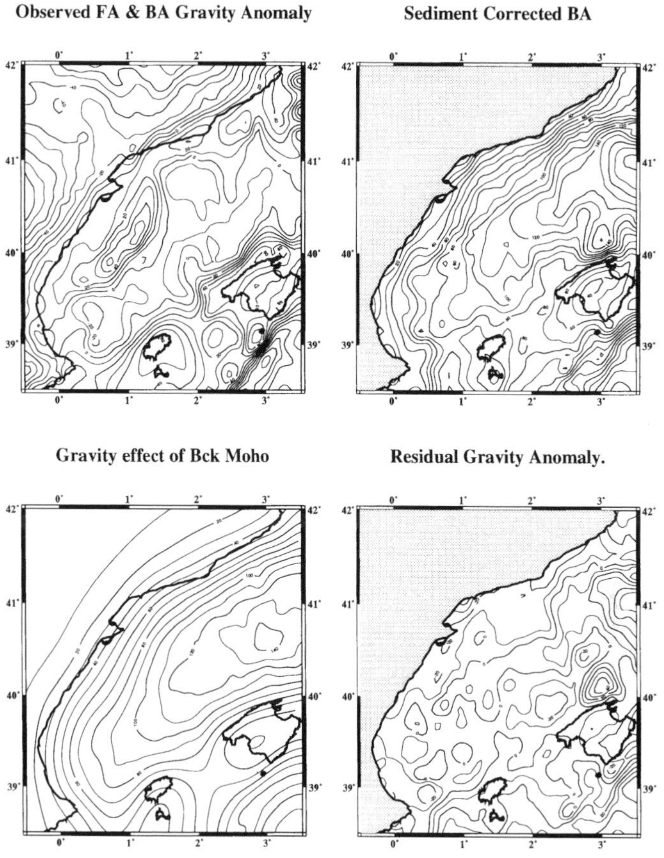

Fig. 7. Comparison of the observed gravity anomaly with the calculated anomaly of the flexed Moho. (top left) Observed gravity anomaly map (Bouguer anomaly on land and Free-Air anomaly at sea). (top right) Bouguer anomaly, corrected at sea for the gravity effect of the Neogene sediments. (bottom) Calculated gravity anomaly of the flexed Moho for values of: Te=0 km, ρ_s=2400 km/m^3, and ρ_m=3220 kg/m^3.

Concluding remarks

Available deep-seismic profiling and gravity data provide a well-constrained image of the present-day crustal geometry of the trough. These data reveal that the area is underlain by thin continental crust – thinned by up to a factor of three in the central part of the basin – and a low-density uppermost mantle. Thermal modelling combined with gravity and geoid data indicates that lithospheric thinning – up to 65 km – continues uninterrupted beneath its eastern margin further to the SE.

Despite advances in knowledge of the present-day structure of the basin, there are still some questions to be solved. Field geology and extensive mapping of its borders reveal that compression and extension were coeval at the first stages of the basin formation. The present-day structure shows a clear crustal and lithospheric asymmetry across the trough. Increasing values of heat flow in a NE–SW direction (Foucher *et al.*, 1992) are in contradiction with crustal thickness variations and observed tectonic subsidence. Likewise, the observed discrepancy between upper and lower crustal thinning and low *P*-wave uppermost-mantle velocities, widely observed in continental extended areas, are also found in the study area. Due to all the above factors it is difficult to apply any plausible basin formation model to the Valencia Trough. Much has yet to be developed in basin modelling approaches to integrate all geological and geophysical data into self-consistent models of basin formation.

References

Banda, E. and Santanach, P. (eds.) (1992a). Geology and geophysics of the Valencia Trough, Western Mediterranean. *Tectonophysics*, **203**, 361 pp.

Banda, E. and Santanach, P. (1992b). The Valencia trough (western Mediterranean): an overview. *Tectonophysics*, **208**: 183–202.

Banda, E., Fernàndez, M. and Torné, M. (1992). The role of the lower crust and upper mantle in extended areas. Mechanisms and consequences. *Ann. Geophysi.*, **10** (I), C81.

Barton, P.J. and Wood, R.J. (1984). Tectonic evolution of the North Sea basin: crustal stretching and subsidence. *Geophys. J. Royal Astron. Soc.*, **79**: 980–1022.

Dañobeitia, J.J., Arguedas, M., Gallart, J., Banda, E. and Makris, J. (1992). Deep crustal structure of the Valencia Trough and its Iberian and Balearic borders from extensive refraction and wide-angle reflection seismic profiling. *Tectonophysics*, **203**: 37–55.

Dewey, J.F., Helman, M.L., Turco, E., Hutton, D.H.W. and Knott, S.D. (1989). Kinematics of the Western Mediterranean. In Coward, M.P., Dietrich, D. and Park, R.G. (eds.). *Geol. Soc., London, Spec. Pub.*, **45**: 265–283.

Doblas, M. and Oyarzun, R. (1990). The late Oligocene–Miocene opening of the North-Balearic Sea (Valencia basin, Western Mediterranean): a working hypothesis involving mantle upwelling and extensional detachment tectonics. *Mar. Geol.*, **94**: 155–163.

Dyment, J. (1990). Some complementary approaches to improve deep seismic reflection studies in sedimentary basin environment: The Celtic Sea basin. In Pinet, B. and Bois, C. (eds.), *The Potential of Deep Seismic Profiling for Hydrocarbon Exploration*. Technip, Paris, 403–423.

Fernàndez, M., Torné, M. and Zeyen, H. (1990). Lithospheric thermal structure of the NE Spain and the North-Balearic basin. *J. Geodynam.*, **12**: 253–267.

Fontboté, J.M., Guimerà, J., Roca, E., Sàbat, F., Santanach, P. and Fernandez-Ortigosa, F. (1990). The Cenozoic geodynamic evolution of the Valencia Trough (Western Mediterranean). *Rev. Soc. Geol. Esp.*, **3**: 249–259.

Foucher, J.P., Mauffret, A., Steckler, M., Brunet, M.F., Maillard, A., Rehault, J.P., Alonso, B., Desegaulx, P., Murillas, J. and Ouillon, G. (1992). Heat flow in the Valencia Trough: geodynamic implications. *Tectonophysics*, **203**: 77–97.

Lanaja, M. (1987). *Contribución de la exploración petrolífera al conocimiento de la geología de España*. IGME, Madrid.

Maillard, A., Mauffret, A., Watts, A.B., Torné, M., Pascal, G., Buhl, P. and Pinet, B. (1992). Tertiary sedimentary history and structure of the Valencia Trough (Western Mediterranean). *Tectonophysics*, **203**: 57–76.

Pascal, G., Torné, M., Buhl, P., Watts, A.B. and Mauffret, A. (1992). Crustal and velocity structure of the Valencia Trough (Western Mediterranean). Part II: Detailed interpretation of 5 expanded spread profiles. *Tectonophysics*, **203**: 21–36.

Roca, E. and Desegaulx, P. (1992). Geological evolution and vertical movement analysis of the Valencia Trough area (Western Mediterranean). *Mar. Petrol. Geol.*, **9**: 167–185.

Torné, M., Pascal, G., Buhl, P., Watts, A.B. and Mauffret, A. (1992). Crustal and velocity structure of the Valencia Trough (Western Mediterranean). Part I: A combined refraction/wide angle reflection and near vertical reflection study. *Tectonophysics*, **203**: 1–20.

Vegas, R. (1992). The Valencia Trough and the origin of the Western Mediterranean basins. *Tectonophysics*, **203**: 249–261.

Watts, A.B. (1988). Gravity anomalies, crustal structure and flexure of the lithosphere at the Baltimore Canyon Trough. *Earth Planet. Sci. Lett.*, **89**: 221–238.

Watts, A.B. and Torné, M. (1992a). Subsidence history, crustal structure and thermal evolution of the Valencia Trough: a young extensional basin in the Western Mediterranean. *J. Geophys. Res*, **97**: 20021–20041.

Watts, A.B. and Torné, M. (1992b). Crustal structure and the mechanical properties of extended continental lithosphere in the Valencia Trough (Western Mediterranean). *J. Geol. Soc., Lond.*, **149**: 813–827.

Watts, A.B., Torné, M., Buhl, P., Mauffret, A., Pascal, G. and Pinet, B. (1990). Evidence for deep reflections from extended continental crust beneath the Valencia Trough, Western Mediterranean. *Nature*, **348** (6302): 631–635.

E3 Depositional sequences in the Gulf of Valencia Tertiary basin

W. MARTÍNEZ DEL OLMO

Abstract

Well-logs and seismic lines from hydrocarbon exploration of the Gulf of Valencia can be analysed in the light of the concepts of sequence stratigraphy. Two major tecto-sedimentary stages are recognized: an early (Paleogene) continental molasse stage, which is correlative with the compressional deformation of the Iberian and Pyrenean mountain belts, and a later (Neogene) stage, marking the establishment of a marine foreland to the Betic Cordillera. Internal discontinuities in the continental molasse are results of changes in position and deformational kinematics of the compressional mountain fronts from which the sediments were derived. In the second stage, eight marine sequences are recognized. Seven of these contain only transgressive and highstand systems tracts. The eighth starts with a lowstand system tract, which is correlative with the Messinian unconformity. Detailed study of the sequences shows that the 'Messinian Crisis' was a brief climatic crisis, which was followed by a eustatic fall of sea level. During the climatic–eustatic events the Mediterranean Sea generally maintained a geometry and water depth very similar to the present.

Introduction

The Gulf of Valencia Basin became a geographically and geologically distinct entity with the tectonic emergence of the Balearic Islands during the Middle Miocene.

The present SW–NE orientation of the Basin (Fig. 1) is subparallel to the trend of the Betic Orogen (Lower Miocene), slightly oblique to the trend of the Catalan Coast Range (Upper Eocene), and perpendicular to the trend of the Iberian Range structures (Oligocene). This pattern, along with the decreasing degree of deformation of the Lower–Middle Miocene towards the NW, suggest that the Gulf of Valencia Basin was the foreland basin of the Betic Orogen. This is why the basin shows strong similarities with the Guadalquivir–Gulf of Càdiz Basin.

From the early days of hydrocarbon exploration, three Neogene Groups have been distinguished (Alcanar, Castellon and Ebro) on the basis of their lithological contrast, their seismic attributes and the erosive Messinian unconformity on the continental platform (Stokinger, 1976; Martínez, 1978; García Siñeriz *et al.*, 1978; Soler *et al.*, 1980–83; Clavell 1991; Martínez *et al.*, 1991) and this nomenclature should be continued.

Because the Alcanar Group is very thin and poorly defined seismically, it was not distinguished as early as the Castellon (sandstone and shale Formations), and the Ebro (sand and clay Formations) (Shell, 1973, 1974).

Well logs of the Neogene sediments have recently been interpreted (Martínez *et al.*, 1991) using the sequence stratigraphy concepts and methodology of Vail (1987) and Galloway (1989).

When sequences are thin, or lithological and acoustic impedance contrasts are not sufficient, seismic profiles often do not show evidence of the sequence boundaries, and it may be the system tracts themselves that can be recognized. In these situations, well-logs, particularly gamma-ray analysis, and hole-to-hole correlations may allow the recognition and mapping of transgressive and regressive parasequences, and the identification of the Lowstand System Tracts (LST), Transgressive Systems Tracts (TST) and Highstand Systems Tracts (HST) that make up the Depositional Sequences (Vail *et al.*, 1989; Van Waagoner *et al.*, 1988; Sangree *et al.*, 1988; Martínez, 1986, 1990; Martinez *et al.*, 1991).

Stratigraphic interpretation

The Tertiary sediments of the Gulf of Valencia can be subdivided into a Paleogene 'Red Complex' unit of continental facies, and a Neogene unit of marine facies, which has been subdivided into the three groups: Alcanar, Castellon and Ebro.

The main characteristics of these two units, and their general distribution, are illustrated in Fig. 2. A number of specific points can be noted:

1. The only intra-Neogene erosional unconformity visible in the seismic profiles is correlative to the Messinian–Lower Pliocene LST (M_8–P_1).
2. The other Sequences of the Neogene (M_1–M_7) only show Systems Tracts of the TST and HST types. If Lowstand Systems Tracts are present, they were strongly condensed or were deposited SE of the basin axis in what are now the

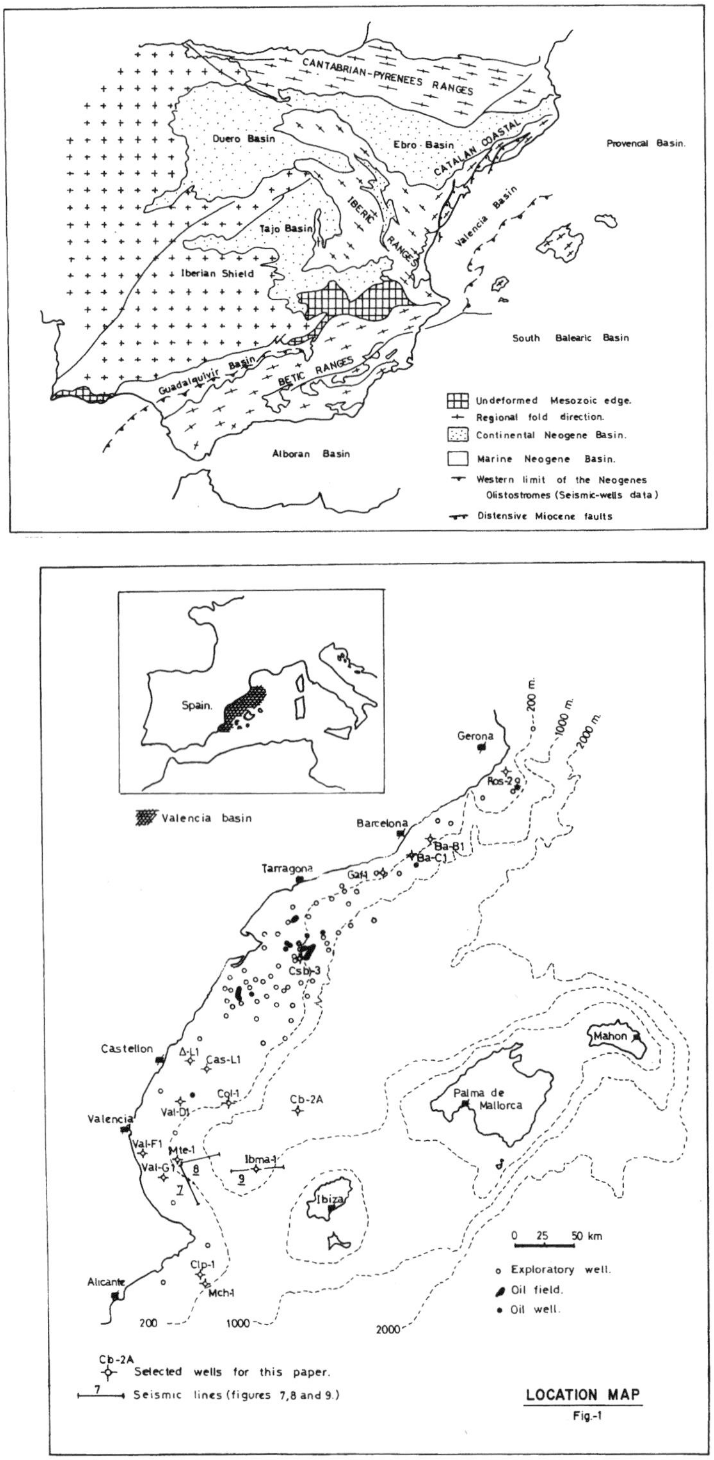

Fig. 1. General location of Valencia Basin, and outline map of Basin, with location of exploratory wells, oil fields, oil wells, along with wells and seismic lines further illustrated in this chapter.

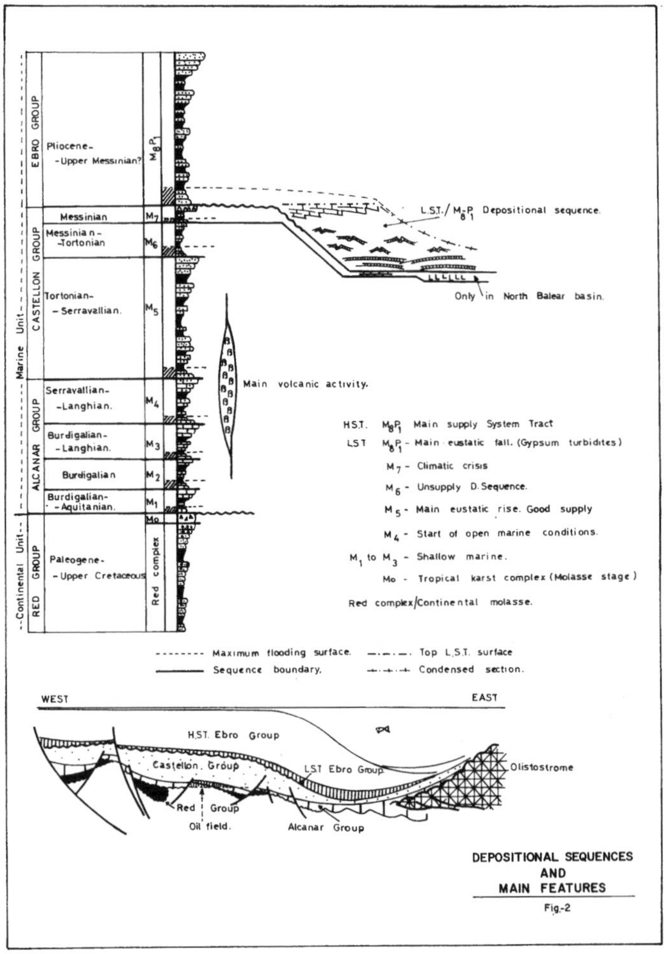

Fig. 2. The main depositional sequences and stratigraphic groups recognized in the Valencia Basin.

Balearic Islands, the Provencal Basin or the South Balearic Basin.

3. Late Miocene and Pliocene sediments make up 90% of the Basin fill. Apart from olistostrome sediment, this sedimentary fill was derived from the passive margin of the peninsula to the W–NW.
4. There is a clear unconformity between the Neogene unit, and the Mesozoic and Palaeogene. This unconformity separates the two main tecto-sedimentary stages.
5. Eight depositional sequences have been distinguished on the basis of their systems tracts and well-to-well correlation.

   If there are other sequences that correlate with episodes on the global curve (Haq *et al.*, 1987), they are too local or too thin to be recognized.
6. Lack of good biozonation, or any other chronostratigraphic control, means that the stage and system attribution on Fig. 2 is general and approximate.
7. For technical reasons, well logs are not available for the Upper Pliocene and Quaternary sediments, so they are excluded from consideration.

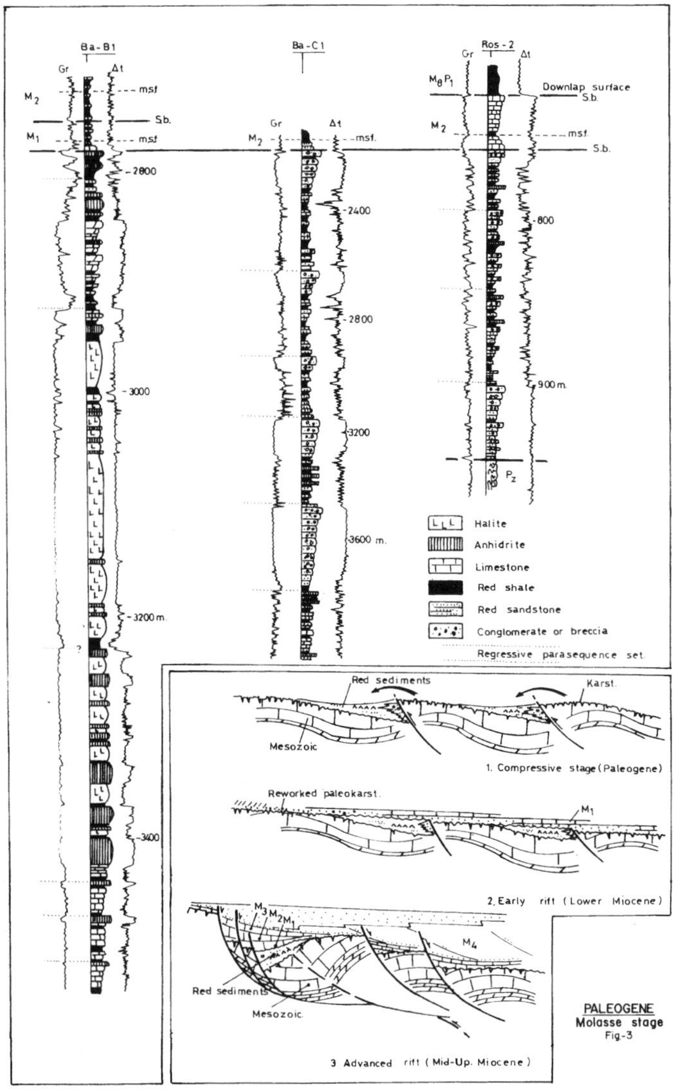

Fig. 3. Typical wells illustrating the Paleogene Molasse Stage, and cross-sections illustrating the evolution of the Red and Alcanar Groups.

The 'Red Complex' and the M_0 deposits

A relatively small but significant number of exploration wells have penetrated thick continental sediments; these are generally red, and located above Mesozoic and Paleozoic rocks and below Neogene marine sediments (Fig. 3). In a few wells, ages have been determined for these red sediments, and they range from Upper Cretaceous to Oligocene.

The unconformity between this 'Red Complex' and the marine facies of the Neogene, the correlation with sediments outcropping in the Iberian and Catalan Coast Ranges, and the distinctly prograding geometry of its lacustrine and fluvial parasequences suggest that these are syntectonic sediments of continental, molasse type. They appear to have formed during compressional phases of the Iberian and Catalan Coastal Ranges (Guimera, 1983, 1987; Anadon *et al.*, 1982, 1985).

During the compressional phases, many local highs were subjected to erosion and subaerial karstification that formed a widespread weathering surface, M_0 (Figs. 2, 3) marked by the Alcanar Breccia (Soler *et al.*, 1980–83). Reworking of this breccia, during Lower Miocene episodes of marine submergence, generated a larger and more-diverse complexity of facies than the originally simple karstic profile, which can still be recognized in many wells (Martinez *et al.*, 1983). Miocene reworking has led to misinterpretation of the origin of the Alcanar Breccia.

The erosion of the structural highs, and the simultaneous filling of the adjacent synclines by molasse sediments, produced a peneplane during the Oligocene, such that there was very little relief at the time of the Miocene marine transgression. The hydrocarbon fields of the Gulf of Valencia have been controlled by the extensional rejuvenation of the peneplane (Fig. 2).

In basin-margin situations, and in the case of early horsts not covered by the marine transgression of the Miocene, the 'Red Complex' may contain thin continental sediments of Lower Miocene age, similar to those recognized in outcrops on the Catalan margin (Agusti *et al.*, 1980; Cabrera *et al.*, 1991). However, these sediments were formed under an extensional tectonic regime and should show an agradational pattern different from the prograding pattern of the molasse sediment formed under a compressional regime.

Some wells drilled through the 'Red Complex' show two, three and up to four stacked regressive parasequences (Fig. 3). These internal sedimentary parasequences are thought to be mainly due to changes reflecting movements of local syntectonic unconformities similar to the changes seen in the compressional fronts of the Iberian and Catalan Coastal Ranges (Riba, 1989; Gonzalez *et al.*, 1988).

The change to extension in the Neogene, which resulted in the present Gulf of Valencia Basin, has also resulted in the preservation of the 'Red Complex', found within the young half-grabens. It has also produced the regional seismic unconformity involving the Lower Miocene sediments (Fig. 3).

The Alcanar Group

The Alcanar Group consists of marly and carbonate facies of Early and Middle Miocene age. It can be divided into four depositional sequences.

The thickness of the Alcanar Group can reach as much as 400–500 m in the depocentres of the early half-grabens. However, the group is normally poorly defined seismically, due to the absence of contrasts in seismic impedance, and it is often thin. Because of this, the most important approach is sequence interpretation based on the well-logs (Martínez, 1990).

Each of the sequences extends further than the previous one and they were deposited on an external platform that was sometimes anoxic, and subject to tectonic subsidence that was compensated by the production of carbonate sediment (Fig. 4).

The Transgressive and Regressive Systems Tracts show a notable facies convergence. The TSTs of the earlier three sequences (M_1, M_2 and M_3) start in the western margin with significant reworking during onlap of the M_0 karstic surface; these deposits have been

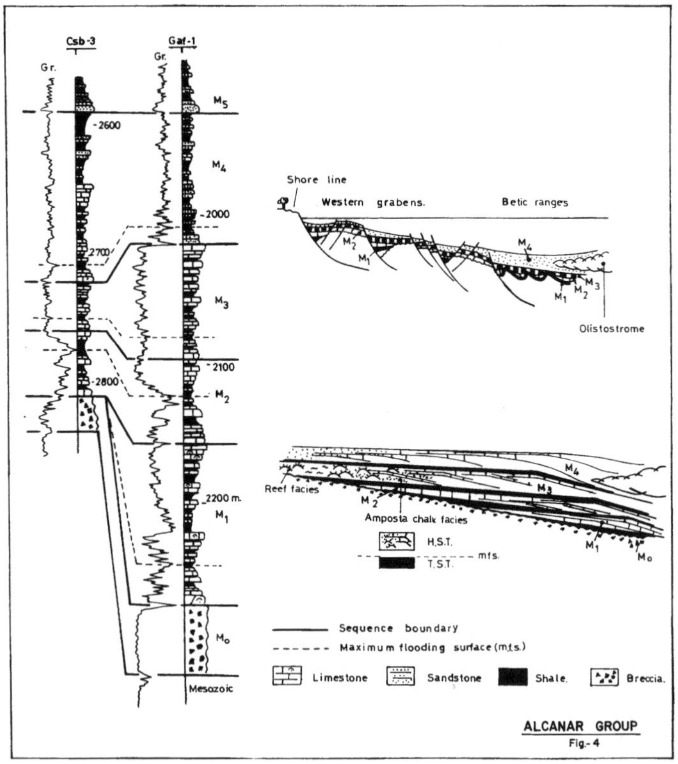

Fig. 4. Typical wells through the Alcanar Group, with cross-sections illustrating the structural geometry and sequence stratigraphy.

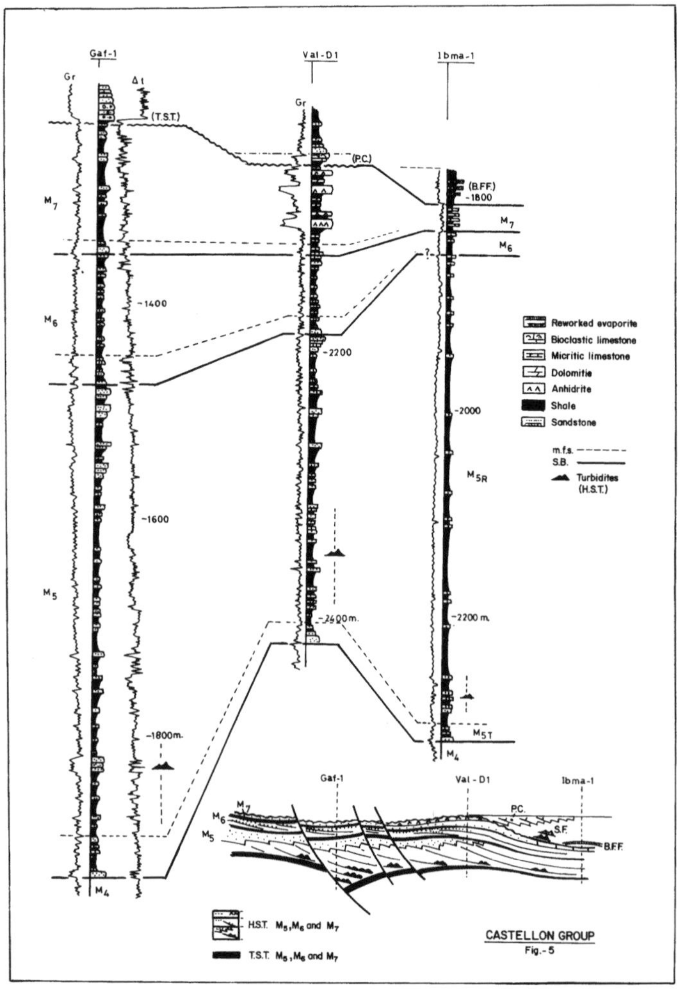

Fig. 5. Typical wells through the Castellon Group, with a cross-section illustrating the sequence structure.

called the Alcanar Conglomerate (Soler *et al.*, 1980–83). The regressive HSTs of the three earlier sequences (M_1, M_2 and M_3) appear to consist of smooth carbonate ramps, at least one of which, the Upper Burdigalian–Langhian one (M_3), developed a reef belt whose proximal talus facies is known as the Amposta Chalk (Fig. 4). These facies outcrop on land, in the Catalan margin (Permanyer, 1982; Agusti *et al.*, 1991).

The Langhian–Serravallian sequence (M_4) began with a particularly extensive onlap marking the onset of open marine conditions, and the later development of a platform–talus–basin system deriving from a siliciclastic coastal belt (HST).

During Langhian–Serravallian times, the new Neogene extensional regime developed, and this was accompanied by an episode of calc-alkaline submarine volcanism. The distal sediments of this sequence also contain the olistostromes derived from the Balearic and Betic front.

In the present axial zone of the Gulf of Valencia, the compressional Betic front is visible in the seismic profiles, with anticlines and synclines oriented SW–NE. The M_3 sequence includes the Neogene sediments deformed by this compression into synclines and thrusts. Similar relationships can be seen on top of the olistostrome unit, where small structural basins include deformed Lower–Middle Miocene sediments (Martinez *et al.*, 1979).

In the Balearic Archipelago, the last compressional phase is Langhian in age (Alvaro *et al.*, 1984; Pomar Goma, 1982; Ramos-Guerreo *et al.*, 1989). This adds support to the correlation of the submarine Mediterranean compressional front with the Betic–Balearic system.

The major gravitational olistostromes appear to have formed in response to the uplift of the Balearic horst. This episode was coeval with the greatest volcanic activity and the major extensional phase of the foreland. Flexuring appears to have been related to crustal thickening of the Betic system.

The Castellon Group

The Castellon Group represents most of the Middle Miocene and all of the Upper Miocene, and three depositional sequences (M_5, M_6 and M_7) have been recognized in it.

This group represents the development of the siliciclastic facies during the peak of tectonic extension. In spite of the strong extension, large amounts of sediment from the neighbouring continent entered the 'accomodation' space available. This strongly uncompensated HST (M_5) provided one of the best examples of an HST with well-developed turbidites (Fig. 5).

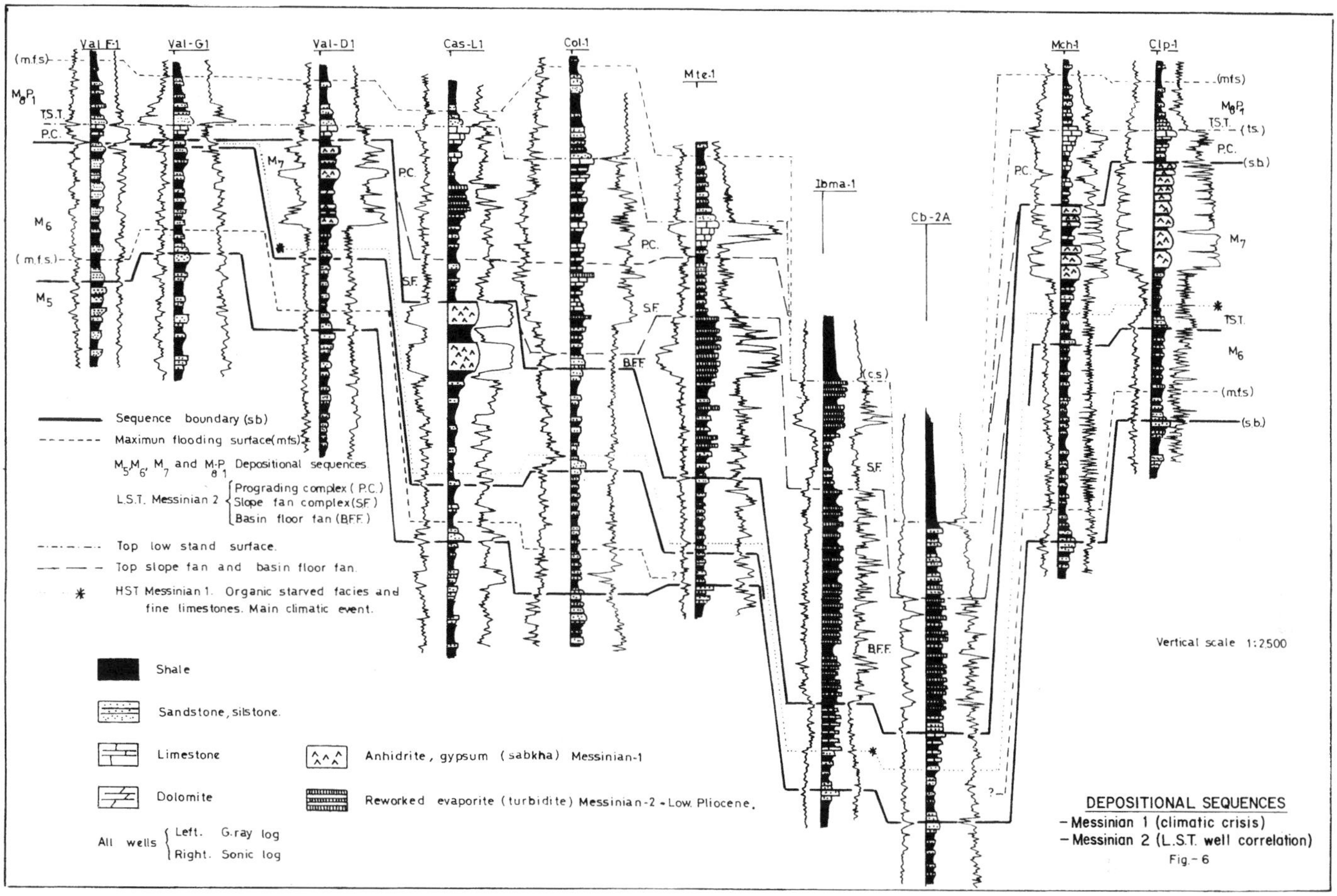

Fig. 6. Correlated wells through the Castellon and Ebro Groups with their interpretation in terms of sequences and relative sea-level changes.

The migration of deep facies towards the W–NW, relative to those of the previous Alcanar Group, appears to have taken place because of a distinct rise of relative sea level. This was not accompanied by an aggressive migration of the coastline, because on land, in the Catalan Coastal system, outcrops of this age have a continental facies (Barnolas *et al.*, 1983; Cabrera *et al.*, 1991) and overly marine or continental sediments of the Alcanar Group. It is therefore clear that the coastline was controlled by steep, faulted slopes which prevented significant penetration of the sea during the TST of this sequence. In the Gulf of Valencia, many extensional faults were strongly active during the sedimentation of this sequence (e.g. the Amposta oil field). This represents the continuation of the extension that began during the Alcanar Group.

The Depositional Sequence M_6 is geometrically defined by the platform–talus–basin profile inherited from previous events. This sequence is particularly shaly, implying a notable decrease in the amount of sediment supplied at its margin. This relative starvation may have been due to a decrease in the extension, or to the beginning of the climatic crisis of the Messinian. The latter possibility seems more reasonable, because many faults are still active, emerging at the contemporaneous sea-bed, and, in addition, it seems unlikely that erosion of the continent would decrease so rapidly for tectonic reasons alone (Fig. 5).

The Depositional Sequence M_7 (Messinian s.s.) started with a very thin and shaly TST, which was then followed by organic laminites, which generally precede the evaporites at the top of its HST. The extensional activity continued, especially at the NW margin of the basin, where local depocentres of the HST formed. The greatest thickness of the evaporite sabkha sediments that characterize the Messinian platform occur in these depocentres. The rapid development of the evaporite facies (Figs. 5 and 6) demonstrates the strength of the climatic crisis that characterized this episode.

The seismic and well information for this HST allows the reconstruction of a depositional model from margin platform to basin:

1. In the active grabens of the margin, a largely shaly facies accumulated. Although evaporite sediments were probably originally quite thick here, they have been generally removed by the Messinian unconformity (Figs. 5 and 6).
2. On the topset of the smooth platform, organic laminites and evaporite sabkha sediments formed, marking the extreme of the Messinian climatic crisis.
3. In the deeper areas, fine calcareous mudstones formed with *Globigerina acostaensis* and *G. margaritae* (Granados, 1978).

Messinian-1 salt deposits, present in the Provencal and Sub-Balearic Basins are assigned to this M_7 Sequence, and indicate the maximum degree of aridity preserved. The salt represents super-saturation in a significant water depth, and, as we shall see below, the alternating gypsum and pelagic clay sedimentation (Messinian-2) is interpreted as dominated by turbidity current and processes of the LST of Sequence M_8–P_1.

The Ebro Group

This Group comprises a fully developed depositional sequence (LST–TST and HST) that extends from the Messinian (*G. margaritae*) to the Upper Pliocene (*G. truncatulinoides*).

The quality of the seismic profiles, and the interpretation of the deposits using the methods and concepts of sequence stratigraphy, allow a very different interpretation of the 'Messinian Crisis' (Martinez, 1986; Martinez *et al.*, 1991) from that produced in the 1970s from deep sea drilling data (Hsu *et al.*, 1978; Ryan *et al.*, 1973, 1978).

This different interpretation satisfactorily solves the main problems inherent in the generally accepted theory involving the total desiccation on the Mediterranean Basin:

1. The absence of alluvial or fluvial sediments during the desiccation episode. How did these sediments escape from an internally draining trap of these dimensions?
2. The lack of a reasonable tectonic mechanism to explain the regular openings and closings of the Gibraltar Falls that have been invoked to account for the successions of alternating gypsum and pelagic clays that are a feature of the Messinian-2 of the deep Mediterranean.
3. The lack of evidence for erosion during sub-aerial episodes of these soluble sediments, or during the catastrophic floods from the Atlantic. In fact, Messinian-2 seismic facies are characteristically flat, parallel layering. Even more surprising, in spite of the 'Pliocene Revolution' that has been suggested, is the seismic conformity visible between Messinian-2 and the Pliocene in the deep water of the Gulf of Valencia.

A profile from the platform to the basin at the time of the Messinian eustatic crisis shows a gradational change from an erosional unconformity to a distinctly conformable sequence. This depositional sequence boundary is regarded as a Lowstand System Tract separating the Ebro Group from the Castellon Group, and it characterizes the global low sea-level event that has traditionally been regarded as responsible for the desertization of the Mediterranean. The deposits of the lower part of the Ebro Group are interpreted as belonging to a Lowstand Systems Tract (Figs. 6 and 11) in which the Prograding Complex (PC), the Slope Fan Complex (SF) and the Basin Floor Fan (BFF) of Vail's model (Vail, 1978; Vail *et al.*, 1988) can be recognized. Accepting this, the Messinian-2 cycles previously argued to result from desertization followed by flooding from the Gibraltar Falls are regarded as a sequence of turbidites accumulating in a continuously marine Mediterranean Basin.

The Basin Floor Fan Complex which initiated this Lowstand System Tract started with thin sandy levels at the base of the prograding parasequence, and these passed upwards into the alternating evaporites and claystones already discussed. Individual (not amalgamated) turbidites and fine contourites appear to form the predominant sediments in this turbidity fan.

The Slope Fan Complex contains thin channelized sandstone–claystone sequences as well as gypsum–clay sequences. This Slope Fan extended further than the previous Basin Floor Fan, and overlies material as high as the middle-high talus facies of the preceding depositional sequence (M_7, Messinian-1).

The prograding complex is characterized by shallowing upward sequences which are siliciclastic below, and become more carbonate-rich upwards. In its more proximal part, this complex is very thin, but in its distal part it contains occasional levels of detrital gypsum and finishes with a condensed sequence on its more distal segment. The most proximal prograding complex probably corresponds to the terminal complex (Esteban *et al.*, 1977) and to the post-evaporite carbonates recognized in the province of Almería (Megias, 1983). This interpretation, based mainly on the well logs, is also supported by the seismic profiles in the areas near the wells that have been selected for this paper (Figs. 1, 7, 8, 9 and 10).

The Transgressive System Tract overlies, as expected, the prograding complex and passes upwards into condensed levels, which overlie the deeper basin floor and slope fan complexes (Figs. 6, 10 and 11).

The subsequent Highstand System Tract is very thick and largely forms the present Gulf of Valencia platform (Soler *et al.*, 1980; Martinez *et al.*, 1991; Johns *et al.*, 1989). The high rate of progradation is typical of a lack of balance between subsidence and sediment supply. Because of this, the facies distribution is very similar to the one described for the depositional sequence M_5. In many wells presented here (Fig. 11), important Highstand System Tract turbidites, located in the distal foreset toe area, are again distinguished (Martínez, 1986; Martínez *et al.*, 1991).

Neogene stratigraphic model

The seismic profiles and data from exploration wells can be integrated by projecting them onto a transverse section to provide a sedimentary model for the Neogene of the Gulf of Valencia. This model (Fig. 12) has been drawn up to exclude the widespread synsedimentary deformation which differentiated local depocentres and highs and extended into the North Balearic and Provencal Basins, but which were not deformed by the Betic compressional phase.

The division of the sediments into the Alcanar, Castellon and Ebro Groups requires the introduction of new formational nomenclature, which has been extended to all the System Tracts identified, using old names where possible (Fig. 12).

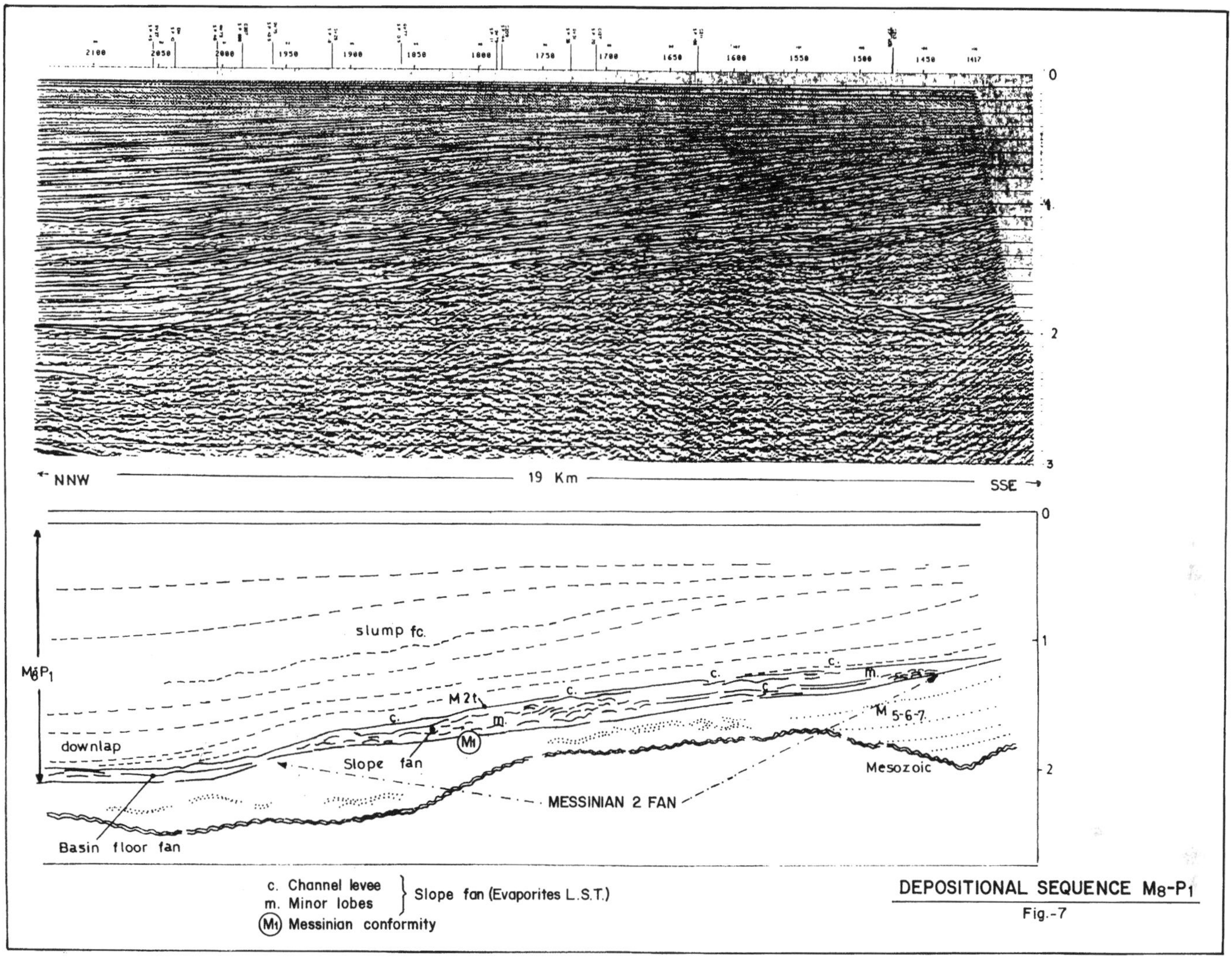

Fig. 7. Seismic profile (located at 7 on Fig. 1) through depositional sequence M_8–P_1, with interpretation.

Conclusions

The Tertiary sediments of the Gulf of Valencia Basin have been divided into a Paleogene molasse stage, and a Neogene marine stage that filled a foreland basin NW of the Betic Ranges.

The sparse information available for the Paleogene molasse stage does not allow detailed stratigraphic description or correlation with the equivalent sediments outcropping on land, on the basin margin.

The marine foreland stage included at least eight depositional sequences, and the main tectonic, climatic, eustatic and sedimentary features of these are outlined, as follows:

Aquitanian–Burdigalian. Early extension to the NW of the foreland basin, moderate sea-level rise, and restricted shallow marine carbonates with anoxic episodes.

Langhian–Serravallian. Important rifting in the W–NW margin, with volcanic activity. Last compressional deformation in the Betic ranges. Uplift of the Balearic Islands, and emplacement of the gravity slide olistostromes. First open marine facies due to a moderate sea level rise.

Tortonian. Major sea-level rise enhanced by thermal flexuring. High sediment supply, rapid progradation and the first deep-water facies over the Spanish margin of the basin.

Upper Tortonian – Messinian. Beginning of the climatic crisis. Low sediment supply on the platform.

Messinian. Climatic crisis: sabkhas over the Peninsula and Balearic platforms, starved limestones and marls in the Valencia sub-basin and salt deposits due to supersaturated waters in the deep Mediterranean Sea.

Upper Messinian – Lower Pliocene. Major sea-level fall. Gypsum turbidites. Erosional unconformity over the previous platforms and conformable in the deep-water

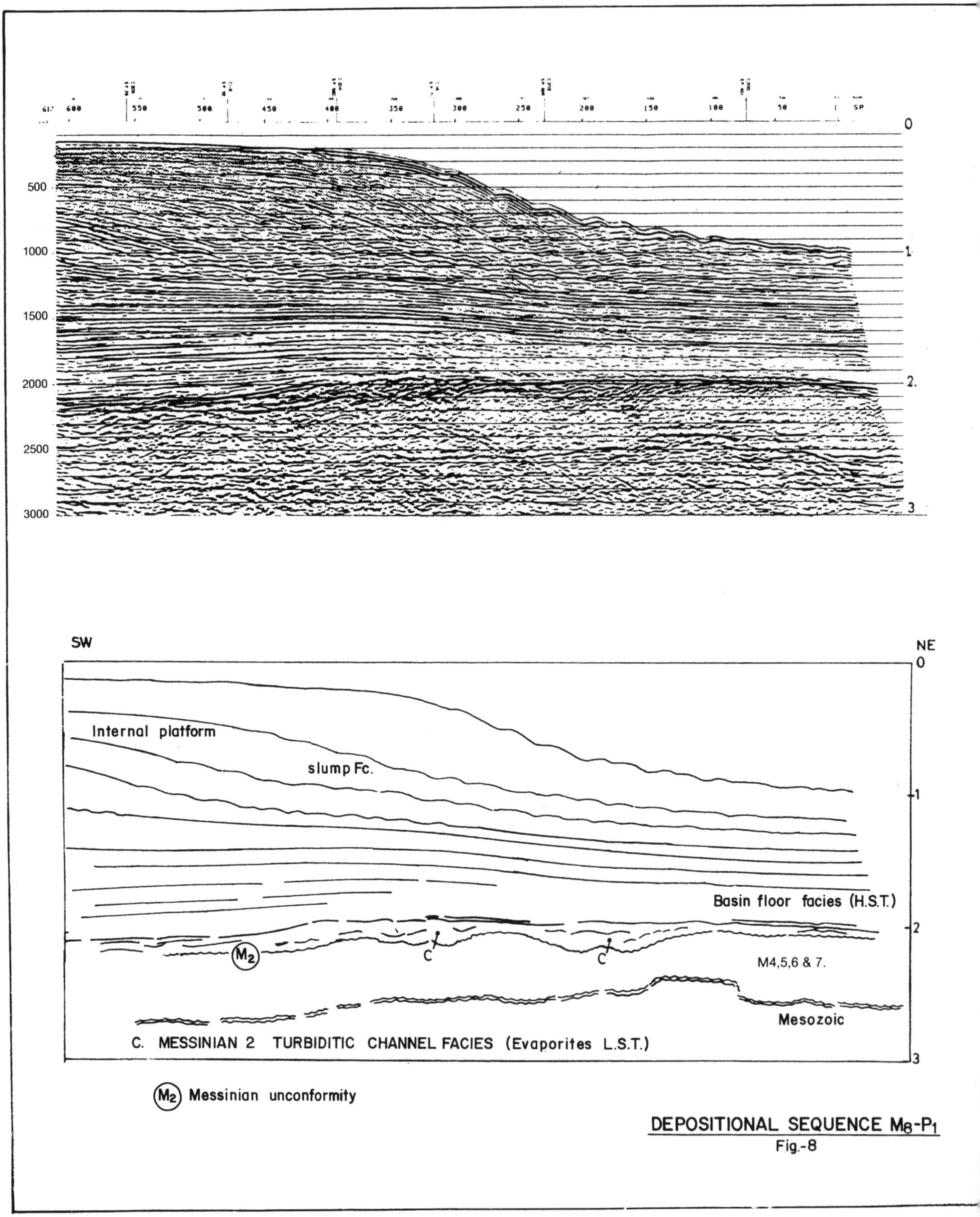

Fig. 8. Seismic profile (located at 8 on Fig. 1) through depositional sequence M_8–P_1, with interpretation.

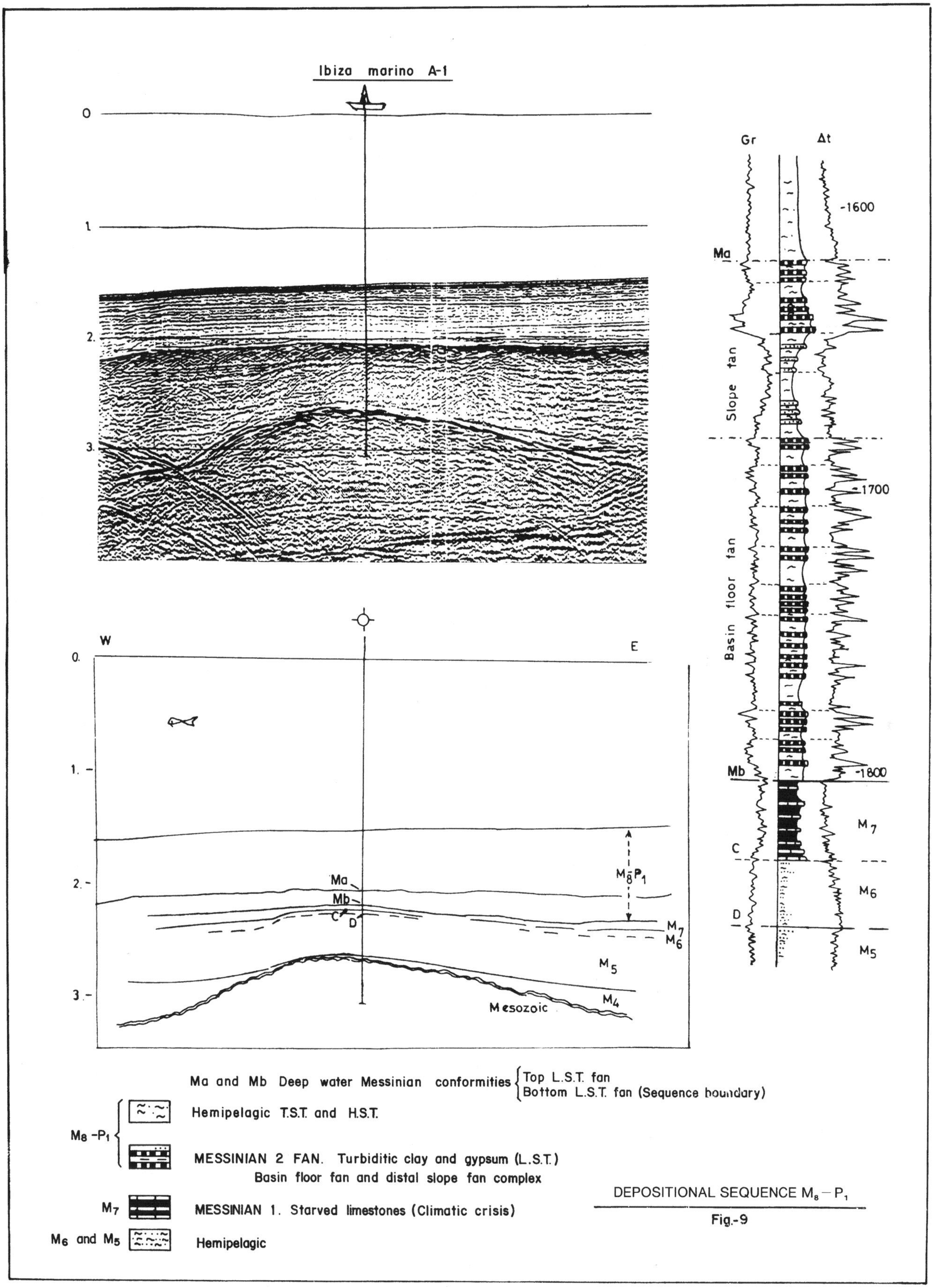

Fig. 9. Seismic profile (located at 9 on Fig. 1) through depositional sequence M_8–P_1, with interpretation and summary well succession.

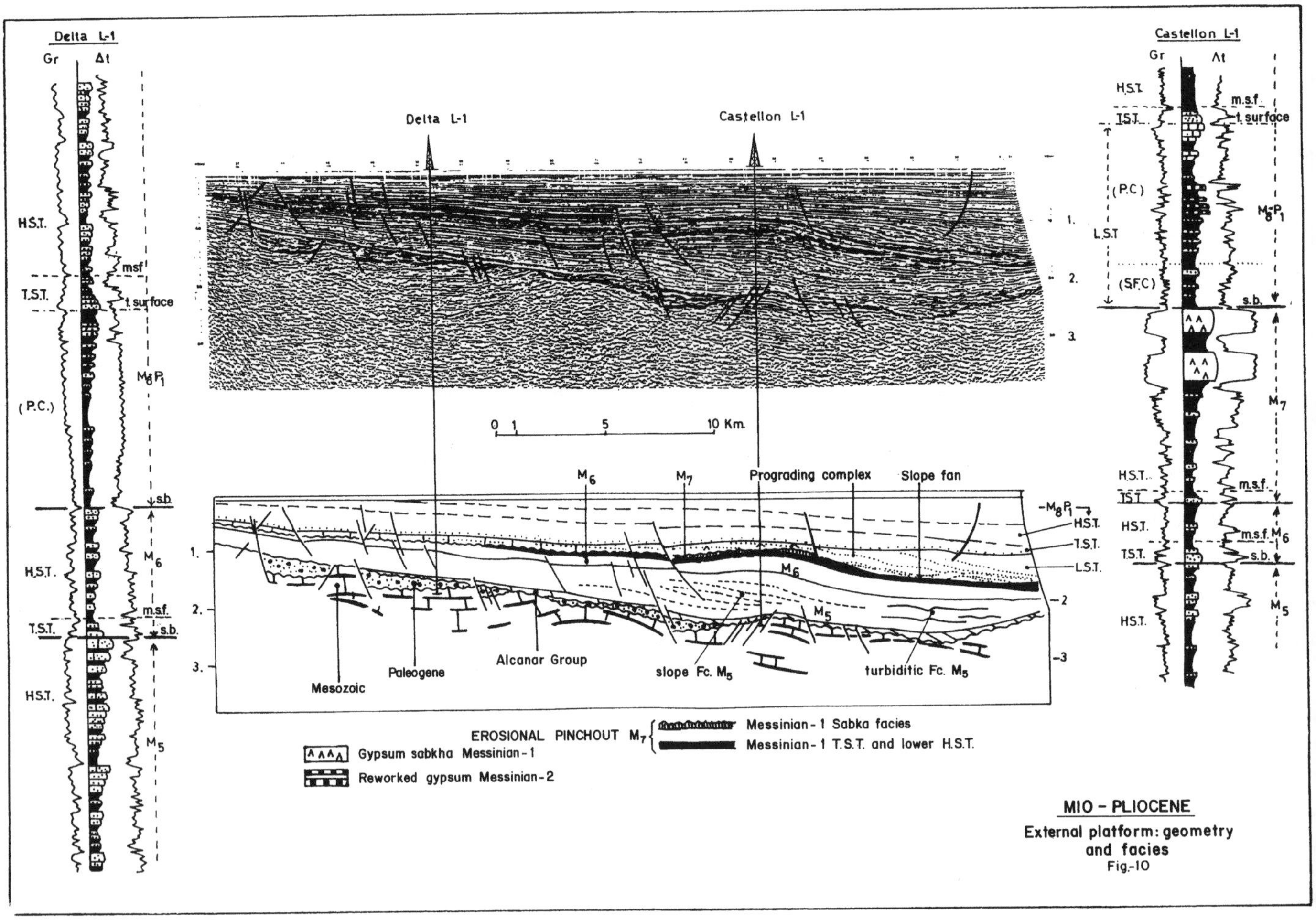

Fig. 10. Seismic profile (located at 10 on Fig. 1) through sequences M_5 to M_8–P_1, with interpretation and two well successions interpreted in terms of relative sea-level changes.

basin. The interpreted deep-water Messinian Unconformity is the seismic expression of the erosional and sedimentary topography of the Messinian LST below and above the gypsum turbiditic system.

Lower Pliocene. Main eustatic sea-level rise.

Pliocene–present-day. Strong sediment supply overwhelms subsidence and produces rapid progradation.

References

Agustí, J. and Cabrera, L. (1980). Nuevos datos sobre la biozonación del Burdigaliense continental en la Cuenca del Vallés-Panadés. *Act. Geol. Hispánica*, **15**(3): 81–84.

Agustí, J., Cabrera; LL., Calvet, F., Macpherson, I., De Porta, J. and Ramos-Guerrero, E. (1991). Registro sedimentario Mioceno en las zonas emergidas del sector central del margen Catalán *I. Congr. Grupo Español del Terciario. Vic.*

Alvaro, M., Capote, R. and Vegas, R. (1979). Un modelo de evolución geotectónica para la Cadena Celtibérica. *Acta Geol. Hispánica*, **14**: 171–177.

Alvaro, M., Barnolas, A., Del Olmo, P., Ramírez del Pozo, J. and Simó, A. (1984). El Neógeno de Mallorca: caracterización sedimentológica y bioestratigráfica. *Bol. Geol. Mineral*, **95**: 3–25.

Anadón, P., Colombo, F., Esteban, M., Marzo, M., Robles, S., Santanach, P. and Solé Sugrañes, L. (1982). Evolución tectoestratigráfica de los Catalánides. *Acta Geol. Hispánica*, **14**: 242–270.

Anadón, P., Cabrera, L., Guimerá, J. and Santanach, P. (1985). Paleogene strike–slip deformation and sedimentation along the south-eastern margin of the Ebro Basin. In Biddle, K. and Christie-Blick, N. (eds.). *Soc. Econ. Paleontol. Mineral. Spec. Publ.*, **37**: 303–318.

Barnolas, A., Clavet, F., Marzo, M. and Torrent, J. (1983). Sedimentología de las secuencias deposicionales del Mioceno del Camp de Tarragona. *X Congr. Español de Sedimentología. Menorca.*, **7**: 28–35.

Cabrera, L., Calvet, F., Guimerá, J. and Permanyer, A. (1991). El registro sedimentario Miocénico en los semigrabens del Vallés–Penedés y del Camp: organización secuencial y relaciones tectónica–sedimentación. *I Congr. Grupo Español del Terciario. Vic.*

Calvet, F., Permanyer, A. and Vaquer, R. (1983). El paleokarst del contacto Mesozoico-Mioceno en el Penedés y Camp de Tarragona. *X Congr. Español de Sedimentología. Menorca*, 1.73–1.75.

Clavell, E. (1991). Geologia del Petroli de les conques Terciaries de Catalunya. Doctoral Thesis. Universidad de Barcelona.

Esteban, M., Calvet, F., Dabrio, C. Barón, A., Giner, J., Pomar, L. and Salas, R. (1977). Aberrant features of the Messinian coral reefs, Spain. *3rd Messinian Seminar. Málaga.*

Galloway, W.E. (1989). Genetic stratigraphic sequences in basin analysis: I:

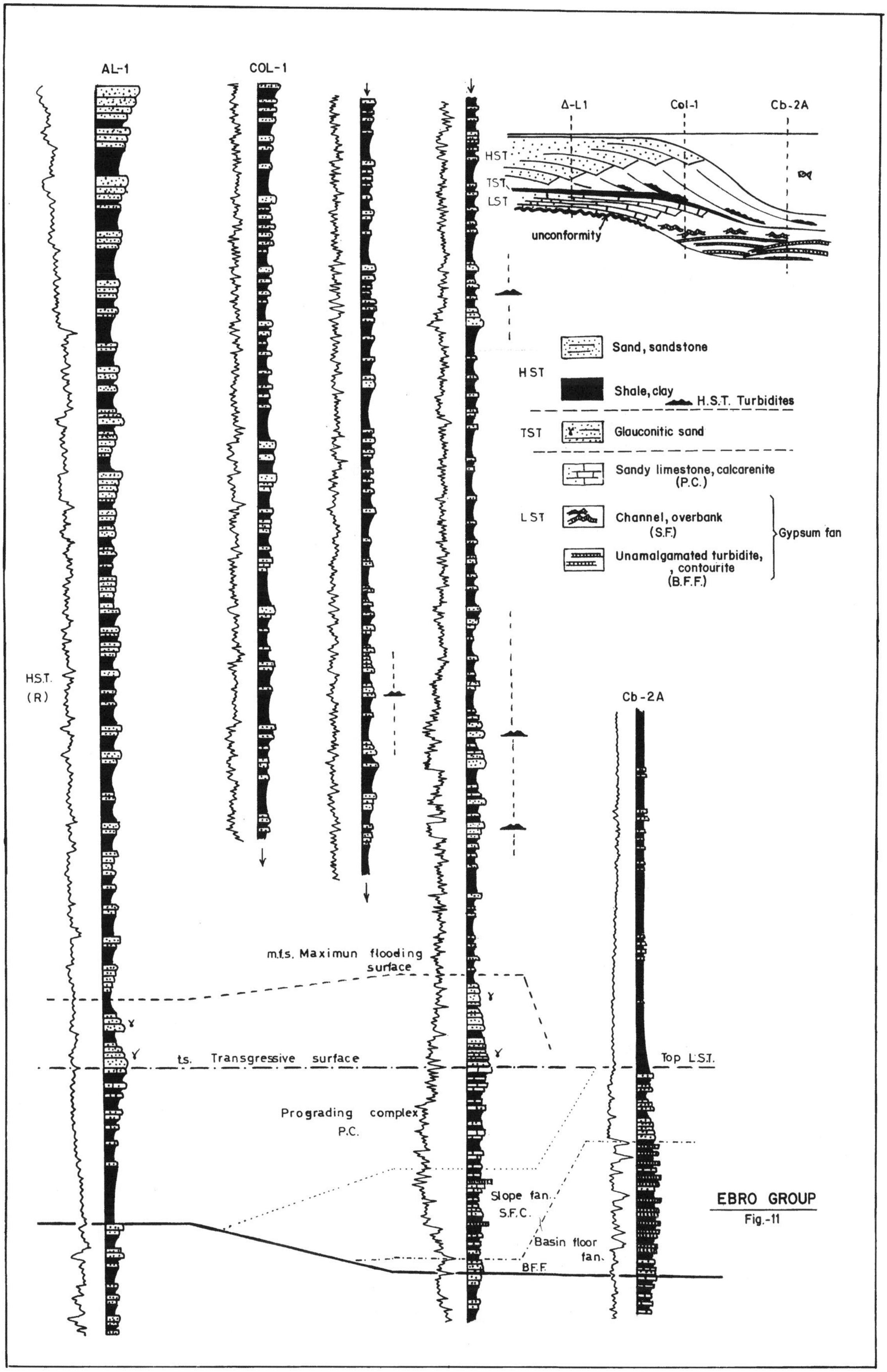

Fig. 11. Wells through the Ebro Group, with interpretations of relative sea-level changes, and cross-section illustrating the postulated relationships.

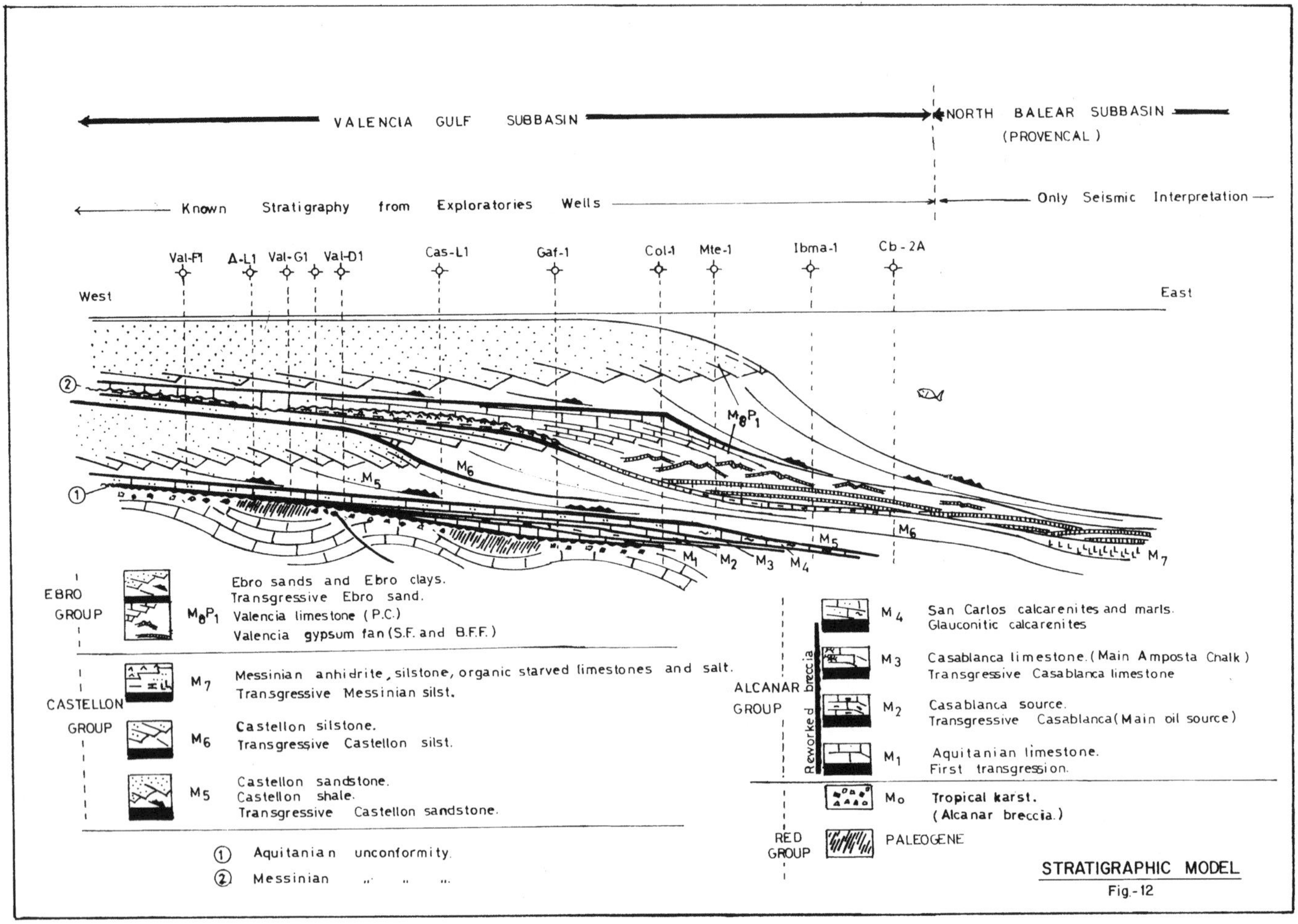

Fig. 12. Sedimentary model for the Neogene of the Gulf of Valencia.

architecture and genesis of flooding-surface bounded depositional units. *Am. Assoc. Petrol. Geol. Bull.*, **73** (2): 125–142.

García-Siñeriz, B., Querol, R., Castillo, F. and Fernández, J.R. (1978). A new hydrocarbon province in the Western Mediterranean. *Proceedings of the 10th World Petroleum Congress, Bucharest*, **4**: 1–4.

González, A., Pardo, G. and Villena, J. (1988). El análisis tectosedimentario como instrumento de correlación entre cuenca. *II Congr. Geol. España, S.G.E. Granada. Simposios*: 175–184.

Granados, L. (1978). Informe micropaleontológico del sondeo Ibiza Marino An-1. ENIEPSA. Internal memo.

Guimerá, J. (1983) Evolution de la deformation alpine dans le NE de la Chaîne Iberique et dans la Chaîne Côtier Catalane. *C. R. Acad. Sci., Paris.*, **279**: 413–420.

Guimerá, J. (1984). Paleogene evolution of deformation in the northeast Iberian Peninsula. *Geol. Mag.*, **121** (3): 413–420.

Hsü, K.I., Montadert, L., *et al.* (1978). *Initial report of the deep sea drilling project*. Washington DC. 42. Part I.

Johns, D.R., Herber, M.A. and Schwinder, M.N. (1989). Deposition sequences in the Castellon area, offshore northeast Spain. In Bally, A.W. (ed.). *AAPG*, **27**: 181–184.

López Blanco, M. and Marzo, M. (1992). Secuencias Deposicionales de tercer y cuarto orden en los abanicos costeros de Sant Llorenç del Munt y Montserrat (Eoceno, Cuenca de Antepais Surpirenaica, NE de España). *III Congr. Geol. España y VIII Congr. Latinoamericano de Geología. Salamanca 1991*, **2**: 123–129.

Magne, J. (1979). Etudes microstratigraphiques sur le Néogéne de le Mediteran Nord-occidentale. Thesis. Université Paul Sabatier. Toulouse. 259 pp.

Maldonado, A., Got. H., Monaco, A., O'Conell, S. and Mirabile, L. (1985). Valencia fan (Northwestern Mediterranean): distal depositional fan variant. *Mar. Geol.*, **62**: 295–319.

Martínez del Olmo, W. (1978). *Estratigrafía y estructura del Golfo de Valencia*. Coloquio Internacional Tectónica de Placas. Madrid.

Martínez del Olmo, W. (1986). Estratigrafía y estructuración del Golfo de Valencia desde líneas sísmicas y pozos. *Conferencia (Abstract). X Congreso Español de Sedimentología. Barcelona.*

Martínez del Olmo, W. (1990). Secuencias deposicionales a través de diagrafías de pozo. Cursos de Doctorado. Universidad de Madrid.

Martínez del Olmo, W. (1995). Geología del subsuelo aplicada a la interpretación stratigráfica y estructural: Dominio Prebético, Ibérico y Mediterráneo. Doctoral Thesis, Univ. de Madrid (in prep.)

Martínez del Olmo, W. and Esteban, M. (1983). Paleokarst development (Western Mediterranean). In P.A. Scholle, D.G. Bebout, C.M. Moore (eds.), *Carbonate depositional environments. Memoir AAPG*, **30**: 93–95.

Martínez del Olmo, W. and Garrido-Megías, A. (1991). La crisis Messiniense a partir de los datos del Golfo de Valencia (nueva hipótesis). *I Congr. Grupo Español del Terciario. Vic.*

Martínez del Olmo, W. and Gentou Penedes, R. (1979). Informe final de los permisos Ibiza Marion A, B y Cabriel, A, D. Empresa Nacional de Investigación y Explotación de Petróleos (ENIEPSA). Internal memo.

Martínez del Olmo, W. and Jurado, M.J. (1991). El Neogeno de la cuenca del Mar Menor (Murcia) a partir de datos del subsuelo. *I Congr. Grupo Español del Terciario. Vic.*

Martínez del Olmo, W., Murillas Angoiti, J. and Fernández Ortigosa, F. (1991). Los ciclos eustático-sedimentarios del Neogeno en el Golfo de Valencia (Mediterráneo Occidental). *I Congr. Grupo Español Terciario. Vic.*

Megías, A.G. (1983). Relación espacio-temporal entre arrecifes y evaporitas de las cuencas neógenas de Almería y Sorbas. *X Congreso Español de Sedimentología. Menorca*, **2**: 33–36.

Megías, A.G., Leret, G., Martínez, W. and Soler, R. (1980–83). La sedimentación neógena en las Béticas: análisis tectosedimentario. Congr. Grupo Español de Sedimentología. Salamanca (1980). *Mediterránea Ser. Geol.*, **1**: 83–103 (1983).

Permanyer, A. (1982). Sedimentologia i diagénesi dels esculls miocenics de la Conca del Penedés. Thesis, Univ. Barcelona. 545 pp.

Pomar Gomá, L. (1982). La evolución tectosedimentaria de las Baleares: Análisis crítico. *Acta Geol. Hispánica*, **14**: 239–310.

Ramos-Guerrero, E., Rodríguez Perea, A., Sabat, F. and Serra-Kiel, I. (1989). Cenozoic tectosedimentary evolution of Mallorca island. *Geodinám. Acta.*, **3** (1): 53–72.

Riba, O. (1989). Las discordancias sintectónicas como elementos de análisis de cuencas. In Arche, A. (ed.) *Sedimentología.* C.S.I.C., **2**: 489–522.

Ryan, W.B.F. and Cita, M.B. (1978). The nature and distribution of Messinian erosional surface: Indication of a several kilometer-deep Mediterranean in the Miocene. *Mar. Geol.*, **27**: 193–230.

Ryan, W.B.F., Hsü, K.I., *et al.* (1973). *Initial report of the deep sea drilling project.* Washington DC. Vol. 13.

Sangree, J.B., Vail, P.R. and Sneider, R.M. (1988). Evolution of facies interpretation of the shelf-slope; application of the new eustatic framework to the Gulf of Mexico. *20th Annual Offshore Technology Conference OIC.* Paper 5695.

Shell España. *Informes internos (1973–1974). Anónimos.*

Soler, R., Martínez, W., Megías, A.G. and Abeger, J.A. (1980). Rasgos básicos del neógeno del Mediterráneo Español. Congr. Grupo Español de Sedimentología. Salamanca (1980). *Mediterránea. Ser. Geol.*, **1**: 71–82 (1983).

Stoeckinger, W.T. (1976). Valencia gulf offer dead-line nears. *Oil Gas J.*, **29**: 197–204.

Stampfli, G.M. and Höcker, C.F.W. (1989). Messinian paleorelief from a 3D seismic survey in the Tarraco Concession area (Spanish Mediterranean Sea). *Geol. Minjb.* **68**: 201–210.

Vail, P.R. (1987). Seismic stratigraphy interpretation procedure. Am. Assoc. Petrol. Geol. In Bally, A.W. (ed.), *Atlas of Seismic Stratigraphy.* **27** (11): 11.

Vail, P.R. and Sangree, J.B. (1988). Sequence stratigraphy workbook, fundamentals of sequence stratigraphy. *AAPG Annual Convention Short Course: Sequence stratigraphy interpretation of seismic, wells and outcrop data.* Houston. Texas.

Van Waggoner, J.C., Posamentier, H.W., Mitchum, R.M. (Jr.), Vail, P.R., Sarg, J.F., Loutit, T.S. and Hardenbol, J. (1988). An overview of the fundamentals of sequence stratigraphy and key definitions. In Wilgus *et al.* (eds.) *SEMP. Spec. Publ.*, **42**: 39–46.

E4 Neogene basins in the Eastern Iberian Range

P. ANADÓN AND E. MOISSENET

Abstract

The structural setting and geological history of the eastern Iberian range and the Neogene basins are outlined. The following Neogene basins and structures are then reviewed: Calatayud–Daroca, Teruel, Serrion with Rubielos and Mijares, Eastern Maestrat, Alcora–Ribeselbes.

Introduction

During the Paleogene, the convergence of Iberia and Eurasia produced several major compressive structures in the inner part of the Iberian Plate, including the Iberian Chain. This NE–SW intraplate chain resulted from a tectonic inversion of Mesozoic extensional basins (Alvaro *et al.*, 1979). The structure of this chain includes a Hercynian basement and a thick Mesozoic cover. A thick packet of Middle–Late Triassic mudstones and evaporites constitutes a detachment level which caused different structural responses in the Hercynian basement with its Permian–Early Triassic tegument, and the Jurassic–Cretaceous cover (Richter & Teichmüller, 1933; Viallard, 1989; Guimerà & Alvaro, 1990; Roca, 1992). The basement faults controlled the history of the Mesozoic basins (Alvaro *et al.*, 1979).

The complex structure of this chain is determined by the diverse orientation of the basement faults in relation to the Paleogene compression direction. A transpressive strike–slip motion in the basement faults and a complex array of folds and thrusts in the cover were produced when the orientation of the basement faults was oblique to the regional shortening direction (N–S approximately). In these areas, the cover structures display a double vergence: to the NE in the northern part of the chain and to the SW in the southern part (Riba & Rios, 1962). When the orientation of the basement faults was perpendicular to the regional shortening, only complex thrust arrays were produced in the cover. Striking examples of these thrust arrangements are the Linking Zone between the Iberian Chain and the Catalan Coastal Range (Guimerà, 1988) and the Cameros–Demanda Range (Guimerà & Alvaro, 1990). These zones are located in the margins of the main Mesozoic basins (Maestrat Basin and Cameros Basin).

The compressive deformation in the eastern Iberian Chain took place between the Late Eocene and the Late Oligocene. In some sectors, it seems that the compressive episode ended during the Early–Middle Miocene (Calatayud Basin; Colomer & Santanach, 1988) or during the Middle–Late? Miocene in the northern part of the Cameros–Demanda Range (Riba *et al.*, 1983). During the Eocene and the Oligocene a set of basins linked to the compressive structures was formed close to the Ebro Basin. These basins show abundant syntectonic structural and intraformational arrangements: progressive unconformities and supratenuous synclines (Anadón *et al.*, 1986; Guimerà, 1988).

During the Oligocene–Miocene transition the compression changed to extensional (rifting) processes that continued through the Neogene (Simón, 1982, 1986; Guimerà, 1984, 1988; Moissenet, 1989). As a result of this Neogene rifting, numerous grabens were formed in the Eastern Iberian Chain. This graben system was also related to a generalized rifting that formed the Valencia Trough (Roca & Guimerà, 1992). The structures generated during the rifting process were formed either in tensional or transtensional situations, and were superimposed on the Paleogene compressive structures. Apart from the NW Iberian Chain basins, where the faults are parallel to the Paleogene structures (NW–SE), the main Neogene extensional faults are oblique to the Paleogene structures (Teruel–Mira Graben) and they are parallel to the Mediterranean coast (NNE–SSW).

In the southern Iberian Chain, the Jurassic–Cretaceous cover overlies a thick Late Triassic sequence formed by mudstones and evaporites. In this zone the Neogene faulting was accompanied by important diapiric extrusions (Ortí, 1981; Moissenet, 1985). Some diapiric bands are located in the axial parts of the grabens or linked to the graben bounding faults (e.g. Teruel Graben). However, in some zones, the diapiric extrusions display an irregular form and occur at the bottom of irregular depressions (e.g. Cabriel Basin). In some Neogene basins the diapiric movements initiated lakes by damming, after the diapiric extrusions (Moissenet, 1985).

The infill of the large Neogene basins of the Iberian Chain (Calatayud–Daroca, Teruel) display an almost complete stratigraphic sequence which ranges from the Early Miocene to the Pleistocene (Mein *et al.*, 1990; Moissenet, 1989). The small grabens, however, were formed in two different periods (Fig. 1). During the first period, which extended from the Late Oligocene–Early Mio-

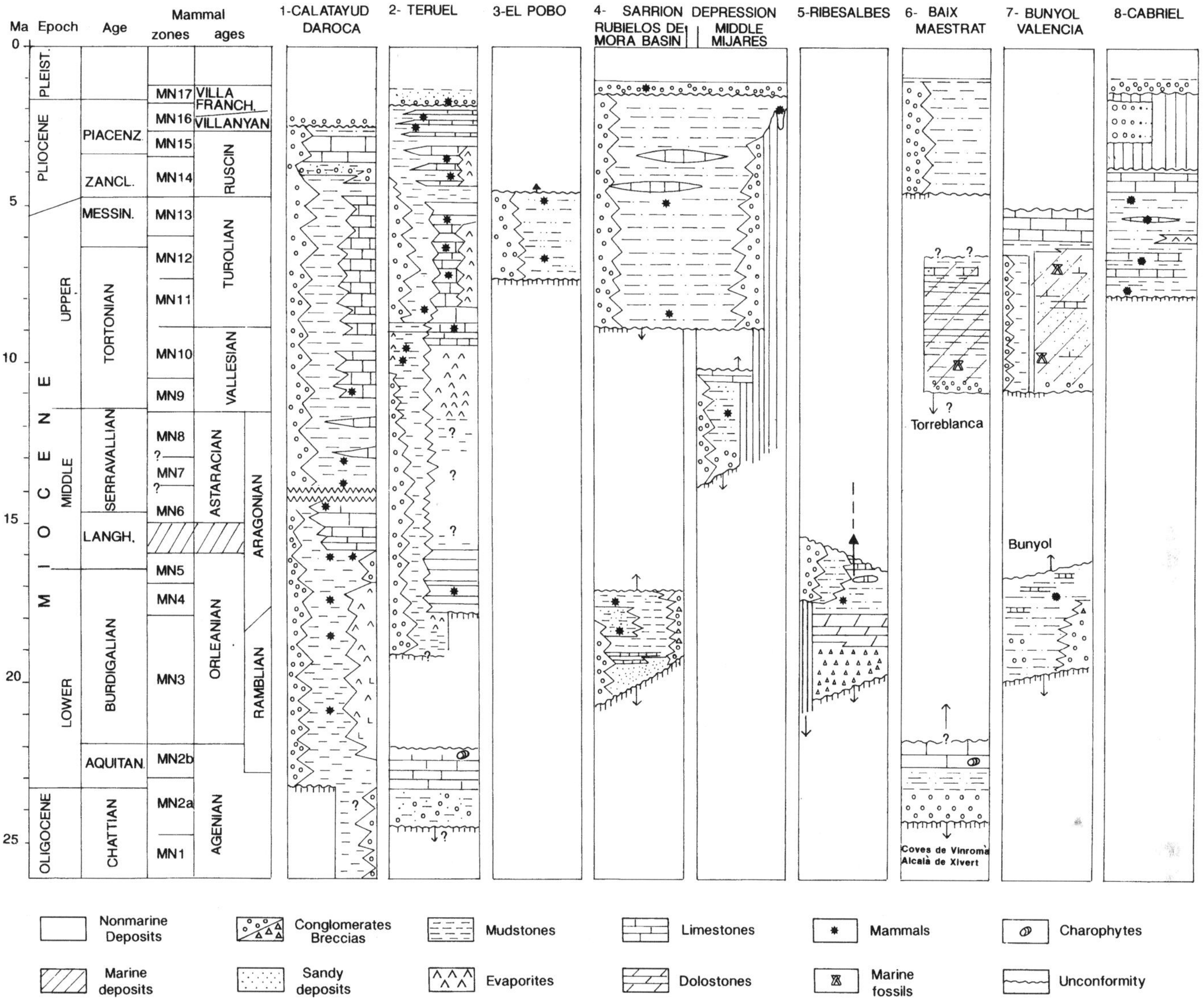

Fig. 1. Synthetic stratigraphy of the Neogene basins of the Iberian Chain. The basin-fill sequences are explained in the text. See Alberti & Pailhé (1983) for additional data on the El Pobo Basin.

cene to the Middle Miocene, the Rubielos de Mora Graben and some basins in the Maestrat, including the Ribesalbes Basin, were formed. During the second period, some graben developed in the Eastern Maestrat by new faulting or reactivation of ancient faults. The El Pobo Basin also formed in this second period (Alberti & Pailhé, 1983). In this period wide and shallow basins (depressions) were also initiated by flexural deformations of the crust (e.g. Sarrión Depression, Cabriel Basin). The sedimentary sequences formed in these depressions onlap the sequences formed during the first period (e.g. Sarrión Depression).

Neogene basins (Fig. 2)

The Calatayud–Daroca basin

The Calatayud–Daroca basin is located in the central-northern part of the Iberian Range. This Neogene basin is bounded by NW–SE faults, and has a complex structure (Julivert, 1954; Colomer & Santanach, 1988).

In the NW area (Calatayud zone), the basin displays a graben structure, and some of the faults are overlain by Early Miocene deposits (Julivert, 1954). In the NE margin of this zone, the faults have significant slip and constitute an 'en échelon' fault system, whereas in the SW margin the faults have minor slips (Julivert, 1954).

In the central sector of the basin (Daroca zone), the SW margin is formed by the Jiloca fault, whereas in the NE margin the Late Neogene sediments onlap the basin margin formed of Hercynian rocks. During the Paleogene the strike–slip motion of the Jiloca fault led to the development of fault-gauge. It seems that during the Early Miocene, after the beginning of the Miocene sedimentation in this area, the Jiloca fault was reactivated, producing basement slices with low-angle thrusts over the Early Miocene alluvial deposits

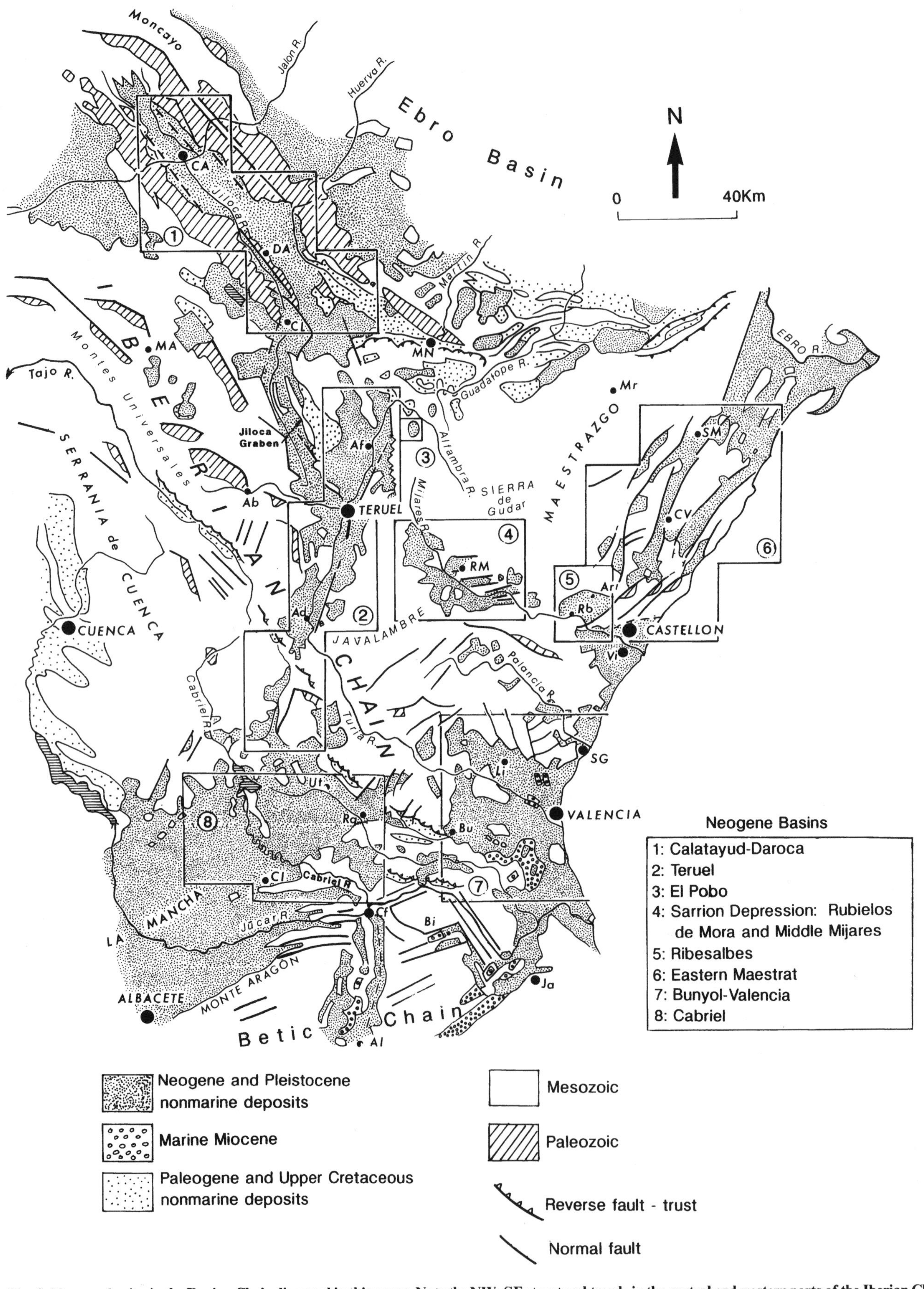

Fig. 2. Neogene basins in the Iberian Chain discussed in this paper. Note the NW–SE structural trends in the central and western parts of the Iberian Chain, the table-land zone of el Maestrat (Maestrazgo), and the NNE–SSW trends of the eastern Maestrat grabens. To the south of Valencia observe the NE–SW structures of the Betic Chain.

(Julivert, 1954). In this zone, the basin does not display the typical graben features associated with extensional kinetics. The SW margin, at least, was linked to a transpressive strike–slip fault (Colomer & Santanach, 1988). During the Pliocene an extensional phase occurred in this area (IGME, 1983) and the Jiloca fault acted as a normal fault. This motion led to the development of the Jiloca semi-graben to the S and to the SW of the Calatayud–Daroca Basin (Moissenet, 1989).

The SE area of the Calatayud–Daroca Basin (Navarrete–Bañón zone) displays a complex pattern, and the ancient SW margin is now connected to the Jiloca Graben. In the eastern basin margin, the base of the Neogene basin fill unconformably overlies the clastic sequences of the Paleogene Montalban Basin.

The basin fill deposits of the Calatayud–Daroca Basin have complex facies relationships. Three main areas which show different basin-fill sequences may be considered. In the Calatayud area (Jalón Basin; Aguirre *et al.*, 1974) the basin-margin facies consist of alluvial conglomerates and breccia deposits that pass basinwards into sandstone- and mudstone-dominated successions. These detrital facies pass into evaporite successions that form the lower part of the basin-fill sequence which crops out in the centre of the basin. The evaporite deposits comprise mainly nodular, alabastrine gypsum, and minor halite and epsomite. A mudstone- and limestone-dominated sequence overlies the evaporite deposits in the central part of the Calatayud area. The Miocene sequence in this area, over 400 m thick, ranges from Early to Middle Miocene in age (De Bruijn, 1967; Aguirre *et al.*, 1974).

In the central part of the basin (Daroca area) the lower part of the Neogene sequence is formed of alluvial deposits, up to 200 m thick. The alluvial deposits consist of conglomerates and red mudstones and sandstones. This sequence is overlain by an alternation of mudstones and lacustrine limestones (Crusafont *et al.*, 1954). The basin-fill sequence in this zone, up to 350 m thick, ranges from the Early?–Middle Miocene to the Early Pliocene. In this zone, the Aragonian mammal stage stratotype (Daams *et al.*, 1977) has been established.

In the SE part of the basin (Navarrete–Bañon area) the Neogene deposits show a relationship similar to that of the Calatayud area. The alluvial coarse detrital deposits in the SW margin of the basin pass laterally to mudstone-dominated deposits and to evaporites (Navarrete Gypsum). A limestone unit overlies the evaporites in the central part of the basin. The basin-fill sequence in this area ranges from the Early to Late Miocene. In this area, near Navarrete, the Ramblian mammal stage stratotype (Daams *et al.*, 1987) has been established.

The Teruel Graben

The Teruel Graben should be regarded as a semi-graben in which several sectors or subbasins may be distinguished (from north to south: Alfambra, Teruel, Ademuz and Moya–Mira). Each basin is bounded by en échelon NNE–SSW normal faults. The hanging-wall subsidence is westwards of the fault, the footwall uplift eastwards of the fault. The NNE–SSW Teruel Graben is 15 km wide and 100 km long, and dissects the NW–SE Paleogene compressional structures of the Iberian Chain. This graben is connected to the Neogene Calatayud–Daroca Graben by the Jiloca 'corredor'. The Teruel Graben fill sequence ranges from the Early Miocene to the Late Pliocene (Moissenet, 1989). To the south of Ademuz (Mira–Moya area) the Neogene extensional phase also led to the opening of a Triassic-floored graben which is the result of extension combined with diapirism.

The Neogene deposits of the Teruel Graben accumulated in an internal drainage setting. These deposits are characterized by frequent lateral and vertical facies changes. In the basin margin the Neogene sequences are formed by alluvial fan red conglomerates grading laterally towards inner basin areas, to red, fine-grained detrital deposits (mudflat and flood basin deposits) and/or to shallow lacustrine carbonates and evaporites. Locally deeper lacustrine carbonates and evaporites (the Libros Gypsum sequence) are present (Anadón *et al.*, 1992). The thick, red, coarse-grained deposits in the basin margin correlate with thick limestone-dominated successions in the inner-basin zones. In between these two settings the facies changes result in the superposition of detrital carbonate and gypsum deposits. A common sequence is formed of red mudstones at the base, calcareous paleosols above and lacustrine limestones on top. This type of sequence originated during the Early–Middle Miocene (Libros), Turolian, Late Miocene (Los Mansuetos) and Pliocene (Celadas, Orrios and Villalba Alta). These sequences (Fig. 3) may be interpreted as forming by an attenuation of the subsidence and a correlative lacustrine flooding (tectonic and climatic controls). The superposition of these sequences has been controlled by tectonics and climate, which also led to the migration of the depocentres.

Villafranchian (Plio-Pleistocene) alluvial fan deposits lie unconformably on top of the graben-fill deposits. The Villafranchian deposits are related to the erosion of the surrounding uplifted blocks due to the reactivation of the faults at the graben margin and to climate changes. They are also related to the beginning of external drainage of the graben (Moissenet, 1982a, b).

The Teruel Graben can be regarded as a reference basin for the Mediterranean non-marine Late Neogene. The Turolian (Latest Miocene) mammal stage has been defined on the fauna of the mammal sites of Concud and Los Mansuetos. The recent discovery of a large number of Early and Middle Pliocene mammal sites in the centre and northern areas of the Teruel Graben has allowed the definition of the Alfambrian as a potential new mammal stage stratotype for the Mediterranean Neogene (Mein *et al.*, 1990).

The Sarrion depression and the Rubielos and Mijares basins

In the area around the Middle Mijares River, extensive Neogene outcrops occur in a large monoclinal depression located between the Jurassic block of Javalambre to the south and the Cretaceous cliffs of the Sierra de Mora to the north. The Neogene deposits may be grouped into several assemblages or sequences which originated in different tectonostratigraphic settings.

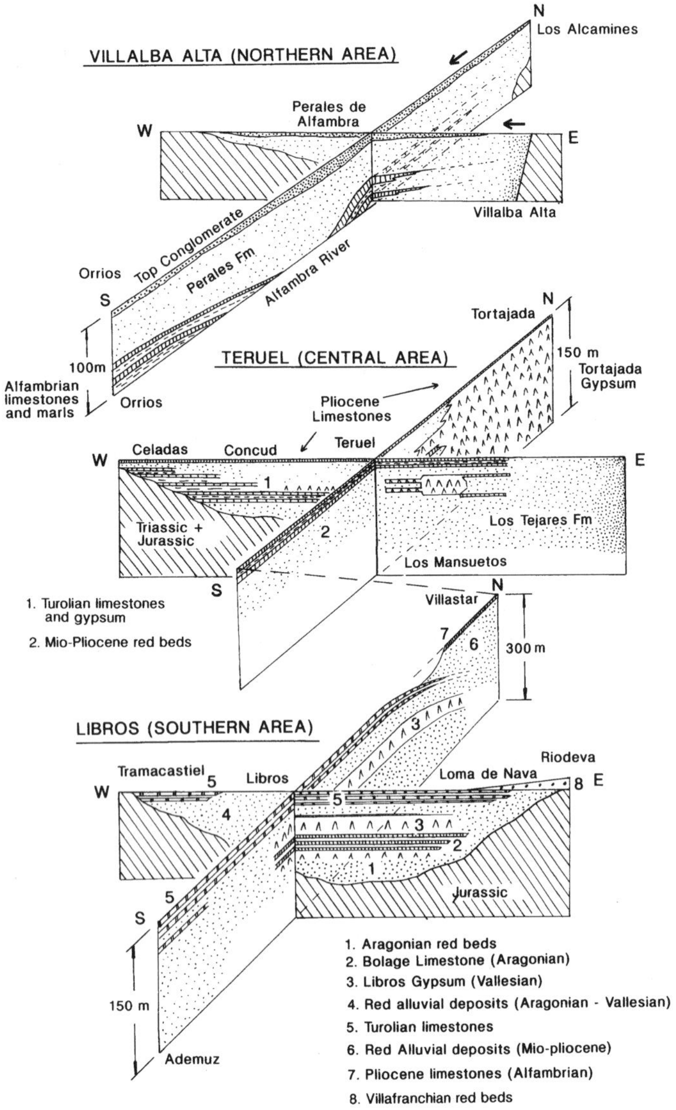

Fig. 3. Schematic facies relationships of the northern, central and southern areas of the Teruel Graben.

The Early Neogene Assemblage

The Early Neogene Assemblage was formed during an early phase of graben development. This assemblage comprises the Early–Middle Miocene deposits of the Rubielos de Mora Basin and the Middle Miocene deposits of the Middle Mijares grabens.

The Rubielos de Mora Basin The Rubielos de Mora Basin is located in a graben 10 km long by 3 km wide bounded by ENE–WSW-striking normal faults. The structure of the basin fill is formed by an overall syncline which resulted from the activity of the normal faults which bound the Rubielos Basin. A vertical slip of 1500 m for the southern fault has been deduced (Guimerà, 1990). Two fossil mammal sites in the middle–upper part of the basin-fill sequence have been attributed to the Early–Middle Miocene (Crusafont *et al.*, 1966; De Bruijn & Moltzer, 1974).

Three main stratigraphic units have been established in the Rubielos de Mora basin fill, which is over 800 m thick. The main depositional features of these units have been described by Anadón *et al.* (1988a, b, 1989, 1991). The Lower Unit records an early stage of the basin evolution dominated by alluvial sedimentation. This unit, from 75 to 300 m thick, is formed by sandstones with minor interbedded mudstones. The bulk of this unit is formed of alluvial deposits. The Middle Unit, up to 100 m thick, consists mainly of lacustrine deposits. In the eastern part of the basin, this unit is formed of bioclastic limestones with interbedded grey mudstones, sandstones and lignites. In the western part of the basin, fluvial mudstones and sandstones also occur. The Upper Unit marks the maximum extent of the open lacustrine facies. This unit is up to 400 m thick and displays a large variety of facies. The bulk of the fluvial inputs were fed to the lacustrine basin from the graben ends and the detrital material was axially distributed. The depositional framework of the Upper Unit was a perennial meromictic lake influenced by contributions from surrounding alluvial systems. The deposits of this unit are characterized by a striking asymmetric facies distribution and by a hierarchical sequential arrangement. The open lacustrine facies comprise laminated and massive mudstones, carbonate–clay varve-like rhythmites and oil shales. The marginal lacustrine facies include shallow limestones, sandstones and mudstones.

The Rubielos de Mora basin fill sequence is unconformably overlain by Late Miocene alluvial deposits of the Upper Neogene Assemblage of the Mijares Depression.

The Middle Miocene deposits of the Mijares grabens The Middle Mijares grabens were formed during the Middle Miocene. They are bounded by ENE–WSW faults. The basin-fill deposits of these grabens range from 10 to 200 m in thickness and comprise alluvial and fluvio-lacustrine facies (Paricio, 1985). The alluvial facies are formed by coarse conglomerates, cross-bedded sandstones and minor mudstones. The fluvio-lacustrine facies, which generally overlie the alluvial facies, consists of travertinic limestones, sandstones and thin coal beds. The graben fill has been dated as Middle Miocene using fossil micromammal faunas from two sites (Paricio, 1985).

The Late Neogene assemblage: The Sarrion Depression infill

The Late Neogene assemblage, which extends over a large area (Sarrion Depression), comprises Late Miocene to Pleistocene alluvial and palustrine deposits. These deposits lie unconformably over the Early Neogene assemblage and the Mesozoic rocks. The Late Neogene assemblage comprises two units. The Lower Unit is made up of red mudstones and conglomerates of alluvial origin and lacustrine travertines. This unit is Late Miocene (= Turolian) in age (Moissenet, 1982a, b). The upper unit comprises two subunits, which are related to the evolution of the Sierra de Javalambre piedmont. The lower subunit, or Sarrion Formation, is formed of red mudstones and interbedded breccias. Calcareous palaeosols are also present. These materials overlie a karst filled with fossiliferous breccias which contain a Late Pliocene (MN 16 zone) fossil mammal fauna (Adrover, 1974). The uppermost deposits of the Late Neogene assemblage constitute the Puebla de Valverde Formation, up to 30 m thick. This formation consists of breccia deposits with interbedded red sandstones. In the upper part of the La Puebla de Valverde section, a red carbonate-enriched sandy mudstone bed contains a fossil mammal fauna of the Middle Villafranchian (Gautier & Heintz, 1974), equivalent to the MN 17 zone of Mein (1975).

The Eastern Maestrat graben system

The Eastern Maestrat is characterized by a complex structure of horsts and grabens, 70 km long and 30 km wide, which is located between the Lower Ebro River area and the Castelló–Ribesalbes area. Horsts and grabens are bounded by NE–SW extensional faults. In the Eastern Maestrat the overall structure is characterized by a graben and semi-graben complex system with Neogene non-marine basin fills. In the Western Maestrat semi-grabens bounded by eastern faults are predominant. To the SW of the Maestrat graben system, the Ribesalbes–Alcora Basin is also bounded by NE–SW faults (see below).

The Tertiary deposits in the Eastern Maestrat comprise two main sequences (Anadón *et al.*, 1990). The lower sequence lies unconformably on the Mesozoic basement and has been deformed by the later horst and graben development. The Late Tertiary sequence is located within the grabens forming the upper graben-fill deposits. The Early Tertiary sequence is composed of two main units. The Lower Unit is formed by coarse conglomerates up to 100 m thick. In the upper part of the unit the conglomerates alternate with sandstones and mudstones. Some minor, thin interbedded lignites also occur. The generally fining-upward succession of the Lower Unit was formed in alluvial fan environments (Anadón *et al.*, 1990). The Upper Unit is formed by lacustrine mudstones and marls up 50 m thick. The limestones are thin bedded, display slumped horizons in some places and contain abundant fossils: lacustrine gastropods, charophytes and ostracodes. The charophytes found in the limestones and marls of the Upper Unit indicate that they were probably deposited during the Early Miocene, although a Late Oligocene age cannot be discounted (Anadón, Moissenet and Feist in IGME, 1985). In the Coves de Vinromà – Alcalá de Chivert area (CV in Fig. 1), there is a unit intermediate between the two above mentioned ones. This unit consists of blue mudstones and marls with interbedded coals (Suarez *et al.*, 1983). In some places, individual coal beds attain 2.5 m in thickness. The Early Tertiary sequence was formed in an early semi-graben system, probably during the Miocene (Simón, 1982). The overall megasequence probably indicates a diminution of the fault motion or the erosion of the surrounding relief, although a back faulting mechanism cannot be excluded.

The Late Tertiary sequence, up to 150 m thick, is mainly formed by alluvial deposits which include proximal alluvial fan conglomerates and distal alluvial, finer-grained facies. These distal deposits comprise red to yellow mudstones with interbedded conglomerates. This sequence may be dated as Late Neogene and Early Pleistocene.

In the Eastern Maestrat coastal plain near Torreblanca a shallow well has recorded a marine sequence probably Tortonian in age (Acuña, 1982). The marine deposits, which overlie calcareous conglomerates, consist of limestones, sandstones and mudstones. This succession may be correlated with similar sequences in the Valencia area that are related to the Tortonian transgression.

The Alcora–Ribesalbes Basin

The Miocene Alcora–Ribesalbes Basin is a complex graben bounded by ENE–WSW faults. To the SW, in the Ribesalbes area, the Neogene deposits unconformably overlie Cretaceous limestones whereas in the NE part of the graben, in the Alcora area, the latest Neogene deposits overlie Triassic and Cretaceous rocks. Two main sequences have been established in the Alcora–Ribesalbes Basin (Agustí *et al.*, 1988; Anadón *et al.*, 1989).

The Ribesalbes Sequence

This lower sequence is formed by alluvial and lacustrine deposits, up to 600 m thick, which were formed throughout the Early–Middle Miocene. The most complete succession of this sequence may be observed in the outcrops close to the village of Ribesalbes, where the Neogene sequence consists of five main units (Anadón, 1983; Agustí *et al.*, 1988; Anadón *et al.*, 1989). The lower part of the Ribesalbes section is formed of 300 m of alluvial conglomerates (unit A). The clasts are heterometric, angular and form a framework. Thin interbedded sandstones and mudstones also occur. Overlying these conglomerates, a lacustrine carbonate unit (B), up to 100 m thick, is formed of finely stratified and laminated dolostones, limestones and mudstones. In general these lacustrine deposits are organic-rich (oil shales). A mudstone-dominated unit (C) overlies the carbonate unit B. Unit C, 90 m thick, consists of yellow and grey mudstones with interbedded dolostone and sandstone beds. A thick olisthostromic unit (D), up to 70 m thick, overlies unit C. The olisthostrome is formed of large blocks of Mesozoic carbonate rocks, each up to several metres thick. The upper part of the Ribesalbes section consists of micritic limestones up to 20 m thick (unit E). The limestones are thin bedded and laminated and contain gastropod and charophyte bioclasts. Near Araya, 4 km to the North of Ribesalbes, several mammal sites in a mudstone sequence correlated to the C unit contain fossil mammal faunas attributed to the MN 4 zone of Mein (1975) (Middle Aragonian = Middle Miocene, Agustí *et al.*, 1988).

The Alcora Sequence

This upper sequence consists of alluvial deposits up to 200 m thick, which probably formed during the Middle and Late? Miocene. In the inner part of the basin the Alcora Sequence conformably overlies the Ribesalbes sequence (Agustí *et al.*, 1988), whereas in the NW basin margin the upper beds of the Alcora sequence onlap the Mesozoic rocks. The Alcora Sequence, in the NW part of the basin, is mainly formed of conglomerates with thin interbedded sandstones and mudstones. Basinwards, to the SE, the conglomerates gradually change to a red mudstone-dominated sequence, with minor interbedded sandstones.

The Neogene sequences in the Coastal Valencian Depression

The Neogene deposits in the western part of the Valencian Depression (Bunyol area) comprise two main units. The lower unit is formed by a complex sequence of conglomerates, sandstones and mudstones of alluvial origin. Near Bunyol there is a lacustrine interbedded sequence formed by limestones, mudstones, clays and thin coals which have yielded an Early Miocene mammal fossil fauna (Crusafont & Truyols, 1957; Adrover, 1968; Daams, 1976) which may be ascribed to the MN 4 zone of Mein (1975). The upper unit in the Bunyol area is formed by conglomerates, sandstones and mudstones that laterally change to – and are overlain by – algal–travertinic limestones and yellow to red mudstones. This unit unconformably overlies the lower unit in the Bunyol area.

In the eastern part of the Valencian Depression, the Neogene deposits comprise two units. The lower unit is formed by calcareous sandstones and mudstones with interbedded limestones. Biostromic accumulations of oysters are frequent. These deposits, which were formed in littoral marine environments, are attributed to the Tortonian (Usera, 1972, 1974). The upper unit, which crops out mainly in the Lliria–Burjassot area, is formed by massive limestones which contain oncoids and lacustrine gastropods. Pedogenic features, typical of limestones originated in palustrine (shallow lacustrine) settings also occur. Near the ancient basin margins these deposits alternate with mudstones and conglomerates. This unit has been attributed to the Late Miocene because of its stratigraphic relationships. The upper unit in the eastern Valencian Depression may be correlated with the upper part of the Neogene sequence in the Bunyol area. The Tortonian transgression did not attain the western part of the Valencia Depression.

The Cabriel Basin

In the Southern Iberian Chain, the Neogene extensional phase caused the opening of Triassic floored depressions which were the result of extension combined with diapirism, and diapirism and dissolution of the evaporite-bearing Triassic rocks (Moissenet, 1985). The Cabriel Basin is one of these depressions. The basin fill is Late Miocene to Early Pliocene in age (Mein *et al.*, 1978; Opdyke *et al.*, 1989). It is composed of over 400 m of terrigenous alluvial deposits with minor, shallow lacustrine carbonates and gypsum (Assens *et al.*, 1973; Robles *et al.*, 1974; Anadón, 1985). The alluvial deposits, formed by conglomerates, sandstones and mudstones predominate in the marginal areas of the NE part of the basin. These deposits pass basinwards into alluvial mudstones and sandstones with interbedded carbonate and gypsum lacustrine deposits. The gypsum deposits comprise lenticular, selenitic and clastic facies which indicate accumulation in shallow lakes (Anadón *et al.*, 1992). The upper part of the Cabriel basin fill in the Cabriel River area comprises a lacustrine limestone sequence, up to 30 m thick, which was called the Mirador Member by Robles (1970). This unit has been correlated with the Rio Jucar Formation of the Jucar Basin (Robles, 1970; Robles *et al.*, 1974). The Jucar Basin is not strictly related to the history of the Iberian Chain and will not, therefore, be considered in this chapter.

References

Acuña, J.D. (1982). Algunas regularidades tafonómicas y paleoecológicas en una secuencia tortoniense de Torreblanca (Castellón). *Estud. Geol*, **38**, 61–73.

Adrover, R. (1974). Un relleno kárstico plio-pleistoceno en el Cerro de los Espejos de Sarrión (Provincia de Teruel, España). *Acta Geol. Hispánica*, **9**, 142–143.

Adrover, R. (1968). Los primeros micromamíferos de la cuenca valenciana en Buñol (Nota preliminar). *Acta Geol. Hispánica*, **3**, 78–80.

Aguirre, E., Hoyos, M., Mensua, S., Morales, J., Pérez-González, A., Quirantes, J., Sánchez de la Torre, L. and Soria, M.D. (1974). Cuenca del Jalón. *Col. Int. Biostratigrafia continental del Neógeno Superior Quaternario inferior. Libro Guia*, Guía 1.10, 10–48.

Agustí, J., Anadón, P., Ginsburg, L., Mein, P. and Moissenet, E. (1988). Araya et Mira: nouveaux gisements de mammifères dans le Miocène inférieur moyen des Chaines Ibériques orientales et méditerranéennes. Consequences stratigraphiques et structurales. *Paleontol. Evol.*, **22**, 83–101.

Alberti, M.T. and Pailhé, P. (1983). Precisiones sobre la edad de los sedimentos de la Sierra del Pobo (Cadena Ibérica oriental, Provincia de Teruel). *Estud. Geol.*, **39**, 117–119.

Alvaro, M., Capote, R. and Vegas, R. (1979). Un modelo de evolución geotectónica para la Cadena Celtibérica. *Acta Geol. Hispánica*, **14**, 172–177.

Anadón, P. (1983). Características generales de diversas cuencas lacustres terciarias con pizarras bituminosas del NE de la Península Ibérica. *Com. X Congr. Nacional de Sedimentología*, Menorca, **1**, 9–12.

Anadón, P. (1985). Terciario. *Memoria del Mapa Geol. España 1:200 000, 2 ser., Inst. Geol. Minero España. Hoja de Lliria*, **55**, 80–97.

Anadón, P., Cabrera, L., Guimerà, J. and Santanach, P. (1985). Paleogene strike–slip deformation and sedimentation along the southern margin of the Ebro Basin. In *Strike–slip deformation, basin formation and sedimentation* (ed. K.T. Biddle and N. Christie-Blick). *Spec. Publ., Soc. Econ. Paleontol. Mineral.*, **37**, 303–318.

Anadón, P., Cabrera, L., Colombo, F., Marzo, M. and Riba, O. (1986). Syntectonic intraformational unconformities in alluvial fan deposits, eastern Ebro Basin margins (NE Spain). In *Foreland basins* (ed. P. Allen and P. Homewood). *Int. Assoc. Sedimentol., Spec. Publ.*, **8**, 259–271.

Anadón, P., Cabrera, L. and Julià, R. (1988a). Anoxic–oxic cyclical lacustrine sedimentation in the Miocene Rubielos de Mora Basin. In *Lacustrine petroleum source rocks* (ed. A.J. Fleet, K. Kelts and M.R. Talbot). *Geol. Soc. Spec. Publ.*, London, **40**, 353–367.

Anadón, P., Cabrera, L., Inglès, M., Julià, R. and Marzo, M. (1988b). The Miocene lacustrine basin of Rubielos de Mora. *International workshop on 'Lacustrine facies models in rift systems and related natural resources'. Barcelona–Rubielos de Mora. Excursión guidebook*, 32 pp.

Anadón, P., Cabrera, L., Julià, R., Roca, E. and Rossell, L. (1989). Lacustrine oil-shale basins in Tertiary grabens from NE Spain (Western European Rift System). *Paleogeogr. Paleoclimatol. Paleoecol.*, **70**, 7–28.

Anadón, P., Moissenet, E. and Simón, J.L. (1990). The Neogene grabens of the Eastern Iberian Chain (Eastern Spain). In *Iberian Neogene basins: field guidebook*. (Ed. J. Agustí and J. Martinell). *Paleontol. Evol. Spec. Publ.*, **2**, 97–130.

Anadón, P., Cabrera, L., Julià, R. and Marzo, M. (1991). Sequential arrangement and asymmetrical fill in the Miocene Rubielos de Mora Basin (northeast Spain). In *Lacustrine facies analysis* (ed. P. Anadón, L. Cabrera and K. Kelts). *Int. Assoc. Sedimentol. Spec. Publ.*, **13**, 257–275.

Anadón, P., Rosell, L. and Talbot, M. (1992). Carbonate replacement of lacustrine gypsum deposits in two Neogene continental basins, eastern Spain. *Sediment. Geol.*, **78**, 201–216.

Colomer, M. and Santanach, P. (1988). Estructura y evolución de borde suroccidental de la Fosa de Calatayud–Daroca. *Geogaceta*, **4**, 29–31.

Crusafont, M. and Truyols, J. (1957). Descubrimiento del primer yacimiento de mamíferos miocénicos de la cuenca valenciana. *Notas Com. IGME*, **48**, 3–20.

Crusafont, M., Villalta, J.F. and Julivert, M. (1954). Notas para la estratigrafía y paleontología de la Cuenca de Calatayud – Teruel. *Notas y Com. IGME*, **34**, 41–58.

Crusafont, M., Gautier, F. and Ginsburg, L. (1966). Mise en évidence du Vindobonien inférieur continental dans l'Est de la province de Teruel (Espagne). *Com. R. Somm. Soc. Geol. Fr.*, 1966, 30–32.

Daams, R. (1976). Miocene rodents (Mammalia) from Cetina de Aragón (prov. Zaragoza) and Buñol (prov. Valencia), Spain. *Proc. Konnink. Nederland. Akad. Wetenschappen*, **B-79**, 152–182.

Daams, R., Freudenthal, M. and Van de Weerd, A. (1977). Aragonian, a new stage for continental deposits of Miocene age. *Newsl. Stratigr.*, **6**, 42–55.

Daams, R., Freudenthal, M. and Alvarez-Sierra, M. (1987). Ramblian; A new stage for continental deposits of early Miocene age. *Geol. Mijnbouw*, **65**, 297–308.

De Bruijn, H. (1967). Gliridae, Sciuridae y Eomidae (Rodentia, Mammalia) miocenos de Calatayud (provincia de Zaragoza, España) y su relación con la biostratigrafía del área. *Bol. Inst. Geol. Mineral. Esp.*, **78**, 187–373.

De Bruijn, H. and Moltzer, J.G. (1974). The rodents of Rubielos de Mora; first evidence of the existence of different biotopes in the Early Miocene of Eastern Spain. *Proc. Konnink. Nederland. Akad. Wetenschappen*, **B-77**, 129–145.

Gautier, F. and Heintz, E. (1974). Le gisement villafranchien de la Puebla de Valverde (Province de Teruel, Espagne). *Bull. Mus. Nat. Hist. Nat. Paris*, **228**, 113–133.

Gautier, F., Moissenet, E. and Viallard, P. (1972). Contribution à l'étude stratigraphique et tectonique du fossé néogène de Teruel (Chaînes ibériques, Espagne). *Bull. Mus. Natl. Hist. Nat. Paris*, **77**, 179–207.

Guimerà, J. (1984). Paleogene evolution of deformation in the northeastern Iberian Peninsula. *Geol. Mag.*, **121**, 413–420.

Guimerà, J. (1988). Estudi estructural de l'enllaç entre la Serralada Ibèrica i la Serralada Costanera Catalana. PhD Thesis. Universitat de Barcelona, 600 pp.

Guimerà, J. (1990). Formación de una cubeta sinclinal en un contexto extensivo: la cuenca miocena de Rubielos de Mora (Teruel). *Geogaceta*, **8**, 33–35.

Guimerà, J. and Alvaro, M. (1990). Structure et evolution de la compression alpine dans la Chaîne Ibérique et la Chaîne Côtière Catalane. *Bull. Soc. Géol. Fr.*, ser. 8, **6**, 339–348.

IGME (1983). *Mapa Geológico de España E. 1:50000. Hoja n. 465: Daroca.*

IGME (1985). *Mapa Geológico de España E. 1:200000, 2 serie, Hoja n. 48: Vinaròs.* Memoria, 100 pp.

IGME (1991). *Mapa Geológico de España E. 1:200000, 2 serie, Hoja n. 40: Daroca.* Memoria, 239 pp.

Julivert, M. (1954). Observaciones sobre la tectónica de la Depresión de Calatayud. *Arrahona*, **1954**, 3–18.

Mein, P. (1975). Résultats du groupe de travail des vertébrés: Biozonation du Neogène méditerranéen a partir des mammifères. In *Report on activity of RCMNS working groups (1971–1975)* (ed. J. Senes), Bratislava, 78–81.

Mein, P., Moissenet, E. and Truc, G. (1978). Les formations continentales du néogène supérieur des vallées du Jucar et du Cabriel au NE d'Albacete (Espagne). Biostratigraphie et environnement. *Doc. Lab. Géol. Fac. Sci. Lyon*, **72**, 99–147.

Mein, P., Moissenet, E. and Adrover, R. (1990). Biostratigraphie du Néogène supérieur de Teruel. *Paleontol. Evol.*, **23**, 121–139.

Moissenet, E. (1982a). Observations preliminaires sur les piemonts des monts ibériques dans la region de Teruel. *Actes Coll. Montagnes Piemonts*, Toulouse, May 1982.

Moissenet, E. (1982b). Le Villafranchien de la region de Teruel (Espagne). *Actes Coll. Int. 'Le Villafranchien Méditerranéen*, **1**, 229–253.

Moissenet, E. (1984). L'evolution tectonique du fossé néogène de Teruel (Chaînes Ibériques Orientales, Espagne). *C.R. Acad. Sci. Paris*, sér. 2, **299**, 173–178.

Moissenet, E. (1985). Les dépressions tarditectoniques des Chaînes Ibériques méridionales: distension, diapirisme et dépots néogènes associés. *C. R. Acad. Sci. Paris*, sér 2, **300**, 523–528.

Moissenet, E. (1989). Les fossés néogènes de la Chaîne Ibérique: leur évolution dans le temps. *Bull. Soc. Géol. Fr., ser. 8*, **5**, 919–926.

Opdyke, N., Mein, P., Moissenet, E., Pérez-González, A., Lindsay, E. and Petko, M. (1990). The magnetic stratigraphy of the late Miocene sediments of the Cabriel Basin, Spain. In *European Neogene mammal chronology* (ed. E.H. Lindsay). Plenum Press, New York, Nato ASI series A, **180**, 507–514.

Ortí, F. (1981). Diapirismo de materiales triásicos y estructuras de zócalo, en el sector central valenciano. *Estud. Geol.*, **37**, 245–256.

Paricio, J. (1985). La unidad neógena del valle medio del Mijares (Cordillera Ibérica). *Teruel*, **74**, 9–65.

Riba, O. and Rios, J.M. (1962). Observations sur la structure du secteur Sud-Ouest de la Chaîne Ibérique (Espagne). *Mem. Hors Série, Soc. Géol. France. Livre à la Mémoire du Prof. Paul Fallot*, **1**, 275–290.

Riba, O., Reguant, S. and Villena, J. (1983) Ensayo de síntesis estratigráfica y evolutiva de la cuenca terciaria del Ebro. In *Libro Jubilar J.M. Rios. Geología de España.*, **2**, 131–159. IGME, Madrid.

Richter, G. and Teichmüller, R. (1933). Die Entwicklung der Keltiberischen Ketten. *Abh. Ges. Wiss. Göttingen Math. Phys. Kl.*, Ser 3, **7**, 1065–1186.

Robles, F. (1970). Estudio estratigráfico y paleontológico del Neógeno continental de la cuenca del Río Júcar. PhD Thesis, Universidad de Valencia, 275 pp.

Robles, F., Torrens, J., Aguirre, E., Ordoñez, S., Calvo, J.P. and Santos, J.A. (1974). Levante. *Libro Guía, Coloquio Int. 'Bioestratigrafía continetal Neógeno superior y Cuaternario inferior'*, 87–133.

Roca, E. (1992). L'estructura de la Conca Catalano-balear: Paper de la compressió i de la distensió en la seva gènesi. PhD Thesis. Universitat de Barcelona, 330 pp.

Roca, E. and Guimerà, J. (1992). The Neogene structure of the eastern Iberian margin: structural constraints on the crustal evolution of the Valencia trough (Western Mediterranean). *Tectonophysics*, **203**, 203–218.

Simón, J.L. (1982). Compresión y distensión alpinas en la Cadena Ibérica Oriental. PhD Thesis. Universidad de Zaragoza, 504 pp.

Simón, J.L. (1986). Analysis of a gradual change in stress regime (example from the Eastern Iberian Chain, Spain). *Tectonophysics*, **124**, 37–53.

Suárez, J.M., Alonso, T. and Pendas, F. (1983). Los lignitos terciarios de Cuevas de Vinromà, Alcalà de Chivert (Castellón de la Plana). In *Libro Homenaje a Carlos Felgueroso*. Madrid, 139–152.

Usera, J. (1972). Paleogeografía del Mioceno marino en la provincia de Valencia. *Bol. R. Soc. Esp. Hist. Nat.*, **70**, 307–312.

Usera, J. (1974). Microbioestratigrafía del Neogeno marina en la provincia

de Valencia. *Bol. R. Soc. Esp. Hist. Nat.*, **72**, 213–223.
Viallard, P. (1973). Recherches sur le cycle Alpin dans la Chaîne Ibérique Sud-Occidentale. PhD Thesis. Université Paul Sabatier de Toulouse, 445 pp.
Viallard, P. (1989). Décollement de couverture et décollement médio-crustal dans une chaîne intraplaque: variations verticales du style tectonique des Ibérides (Espagne). *Bull. Soc. Géol. France*, ser. 8, **5**, 913–918.

E5 The Tertiary of the Iberian margin of the Ebro basin: sequence stratigraphy

J. VILLENA, G. PARDO, A. PÉREZ, A. MUÑOZ AND A. GONZÁLEZ

Abstract

The general definition and recognition criteria of 'tecto-sedimentary units' (TSUs) are described and discussed. The non-marine sediments of the Iberian margin of the Ebro basin are divided into tecto-sedimentary units. Eight of these units (labelled T_1 to T_8) are described from the area; their ages range from Thanetian (Late Paleocene) to Turolian (Late Miocene).

Introduction

The Tertiary sediments of the Iberian margin of the Ebro basin were deposited in alluvial and lacustrine systems and mostly derived from this same margin. In these continental sediments detrital-clastic and evaporitic facies dominate, so there is a scarcity of fossiliferous beds, and chronostratigraphic and correlation studies using fossils are very difficult.

In contrast to the classical methods of stratigraphy, present methods of basin analysis are often concerned with the division of the basin-fill into 'genetic units'. These units are defined by objective criteria that result from events that are thought to have affected the whole basin (Vera, 1989). We have followed this approach in our study of the southern margin of the Ebro basin using the 'tectosedimentary analysis' of Garrido-Megías (1973, 1982). This method is based on analysis of sequence evolution and the large-scale geometry of sedimentary bodies and results in the division of the basin-fill into tectosedimentary units (TSUs).

The concept of TSU

Garrido-Megías (1982) defined the simple TSU as 'a stratigraphic unit made up by a succession of strata (not necessarily in concordance) deposited within a concrete interval of geological time and under a tectonic and sedimentary dynamic of definite polarity'. He established also that the TSU boundaries are defined by sedimentary breaks of basinal extent.

González *et al.* (1988) and Pardo *et al.* (1989) established that a sedimentary break is a surface recognised on a basinal scale, across which the sedimentary fill undergoes either a sudden jump or a change of direction of trend in its evolution. The scale of evolution involved is that of the upper-order sequences or rhythms in the sequential analysis of Delfaud (1972). The surface can be conformable everywhere, or conformable in the central parts of the basin and correlative to an unconformity of any type at the basin margins.

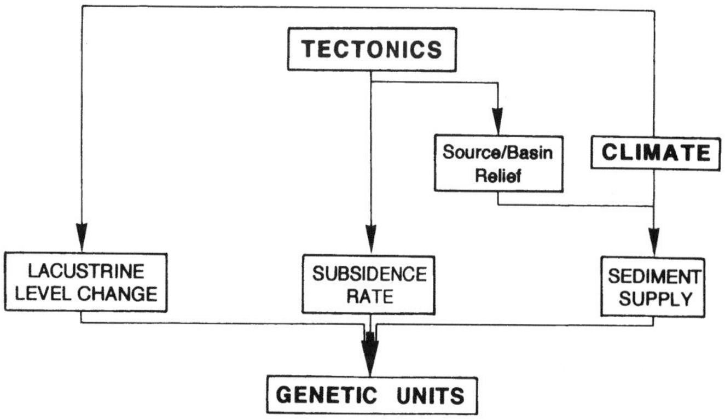

Fig. 1. Variables determining the depositional story and the architecture of sequences.

From a genetic point of view, a sedimentary break in the stratigraphic record indicates an inflection, or a sharp change in rate, in the allocyclic factors that controlled the dynamics of the sedimentation.

Types of sedimentary break

The factors that give rise to the filling of a basin with successive sequences have been analysed by Galloway (1989). According to this author, the variables that give rise to sequences are changes of 1. eustatic sea level, 2. subsidence, and 3. sedimentation rate.

The variables that control the history of basin filling for continental basins isolated from marine influence are shown in Fig. 1. Here, the only allocyclic factors are 1. climate, which influences sediment supply and variation of the lacustrine level, and 2. tectonics, which influence sediment supply and subsidence.

In the case of the Ebro basin which has been endorheic (i.e. land-

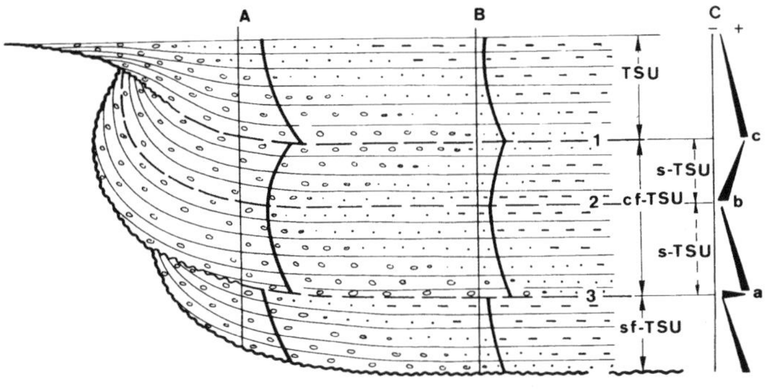

Fig. 2. Sedimentary break types originated at active edges of continental basins when the allocyclic factor conditioning the basin filling is essentially the diastrophic activity. A and B: sequential evolution; C: inferred evolution for the diastrophic activity; 1, 2 and 3: sedimentary breaks types; sf-TSU: simple fundamental tectosedimentary unit; cf-TSU: cyclic fundamental tectosedimentary unit; s-TSU: non-fundamental simple tectosedimentary unit; a and c: relative maxima of the deformation rhythm; b: relative minimum of the deformation rhythm.

locked) since the Oligocene, climatic factors are manifested mainly in facies changes, such as from evaporitic to carbonate lacustrine sedimentation. These lakes were always very shallow (about 2 m in depth), as demonstrated by recent facies analysis (Cabrera, 1983; Salvany, 1989a, b; Pérez *et al.*, 1989). Consequently, oscillations of lake level cannot produce sedimentary changes on the scale of the genetic units (TSU). It is possible that in more central areas of the basin, the climatic factor could modify the succession in a unit in a different way to the one observed in marginal areas of the basin, by controlling the extent of the area of lake deposition.

However, along the Iberian margin of the Ebro basin, there is a clear parallelism between the retreat of alluvial fans and the extension of the lacustrine areas, and also between the progradation of the fans and the shrinking of the lacustrine areas. In fact, along the Iberian margin, the tectonic factor is the main one which conditions the geometry and the polarity of the sedimentary basin-fill. Furthermore, along this margin, the study of the relationships between the active tectonic structures, the vertical changes and the geometrical relationships between the beds shows that a coarsening upwards together with an offlap geometry can be interpreted as evidence for an episode of increasing diastrophism. In contrast, a fining upwards together with an onlap geometry indicates an episode of decreasing diastrophism. Using these observations and interpretations, González *et al.* (1988) and Pardo *et al.* (1989) define three main types of sedimentary break (Fig. 2) as general features of active margins.

Type 1 sedimentary breaks are those related to increasing, then decreasing, changes in diastrophic activity. They are recorded by changes in polarity in the successions from coarsening upwards to fining upwards, and by a simultaneous change in the geometry from offlap to onlap on the active margin. This geometry corresponds to that described by Riba (1976) for syntectonic unconformities; so type 1 sedimentary breaks are conformable surfaces in the centre of the basin, but correlative to syntectonic unconformities on the active margin of the basin and/or on the flanks of the synsedimentary folds located within it.

Type 2 sedimentary breaks are those related to decreasing, then increasing, changes in diastrophic activity. They are recorded by changes in polarity in the successions from fining upwards to coarsening upwards, and a simultaneous change in the geometry from onlap to offlap on the margin of the basin. These sedimentary breaks are conformable surfaces throughout this basin.

Thirdly, if a period of decreasing diastrophic activity is followed by a relatively short period (instantaneous on a geological time scale) of increasing activity, and then by another period of decreasing diastrophic activity, the sedimentary record will consist of two onlapping fining-upward sequences, separated by an unconformity near the active margins of the basin. The upper unit shows its onlap displaced toward the centre of the basin. This type is called a type 3 sedimentary break (Fig. 2).

The TSU *sensu stricto*, or *simple* TSU, is characterised by a vertical sequence variation trend in one direction (coarsening upward or fining upward). In the words of Garrido-Megías (1982), it possesses a 'defined sedimentary polarity'. It is sometimes useful to define a *complex* TSU consisting of two, or more, simple TSUs.

In our studies of the Tertiary continental deposits, we have distinguished a set of TSUs that directly represents tectonic pulses. This fact is demonstrated by the bounding surfaces of the units, which are invariably unconformable at the basin margins, i.e. of types 1 and 3 (González *et al.*, 1988). We have called these *fundamental* TSUs (see Fig. 2).

The fundamental TSU can be simple, with a coarsening-upward or fining-upward sequence; however, most of the Paleogene units that we have studied are complex, since they show a fining-upward, then coarsening-upward sequence.

Thus, we distinguish a cyclic fundamental TSU made up of two simple non-fundamental TSUs, separated by a break of type 2 located at the inflection point in the sequence (Fig. 2).

Let us consider a continental basin, where the sedimentary evolution is controlled entirely by the tectonic factor and the sedimentary system is composed of alluvial fan-playa lake components. Here, the fining-upwards then coarsening-upwards sequence of a cyclic fundamental TSU is equivalent to a cycle of retrogradation of alluvial fans with expansion of the playa-lake followed by progradation of the alluvial fans with shrinkage of the playa-lake system. A TSU of this type is the continuous sedimentation response to diastrophic activity that varies with time. During the deposition of a cyclic fundamental TSU, the deformational rate first decreases, and then increases, between the two relative maxima responsible for the lower and upper sedimentary breaks (Fig. 2). The sedimentary break of type 2 in the inflexion of the cycle is interpreted as the moment at which the deformation rate reaches a minimum, or, in other words, the moment at which the diastrophic activity changes its sense from decreasing to increasing.

Stratigraphy

The application of 'tectosedimentary analysis', as just described, has allowed us to define for the Tertiary rocks of the Ebro

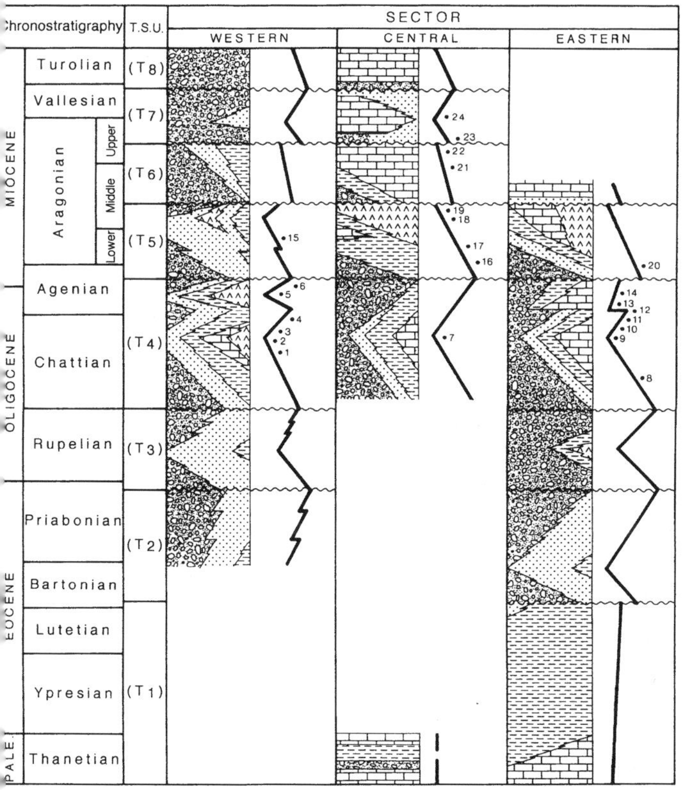

Fig. 3. Stratigraphic logs and sequential evolution of the tectosedimentary units and stratigraphic location of the vertebrate sites in the study area. 1: Arnedo and Bergasa. 2: Autol-0, Autol-1 and Valhondo. 3: Quel-1. 4: Carretil. 5: Islallana. 6: Fuenmayor-2. 7: Las Torcas. 8: Gandesa. 9: Torre del Compte and Mina del Pilar-3. 10: Torrente de Cinca-4, Torrente de Cinca-7, Torrente de Cinca-18 and Fraga w-4. 11: Fraga w-6. 12: Velilla de Cinca-5 and Ballobar-12. 13: Fraga w-11 and Ballobar-21. 14: Torrente de Cinca-68. 15: Los Agudos. 16: Tudela-1. 17: Tudela-2. 18: Monteagudo. 19: Tarazona. 20: Paridera del Cura. 21: Borja. 22: Villanueva de Huerva. 23: Moyuela. 24: El Buste and La Ciesma.

basin eight genetic units, TSUs, bounded by recognisable sedimentary breaks over the whole of the area studied. Their correlation has been established mainly by mapping, and by the similarity of their vertical sequences, as well as by their stratigraphic relationships with the adjacent units. Their chronostratigraphic dating is doubtful in the units without vertebrate sites. Even in the TSUs with vertebrate sites, the sites are usually few and scattered; but if they allow dating, they provide very important evidence for the ages of the TSUs. Three of the TSUs are regarded here as Paleogene, one as from the Paleogene–Neogene transition, and four as Neogene.

These units consist of conglomerates, sandstones, lutites, limestones and/or gypsum. In each unit the relative importance of each of these rock types, as well as their areal and vertical distribution, are determined largely by proximity to the basin margin from which the sediment was derived.

Each of the eight units has a particular vertical evolution, generally complex, but similar over the whole area studied.

The data on chronostratigraphy, lithology and vertical sequence, outcrop area and thickness variations are synthesised in Figs. 3, 4, 5 and 6.

For detailed treatment, the region has been divided into three sectors. The western sector is located NW of the Queiles river, the eastern sector is SE of the Martin river, and the central sector is between the two rivers. To identify more easily the units defined in this chapter, we use the unit terms T1–T8, as in previous works.

We also add below extra detail on certain aspects typical of each unit that have not been adequately portrayed in the figures.

Paleogene: Thanetian–Bartonian (T1)

This unit displays a cyclic sequence (fining upward, then coarsening upward) when complete to the SE of the eastern sector outside our area. Within our area it can only be partially seen (see Figs. 3 and 5).

Its lower boundary is a mappable unconformity above a basement of different Palaeozoic and Mesozoic rocks.

Its chronostratigraphic assignment is based on the presence of *Vidaliella gerundensis* and an association of charophytes (Colombo, 1986; see González, 1989 and Villena *et al.*, 1992).

Paleogene: Bartonian–Priabonian (T2)

When complete, in the eastern sector, this unit presents a cyclic sequence (fining upward, then coarsening upward).

Its lower boundary is a sedimentary break of type 1, also marked by a sudden increase in grain-size. It is locally represented by a low-angle unconformity.

Its chronostratigraphic assignment is based on charophyte associations (Anadón *et al.*, 1981 and Anadón & Feist, 1981), and from the correlations that González (1989) establishes with the stratigraphy outcropping in the Catalán sector of the Ebro basin, where Ferrer (1971) pointed out an association with microforaminifera of late Eocene age (see Villena *et al.*, 1992).

Paleogene: Priabonian – Lower Oligocene (T3)

The lower boundary of this unit is a sedimentary break of type 1 in the western sector, where it is best seen.

Its chronostratigraphic assignment is based on the correlations that have been established with the neighbouring intramontane Iberian basin of Montalbán, where fossil remains have been found (Pérez *et al.*, 1983).

Paleogene–Neogene transition: Upper Oligocene–Agenian (T4)

The vertical sequence of this unit is complex to a variable degree, depending on the sector, but is uniform in any one sector (see Fig. 3). The differences in the sequences can be attributed to the activity of local tectonic structures.

Its lower boundary is a break of type 1.

Its chronostratigraphic assignment is based on vertebrate faunas from numerous sites in this unit (Fig. 3), and is discussed further by Villena *et al.*, (1992).

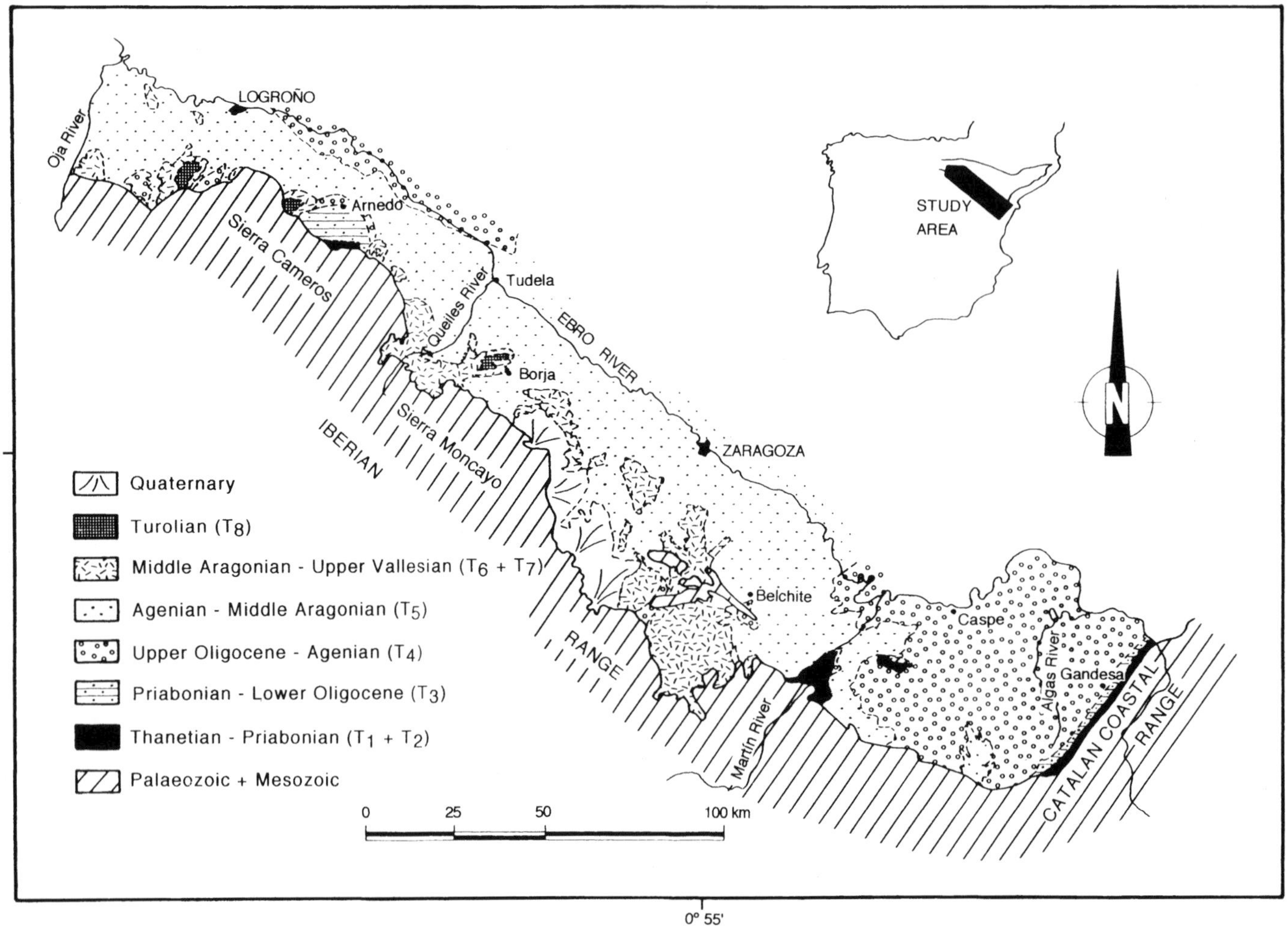

Fig. 4. Location map and cartography of tectosedimentary units.

Neogene: Agenian–Middle Aragonian (T5)

This unit does not have the same vertical sequence in the different sectors (see Fig. 3). This is attributed to differences in the tectonic regime, discussed elsewhere (see Chapter E6).

Its lower boundary is a sedimentary break of type 1.

Its chronostratigraphic assignment is based on the vertebrate faunas of the sites noted in Fig. 3.

Neogene: Middle Aragonian–Upper Aragonian (T6)

In the western sector, this unit is separated from the underlying one by a syntectonic unconformity. In the central sector its lower boundary is an angular unconformity.

Its chronostratigraphic assignment is based on the fossils of the vertebrate sites noted in Fig. 3.

Neogene: Upper Aragonian–Vallesian (T7)

The lower boundary of this unit is a sedimentary break of type 3.

In the central sector, its vertical sequence is cyclic and strongly asymmetrical, with a thin lower hemicycle. This asymmetry is not clear on Fig. 3, because the vertical scale is time, not sediment thickness.

This is the first unit in the central sector of the southern margin of the Ebro basin to contain sediments derived from the north (Pérez *et al.*, 1988).

The chronostratigraphic assignment of this unit is based on the fossils of the vertebrate sites noted on Fig. 3.

Neogene: Turolian (T8)

The lower boundary of this unit is a vertical sequence change from coarsening upward to fining upward, sometimes involving a sharp change in grain-size.

This unit may be equivalent to unit T8 of the Tertiary succession of the Daroca–Calamocha basin (ITGE, 1991), and we have accepted this chronostratigraphic suggestion.

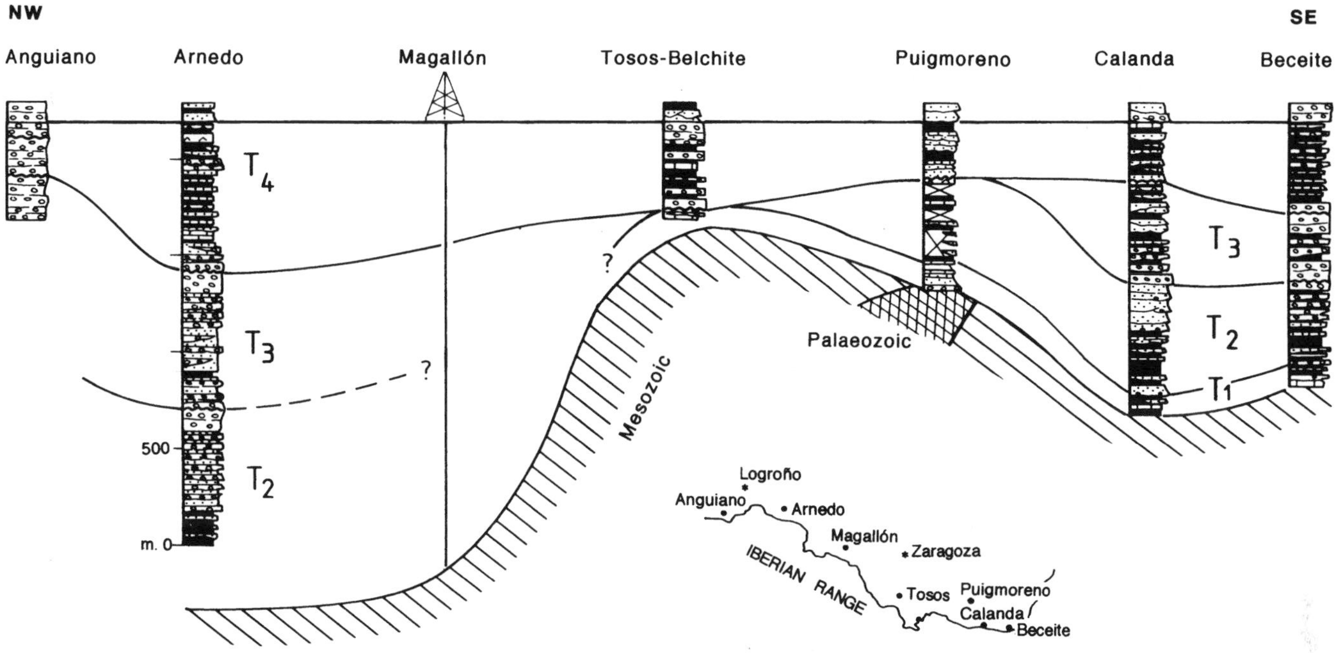

Fig. 5. Representative synthetic stratigraphic sections of the Paleogene tectosedimentary units.

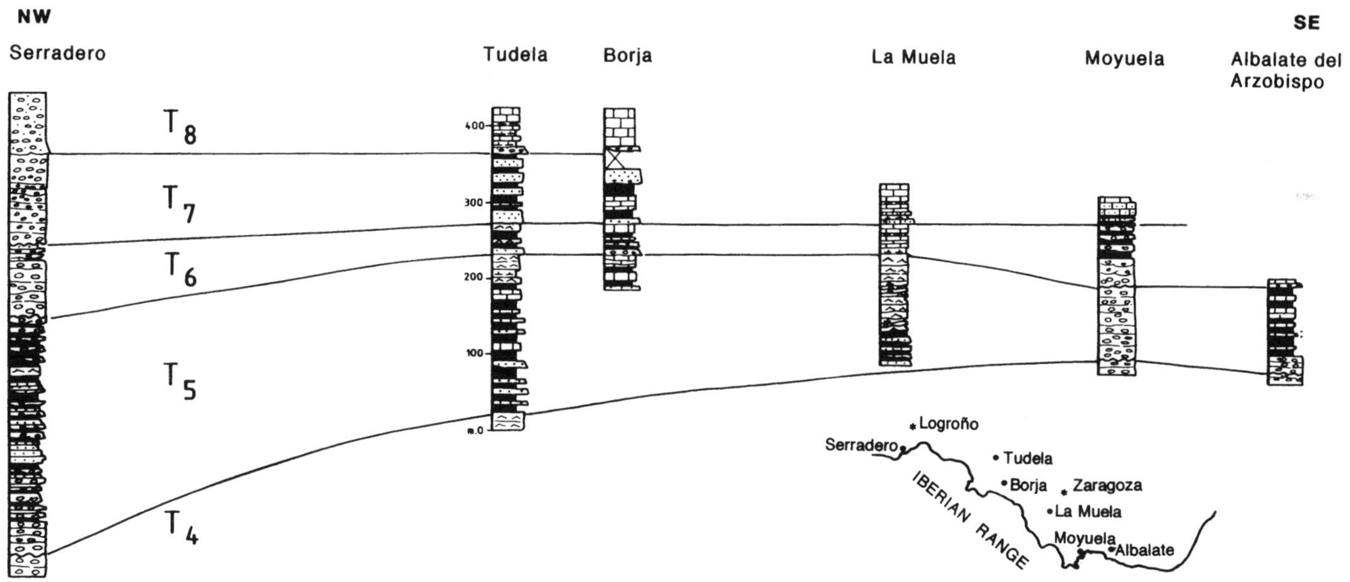

Fig. 6. Representative synthetic stratigraphic sections of the Neogene tectosedimentary units.

Acknowledgements

This work has been supported by project no. PB–89/0342 of DGICYT.

References

Anadón, P. and Feist, M. (1981). Charophites et biostratigraphie du Paléogène inférieur du bassin de l'Èbre oriental. *Paleontographica*, **178** Abt. B: 143–168.

Anadón, P., Cabrera, L., Colombo, F., Marzo, M. and Riba, O. (1981). Estudio estratigráfico y sedimentológico del borde meridional de la Depresión del Ebro entre Alcañíz y Borjas Blancas. (Provincias de Teruel, Zaragoza, Lérida y Tarragona). *Junta de Energía Nuclear*. Internal Report.

Cabrera, L. (1983). Estratigrafía y sedimentología de las formaciones lacustres del tránsito Oligoceno-Mioceno del SE de la Cuenca del Ebro. Unpublished PhD Thesis. Barcelona University. 443 pp.

Colombo, F. (1986). Estratigrafía y sedimentología del Paleógeno continental del borde meridional occidental de los Catalánides (provincia de Tarragona. España). *Cuad. Geol. Ibérica*, **10**: 55–115.

Delfaud, J. (1972). Application de l'analyse séquentielle à l'exploration lithostratigraphique d'un bassin sédimentaire. L'exemple du Jurassique et du Crétacé inferieur de l'Aquitaine. *Mem. B.R.G.M. France*, **77**: 593–611.

Ferrer, J. (1971). El Paleoceno y Eoceno del borde suroriental de la Depresión del Ebro (Cataluña). *Mem. Suisses Paleontol.*, **90**: 1–70.

Galloway, W.E. (1989). Genetic stratigraphic sequences in basin analysis I: architecture and genesis of flooding-surface bounded depositional units. *Am. Assoc. Petrol. Geol. Bull.*, **73**: 125–142.

Garrido-Megías, A. (1973). Estudio geológico y relación entre tectónica y sedimentación del Secundario y Terciario de la vertiente meridional pirenaica en su zona central (provincias de Huesca y Lérida). Unpublished PhD Thesis. Granada University. 395 pp.

Garrido-Megías, A. (1982). Introducción al análisis tectosedimentario: aplicación al estudio dinámico de cuencas. *Actas V Congreso Latinoamericano de Geología*, 385–402. Argentina.

González, A. (1989). Análisis tectosedimentario del Terciario del borde SE de la Depresión del Ebro (sector bajoaragonés) y cubetas ibéricas marginales. Unpublished PhD Thesis. Zaragoza University, 507 pp. 2 vols.

González, A., Pardo, G. and Villena, J. (1988). El análisis tectosedimentario como instrumento de correlación entre cuencas. *II Congreso Geológico de España. Simposios*: 175–184. Granada.

ITGE. (1991). *Mapa Geológico de España*. Escala 1:200000. Hoja no. 40 (Daroca). Serv. Publ. Ministerio de Industria. Madrid.

Pardo, G., Villena, J. and González, A. (1989). Contribución a los conceptos y a la aplicación del análisis tectosedimentario. Rupturas y unidades tectosedimentarias como fundamento de correlaciones estratigráficas. *Rev. Soc. Geol. Esp.*, **2** (3–4): 199–219.

Pérez, A., Pardo, G., Villena, J. and González, A. (1983). Estratigrafía y sedimentología del Paleógeno de la Cubeta de Montalbán (Prov. de Teruel, España). *Bol. R. Soc. Esp. Ha. Nat.*, **81** (3–4): 197–223.

Pérez, A., Muñoz, A., Pardo, G., Arenas, C. and Villena, J. (1988). Características de los sistemas lacustres en la transversal Tarazona-Tudela (sector Navarro-Aragonés de la Cuenca terciaria del Ebro). *II Congreso Geológico de España. Simposios*, 519–527.

Pérez, A., Muñoz, A., Pardo, G. and Villena, J. (1989). Evolución de los sistemas lacustres del margen ibérico de la Depresión del Ebro (sectores central y occidental) durante el Mioceno. *Acta Geol. Hispánica*, **24**: 243–257.

Riba, O. (1976). Syntectonic unconformities of the Alto Cardener, Spanish Pyrenees: a genetic interpretation. *Sediment. Geol.*, **15**: 213–233.

Salvany, J.M. (1989a). Las formaciones evaporíticas del Terciario continental de la Cuenca del Ebro en Navarra y La Rioja. Litoestratigrafía, petrología y sedimentología. Unpublished PhD Thesis. Barcelona University. 397 pp.

Salvany, J.M. (1989b). Aspectos petrológicos y sedimentológicos de los yesos de Ablitas y Monteagudo (Navarra): Mioceno de la Cuenca del Ebro. *Turiaso* IX: 121–146.

Vera, J.A. (1989). División de unidades estratigráficas en el análisis de cuencas (introducción). *Rev. Soc. Geol. Esp.*, **2** (3–4): 169–176.

Villena, J., González, A., Muñoz, A., Pardo, G. and Pérez, A. (1992). Síntesis estratigráfica del Terciario del borde Sur de la Cuenca del Ebro: unidades genéticas. *Acta Geol. Hispánica*, **1–2**: 225–245.

E6 The Tertiary of the Iberian margin of the Ebro basin: paleogeography and tectonic control

J. VILLENA, G. PARDO, A. PÉREZ, A. MUÑOZ AND A. GONZÁLEZ

Abstract

The major structures of the Iberian margin of the Ebro basin are described, and their regional kinematics are discussed. Two types of alluvial fan system are distinguished. The paleogeographic and tectonic evolution of the margin is analysed in terms of the eight tecto-sedimentary units defined in the previous chapter. Rates of sedimentation averaging 10 cm/1000 years for the Paleogene dropped to 7 cm/1000 years for the Neogene.

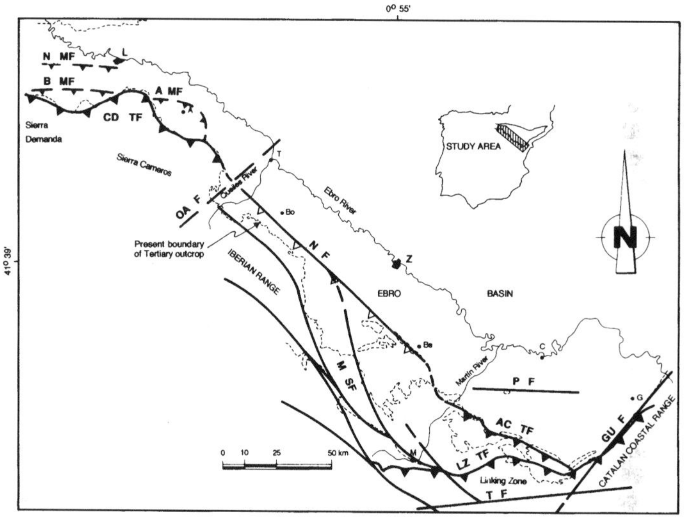

Fig. 1. Major tectonic structures of the Iberian margin of the Ebro basin. N MF: Nájera monoclinal fold. B MF: Baños de Río Tobía monoclinal fold. A MF: Arnedo monoclinal fold. CD TF: Cameros–Demanda thrust front. OA F: Odemira–Avila fault. N F: North Iberian fault. M SF: Montalbán set of faults. P F: Puigmoreno fault. AC TF: Albalate–Calanda thrust front. LZ TF: Linking zone thrust front. T F: Tarragona fault. GU F: Gandesa-Ulldemolins fault. L: Logroño. A: Arnedo. T: Tudela. Bo: Borja. Z: Zaragoza. Be: Belchite. M: Montalbán. C: Caspe. G: Gandesa.

Introduction

The sedimentation of the Tertiary of the Iberian margin of the Ebro basin was mainly controlled by tectonic activity. Tectonic activity controlled the horizontal and vertical evolution of the sedimentary fill (see Chapter E5), the geographic location and orientation of the thresholds and the depositional areas, the extent and relief of the source areas, and also certain aspects of the facies of the deposits.

Because of the controlling importance of tectonic activity, we start this chapter with an outline of the main tectonic structures that were active under general tectonic compression throughout the Tertiary. Some of these structures probably originated as strike–slip faults during the Late Paleozoic, and then controlled the Mesozoic sedimentary history of the Iberian Range (Alvaro *et al.*, 1979).

We first describe the most important structures (Fig. 1).

In the western sector, to the NW of the Queiles river, the major structures (Muñoz, 1992) are the thrust front of the Sierra de Cameros–Demanda, the NE–SW prolongation of the Odemira–Avila basement-cutting fault, and the Arnedo, Baños de Río Tobía and Nájera monoclinal folds. There is geophysical evidence that these monoclinal folds overlie blind thrusts in the pre-Tertiary substratum. Apart from the NE–SW faulting, the dominant structures in this sector appear, on seismic grounds to be low-angle, thrust sheets, some with a throw of up to 25 km (in the central sector of the Sierra de Cameros; Casas, 1992).

In the central sector, between the Queiles and Martin rivers, the major structures are NW–SE-trending faults, involving the basement. The most important of these faults are the North Iberian, the Ateca–Castellón (Viallard, 1979, 1980) and the Montalbán complex (Pérez, 1989). The sedimentary history of the intramontane Iberian basins provides evidence of the dextral strike–slip behaviour of the above faults during the Paleogene (Pardo *et al.*, 1984). There is also evidence that the North Iberian fault was active as the thrust front of the Iberian Range (Klimowitz, 1991).

The eastern sector, SE of the Martin river, is structurally complex. It is the zone of transition where the NW–SE faults that continue those of the central sector, just described, meet other basement-cutting faults that trend NE–SW, like those of Gandesa–Ulldemolins (Anadón *et al.*, 1985). E–W trending, basement-cutting faults, such as the Tarragona (Salas, 1987) and Puigmoreno faults, also occur, and thrust fronts, such as those of the linking zone and the Albalate–Calanda area, are also present.

Two kinematic components appear to dominate the regional tectonics:

(a) the convergent motion of the European and African lithosphere plates that produced a N–S compressional component in the Iberian plate (Guimerá, 1984, 1988).
(b) the E–W extensional regime that produced widespread rifting from the Rhine graben to the Alborán Sea (Casas, 1992).

The N–S compression began at the end of the Cretaceous times, and continued until the Late Miocene in some sectors of the Iberian Range (Sierra de Cameros–Demanda; Muñoz *et al.*, 1993). The extensional rifting began in the Miocene in the Catalan region, and spread westwards during the Miocene (Simón, 1984).

The geographical arrangement of the tectonic structures has had a major influence on the sedimentary facies associations by determining the location and magnitude of the source areas for the alluvial fan systems that controlled the deposition of most of the Tertiary sediments.

Two basic types of alluvial fan system can be distinguished in our area of study:

1. small fans, less than 15 km apex to toe, mass-flow deposits important proximally, minor development of the mid fan sector, aqueous sheet flow processes important;
2. larger fans, more than 20 km and sometimes 100 km apex to toe, all fan sectors well developed, particularly the distal one, dominance of fluvial processes.

Very commonly the small fans would be regarded as typical of arid climates, whereas the larger fans would be regarded as typical of humid climates (Figs. 3 to 10). But this traditional climatic assignment is not valid in our case, because the two fan types coexist too closely, in both space and time, to be explained in terms of different climates. We prefer to designate these fan types as low and high transport-efficiency types respectively (Colombo, 1989). The differences result from the dimensions and slope of the source or drainage area. The fans of low transport efficiency were related to tectonic structures developed over relatively short distances, such as folds and high-angle reverse faults, propagated from the Mesozoic cover. Those of high transport efficiency were related to larger active fronts, generally aligned with principal structures coming from the basement.

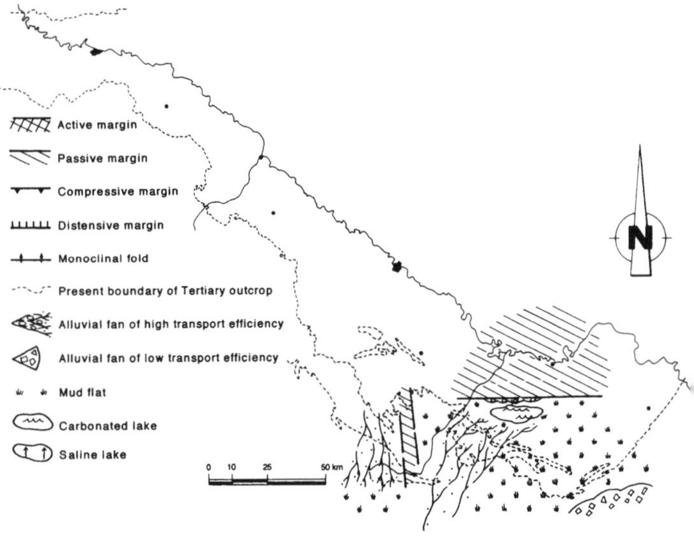

Fig. 2. Palaeogeographic sketch of Thanetian–Bartonian (TSU T1).

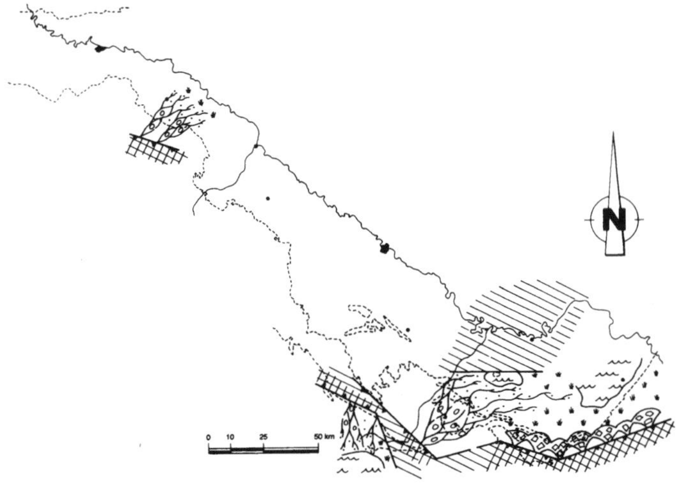

Fig. 3. Palaeogeographic sketch of Bartonian–Priabonian (TSU T2). See Fig. 2 for key.

Palaeogeography and tecto-sedimentary evolution

Paleogene

The three tecto-sedimentary units (TSUs) in the Paleogene (see Chapter E5) have limited outcrop areas, so the palaeogeographical reconstructions are also limited. All these TSUs were separated by compressive diastrophic maxima that produced sedimentary breaks of type 1. The TSUs T2 and T3 show cyclic evolution, with first retrogradation, then progradation, of the alluvial systems, suggesting phases of decreasing diastrophic activity, followed by phases of increasing activity.

At the beginning of the Paleogene (Fig. 2), the margin of the basin in the eastern sector, which is the only one with outcrops, was located at least 50 km south of the present outcrop margin on which the distal sector of an alluvial fan system of high transport efficiency is recognized. Towards the west, this system is bounded by a roughly N–S threshold that separated the Ebro basin from the Montalbán basin, in which the paleocurrents flowed south.

The diastrophic maxima that separate the Paleogene TSUs developed the structure of the southern margin by activity of basement structures there, such as the Montalbán and Tarragona structures. As a result the margin was displaced progressively towards the north. In the early Bartonian (boundary of TSUs T1 and T2), it was located 20 km south of the present outcrop margin, roughly coinciding with the present thrust front of the linking zone (Fig. 3). This thrust front was formed in the Priabonian (boundary

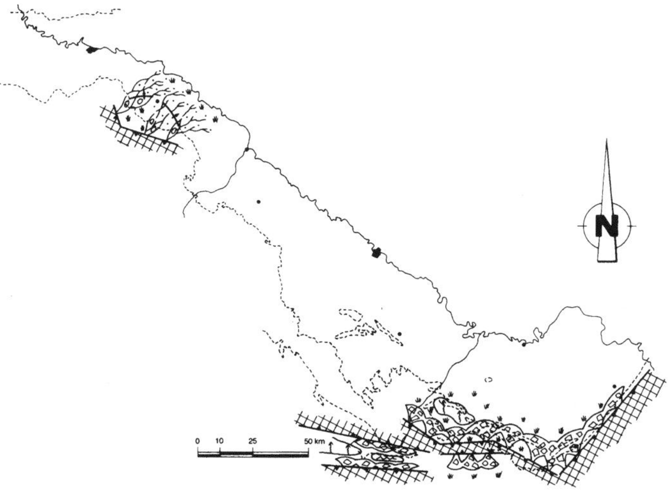

Fig. 4. Palaeogeographic sketch of Priabonian–Early Oligocene (TSU T3). See Fig. 2 for key.

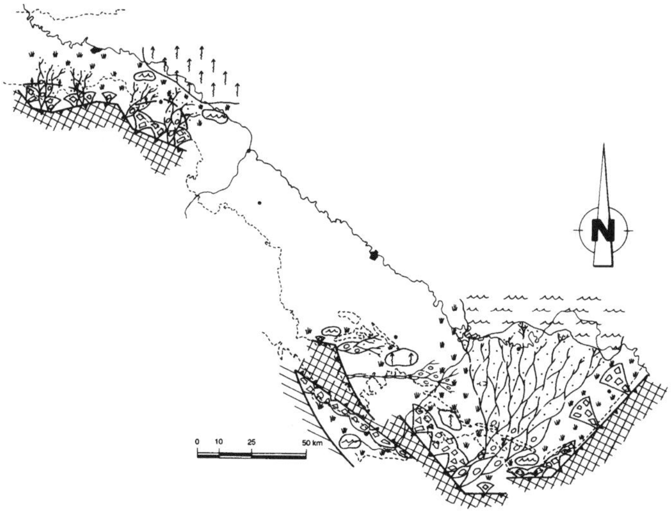

Fig. 5. Palaeogeographic sketch of Late Oligocene (early stage of TSU T4). See Fig. 2 for key.

of TSUs T2 and T3) at the same time that the Catalan margin developed its NE–SW arrangement (Fig. 4).

In the eastern sector, during the Thanetian–Bartonian (T1) and the Bartonian–Priabonian (T2), the Puigmoreno fault formed, at least partly, the northern margin of the basin. This margin was passive, but caused the eastward orientation of the alluvial fan systems that originated from the active southern margin (Fig. 3).

In the western sector, the earliest known sediments are from the end of the Bartonian. There is evidence for an active NW–SE front, corresponding to a segment of the Cameros–Demanda thrust (Figs. 3 and 4). There is some evidence of a dextral component of movement along this front. In the Early Oligocene (TSU T3) the Arnedo thrust sheet became active.

The Paleogene–Neogene transition (Late Oligocene–Agenian TSU T4)

The diastrophic maximum of the Early–Late Oligocene boundary developed an *en echelon* margin, orientated NW–SE between the eastern and central sectors (Fig. 5). Although TSU T4 records a general cycle in tectonic activity, the later hemicycle of increasing diastrophism is developed differently in each of the sectors. Below, we outline the main features observed in this later hemicycle.

In the eastern sector two thresholds developed (Fig. 6), one coinciding with the E–W Puigmoreno basement fault (and the Puigmoreno monoclinal fold), and the other more to the south, coinciding, in part, with the present Albalate–Calanda thrust front. This second threshold acted as both a local source area and as a barrier for the alluvial fans coming from the thrust front of the linking zone. This caused the anomalous and complex evolution of the upper part of the unit (González, 1989; see Chapter E5, Fig. 3).

In the central sector, small thrusts trending E–W developed, and they appear to have been associated with the NW–SE *en echelon* active margin, and some north-verging folds, such as the Aguilón and Mezalocha–Belchite anticlines. These folds may have been surface expressions of movement of the North Iberian fault.

In the western sector, during the later hemicycle of TSU T4, there was a complex sequence of tectonic activity with two intervals of increasing diastrophism during which the Cameros–Demanda thrust front moved about 8 km to the north. During the first interval the Baños de Río Tobía monoclinal fold was active. The two intervals were separated by an intermediate interval of decreasing diastrophism (Muñoz, 1992; see also Chapter E5) during which lacustrine deposition in this sector reached its maximum Tertiary extent.

The increasing diastrophism of the end of this unit is marked by a maximum in compressional activity.

Neogene

During the Late Agenian, at the beginning of TSU T5, the southern margin of the Ebro basin continued to be affected by compressive tectonic activity, but of decreasing strength. In the eastern sector, the present Iberian margin acted for the first time as a source area for the sediments of the Ebro basin (Fig. 7). But starting from this time, there was an important change in the geodynamic pattern of the different sectors of the Iberian Range.

In the western sector, the compressional regime continued, as shown by the sedimentary breaks of type 1 between the Neogene TSU (see Chapter E5). The compression is also shown by thrusting along the Cameros–Demanda thrust front in all the units except the highest, Turolian, which oversteps the thrusting. In other words, until the middle Aragonian, the Cameros–Demanda thrust front advanced progressively to the north, a distance of up to 10 km, arriving at more or less its present position. At the same time, the Baños de Río Tobía and Nájera monoclinal folds were also active, and continued so until the Late Aragonian (Figs. 7 and 8). After that, there were only minor horizontal movements, but important vertical movements, on the Cameros–Demanda thrust front, until the Turolian, by which time it became inactive.

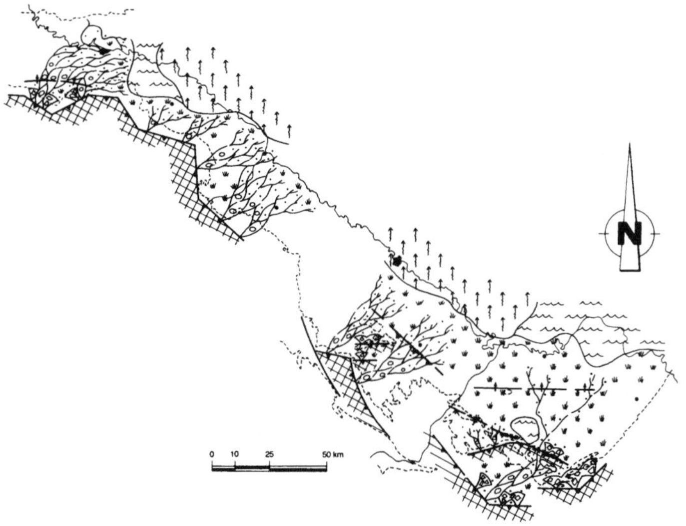

Fig. 6. Palaeogeographic sketch of Late Oligocene–Agenian (late stage of TSU T4). See Fig. 2 for key.

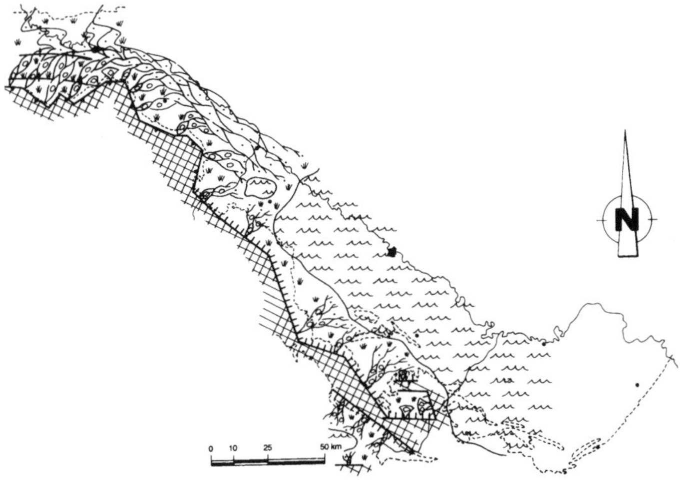

Fig. 8. Palaeogeographic sketch of Middle Aragonian–Late Aragonian (TSU T6). See Fig. 2 for key.

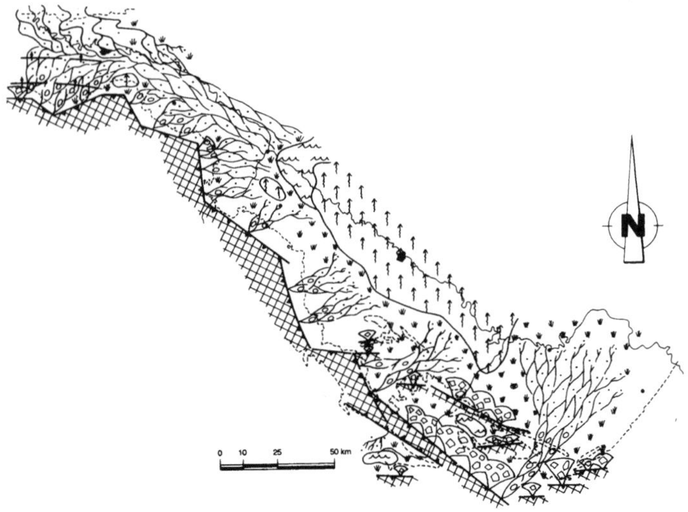

Fig. 7. Palaeogeographic sketch of Agenian–Middle Aragonian (TSU T5). See Fig. 2 for key.

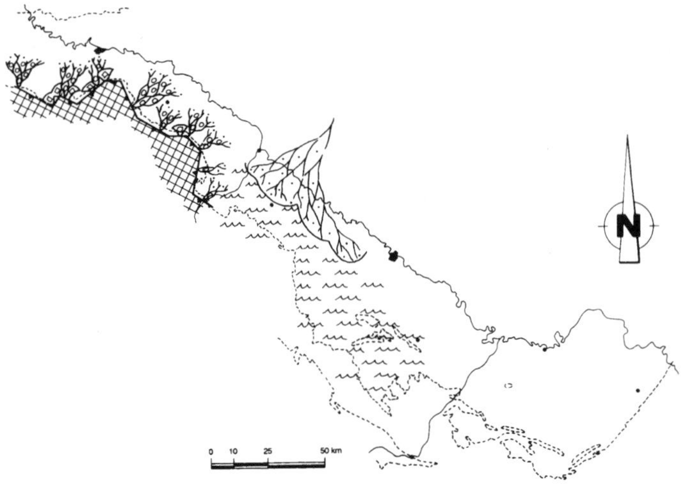

Fig. 9. Palaeogeographic sketch of Late Aragonian–Vallesian (TSU T7). See Fig. 2 for key.

In contrast, the eastern and central sectors were dominated by an extensional regime (Simón, 1984; Guimerá, 1988), and this is demonstrated by the sedimentary breaks of type 3 in the Neogene TSUs, and the way these units overlap the margin. In the central sector, during this time, the alluvial fans decreased in size, in a succession of stages, and the lacustrine systems moved southwards progressively. These lacustrine systems extended across the present margin during the Late Aragonian–Turolian, with the almost complete disappearance of alluvial systems from the Iberian margin (Figs. 9 and 10). At the same time, alluvial systems from the north introduced large quantities of clastic sediments close to the Iberian margin, and this produced a fining to coarsening grain-size cycle in TSU T7 in the central sector.

As a generalization, the northern margin of the Iberian Range became divided, in the Early Miocene, into two very different structural domains (Casas, 1992). Muñoz (1992) suggests that the boundary between the domains was located where Guiraud and Seguret (1985) placed the important NE–SW basement structure known as the Odemira–Avila fault, and this coincides with the post-Pontian monoclinal fold recognized by Bomer (1954). To the west of this structural zone, the compressional regime continued, while to the east, extension was dominant and may be related to the general regional rifting mentioned above.

Concluding remarks

We will finally consider some aspects of the relationship between tectonics and sedimentation, particularly subsidence and sedimentation rate.

A first point is the notable difference in Paleogene sediment

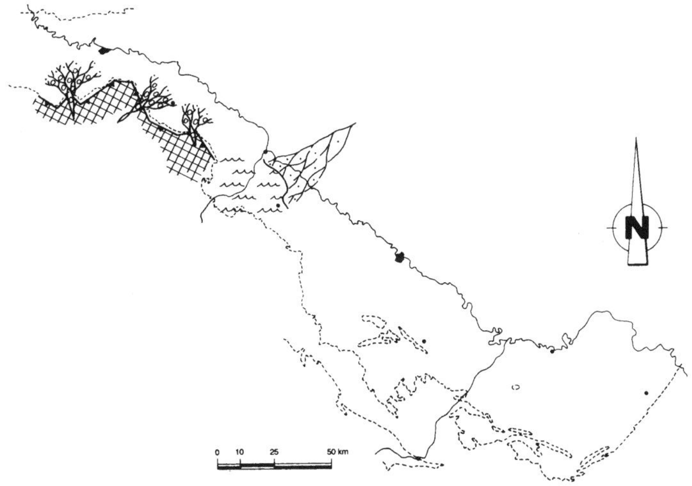

Fig. 10. Palaeogeographic sketch of Turolian (TSU T8). See Fig. 2 for key.

thickness between the three sectors into which we have divided our study area. The Paleogene of the central sector is significantly thinner (see Chapter E5), so that in a NW–SE section the central sector appears as a threshold between two more strongly subsiding domains, recognised by earlier authors (Riba, 1964; Riba *et al.*, 1983). At present we do not have sufficient subsurface information to allow us to understand the reasons for this differential subsidence. However, in the central sector, although the outcropping Paleogene stratigraphy corresponds to that south of the North Iberian fault, subsurface data (from the Zaragoza and Magallón boreholes: IGME, 1987; see Chapter E5) show that the thickness increases considerably. This suggests that the North Iberian fault may have acted as a thrust front, and that our surface observations in the central sector may have been south of this front, in a southern piggy-back basin, while the Zaragoza and Magallón boreholes might be in the footwall block, north of the fault front. Both the western and the eastern sectors are located north of this fault, moreover, and the subsidence in these sectors may have involved a continuation of Mesozoic movements of the NE–SW fault systems, such as those of Odemira–Avila, Gandesa, etc.

In Neogene times, despite the different tectonic arrangement observed between the western and central–eastern sectors, sediment thicknesses do not vary strongly across the area being reviewed, although the western sector shows slightly greater thicknesses.

If we focus on the western sector, it provides us with a record that is relatively complete, and was formed in a relatively continuous compressional tectonic regime. Here, the Paleogene units have the best development of sediments representing increasing diastrophic activity (coarsening-upward sequence of TSUs), whereas the Neogene units have the best development of sediments representing decreasing diastrophism (fining-upward sequence of TSUs).

Overall averaging of sedimentation rates through the Tertiary gives rates of more than 10 cm/1000 years for the Paleogene, and less than 7 cm/1000 years for the Neogene. This further supports the earlier evidence for a general decrease in diastrophic activity.

To sum up, throughout the greater part of the Tertiary, the Iberian margin of the Ebro basin was an active margin, and it would not be completely wrong to speak of the basin as an Iberian foreland in Paleogene times. In the western sector, this persisted until the Late Miocene.

Acknowledgements

This work has been supported by projects nos. PB–89/0342 and PB–89/0344 of DGICYT.

References

Alvaro, M., Capote, R. and Vegas, R. (1979). Un modelo de evolución geotectónica para la Cadena Celtibérica. *Acta Geol. Hispánica*, **14**: 172–177.

Anadón, P., Cabrera, L., Guimerá, J. and Santanach, P. (1985). Paleogene strike–slip deformation and sedimentation along the southeastern margin of the Ebro basin. In Biddle, K.T. and Christie-Blick, N., eds. *Strike–slip deformation, basin formation and sedimentation, S.E.P.M. Spec. Publ.*, **37**: 303–318.

Bomer, B. (1954). Trois aspects du contact entre Monts Celtibériques occidentaux et Bassin de l'Ebre. *Bull. Assoc. Geogr. Fr.*, **239–240**: 35–41.

Casas, A.M. (1992). *El frente Norte de las Sierras de Cameros: estructuras cabalgantes y campo de esfuerzos*, ed. Instituto de Estudios Riojanos. Logroño. *Zubía*, **4**. 220 pp.

Colombo, F. (1989). Abanicos aluviales. In Arche, A., ed. *Sedimentología: colección nuevas tendencias*, **1**: 143–218. C.S.I.C. Madrid.

González, A. (1989). Análisis tectosedimentario del Terciario del borde SE de la Depresión del Ebro (sector bajoaragonés) y cubetas ibéricas marginales. Unpublished PhD Thesis. Zaragoza University. 507 pp.

Guimerá, J. (1984). Paleogene evolution of deformation in the northeastern Iberian Peninsula. *Geol. Mag.*, **121**: 413–420.

Guimerá, J. (1988). Estudi estructural de l'enllac entre la Serralada Iberica i la Serralada Costanera Catalana. Unpublished PhD Thesis. Barcelona University. 600 pp.

Guiraud, M. and Seguret, M. (1985). Releasing solitary overstep model for the Late Jurassic–Early Cretaceous (Wealdian) Soria strike–slip Basin (North Spain). In Biddle, K.T. and Christie-Blick, N., eds., *Strike–slip deformation, basin formation and sedimentation. S.E.P.M. Spec. Publ.*, **37**: 159–175.

IGME (1987). *Contribución de la exploración petrolífera al conocimiento de la geología de España*. Instituto Geológico y Minero de España. Madrid. 465 pp. 17 figs.

Klimowitz, J. (1991). Notas sobre la estratigrafía y estructura del terciario inferior en el subsuelo del sector central de la cuenca del Ebro. *I Congreso del G.E.T.* Comunicaciones: 174–177. Vic.

Muñoz, A. (1992). *Análisis tectosedimentario del Terciario del sector occidental de la Cuenca del Ebro (Comunidad de La Rioja)*, ed. Instituto de Estudios Riojanos. Logroño. *Ciencias de la Tierra*, **15**. 347 pp.

Muñoz, A., Pardo, G. and Villena, J. (1992). Evolución paleogeográfica de los conglomerados miocenos adosados al borde norte de la Sierra de Cameros (La Rioja). *Acta Geol. Hispánica*, **1–2**: 3–14.

Pardo, G., Villena, J., Pérez, A. and González, A. (1984). El Paleógeno de los márgenes del umbral de Montalbán: relación tectónica–sedimentación. *Publ. Geol.*, **20**: 355–363. Barcelona.

Pérez, A. (1989). Estratigrafía y sedimentología del Terciario de borde meridional de la Depresión del Ebro (sector riojano-aragonés) y cubetas de Muniesa y Montalbán. Unpublished PhD Thesis. Zaragoza University. 474 pp.

Riba, O. (1964). Estructura sedimentaria del Terciario continental de la Depresión del Ebro en su parte Riojana y Navarra. *Aport. Esp. a Congr. Geogr. Int.*, 127–138. UK.

Riba, O., Reguant, S. and Villena, J. (1983). Ensayo de síntesis estratigráfica y evolutiva de la Cuenca terciaria del Ebro. *Libro jubilar J.M. Ríos. Geología de España*, **2**: 131–159.

Salas, R. (1987). El Malm i el Cretaci inferior entre el Massís de Garraf i la Serra d' Espadá. Anàlisi de conca. Unpublished PhD Thesis. Barcelona University. 345 pp.

Simón, J.L. (1984). *Compresión y distensión alpinas en la Cadena Ibérica oriental*, ed. Instituto de Estudios Turolenses de la Excma. Diputación de Teruel. 269 pp.

Viallard, P. (1979). La Chaîne Ibérique: zone de cisaillement intracontinental pendant la tectogenèse alpine. *C.R. Acad. Sci. Paris.*, **289** (serie D): 65–68.

Viallard, P. (1980). Les Ibérides (Chaînes Ibériques y Catalane): interprétatión de la fracturatión majeure fini-oligocène. *C.R. Acad. Sci. Paris*, **291** (serie D): 873–876.

E7 Stratigraphy of Paleogene deposits in the SE margin of the Catalan basin (St. Feliu de Codines – St. Llorenç del Munt sector, NE Ebro basin)

J. CAPDEVILA, E. MAESTRO-MAIDEU, E. REMACHA AND J. SERRA ROIG

Abstract

Three depositional sequences in the sense of Vail (1987) are defined in the Middle and Upper Eocene deposits of the NE Ebro Basin. These sequences are correlated along the SE border of the basin in the area of deposition of the deltaic complexes of St. Llorenç del Munt, Gallifa and St. Feliu de Codines. The sedimentation took place during a period of block-differentiation of the basin margin, which induced thickness variations in the depositional areas. Although the tectonic-sedimentation evolution model that we propose shows the control of the paleogeography by tectonics, the sequence stratigraphy of the alluvial and deltaic complexes is influenced by the relative sea-level changes.

Introduction

The area studied is located on the SE side of the Catalan Central Depression (NE Ebro basin), close to the Catalan Coastal Ranges (Fig. 1). During Paleogene times, the NE margin of the Ebro Basin was filled with predominantly clastic terrigenous deposits which formed several deltaic complexes. The purpose of this chapter is to analyse and compare the sequence stratigraphy and the facies distribution patterns of the Eocene St. Llorenç del Munt and St. Feliu de Codines–Gallifa deltaic complexes, and find the relationships between tectonics and sedimentation for this side of the NE Ebro basin margin.

The strike–slip tectonics of the NE margin of the Ebro basin was contemporaneous with the sedimentation of the Paleogene deposits. Thus, the facies distribution of these materials near the basin margin is strongly controlled by the evolution of thrusting and fault development (Anadón *et al.*, 1985). This aspect is displayed by the Paleogene deltaic systems.

Tectonic setting

The Catalan Coastal Range is an orogenic chain with an Hercynian basement unconformably overlain by a Mesozoic cover. The main structural features are of Tertiary age. They are defined by vertical, basement-involved, strike–slip faults that constitute a right-stepping, en echelon array (Anadón *et al.*, 1985) (Fig. 1).

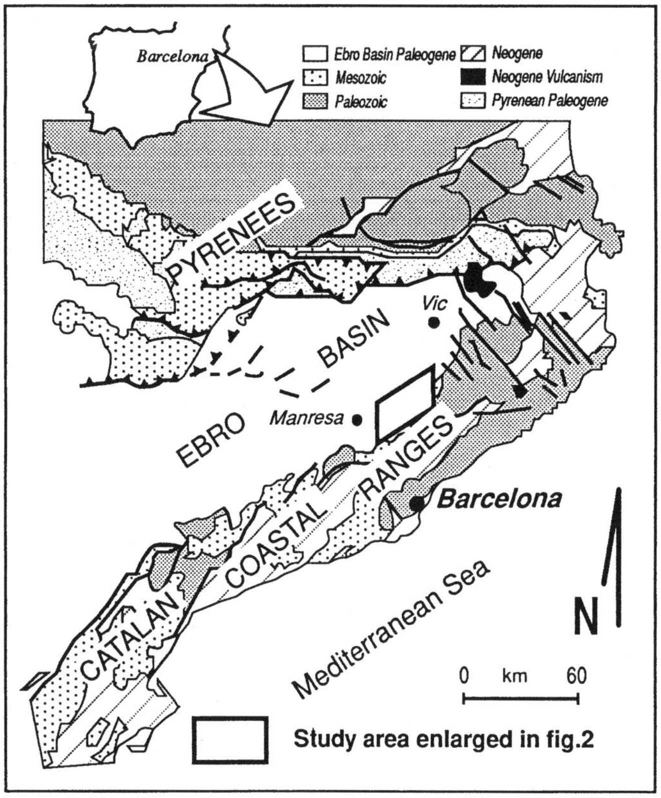

Fig. 1. Geological sketch map of Catalonia. The view of this figure shows the Eastern side of the Ebro basin, and the main units of the basin margins, the Pyrenees to the North and the Catalan Coastal Ranges to the SE. The enclosed area corresponds to the study area, shown and an enlarged version is given in Fig. 2.

These faults are oriented NE–SW and ENE–WSW, with a sinistral or sinistral-reverse sense of movement (Guimera, 1984). There are also some transverse, strike–slip, SE–NW- and SSE–NNW-directed, basement faults, consistent with the movement of the first array, which is the principal fault set (Figs. 1, 2). During Paleogene times, these transverse faults produced a block-differentiation of the basin margin. Each one of the blocks was a differentiated sedimentary domain (Fig. 2).

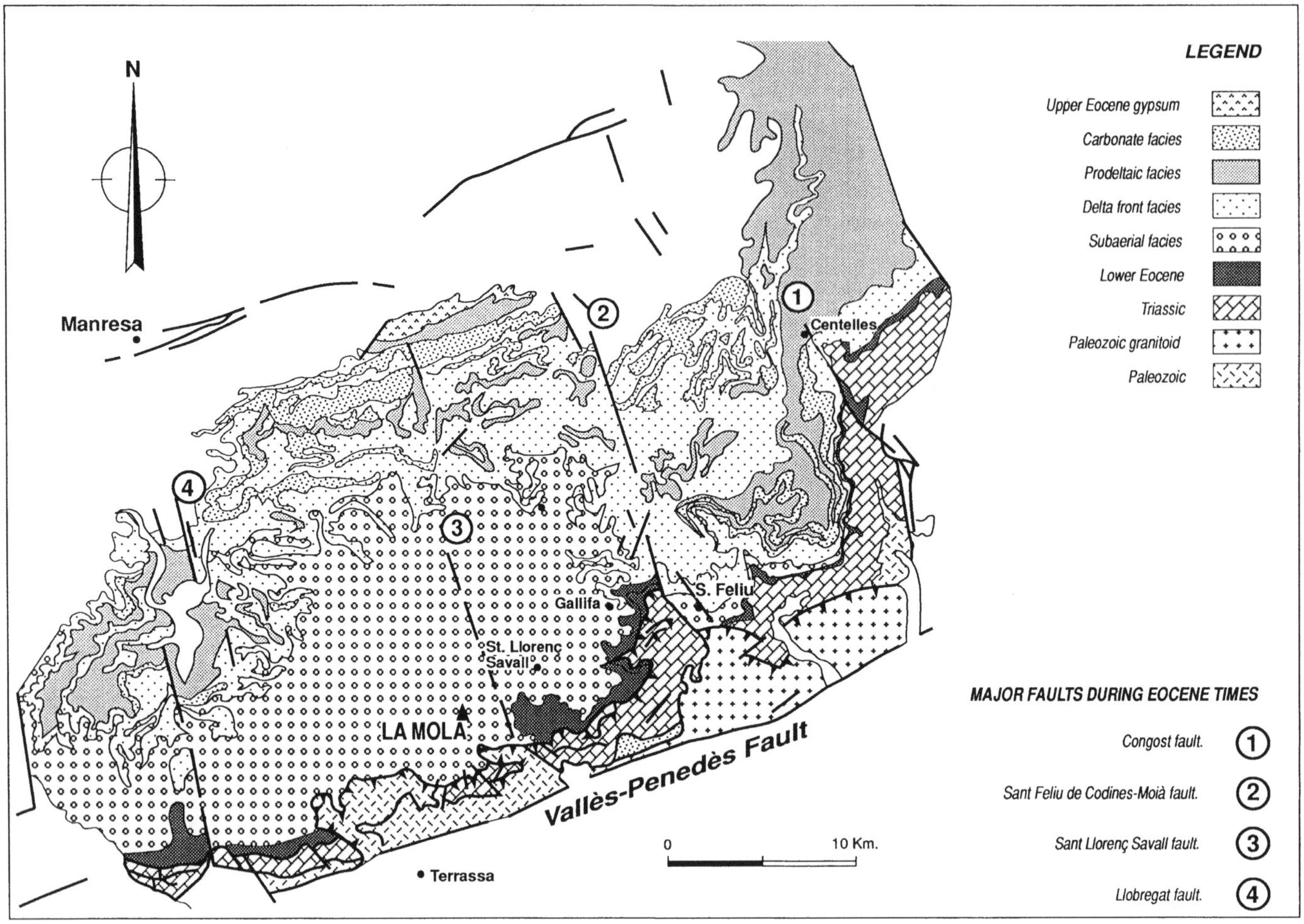

Fig. 2. Geological sketch map of the study area, showing the different alluvial and deltaic facies associations and their distribution. This map also shows the main features of the basin margin and the major normal basement faults, which led to the syn-sedimentary block-compartmentation of the basin margin during Eocene times.

The main tectonic structures in the area studied are controlled by the El Vallès–Penedès fault (Fig. 2). They are Triassic and Paleozoic basement thrust sheets, which overthrust the Mesozoic cover and even the Paleogene deposits of the Ebro foreland basin (Fig. 2), with a S–N propagation (Anadón *et al.*, 1985). There are also some transverse, SE–NW- and SSE–NNW-directed, basement faults (Fig. 2).

General stratigraphy of the Paleogene deposits

The Paleogene sediments of the NE margin of the Ebro basin are mainly composed of alluvial and deltaic deposits (Fig. 2), and are arranged into a transgressive–regressive megacycle. The sedimentation of the alluvial deposits near the basin margin occurred from Paleocene to latest Paleogene, whereas the deltaic deposits are more basinal and Middle–Late Eocene. A synthesis of the nomenclature used by several authors is shown in Fig. 3. From base to top, a general section of the Paleogene is:

Paleocene. The Paleocene sediments lie uncomformably over the Mesozoic substratum, and are composed of red siltstones, with

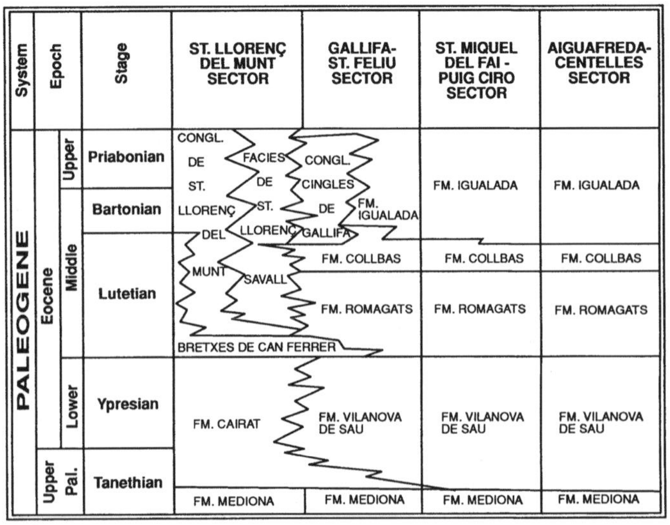

Fig. 3. Paleogene stratigraphy of St. Llorenç del Munt–St. Feliu de Codines area, from the most important work of the last 25 years (see references for more information).

some sandstone and conglomerate layers, and abundant Microcodium. These deposits have a high carbonate content, including caliche nodules, calcareous crusts and lacustrine–palustrine limestones. At the top of the alluvial facies there is a condensed paleosol layer, a sign of an important sedimentary break (Anadón & Marzo, 1986).

Early Eocene. The early Eocene sediments of the study area are composed of continental red sandstones, mudstones and conglomerates, forming part of several coalescent alluvial fans. Overlying these deposits there are breccia alluvial fans. Some alluvial fans extended into the basin for a considerable distance and received a great deal of rounded, lithologically varied clasts (e.g., the base of the St. Llorenç del Munt deltaic complex). The other alluvial fans are characterized by angular clasts fed by local basement thrust sheets (e.g. the base of the St. Feliu de Codines deltaic complex).

Middle Eocene. In this period the conglomerate deltaic deposition started both in the St. Llorenç del Munt and St. Feliu de Codines areas. The petrologic content of the two fan deltaic units is different. The St. Llorenç del Munt depositional system contains cobbles of basement and lower Triassic, and even Paleocene clasts, whereas the St. Feliu de Codines deltaic complex encloses cobbles exclusively of the basement thrust sheets, composed of igneous and metamorphic rocks. Both depositional systems display a similar distribution pattern, consisting of reddish continental deposits in the proximal parts, marine sandstones and conglomerates in the deltaic front, sometimes including subordinate fringing reefs, and a prodeltaic platform that extends over the Catalan basin.

Middle–Late Eocene. Passing upwards into a more fully marine context, the deposition of the middle and late conglomerate fan delta units occurs in St. Llorenç del Munt. In the St. Feliu de Codines area, the deposition of the upper fan delta unit occurred, equivalent in time to the middle fan delta unit of St. Llorenç del Munt. At this time the Gallifa fan delta unit developed, situated between St. Llorenç del Munt and St. Feliu de Codines.

Stratigraphy of the deltaic complexes

Three genetically related stratigraphic units are defined here. Each one is characterized by alluvial to deltaic successions, including proximal to distal depositional environments. These are defined as depositional sequences in the sense of Vail (1987). The locations of sedimentary sections through these deposits are shown in Fig. 4.

Rellinars sequence

The Rellinars sequence ranges in age from Upper Lutetian to Bartonian. The sequence consists mainly of 160–180 m of alluvial deposits that grade up into deltaic sediments, and 100 m of prodelta marls with fine-grained sandstones and siltstones in the distal environments (Figs. 4 and 5). In the St. Llorenç del Munt deltaic complex this sequence consists of conglomeratic proximal deltaic sediments, showing reduced delta front and prodelta sediments, deposited in a lagoonal or shallow shelf environment (Fig. 6). Basinward these deposits merge into shelf facies (Fig. 7).

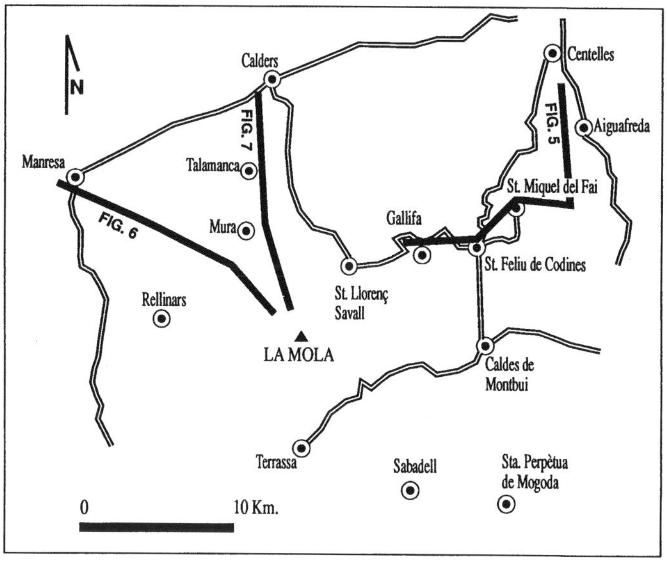

Fig. 4. Map of the cross-sections shown in Figs. 5, 6 and 7.

The lower boundary of the Rellinars sequence is an erosional uncomformity over the Lower Eocene continental materials (Fig. 5) that has been recognized both in St Llorenç del Munt and St. Feliu de Codines areas, but it has not been identified in the Gallifa area, which is the intermediate zone. The differentiated systems tracts are:

LST

The LST (lowstand systems tract) is formed in the St. Feliu de Codines area by about 80 m of continental alluvial deposits, basically with an agradational parasequence stacking pattern. As mentioned above, the lower boundary is an erosional unconformity surface over the Lower Eocene continental materials; the upper boundary is marked by a metre-thick level of channel abandonment facies, with abundant root marks and caliche nodules. The upper boundary erosive surface seems to be overlain uncomformabily by the TST materials. Therefore, we cannot discount the possibility that this unit could constitute an entire sequence on its own. On the other hand, we can not assume this, because all the outcrops of this unit are of proximal facies.

We have recognized in this unit several alluvial depositional systems (the St. Feliu de Codines alluvial complex and the Gallifa alluvial fan) which developed contemporaneously (Capdevila, 1992). There are some differences between the two alluvial systems: The St. Feliu de Codines alluvial system consists of several coalescent breccia alluvial bodies, which consist of layers of igneous and metamorphic angular heterometric clasts alternating with metre-thick levels of arkosic, very coarse-grained sandstone, with root marks. The main sedimentation events were sheetfloods. The Gallifa alluvial system is only characterized by one alluvial body, well developed, with alluvial plain facies eroded by channels filled

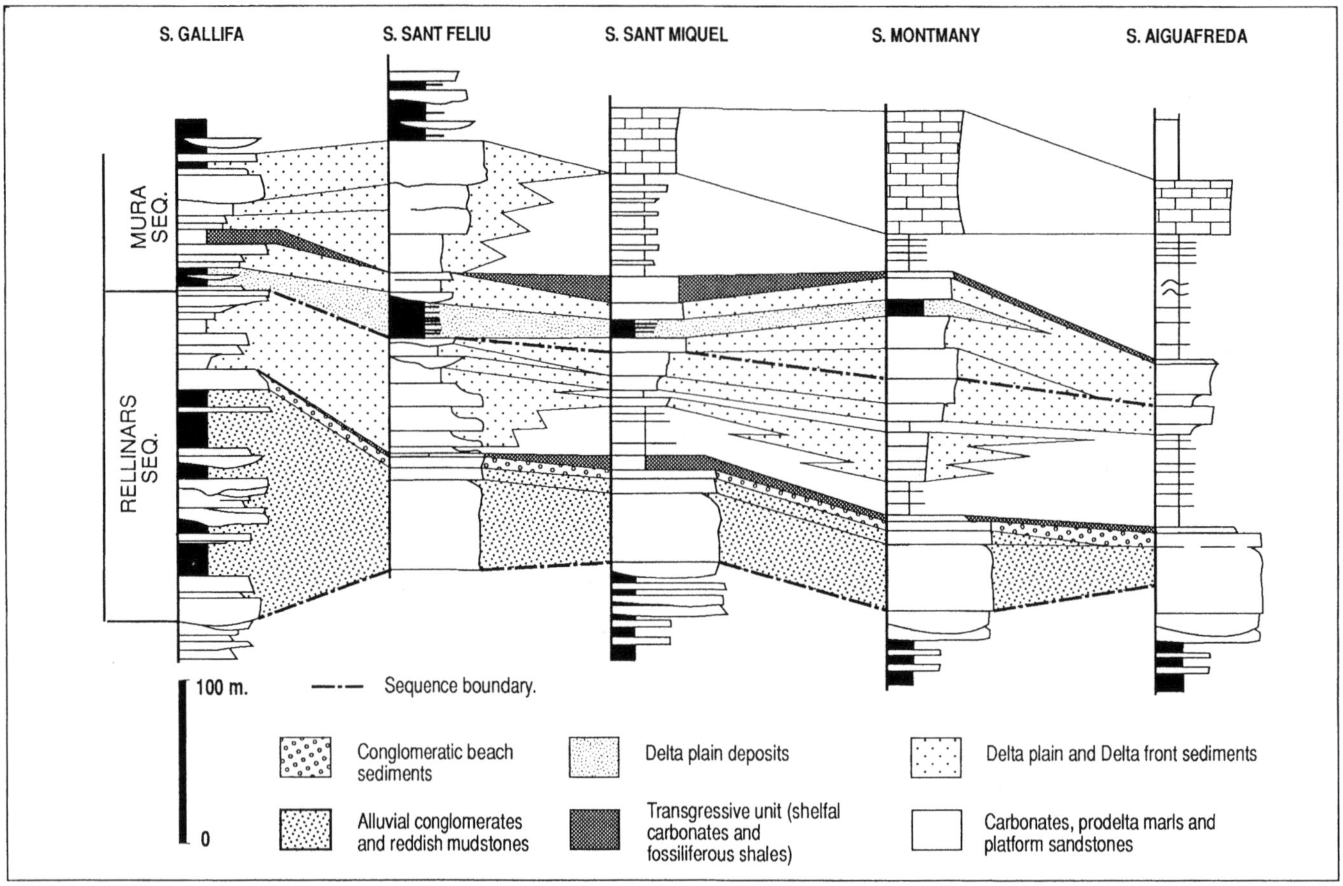

Fig. 5. Synthetic correlation fence of the St. Feliu de Codines Deltaic Complex. Note the parasequence stacking patterns of the two upper sequences, which indicates the different system tracts.

with well-sorted conglomerate, which displays a lithological composition very similar to that of the St. Llorenç deltaic complex (Maestro, 1987, 1991).

In St. Llorenç del Munt, the LST is poorly exposed, but we have recognized a prograding deltaic lowermost unit that consists of delta-plain crevasse couplets, thin delta-front distributary mouth bars and upper prodelta grey mudstones (Fig. 6). This lowstand prograding complex is thin and only represented in the southwestern area of St. Llorenç del Munt.

TST

The TST (transgressive systems tract) is the first depositional unit that extends in continuity across both sectors of the studied basin margin (Figs. 5, 8).

In the St. Feliu de Codines depositional system the TST is characterized by two main intervals. The lower part is composed of retrograding cycles of middle to distal alluvial fan facies cycles. These facies pass laterally to conglomeratic foreshore and nearshore facies. In the Gallifa area the LST consists of distal alluvial fan deposits with similar features to that of the alluvial units of St. Feliu de Codines, overlain by interdistributary bay sandstone and conglomeratic deposits.

The upper part consists of a relatively thin layer of fine-grained sandstones and siltstones deposited in a distal shoreface–offshore environment, with a high content of marine fauna (mainly foraminifera and bryozoa). In the most distal sectors, this interval of TST displays a high content of glauconite.

In the St. Llorenç del Munt deltaic complex, particularly in proximal areas a transgressive conglomerate unit rich in limestone cobbles can be distinguished. The conglomerates are strongly lithified by carbonate cements. These conglomerates pass basinward into shelfal carbonate deposits (Figs. 6, 7).

HST

The HST (highstand systems tract) consists of a prograding unit composed of conglomerate alluvial facies in the proximal areas. These merge to delta-plain deposits and rapidly pass to delta-front facies to the NE (Fig. 5). To the W, the HST is mainly composed of large conglomerate submarine channels. In front of these conglomerate distributary channels there are tabular sandstone bodies with large-scale and low-angle oblique stratification, deposited in proximal stream-mouth bar facies. In marginal positions of the deltaic system, the sedimentation in the nearshore environments was dominated by wave and storm processes. Pro-

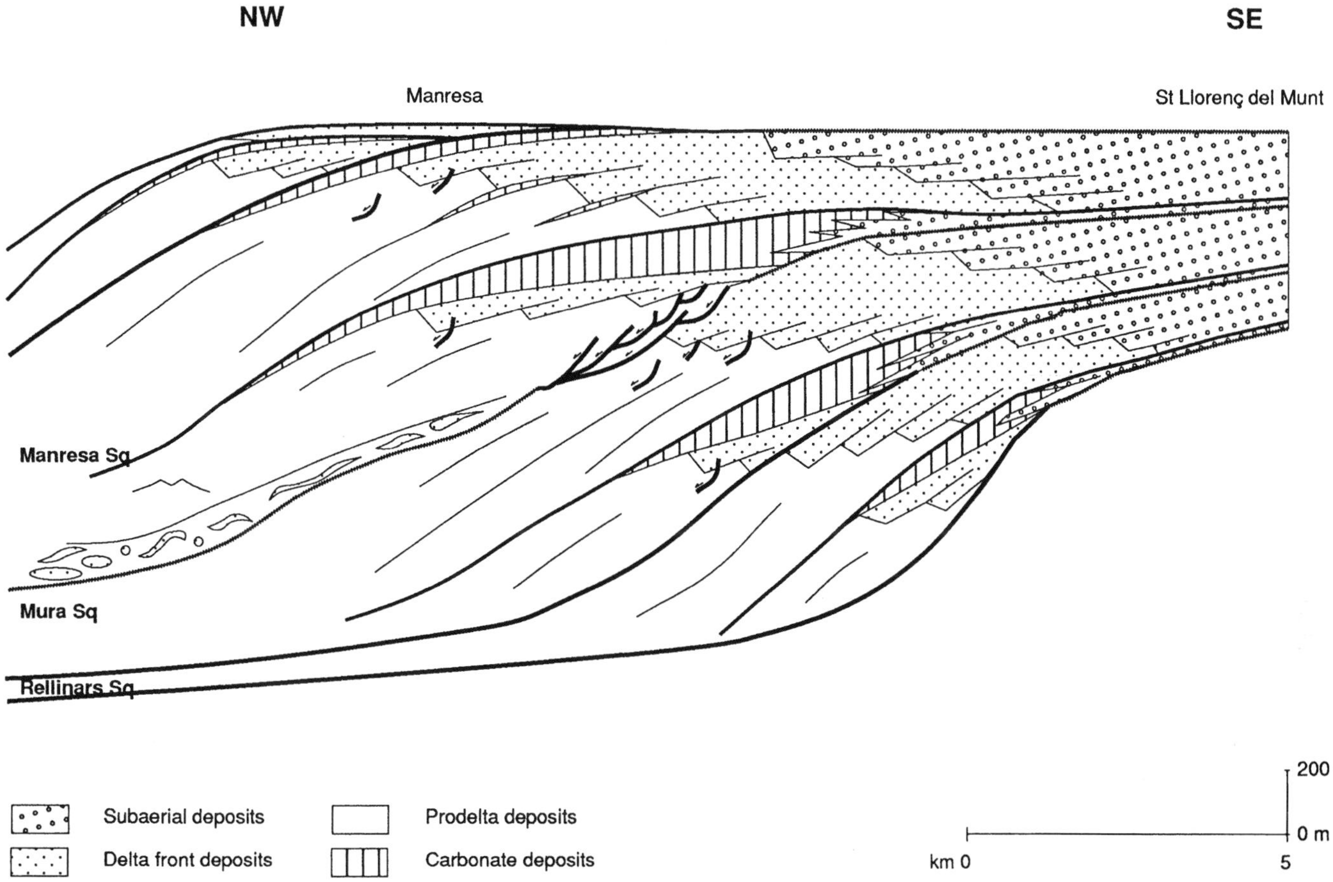

Fig. 6. Synthetic NW–SE cross section of the depositional sequences exposed in the St. Llorenç del Munt Deltaic Complex. This figure shows the main facies associations.

gressively to the W the HST deposits pass to delta-plain environments related to the St. Llorenç del Munt deltaic complex (Figs. 7, 8).

In this area, the HST consists of a progradational conglomerate body in the proximal segment, which is well exposed in the St. Llorenç del Munt area (Figs. 6, 7). It merges into reddish delta plain deposits, which are fully developed in the St. Feliu de Codines–St. Llorenç del Munt transition zone (Fig. 8). These were deposited in a temporally flooded and plant-rich closed-bay environment, with conglomerate channels and reddish sandstones and siltstones in the interchannel areas. To the N and NW the conglomerate deposits pass into prograding conglomerate distributary mouth-bar deposits, and to poorly exposed prodelta sediments (Figs. 6, 7).

Mura Sequence

The Mura Sequence is Bartonian in age. This sequence is mainly composed of deltaic sediments, and was deposited in St. Llorenç del Munt and St. Feliu de Codines deltaic depositional systems. It is 150–200 m thick in its delta-front segment, located near St. Miquel del Fai. A distinctive feature of this sequence is the presence of fringing and barrier reefs, and carbonate platforms (Taberner, 1982). In the Gallifa sector the Mura sequence is mainly represented by a thick continental succession. In the upper parts of the western sector it was influenced by the St. Llorenç del Munt deltaic complex (Capdevila, 1992). In the distal facies of this deltaic sequence we have identified sedimentary instability processes, such as slumping, intraformational unconformities and frequent ball and pillow structures (Maestro, 1987, 1991) (Figs. 5, 6, 7).

The lower boundary of this sequence is of type II. The upper limit is a type I sequence boundary (Maestro, 1991) marked by an erosional surface clearly distinctive over both continental and delta front facies in the St. Llorenç del Munt sector (Figs. 6, 7).

LST

The lowstand systems tract is characterized by interdistributary bay deposits with an agradational–progradational parasequence stacking pattern, which onlaps over the lower boundary (Fig. 5). These deposits are overlain by delta-plain sediments, which pass distally to delta-front deposits with distributary channel and stream-mouth bar facies. Over this level there is a relatively thick succession of delta-front deposits, with stream-mouth bar abandonment facies marked by the presence of thin carbonate layers of fringing reef facies. These carbonate layers show the

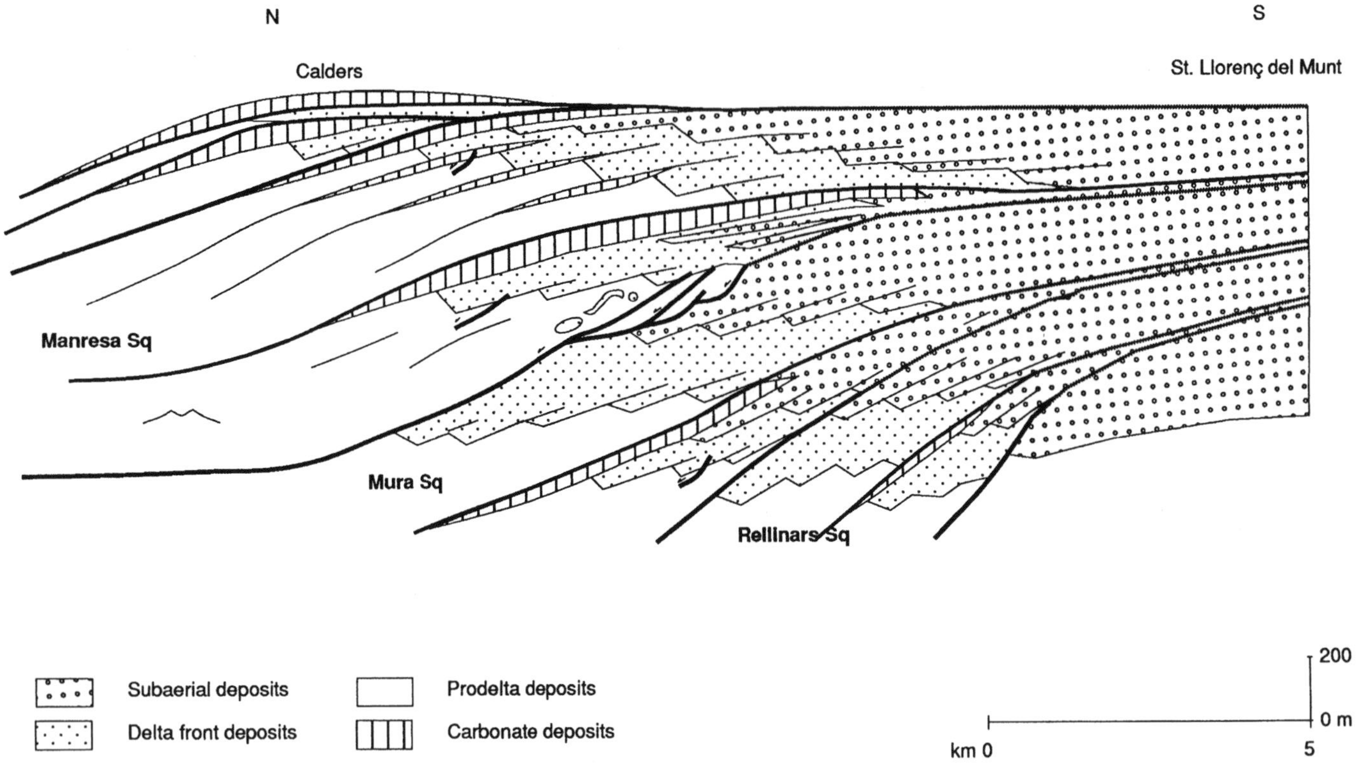

Fig. 7. Synthetic N–S cross section of the depositional sequences in St. Llorenç del Munt, showing the main facies associations. In this section we can observe that subaerial delta plain deposits prograde much more over the Ebro basin than in the previous cross section (Fig. 6). This is due to the activity of the Eocene block-compartmentation of the basin margin.

agradational–retrogradational parasequence stacking pattern of the upper LST.

In the St. Llorenç del Munt area, only the prodeltaic deposits, with a thick succession of grey mudstones and claystones are exposed.

TST

The TST consists of a shallow carbonate platform that overlies the entire LST of this sequence (Fig. 5). Locally, this carbonate platform incorporates some bioconstructions. At the top of the TST there is a condensed section marked by the existence of a hardground layer.

In the Gallifa sector we have not identified the carbonate platform. However, the TST in this sector is marked by the presence of the correlative hardground layer between the delta front sediments (Fig. 8).

In St. Llorenç del Munt, a second carbonate-rich conglomerate unit is identified in the proximal parts. In the intermediate area of the basin, at the NW side of the deltaic complex, the TST consists of retrogressive, metric to decametric, carbonate marine cycles overlying the LST. This unit basinward interfingers with coralgal patch reefs growing on the wedge of conglomerates (Figs. 6, 7).

HST

The HST consists of a thick succession of delta-plain deposits alternating with braided conglomerate distributary channels that merge into a prograding succession of fine-grained sandstones with thin fringing-reef carbonate layers, pertaining to the delta-front facies. Distally to the E–NE, the HST is composed of a thick outer platform succession with sedimentary instabilities. In its uppermost part, the HST includes a large prograding barrier-reef (Fig. 5).

In the Gallifa sector the HST consists of a delta-front succession overlain by a thick succession of delta-plain materials. In this sector, the HST is strongly affected by the influence of the St. Llorenç del Munt deltaic complex.

In the St. Llorenç del Munt sector, the HST is very similar to that of the St. Feliu de Codines sector. The proximal areas consist of a progradational conglomerate sequence. The N and NE reddish delta-plain deposits (St. Feliu de Codines–St. Llorenç del Munt transition zone) developed to the NW, shallowing and coarsening up distributary mouth bar sequences, with a conglomerate bedset in the uppermost part. These deltaic sequences merge into wave and storm reworked distal deltaic facies. In the shoreface environments, reefal limestones develop over the paleotopography of the mouth bar cycles. Fine prodelta deposits built the distal deltaic platform (Figs. 6, 7). Near the faulted area of the Llobregat river, affecting the upper part of the prodelta and the lower part of the delta-front environments, we find some deformational features such as gullies, rotational slides and mudlumps, induced by differential weight and fault movements (Maestro, 1987, 1991).

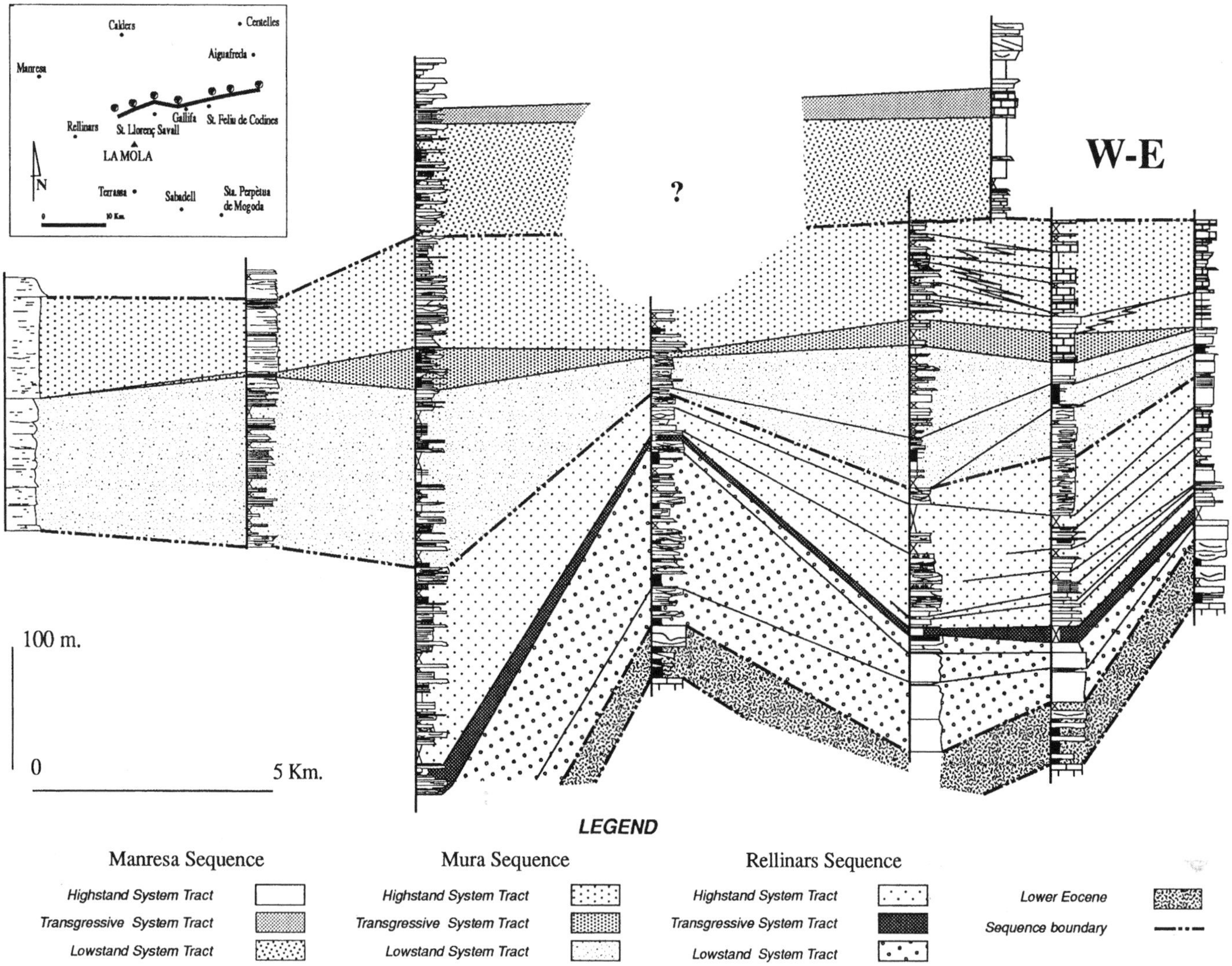

Fig. 8. Stratigraphic correlation fence of the depositional sequences and the related system tracts through the St. Llorenç, Gallifa and St. Feliu deltaic complexes.

Manresa Sequence

The age of this sequence is Late Bartonian–Priabonian, and it consists of sediments deposited in the St. Llorenç del Munt deltaic complex and in the Gallifa deltaic system.

The erosive unconformity that separates the Manresa sequence from the Mura sequence is clearly marked in the western marine areas by kilometre-scale sedimentary breaks that prove that erosion occurred in the uppermost part of the HST deposits of the Mura sequence. In the continental areas we recognize an erosional surface.

LST

The LST of this sequence contains the deepest deposits identified in the deltaic complex, and consists of grey shales and silts with channel-shaped sandstone bedsets interpreted as type III turbidites (Mutti, 1985). These slope deposits are thin, and pass landwards into prodelta and delta-front prograding facies. These sediments together represent the lowstand and wedge prograding complex (Vail, 1987). The sedimentary instabilities form part of the LST and distally they include the type III turbidites (Figs. 6, 7).

TST

In the continental areas the TST consists of a third white conglomerate layer. In the marine areas, two calcareous wedges, which onlap the LST deltaic sediments, can be identified. In the transition from continental to marine areas some patch reef facies were deposited, overlying calcareous conglomerate beds (Figs. 6, 7, 8).

HST

As in the previous sequence, in the proximal areas, the HST consists of a thick, progradational conglomerate sequence. The reddish delta-plain environments were sited in the NE part of the fan. The delta-front deposits consist of several shallowing, coarsening distributary mouth bar sequences prograding over prodeltaic facies. Reefal limestones, or their storm-reworked calcarenite equivalent sediments form the topmost deposits of the HST. They were deposited when the deltaic sedimentation was lowered during the modification of the basin in the studied area. They represent the late Highstand Systems Tract (Vail, 1987).

Conclusions

In this work we have defined four depositional sequences or third-order cycles in the sense of Vail (1987) in the Middle and Upper Eocene deposits of the SE Catalan Basin. These sequences are correlated along the whole border of the basin, in the area of deposition of the deltaic complexes of St. Llorenç del Munt, Gallifa and St. Feliu de Codines. The sedimentation took place during a period of block-differentiation of the basin margin (Fig. 2), which induced thickness variations in the depositional areas. This fact is displayed in Fig. 8.

References

Anadón, P., Colombo, F., Esteban, M., Marzo, M., Robles, S., Santananch, P. and Sole Sugranyes, Ll. (1979). Evolución tectonoestratigrà-fica de los Catalànides. *Acta Geol. Hisp.*, **14**: 242–270.

Anadón, P., Cabrera, Ll., Guimera, J. and Santanach, P. (1985). Paleogene Strike–slip deformation and sedimentation along the Southeastern margin of the Ebro Basin. *S.E.P.M., Spec. Publ.* **26**: 303–318.

Capdevila, J. (1992). El Complex Fan-Deltaic de St. Feliu de Codines-Gallifa (Eocè del marge SE de la Depressió Central Catalana). Treball de Recerca U.A.B. 100 pp. Unpublished thesis.

Ferrer, J. (1971). El Paleoceno y Eoceno del borde sur-oriental de la Depresión del Ebro (Cataluña). *Mem. Suisses Paleontol.*, **90**: 1–70.

Ferrer, J., Rosell, J. and Reguant, S. (1968). Síntesis litoestratigráfica del Paleógeno del borde oriental de la Depresión del Ebro. *Acta Geol. Hisp.*, **3**: 54–56.

Guimera, J. (1984). Paleogene evolution of deformation in the northeastern Iberian Peninsula. *Geol. Mag.*, **121**: 413–420.

Maestro, E. (1987). Estratigrafia i fàcies del complex deltaic (fan delta) de St. Llorenç del Munt (Eocè mid-superior. Catalunya. Thesis. U.A.B., 306 pp.

Maestro, E. (1991). The deltaic complex of St. Llorenç del Munt (Middle–Upper Eocene), SE Catalan Basin. *Cuad. Geol. Ibérica*, **15**: 73–102.

Mutti, E. (1985). Turbidite systems and their relations to depositional sequences. In *Provenance of Arenites*, ed. G.G. Zuffa, pp. 65–93. Reidel.

Reguant, S. (1967). El Eoceno marino de Vic (Barcelona). *Mem. Inst. Geol. Min.*, **68**, 330 pp.

Riba, O., Reguant, S., Villena, J. (1983). Ensayo de síntesis estratigráfica y evolutiva de la cuenca Terciaria del Ebro. *In Libro jubilar J.M. Ríos, Estudios sobre Geología de España*, ed. I.G.M.E. *Chap. III.3.8*: 131–159.

Santisteban, C. and Taberner, C. (1988). Geometry, structure and geodynamics of a sand wave complex in the Southeast margin of the Eocene Catalan basin, Spain. In *Tide-influenced sedimentary environments and facies*, ed. P.L. De Boer, A. van Gelder and S.D. Nio, pp. 123–138. D. Rerdel Publishing, Dordrecht, 530 pp.

Taberner, C. (1982–83). Evolución ambiental y diagenética de los depósitos del Terciario Inferior (Paleoceno y Eoceno) de la Cuenca de Vic. Thesis, U.B. 1400 pp.

Vail, P.R. (1987). Seismic stratigraphy interpretation procedure. In *Atlas of seismic stratigraphy*, ed. A.W. Bally, *Am. Assoc. Petrol. Geol., Stud. Geol.*, **27–1**: 277–281.

E8 Onshore Neogene record in NE Spain: Vallès–Penedès and El Camp half-grabens (NW Mediterranean)

L. CABRERA AND F. CALVET

Abstract

There are several Late Oligocene – Neogene onshore half-grabens in the so called Catalan continental margin (NE Spain) which are related to the NE–SW and NNE–SSW alignments of major basement faults. The extensional activity of these, and other similar faults in the offshore continental margin zones, dominates the structure of the Catalan continental margin. Extensional half-grabens evolved under the local influence of tectonism and of major paleogeographic, sea-level and paleoclimatic changes. These changes had a major influence on the Neogene sedimentary framework. Of these basins, two of them (the Vallès–Penedès and El Camp half-grabens) display a well-preserved Neogene sedimentary record. They were infilled mainly by Late Oligocene (?) and Early to Late Neogene deposits. The Miocene infill in these half-grabens resulted mainly from sedimentation on Early to Late Miocene alluvial fan systems which became fan deltas when they were affected by late Burdigalian, Langhian and Serravallian marine transgressions and highstand levels. Late Burdigalian to Serravallian terrigenous bay-shelves as well as Langhian coralgal carbonate platforms and Serravallian mixed shelves developed beyond the distal fan delta zones. The analysis of the structure and stratigraphy of these basins has contributed to detailed knowledge of the Miocene evolution of the Catalan continental margin.

Introduction

There are several Late Oligocene–Neogene onshore and offshore half-graben basins in NE Spain, which are bounded by NE–SW and NNE–SSW major extensional faults. These extensional faults, which extend along the so-called Catalan continental margin (NW border of the Valencia Trough), were the major features which determined the Neogene structure of the NE Iberian region and gave rise to a number of extensional basins which evolved under the local influence of tectonic structures and of major paleogeographic, eustatic and paleoclimatic changes. Among these basins, the Vallès–Penedès and El Camp onshore half-grabens display a well-preserved Neogene sedimentary record.

This chapter provides a general description of the Vallès–Penedès and El Camp Neogene infills. Most of these Early to Late Miocene and Pliocene–Pleistocene sedimentary infills consist of terrigenous facies which resulted from non-marine, transitional and marine sedimentation.

One interesting aspect of the onshore Neogene record in the Catalan margin is the possibility of detailed sedimentary analysis, which will contribute to better understanding of the evolution of this Valencia Trough area. The major influence of the changing tectonic regimes on the overall sedimentary framework in the half-grabens is well established, as well as the importance of local and regional controlling features on the sedimentary evolution. From this point of view, this is one of the rare case studies where the sedimentary record of the rifting and thermal subsidence tectonic stages can be studied in detail, and evaluated with paleoclimatic, tectonic and paleogeographic constraints.

Geological setting: The Catalan continental margin

The Catalan continental margin, at the NE Iberian Margin, makes up part of the northwestern edge of the Catalan–Balearic basin, also known as the Valencia Trough (Fig. 1). This basin is located in the Northwestern Mediterranean between the Iberian Peninsula and the Balearic Islands and was formed in a tectonic setting of continuous convergence between Africa and Europe (Pitman & Talwani, 1972; Biju Duval & Montadert, 1977; Rehault *et al.*, 1984; Dewey *et al.*, 1989; Roca, 1992; Roca & Dessegaulx, 1992; see also Chapter E1). While the SE Valencia Trough margins are formed by the NE end of the Betic thrust–fold belt, its NW margins are essentially extensional and correspond to the diverse parts of the Eastern Iberian continental margin (Fontboté *et al.*, 1990; Roca, 1992; see also Chapter E2).

The main Neogene structure of the Catalan continental margin is extensional (Figs. 1 and 2) and was the result of the Late Oligocene (?)–Early Miocene extensional reactivation of NE–SW and NNE–SSW earlier major basement faults (Fontboté, 1954; Anadón *et al.*, 1979, 1985; Guimerà, 1984, 1988; Bartrina *et al.*, 1992). These faults occur along the Catalan continental margin and led to the generation of major half-grabens, horst and tilted blocks, which are observed both in the onshore and offshore zones (Fig. 1). The

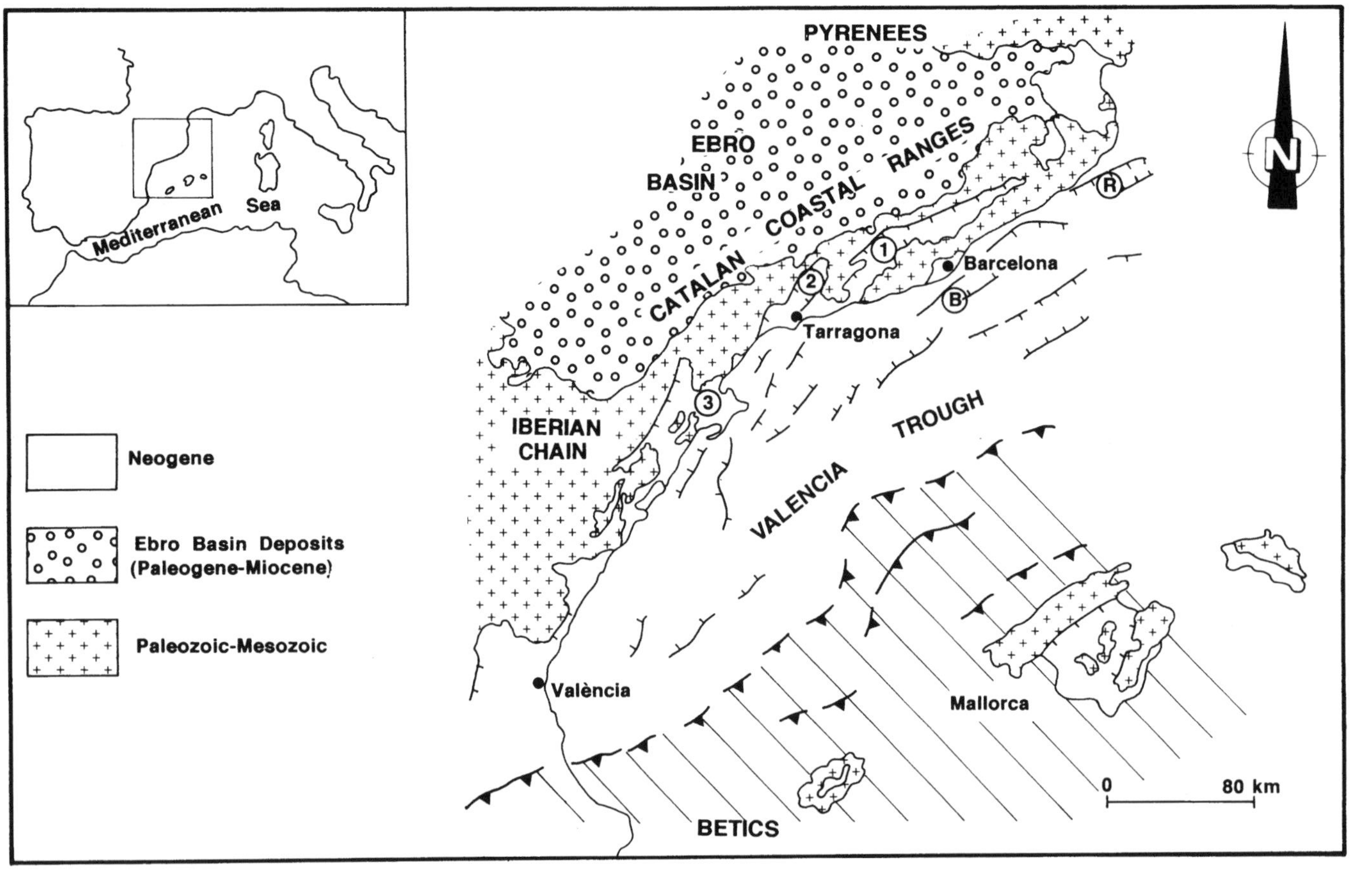

Fig. 1. General structural sketch of the Valencia Trough showing the setting of the onshore half-graben basins in NE Spain (adapted and modified from Fontboté *et al.*, 1990). The orientation of the main onshore half-grabens (1 to 3) corresponds well to that recorded for some of the major basement offshore faults. *Major onshore half-grabens*: 1, Vallès–Penedès; 2, El Camp; 3, Baix Ebre. *Offshore half-grabens*: R, Roses; B, Barcelona.

present coastline is nearly parallel to these major faults. Moreover, other minor NW–SE-oriented faults, transverse in relation to the major faults, also influenced the final margin structure (Fig. 2).

Two major structural evolutionary stages gave rise to the present Catalan margin structure (Figs. 1 and 2; Fontboté *et al.*, 1989; Roca & Guimerà, 1991; Bartrina *et al.*, 1992; Roca, 1992) and to the observed tectono-sedimentary evolution (Figs. 3 to 6; Cabrera and Calvet, 1990; Cabrera *et al.*, 1991; Bartrina *et al.*, 1992):

1. Rifting stage (Late Oligocene (?)–Early Miocene), which resulted in generation of the above-mentioned extensional structures (Figs. 1 and 2). The beginning of the non-marine sedimentary infilling of the half-grabens characterized this evolutionary stage (Figs. 3, 4 and 6).
2. Thermal subsidence stage (Late Burdigalian–Late Neogene) related to the cooling and crustal thinning resulting from the former stage. The final infilling of the early half-grabens and the overlapping of the surrounding horsts and tilted blocks characterized this stage (Figs. 3 to 6), together with a generalized subsidence of the Catalan continental margin. This subsidence was associated with the persistence of activity of some of the major faults which bounded the half-grabens and of other minor faults. A large extension of marine influence on the inner continental margin sedimentation took place during this stage (Figs. 4 to 6).

The onshore half-grabens

The onshore central zones of the Catalan continental margin include the Vallès–Penedès and El Camp half-grabens. Several major structural highs surround the half-grabens: Prades, Bonastre and the Garraf–Montnegre horst complex, which in its turn includes minor fault-bounded depressions (Vilanova, Baix Llobregat). The tectonic depressions show slightly different orientations (ENE–WSW and NE–SW respectively) because of the changes in orientation of the basement faults which led to their generation.

The Vallès–Penedès half-graben is up to 100 km long and 12 to 14 km wide and the Vallès–Penedès master fault, with a vertical slip larger than 3000 m, bounds its NW margin. Activity along this fault took place at least from Earliest Miocene to Late Miocene (Gallart, 1981; Amigó, 1986). Minor faults with up to a few hundred metres of slip make up the present southern boundary of the depression,

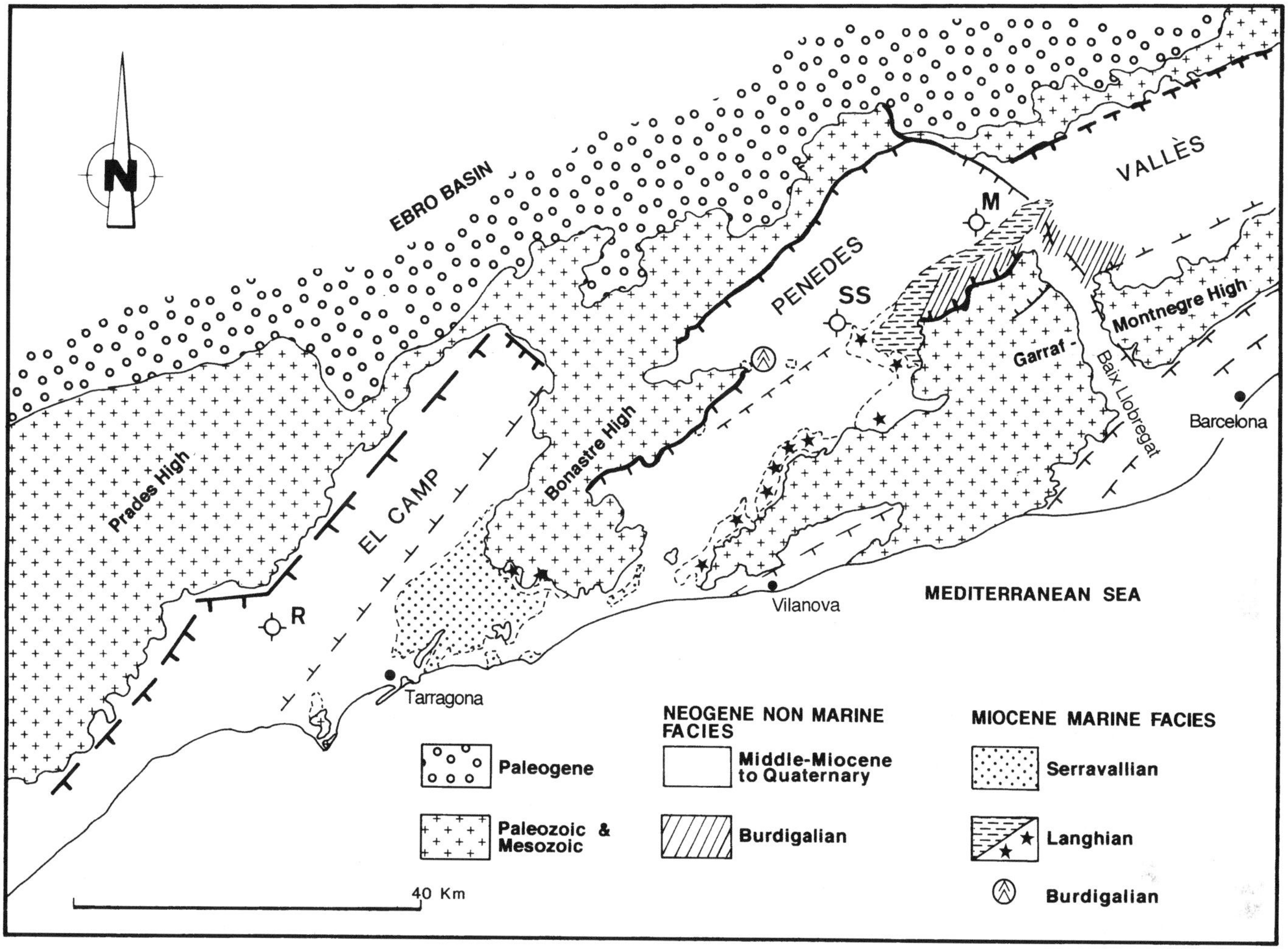

Fig. 2. Geological sketch showing the main structural features of the Late Oligocene (?) – Neogene Vallès–Penedès and El Camp half-grabens (see Fig. 1 for location) and the distribution of the main outcrops. These half-grabens are the best known in the Catalan margin. The basin basement and the surrounding source areas consist of Paleozoic rocks deformed during the Hercynian orogeny, and of Mesozoic and Paleogene rocks affected by the Alpine deformation. In both basins the basement dips gently towards the northwestern basin margin, attaining there the maximum basement depth (3000–1500 m) and basin infill thickness. This thickness decreases clearly towards the southern structural highs (Bonastre, Garraf–Montnegre), where clear southward Middle Miocene onlap and overlap occur. Apart from the major NNE–SSW and NE–SW extensional faults there are other transverse faults which also contributed to the structure of the margin. *Oil exploration wells*: M, Martorell I; SS, Sant Sadurní; R, Reus. Dotted and dashed patterns in the key of the Miocene marine facies correspond to the Serravallian mixed carbonate–terrigenous and the Langhian terrigenous bay-shelf deposits, respectively. Stars correspond to carbonate coralgal platform facies, while encircled angles show the location of the Late Burdigalian coastal sebkha evaporites.

although in most of the southern structural highs of the Garraf–Montnegre horst, these faults were overlapped by latest Burdigalian–Langhian sediments.

The El Camp half-graben is up to 60 km long and 10 to 14 km wide. The El Camp master fault bounds its NW margin with a vertical slip which took place at least from Early Miocene to Quaternary and was larger than 1500 m (Gallart, 1981; Amigó 1986). In contrast to the Vallès–Penedès fault, the El Camp fault is only locally concealed by Pliocene and Quaternary deposits, which attain a noticeable thickness at the foot of the fault. Another NE–SW major fault occurs to the SE of the half-graben. This fault bounded the NW margin of the Bonastre horst and was overlapped by Langhian–Serravallian sediments.

Sedimentation in both tectonic depressions was mainly terrigenous and developed on alluvial fan systems (Figs. 3 to 5). Development of shallow palustrine, marsh-swamp and lacustrine minor systems was sometimes enhanced by the existing geomorphic conditions (Figs. 3 and 4). The alluvial fans evolved into fan deltas when they were affected by Langhian and Lower Serravallian (Figs. 3 and 5) marine transgressions and highstand levels. Moreover, beyond the distal fan delta zones, the marine episodes (Figs. 3 to 5) resulted in a varied array of environments linked to coastal sebkhas (Late Burdigalian), terrigenous bay-shelves (Langhian–Lower Serravallian), coralgal carbonate platforms (Langhian) and mixed terrigenous carbonate shelf systems (Lower Serravallian).

The chronostratigraphy of the outcropping Neogene successions

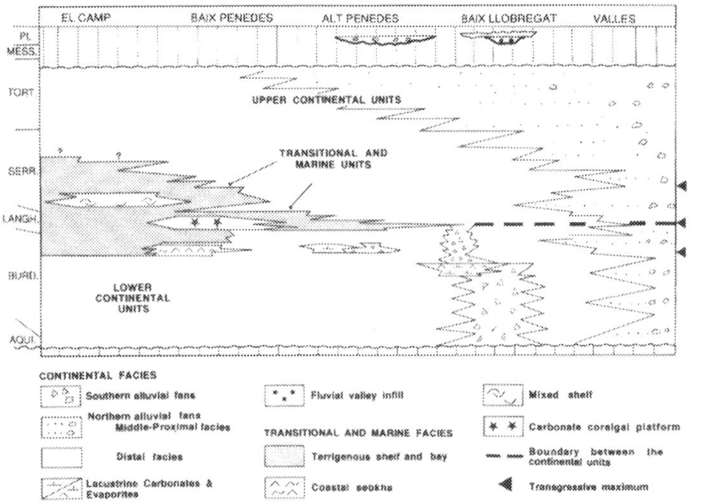

Fig. 3. **Generalized stratigraphic framework of the Neogene onshore sedimentary record in the Central Catalan margin. The relationships between the major continental, transitional and marine facies assemblages show their successive progradations and retreats and record the major structural and paleogeographic events which affected the continental margin. *Basal unconformity (Late Oligocene–Early Miocene)*: An often karstified pre-rift unconformity, associated with a variety of weathering products (scree deposits, paleosoils), underlies the lowermost Neogene basin infill. *Lower continental units. (Early Miocene–Earliest Langhian). Alluvial fan*: Intensely coloured red conglomerates and sandy conglomerates, breccias and muddy sandstones. Northern alluvial fans were areally spread while southern ones were rather more restricted. *Playa-lake* limestones, nodular anhydrite bearing dolostones, and clays and *shallow carbonate lake* limestones, mudstones and brown coal also occur (Cabrera, 1981 a and b; Cabrera *et al.*, 1991). *Transitional and marine units (Late Burdigalian–Early Serravallian). Late Burdigalian evaporitic coastal sebkha* laminated and nodular gypsums (Ortí & Pueyo, 1976) and shallow transitional to bay-shelf mudstones and sandstones. *Langhian coralgal platform.* These carbonate marine sequences include coral–algal reefs which display a discontinuous fringing arrangement and are related to structural highs. Also patch reefs interfingering with terrigenous transitional facies occur (Permanyer, 1982; Calvet *et al.*, in press). *Serravallian mixed carbonate–siliciclastic shelf deposits* displaying a coarsening and shallowing up sequence range from basinal marly sediments and calcisiltites to calcarenite–sandstones and coquina facies. The lower Serravallian successions do not include coral reef facies (Cabrera *et al.*, 1991). *Upper continental units (Langhian–Tortonian). Alluvial fan.* The alluvial deposits included in these units display facies assemblages and sequential arrangements similar to those observed in the lower continental units. Nevertheless, yellowish and reddish brown colour shades are dominant in their fine grained facies. Moreover, the basin depositional framework was noticeably different from that observed in the Early Miocene. The middle and distal parts of these alluvial systems overlie and interfinger with the marine and transitional units (Cabrera *et al.*, 1991). *Messinian erosive surface. Pliocene Units.* This deeply entrenched erosive surface affected both the Pre-Neogene basement and the Neogene sequences. It is overlain by thin alluvial fan conglomerates in the Penedès and El Camp (Gallart, 1981), while scree deposits, fluvial valley infill conglomerates and transitional-bay estuarine, sandy and muddy dominated sequences developed in the Baix Llobregat area (Martinell, 1988).**

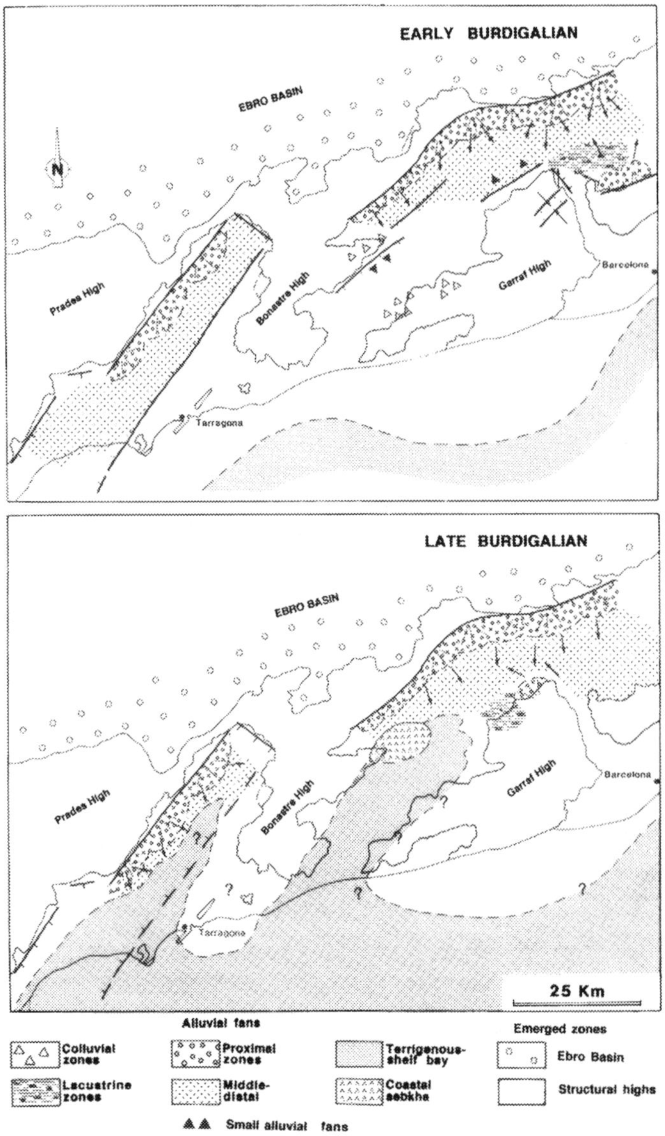

Fig. 4. *Early Miocene paleogeographic evolution of the onshore zones of the central Catalan continental margin. Early Burdigalian.* The Late Oligocene (?)–Early Miocene sedimentary record in the onshore Vallès–Penedès and El Camp half-grabens was largely developed under conditions of complete isolation from the sea. Areally widespread and restricted alluvial fan–lacustrine systems developed, first in narrow subsiding fault bounded troughs and later in a wider setting. Minor shallow carbonate lacustrine and playa-lake deposits developed between the terminal and marginal alluvial fan zones (Cabrera, 1981 a and b). *Late Burdigalian.* Evaporitic coastal sebkha deposits (Ortí and Pueyo, 1976) and sandy and mudstone dominated marine and transitional facies recorded in exploration wells (Sant Sadurní I, Bartrina *et al.*, 1992; Reus I) show the beginning of the marine influences in the half-graben zones. Thick red bed sequences continued being deposited on the alluvial fan systems. Minor shallow carbonate lacustrine and swamp deposits developed distal to the terminal and marginal alluvial fan zones in the Penedès area (Cabrera, 1981 a and b).

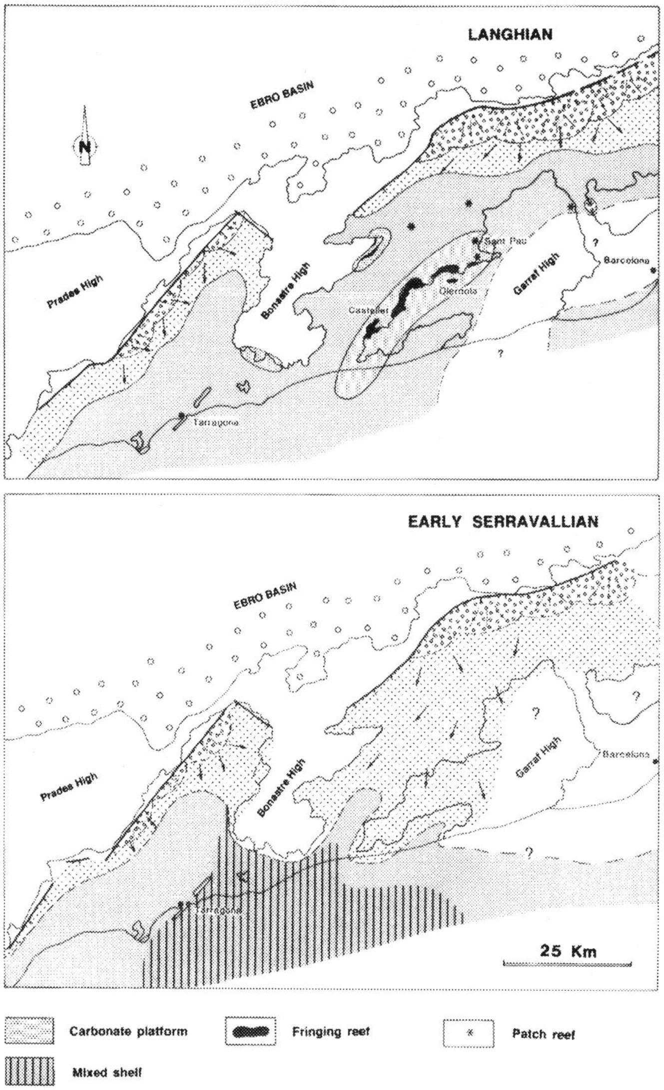

Fig. 5. ***Middle Miocene paleogeographic evolution of the onshore zones of the central Catalan continental margin. Langhian.*** **Langhian successions in the Vallès–Penedès record the development of a variety of alluvial fan systems which developed along the northern borders of the major depressions. In the central half-graben zones, the middle and distal alluvial fan environments graded into fan delta and terrigenous bay zones. These passed laterally in their turn into marine shelf zones. Some of the main locality-sections of the carbonate platform and patch reef assemblages are shown in the sketch.** ***Early–Middle Serravallian.*** **The continental transitional and marine facies assemblages of this age show an overall depositional pattern similar to that observed during Langhian. Nevertheless, no coralgal carbonate platforms developed during this depositional stage. The observed mixed siliciclastic–carbonate shelf deposits include a typical temperate carbonate biota. Later, during the Late Serravallian–Tortonian time span (not represented), alluvial fan environments developed over all the depressions and also over part of the structural highs. Their uppermost Tortonian deposits constitute the youngest pre-Messinian sedimentary record in the onshore zones.**

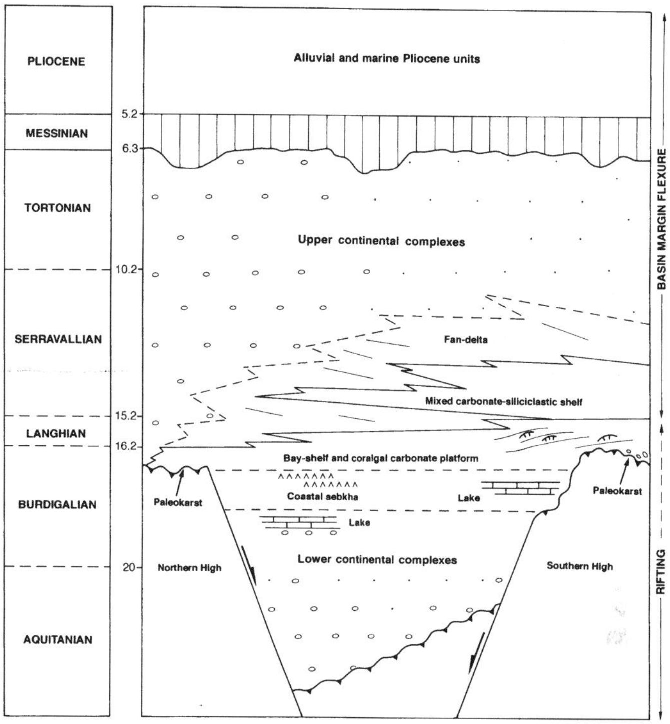

Fig. 6. Conceptual sketch showing the most characteristic elements of the Neogene sedimentary record in the onshore Central Catalan margin.

of both depressions has been well established in the continental successions thanks to fossil mammal assemblages (Crusafont *et al.*, 1955; Agustí *et al.*, 1985) dating from Late Ramblian to Early Turolian (i.e. Burdigalian to Tortonian) and Ruscinian (i.e. Pliocene). The marine successions have been attributed to the Late Burdigalian, Langhian and Lower Serravallian using the calcareous nannoplankton and the planktonic microforaminifera assemblages (Magne, 1979; Salvatorini and Mazzei, in Permanyer, 1982; Porta and Civis, 1990; MacPherson, 1992).

No paleontological data are available to date the earliest subsurface deposits in both half-grabens, which consist essentially of continental deposits. As a consequence an Earliest Miocene (Aquitanian–Early Burdigalian) age for the early deposits in both half-grabens is the most likely although the existence of Late Oligocene deposits cannot be discounted.

Sedimentary infill in the half-graben basins

The Late Oligocene to recent sedimentary evolution of the Catalan continental margin has been taking place on the northwestern border of the Valencia Trough. The major features of the offshore sedimentary record of this border have been established previously (Stoeckinger, 1976; García Siñériz *et al.*, 1978; Soler *et al.*, 1983; Johns *et al.*, 1989; Clavell and Berastegui, 1991; Martínez del Olmo *et al.*, 1991; see also Chapter E3). But the offshore half-grabens which can be recognized along the Catalan margin (i.e. Barcelona and Roses, Fig. 1) differ from the onshore Vallès–Penedès and El Camp tectonic depressions, these ones being located further from the trough centre. Thus, they were influenced differently by the general tectonic and paleogeographic changes which affected the Valencia trough.

Paleoclimatic constraints

Paleoclimate in the Western Mediterranean regions changed during the Neogene from a warm tropical–subtropical regime, with dry and rainy seasons, to a mediterranean regime. The latter regime was finally established during the Pliocene (Suc *et al.*, 1992) and is characterized by a larger seasonal temperature contrast, with a dry summer season.

Palynological assemblages (Bessedik & Cabrera, 1985), fossil plant leaf assemblages (Sanz de Siria, 1981) and crocodile remains (Crusafont *et al.*, 1955; Cabrera, 1979) found in Early Burdigalian to Langhian sections suggest that Late Burdigalian–Langhian paleoclimatic conditions were warm, seasonal and arid tropical–subtropical. This is confirmed by the widespread occurrence of coral reef complexes (linked to the occurrence of an impoverished mangrove assemblage; Bessedik & Cabrera, 1985), as well as by isotopic paleotemperatures, ranging from 15.5 to 24.6 °C (Vázquez *et al.*, 1991). However, the marine paleobiota recorded in the later Lower Serravallian carbonate–siliciclastic shelf suggest that these deposits were already being deposited under more temperate conditions.

A trend of alternating warm versus more temperate and humid versus drier conditions, developed during the Late Serravallian–Tortonian and until the Pliocene. The varied paleomastological fossil assemblages suggest several intra-Miocene changes in the vegetation cover from at least partially forested to more savanna-like landscapes (Agustí *et al.*, 1984).

Tectonic and paleogeographic constraints

The relationships between the faults and the associated basin infills point to the syntectonic development of the latter (Cabrera, 1979, 1981a, b; Cabrera *et al.*, 1991; Bartrina *et al.*, 1992). Moreover, the evolution of the tectonic depressions records the general evolution of the continental margin. During the rifting stages (Late Oligocene (?)–Early Burdigalian), the sedimentary accumulation was closely controlled by major fault activity along the northwestern basin margins as well as by the activity of other less-important faults along the southern margins and across the half-grabens. From the beginning of the thermal subsidence stage (Late Burdigalian) the faults which bounded the southern half-graben margins became inactive and the southern structural highs were gradually onlapped and overlapped by the sedimentary infill. The depositional framework in the half-grabens became more assymmetrical and the Early Miocene facies distribution differed noticeably from those observed during the Middle and Late Miocene.

Both structural and sea-level evolution in the NW Mediterranean, after the early stages (Late Oligocene (?)–Early Burdigalian) of isolation from marine influences, caused the half-grabens located in the inner Catalan continental margin zones to be finally connected to the open sea. From the Late Burdigalian until Serravallian, marine environments were diversely widespread in both tectonic depressions. The encroaching of the marine influence on the inner zones of the continental Catalan margin was the culmination of a general process of spreading marine influence which affected most of the NW Mediterranean continental margins from Earliest Miocene to Langhian. The Late Burdigalian, Langhian and Serravallian marine transgressive episodes were coincident with the above-mentioned generalized ending of fault activity along the southeastern half-graben margins and resulted in a noticeable diversity of depositional systems.

From Early to Late Serravallian the extent of the marine influenced areas during the highstand episodes decreased. From Late Serravallian until Tortonian, sedimentation was predominantly non-marine in both basins, with widespread dominance of alluvial fan environments.

Since sea-level changes which affected the Mediterranean regions influenced the half-graben sedimentary record from Late Burdigalian to present, erosive surfaces linked to Miocene sea-level fall might be potentially developed in the basin infills. However, the only well-preserved major erosive surface in the studied onshore sedimentary record is the Messinian erosive surfaces (Hsü *et al.*, 1978; Ryan & Cita, 1978). The alluvial–fluvial (Gallart, 1981) and/or marine (Martinell, 1988) Lower Pliocene units recorded in the tectonic depressions of the central Catalan continental margin (El Camp, Penedès and Baix Llobregat;) overlie a deeply entrenched erosive surface which affected both basement rocks and the earlier Neogene sequences. Pleistocene erosive surfaces related to glacio-eustatic changes are superimposed on these features.

Main features of the sedimentary system and basin infill organization

From their earlier evolutionary stages (Late Oligocene ?–Aquitanian) until the Late Miocene (Tortonian) both half-grabens were fed largely from their NW margins by medium- to large-sized alluvial fans (up to 12 km in radial extent), which displayed a variety of paleoenvironments. As a consequence, a varied array of alluvial–fluvial paleoenvironments developed in both half-grabens. Both radial and longitudinal drainages are suggested for these major alluvial systems by facies distribution and paleocurrents. Up to 1500 or 3000 m of alluvial fan deposits were deposited at the foot of the active faults which bounded the half-grabens (Cabrera, 1981a, b; Cabrera *et al.*, 1991).

Moreover, the half-grabens were also fed during their early evolutionary stages by local, small alluvial fan systems fed from the evolving structural highs located to the south of the main sedimentary troughs (Cabrera, 1981a, b; Cabrera *et al.*, 1991). These alluvial deposits (up to 200 m thick) do not display a well-developed facies differentiation due to the short fan radial extent (less than 1 km). Thick, matrix- and clast-supported conglomerates and breccias, mudstones and sandstones which show poorly developed sequential trends are the most widespread facies in these alluvial systems.

In the Vallès–Penedès these southern alluvial fans were active until Late Burdigalian (Fig. 3).

Distinctive lacustrine deposits, making up single sequences up to 5 m thick, have only been recorded in the early evolutionary stages of the basins (i.e. Vallès–Penedès, see Figs. 3 and 4). These lakes (evaporitic playa-lakes, carbonate lakes and swamps) were shallow and in some cases probably developed under internal drainage conditions (Cabrera, 1981a, b; Cabrera *et al.*, 1991). Marsh–swamp paleoenvironments developed less frequently. Limestones, dolostones, minor anhydrite beds and minor coal deposits formed in these lakes (Figs. 3, 4 and 6).

In those half-graben zones where non-marine sedimentation was exclusive, a succession of progradational and retrogradational alluvial sequences was deposited and records the successive progradations and retreats of proximal and middle alluvial system zones. In the zones where lacustrine deposits developed, they alternate with terminal alluvial fan mudstones, recording frequent alluvial progradations and retreats. The higher rank megasequences recorded in the alluvial fan systems were probably related to base-level changes linked to changing subsidence–sedimentation rates. Changes of the major fault activity could cause these variations, although paleoclimatic changes also took place in the time interval.

During the Late Burdigalian, the marine influence reached both basins. Nevertheless, the depositional gradient was not large enough in this earlier stage to give rise to a large environmental diversification of the depositional framework. Terrigenous shallow bay successions (up to 100 m thick) and evaporitic coastal sebkha sequences up to 60 m thick (Ortí & Pueyo, 1976) were the most characteristic elements developed during this stage. No obvious major sequences have been recognized in these successions.

In the Langhian and Lower Serravallian, marine influence was persistent and the resulting depositional gradient large enough to result in rather more varied depositional conditions. After initial episodes of transgression and sea level rise, thick fan delta successions (up to 400 m thick) developed (Cabrera *et al.*, 1991). These sequences display several punctuated progradational and retrogradational stages, not necessarily isochronous in all the sectors.

Langhian coralgal carbonate platform successions (up to 60 m thick) developed extensively on the northern edge of the Garraf high. After early evolutionary stages which took place under transgressive and rising sea-level conditions, the carbonate platforms developed a characteristic progradational pattern with sigmoidal basinward clinonoforms (Permanyer, 1982; Calvet *et al.*, in press).

The Serravallian mixed terrigenous–carbonate shelf successions (up to 400 m thick) have been well recorded in the El Camp half-graben where they display a shallowing and coarsening-upward sequential arrangement (Cabrera *et al.*, 1991).

The interrelationship and interfingering between the fan delta–bay systems, the Langhian carbonate platform and the Serravallian mixed shelf are recorded in both the Vallès–Penedès and the El Camp depressions. This relationship terminated with large fan delta progradations, which ended the marine influence on the half-grabens.

Concluding remarks

The onshore half-grabens described here and their associated sedimentary deposits record the Cenozoic (Oligocene–Miocene) stages of the formation of the Catalan continental margin (NW Valencia Trough border).

Examination of the relationship between tectonics and sedimentation in the Vallès–Penedès and El Camp depressions shows the areally extending evolution of the sedimentary areas from an early setting of narrow and elongated sedimentary troughs. This process resulted from waning of the fault activity and was linked to the overlapping of the southern structural highs.

Major alluvial fan systems were highly developed in these basins and dominated sedimentation. Lacustrine systems were minor features in the depositional framework while fan delta, bay, carbonate platform and mixed shelf systems were important during the marine-influenced episodes (Late Burdigalian, Langhian and Late Serravallian). These features are related to the evolution of tectonic subsidence in the half-grabens and to the influence of sea-level changes.

In studies of onshore grabens there are many similarities in the influence on the sedimentary record of both individual structures (for example faults and structural highs), and of sequence evolution. The beginning of the non-marine fault-controlled sedimentation, the overlapping of marginal structural highs and the spreading and retreat of the marine environments in the half-grabens took place during relatively specific time intervals. However, the existence of reliable and relatively refined paleontological data (planktonic foraminifera biozones) demonstrates, in some cases, a noticeable heterochrony of the deposits involved in these processes, and indicates a lack of direct correlation between the different sequence trends.

The basin-infill sequential-analysis approach is potentially applicable to the study of the half-grabens, but a more refined dating is needed to establish definitive conclusions and sequential analysis criteria cannot be used for correlation.

Acknowledgements

The authors thank the scientific editors of this book, Cristino J. Dabrio and Peter F. Friend for their useful suggestions for improving and clarifying the text. This paper has been financed by the CICYT Project. GEO 89–0381 of the Spanish Ministry of Education and Science.

References

Agustí, J., Moya Solà, S. and Gibert, J. (1984). Mammal distribution dynamics in the Eastern Margin of the Iberian Peninsula during the Miocene. *Paléobiologie Continentale, Montpellier*, **XIV** (2): 33–46.

Agustí, J., Cabrera, Ll. and Moya, S. (1985). Sinopsis estratigráfica del Neógeno de la fosa del Vallès-Penedès. *Paleontol. Evol.*, **18**: 57–81.

Amigó, J. (1986). Estructura del massís del Gaià. Relacions estructurals amb les fosses del Penedès i del Camp de Tarragona, PhD Thesis, University of Barcelona, 253 pp.

Anadón, P., Colombo, F., Esteban, M., Marzo, M., Robles, S., Santanach, P.F. and Solé Sugranyes, Ll. (1979–1981). Evolución tectono-estratigráfica de los Catalánides. *Acta Geol. Hisp.*, **14**: 242–270.

Anadón, P., Cabrera, Ll., Guimerà, J. and Santanach, P.F. (1985). Paleogene strike–slip deformation and sedimentation along the south-eastern margin of the Ebro Basin. In K. Biddle and N. Christie-Blick (eds.), *Strike–slip deformation, basin formation and sedimentation*. Special Publication of the *Soc. Econ. Paleont. Mineral.*, **37**: 303–318.

Bartrina, M.T., Cabrera, Ll., Jurado, M.J., Guimerà, J. and Roca, E. (1992). Evolution of the central Catalan margin of the Valancia Trough. *Tectonophysics*, **203** (1–4), 219–247.

Bessedik, M. and Cabrera, Ll. (1985). Le couple récif-mangrove à Sant Pau d'Ordal (Vallès–Penedès), témoin du maximum transgressive en Méditerranée nord occidentale (Burdigalien supérieur–Langhien inferieur). *Newsl. Stratigr.*, **14**: 20–35.

Biju Duval, B. and Montadert, L. (1977). Introduction to the structural history of the Mediterranean basins. In L. Biju Duval and L. Montadert (eds.), *Structural history of Mediterranean basins*. Technip, Paris, pp. 1–12.

Bouvier, J.D., Gevers, E.C.A., Wigley, P.L. and Omann, P.D. (1990). 3-D Seismic interpretation and lateral prediction of the Amposta Marino Field (Spanish Mediterranean Sea). *Geol. Mijnbouw*, **69**: 105–120.

Cabrera, Ll. (1979). Estudio estratigráfico y sedimentológico de los depósitos continentales basales de la depresión del Vallès–Penedès. Masters Degree. University of Barcelona, 361 pp.

Cabrera, Ll. (1981a). Estratigrafía y características sedimentológicas generales de las formaciones continentales del Mioceno inferior de la cuenca del Vallès-Penedès (Barcelona, España). *Estud. Geol.*, **37**: 35–43.

Cabrera, Ll. (1981b). Influencia de la tectónica en la sedimentación continental de la cuenca del Vallès–Penedès (provincia de Barcelona, España) durante el Mioceno inferior. *Acta Geol. Hisp.*, **16**: 163–169.

Cabrera, Ll. and Calvet, F. (1990). Sequential arrangement of the Neogene sedimentary record in the Vallès–Penedès and Valls–Reus half grabens, Iberian margins, NE Spain. In *The Valencia Trough: geology and geophysics. Terra Abstracts*, **2** (2): 110.

Cabrera, L., Calvet, F., Guimerà, J. and Permanyer, A. (1991). El registro sedimentario miocénico en los semigrabens del Vallès–Penedès y de El Camp: organización secuencial y relaciones tectónica sedimentación. In F. Colombo (ed.) Field guide book no. 4 of the I Congreso del Grupo Español del Terciario, Vic, 132 pp.

Calvet, F., Esteban, M. and Permanyer, A. (in press). Langhian reefs of the Penedès–Vallès and El Camp Depressions, NE Spain. In C. Jordan, M. Colgan and M. Esteban (eds.), *Miocene reefs: a global comparison*. Springer Verlag.

Clavell, E. and Berástegui, X. (1991). Petroleum Geology of the Gulf of Valencia. In A.M. Spencer (ed.), *Generation, accumulation and production of Europe's hydrocarbons. Special Publ. Petrol. Geosci.*, **1**, 355–368. Oxford University Press, Oxford.

Crusafont, M., Villalta, J.F. de and Truyols, J. (1955). El Burdigaliense continental de la cuenca del Vallès–Penedès. *Mem. Com. Inst. Geol. Dip. Prov. Barcelona*, **12**: 260 pp.

Dewey, J.F., Helman, M.L., Turco, E., Hutton, D.H.W. and Knott, S.D. (1989). Kinematics of the Western Mediterranean. In M.P. Coward, D. Dietrich and R.G. Park (eds.), *Alpine tectonics. Geol. Soc. Lond. Spec. Publ.*, **45**: 265–283.

Fontboté, J.M. (1954). Las relaciones tectónicas de la depresión del Vallès–Penedès con la Cordillera Prelitoral Catalana y con la Depresión del Ebro. In *Tomo homenaje Prof. E. Hernández Pacheco. R. Soc. Esp. Hist. Nat.*, Madrid, 281–310.

Fontboté, J.M., Guimerà, J., Roca, E., Sàbat, F., Santanach, P. and Fernández Ortigosa, F. (1990). The Cenozoic geodynamic evolution of the Valencia Trough (Western Mediterranean). *Rev. Soc. Geol. Esp.*, **3** (2): 7–18.

Gallart, F. (1981). Neógeno superior y Cuaternario del Penedès (Catalunya, España). *Acta Geol. Hisp.*, **16**: 151–157.

García Siñériz, B., Querol, R., Castillo, F. and Fernández, J.R. (1978). A new hydrocarbon province in the Western Mediterranean. *10th World Petroleum Congress Bucharest, Proc.* **4**: 1–4. Bucharest.

Guimerà, J. (1984). Palaeogene evolution of deformation in the north-eastern Iberian Peninsula. *Geol. Mag.*, **121** (5): 413–420.

Guimerà, J. (1988). Estudi estructural de l'enllaç entre la Serralada Ibèrica i la Serralada Costanera Catalana. PhD Thesis, University of Barcelona, 600 pp.

Hsü, K.J., Montadert, L.C. and Bernouill, D. (eds.) (1978). *Initial reports of the deep sea drilling project*, **42**, 1.

Johns, D.R., Herber, M.A. and Schwander, M.M. (1989). Depositional sequences in the Castellón area, offshore northeast Spain. In A.W. Bally (ed.), *Atlas of seismic stratigraphy*, vol. 3. *AAPG Studies in Geology*, **27**: 181–184.

Macpherson, I. (1992). Paleoecología de los foraminíferos en el Mioceno Medio de la Cuenca del Penedès. Tesis doctoral Universidad de Barcelona. Deptos. de Ecología y Geología Dinámica, Geofísica y Paleontología. 434 pp.

Magné, J. (1979). *Etudes microstratigraphiques sur le Néogène de la Mediterranée nordoccidentale*. Vol. I. *Les basins Néogènes Catalans*. Univ. Paul Sabatier. Toulouse, 260 pp.

Martinell, J. (1988). An overview of the marine Pliocene of NE Spain. *Géol. Méditerr.*, **15** (4): 227–233.

Martínez del Olmo, W. and Esteban, M. (1983). Paleokarst Development (Western Mediterranean). In P.A. Scholle, D.G. Bebout and C.H. Moore (eds.), *Carbonate Depositional Environments. Am. Assoc. Petrol. Geol. Mem.*, **33**: 93–95.

Martínez del Olmo, W., Murillas, J. and Fernández, F. (1991). Los ciclos eustáticos-sedimentarios del Neógeno en el Golfo de Valencia (Mediterráneo occidental). *Abstracts of the I Congreso del Grupo Español del Terciario, Vic*, pp. 206–209.

Ortí, F. and Pueyo, J.J. (1976). Yeso primario y secundario del depósito de Vilobí (provincia de Barcelona, España. *Instituto de Investigaciones Geológicas. Universidad de Barcelona*, **31**: 5–34.

Permanyer, A. (1982). Sedimentologia i diagènesi dels esculls miocènics de la conca del Penedès. PhD Thesis. University of Barcelona. 545 pp.

Pitman, W.C. and Talwani, M. (1972). Sea-floor spreading in the North Atlantic. *Bull. Geol. Soc. Am.*, **82**: 619–646.

Porta, J. de and Civis, J. (1990). Events and correlation in the Neogene of prelitoral Catalonian depression. In *The Valencia Trough: geology and geophysics. Terra Abstracts*, **2** (2): 116–117.

Rehault, J.P., Boillot, G. and Mauffret, A. (1984). The western Mediterranean basin geological evolution. *Mar. Geol.*, **55**: 447–477.

Roca, E. and Desegaulx, P. (1991). Analysis of the geological evolution and vertical movement analysis in the Valencia Trough area, western Mediterranean. *Mar. Petrol. Geol.*, **9**: 167–184.

Roca, E. and Guimerà, J. (1991). The Neogene structure of the eastern Iberian margin: structural constraints on the crustal evolution of the Valencia Trough (western Mediterranean). *Tectonophysics*, **203**: 203–218.

Ryan, W. and Cita, M.B. (1978). The nature and distribution of Messinian erosional surfaces. Indicators of a several kilometer deep Mediterranean in the Miocene. *Mar. Geol.*, **27**: 193–230.

Sanz de Siria, A. (1981). La flora Burdigaliense de los alrededores de Martorell (Barcelona). *Paleontol. Evol., Sabadell*, **16**: 3–13.

Soler, R., Martínez del Olmo, W., Megías, A.G. and Abeger, J.A. (1983). Rasgos básicos del Neógeno del Mediterráneo Español. *Mediterranea*, **1**: 71–82.

Stoeckinger, W. (1976). Valencia Gulf offer deadline nears. *Oil Gas J.*, **29**: 197–204.

Suc, J.P., Clauzon, G., Bessedik, M., Leroy, S., Zheng, Z., Drivaliari, A., Roiron, P., Ambert, P., Martinell, J., Domenech, R., Matias, I., Julià and Anglada, R. (1992). Neogene and Lower Pleistocene in Southern France and Northeastern Spain. Mediterranean environments and climate. *Cah. Micropaleontol.*, **7** (1–2): 165–187.

Vázquez, A., Zamarreño, de Porta, J. and Plana, F. (1991). La composición isotópica y los elementos traza de *Amussiopecten baranensis* (Pectinidae) como indicadores paleoambientales, en el Langhiense catalán. *Rev. Soc. Geol. Esp.*, **4** (3–4): 215–227.

E9 The Paleogene basin of the Eastern Pyrenees

J.M. COSTA, E. MAESTRO-MAIDEU AND CH. BETZLER

Abstract

We have distinguished 10 depositional sequences in the Eocene succession of the eastern South Pyrenean Basin (Ilerdian to middle Lutetian). The lowermost sequences (Puig Aguiló and Fòrnols sequences) consist of shelfal carbonate facies (Alveolina limestones) in the eastern and western areas of the basin. In the central area, they consist of offshore marls. The Sagnari and Corones sequences are characterized by fluvio-deltaic deposits and shelfal carbonates. The Cadí, Armàncies and Vall del Bac sequences consist of resedimented carbonates and offshore marls in the central parts of the Basin, and of carbonate platforms (Assilina limestones) in the marginal areas. The Vall del Bac sequence exposes a deltaic platform in the eastern area. The uppermost sequences (Campdevànol, Ripoll and Josa sequences) consist of deltaic platforms in the eastern area and of turbiditic systems in the western area.

The boundary between the Vall del Bac and Campdevànol sequences is related to an important paleogeographic change in the basin. The depositional systems of the sequences below this level display the proximal environments in the two terminations of the Basin, and the distal environments in the central areas. The depositional systems of the sequences above this level display the proximal environments in the eastern areas of the Basin, and the distal ones in the central and western areas.

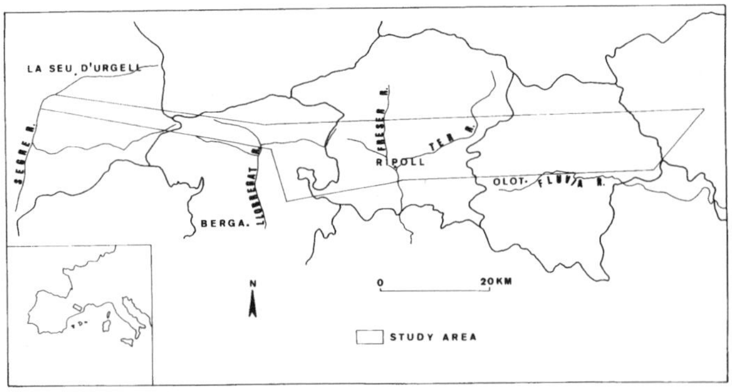

Fig. 1. Geographical setting of the study area.

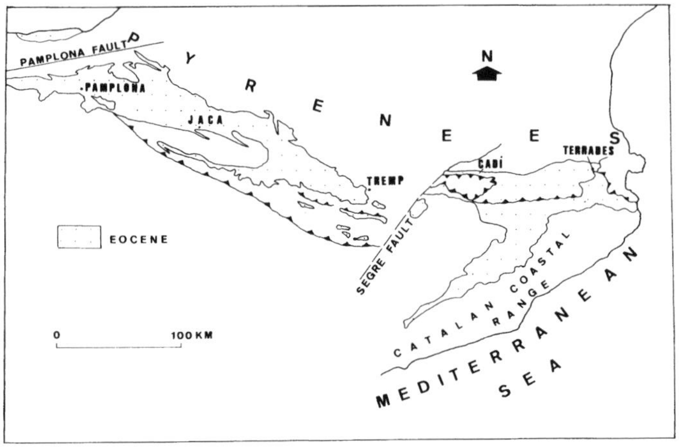

Fig. 2. Eocene sedimentary basins in the Pyrenees and in the Ebro Depression (*sensu* Rosell, 1988).

Geographical and geological setting

The area studied corresponds to a narrow band of the Eastern Pyrenees, oriented from E to W, in the NE of Catalonia (Fig. 1). It is bounded by the Segre river to the W and by the Mediterranean Sea to the E.

During the Paleocene and the Eocene, the Prepyrenees and the Ebro Depression formed part of the same sedimentary basin. The basin was opened to the W and connected to the Atlantic. To the N, the basin was bounded by the north-Pyrenean thrusts, which affected Paleozoic and Mesozoic rocks. They consist of thrust sheets that formed successively in time (Paleocene to Oligocene; Puigdefábregas and Souquet, 1986), and progressively displaced the foreland basin southwards.

At the beginning of the Ilerdian, the Pyrenean Basin was subdivided into three sub-basins by the Segre and the Pamplona tectonic structures (Rosell, 1988): an Eastern Basin (Terrades–Cadí Basin), a Central Basin (Tremp–Jaca–Pamplona Basin), and a Western Basin (Basque Basin) (Fig. 2).

In this study we describe the deposits which filled the Eastern Basin (Fig. 3) from the Ilerdian to the middle Lutetian. The

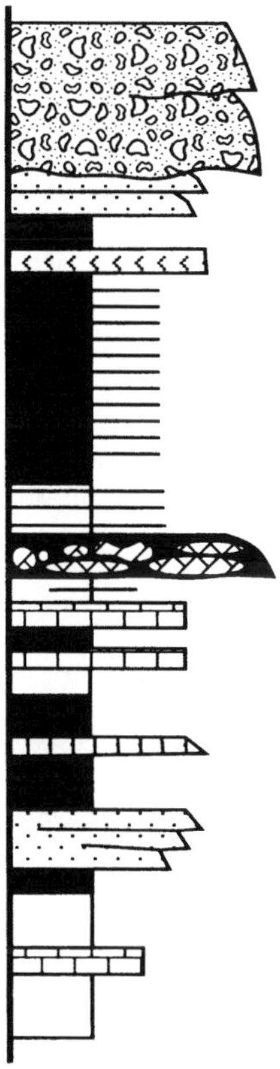

GICH (1969) PALLI (1972)	ESTEVEZ (1973)	BUSQUETS (1981)	PUIGDEFABREGAS et al (1986)	COSTA (1985) MAESTRO (1985)	BETZLER (1989)	THIS WORK
Bellmunt Fm	Tramo rojo intermedio	Bellmunt Fm	Bellmunt Sq			Uppermost Sqs
Vallfogona Fm	Vallfogona Fm	Cal Bernat Fm	Beuda Sq	Sistema		Josa Sq
		Guixos de Campdevànol				
Campdevànol Fm		Vallfogona Fm	Campdevànol Sq	Josa Bagà Gombrèn	Josa Sq	Ripoll Sq
						Campdevànol Sq
Fm Armancies / Terrades	Armancies Fm	Fm Armancies / La Penya	Armancies Sq	Armancies Sq	Oden B Sq	Vall del Bac Sq
						Armancies Sq
Corones Fm	Corones Fm	Corones Fm	Corones Sq	Cadí Sq		Cadí Sq
				Corones Sq	Oden A Sq	Corones Sq
				Sagnari Sq	Alinyà Sq	Sagnari Sq
Sagnari Fm	Sagnari Fm	Sagnari Fm	Cadí Sq	Ager 2 Sq	Llimiana Sq	Puig Aguiló Sq
				Ager 1 Sq	Ager Sq	Fòrnols Sq

Fig. 3. Table of the most usual lithostratigraphic and unconformity-bounded units in the Eocene deposits of the Eastern South Pyrenees.

sedimentation pattern of this basin agrees with that described by Rosell (1988). Rosell considers the establishment of the sedimentation of an extensive carbonate ramp during the Ilerdian, in the Pyrenean and the Ebro Basins. Afterwards, the Eastern Basin was tectonically divided from the Central Basin. The two basins contain different depositional systems, bounded in space by the Segre tectonic threshold. During the rest of the Ilerdian, a succession of deltaic systems developed in the two sub-basins. Sedimentation of the deltas concluded with the establishment of a new stage where carbonate sedimentation prevailed (Cuisian). The last stage described by Rosell encompasses the lower and middle Lutetian. It consists of deltaic depositional systems in the proximal areas of each basin. The erosion of the deltas, during periods of eustatic falls, formed turbiditic systems in the distal areas of the basins.

The sedimentation in the Eastern basin ended with the progressive establishment of fluvio-deltaic, and finally, fluvio-alluvial systems, which are not described in detail in this chapter.

Structural setting

The Eastern Pre-Pyrenees consist of two structural units (Fig. 4): a lower unit, the Cadí Unit (Muñoz, 1985), and an upper unit exposed at the eastern and western terminations of the basin: to the E, the Figueres–Montgrí Unit (Solé *et al.*, 1955; Estévez, 1973), and to the W, the Pedraforca Unit (Seguret, 1972; Vergés and Martínez, 1988).

The Cadí Unit consists of thrust sheets with post-Silurian (Devonian–Carboniferous) deposits, which are covered by deposits of Stefano-Permian, upper Cretaceous, Paleocene and Eocene. This unit is bounded by the Serra Cavallera thrust, in the N, and the Vallfogona thrust, in the S. This last thrust differentiates the Eastern Pyrenean basin from the Ebro basin (Fig. 4). The Upper Unit consists mainly of Mesozoic deposits.

The deposits studied below belong to the Cadí Unit.

Stratigraphy

The first lithostratigraphic studies of the Eastern South Pyrenees were presented by Kromm (1966, 1967a, 1967b, 1968a, 1968b, 1968c, 1969), Gich (1969, 1972), Solé Sugrañes (1970), Pallí (1972), Estévez (1973) and Busquets (1981).

Recently, a variety of sequence stratigraphic interpretations have been presented from this area (Maestro, 1985; Costa, 1985, 1989a, 1989b; Rosell and Costa, 1989; Costa and Maestro, 1989; Betzler,

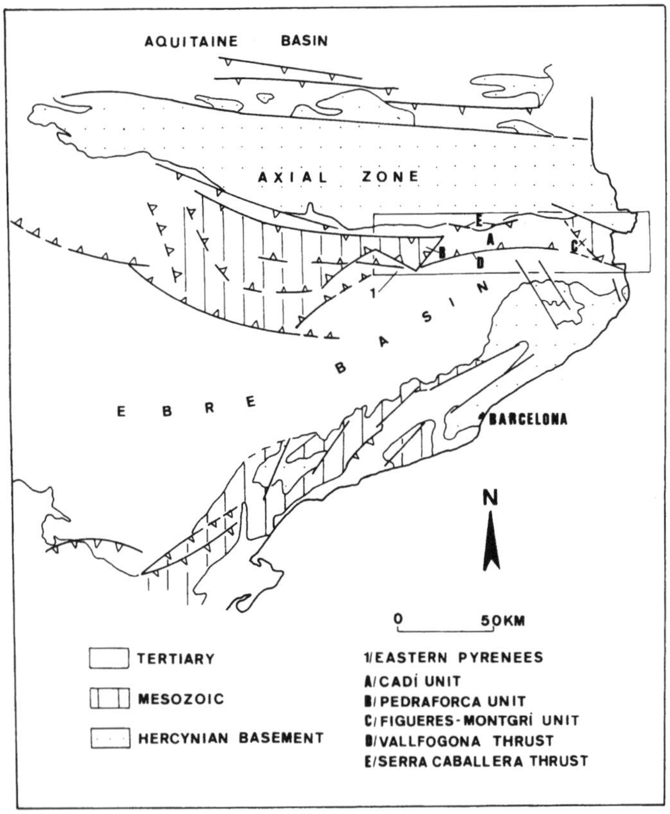

Fig. 4. Structural sketch-map of Catalonia, showing the main structural units of the Eastern Pyrenees.

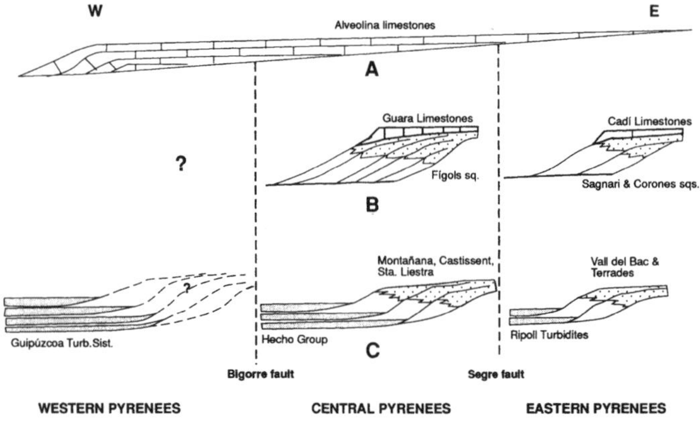

Fig. 5. E–W sedimentologic evolution of the lower and middle Eocene of the Southern Pyrenees (Rosell, 1988). A, Carbonate ramp (Alveolina limestones): Fòrnols and Puig Aguiló sequences in the Eastern Pyrenees. B, Siliciclastic platforms: Sagnari and Corones sequences. Carbonate platforms: Cadí and Armàncies sequences. C, Deltaic platforms and their related Turbiditic systems: Vall del Bac, Campdevànol, Ripoll and Josa sequences.

1989a, 1989b; Luterbacher *et al.*, 1991; Maestro *et al.*, 1991). Furthermore, Puigdefábregas and Souquet (1986) defined a series of depositional sequences related to the progressive emplacement of the Cadí and Pedraforca Units.

Fig. 5 shows the correlation of the lithostratigraphic units with the depositional sequences.

In this chapter we describe the depositional sequences (*sensu* Vail, 1987) characterized for the basin. The data used are synthesized and updated from Maestro (1985), Costa (1985), Costa (1989a), Betzler (1989a, b) and Luterbacher *et al.* (1991).

Depositional sequences

The Eocene succession of this Basin is more than 3000 m thick, and ranges in age from Ilerdian to middle Lutetian. We have distinguished and named 10 depositional sequences (*sensu* Vail, 1987) (Fig. 6).

The lowermost sequences (Puig Aguiló and Fòrnols) consist of shelfal carbonate facies (Alveolina limestones) in the eastern and western ends of the basin. In the central area, they consist of offshore marls. The following Sagnari and Corones sequences are characterized, throughout the basin, by fluvio-deltaic deposits and expose shelfal carbonates in the middle and upper parts of the sequences. The three following sequences (Cadí, Armàncies and Vall del Bac) consist of resedimented carbonates and offshore marls in the central parts of the basin. In the marginal areas they consist of carbonate platform sediments (Assilina limestones), except for the Vall del Bac sequence, which exposes a deltaic platform in the Eastern end. The uppermost sequences (Campdevànol, Ripoll and Josa) consist of deltaic platform sediments in the eastern area (Vall del Bac–Terrades Platform Complex, *sensu* Costa, 1989a) and of turbiditic systems in the western area (Ripoll Turbiditic Complex, *sensu* Costa, 1989a).

On the basis of biostratigraphic dating (Kromm, 1969; Gich, 1969; Pallí, 1972; Rosell *et al.*, 1973) and the position of the Ripoll Turbiditic Complex in the Eocene succession of the Pyrenees, the age of the boundary between the Vall del Bac and Campdevànol sequences can be estimated as 49.5 Ma. This boundary would then correspond to the boundary of two second-order global sequences (Haq *et al.*, 1987, 1988). An important paleogeographic change occurred in the Basin, at this boundary. The depositional systems of the sequences below the boundary display proximal environments at the two ends of the Basin, and the distal environments in the central areas. This set of sequences exposes a predominance of carbonate sediments. The depositional systems of the sequences above the boundary display proximal environments in the eastern areas of the Basin, and distal ones in the central and western areas. Carbonate sediments are rare, and the siliciclastic deposits prevail.

Fòrnols Sequence (Fig. 7, number 1)

The lower part of this sequence consists of siliciclastic deposits (uppermost part of the Garumnian facies). The middle and upper parts consist of marine carbonate deposits. It has a thickness of 145 m in the western areas, 150 m in the central areas and 175 m in the eastern areas. The age of this sequence is early Ilerdian (*Alveolina cucumiformis*, *Alveolina ellipsoidalis* and *Alveolina moussoulensis* zones; Betzler, 1989a). The basal unconformity consists of a karstified horizon, which is developed on the lacustrine limestones of the middle part of the Garumnian facies. In the eastern area, the

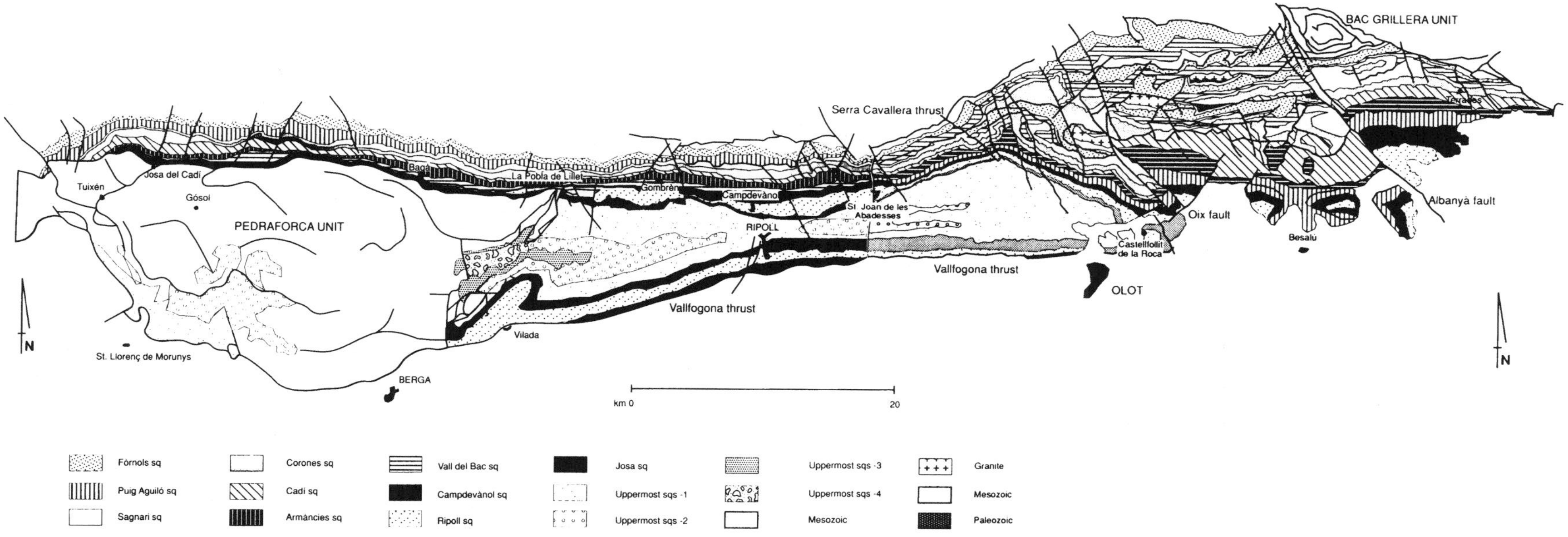

Fig. 6. Geologic map of the Eocene depositional sequences in the Eastern Pyrenees. The map is based on previous geologic maps by Costa (1985), Maestro (1985), Losantos *et al.* (1989) and Costa (1989).

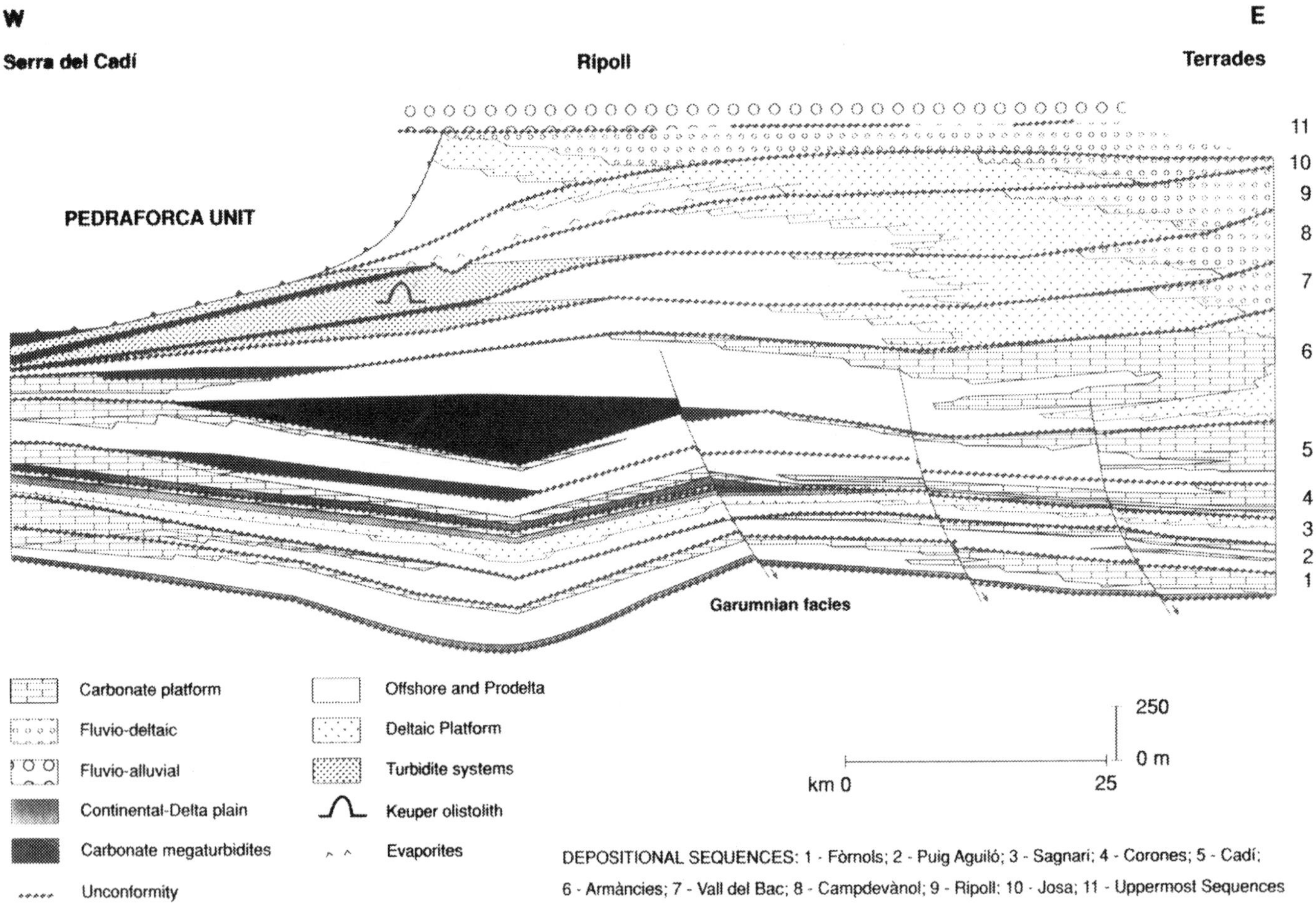

Fig. 7. E–W cross-section of the Eocene depositional sequences in the Eastern South Pyrenees.

unconformity is located at the base of conglomeratic bodies in the upper part of the Garumnian facies.

The lower part of the sequence (Fig. 7) consists of siliciclastic sediments deposited in coastal mud plain environments, which represent the lowstand systems tract (LST) of the sequence (*sensu* Van Wagooner *et al.*, 1990). In the upper part, the maximum regression surface (Betzler, 1989a) consists of a succession of edaphic, dolomitic and brecciated horizons. These deposits are overlain by offshore limestones and marls indicating a marine transgression trend. The maximum flooding surface (Betzler, 1989a) exposes coralgal reefs. This carbonate ramp depositional system represents the transgressive systems tract (TST) (*sensu* Van Wagooner *et al.*, 1990). Overlying this surface, carbonate deposits are developed in a marine platform with prograding geometries. They represent the highstand systems tract (HST) of this depositional sequence (*sensu* Van Wagooner *et al.*, 1990).

Puig Aguiló Sequence (Fig. 7, number 2)

This consists of an alternation of limestones and marls, with carbonate platform deposits in the distal ends of the Basin (Terrades and Cadí areas). This sequence has a thickness of 200 m in the western area, 175 to 260 m in the central areas and 150 m in the eastern area. The age of the sequence is early to middle Ilerdian (*Alveolina moussolensis* and *Alveolina corbarica* biozones; Betzler, 1989a). The lower boundary is reflected by a sharp contact between Nummulites and Alveolina mudstones and wackestones (uppermost part of the Fòrnols sequence), and the mixed siliciclastic–carbonate facies at the base of the Puig Aguiló sequence, which often consist of alveolina bars. These facies correspond to shoals and bioclastic lagoon deposits, and represent the LST of this sequence.

The TST consist of a succession of shallowing cycles, with agradational to retrograding-upward patterns. Finally, the HST exposes prograding carbonate banks and regressive cycles of tidal flats.

To the centre of the Basin, these deposits grade into offshore facies (Fig. 7).

Sagnari Sequence (Fig. 7, number 3)

The Sagnari sequence consists of siliciclastic deposits of deltaic origin. It has a thickness of 90 m at both ends of the Basin and 150 m in the centre of the Basin. The age of the sequence is middle to upper Ilerdian (*Alveolina corbarica* and *Alveolina trempina* zones; Betzler, 1989a). The basal unconformity is reflected in the contact between the marly-calcareous facies of the uppermost part of the Puig Aguiló sequence, and the siliciclastics at the base of this sequence.

The LST consists of sandstones and marls, organized in tidally influenced prodelta and delta-front cycles. The stratigraphic succession of this tract has a progradational pattern. The TST is characterized by an upward-retrogradational trend. At the top of the TST, some bioclastic- and oolite-rich carbonate levels are developed. The delta-front cycles of the uppermost tract (HST) of the sequence show a progradational pattern. The delta-front facies progressively pass into delta-plain facies, with red shales, sandstones and limestones (Fig. 7).

Corones Sequence (Fig. 7, number 4)

The lower part of the sequence (red siliciclastic deposits) is only exposed in the western and central areas of the Basin. The middle and upper parts of the sequence, which consist of marls and limestones, are developed from the west to the east (Fig. 7). The sequence has a thickness of 140 m, except in the eastern area, where it is 60 m. The age of this sequence is late Ilerdian to middle Cuisian (*Alveolina trempina* and *Alveolina oblonga* biozones; Betzler, 1989a).

The basal boundary of this sequence is placed in the red sediments of the Corones Fm (Gich, 1969; Fig. 5), at the top of limestones (Figs. 5 and 7). In the eastern area (Terrades zone), the reddish facies are not exposed, and the boundary between the two sequences is placed at the top of the delta-front cycles of the Sagnari sequence (Fig. 7).

We distinguish three levels in the Corones sequence. The lower level contains red siliciclastics of a coastal fluvial plain facies association, with edaphic horizons in the uppermost part. It forms the LST of the sequence. The TST, mainly calcareous, consists of shallowing cycles and a tidal plain facies association. It is characterized by abundant algal laminations, chert nodules and hydrocarbons. There are some occasional slumped levels. The HST consists of an alternation of marls and marly-limestones. They are interpreted as carbonate ramp deposits. (Fig. 7).

During the deposition of the TST and the HST of this sequence, the tectonic subsidence of the basin progressively increased. This is reflected by the gradually increasing trend to distal sedimentary environments upwards through the stratigraphic succession.

Cadí Sequence (Fig. 7, number 5)

The Cadí sequence consists of marls and limestones. Its thickness ranges from 100 to 230 m. The age of this sequence is middle Cuisian (*Alveolina oblonga* and *Alveolina laxispira* biozones; Betzler, 1989a). At the two ends of the basin, the lower boundary separates the marls and marlstones of the uppermost part of the Corones sequence from the marly facies of the lowermost part of the Cadí sequence. In the central areas this boundary is noted at the base of some resedimented carbonate bodies that characterize the lower part of this sequence.

The LST, in the central areas of the basin, consists of resedimented Corones-sequence material (Fig. 7). It consists of slumped levels, olistoliths and, locally, megaturbidites (Bagà area). In the proximal areas of the Basin, the LST and the TST consist of shallowing cycles of marls and marly-limestones of external carbonate ramp and platform environments. The HST consists of a succession of shallowing cycles of bioclastic limestones (shoals and bars with *Assilinas*). In the centre of the basin the deposits are mainly marls (Fig. 7).

Armàncies Sequence (Fig. 7, number 6)

The Armàncies sequence consists of carbonate, mixed and siliciclastic deposits, with a predominance of limestones. Its thickness ranges from 300 m in the eastern area to 100 m in the western one. The age of the sequence is late Cuisian (*Alveiolina laxispira* and *Alveolina major* biozones; Betzler 1989a).

The lower boundary is exposed as the contact between the carbonate deposits of the Cadí sequence and the mixed deposits of the Armàncies sequence. In the central and western areas, this boundary is placed at the base of some resedimented carbonates of the lowermost part of this sequence.

The LST consists of mixed deltaic facies cycles (eastern area) and of carbonate platform deposits (western area). These cycles have a prograding pattern. The TST and the HST are represented by agradational cycles of *Assilina* limestones (western area) or mixed carbonate-siliciclastic deposits (eastern area). In the central part of the Basin, the LST consists of an alternation of offshore marls and marly-limestones and resedimented carbonates. There are turbiditic sandstone layers interbedded with the marls. The resedimented carbonates consist of five megaturbidites, each approximately 20–30 m thick. The TST and the HST are represented by shallowing cycles of marls and marly-limestones (Fig. 7).

Vall del Bac Sequence (Fig. 7, number 7)

In the eastern area, the Vall del Bac sequence consists of siliciclastic deposits, with a thickness of between 250 and 300 m. In the central areas, the sequence consists of 150 m of marls, marly-limestones, silts and fine-grained sandstones. Finally, in the western areas, some carbonate deposits are exposed in the lower part. There, the upper part of the sequence exposes a marly succession. Its thickness reaches 150 m. The age of this sequence is late Cuisian (*Alveolina major* biozone; Betzler, 1989a).

The eastern areas expose deltaic platform deposits (Platform 1 of the Vall del Bac–Terrades Complex; Costa, 1989a). Its lower boundary is marked by a sharp contrast between the mixed carbonate–siliciclastic facies of the Armàncies sequence and the siliciclastic facies (sandstones and conglomerates) of the base of the Vall del Bac sequence. The lower stretch of the sequence consists of delta-front cycles with progradational patterns, and is interpreted as the LST. The middle part exposes retrogradational delta-front cycles, and represents the TST. The upper tract (HST) consists of the same kind of cycles with a progradational pattern. The delta cycles pass upwards into delta-plain facies. To the west these facies progressively become more distal (Fig. 7). So, in the central areas of the Basin, they consist of fine-grained deposits which represent the prodelta facies associations (Fig. 7). In these zones, the lower boundary of the sequence is difficult to distinguish, due to its similarity with the facies of the Armàncies sequence. These prodeltaic facies of the Vall del Bac sequence have a high content of siliciclastics. In the western area of the Basin, the base of the sequence consists of the resedimented carbonates of Les Balmes (Maestro, 1985) (Fig. 7). They consist of carbonate breccias and poorly differentiated megaturbidites, which are formed by limestones of the uppermost part of the Armàncies sequence.

Campdevànol, Ripoll and Josa Sequences (Fig. 7, numbers 8 to 10)

The descriptions of these three sequences are included in the same section because all of them show the same facies distribution pattern in the Basin. Each one consists of systems of deltaic platforms in the East, and turbiditic systems in the West (Costa, 1989a). The base of each sequence is bounded by a type-1 unconformity (*sensu* Vail, 1987). These sediments range in age from the late Cuisian to the middle Lutetian.

The Campdevànol sequence consists of the first of the turbiditic systems, in the western areas. Its thickness ranges from 350 m in the central–western sector to 20 m in the westernmost area. The turbiditic facies overly the prodelta deposits of the Vall del Bac sequence. The LST of the sequence consists of this first turbiditic system and the slope facies which are developed over the turbiditic system evolving (from west to east) to prodelta and delta-front facies. The TST and the HST are only exposed in the deltaic platform deposits, with a thickness ranging from 120 to 350 m. The TST is represented by hemipelagic marls developed in the central–eastern areas. The HST consists of deltaic facies showing progradational patterns. These deposits evolved from delta-plain facies to delta-front facies and finally, to prodelta facies, from east to west (Fig. 7).

The Ripoll sequence consists of a second turbiditic system. The thickness of this system ranges from 600 m, in the central–western areas, to 20 m in the western area. This sequence contains, also, a new deltaic system, developed in the eastern areas, which thickens from 250 to 350 m. In the base of the turbiditic system an interval of carbonate tubidites is developed, which separates this system from the underlying turbiditic system. Overlying the carbonate turbidites, from the central–western areas to the East, are developed slope facies, which pass progressively to prodelta and delta-front facies. The sediments consisting of turbidites, the slope facies and its deltaic equivalents represent the LST of this sequence. The TST is represented by marls and limestones, and the HST consists of delta-plain facies, in the eastern zone, which evolved to delta-front and prodelta facies (Fig. 7).

The Josa sequence consists of the third turbiditic system and the third deltaic system. Its thickness ranges from 250 m, in the east, to 200 m, in the west. In the base of the turbiditic system two carbonate megaturbidites are developed, which separate this system from the underlying sequence (Fig. 7). In the central areas and in the base of the deltaic system, the evaporites of the Vallfogona Fm (Gich, 1969; fig. 5) are developed, separating this sequence from the underlying one. Over the turbidites and the evaporites, slope facies are exposed which evolve, to the east, to prodelta and delta-front facies. This group of deposits is interpreted as the LST of the sequence. The TST is represented by carbonate reefal facies. The HST consists of a succession of delta-front cycles, which evolved to the east and to the top, to delta-plain deposits that belong to the Bellmunt Fm (Gich, 1969).

Uppermost Sequences (Fig. 7, number 11)

After the sedimentation of the previous sequence, fluvio-deltaic systems were progressively established in the Basin. They prograded, in a general sense, from east to west (Fig. 7). The delta-plain facies of these systems are included in the Bellmunt Fm (Gich, 1969), whereas the delta-front facies of these systems are included in the Cal Bernat Fm (Busquets, 1981; Fig. 5). The sedimentation in the Basin ended with the development of some fluvio-alluvial fan systems. The uppermost alluvial fan systems fossilized the Pedraforce nappe (Martínez *et al.*, 1988). A more detailed study of this group of sediments will allow their division into lower-rank depositional sequences.

Conclusions

The conclusions of this study are summarized in Fig. 7, which shows a diagram of the Basin. We have noted the surfaces which bound the depositional sequences, and the evolution of the minor-rank units (systems tracts) in space and time.

References

Betzler, Ch. (1989a). The upper Paleocene to middle Eocene between the rio Segre and the rio Llobregat (Eastern south Pyrenees): facies stratigraphy and structural evolution. *Tübinger Geowissenschaftliche Arbeiten*, **2**: 131 pp.

Betzler, Ch. (1989b). A carbonate complex in an active foreland basin. The Paleogene of the Sierra de Port del Comte and the Sierra del Cadí (southern Pyrenees). *Geodinam. Acta*, **3**(2): 207–220.

Costa, J.M. (1985). Estratigrafia fisica i fàcies del Paleogè Pre-pirinenc entre els rius Gréixer i Arija. Tesi de llicenciatura, 123 pp. Universitat Autònoma Barcelona.

Costa, J.M. (1989a). Turbidites de Ripoll. Relació amb llurs plataformes. Thesis doctoral, 154 pp. Universitat Autònoma Barcelona.

Costa, J.M. (1989b). El complejo turbidítico de Ripoll. Facies. *Simposios XII Congr. Esp. Sediment., Bilbao*: 121–127.

Costa, J.M. and Maestro, E. (1989). Las secuencias deposicionales del Eoceno del Prepirineo Oriental. *Geogaceta*, **6**: 66–68.

Estévez, A. (1973). La vertiente meridional del Pirineo Catalán al norte del curso medio del rio Fluvià. Tesis doctoral. Univ. Granada, **44**: 514 pp.

Gich, M. (1969). Las unidades litoestratigráficas del Ripollès oriental (prov. Gerona y Barcelona). *Acta Geol. Hisp*, **IV**: 5–8.

Gich, M. (1972). Estudio geológico del Eoceno prepirenaico del Ripollés oriental. Doctoral thesis. Univ. Barcelona.

Haq, B.U., Hardenbol, J. and Vail, P.R. (1987). Chronology of fluctuating sea levels since the Triassic. *Am. Assoc. Adv. Sci.*, **235**: 1156–1166.

Haq, B.U., Hardenbol, J. and Vail, P.R. (1988). Mesozoic and Cenozoic chronostratigraphy and cycles of sea-level change. In (C.K. Wilgus *et al.*, eds.): *Sea level change: an integrated approach, Soc. Econ. Paleontol. and Mineral. Spec. Publ.*, **42**: 71–108.

Kromm, F. (1966). L'age et les conditions de sédimentation des couches rouges de l'Eocène terminal entre l'Ampurdan et le rio Ter (prov. de Gerone, Espagne). *Actes Soc. Linn. Bordeaux*, **104**, B(3).

Kromm, F. (1967a). Caractères géologiques principaux de la region d'Olot-Besalú (prov. Gerone, Espagne). *Actes Soc. Linn. Bordeaux* **104**, B(17): 4–18.

Kromm, F. (1967b). Le flysch de Vallfogona et son contexte paléogéographique (prov. de Gerone). *Actes Soc. Linn. Bordeaux*, **104**, B(3): 4–9.

Kromm, F. (1968a). Notice explicative d'une carte à 1:100000 de formations eocènes de la zone prepyrénéenne (prov. Barcelone et Gerone, Espagne). *Actes Soc. Linn. Bordeaux*, **105**, B–8.

Kromm, F. (1968b). Stratigraphie comparée des formations eocènes du revers sud des Pyrénées et de la Cordillière Prélittorale Catalane (prov. Gerone et Barcelona). *Actes Soc. Linn. Bordeaux*, **105**, B (2).

Kromm, F. (1968c). Stratigraphie résumée de l'Eocène du versant sud des Pyrénées orientales de la zone prépyrénéennes (prov. Barcelone et Gerone). *Ext. C. R. Som. Sceanc. Soc. Geol. France*, 7.

Kromm, F. (1969). Répartition des faciès et pósition stratigraphique des formations ilerdiennes en Catalogne orientale. *Coll. sur l'Eocène, Paris, Mai 1968, vol. 3, Mem. Bur. Rech. Geol. Mineral.*, **69**, 209–217.

Luterbacher, H.P., Eichenseer, H., Betzler, Ch. and Hurk, A. van den (1991). Carbonate–siliciclastic depositional systems in the Paleogene of the south-Pyrenéan foreland basin: a sequence-stratigraphic approach. In McDonald, D. (ed.), *Sea-level changes at active plate margins. I.A.S. Spec. Publ.*, **12**: 391–407.

Maestro, E. (1985). Estratigrafia física i fàcies del Paleogè de la unitat Cadí-Ripoll entre els rius Segre i Gréixer. Tesi de Llicenciatura. Univ. Autònoma de Barcelona: 135 pp.

Maestro, E., Costa, J.M., Betzler, Ch. and van den Hurk, A. (1991). Aplicación de la estratigrafía secuencial al margen activo de la cuenca de antepaís surpirenaica Oriental (Unidad Cadí–Ripoll: zona Figueres, zona Ripoll, zona Port del Comte). Sistemas deposicionales carbonatados y siliciclásticos, espacio de acomodación, énfasis y oclusión de discontinuidades sedimentarias. In F. Colombo (ed.), *Guidebook, I Congr. Grupo Español Terciario* 10: 152 pp.

Martínez, A., Vergés, J. and Muñoz, J.A. (1988). Secuencias de progadación del sistema de cabalgamientos de la terminación oriental del manto del Pedraforca y relación con los conglomerados sinorogénicos. *Acta Geol. Hisp.*, **23**(2): 119–27.

Pallí, Ll. (1972). Estratigrafía del Paleogeno del Empordà y zonas limítrofes. *Publ. Geol. Univ. Autónoma Barcelona*, **1**: 328 pp.

Puigdefábregas, C. and Souquet, P. (1986). Tectosedimentary cycles and depositional sequences of the Mesozoic and Tertiary from the Pyrenees. *Tectonophysics*, **129**: 173–203.

Rosell, J. (1988). Ensayo de síntesis del Eoceno surpirenaico: el fenómeno turbidítico. *Rev. Soc. Geol. Esp.*, **1**: 3–4.

Rosell, J. and Costa, J.M. (1989). Secuencias deposicionales en las turbiditas de Ripoll. In *Simposios. XII Congr. Esp. Sediment. Bilbao. Septiembre 1989*: 115–120.

Rosell, J., Ferrer, J. and Luterbacher, H.P. (1973). El Paleogeno marino del noroeste de España. In E. Perconig (ed.) *XIII Col. Eur. Micropaleontol. ENADIMSA, Madrid*: 29–61.

Séguret, M. (1972). Étude tectonique des nappes et sèries décollées de la partie centrale du versant sud des Pyrenées. Caractère synsédimentaire, rôle de la compression et de la gravité. These doctoral. Univ. Montpellier. *Publ. USTELA, sèrie geol struct.*, **2**: 210 pp.

Solé, F., Fontboté, J., Masachs, V. and Virgili, C. (1955). Continuidad de las escamas de corrimiento del Ampurdán entre Figueres y el macizo del Montgrí. *Hom. Dr. F. Pardillo. Univ. Barcelona*: 10 pp.

Solé Sugrañes, L. (1970). Estudio del Prepirineo entre los rios Segre y Llobregat. Tesi doctoral, Univ. Barcelona: 495 pp.

Vail, P.R. (1987). Seismic stratigraphy interpretation procedure. In Bally, A.W. (ed.) *Atlas of seismic stratigraphy. Am. Assoc. Petrol. Geol., Studies in geology*, **27**(1), 277–281.

Van Waggoner, J.C., Mitchum, R.M., Campion, K.M. and Rahmanian, V.D. (1990). Siliciclastic sequence stratigraphy in well logs, cores and outcrops. *Am. Assoc. Petrol. Geol. Method. Explor. Ser.*, **7**: 55 pp.

Vergés, J. and Martínez, A. (1988). Corte compensado del Pirineo oriental: geometría de las cuencas de antepaís y edades de emplazamiento de los mantos de corrimiento. *Acta Geol. Hisp.* **23**: 95–105.

E10 The Neogene Cerdanya and Seu d'Urgell intramontane basins (Eastern Pyrenees)

E. ROCA

Abstract

The Cerdanya, Seu d'Urgell, Conflent and Roselló basins began to form in the Miocene, as a result of the dextral slip of the La Tet and La Tec faults that extend ENE and WSW across the eastern Axial Pyrenees. The first stage of sedimentation was followed by an episode of basin inversion, which was, in turn, followed by a second stage of sedimentation in the Late Miocene (Late Messinian) and Pliocene. Plant material fossilised during the sedimentation provides evidence of the development of the present Mediterranean climate with warm and dry summers and cold and damp winters.

Introduction

The present relief of the Eastern Axial Pyrenees is characterized by wide ENE–WSW oriented valleys and depressions (Cerdanya, Conflent, Rosselló) along which the main rivers of this area run (Segre, La Tec and La Tet rivers). Partially modified by the Quaternary glacial processes, this morphology was caused by the development, during the Neogene, of a horst and basin system which is bounded by E–W and ENE–WSW faults (Fig. 1). The most prominent are the ENE–WSW La Tet and La Tec fault systems, which are formed by NE–SW right-stepping en echelon faults and E–W faults that developed at the end of its western block and in the overstep zones of the NE–SW faults.

The Cerdanya, Seu d'Urgell and the Conflent basins are related to the La Tet fault (Fig. 1) and formed on the NW side of this fault, where major extensional E–W faults developed (Fig. 2). From the tectonic structures in these basins and the stratigraphic features of the basin infill, the Neogene tectonic evolution of the Eastern Pyrenees has been divided into two main periods (Roca and Santanach, 1986; Cabrera *et al.*, 1988).

The first stage, developed during the Miocene, is characterized by strong tectonic activity that originated an intense fracturing of the Eastern Pyrenees basement rocks. During this period, the main Neogene basins (Seu d'Urgell, Cerdanya, Conflent and Rosselló) formed mainly as a result of the dextral slip of the ENE–WSW La Tet and La Tec basement faults, which created extensional conditions in the SW fault terminations (e.g. Seu d'Urgell and Cerdanya basins) and in the overstep zones (e.g. Conflent basin) (Fig. 2).

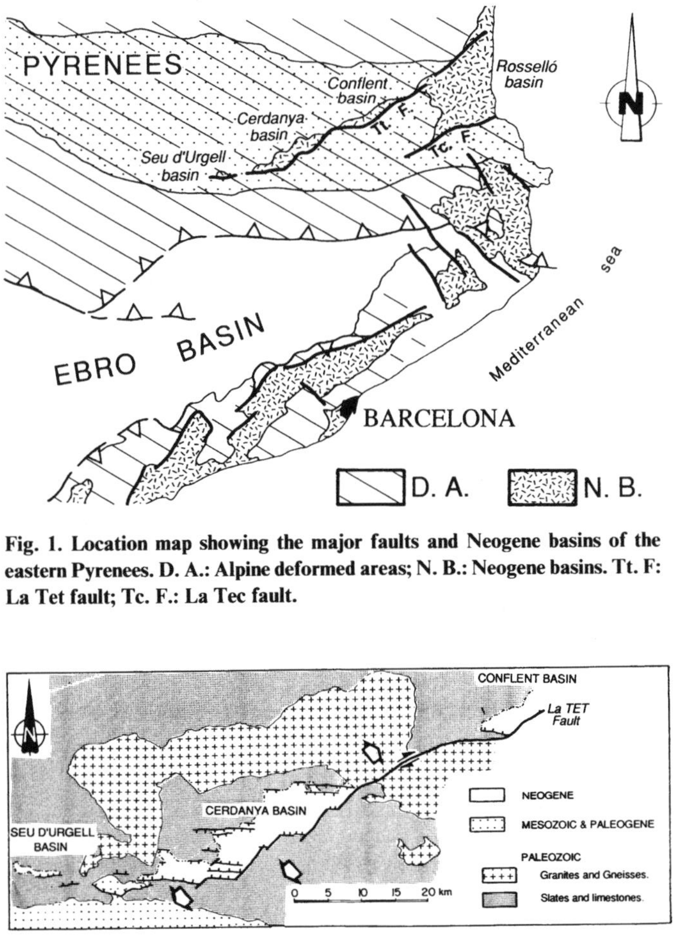

Fig. 1. Location map showing the major faults and Neogene basins of the eastern Pyrenees. D. A.: Alpine deformed areas; N. B.: Neogene basins. Tt. F: La Tet fault; Tc. F.: La Tec fault.

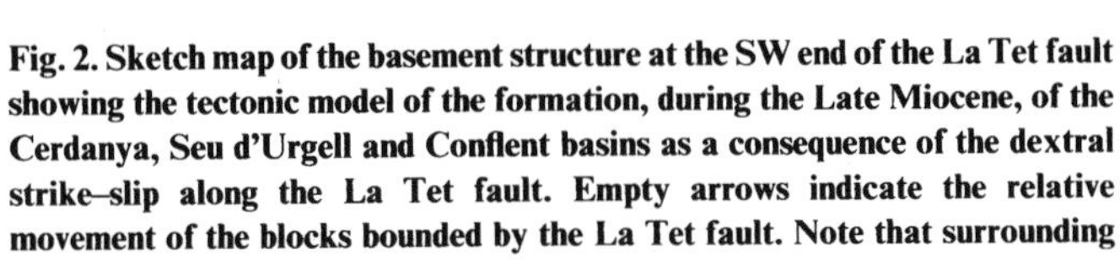

Fig. 2. Sketch map of the basement structure at the SW end of the La Tet fault showing the tectonic model of the formation, during the Late Miocene, of the Cerdanya, Seu d'Urgell and Conflent basins as a consequence of the dextral strike–slip along the La Tet fault. Empty arrows indicate the relative movement of the blocks bounded by the La Tet fault. Note that surrounding source areas of the Cerdanya basin are also indicated.

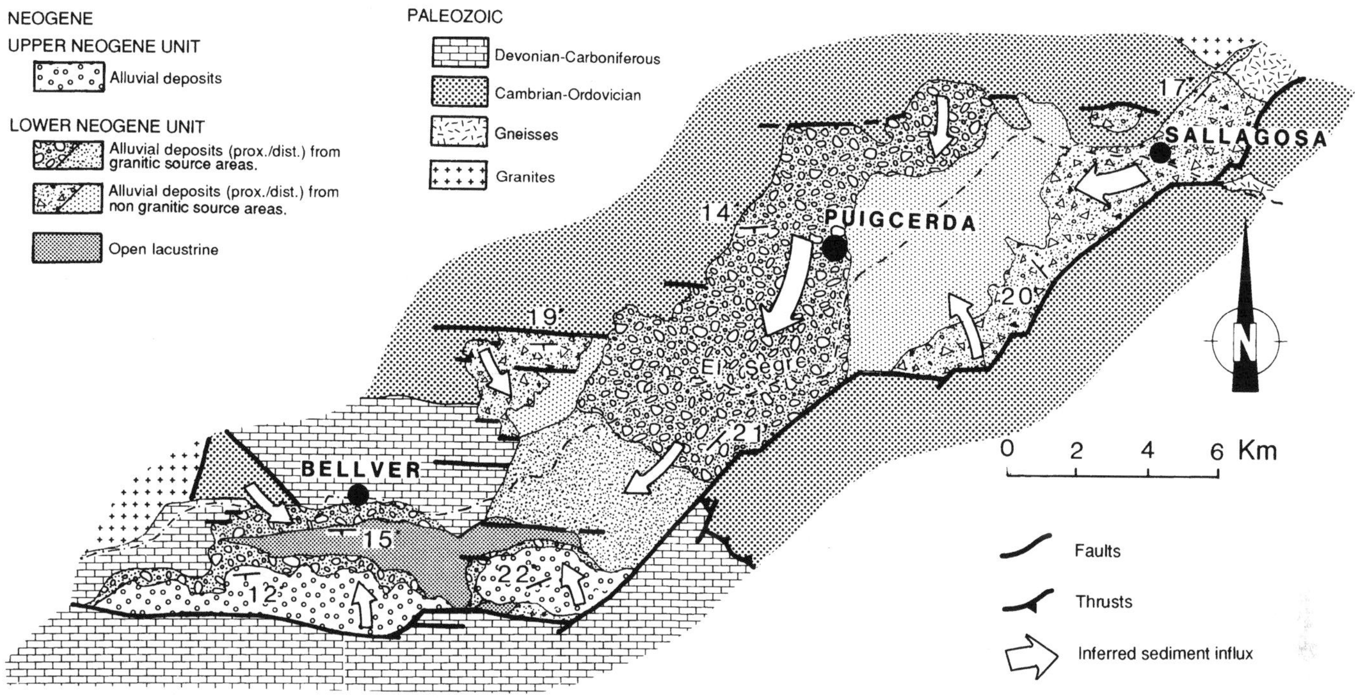

Fig. 3. Simplified geological map of the Cerdanya basin, without the Quaternary deposits, showing the major faults and the main depositional facies (based on Roca, 1986). The numbers on the map refer to dip of strata on the nearest strike and dip symbol.

The second stage began in the Late Messianian after a short period in which the basin infills were inverted and eroded as a consequence of the development of compressional tectonic structures inside the basins (Calvet, 1985; Pous *et al.*, 1986; Cabrera *et al.*, 1988). During this Late Messinian–Pliocene stage, the tectonic activity attenuated in such a way that the deformation concentrated close to the major faults whose movement became normal, giving rise to the southwards tilting of the basin infills and the reactivation of the subsidence in the previously eroded basins.

During these two tectonic stages, the paleoclimatological conditions, which also influenced the main sedimentological features of the infill of the Neogene basins, were not constant. Effectively, macroflora and palynological studies carried out in these basins (Sanz de Siria, 1978; Suc, 1984; Bessedik, 1985) conclude that while during the Late Miocene warm temperate and damp conditions were dominant, during the Pliocene the climatology became typical Mediterranean with warm and dry summers and cold and relatively damp winters.

The Cerdanya basin

The Cerdanya basin is an ENE–WSW-orientated elongate half-graben, located in the northwestern block of the La Tet fault at its southern horsetail termination (Figs. 2 and 3). Its formation and later evolution during the Late Miocene were conditioned by the dextral slip of the La Tet fault and by the normal slip of the E–W faults, which absorbed the strike slip of that fault (Cabrera *et al.*, 1988). The structure of the Cerdanya basin is due, therefore, to subvertical NE–SW faults that form its SE boundary and to the presence, within the basin, of E–W normal faults that divided the basin into different SE to E tilted blocks.

This structure confers a clear asymmetry between the S–SE and the N basin margins (Fig. 3). The S and SE margins are sharp and characterized by E–W and NE–SW-striking faults which separate the Neogene basin fill from the Hercynian basement rocks. In contrast, the northern basin margin is more irregular and the contact between the basin infill and the basement, although controlled by E–W faults, is basically an unconformity.

The Neogene infill of the Cerdanya basin basically consists of 400 to 1000 m thick siliciclastic (Pous *et al.*, 1986; Cabrera *et al.*, 1988), muddy, sandy and conglomerate alluvial fan to fluvial sequences (Figs. 3 and 4). Diatomites and thin lignite seams linked respectively to open and marginal lacustrine paleoenvironments are also recorded.

Two main stratigraphic units have been identified in the Cerdanya basin (Roca and Santanach, 1986; Agustí and Roca, 1987): the Lower Neogene Unit, dated as Middle–Late Miocene (Vallesian) by Déperet and Rérolle (1885) and Golpe Posse (1981), and the Upper Neogene Unit dated as Latest Miocene (Turolian) – Pliocene (?) (Agustí and Roca, 1987). These units reflect the two tectonic periods recognized in the Neogene evolution of the Eastern Pyrenees (Cabrera *et al.*, 1988). Thus, the sedimentation of the Lower Unit was controlled by dextral slip on the La Tet fault which created the basin, whereas the Upper Unit records the final evolution of the basin under a generally extensional regime.

The geometric relationship between both units changes depend-

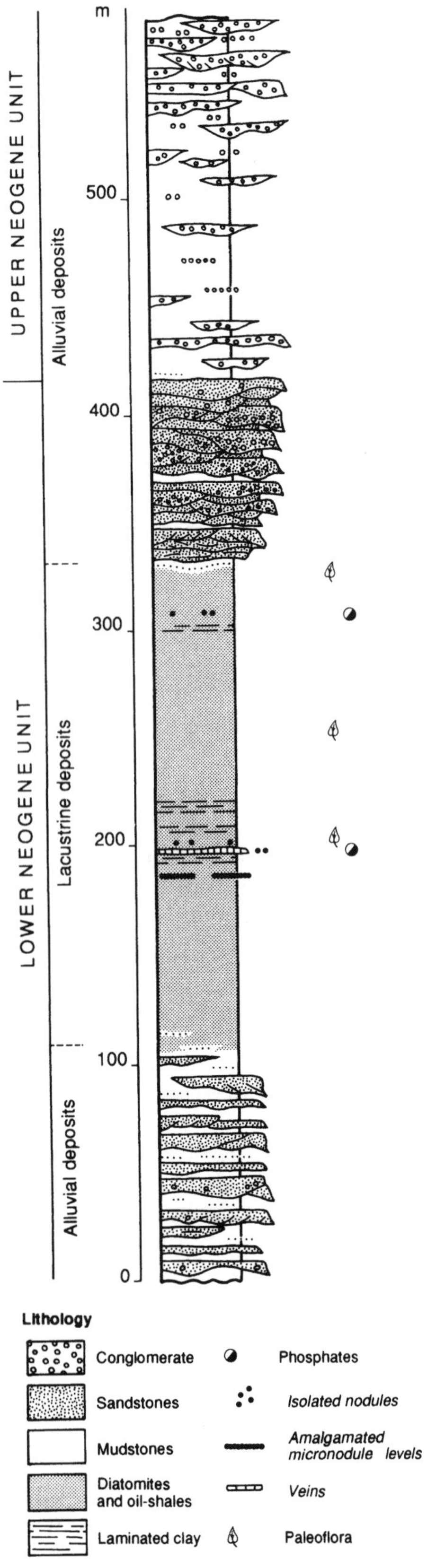

Fig. 4. Stratigraphic log of the outcropping western Cerdanya basin infill (Roca, 1992a).

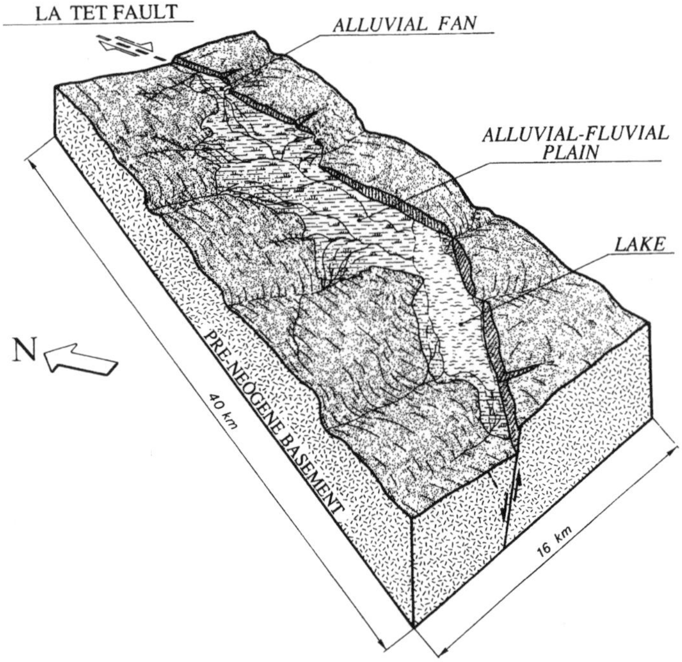

Fig. 5. Block diagram of the Cerdanya basin showing the general palaeoenvironmental arrangement during the Lower Neogene Unit deposition (Late Miocene–Vallesian).

ing on the development of the local small- to medium-scale tectonic structures (mainly NW–SE folds originated by the dextral slip of the La Tet fault) that only affect the Lower Neogene Unit. In this way, the contact between both units is unconformable where the Lower Unit was folded and conformable where no folding took place.

Later on, during the Quaternary, the Cerdanya basin has recorded several glacial pulsations (Panzer, 1932; Gourinard, 1971a, b) reflected by the sedimentation of widespread glacial, fluvio-glacial and fluvial deposits that unconformably overly the Neogene basin infill and the Hercynian basement rocks.

Lower Neogene unit

The successions of this unit constitute most of the Neogene that nowadays crops out in the Cerdanya basin, except in the southernmost basin areas, where they are overlain and concealed by the Upper Neogene Unit sequences (Fig. 3). According to the surface and geoelectrical data (Pous *et al.*, 1986) the changing thickness (400–800 m) of this unit is strongly controlled by the differential normal slip of the E–W normal faults that compartmentalized the basin.

The sequences of this unit were deposited in a variety of depositional settings ranging from talus-scree, to alluvial fan, distributary fluvial plain, marginal shallow lacustrine (deltaic) and open, relatively deep, lacustrine environments. In general, the alluvial fan environments developed largely in the northern basin areas, while fluvial and lacustrine settings were more widespread in the central and southern basin areas (Fig. 5).

The spatial distribution as well as the lithological and sedimentological features of the alluvial fans are closely linked to the source areas (Roca, 1986; Cabrera *et al.*, 1988). In this way, two types of alluvial fan assemblages have been distinguished: (i) alluvial fans fed from slate (or limestone) source areas, in which the main terrigenous deposits consist of lenticular bodies of massive conglomerates and massive red mudstones; and (ii) alluvial fans fed from granitic source areas, consisting of pale-brown lenticular sand beds and planar- and trough-cross-stratified conglomeratic sandstones.

Towards the central basin areas, the proximal alluvial facies grade laterally into channelled alluvial facies assemblages and to marginal lacustrine facies consisting of massive grey mudstones and siltstones with interbedded lenticular and laterally extensive sheet sandstones. In the sequences of distal–marginal fluvial plain, interbeds of thin lignite and grey silt with paleoflora (Bessedik, 1985; Baltuille *et al.*, 1992; Julià, 1992) and fish remains (Julià, 1992) occur, recording the development of shallow lacustrine to paludine zones.

The open lacustrine facies are largely developed in the western Cerdanya basin, although subsurface data suggest that they may be present in other areas. They consist of up to 250 m thick fine blue-grey to pale-grey organic mudstones and diatomites with interbedded ostracod-rich laminae (Fig. 4). Early diagenetic Ca–Fe phosphates (widespread single or amalgamated nodules and very scarce thin veins) occur at several levels in these deposits (Bech and Vallejo, 1977; Roca *et al.*, 1987; Anadón *et al.*, 1989) which yield insects (Vilalta and Crusafont, 1945; Villalta, 1962) and a very rich paleoflora (Villalta and Crusafont, 1945; Menéndez Amor, 1955; Alvarez and Golpe Posse, 1981; Barrón, 1992).

According to surface and subsurface interpretations, the Lower Neogene Unit shows a megasequential arrangement consisting of two major thickening and coarsening-upward sequences capped by small fining-upward sequences (Fig. 6). The upper thickening and coarsening-upward sequence records a sharp progradation of the alluvial fan delta systems into the deep lacustrine areas (Margalef, 1957; Margalef and Marrasé, 1985), which initiated a generalized decrease of lake depth, and finally, their disappearance (Julià, 1984; Roca, 1986).

Upper Neogene Unit

Overlying the Lower Neogene Unit deposits, the sequences of the Upper Neogene Unit are preserved only in the southernmost areas of the Cerdanya basin (Fig. 3). In these areas, the Upper Unit is tilted 15–20° towards the south and is mainly formed by terrigenous sequences deposited in alluvial depositional settings (Roca, 1986). Lithologically, this unit consists of lenticular bodies of poorly-sorted conglomerates and breccias with massive red mudstone interbeds, which grade laterally, towards the central basin areas, to massive grey mudstones with thin conglomerate interbeds and lignite seams.

Unlike the Lower Neogene Unit, the conglomerate facies of the Upper Neogene Unit are essentially made up of carbonate pebbles.

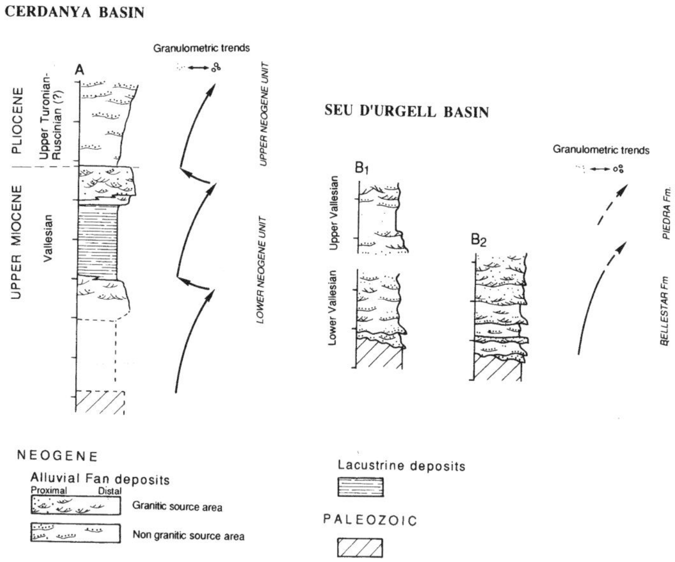

Fig. 6. Synthetic stratigraphic logs of the Cerdanya and Seu d'Urgell basins showing the major megasequential trends of their fill. The lower part of the Cerdanya basin stratigraphical log (A), marked in dashed lines, has been inferred from geoelectric profiles (Pous *et al.*, 1986). Location of the logs: A – Western sector of the Cerdanya basin; B_1 – Bellestar subbasin (western half-graben of the Seu d'Urgell basin); B_2 – Western sector of the Alàs subbasin (eastern half-graben of the Seu d'Urgell basin).

This lithological change indicates that the source area of the Upper Neogene Unit alluvial sediments was located southwards, where the basement mainly consists of Devonian, Mesozoic and Paleogene carbonate rocks. In this way, sedimentological studies (Roca, 1986) report that the sequences of the Upper Neogene Unit correspond to a northwards prograding alluvial fan system, situated near to the southern basin marginal faults.

The thickness of the Upper Neogene Unit is quite constant at about 250 m. Like the Lower Neogene Unit, this unit is arranged in a thickening and coarsening-upward megasequence (Fig. 6) that records, in its uppermost parts, the final infill processes of the Cerdanya basin.

The Seu d'Urgell basin

The Seu d'Urgell basin is the westernmost Neogene basin of the Eastern Pyrenees (Fig. 1). Like the Cerdanya basin, its formation and evolution during the Late Miocene was conditioned by the dextral slip of the La Tet fault that originated the normal slip of the E–W faults located in its southern termination. Located at the western end of this system of E–W normal faults that developed westwards of the Cerdanya basin (Fig. 2), the Seu d'Urgell basin is a small and E–W elongate graben (Fig. 7).

Separated by a prominent NE–SW high, two subbasins have been identified in the Seu d'Urgell basin: the E–W-oriented Alàs subbasin in the east, and the NE–SW-oriented Bellestar subbasin in the west (Fig. 7). The structure of these two subbasins is quite different. Whereas the structure of the Alàs subbasin is mainly due

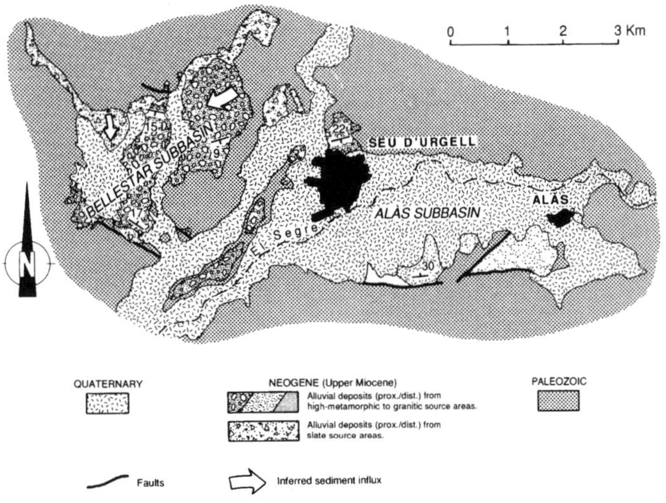

Fig. 7. Simplified geological map of the Seu d'Urgell basin showing the major faults and the main depositional facies. The numbers on the map refer to dip of strata on the nearest strike and dip symbol.

to E–W normal faults that generally dip to the north, the structure of the Bellestar subbasin is more complex and, to a great extent, controlled by NE–SW and NW–SE normal faults. In both subbasins, the faults, which are difficult to recognize at the surface, have a larger slip in the southern parts of the basin, conferring some asymmetry between the north and south subbasin margins (Roca, 1992b).

Anyway, the present morphology of the Seu d'Urgell basin is very complex (Fig. 7), as a result of the small dip of the E–W normal faults located in the northern basin areas, the strong Quaternary erosion, the onlapping character of the upper Neogene infill sequences and the presence of a complex pre-Late Miocene paleotopography.

Less well known than the Cerdanya basin, the Neogene infill of the Seu d'Urgell basin (Fig. 6) has a total thickness of less than 250 m of siliciclastic sequences, mainly deposited from proximal to distal alluvial fan settings (Agustí *et al.*, 1979). These sequences, dated as Middle–Late Miocene (Vallesian) by mammalian assemblages (Chevalier, 1909; Crusafont and Golpe Posse, 1974; Agustí *et al.*, 1979), have been grouped into two members.

The lower member, which unconformably overlies the Hercynian basement, has a very variable thickness (0–15 m) and records strong tectonic activity of the normal faults that compartmentalized the basin. It broadly consists of red mudstones and breccias deposited in colluvial and local small alluvial fans (Agustí *et al.*, 1979).

The upper member is thicker (70–240 m) and also consists of alluvial sequences. It includes the two lithostratigraphic units differentiated in the basin fill (Hartevelt, 1970; Agustí *et al.*, 1979): the Bellestar Fm., which corresponds to the proximal alluvial fan sequences, and the Piedra Fm., which records very distal or marginal, poorly drained areas of alluvial fans. In this member, generally, the proximal alluvial fan depositional settings are widespread in the Bellestar subbasin and in the northern and western areas of the Alàs subbasin, whereas the distal and marginal alluvial settings are developed in the southern and eastern areas of the Alàs subbasin (Fig. 7). Deposited in a structural framework with only minor fault activity, all these upper member sequences clearly extend through time over their substratum, onlapping both the margins and the minor horsts within the Seu d'Urgell basin.

As in the Cerdanya basin, the spatial distribution as well as the lithological and sedimentological features of the alluvial fans are closely linked to the source area (Roca, 1992b). Thus, two types of alluvial fan assemblage have been distinguished: (i) alluvial fans fed from nearby slate source areas, consisting of red mudstones and lenticular bodies of massive slate conglomerates; and (ii) alluvial fans fed from distant high-metamorphic to granitic source areas, in which the main terrigenous deposits consist of grey and yellow quartzitic conglomerates and planar- and trough-cross-stratified arkosic sandstones. In general, the deposits of the first alluvial fan assemblage are only well developed in the lowermost sequences of the upper member, whereas those of the second assemblage are widespread in the whole upper member sequences.

All these alluvial proximal facies grade laterally southwards into channelled alluvial fan facies assemblages of terrigenous sequences of red and grey sandstones and mudstones that, locally, include thin conglomerate levels (Agustí *et al.*, 1979). In the poorly drained areas of the distal or marginal alluvial fan sequences, lignite seams and grey silt levels with paleoflora (Sanz de Siria, 1980; Baltuille *et al.*, 1992) record the development of shallow ponds and swampy zones.

From the spatial distribution of these deposits, the lithological composition of the pebbles and the paleocurrent measurements, it has been inferred that the sequences of the upper member were deposited in various alluvial fans developed in the northern basin margins that prograded southwards. This southward progradation of the alluvial fans is clearly recorded by the coarsening-upward megasequence arrangement of this member (Fig. 6).

Acknowledgements

I would like to thank Lluis Cabrera and Carles Martín for critically reading a first draft of the manuscript and making helpful suggestions. I also wish to thank Peter Friend for his useful comments which improved the final version of this paper. This work is a contribution to the project PB91–0252 founded by the Comisión Interministerial de Ciencia y Tecnología (CICYT).

References

Agustí, J. and Roca, E. (1987). Síntesis biostratigráfica de la fosa de la Cerdanya (Pirineos orientales). *Estud. Geol.*, **43**, 521–529.

Agustí, J., Gibert, J., Moya, S. and Cabrera, L. (1979). Roedores e insectivoros (Mammalia) del Mioceno superior de la Seu d'Urgell (Cataluña, España). *Acta Geol. Hisp.*, **14**, 362–369.

Alvarez, C. and Golpe Posse, J.M. (1981). Sobre la paleobiología de la cuenca de Cerdanya (depresiones pirenaicas). *Bol. R. Soc. Esp. Hist. Nat. (Geol)*, **79**, 31–44.

Anadón, P., Cabrera, L., Julià, R., Roca, E. and Rosell, L. (1989). Lacustrine oil–shale basins in Tertiary grabens from NE Spain

(Western European rift system). *Palaeogeogr. Palaeoclimatol. Palaeoecol.*, **70**, 7–28.

Baltuille, J.M., Becker-Platen, J.D., Benda, L. and Ivanovic Calzaga, Y. (1992). A contribution to the subdivision of the Neogene in Spain using palynology. *Newsl. Stratigr.*, **27**, 41–57.

Barrón, E. (1992). Relaciones paleobotánicas de la flora del Mioceno superior de la Cerdanya (Lleida). *VIII Jornadas de Paleontologia, Barcelona 1992, Resúmenes*, 15–17.

Bech, J. and Vallejo, V.R. (1977). Contribución al conocimiento de la anapaita de la Cerdanya. *Acta Geol. Hisp.*, **12**, 113–116.

Bessedik, M. (1985). Reconstitution des environnements miocènes des régions nordouest mediterranéennes à partir de la Palynologie. PhD Thesis, Univ. Montpellier, 162 pp.

Cabrera, Ll., Roca, E. and Santanach, P. (1988). Basin formation at the end of a strike–slip fault: the Cerdanya Basin (eastern Pyrenees). *J. Geol. Soc., Lond.*, **145**, 261–268.

Calvet, M. (1985). Néotectonique et mise en place des reliefs dans l'Est des Pyrénées: l'example du horst des Albères. *Rev. Géol. Dynam. Géogr. Phys.*, **26**, 119–130.

Chevalier, M. (1909). Note sur la 'cuencita' de la Seo de Urgel. *Bull. Soc. Géol. France*, Sèrie IV, **9**, 158–178.

Crusafont, M. and Golpe Posse, J.M. (1974). El nuevo yacimiento vallesiense de Ballestar (nota preliminar). *Bol. R. Soc. Esp. Hist. Nat. (Geol.)*, **72**, 67–73.

Depéret, Ch. and Rérolle, L. (1885). Note sur la géologie et sur les mammifères fossiles du bassin lacustre Miocène superieur de la Cerdagne. *Bull. Soc. Géol. France*, **13**, 488–506.

Golpe Posse, J.M. (1981). Los mamíferos de las Cuencas de Cerdanya y Seu d'Urgell (depresiones pirenaicas) y sus yacimientos: Vallesiense Medio-Superior. *Bol. Geol. Mineral.*, **92**, 91–100.

Gourinard, Y. (1971a). Détermination cartographique et géophysique de la position des failles bordières du fossé néogène de Cerdagne (Pyrénées-Orientales franco-espagnoles). *96 Congr. Nat. Soc. Sav., Toulouse, Sciences*, **2**, 245–263.

Gourinard, Y. (1971b). Les moraines de la basse vallée du Carol entre Latour et Puigcerdà (Pyrénées orientales franco-espagnoles). *C. R. Acad. Sci. Paris*, **272**, 3112–3115.

Hartevelt, J. (1970). Geology of the upper Segre and Valira valleys, Central Pyrenees, Andorra, Spain. *Leidse Geol. Meded.*, **45**, 167–236.

Julià, R. (1984). Síntesis geológica de la Cerdanya. Girona. In *El borde mediterráneo español: evolución del orógeno bético y geodinámica de las depresiones neógenas*. C.S.I.C., Granada, 95–98.

Julià, R. (1992). The Neogene lacustrine deposits from the Cerdanya intramontane basin, (eastern Pyrenees). In J. Catalán and J. Ll. Pretus (eds.), *Mid-Congress Excursions. XXV SIL Internatl. Congr., Barcelona 1992*, 13-1–13-7.

Margalef, R. (1957). Paleoecologia del lago de la Cerdaña. *Publ. Inst. Biol. Aplicada*, **25**, 131–137.

Margalef, R. and Marrasé, C. (1985). On the Paleoecology of the Miocene lake of La Cerdanya (Pyrenees) and a multiple event of massive phosphate precipitation. *4th Int. Symp. Palaeolimnol., Ossiac 1985; abstracts*, 56.

Menéndez Amor, J. (1955). *La depresión ceretana española y sus vegetales fósiles*. Mem. Real. Acad. Cienc. Exact. Fis. Nat. Madrid, **18**, 344 pp.

Panzer, W. (1932). Die eiszeitlichen Endmoränen von Puigcerdá (ostpyrenäen). *Zeitschrift für Gletscherkunge. An.. Glaciol.*, **20**, 411–421.

Pous, J., Julià, R. and Solé Sugrañes, L. (1986). Cerdanya basin geometry and its implication on the Neogene evolution of the Eastern Pyrenees. *Tectonophysics*, **129**, 355–365.

Roca, E. (1986). Estudi geològic de la fossa de la Cerdanya. Thesis. Univ. Barcelona, 109 pp.

Roca, E. (1992a). La fossa de la Cerdanya. In J. Guimerà (ed.), *Història natural dels països Catalans. 2. Geologia (II)*. Enciclopèdia Catalana S.A., Barcelona, 301–305.

Roca, E. (1992b). La fossa de la Seu d'Urgell. In J. Guimerà (ed.), *Història natural dels països Catalans. 2. Geologia (II)*. Enciclopèdia Catalana S.A., Barcelona, 305–306.

Roca, E. and Santanach, P. (1986). Génesis y evolución de la fosa de la Cerdanya (Pirineos orientales). *Geogaceta*, **1**, 37–38.

Roca, E., Julià, R., Cabrera, Ll. and Anadón, P. (1987). Late Miocene lacustrine diatomites and early diagenetic phosphates from the Cerdanya basin (eastern Pyrenees). *Terra cognita*, **7**, 222.

Sanz de Siria, A. (1978). La flora miocénica de las cuencas pirenaicas catalanas. *Bul. Inf. Inst. Paleontol., Sabadell*, **10**, 52–62.

Sanz de Siria, A. (1980). Estudio sistemático y paleontológico de la flora miocénica de la cuenca de la Seu d'Urgell. *Paleontol. Evol.*, **15**, 3–29.

Suc, J.P. (1984). Origin and evolution of the Mediterranean vegetation and climate in Europe. *Nature*, **307**, 429–432.

Villalta, J.F. (1962). Dos coleópteros fósiles procedentes de la depresión de la Cerdaña (Lérida). *Estud. Geol.*, **18**, 105–109.

Villalta, J.F. and Crusafont, M. (1945). La flora miocénica de la depresión de Bellver. *Ilerda*, **3**, 339–353.

E11 Eocene–Oligocene thrusting and basin configuration in the eastern and central Pyrenees (Spain)

J. VERGÉS AND D.W. BURBANK

Abstract

Outcrop, borehole, seismic, paleontologic and paleomagnetic data have been combined in order to define the geometric evolution of the eastern and central Pyrenean fold-and-thrust belt and its adjacent foreland basin during Eocene and Oligocene times. Restoration of balanced sections within a concise temporal framework forms the basis for more reliable and detailed palinspastic reconstructions than previously attained for this part of the Pyrenean orogeny.

Introduction

The eastern and central part of the southern Pyrenees and related foreland Ebro basin display an unusually well-preserved geological record of the tectonic evolution of the chain. Numerous aspects of the compressional history of the south-eastern and south-central Pyrenees have been recently delineated and provided an initial basis for detailed reconstructions of the kinematic development of the range. Within the adjacent foreland and piggyback basins, extensive paleomagnetostratigraphic sections provide good age control for the syntectonic sediments of the foreland basins.

The goal of this chapter is the construction of palinspastic maps ranging in age from the Early Eocene to the Early Oligocene showing the tectonic evolution of the eastern and central southern Pyrenees. These maps are constrained by both structural and temporal data sets. Two balanced and restored cross-sections extending from the undeformed foreland strata to the inner part of the chain provide the main structural framework. The cross-sections are constructed using geological surface information and the available subsurface data, including seismic lines, oil-well data and potash-well data.

When considered sequentially in the context of the detailed temporal control of the syntectonic sedimentary record, these maps provide a three-dimensional reconstruction of the evolution of the southern Pyrenees and stand in contrast to the two-dimensional image that is typically derived from single cross-sections.

Geological setting

Shortening related to north-dipping subduction of Iberia under Europe caused the Pyrenean orogeny (e.g., Muñoz, 1992) and the formation of the piggyback foreland basins (Fig. 1). The Ebro foreland basin constitutes the latest stage of the foreland evolution in the southern Pyrenees. The eastern and central Pyrenees display contrasting tectonic styles. The eastern part is characterized by widespread deformation in the foreland basin and by the stacking of the 'upper' and 'lower' Pedraforca thrust sheets, primarily involving Mesozoic cover rocks, over the Cadí thrust sheet, which comprises both basement and cover rocks (Fig. 2). The central part of the Pyrenees is characterized by an extensive thrust sheet (South Central Unit, SCU, of Séguret, 1972) comprising mainly Mesozoic strata. This SCU is formed by three major structural units that are equivalent to the cover thrust sheets in the eastern Pyrenees (Vergés *et al.*, 1992). The boundary between the eastern and central Pyrenees comprises a set of oblique thrusts, the Segre oblique thrust zone, which represents a long-lived hanging wall structure that both pre-dates and is coeval with southward emplacement of the SCU.

Structural reconstructions

Description of the cross-sections

Two balanced and restored cross-sections (see Fig. 2 for location) were constructed in order to calculate the absolute amounts of shortening and the position of the traces of the main thrusts. The eastern cross-section, across the southeastern Pyrenees, extends to the southern margin of the foreland basin, as represented by the Catalan Coastal Ranges. The western cross-section, across the eastern termination of the SCU, reaches the undeformed foreland Ebro strata. Both cross-sections are constructed across oblique structures, but they are parallel to the tectonic transport direction (Martínez *et al.*, 1988). Both cross-sections were constructed using the length-line balancing method and including the available drill hole and seismic data. The timing and magnitude of advance of the traces of the thrusts form the basis

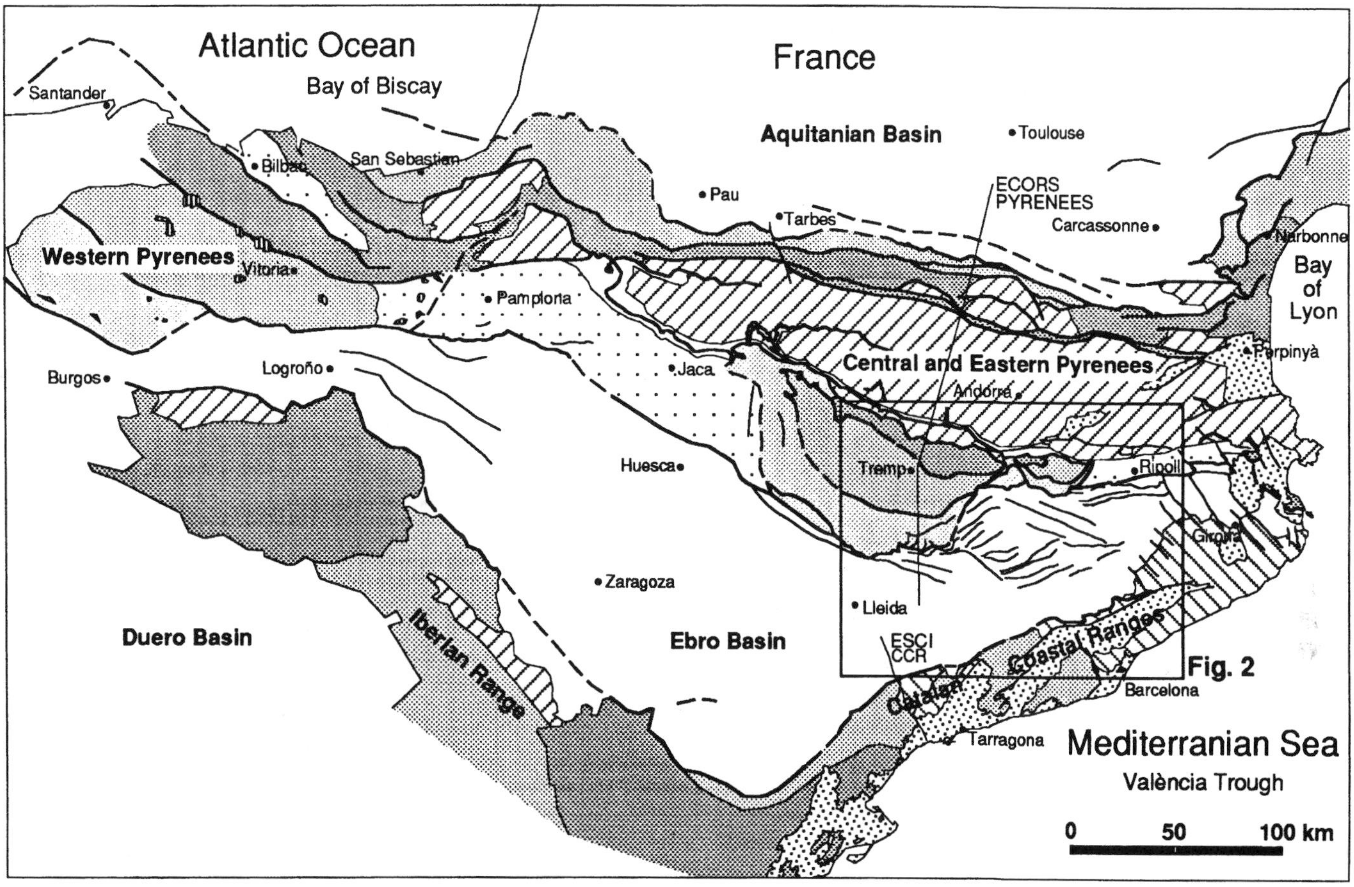

Fig. 1. Simplified geologic map of the Pyrenees, Iberian Range, and Catalan Coastal Range and of the adjacent foreland and piggyback basins. Box shows the location of Figure 2.

for the palinspastic maps. Note that the magnitude of advance of a thrust in map view is often smaller than the shortening of the thrust sheet due to internal deformation. However, when there is an efficient décollement level at depth, shortening and advance of the thrust traces will be nearly equal.

Eastern balanced and restored cross-section

Structure

Beginning in the south, the deformed foreland Ebro basin shows undeformed strata at depth and a set of detached folds above the Cardona salt level (Fig. 3). These salts represent the main décollement level of the eastern and eastern-central foreland fold-and-thrust belt (Ramírez & Riba, 1975; Vergés *et al.*, 1992). The Oló anticline and thrust represent the southernmost structure of the foreland and the tip line of the Pyrenean thrust system in this traverse (Vergés *et al.*, 1992). The Puig-reig anticline represents the transition between the foreland-basin structures and the Pyrenean thrust sheets, and it corresponds to a duplex structure as imaged by seismic lines.

The Pyrenean thrust sheets form a succession of three piggyback tectonic units (Muñoz *et al.*, 1986; Vergés & Martínez, 1988). The stratigraphy of each thrust sheet is depicted in the balanced and restored sections (Fig. 3). The Cadí thrust sheet is the structurally lowest, youngest and southernmost, whereas the upper Pedraforca thrust sheet is the structurally highest, was emplaced earliest, and is situated farther to the north (Fig. 3). The lower Pedraforca thrust sheet is located on top of the underlying Cadí thrust sheet and is folded into a syncline. Its internal structure comprises a set of imbricate thrusts, separating different stratigraphic units (Vergés & Martínez, 1988). The present geometry of the upper Pedraforca thrust sheet, equivalent to that of the Bóixols thrust sheet (Berastegui *et al.*, 1989) to the west, results from a tectonic inversion of pre-existing Lower Cretaceous extensional faults (Vergés & Martínez, 1988).

The pin-point of the restored, eastern cross-section (Fig. 3) is fixed on the undeformed foreland coinciding with the Castellfullit oil-well (Fig. 2). Four different horizontal levels during the time of deposition have been chosen as reference lines for the construction of the restored cross-section. From south to north, they are the top of the Cardona evaporites and their equivalent the transition levels to the north (after Riba, 1973); the top of the Beuda evaporites for the Cadí thrust sheet; the top of the Garumnian red beds for the Cadí–lower Pedraforca thrust sheets; and finally the Cenomanian limestones for the lower–upper Pedraforca thrust sheets.

The total shortening involved in this traverse from the pin-point

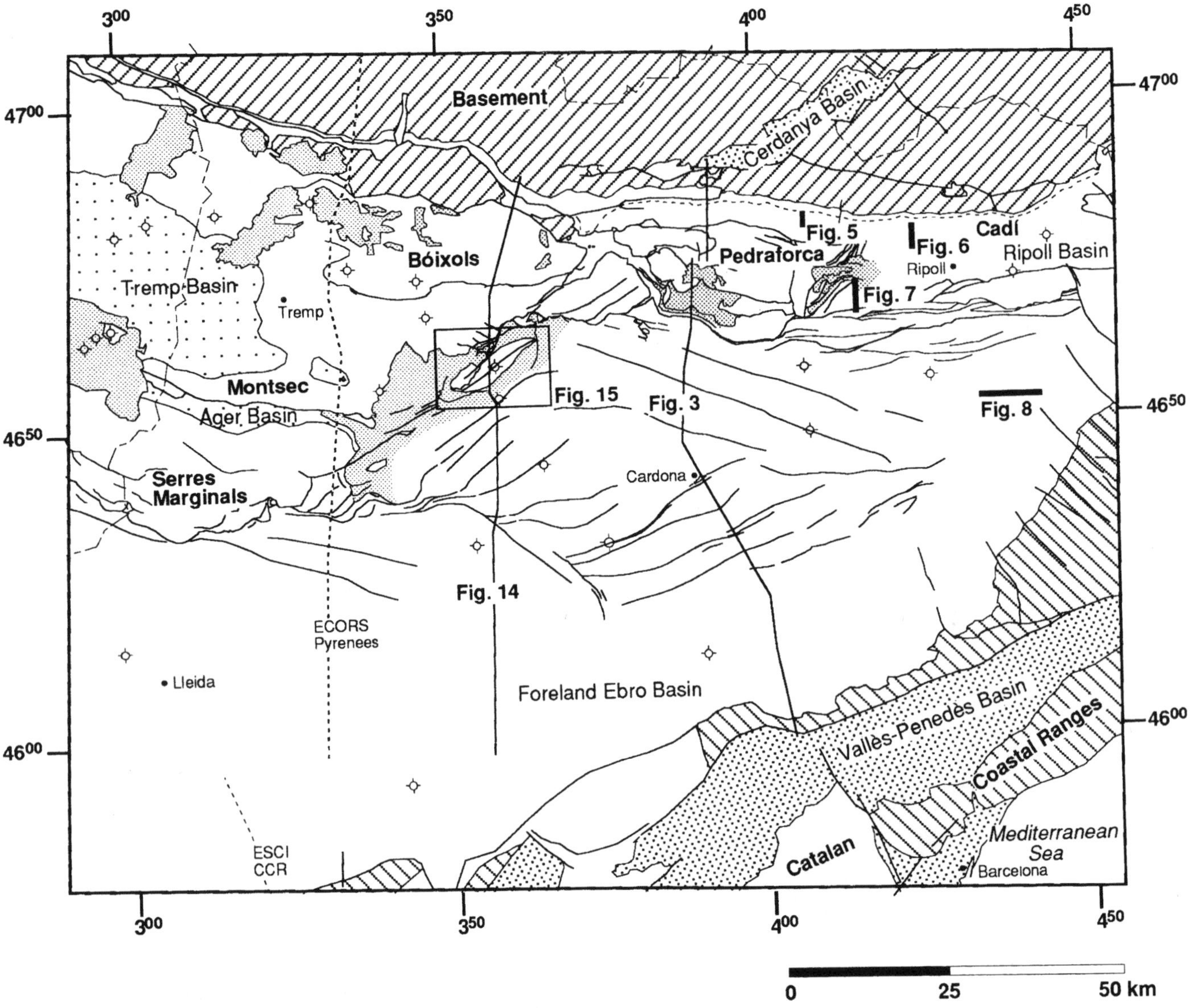

Fig. 2. Map of the central and eastern Pyrenees showng the position of the structural cross-sections and magnetic sections (marked with Figure numbers), well locations (Lanaja, 1987), the ECORS profile, and the major tectonic elements.

to the northern outcrop of Jurassic rocks in the upper Pedraforca thrust sheet is 64 km. We can partition this into 6 km for the upper Pedraforca thrust sheet, 36.2 km for the lower Pedraforca thrust sheet, 16.2 km for the frontal Pyrenean blind and emergent thrust and finally 5.6 km for the deformed foreland. The lower Pedraforca thrust sheet displays 27.4 km of translation over the basal thrust and 8.8 km of internal shortening related to internal imbrication.

Timing

Time control for the complete motion of the different structures and for the related foreland strata (Fig. 4) along this cross-section has been derived from four paleomagnetic sections (Fig. 2). Three sections are located to the north (Bagà; Fig. 5), east (Gombrèn; Fig. 6), and southeast (St. Jaume; Fig. 7) of the present limits of the Pedraforca thrust sheets. The fourth section (Vic; Fig. 8) is situated in the eastern Ebro basin and provides time control on the younger, more distal foreland strata. The magnetic sections are constructed from multiple, superposed magnetic sites (for fuller description, see Burbank *et al.*, 1992a, 1992b) and are correlated with the new magnetic polarity time scale of Cande & Kent (1992). In comparison with many previous time scales (e.g. Berggren *et al.*, 1985), the Oligo-Miocene boundary in the Cande and Kent time scale is the same age, but the upper and lower boundaries of the Eocene epoch are both about 2.7 Ma younger. Whereas estimates of the absolute age of events within the Eocene will thus be affected, derived rates will be largely unchanged. The magnetostratigraphies show that (a) the Bagà section, which encompasses the Cadí and Corones sequences (Fig. 9), spans from ~56 to 51.5 Ma (Fig. 10); (b) the Gombrèn section (Fig. 11), which comprises the Armàncies to Bellmunt sequences, spans 51.5 to ~43 Ma; (c) the St. Jaume

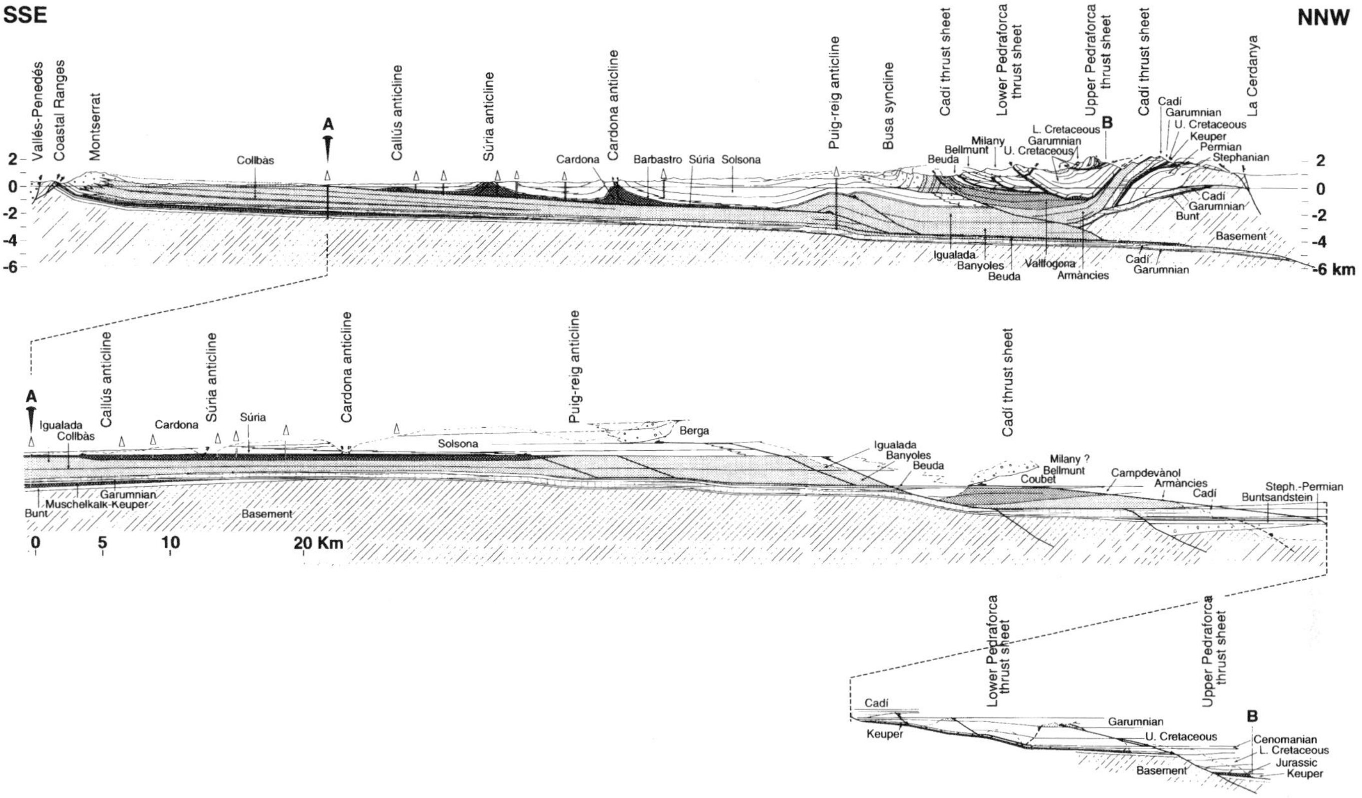

Fig. 3. Upper section: NNW–SSE balanced structural cross-section in the eastern Pyrenees (see Figure 2 for location) from the southernmost Axial Zone, across the cover thrust sheets and the Ebro foreland basin to the Catalan Coastal Range. The deep structure and the depth of the cover–basement contact are based on available seismic information and oil- and potash-well data. The main décollement levels of the Pyrenean thrust system, from N to S, correspond to the Keuper, the top of the Early Eocene limestones, the Beuda and Cardona evaporites.

Lower section: Restored cross-section between the points A (pin point) and B (northernmost outcrop of Jurassic rocks in the Upper Pedrafroca thrust sheet) on the upper section. From S to N, the top of the Cardona and Beuda evaporites, the top of the Garumnian red beds, and the base of the Cenomanian have been taken as sub-horizontal reference levels in the restoration. Between the Beuda and Cardona evaporites, the Banyoles and Igualada marls wedge to the N (Pyrenees) and to the S (Catalan Coastal Ranges).

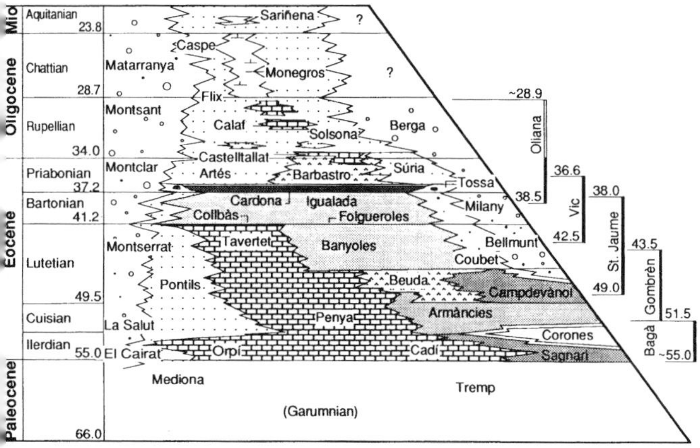

Fig. 4. N–S stratigraphic panel depicting the temporal positions of the syntectonic foreland strata and the dated magnetic sections related to them.

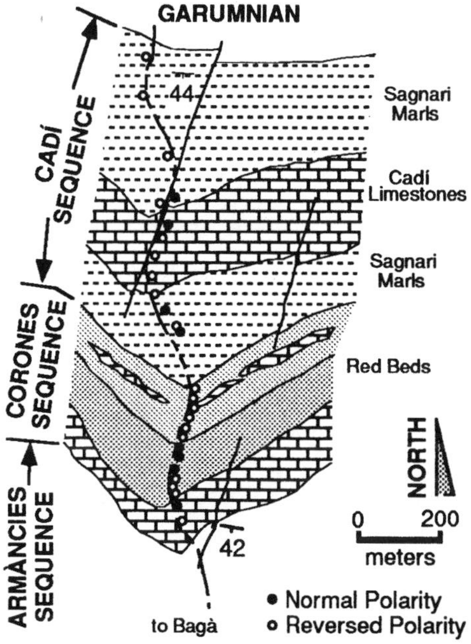

Fig. 5. Location map of the Bagà magnetic section. See Figure 2 for location with respect to the Pedraforca thrust sheets.

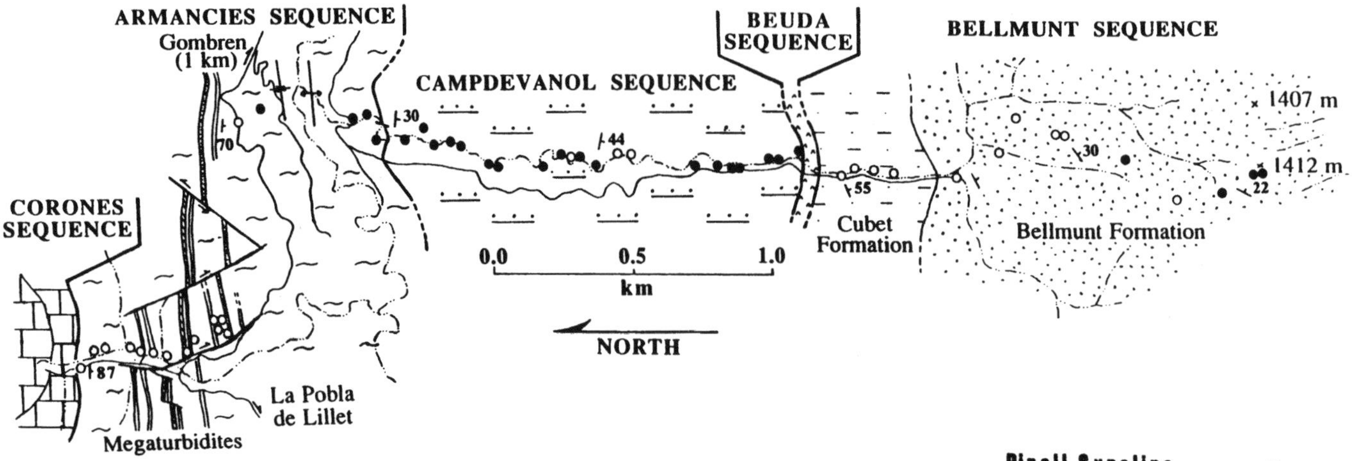

Fig. 6. Location map of the Gombrèn magnetic section in the northern limb of the Ripoll syncline. See Figure 2 for location with respect to the Pedraforca thrust sheets.

section (Fig. 12), which encompasses the Campdevànol to Milany sequences, ranges from ~49.5 to ~38 Ma; and (d) the Vic section, which includes the Bellmunt to Solsona sequences (Fig. 13), ranges in age from ~43 Ma to ~36.5 Ma.

Based on these age constraints (Fig. 10), related paleontological data, and the deformational inferences that can be derived from syntectonic strata associated with individual structures, the timing of the deformation from south to north is as follows:

The initiation of growth of the Cardona anticline occurred in the lowermost part of the early Oligocene.

Multiple phases of motion can be defined for the Vallfogona thrust (southern boundary of the Cadí thrust sheet). The earliest phase of motion corresponds with the base of the Bellmunt red beds at ~43.5 Ma (Fig. 10) and is marked by a progressive unconformity. Continued or later motion resulted in folding of the Ripoll syncline about an E–W axis between ~41 and 40 Ma, and a large progressive unconformity within the Solsona sequence indicates the final stage of Vallfogona thrusting between 36 and perhaps as young as 30 Ma.

Break-back imbricates of the Lower Pedraforca thrust sheets developed during Lutetian and Bartonian times (Martínez *et al.*, 1988) between 44 and 38 Ma (Burbank *et al.*, 1992b).

The initiation of the southward translation of the lower-Pedraforca thrust sheet along its basal detachment is documented by the deposition of the Corones Fm. (Puigdefábregas *et al.*, 1986) between 54 and 51.5 Ma (Burbank *et al.*, 1992b).

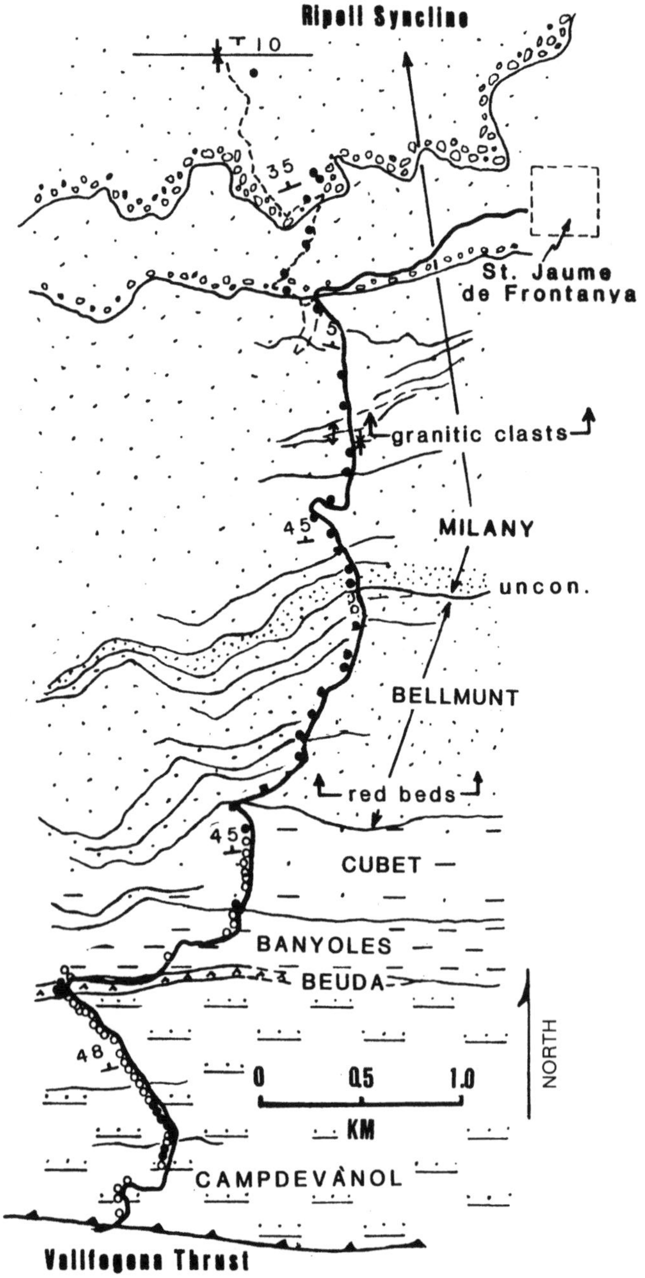

Fig. 7. Location map of the St. Jaume de Frontanyà magnetic section in the southern limb of the Ripoll syncline. See Figure 2 for location with respect to the Pedraforca thrust sheets.

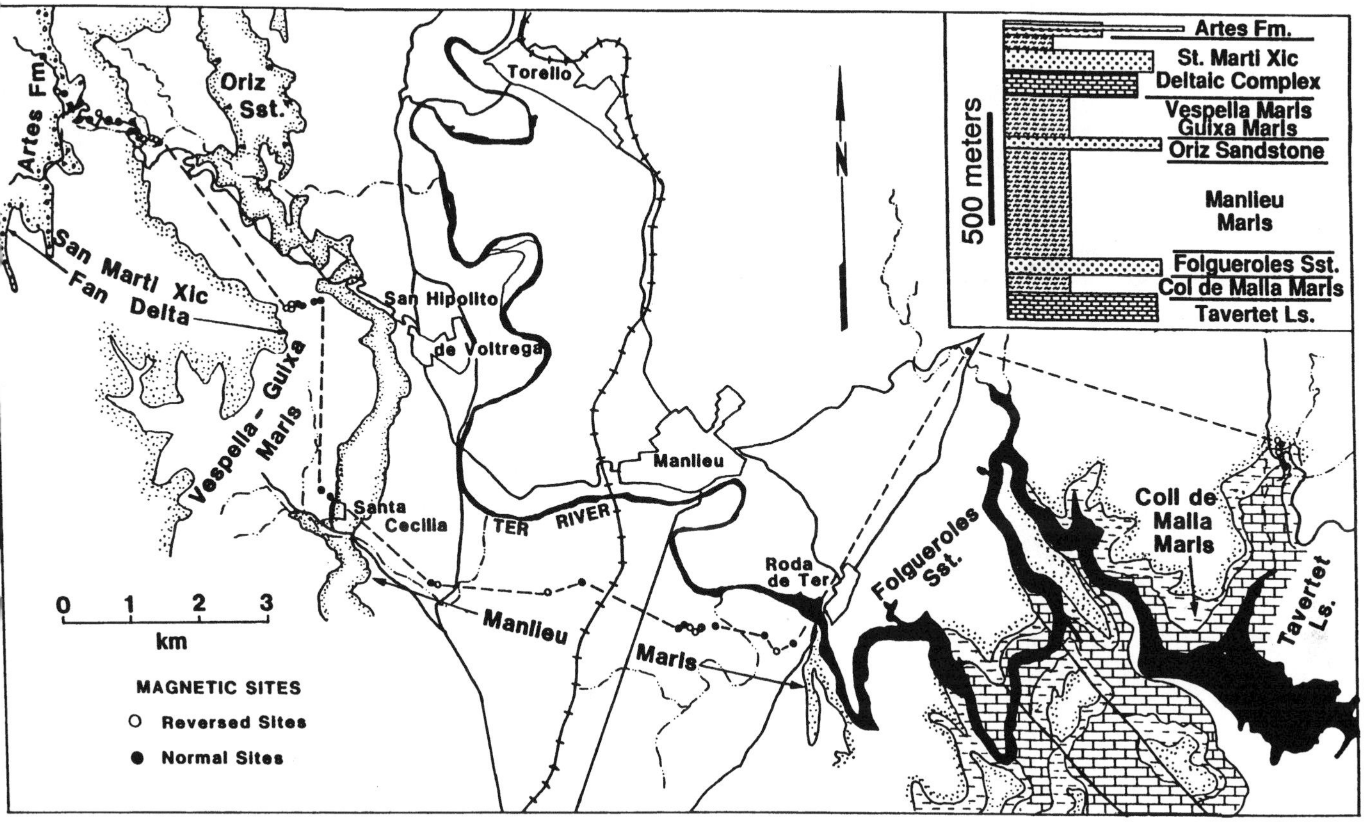

Fig. 8. Location map of the Vic magnetic section in the eastern Ebro foreland to the south of the cover thrust sheets. See Figure 2 for location.

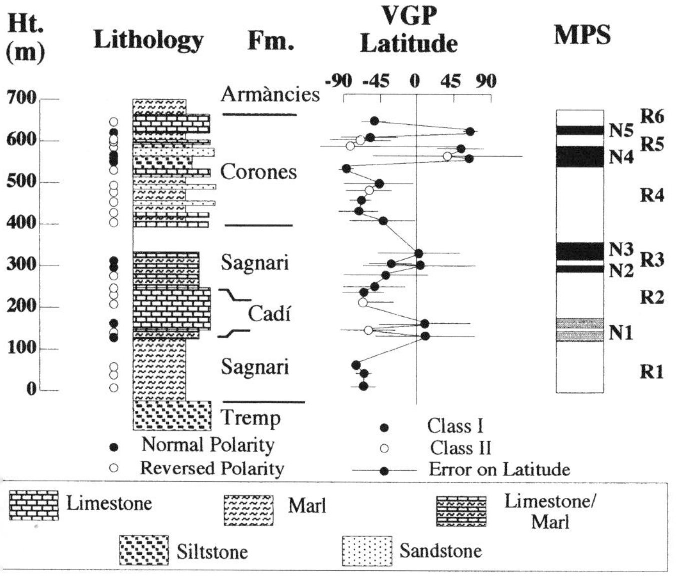

Fig. 9. Magnetic polarity stratigraphy for the 700-m-thick Bagà section, where 33 sites were collected in the Cadí and Corones sequences. Sites yielding low latitude virtual geomagnetic poles with high uncertainties are not included in the magnetic zonation. Normal (N) and Reversed (R) intervals are distinguished in the Magnetic Polarity Sequence (MPS).

The upper Pedraforca thrust sheet was emplaced during the upper Garumnian deposition in Palaeocene times (Vergés & Martínez, 1988).

Western balanced and restored cross-section

Structure

The Sanaüja backthrusted anticline represents the southernmost structure and tip line in this traverse (Fig. 14). Its position is in close proximity to the southern limit of the Cardona salt décollement level (Vergés *et al.*, 1992). The Oliana anticline represents the northern and major structure of the foreland. It constitutes a duplex structure at depth overthrusting the undeformed foreland strata, as defined by seismic lines and well-data (Vergés & Muñoz, 1990; Burbank *et al.*, 1992a). North of the Oliana anticline, the Pyrenean thrust sheets are represented from south to north by the Serres Marginals, Montsec and Bóixols thrust sheets (Fig. 14) that are equivalent to the lower and upper Pedraforca units (Vergés *et al.*, 1992). The Montsec thrust sheet is cut by north-dipping normal faults that could have accommodated the change in thickness of upper Cretaceous strata between the Montsec hanging wall and footwall. The Serres Marginals, Montsec and Bóixols thrust sheets display a more complete geological record in this cross-section than in the eastern cross-section, where southward tilting during emplacement of basement thrust units led to erosion of the

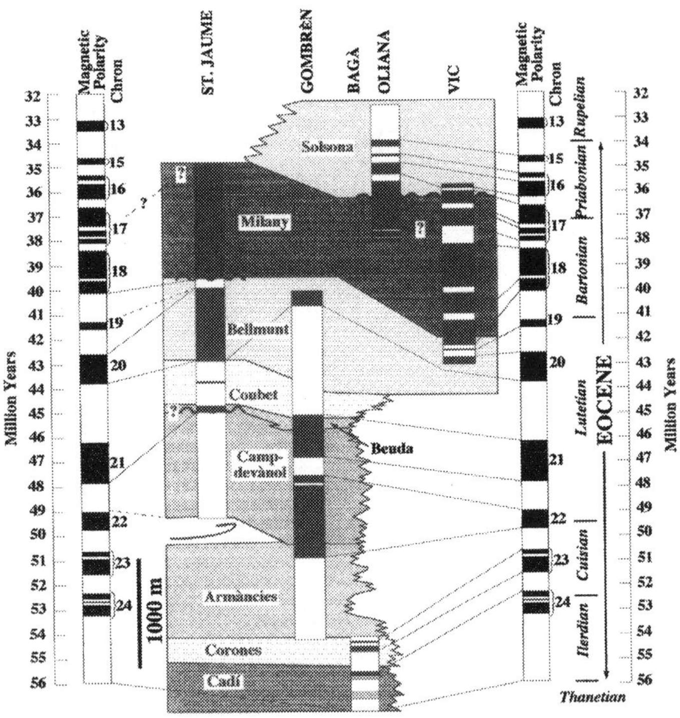

Fig. 10. Correlation of the magnetic polarity stratigraphies of the eastern and central Pyrenean foreland with the magnetic polarity time scale of Cande and Kent (1992). Major depositional sequences (Puigdefábregas *et al.*, 1986) are depicted in their appropriate stratigraphic range for each section. Note that the correlation of the Campdevànol strata between the Gombrèn and St. Jaume de Frontanyà section is problematic. These correlations are guided by relevant faunal data which permit little flexibility in where they must fit into the magnetic time scale. Despite the fact that the St. Jaume section is truncated by the Vallfogona thrust at its base, the magnetic patterns do not match well and appear to require large changes in sediment-accumulation rates across the rather small separation between the sections. Alternatively, weakly magnetized strata and a hard-to-remove normal overprint may account for the apparent mismatch.

hindward part of the upper Pedraforca (Bóxols-equivalent) thrust sheet.

The restored cross-section is fixed on the undeformed foreland coinciding with the Guissona oil-well (Fig. 14). Three different horizontal levels have been chosen as reference lines. From south to north, they are the top of the Cardona evaporites; the top of the Garumnian red beds for the Serres Marginals and Montsec thrust sheets; and the base of the Cenomanian limestones for the Montsec and Bóixols thrust sheets. The Cardona evaporites have been areally restored. The location of the lower footwall ramp is controlled by the restoration of the middle and upper Eocene marls in the Oliana duplex.

The total shortening involved in this western traverse from the pin point to the northern outcrop of Jurassic rocks in the Bóixols thrust sheet is 60.7 km. These can be subdivided in to 7.2 km for the Bóixols thrust sheet, 3.4 km for the Montsec thrust sheet, 27.6 km for the basal thrust of the SCU, 1.1 km for the reactivation within

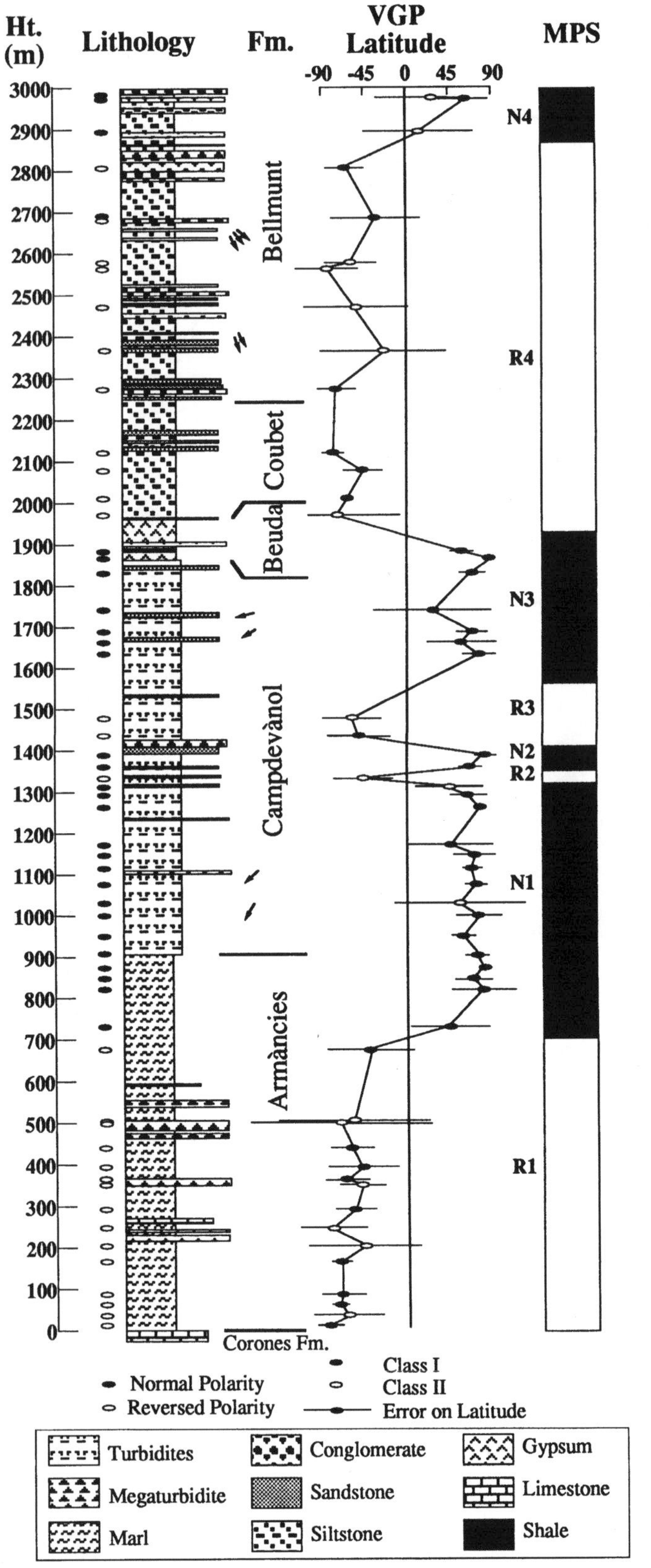

Fig. 11. Magnetic polarity stratigraphy for the 3000-m-thick Gombrèn section, where 94 sites were collected in the Armàncies to Bellmunt sequences. Normal (N) and Reversed (R) intervals are distinguished in the Magnetic Polarity Sequence (MPS).

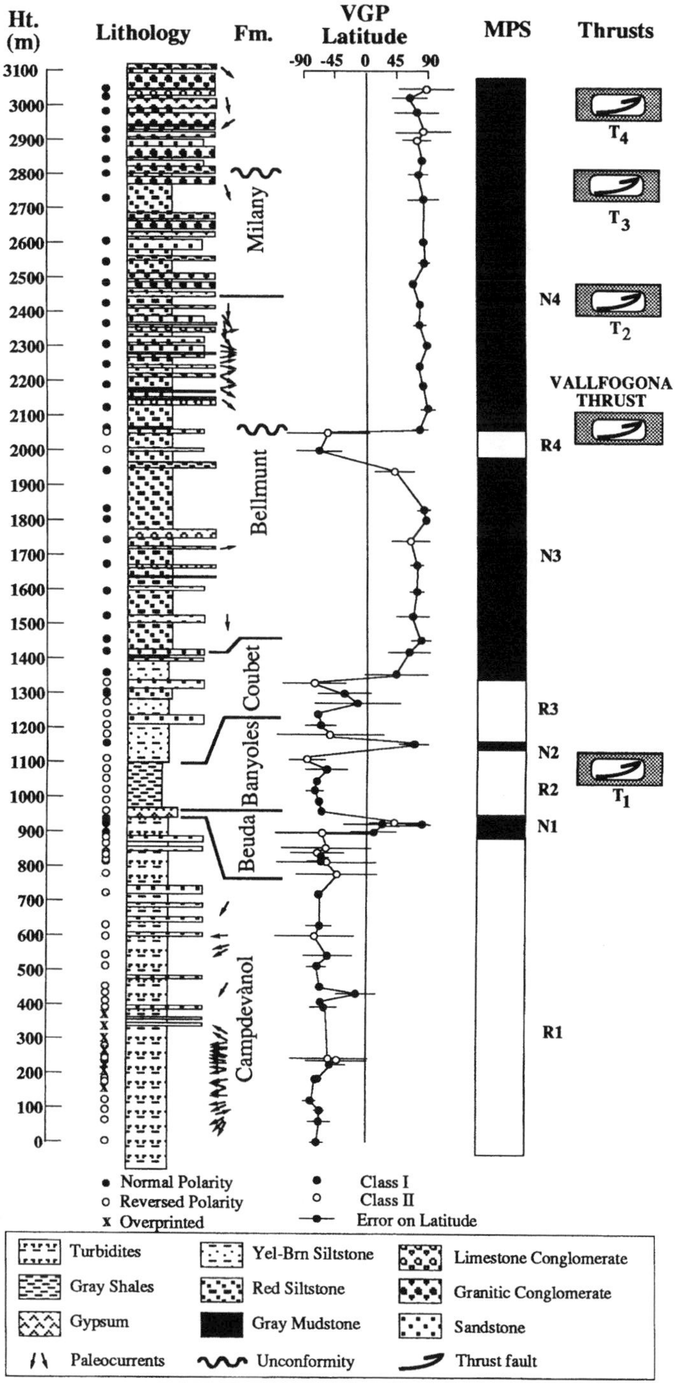

Fig. 12. Magnetic polarity stratigraphy for the 3100-m-thick St. Jaume de Frontanyà section, where 93 sites were collected in the Campdevànol to Milany sequences. Normal (N) and Reversed (R) intervals are distinguished in the Magnetic Polarity Sequence (MPS).

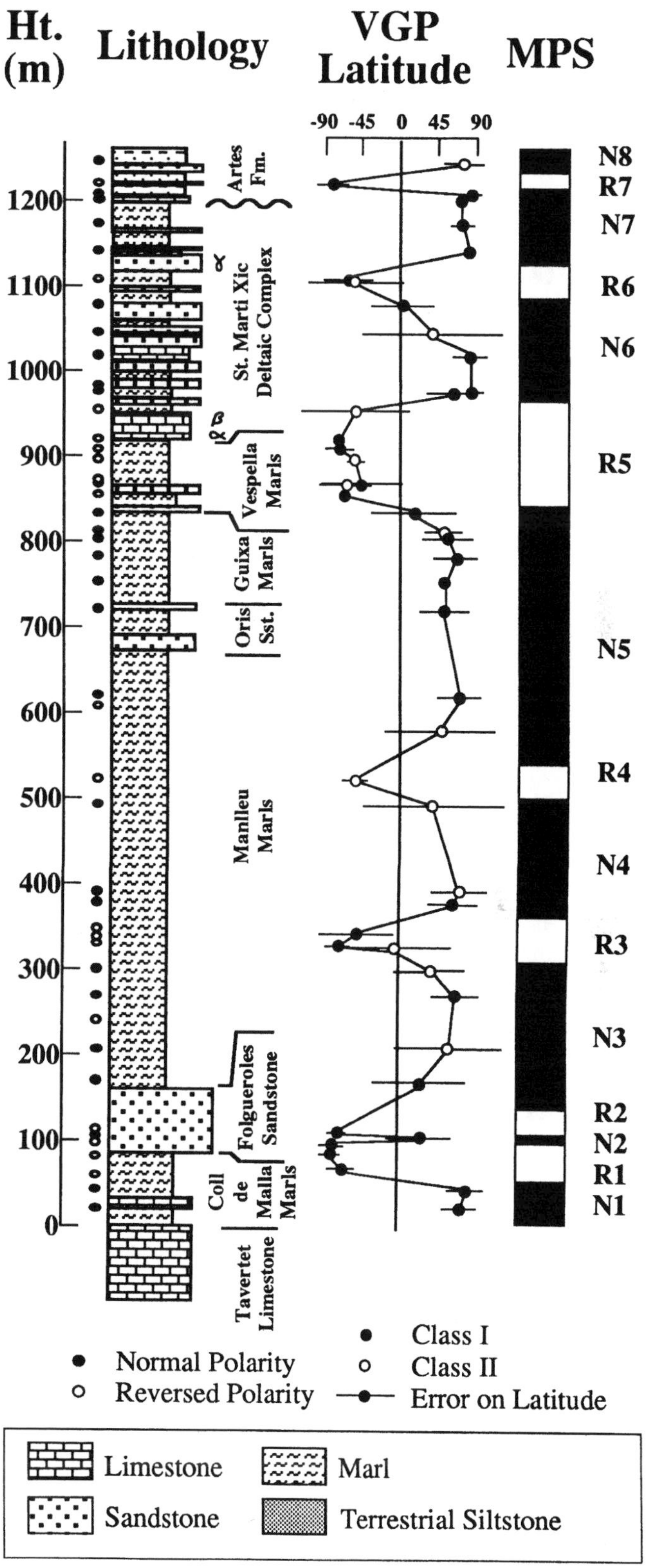

Fig. 13. Magnetic polarity stratigraphy for the 1250-m-thick Vic section, where 72 sites were collected in the Bellmunt to Solsona sequences. Normal (N) and Reversed (R) intervals are distinguished in the Magnetic Polarity Sequence (MPS).

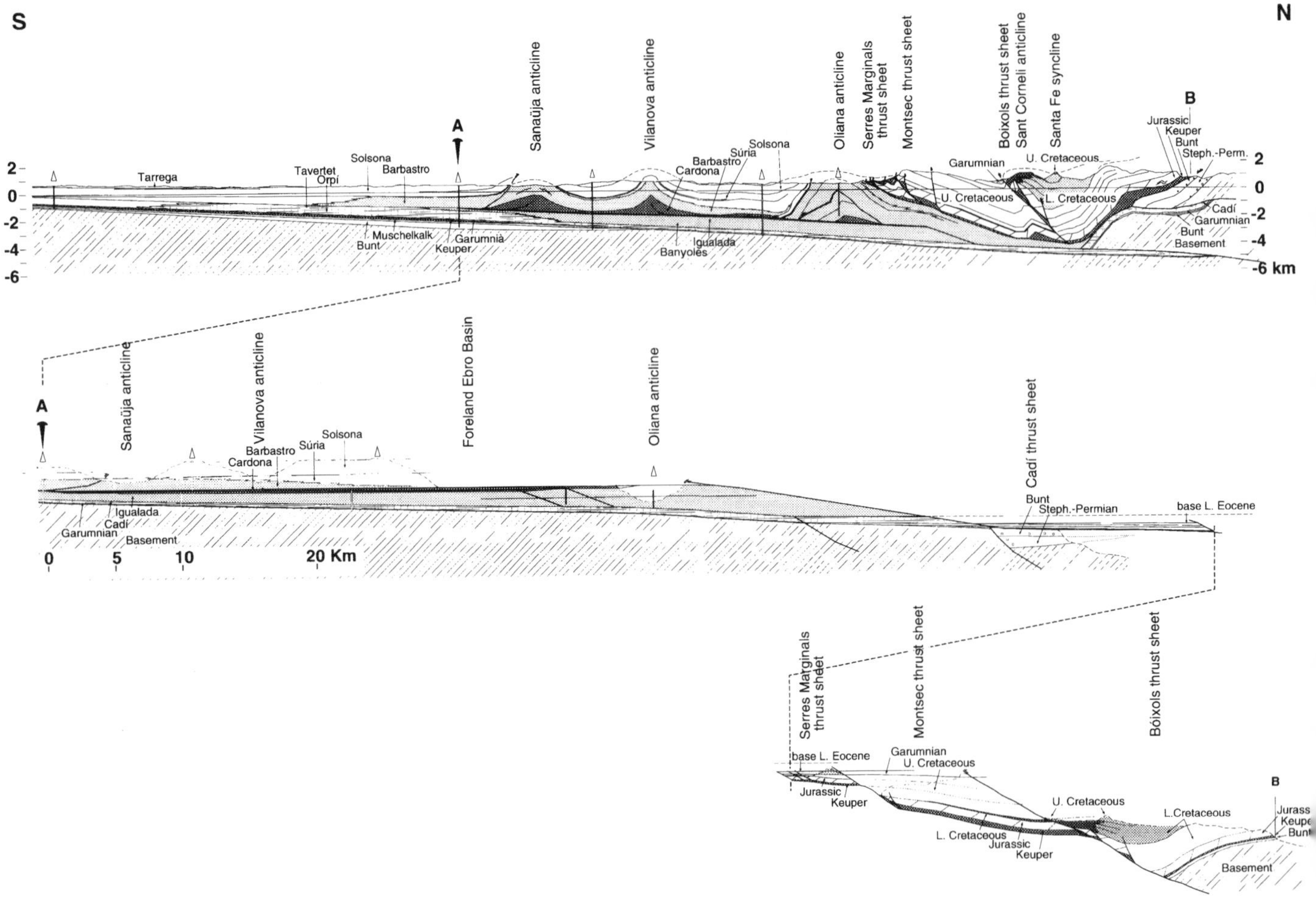

Fig. 14. Upper section: N–S balanced structural cross-section along the eastern part of the central Pyrenees (see Figure 2 for location). The Cadí thrust sheet is not represented in this cross-section, but some shortening (18 km at least) is taken along a detachment located within the Keuper (compare with the restored cross-section). The northern side of the cover thrust sheets was tilted during the emplacement of basement tectonic units over the basal Pyrenean thrust. Notice that if we restored the southern tilting of the upper Pedrafroca thrust sheet, it shows similar characteristics to the frontal part of the Bóixols thrust sheet.

Lower section: Restored cross-section between the points A (pin point) and B (northernmost outcrop of Jurassic rocks of the Bóixols thrust sheet) on the upper section, using the same sub-horizontal reference levels as in Figure 3. The Bóixols thrust sheet has been restored during Cenomanian times (modified from Berastegui *et al.*, 1989). Shortening during the Middle Eocene is represented by the displacement to the south of the Serres Marginals, Montsec and Bóixols thrust sheets along the basal thrust (represented by the present Serres Marginals thrust). A minimum amount of 18 km of shortening has been calculated from the palinspastic map reconstructions.

the Serres Marginals thrust sheet, and 21.4 km for the Oliana duplex and folding and thrusting in the foreland.

Timing

Magnetostratigraphic studies in the vicinity of Oliana (Fig. 15) have been used to constrain the mid-to-late Eocene deformational history (Burbank *et al.*, 1992a). The Oliana magnetic record (Fig. 16) extends from ~39 to 34 Ma (Fig. 10). Based on these age constraints, paleontological data from Oliana and elsewhere, and the deformational inferences that can be derived from syntectonic strata associated with individual structures, the timing of the deformation from south to north is as follows:

The Cardona evaporites was deposited ~37.2 Ma at the beginning of the Priabonian. Development of the Sanaüja backthrust and anticline and of the Vilanova anticline occurred following deposition of the Barbastro evaporites and Súria Formation (Figs. 4 and 14).

The Oliana anticline and duplex developed in a piggyback thrusting sequence between 36.5 and 34 Ma at the same time that the adjacent Serres Marginals and Montsec thrusts moved in a breakback sequence.

A minimum of 18 km of shortening along the basal thrust of the SCU and 3 km along the Montsec thrust occurred during early and middle Eocene times, but is not represented by syntectonic deposits in this section.

During the latest Cretaceous and early Paleocene, the Bóixols thrust inverted an Early Cretaceous extensional basin.

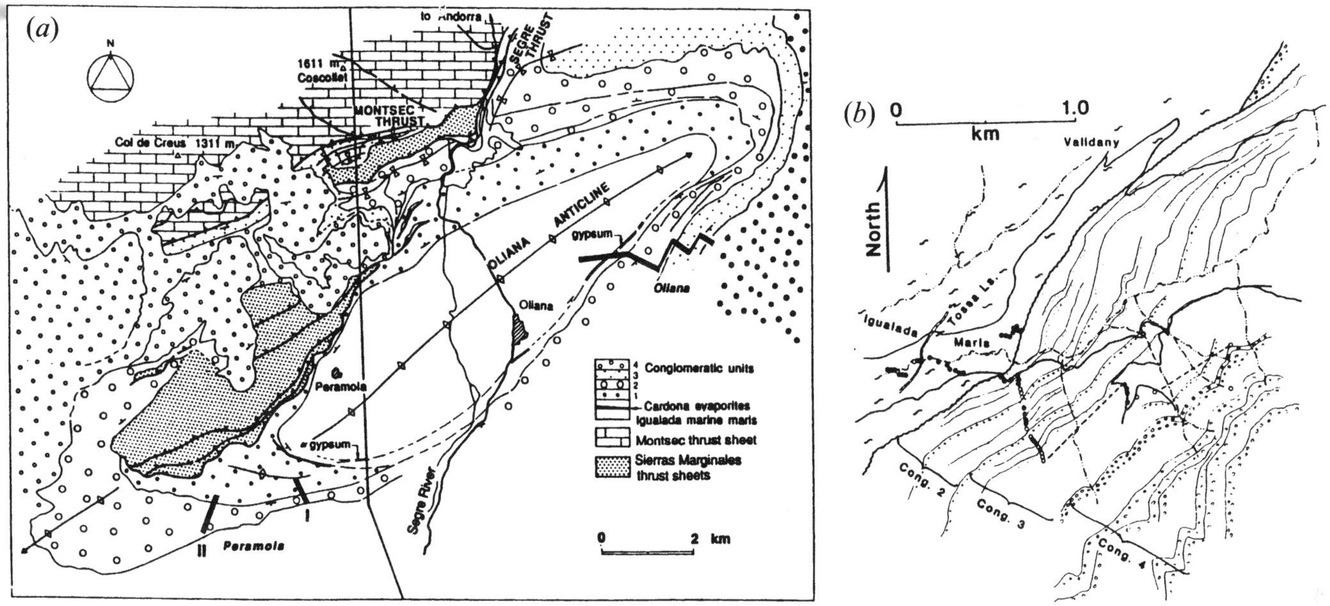

Fig. 15. A. Geologic map of the Oliana anticline, adjacent thrust sheets, and related sedimentary units. (See Figure 2 for location.) The location of the Oliana magnetic section is shown by the solid black line on the east limb of the anticline. Two other sections (Peramola I and II) were sampled to the SW. For more details, see Burbank *et al.*, 1992. B. Detailed site location map for the Oliana magnetic section showing normally (filled circles) and reversely (open circles) magnetized sites in the syntectonic deposits.

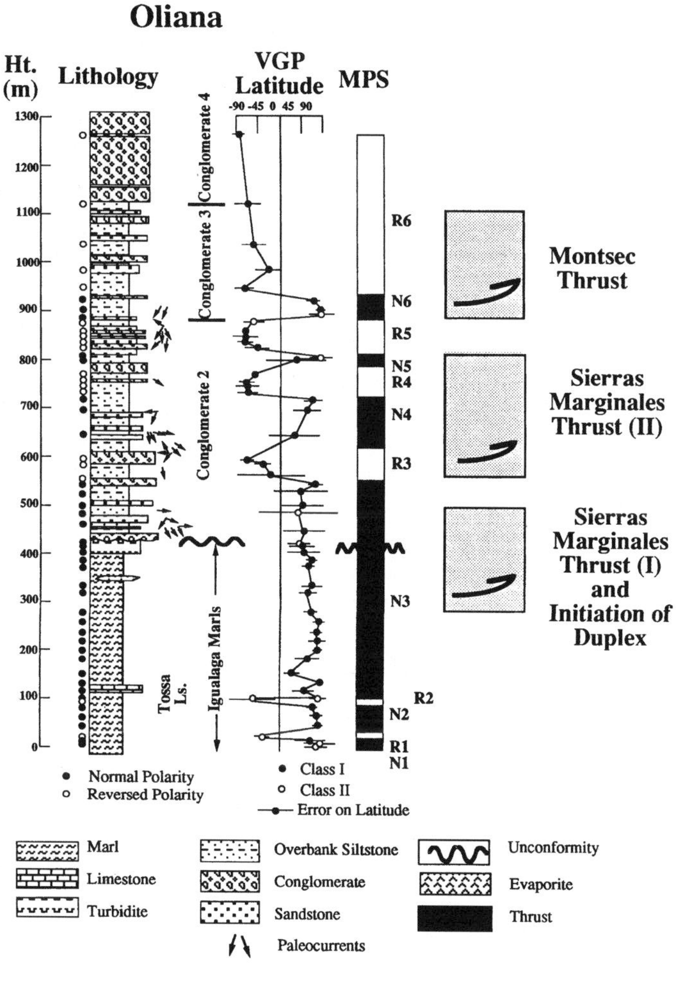

Palinspastic maps

Four different palinspastic reconstructions have been made in order to combine the cross-sections, and the magnetic and stratigraphic data. The maps include cross-section information in two dimensions (2D), thrusting time-control, the location and extension of the foreland evaporitic basins, and oil-well data in order to generate a 3D restoration.

Restoration between 55 Ma and 51 Ma (Ypresian)

Bóixols and upper Pedraforca thrust sheets were emplaced during the late Cretaceous (Simó & Puigdefábregas, 1985) and were fossilized by upper Garumnian red beds (Vergés & Martínez, 1988) prior to 55 Ma. The initial interval of Montsec thrusting occurred during Ilerdian and Cuisian time (Mutti *et al.*, 1985), folds developed within the Serres Marginals thrust sheet, and deposition occurred both within piggyback basins and the Ebro foreland basin (Fig. 17). At the close of this interval, motion of the Pedraforca thrust sheet commenced.

Restoration between 51 Ma (Late Ypresian) and 47 Ma (Middle Lutetian)

During this interval, the tectonic style evolved from detached folds to thrusts over the Keuper (main décollement level during that time), and complete overthrusting of the Mesozoic

Fig. 16. Magnetic polarity stratigraphy for the 1300-m-thick Oliana section, where 73 sites were collected in the Milany to Solsona sequences. Normal (N) and Reversed (R) intervals are distinguished in the Magnetic Polarity Sequence (MPS).

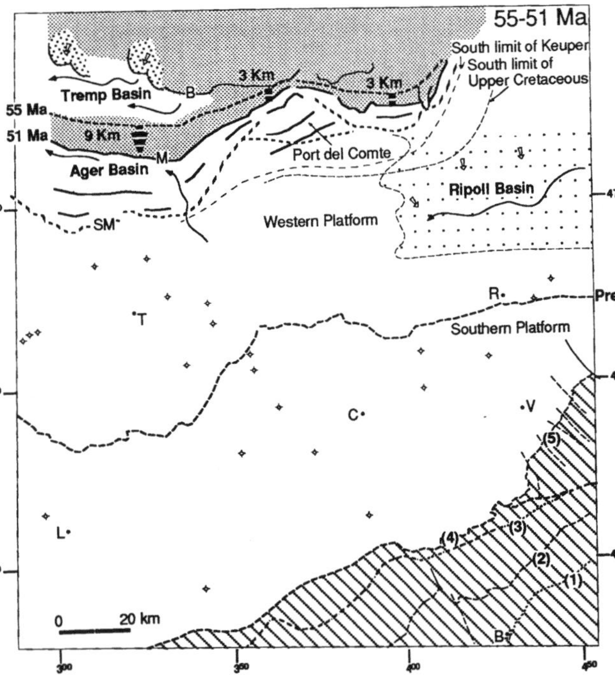

Fig. 17. Palinspastic reconstruction of the eastern and central Pyrenean foreland and adjacent thrusts during early Eocene (Ypresian) time. In this and subsequent palinspastic reconstructions, areas of positive relief in the hinterland thrust sheets in the Pyrenees are shown by shading and in the Catalan Coastal Ranges by hachured regions. Shortening within the Catalan Coastal Ranges is not represented here. Coarse-grained sedimentary fans are depicted as stippled lobes, and palaeocurrent directions are shown by small arrows. Important geographic limits of depositional units are shown by thin lines. The Ripoll trough has been represented during the Armàncies talus deposition. Thick dashed lines and numbers show the reconstructed direction and amount of displacement during the time interval. The amount of thrust trace displacement (in km) is shown for calibrated thrusts, and the starting and ending positions of the thrust traces are shown. The axes of active folds are shown by solid dark lines when they are active. The present position of the thrust front, shown by a dashed line, serves as a reference frame for all reconstructions. Four N–S reference lines display the traces of restored sections (cross-section labels from Verges, 1993). Bóixols (B), Montsec (M), Pedraforca (P), Segre (S), Serres Marginals (SM) and Vallfogona (V) thrust traces are marked in the different palinspastic maps. The positions of drill holes in the foreland, and the various key boundaries in the Catalan Coastal Ranges are shown: (1) Present coastal shoreline; (2) Present southeastern boundary of Vallès–Penedès Neogene Basin; (3) Present position of the north-fault-boundary of the Vallès–Penedès Neogene Basin; (4) Present position of the frontal thrust, except where cut by a Neogene extensional fault in the NE; (5) Present southern boundary of the Ebro foreland basin; and (6) southwesternmost extension of Eocene marine strata. As shown by the wedge-shaped stratal geometries in the Ager basin and by the terrestrial deposits of the Corones Formation in the east, differential shortening occurs along the Montsec thrust and its eastern equivalents at this time. Due to larger displacements in its western and frontal parts, the pre-deformational trace of the Montsec is straighter than today. Within the carbonate platform of the Serres Marginals thrust sheet, folds develop parallel to the Montsec thrust front. Those in the Port del Comte region (PC), which can be linked to the Serres Marginals, are later truncated by the basal Serres Marginals thrust. The Tremp piggyback basin is fed by alluvial cones from the north and distributes detritus to the turbiditic systems of the Jaca basin. In the east, the Ripoll basin is infilled with carbonates, marls, and red beds primarily derived from the north. B = Barcelona; C = Cardona; L = Lleida; R = Ripoll; T = Tremp; V = Vic.

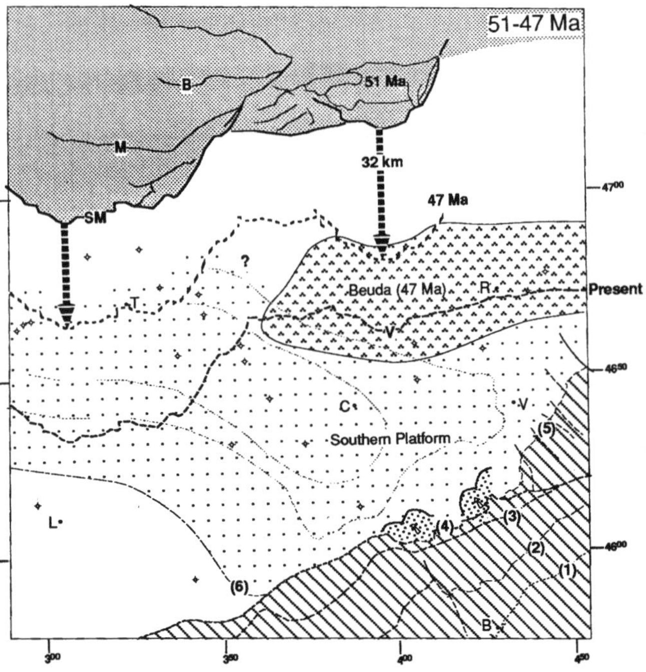

Fig. 18. Palinspastic map for the middle Lutetian (47 Ma) in the eastern and central Pyrenees, during deposition of Beuda evaporites. Although no accurate reconstruction of shortening in the central Pyrenees is available, the frontal parts of the Serres Marginals (SM) were advancing under water, whereas ~32 km of thrust advance can be documented in the eastern Pyrenees where rapid deepening occurred in front of the advancing thrust load (since 55 Ma). As the related depocentre migrated southwards, the Beuda evaporites were deposited above the Campdevànol turbidites, and an extensive carbonate platform extended westward and southward across much of the foreland. The Beuda strata form an important décollement for the next stage of thrusting. Coarse fans (SE: San Esteban) continued to infill the Tremp piggyback basin. Farther south, incipient uplift and fan-delta deposition occurred at St. Llorenç del Munt (SL) and Montserrat (M) in the Catalan Coastal Range.

section across Eocene rocks can be well documented by syntectonic strata (Fig. 18) located in both the hangingwall and footwall of the basal thrust of the lower Pedraforca thrust sheet which moved ~32 km to the south following earlier Corones deposition. As the foreland depocentre migrated in front of the advancing thrusts (Puigdefábregas *et al.*, 1986), a rapidly subsiding trough was filled with marls (Armàncies) and turbidites (Campdevànol) and was succeeded by the Beuda evaporites, which attained thicknesses of up to 2 km (Martínez *et al.*, 1989).

Restoration between 47 Ma (Middle Lutetian) and 34.4 Ma (Latest Priabonian)

At 47 Ma the frontal thrust of the Pedraforca unit reached the southernmost position partially overlying the Beuda evaporites (Puigdefábregas *et al.*, 1986; Martínez *et al.*, 1988). The future Vallfogona thrust (the southern boundary of the Cadí thrust sheet) was located 19 km north of its present eastern position and 25 km in its western segment (Fig. 19). Motion was intiated along the

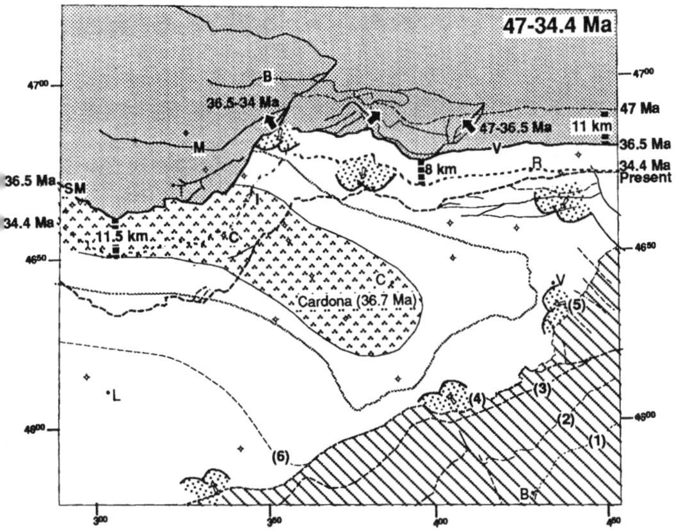

Fig. 19. Palinspastic reconstruction for early Priabonian time (37.2 Ma) during deposition of Cardona marine salts. The Isona (I) and Comiols (C) wells provide important constraints on the shortening along the western traverse. The Pedraforca thrust sheet was blocked at ~47 Ma and subsequently fossilized by terrestrial deposits, during its passive motion on top of the Vallfogona thrust. Vallfogona thrusting initiated in the eastern Pyrenees at ~46 Ma, and 11 km of displacement occurred in the succeeding 10 my, synchronously with breakback thrusting in the Pedraforca thrust sheet (indicated by heavy NE- and NW-facing arrows) from 47 to 36.5 Ma. During the following 2 my, 8 km of further southwards thrust advance occurred in the east, while a breakback sequence developed at Oliana and the Serres Marginals frontal thrust trace migrated 11 km to the south. Several conglomeratic fans fed detritus to the northern foreland basin (C: Collegats; O: Oliana; SL: Sant Llorenç de Morunys; SM: San Marti Xic), whereas deformation along the Catalan Coastal Ranges led to significant influxes of coarse clastic materials (V: Vic; M: Montserrat; L: Llena). The centre of the foreland was dominated by evaporitic deposition (Cardona evaporites; dark shading) and by related gypsiferous deposits (thin continuous line) during earliest Priabonian times, and was largely succeeded by terrestrial deposition later in the Priabonian and Oligocene.

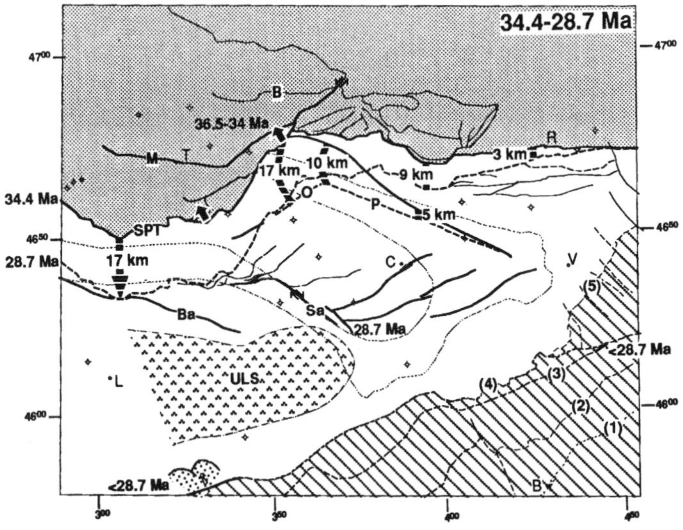

Fig. 20. Palinspastic reconstruction for the early Oligocene. Conglomeratic deposition dominated the Pyrenean deformed margin of the basin, whereas lacustrine deposition in a closed basin occurred in the central foreland (Anoia lacustrine system; Anadón *et al.*, 1989), and coarse fans continued to enter the basin from the southwestern Catalan Coastal Rangers (L: Llena). Highly differential shortening occurred along the Pyrenean front. A minimum of 17 km of southerly thrusting advance took place in the frontal and oblique margins of the South Central Unit. Farther east, the magnitude of shortening diminished and was partitioned differently, such that the Puig-reig (P) and Oliana (O) anticline experienced a scissors-like transport at the same time that shortening along the Vallfogona thrust also decreased to the E. Most of the deformation in the Catalan Coastal Ranges had migrated farther southwest along the range front by the early Oligocene. Ol: Oló; Sa: Sanaüja; Ba: Barbastro–Balaguer; O: Oliana; P: Puig-reig; C: Campins.

Vallfogona thrust at ~43.5 Ma (base of the Bellmunt Fm). The Pedraforca unit was deformed by a breakback thrusting sequence (Martínez *et al.*, 1988) from 47 Ma to 36.5 Ma (Burbank *et al.*, 1992b). Clockwise rotation of the Pedraforca thrust sheet is documented at this time (Keller, 1992). The stacking of basement thrust sheets below the cover–basement boundary was initiated at this time. Farther to the west, the Serres Marginals thrust moved 11.5 km to the south during this interval. By the close of the Eocene epoch, the Cardona evaporites had been deposited and the transition to wholly terrestrial deposition had begun (Busquets *et al.*, 1985).

Restoration between 34.4 Ma (Priabonian/Early Oligocene) and 28.7 Ma (Rupelian/Chattian)

The latest significant movements in the eastern and east–central Pyrenees occurred in the early Oligocene (Fig. 20). The Cardona and Súria anticlines deformed lower Oligocene strata (Sáez & Riba, 1986) in the eastern foreland Ebro basin, whereas the Barbastro–Balaguer anticline, the tip line of the Pyrenean thrust system (Williams, 1985; Muñoz, 1992), developed primarily in the late Oligocene (Pardo & Villena, 1979; Riba *et al.*, 1983). In a scissors-like motion, the Puig-reig anticline experienced strong differential shortening, supported by rotation data (Keller, 1992), such that both shortening and rotation increased towards the west. North of this, limited shortening occurred along the Vallfogona thrust, whereas to the south, numerous detached folds developed in the central foreland. Conglomeratic deposition predominated along the northern foreland margin throughout this interval, whereas closed-basin lakes characterized the central foreland.

Conclusions

The new synthesis of structural, stratigraphic, and chronologic data permits the development of more precise and detailed reconstructions of the deformational and depositional history of the central and eastern Pyrenees than have previously been attainable. Key aspects of this synthesis include:

The geometry of presently preserved thrust sheets is closely controlled by the distribution of evaporitic units which localize basal detachments and by structures, such as extensional faults, inherited from earlier episodes of deformation.

High rates of subsidence during the early stage of the foreland basin evolution in the east (deep and restricted marine conditions) correspond with high rates of thrusting during the emplacement of the Pedraforca thrust sheet.

Shortening calculated in the cover and cover–basement thrust sheets decreases from west to east. This could be due to the absence of structurally high thrust sheets involving Mesozoic strata in the eastern area or to a lack of good information. Nevertheless, the end of deformation shown by the eastern foreland structures suggests an earlier termination of deformation in the east than in the west. The most likely way to have greater and more prolonged shortening in the eastern Pyrenees, which is not recorded in the foreland structures, would be the activity of large out-of-sequence thrusts affecting the inner part of the chain. The Ribes–Camprodon thrust is one of these faults (Muñoz, 1985) but its activity is dated as old as 38 Ma (Burbank *et al.*, 1992b).

The evolution of the eastern and central foreland basin is controlled by a shift from restricted marine sedimentation (Ripoll basin) to broader and shallower marine sedimentation in the Ebro basin during the middle Lutetian (46.5 Ma; chron 21). This shift coincides with the sudden input of conglomerates in the southern margin of the basin, suggesting an increase of the tectonic activity in this margin. The double-wedge geometry of the basin at that moment (see the restored eastern cross-section) suggests an active southern margin with a double flexure caused by the load of both the Pyrenees and the Catalan Coastal Ranges.

By the late Eocene, the Ebro basin became increasingly restricted by the encroachment of thrust sheets from both the north and south. High hinterland topography due to growth of the antiformal stack of basement slices and concurrent high rates of denudation may have generated an enhanced flux of detrital sediments to the foreland. Regardless of cause, previously persistent marine deposition was supplanted by terrestrial deposition during the early Priabonian (younger than 37.2 Ma).

Due to the absence of sediments younger than Middle Oligocene in the eastern foreland Ebro basin, the exact age of the end of the deformation in the eastern Pyrenees is unknown. Deformation in the southeastern Pyrenees subsequent to this time, however, represents less than 1 km of total shortening. Some deformation is still active during the early part of the late Oligocene in the Catalan Coastal Ranges, such as in Campins (Anadón, 1986), east of the eastern cross-section and in la Llena (Colombo & Vergés, 1992), southwest of the western cross-section. The end of shortening in the eastern part of the Pyrenees and Catalan Coastal Ranges is coeval with the Late Oligocene initiation of extension in the València trough (Riba *et al.*, 1983; Fontboté *et al.*, 1990).

Acknowledgments

This research was supported by grants to DWB from the National Science Foundation EAR-8517482, 8816181, and 9018951. Acknowledgment is gratefully made to the donors of the Petroleum Research Fund, administered by the American Chemical Society (grants 17625, 20591, and 23881 to D.W.B.). Support for JV from the Servei Geològic de Catalunya, Subprograma de Perfeccionamiento para Doctores y Tecnólogos (1990) and DGICYT projects PB91–0252 and PB91–0805 facilitated this research.

References

Anadón, P. (1986). Las facies lacustres del Oligoceno de Campins (Vallés oriental, provincia de Barcelona). *Cuad. Geol. Ibérica*, **10**: 271–294.

Anadón, P., Cabrera, Ll., Colldeforns, B. and Sáez, A. (1989). Los sistemas lacustres del Eoceno superior y Oligoceno del sector oriental de la cuenca del Ebro. *Acta Geol. Hisp.*, **24**: 205–230.

Berastegui, X., García, J.M. and Losantos, M. (1989). Structure and sedimentary evolution of the Organyà basin (Central South Pyrenean Unit, Spain) during the Lower Cretaceous. *Bull. Soc. Géol. Fr.*, **8**, VI(2): 251–264.

Berggren, W.A., Kent, D.V., Flynn, J.J. and Van Couvering, J.A. (1985). Cenozoic geochronology. *Geol. Soc. Am. Bull.*, **96**: 1407–1418.

Burbank, D.W., Vergés, J., Muñoz, J.A. and Bentham, P. (1992a). Coeval hindward- and forward-imbricating thrusting in the Central Southern Pyrenees, Spain: timing and rates of shortening and deposition. *Geol. Soc. Am. Bull.*, **104**: 3–17.

Burbank, D.W., Puigdefàbregas, C. and Muñoz, J.A. (1992b). The chronology of the Eocene tectonic and stratigraphic development of the eastern Pyrenean Foreland Basin, NE Spain. *Geol. Soc. Am. Bull.*, **104**: 1101–1120.

Busquets, P., Ortí, F., Pueyo, J.J., Riba, O., Rosell, L., Sáez, A., Salas, R. and Taberner, C. (1985). Evaporite deposition and diagenesis in the saline (potash) Catalan basin, upper Eocene, *Exc. Guide-book 6th European Regional Meeting. Lleida, Spain*: 11–59.

Cande, S.C. and Kent, D.V. (1992). A new geomagnetic polarity time-scale for the late Cretaceous and Cenozoic *J. Geophys. Res.*, **97** (B10): 13917–13951.

Columbo, F. and Vergés, J. (1992). Geometría del margen SE de la Cuenca del Ebro: discordancias progresivas en la Grupo Scala Dei, Serra de la Llena. *Acta Geol. Hisp.*, **27** (1–2): 33–54.

Fontboté, J.M., Guimerà, J., Roca, E., Sàbat, F., Santanach, P. and Férnandez-Ortigosa, F. (1990). The Cenozoic evolution of the València trough (western Mediterranean). *Rev. Soc. Geol. Esp.* **3** (3–4): 249–259.

Keller, P. (1992). Paläomagnetische und strukturgeologische Untersuchungen als Beitrag zur Tektogenese der SE-Pyrenäen. PhD, ETH Zürich: 113 pp.

Lanaja, (1987). Contribución de la exploración petrolífera al conocimiento de la geología de España. *Instituto Geológico y Minero de España*, 465 pp.

Martínez, A., Vergés, J. and Muñoz, J.A. (1988). Secuencias de propagación del sistema de cabalgamientos de la terminación oriental del manto del Pedraforca y relación con los conglomerados sinorogénicos. *Acta Geol. Hisp.*, **23** (2): 119–128.

Martínez, A., Vergés, J., Clavell, E. and Kennedy, J. (1989). Stratigraphic framework of the thrust geometry and structural inversion in the southeastern Pyrenees: La Garrotxa area. *Geodinam. Acta*, **3** (3): 185–194.

Muñoz, J.A. (1985). Estructura alpina i herciniana a la vora sud de la zona axial del Pirineu oriental. PhD, Univ. of Barcelona, 305 pp.

Muñoz, J.A. (1992). Evolution of a continental collision belt: ECORS–Pyrenees crustal balanced cross-section. In *Thrust tectonics* (ed. K. McClay): 235–246. Chapman and Hall, London.

Muñoz, J.A., Martínez, A. and Vergés, J. (1986). Thrust sequences in the eastern Spanish Pyrenees. *J. Struct. Geol.*, **8** (3/4): 399–405.

Mutti, E., Rosell, J., Allen, G.P., Fonnesu, F. and Sgavetti, M. (1985). The Eocene Baronia tide dominated delta-shelf system in the Ager basin. *Exc. Guide-book 6th European Regional Meeting. Lerida, Spain*: 579–600.

Pardo, G. and Villena, J. (1979). Aportación a la geologia de la región de Barbastro. *Acta Geol. Hispànica. Homenatge a Lluís Solé i Sabarís*, **14**: 289–292.

Puigdefábregas, C., Muñoz, J.A. and Marzo, M. (1986). Thrust belt development in the Eastern Pyrenees and related depositional sequences in the southern foreland basin. In *Foreland basins* (ed. P.A. Allen and P. Homewood). *Spec. Publ. Int. Assoc. Sedimentol.*, **8**: 22.

Ramírez, A. and Riba, O. (1975). Bassin potassique catalan et mines de Cardona. *IX Congres Internatl. de Sédimentologie, Nice 1975. Livret-guide Ex.* 20: 49–58.

Riba, O. (1973). Las discordancias sintectónicas del Alto Cardener (Prepirineo catalán), ensayo de interpretación evolutiva. *Acta Geol. Hisp.*, **8** (3): 90–99.

Riba, O., Reguant, S. and Villena, J. (1983). Ensayo de síntesis estratigrafica y evolutiva de la cuenca terciaria del Ebro. Libro Jubilar J.M. Rios. *Geol. Esp.*, **II**: 131–159.

Sáez, A. and Riba, O. (1986). Depósitos aluviales y lacustres paleogenos del margen pirenaico catalán de la cuenca del Ebro. *Libro guia Exc. XI Congreso Español de Sedimentologia. Barcelona*: 6.1–6.29.

Séguret, M. (1972). Étude tectonique des nappes et séries décollées de la partie centrale du versant sud des Pyrénées. *Pub. USTELA, sér. Geol. Struct, no. 2, Montpellier*: 1–155.

Simó, A. and Puigdefábregas, C. (1985). Transition from shelf to basin on an active slope, upper Cretaceous, Tremp area, southern Pyrenees. *Exc. Guide-book 6th European Regional Meeting. Lleida, Spain*: 63–108.

Vergés, J. and Martínez, A. (1988). Corte compensado del Pirineo oriental: geometria de las cuencas de antepaís y edades de emplazamiento de los mantos de corrimiento. *Acta Geol. Hisp.*, **23** (2): 95–106.

Vergés, J. and Muñoz, J.A. (1990). Thrust sequences in the Southern Central Pyrenees. *Bull. Soc. Géol. Fr.*, **8**, VI(2): 265–271.

Vergés, J., Muñoz, J.A. and Martínez, A. (1992). South Pyrenean fold-and-thrust belt: Role of foreland evaporitic levels in thrust geometry. In *Thrust tectonics* (ed. K. McClay): 255–264. Chapman and Hall, London.

Williams, G.D. (1985). Thrust tectonics in the south central Pyrenees. *J. Struct. Geol.*, **7** (1): 11–17.

E12 The Late Eocene–Early Oligocene deposits of the NE Ebro Basin, West of the Segre River

E. MAESTRO-MAIDEU AND J. SERRA ROIG

Abstract

The detailed study of the sediments that fill the depositional complex of Late Eocene and Early Oligocene age, W of the Segre thrust, has allowed the recognition of five major depositional cycles. These cycles are bounded by angular unconformities, particularly in the borders of the basins, and by sharp changes of facies in the more central areas, where the cycles tend to be conformable. The major cycles consist of minor cycles reflecting relative changes of basinal base level. The major cycles are regarded as depositional sequences containing both continental and marine deposits.

The five major cycles are considered to be responses to eustatic sea-level changes. However, tectonism also operated during the sedimentation of these units. The displacement of the Pyrenean thrust sheets enhanced the angular unconformities in the conglomerates that border the thrusts. The emplacement of these thrust sheets caused a paleogeographic change in the orientation of the Basin. During the Middle and Late Eocene, continental deposits formed in the south, while marine conditions existed in the north (Oliana area). In the latest Eocene, and even more clearly in the Early Oligocene, proximal deposits, sourced in the Pyrenees, formed in the north, whereas distal fluviatile and lacustrine sediments formed in the south, towards the centre of the Ebro Basin.

Geological setting

The area described is in the NW Catalan basin, the NE sector of the Ebro basin, W of the Segre river (Fig. 1). It is close to the thrust units of the Pyrenees. The sediments studied were deposited during the Late Eocene and Early Oligocene, and they have been affected by the southward emplacement of the Pyrenean thrusts.

During the Late Eocene (Bartonian to upper Priabonian), fluvial and lacustrine facies were deposited at the southern end of the basin studied, and marine deposits of deltaic origin accumulated at the northern end (Fig. 1). In the latest Eocene and the Early Oligocene, an important paleogeographic change took place, during the period of emplacement of the Pyrenean thrust sheets. Thick conglomerate alluvial fans were built in the N, bordering the thrust sheets. These coarse deposits passed into sandy fluvial systems in southern areas, and then passed further downstream into the deposits of mudplains and lakes.

Tectonic setting

The study area is bounded to the N by the Mesozoic carbonates of the Montsec thrust sheet, to the W by the Sierras Marginales, and to the S by the Artesa thrust and the Seró fault (Fig. 1). The Seró fault extends into the northern flank of the Barbastro–Balaguer anticline. Towards the E, the thrusts change their W–E alignment to a SW–NE alignment and converge to form the Segre thrust (Vergés and Muñoz, 1991). In the Segre area, these thrusts have functioned as oblique ramps with an eastwards vergency (Cámara and Klimovitz, 1985).

The Montsec thrust unit consists of a thick succession of Late Triassic, Jurassic, Cretaceous and Early Tertiary deposits.

The Sierras Marginales consist of an imbricate fan thrust system (Martínez-Peña and Pocoví, 1988; Vergés and Muñoz, 1991) that has deformed a Mesozoic and Early Tertiary succession which becomes reduced progressively towards the S. The northernmost unit of the Sierras Marginales is named Sant Mamet (Fig. 1). It has a dome morphology, and is covered at its eastern end by latest Eocene and Early Oligocene conglomerates. This structure continues as a SW–NE trending anticline as far as the Serra de Peramola, although it is only represented by a transpressive thrust in Montmagastre and Bellfort (Fig. 1). In the Peramola unit a succession of imbricate thrusts is exposed (Fig. 1). All of these thrusts had an eastward component of movement. Their emplacement began in the W, where the latest Eocene conglomerates cover the St. Mamet unit, and were only folded by the southward emplacement of the St. Mamet thrust. Later, the emplacement affected the easternmost areas, where conglomerates of the same age were thrust as the Peramola unit moved to the E.

To the S of the St. Mamet dome, some thin imbricate thrust sheets are exposed. Although the front of these thrust sheets has a predominantly southward vergence and W–E strike, strikes tend to be SW–NE in the E, and NW–SE in the W, displaying oblique ramp characteristics. The Salgar unit forms the southwestern end of the

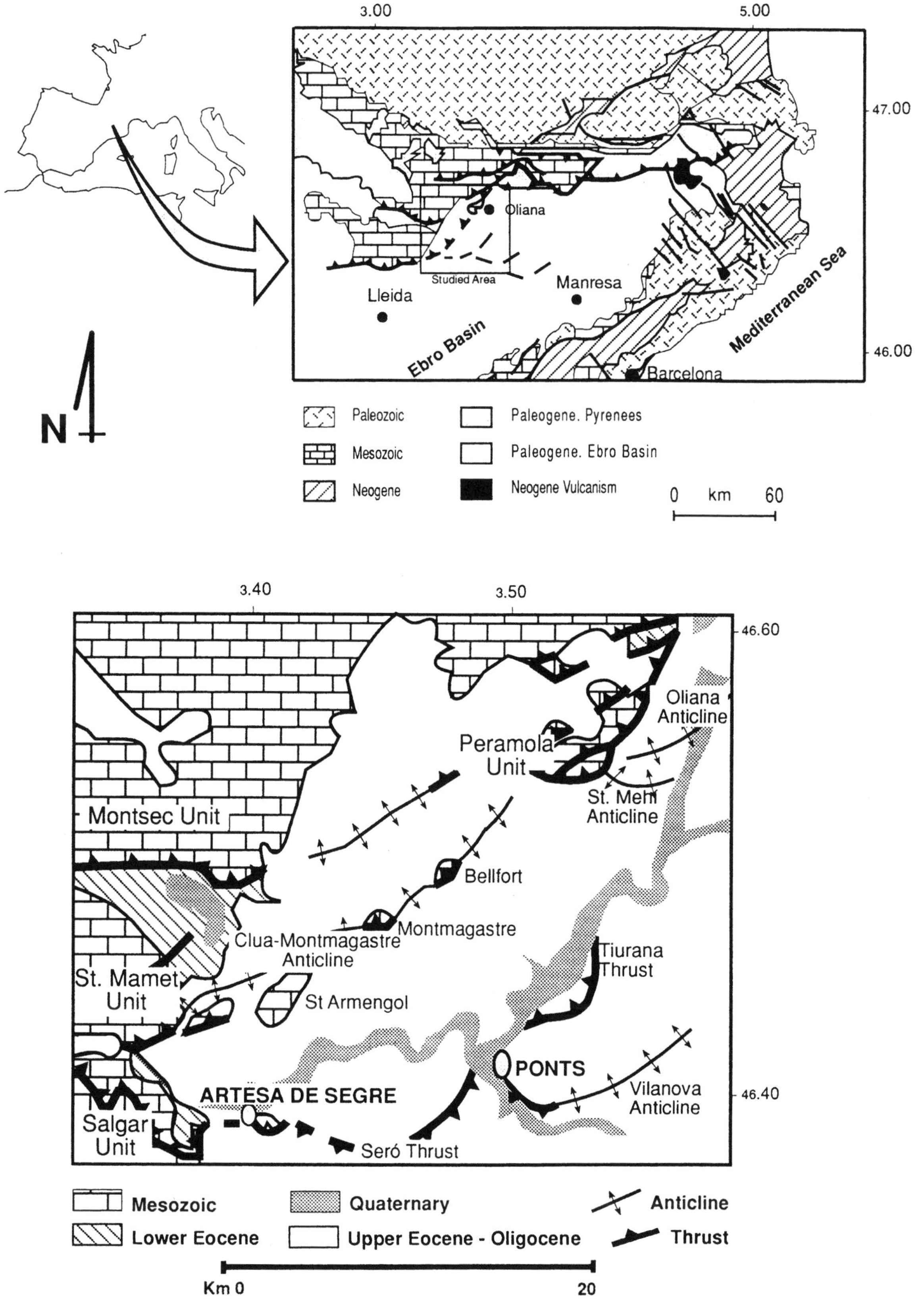

Fig. 1. Geological setting of the studied area in the zone of contact of the Ebro Basin with the south-Pyrenean thrust sheets. The expanded map shows the mainly tectonic units and features of this area.

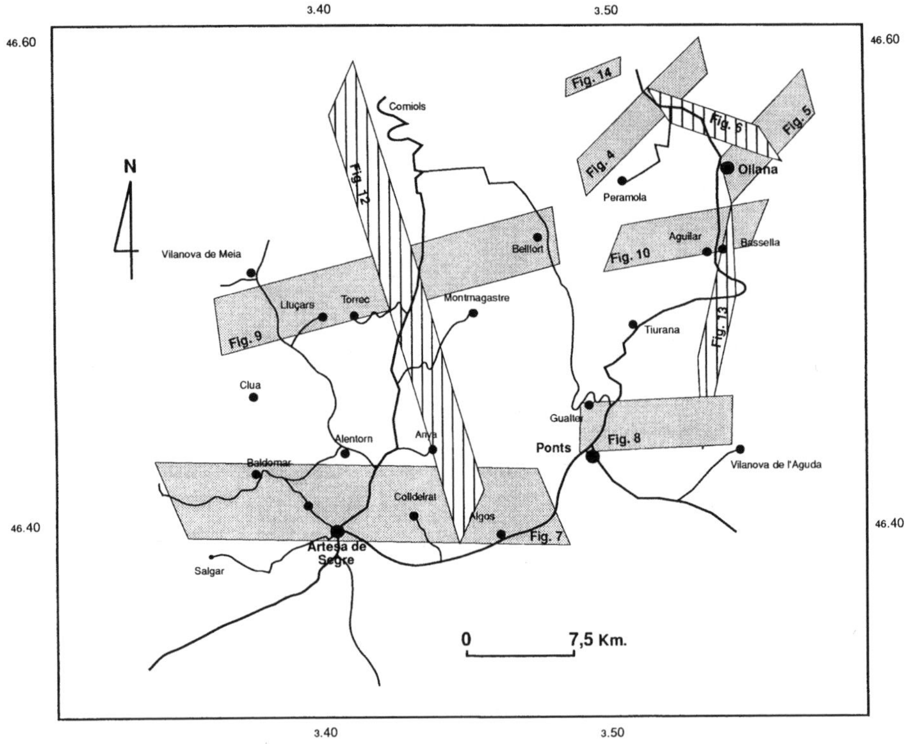

Fig. 2. Map showing the location of the correlation diagrams that constitute the following figures.

study area (Fig. 1). The thrust is elevated in the NW, where unconformable Late Eocene conglomerates onlap the Early Tertiary deposits. In the SE, the Salgar unit thrusts Late Eocene conglomerates and evaporites. The Artesa thrust forms a small outcrop (Fig. 1), apparently related to the Sierra Marginales thrusts. The Seró thrust is the result of movement of the Artesa thrust against Late Eocene evaporites, and it has a northerly vergence.

The Oliana anticline, striking SW–NE, has been developed in front of the Peramola unit (Fig. 1) and is parallel to the Segre thrust. The anticline folds Middle–Late Eocene deposits. Vergés and Muñoz (1990) have intrepreted it as a superimposition of three structural sheets of marls and evaporites forming a foreland dipping duplex. The St. Mehí anticline, oriented WNW–ESE, is exposed S of the Oliana anticline (Fig. 1), and folds latest Eocene conglomerates.

The Barbastro–Balaguer anticline is a WNW–ESE structure which continues the line of the southern Pyrenean frontal thrusts. The core of the anticline consists mainly of Late Eocene evaporites.

Depositional cycles

Introduction

Our analysis of the deposits is based on their cyclicity, which is independent of whether the deposits are continental or marine. The basic data are presented as correlation fences, located on Fig. 2.

In this study, the only marine deposits analysed are those exposed in the northeastern area, near Oliana. They are of Middle to Late Eocene age, and can be described as depositional sequences (*sensu* Vail, 1987).

Hinterland deposits are extensive. Three orders of sedimentary cycles can be distinguished (Fig. 3). The cycles are bounded by unconformities near the borders of the basin, and tend to be conformable in the centre of the basin. The major cycles correspond to composite sequences (Mitchum and Van Wagoner, 1991), since they are formed by minor-order cycles. Despite the lack of chronostratigraphic data, we feel we can distinguish them as third-order cycles. The Late Eocene continental cycles can be correlated with some of the marine sequences. The minor-order cycles can be distinguished as fourth and fifth order, and consist of sequences deposited during high-frequency base-level oscillations.

Hinterland sequence stratigraphy

The sequence stratigraphic analysis in hinterland deposits is based on interpretations of cycles of relative change of base level in the basin, with minor or no relationship to sea level. Instead, the base-level changes were due to changes in climate or tectonism that together controlled the type and quantity of sediment that filled the accommodation space determined by the tectonism.

The stratigraphic expression of the base-level cycles, first described by Wheeler (1964) as base-level transit cycles, comprises different magnitudes of unconformity-bounded sequences called hinterland sequences (Vail *et al.*, 1977) or Mediterranean sequences (Uliana and Legarreta, 1988).

Erosion or non-sedimentation predominated during relative fall

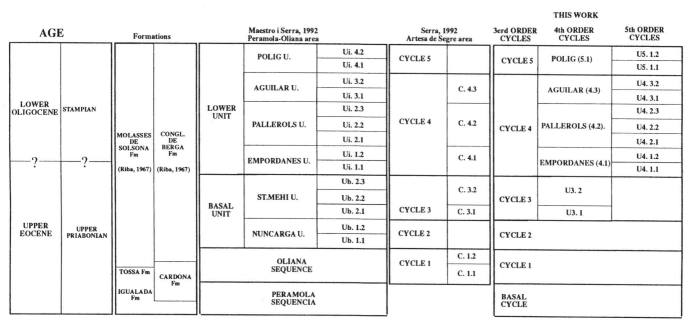

AGE		Formations		Maestro i Serra, 1992 Peramola-Oliana area			Serra, 1992 Artesa de Segre area		THIS WORK 3erd ORDER CYCLES	THIS WORK 4th ORDER CYCLES	THIS WORK 5th ORDER CYCLES
LOWER OLIGOCENE	STAMPIAN	MOLASSES DE SOLSONA Fm (Riba, 1967)	CONGL. DE BERGA Fm (Riba, 1967)	LOWER UNIT	POLIG U.	Ui. 4.2	CYCLE 5		CYCLE 5	POLIG (5.1)	U5. 1.2
						Ui. 4.1					U5. 1.1
					AGUILAR U.	Ui. 3.2	CYCLE 4	C. 4.3	CYCLE 4	AGUILAR (4.3)	U4. 3.2
						Ui. 3.1					U4. 3.1
					PALLEROLS U.	Ui. 2.3		C. 4.2		PALLEROLS (4.2).	U4. 2.3
						Ui. 2.2					U4. 2.2
						Ui. 2.1					U4. 2.1
					EMPORDANES U.	Ui. 1.2		C. 4.1		EMPORDANES (4.1)	U4. 1.2
?——	?——					Ui. 1.1					U4. 1.1
UPPER EOCENE	UPPER PRIABONIAN			BASAL UNIT	ST.MEHI U.	Ub. 2.3	CYCLE 3	C. 3.2	CYCLE 3	U3. 2	
						Ub. 2.2					
						Ub. 2.1		C. 3.1		U3. 1	
					NUNCARGA U.	Ub. 1.2	CYCLE 2		CYCLE 2		
						Ub. 1.1					
		TOSSA Fm	CARDONA Fm	OLIANA SEQUENCE			CYCLE 1	C. 1.2	CYCLE 1		
								C. 1.1			
		IGUALADA Fm		PERAMOLA SEQUENCIA					BASAL CYCLE		

Fig. 3. Table of the cycles described in this work, and its correlation with some existing lithostratigraphic units.

in base level (Wheeler, 1964). At the end of the relative fall, sedimentation of the lowstand deposits begins in some low-lying places, while other places are areas of non-deposition or further erosional cutting-back. The usual sediments are gravels and coarse sand, deposited by braided rivers (Uliana and Legarreta, 1988; Shanley and McCabe, 1989), and these tend to pass into lake deposits in the centre of the basin.

During relative rises of base level, depositional cyclic phases (Wheeler, 1964) were formed. Highstand deposits started to accumulate before the inflexion points of the curves of rising base level. At these points, accommodation space increased, and the supply of coarse sediment decreased. The areas where erosion had prevailed were filled by distal-type sediments (Uliana and Legarreta, 1988). Meandering and anastomosing fluvial systems were developed, with channel-fill and overbank deposits (Uliana and Legarreta, 1988; Shanley and McCabe, 1989). Some coarser meandering river deposits can be found in what were the most marginal areas, which under lowstand conditions could not accommodate sediments (Uliana and Legarreta, 1988). Alluvial fans were developed during highstand conditions in areas of high relief.

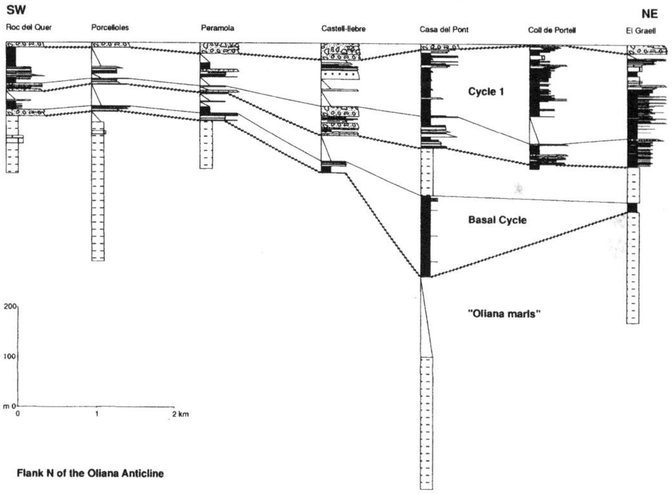

Fig. 4. Correlation panel of the upper Eocene marine deposits in the northern flank of the Oliana anticline. The 'Oliana marls' is a lithostratigraphic unit, not studied in the present work. Overlaying these marls is found the basal cycle, only recognized in marine sediments (Oliana area).

Depositional cycles

The lowest stratigraphic unit, not analysed in this work, is named the Oliana Marls, and is only exposed in the core of the Oliana anticline. It consists of a thick offshore marl succession dating from upper Bartonian to lower Priabonian (Caus, 1973). It represents the distal facies of some eastern deltaic systems (Figs. 4, 5 and 6).

Basal Cycle

The basal cycle consists only of marine deposits occurring in the Oliana area (Figs. 4, 5 and 6). It is interpreted as a depositional sequence, exposed in front of the St. Mamet–Peramola thrust alignment. These deposits have no known equivalent in continental areas, and were deposited during the sedimentation of the Cardona or Barbastro evaporites.

The cycle is of middle to late Priabonian age (Caus, 1970), and was called the Peramola sequence by Maestro (1989). The lowstand deposits consist of a prograding fan-delta system, which developed from SW to NE. In the SW, fluvial conglomerates are exposed whereas to the NE a delta-front cycle was formed, which evolved upwards into prodelta-slope mudstones with occasional thin-bedded turbidities (Fig. 4). Above the deltaic sandstone, some patch

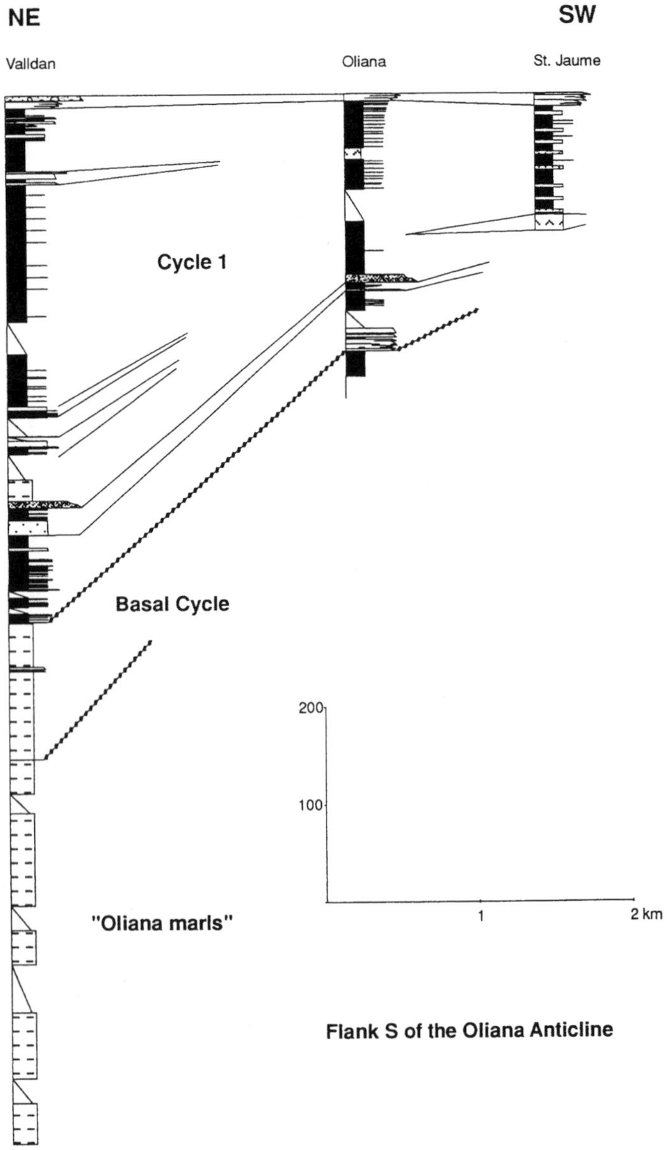

Fig. 5. Correlation panel of the Upper Eocene marine deposits in the southern flank of the Oliana anticline.

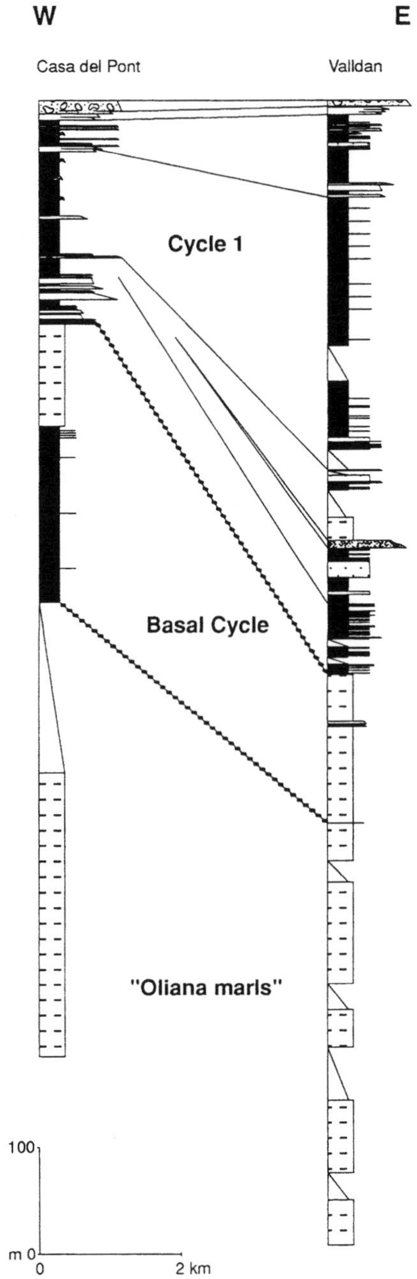

Fig. 6. Correlation of the thickest successions in the northern and southern flanks of the Oliana anticline. The proximal facies are exposed in the northern flank. The southern flank, more subsident, consists of deeper facies associations.

reefs were formed. They constitute the transgressive systems tract, and the highstand deposits consist of fine-grained deltaic facies that passed, from SW to NE, into offshore marls.

Cycle 1

This cycle is exposed as marine deposits in the Oliana area (Figs. 4, 5 and 6), and as continental deposits in the Artesa de Segre area (Fig. 7). It forms a depositional sequence that is only exposed in front of the St. Mamet–Peramola thrust alignment (Fig. 1).

In the Oliana area this cycle was called the Oliana sequence by Maestro (1989). It lies unconformably on the Peramola sequence (Figs. 4, 5 and 6), and is late Priabonian in age (Caus, 1970). The lowstand deposits were formed, in the NW, as alluvial fans that developed as braid deltas (Fig. 4). The alluvial conglomerate material was derived from the erosion of the Mesozoic thrust sheets to the WNW. The stream mouth area of the deltas is characterized by scour-and-fill facies, which may indicate a rapid deepening of the basin to the ESE (Fig. 4). In the areas that were deep, the lowstand deposits are represented by grey clay with turbidite sandstones and occasional slumps (Fig. 5). These slumps and a calcareous megatur-

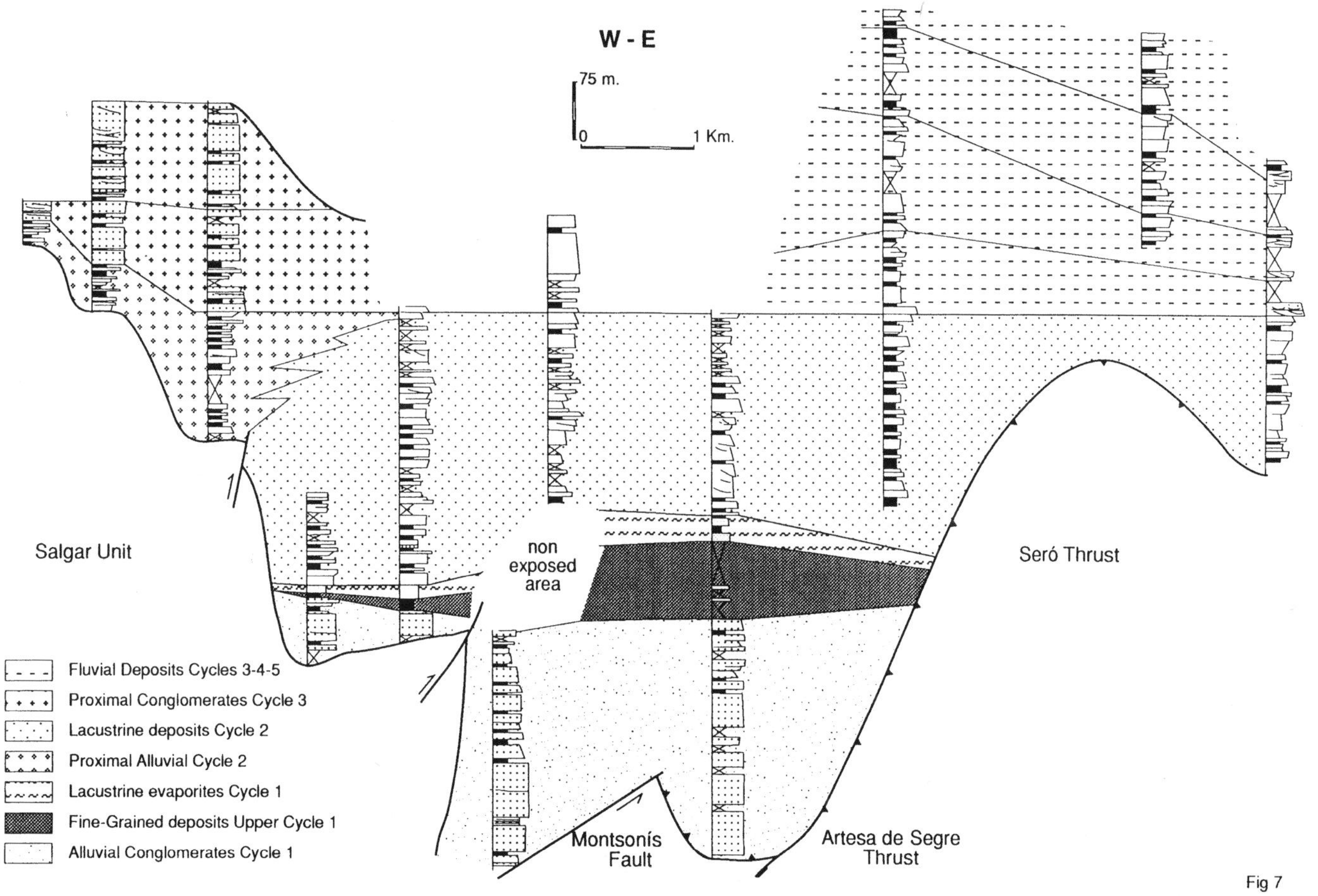

Fig. 7. Correlation panel of the alluvial, fluvial and lacustrine deposits of the upper Eocene and lower Oligocene deposits in the southwestern area of the studied zone, near Artesa de Segre. The main facies associations observed are represented inside each cycle.

bidite appear to reflect emplacement of the thrust sheets during the sedimentation.

The transgressive systems tract is represented by a fossil-rich, pelitic carbonate unit, with reefal facies. The highstand systems tract consists of a coarsening- and thickening-upwards fan delta sequence, with conglomerates in the WNW, and delta-front, pro-delta and slope deposits in the ESE (Figs. 4 and 6). At the base of the river mouth-bar sequence, and in the SSE, a SW–NE-oriented wedge of evaporites is found (Fig. 5). The upper boundary of the Oliana sequence is an erosive unconformity.

In the southernmost part of the study area, near Artesa de Segre, a fining-upwards megasequence lies unconformably over the Early Tertiary deposits of the Salgar and Artesa thrusts (Fig. 7). This sequence filled a subsiding depression created by the tectonics. The succession, from base to top, consists of breccias, conglomerates that progressively pass upwards from alluvial-fan to fluvial, and, finally red mudstones with some sandstone layers of fluvio-lacustrine origin. The top of the sequence consists of lacustrine evaporites, as in the Oliana area, and a fan-delta mouth-bar cycle. It is thought to be middle to late Priabonian in age (Serra Roig, 1992). In the continental succession, two lower-order base-level cycles can be distinguished (1.1 and 1.2), with coarser conglomerates at the base of each.

Cycle 2

The second cycle is developed entirely as continental sediments in the Artesa de Segre area, in front of the St. Mamet–Peramola unit (Fig. 1). It consists mainly of lacustrine sediments deposited in structural depressions (Artesa de Segre, Vilanova de l'Aguda, Oliana) during the late Priabonian (Sáez, 1987). Lateral to these relatively fine-grained deposits are alluvial fan and fluvial facies, derived from the northern Mesozoic thrust sheets.

The base of the sequence consists of coarse-grained sandstone and conglomerates (Figs. 8 and 13). These deposits were followed by grey silts and marls (Figs. 7 and 8). Finally, highstand deposits are represented by fine-grained detrital facies that formed in delta and lake depositional environments, and that show a progradational pattern (Figs. 7, 8, 12 and 13). These deltaic deposits are partly products of supply from anastomosing and meandering fluvial systems in the NE, but were mainly derived from alluvial fans to the W. The fluvio-alluvial streams developed channels and mouth lobes of Gilbert-type (Serra Roig, 1992). The upper bound-

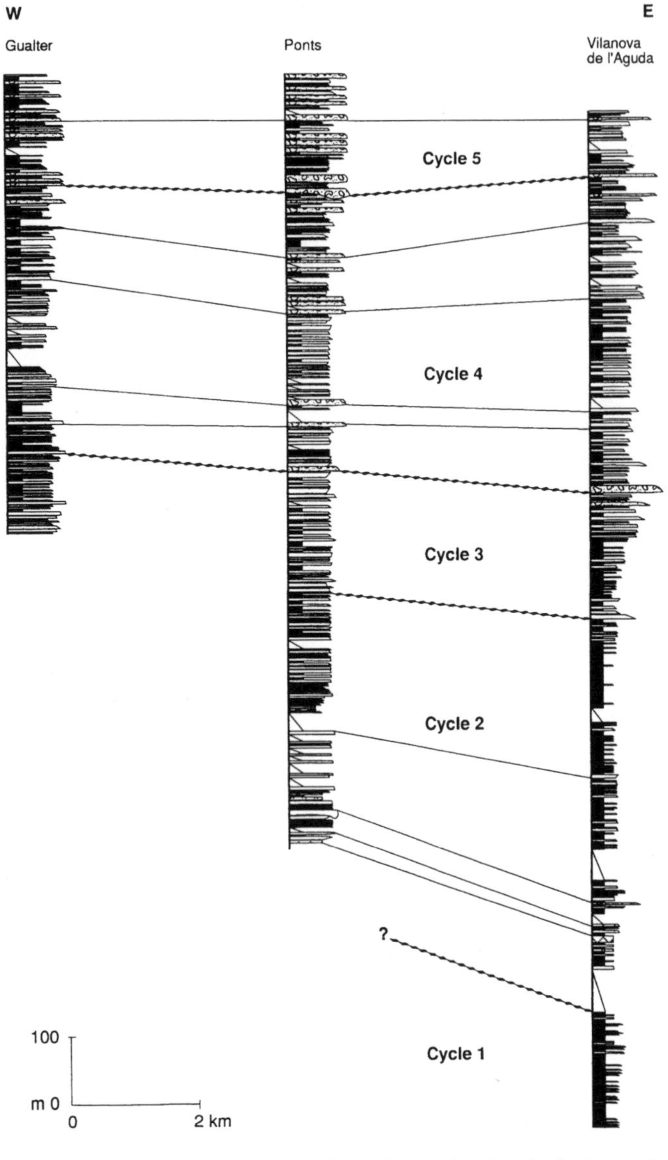

Fig. 8. Correlation panel of the fluvial and lacustrine deposits in the south-eastern sector of the studied area. The facies represented show great similarity with the facies associations in Fig. 7.

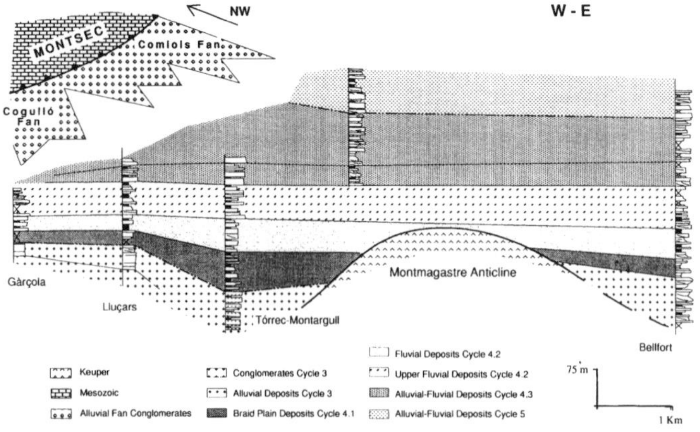

Fig. 9. East–West panel of the northwestern sector of the area, north of the St. Mamet–Montmagastre–Bellfort–Peramola tectonic alignment. The different cycles present are represented with their main facies evocations. Cycles 1 and 2 were not exposed north of the alignment.

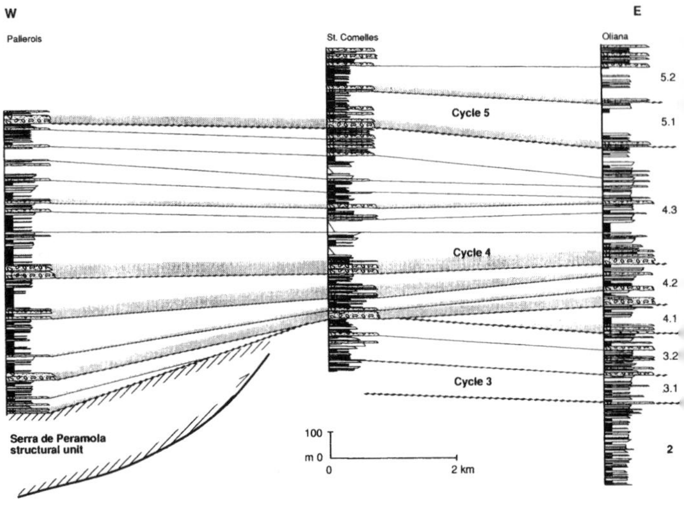

Fig. 10. East–west panel of the northeastern sector of the area east of the Peramola Structural Unit, and south of the St. Mehí anticline. Near the anticline a progressive unconformity is exposed, with on-lap of the deposits of Cycle 4 over the conglomerates of Cycle 3.

ary of the second cycle is a sharp change of facies, marked by the incoming of coarser alluvial and fluvial deposits (Fig. 12 and 13).

The movement of the Oliana anticline resulted in a difference in subsidence rate, between the northwestern limb where conglomerates predominate (and the rate appears to have been low), and the southeastern limb, where shales and sandstones predominate (and the rate appears to have been negligible) (Figs. 10 and 13).

Cycle 3

The third cycle is exposed as alluvial fan deposits in the margins of the basin, near the Prepyrenean thrust sheets (Figs. 7, 9 and 10). The deposits passed into fluvial deposits in areas further from the thrust sheets. All of the deposits of this cycle were continental. They date from the Late Priabonian to basal Oligocene (Riba *et al.*, 1975; Sáez, 1987; Agustí *et al.*, 1987; Anadón *et al.*, 1989a, 1989b, 1989c; Burbank *et al.*, 1992).

The alluvial fans were formed mainly of carbonate conglomerates derived from the Mesozoic and Early Tertiary carbonates. In the distal areas, where the fluvial deposits prevail, there is a high proportion of quartz and Paleozoic clasts, derived from northernmost areas.

This cycle, described as the Basal Unit (Maestro and Serra Roig, 1992), or as Cycle 3 (Serra Roig, 1992), can be divided into two unconformity-bounded minor-order cycles, and these are related to changes of base level of the basin, in areas of alluvial deposition (Figs. 3, 7, 9 and 10). These subunits consist of basal breccias, followed by conglomerates and coarse sands with prograding

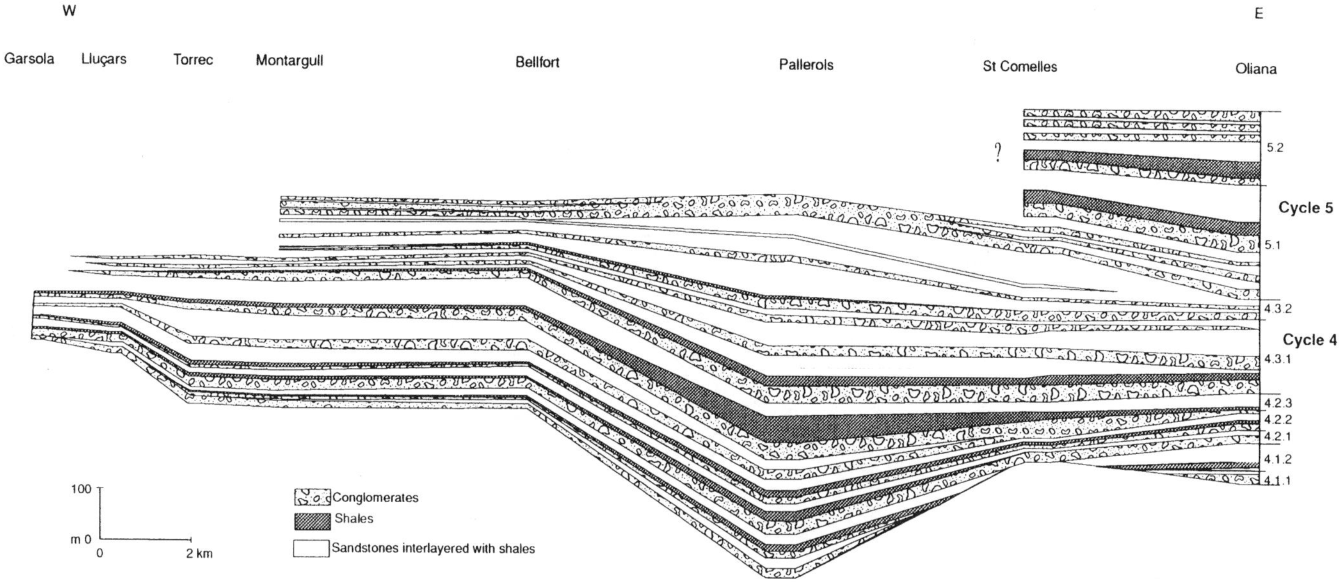

Fig. 11. Correlation diagram of the panels in Figs. 9 and 10.

geometries. In areas of fluvial sedimentation, where the sediments are conformable, the subunits are distinguished by a sharp upward change to coarser deposits.

Some of the alluvial fans are exposed in front of the St. Mamet–Peramola thrust, and have been affected by its emplacement. The thrust has folded the alluvial fan deposits of the western areas (Mosquera alluvial fan and Montmagastre braid plain), and thrust them in the eastern area (Peramola alluvial fan). Other alluvial complexes were deposited in northern structural depressions, and lie unconformably on the St. Mamet–Peramola unit (the Clue, Cogulló and Tórrec alluvial fans) (Fig. 9).

As in the second cycle, the movement of the Oliana anticline induced differential subsidence on its two flanks, but passing upwards from the second to the third cycle, the difference becomes less, implying decrease in the anticlinal activity.

Cycle 4

The base of this cycle is: 1) an angular unconformity in the Tertiary deposits in the area of the Comiols alluvial fan (Fig. 12); 2) an angular unconformity between conglomerates and the Mesozoic carbonates of the Peramola thrust sheet (Fig. 10); and 3) a progressive unconformity against the St. Mehí anticline (Fig. 10). In the centre of the basin, the basal surface corresponds to a conformable surface defined by a sharp change of facies (Figs. 8, 12 and 13). This cycle is of Early Oligocene (Stampian) age, based on biostratigraphical data (Riba *et al.*, 1975; Sáez, 1987; Agustí *et al.*, 1987; Anadón *et al.*, 1989a, 1989b, 1989c; Burbank *et al.*, 1992).

In the northern marginal areas, there is an important development of alluvial conglomerates (Comiols, Fig. 12; El Corb, St. Honorst, Fig. 14), which progressively change to fluvial deposits in the rest of the basin (Figs. 11, 12 and 13). These fluvial deposits were formed in braided, anastomosing and meandering rivers, and

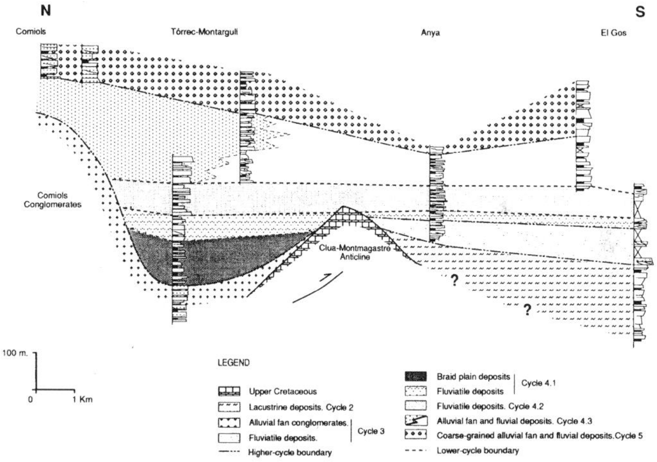

Fig. 12. North–south correlation fence in the western studied area, from Comiols, over the Montsec Unit, to the Seró thrust. The Clua–Montmagastre anticline is shown in the middle of the diagram.

include also overbank and local palustrine and lacustrine sediments.

The St. Mamet–Peramola alignment acted as a topographic barrier, and was only covered by the uppermost deposits of this unit in the area of the Montmagastre anticline, and by the deposits of Cycle 5 in the area of the Peramola thrust.

Three minor-order cycles, bounded by angular, erosional unconformities, are distinguished in the study area, and are called, from bottom to top, the Ca l'Empordanès, Pallerols and Aguilar units (Maestro and Serra Roig, 1992). In these cycles, there are some even lower-order cycles due to base-level changes in the basin (Figs. 3, 10 and 11). In the Ca l'Empordanès unit, two minor cycles can be distinguished (4.1.1, 4.1.2); in the Pallerols unit, there are three

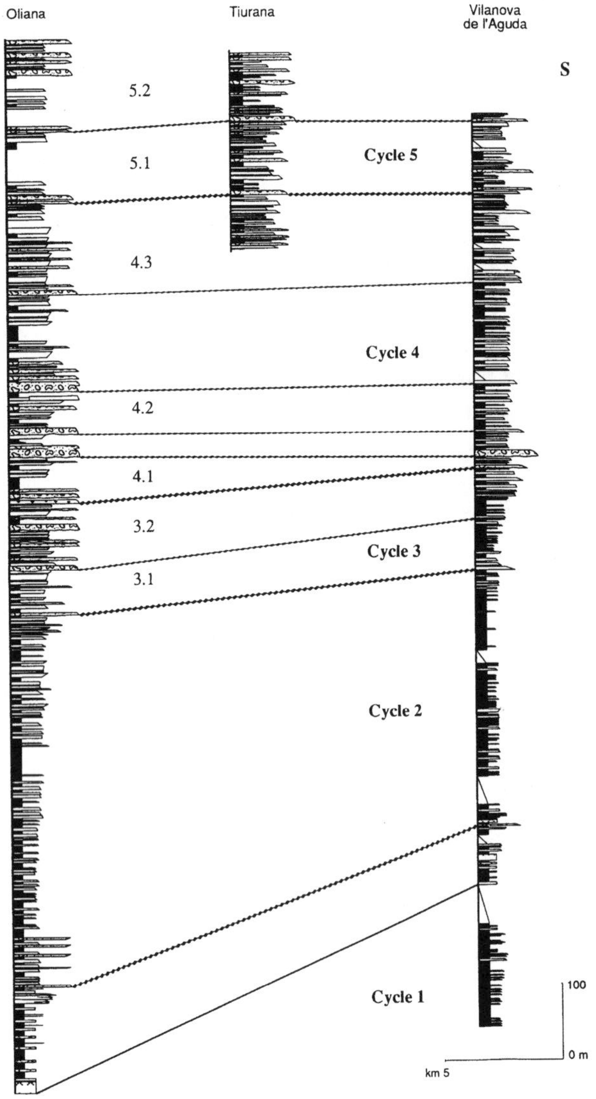

Fig. 13. North–south correlation fence in the eastern studied zone, from the area of the Oliana anticline to the Vilanova de l'Aguda anticline. The deposits of Cycle 1 are marine in the Oliana succession, and lacustrine in the South, in Vilanova de l'Aguda. The above cycles are composed entirely of continental deposits. Numbered subdivisions of the cycles are shown (e.g. 3.1).

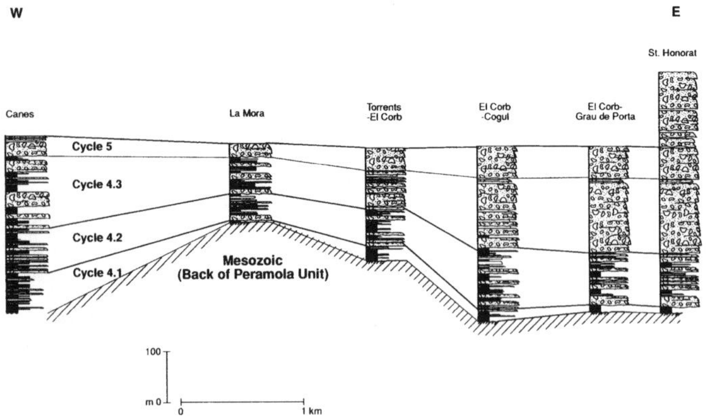

Fig. 14. East–west panel of the lower Oligocene proximal conglomerates deposited behind the Peramola Structural Unit, and in front of the Montsec Unit. These sediments were deposited during the south and southeastern emplacement of the tectonic units, and progressively fossilize them. The uppermost deposits of Cycle 5 are not affected by tectonic movements. They surpass, in some places, the topographic high of the Peramola Mesozoic Unit.

minor cycles (4.2.1, 4.2.2, 4.2.3); finally, in the Aguilar unit, there are two base-level cycles (4.3.1, 4.3.2). The cycles have three distinct systems tracts: the lowstand deposits consist of braided river conglomerates; the rising base-level deposits consist of alternating red shales and caliche horizons; the highstand deposits consist of alternating shales and sandstones of meandering fluvial origin. The deposits of the Ca l'Empordanès and Pallerols units onlap the folded deposits of the third-order cycle in the area of the St. Mehí anticline and the Mesozoic sediments of the Peramola thrust sheet (Fig. 10). The emplacement of this thrust sheet occurred before the fourth cycle, but the southerly and easterly advancing movements of the thrust sheet and its frontal anticline continued during the deposition of the different units of Cycle 4.

The conglomerates of Cycle 4 are polymict, with clasts of Mesozoic and Lower Tertiary carbonates, quartz and Paleozoic sediments. Fluvial facies are more fully developed in Cycle 4 than in Cycle 3.

Cycle 5

This last cycle, although exposed in many parts of the study area, is incomplete because of subsequent erosion. It is thought to be Stampian in age (Riba *et al.*, 1975).

The proximal facies of the cycle is well exposed in Comiols (Fig. 12), El Corb and St. Honorat (Fig. 14), in the northern part of the study area, near the Mesozoic outcrops of the Montsec unit. These sediments extend over the St. Mamet–Peramola tectonic alignment at many points. In front of this, well-developed meandering and anastomosing fluvial sediments were deposited (Figs. 8 to 13). In the Montmagastre–Bellfort area, fluvial deposits are exposed north of the thrust structure (the Montargull depression), and distal facies are exposed south of the thrust. These distal deposits were derived both from the Comiols or El Corb–St. Honorat structures and from the northeastern fluvial systems of the Berga conglomerates (*sensu* Riba, 1973, 1976).

South of the proximal area, the fifth cycle is incompletely exposed, but is represented by fluvial deposits sharply distinguished from the fourth cycle deposits below (Figs. 10 to 13) by the appearance of conglomerates and microconglomerates. In the Peramola area, the fifth cycle is known as the Polig unit (Fig. 3).

The base of the cycle consists of a thick unit of conglomerates, microconglomerates and coarse sandstones. The grain-size of the deposits of this cycle is coarser than that of the lower cycles, due to the progradation of the alluvial and fluvial systems towards the centre of the basin. Two minor-order cycles can be distinguished in this area (5.1, 5.2), showing sedimentary systems tracts related to changes of base-level. The fluvio-alluvial facies predominate in the western areas, while finer-grained facies are well developed in the eastern areas. These sediments represent the last molasse sedimentation preserved in the N of the Catalan basin.

Conclusion

Five major depositional facies are differentiated in the sediments that fill the depositional basin complex of the Late Eocene and Early Oligocene, W of the Segre thrust. These cycles are bounded by angular unconformities, which are enhanced in the borders of the basin, and by sharp changes of facies in central areas, where the cycles tend to be conformable. These cycles are related to relative changes of base level in the basin. Minor and major cycles can be distinguished and the major cycles are regarded as depositional sequences that include both continental and marine deposits. The five major facies are thought to be controlled by eustatic changes of sea-level. During all this sedimentation, tectonism had a decisive effect on the paleogeography of the basin.

The emplacement of the Pyrenean thrust sheets changed the paleogeographic orientation of the basin. During the Middle and Late Eocene, continental deposits formed in the S, while conditions were marine in the N (Oliana area). In the latest Eocene, and even more in the Early Oligocene, proximal deposits were derived from the Pyrenees in the N, while, in the S, towards the centre of the Ebro basin, distal fluviatile and lacustrine sedimentation took place.

The displacement of the Pyrenean thrust sheets towards the S and SE caused the deposition of alluvial fans in the margins of the basin during highstand base-level conditions, and also caused the formation of subsiding areas with lacustrine and deltaic deposits, related to the alluvial fans and fluvial systems. The emplacement of the thrust sheets enhanced the angular unconformities that occur in the associated conglomerates.

The main tectonic events occur over time intervals that surpass the duration of the individual cycles. Thus, the eastward emplacement of the Sierras Marginales units (Salgar and Artesa units) took place during the deposition of Cycle 1 and part of Cycle 2. The movement of the St. Mehí anticline occurred during the deposition of both Cycles 2 and 3. The emplacement of the structural units of St. Mamet, Peramola, Bellfort, Montmagastre and the Clud–Montmagastre anticline took place during both Cycles 2 and 3. The sedimentation above this structural alignment began in Cycle 3 only near the St. Mamet dome, but became extensive in Cycles 4 and 5.

References

Agustí, J., Anadón, P., Arbiol, S., Cabrera, Ll., Colombo, F. and Sáez, A. (1987). Biostratigraphical characteristics of the Oligocene sequences of North-eastern Spain (Ebro and Campins basins). *Müncher Geowiss. Abh.*, **10**: 35–42.

Anadón, P., Cabrera, Ll., Colldeforns, B., Colombo, F., Cuevas, J.L. and Marzo, M. (1989a). Alluvial fan evolution in the SE Ebro basin: response to tectonics and lacustrine base level changes. *4th Int. Conf. Fluvial Sediment. Exc. Guidebook, 9. Publ. Servei Geol. Catalunya*, p. 91.

Anadón, P., Marzo, M., Riba, O., Sáez, A. and Vergés, J. (1989b). Fan delta deposits and syntectonic unconformities in alluvial fan conglomerates of the Ebro basin. *4th. Int. Conf. Fluvial Sediment. Exc. Guidebook, Publ. Sevei Geol. Catalunya*, p. 100.

Anadón, P., Cabrera, Ll., Colldeforns, B. and Sáez, A. (1989c). Los sistemas lacustres del Eoceno superior y Oligoceno del sector oriental de la cuenca del Ebro. *Acta Geol. Hisp.*, **24** (3–4): 205–230.

Burbank, D.G., Vergés, J., Muñoz, J.A. and Bentham, P. (1992). Coeval hindward- and forward-imbricating thrusting in the south-central Pyrenees, Spain: timing and rates of shortening and deposition. *Geol. Soc. Am. Bull.*, **104**: 3–17.

Cámara, P. and Klimowitz, J. (1985). Interpretación geodinámica de la vertiente centro-occidental surpirenaica (cuencas de Jaca-Tremp). *Estud. Geol.*, **41**: 391–404.

Caus, E. (1971). Bioestratigrafía y micropaleontología del Eoceno medio y superior del Prepirineo Catalán. Doctoral thesis. Universidad Autónoma de Barcelona. 187 pp.

Caus, E. (1973). Aportaciones al conocimiento del Eoceno del anticlinal de Oliana (prov. de Lérida). *Acta Geol. Hisp.*, **8**: 7–10.

Maestro, E. and Serra Roig, J. (1992). Ciclicidad en los sedimentos continentales del Eoceno final y del Oligoceno inferior al W del rio Segre (NE de la cuenca del Ebro). *Rev. Soc. Geol. Esp.*, **5**: 117–135.

Maestro-Maideu, E. (1989). Las secuencias deposicionales del Eoceno superior de Peramola (Anticlinal de Oliana. Catalunya). *XII Congr. Nac. Sedimentol.*, Bilbao, 207–210.

Martínez Peña, M.B. and Pocoví, A. (1988). El amortiguamiento frontal de la cobertera surpirenaica y su relación con el anticlinal de Barbastro–Balaguer. *Acta Geol. Hisp.*, **23**: 81–94.

Mitchum, R.M. Jr. and van Wagoner, J.C. (1991). High-frequency sequences and their stacking patterns: sequence-stratigraphic evidence of high-frequency eustatic cycles. *Sediment. Geol.*, **70**: 131–160.

Pocoví, A. (1978). Estudio geológico de las Sierras Marginales Catalanas. Doctoral thesis. Univ. de Barcelona.

Riba, O. (1973). Las discordancias sintectónicas del Alto Cardener (Prepirineo Catalán). Ensayo de interpretación evolutiva. *Acta Geol. Hisp.*, **8**: 90–99.

Riba, O. (1976). Syntectonic unconformities of the Alto Cardener, Spanish Pyrenees: a genetic interpretation, *Sediment. Geol.*, **15**: 213–233.

Riba, O., Ramírez del Pozo, J. and Maldonado, A. (1975). *Mapa y memoria explicativa de la Hoja 329 (34–13–Ponts), del mapa geológico nacional a escala 1:50 000*, IGME.

Sáez, A. (1987). Estratigrafía y sedimentología de las formaciones lacustres del tránsito Eoceno–Oligoceno del NE de la cuenca del Ebro. Doctoral thesis, Univ. Barcelona. 353 pp.

Serra Roig, J. (1992). Els sediments alluvials, fluvials i lacustres de l'Eocè superior i l'Oligocè al sector W del riu Segre. Zona NE de la conca de l'Ebre. Research Work. Universitat Autònoma de Barcelona, 93 pp. (unpubl.)

Uliana, M.A. and Legarreta, L. (1988). Introducción a la estratigrafía secuencial. Análisis de discontinuidades estratigráficas. In *Introducción a la estratigrafía secuencial y discordancias interregionales* (G. González, C. Guilsano, L. Legarreta, A. Ricardi y M.A. Uliana, eds.). *Assoc. Geol. Argentina. Inst. Argentino del Petról.*, 56 pp.

Vail, P.R. (1987). Seismic stratigraphy interpretation procedure. In: *Atlas of seismic stratigraphy* (Bally, A.W., ed.). *Assoc. Petrol. Geol., Stud. Geol.*, **27**(1): 277–281.

Vergés, J. and Muñoz, J.A. (1990). Thrust sequences in the southern central Pyrenees. *Bull. Soc. Geol. Fr.*, **8** (VI–2): 265–271.

Wheeler, H.E. (1964). Baselevel Transit cycle. *Kansas Geol. Surv. Bull.*, **169**: 623–630.

E13 Chronology of Eocene foreland basin evolution along the western oblique margin of the South–Central Pyrenees

P. BENTHAM AND D.W. BURBANK

Abstract

Recently developed magnetic polarity stratigraphies in the western South–Central Unit provide a more-precise temporal database for the analysis of the depositional and deformational history of the southern Pyrenean foreland basin. When combined with lithostratigraphic and structural data, the eight new magnetic sections along the Isabena and Esera valleys and in the Ainsa Basin help define the early stages of development of the Eocene foreland and illustrate the important role played by growing structures, such as the Mediano Anticline, in controlling depositional environments and patterns of subsidence.

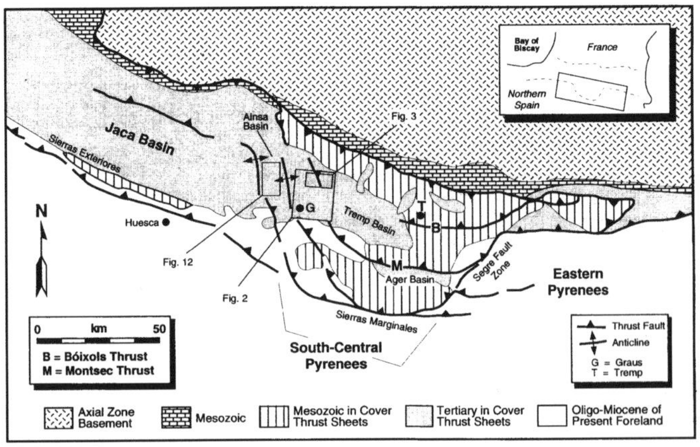

Fig. 1. Map of the Pyrenees and adjacent foreland basin showing the study areas, and the locations of subsequent figures.

Introduction

In an attempt to develop more precise chronological control for the depositional and deformational history of the central part of the South Pyrenean fold-and-thrust belt and its related foreland-basin deposits, several new magnetic polarity stratigraphies have been developed within the syntectonic sedimentary succession. The temporal information derived from these studies permits more detailed correlation between sections and more reliable analysis of the timing, sequencing and rates of sedimentary and tectonic processes. This work has been focused on the western part of the South–Central Unit (Séguret, 1972) and encompasses the western Tremp–Graus and Ainsa basins (Fig. 1). Whereas the majority of these studies has been concerned with late Eocene and Oligocene deposition, part of the studied record begins in the early Eocene. We report here the data and the location of each magnetostratigraphic section, the nature of the magnetic record from these sites, the chronologic significance of each section, and some of the geologic conclusions drawn from these chronologic data.

The methodology utilized here conforms to that described by Burbank *et al.* (1992), whereby three to five specimens were collected at each site and were thermally demagnetized through three to six steps in order to determine a characteristic remanence direction and polarity for each site. The resulting reversal stratigraphies were correlated to the magnetic polarity time scale of Harland *et al.* (1990), and the temporal inferences based on these correlations were used to enhance the interpretation of the geologic history of this part of the southern Pyrenees.

Description of the studied sections

Upper Isabena and Esera River sections

The lower Eocene strata of the Isabena and Esera valleys of the northwestern Tremp–Graus basin (Figs. 2 and 3) record the initiation of the primary phase of the Pyrenean orogeny (Muñoz, 1991). The Roda, St. Esteban, and Campanue fan deltas represent three upward-coarsening, southward-prograding cycles in the Isabena valley and attest to nearby tectonic activity. The Foradada tear fault, active during emplacement of the Cotiella thrust sheet, disrupts a portion of the lower Eocene marine sequence adjacent to the Esera valley. Striking changes in stratigraphic thicknesses between these valleys indicate strong differential subsidence that resulted from either down-to-the-west normal faulting (Cuevas *et al.*, 1985) or from relative uplift of the margin of an inset, more easterly thrust sheet during a regime of regional subsidence. Detailed nummulite biozonations for the lower Eocene of both the

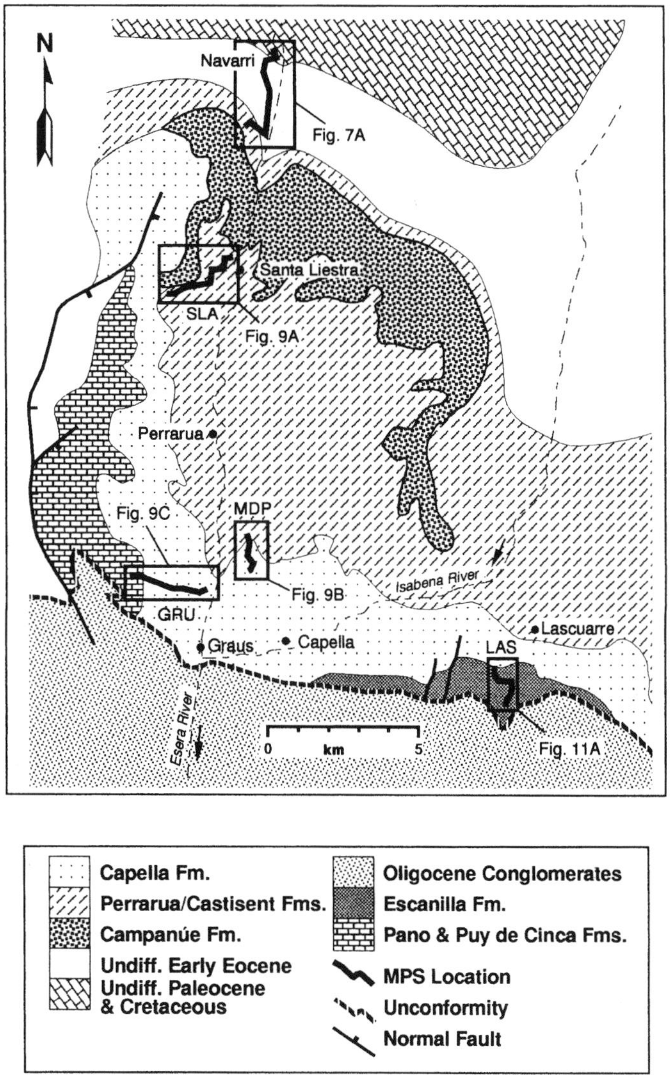

Fig. 2. Simplified geologic map of the western Tremp–Graus Basin showing the location of five of the sections discussed in the text. SLA: Santa Liestra; MDP: Meson de Pascual; GRU: Grustan; LAS: Lascuarre.

Esera and Isabena valleys (Puigdefábregas *et al.*, 1989) have revealed some intriguing problems. For example, the Murillo Limestone can be traced continuously between the two valleys (Fig. 3), but in the Esera valley it contains an Ilerdian fauna, whereas in the Isabena valley it appears to be characterized by a Cuisian fauna.

Two sections were measured in the Isabena valley (Fig. 3). The lower (Roda) section commences at the top of the Alveolina limestone and terminates in the base of the St. Esteban fan-delta (Fig. 4A and 5A). The second section (Esplans section) extends from slightly below the Murillo limestone to the Castisent Formation (Figs. 4B and 5B).

Across a total of ~900 m of measured section, 72 magnetic sites were placed (Fig. 5). Given the known Cuisian age of the middle portion of the Roda section (Puigdefábregas *et al.*, 1989), the magnetic polarity stratigraphy (MPS) can be correlated with the magnetic polarity time scale (MPTS) with considerable confidence (Fig. 6). According to this correlation, the top of the Alveolina limestone dates from >55.5 Ma, the La Pobla Limestone dates from 54.6 Ma, the Murillo Limestone dates from 53.3 Ma, and the base of the St. Estaban Formation is dated at ~52.8 Ma. The mean rate of compacted sediment accumulation during this time was 18 cm/ky.

The section in the upper Esera valley (Fig. 7) begins at the base of the Alveolina Limestone of Ilerdian age and continues to the base of

Fig. 3. Simplified geologic map of the lower Eocene stratigraphy between the Esera and Isabena valleys, showing the major lithostratigraphic units that can be physically traced between the studied sections. The location of the three magnetostratigraphic sections within the northwestern Tremp basin are shown for reference: A = Navarri; B = Esplans; and C = Roda.

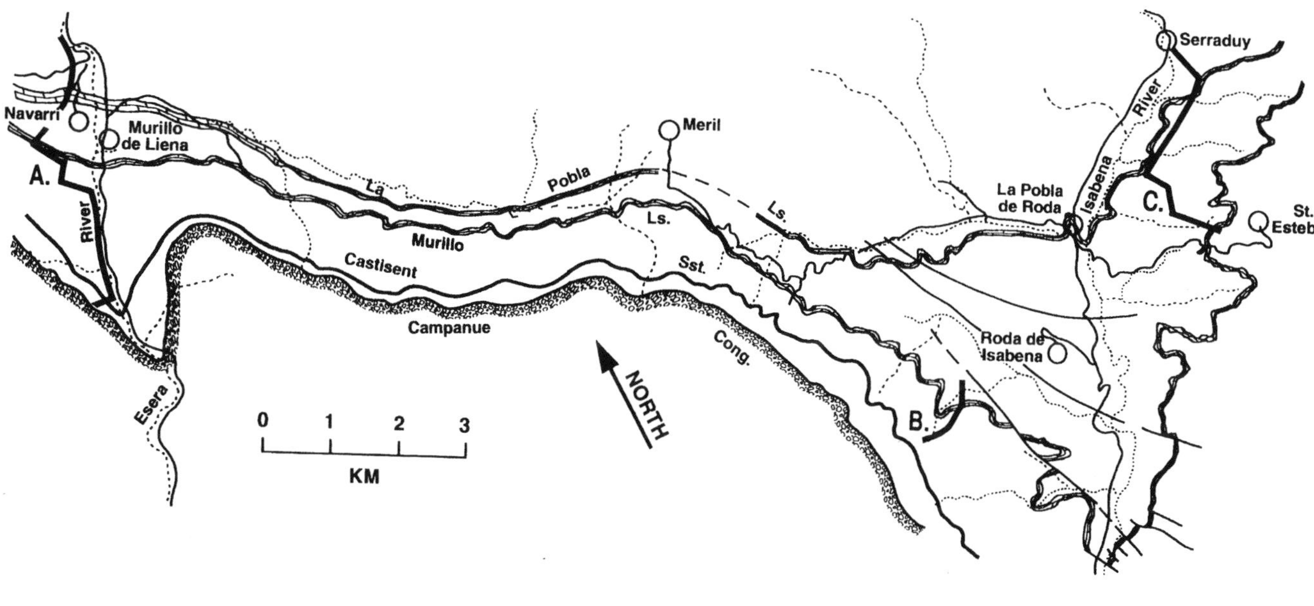

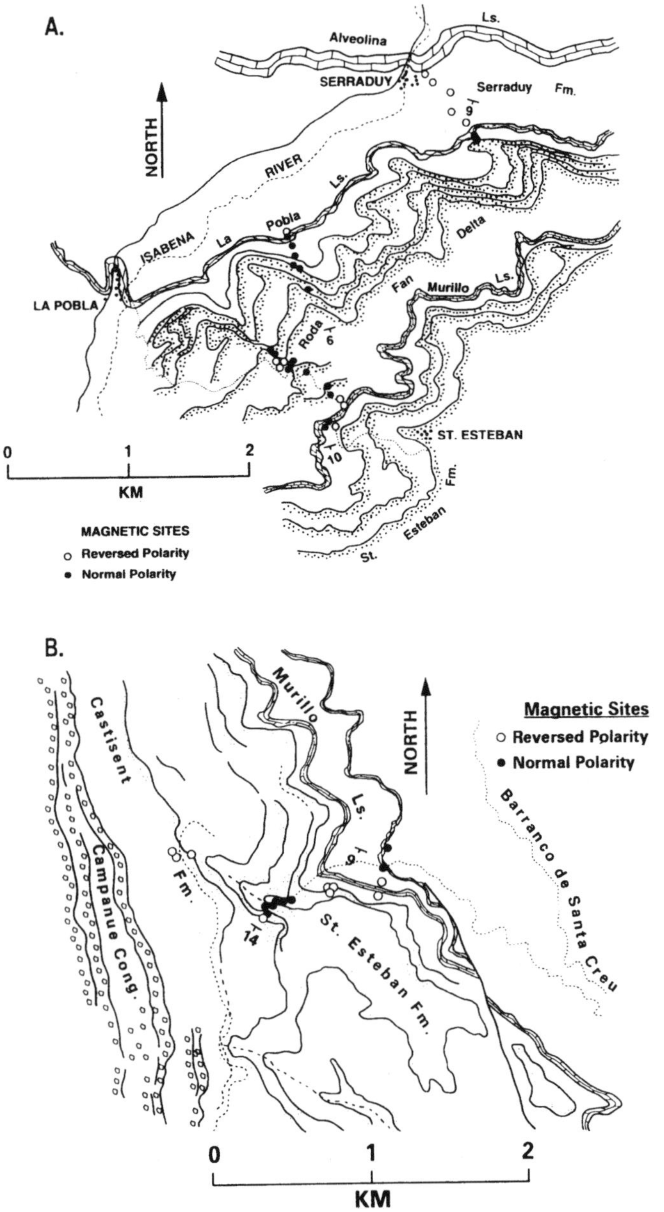

Fig. 4. A. Location map for magnetic samples in the Roda section in the Isabena valley. B. Location map for the Esplans section in the Isabena valley from the Murillo Limestone to the basal part of the Castissent Formation.

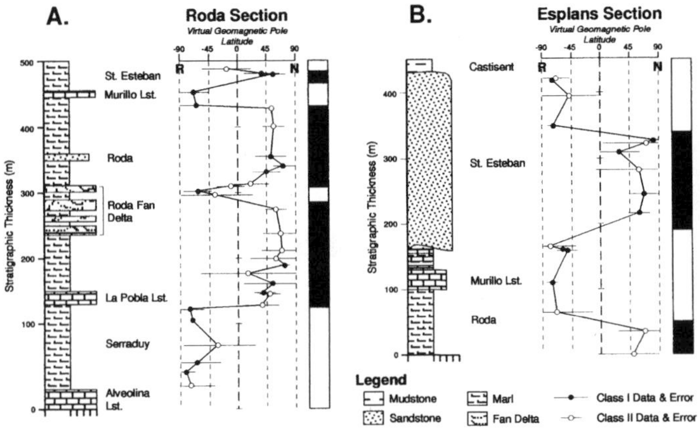

Fig. 5. A. Magnetic polarity stratigraphy and lithostratigraphy for the Roda section, upper Isabena valley. B. Magnetic polarity stratigraphy and lithostratigraphy for the Esplans section, upper Isabena valley. This is correlated with the Roda section on the basis of the Murillo Limestone.

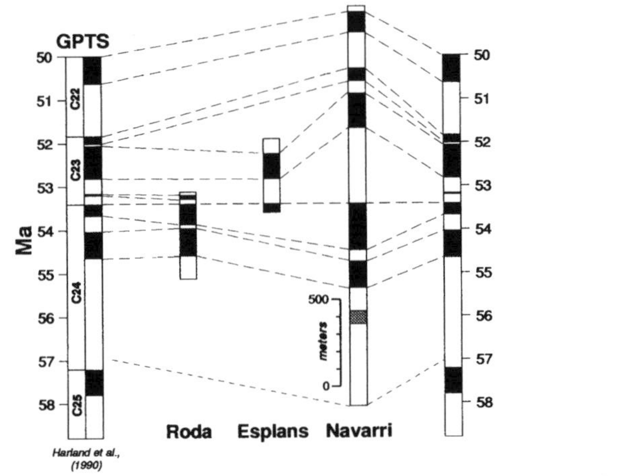

Fig. 6. Correlation of local magnetic stratigraphies in the upper Esera and Isabena valleys with the magnetic polarity time scale of Harland *et al.* (1990).

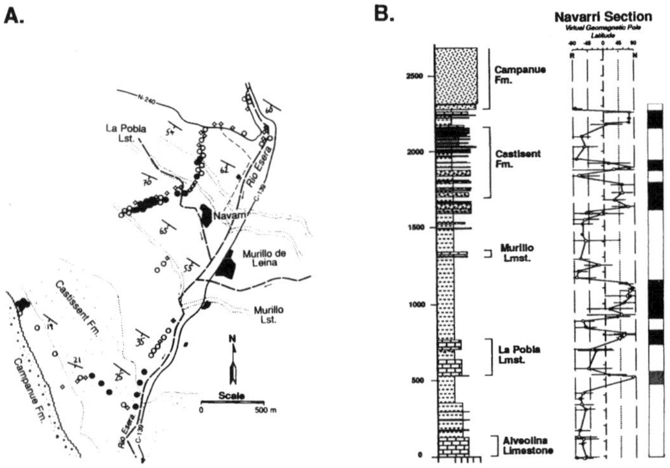

Fig. 7. A Location map for the magnetic sites in lower Eocene strata of the upper Esera valley. B. Magnetic polarity stratigraphy and lithostratigraphy for the upper Esera valley.

the massive Campanue conglomerates. The measured section represents a thickness of 2300 m and contains 97 sampling sites. Following correlation with the GPTS (Fig. 6), the Esera section is interpreted to span from ~57 Ma (base of chron 24r: Alveolina Limestone) to ~50.0 Ma (base of chron 21r: base of the Campanue conglomerate). Mean rates of compacted deposition average 33 cm/ky over this interval.

The much-studied Roda fan delta in the Isabena valley spans an interval from ~54.3 to 53.7 Ma. The St. Esteban fan delta ranges in age in the Isabena valley from 52.8 to 52.1 Ma. Coarse sandstone facies of the St. Esteban fan delta are best developed immediately above the Roda section and are progressively less developed in the

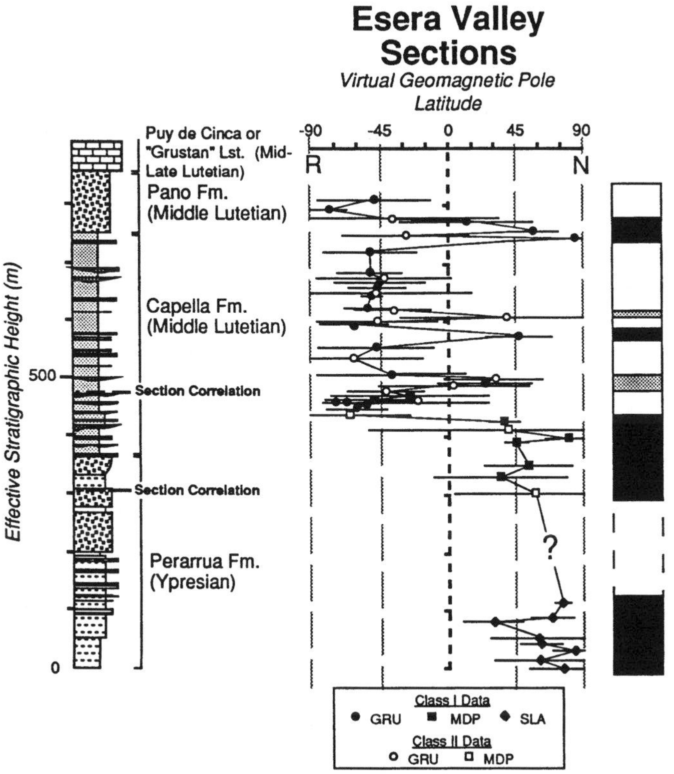

Fig. 8. Composite magnetic polarity stratigraphy for the lower Esera valley based on sections at Santa Liestra, Meson de Pascual, and Grustan.

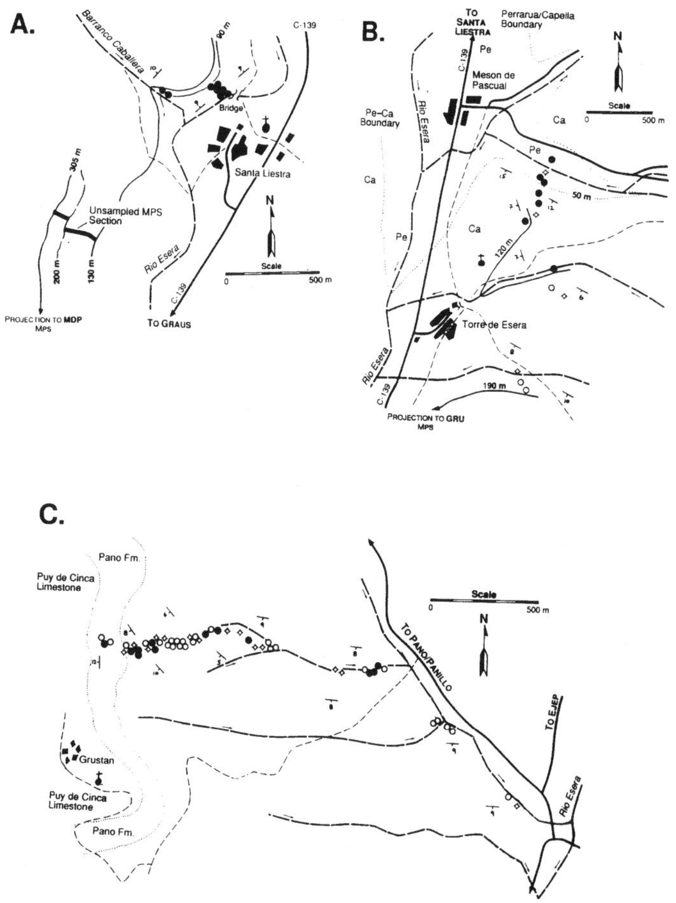

Fig. 9. Location maps for the magnetic sections at: A. Santa Liestra; B. Meson de Pascual; and C. Grustan.

Esplans and Esera sections. In contrast, the conglomeratic facies of the Campanue fan delta is strongly developed in both the Isabena and Esera valleys. Its base occurs shortly after the end of chron 22 ($\sim$50.0 Ma).

The Roda fan delta is interpreted to reflect an early phase of emplacement of the Cotiella thrust sheet in the central Pyrenees (Puigdefábergas *et al.*, 1991). To the west of the Esera section (Fig. 1), the Foradada tear fault, corresponding to the western lateral margin of the Montsec/Cotiella thrusts, cuts all of the Eocene strata older than the Castisent Formation, which is $\sim$52 my/old. Despite the contrast in mean rates of accumulation between valleys, the marl-dominated successions in each are associated with more rapid subsidence and accumulation, whereas the St. Esteban fan and the Castisent Formation are characterized by some of the slowest subsidence rates. This is in agreement with the concept of trapping coarse material in more proximal positions during times of rapid subsidence (Heller *et al.*, 1988; Angevine *et al.*, 1990) and with the slow aggradation with considerable pedogenesis inferred for the Castisent Formation.

Lower Esera and Isabena valleys

In the lower Esera valley (Fig. 2), magnetic polarity studies have been completed from the top of the Campanue fan delta to the middle of the Pano Formation below the base of the Grustan Limestone (Bentham, 1992). A composite MPS (Fig. 8) has been synthesized from the results of three spatially separated sections (Fig. 9) that encompass 71 sites and 900 m of strata. Correlation of the composite Esera MPS with the GPTS (Fig. 10) is based in part on the correlation of the reversal pattern of the strata beneath the Campanue fan with the GPTS (Figs. 6 and 7) and on upper age limits that can be derived from other associated sections which indicate that the top of this section is older than the Escanilla limestone (Bentham, 1992). Accordingly, the top of the Campanue fan dates from $\sim$48.2 Ma and the middle Pano Formation is $\sim$42.8 Ma. Compacted accumulation rates for the Pano Formation ($\sim$16 cm/ky) are only half the rate calculated for the underlying Perarrua Formation. This decrease in rates is interpreted as resulting from initial growth of the Mediano anticline during the latter phases of deposition.

In the lower Isabena river valley, the MPS at Lascuarre (Fig. 11) begins at the Escanilla Limestone, for which biostratigraphic data suggest a late Lutetian to early Bartonian age, spans the lower and middle members of the Escanilla Formation, and comprises 37 magnetic sites. Following correlation with the GPTS (Fig. 10), the top of the Escanilla Limestone is dated at $\sim$42.7 Ma, and the middle of the Escanilla Formation (top of the sampled section) dates from $\sim$39 Ma. Both the temporal data and the common

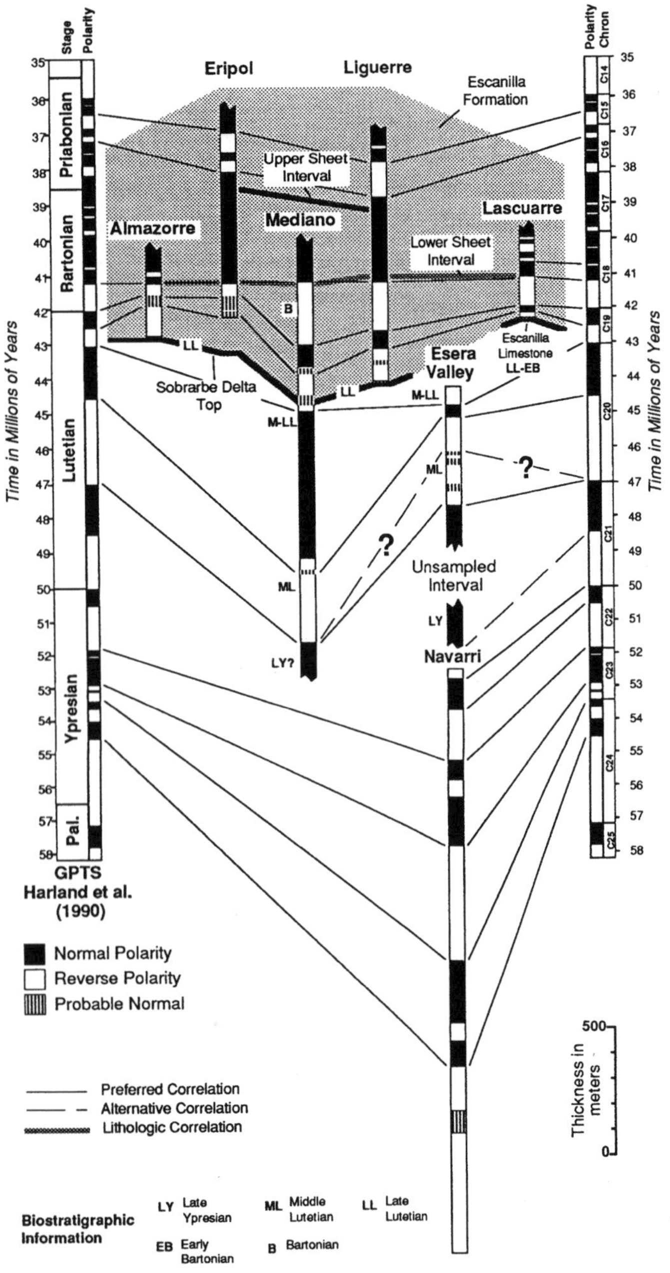

Fig. 10. Correlation of the local magnetic polarity stratigraphies (MPSs) from the lower Esera and Isabena valleys and from the Ainsa basin with the magnetic polarity timescale (MPTS).

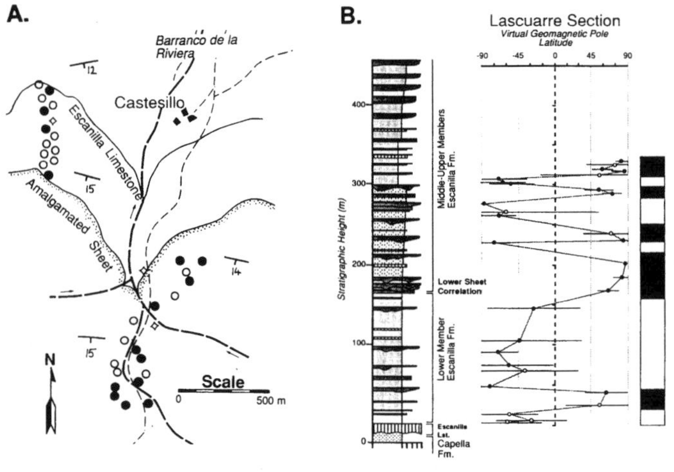

Fig. 11. A. Location map for the Lascuarre section across the Escanilla Formation in the lower Isabena River valley. B. Magnetic polarity stratigraphy of the Lascuarre section.

occurrence of lacustrine strata and abundant paleosol development (Bentham, 1992) indicate that slow sediment accumulation (~10 cm/ky) persisted through this interval, probably as a result of continued growth of the Mediano anticline.

Ainsa Basin

Four magnetic sections have been developed in the Ainsa Basin (Fig. 1). Three of these are concentrated entirely within the Escanilla Formation in the Buil syncline (Fig. 12). The long Mediano section (Fig. 13) encompasses 92 magnetic sites in ~1700 m of predominantly Lutetian and Bartonian strata (Fig. 14A). Based on correlation with the GPTS (Fig. 10) the Hecho Group turbidites and succeeding marls extend from earlier than 47 Ma to 43.3 Ma. The Sobrarbe delta spans 42.9 to 43.3 Ma, and the delta top represents the local transition from marine to terrestrial conditions during late Lutetian times. Growth of the Mediano anticline during deposition of the basal 600 m of strata of the Mediano section is illustrated by the fanning geometry of beds and resedimented carbonates and breccias on the flank of the anticline (Fig. 14B). Although sediment-accumulation rates vary as a function of distance from the growing anticline, overall rates in the Ainsa Basin accelerate from ~15 cm/ky prior to 44.5 Ma to ~40 cm/ky during chron 20n (~43.0–44.5 Ma) at the same time as they decelerate in the western Tremp–Graus basin as shown by the composite Esera and Mediano sections (Figs. 10 and 15).

The three Escanilla sections (Figs. 16 and 17) at Liguerre, Eripol, and Almazorre provide time control across the Ainsa Basin for much of the interval of Escanilla deposition. Correlations among the sections are based on traceable units, like the amalgamated sheet sandstone or the Sobrarbe delta (Figs. 16 and 18), and on the magnetic polarity patterns. The time-constrained depiction of Escanilla depositional geometries and rates from the Ainsa and western Tremp–Graus basins that is attained from these sections (Fig. 18) clearly shows the influence of synsedimentary deformation of the Mediano and Boltaña anticlines. In particular, the most rapid accumulation occurs furthest from the anticlines and coeval depositional units taper towards them (Fig. 18). Lithostratigraphic data (Bentham, 1992) indicate that depositional systems were focused along the core of the Buil syncline by this deformation and that the fluvial systems were deflected away from the growing structures.

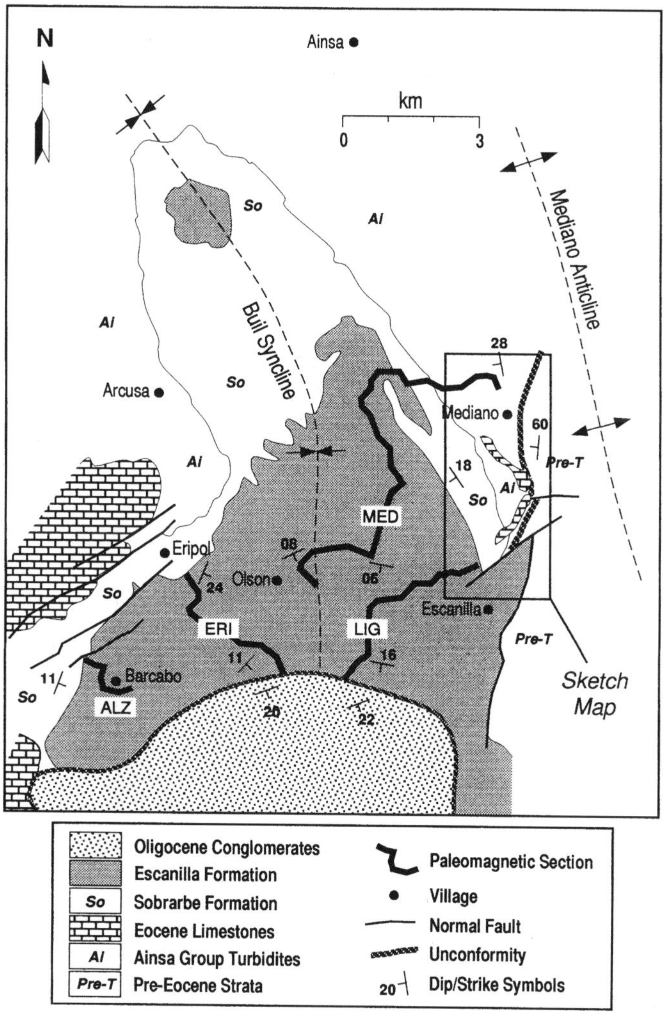

Fig. 12. Simplified geologic map of the Ainsa Basin, showing the locations of the four magnetic sections. Three of these are located entirely within the Escanilla Formation (ALZ: Almazorre; ERI: Eripol; and LIG: Liguerre), whereas the Mediano section (MED) extends from the flank of the Mediano anticline, through the Hecho Group turbidites, and into the Escanilla Formation.

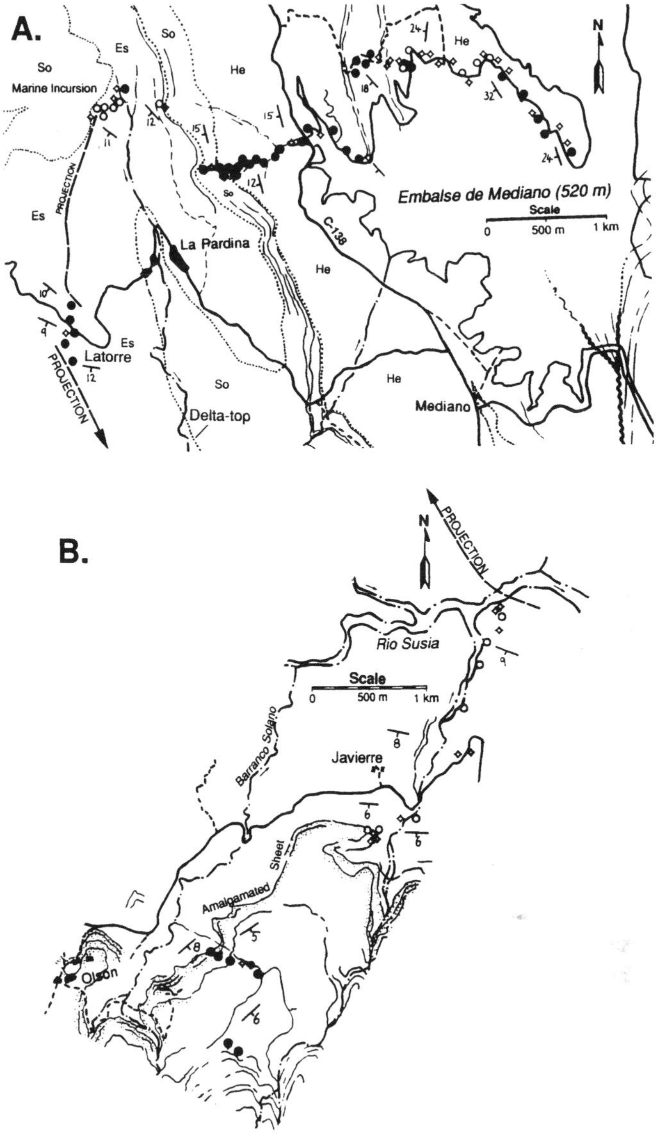

Fig. 13. Site location map for the Mediano magnetic section (in two subsections, A and B), extending from the flank of the Mediano anticline to the core of the Buil syncline in the Escanilla Formation.

Middle to late Eocene paleogeography of the Ainsa and western Tremp–Graus Basins

Based on the regional magnetostratigraphic correlations and lithostratigraphic analysis within the Ainsa and Tremp–Graus basins, the tectono-stratigraphic evolution of the region along the western oblique margin of the South–Central Unit (SCU) can be reconstructed. Four sequential reconstructions will be presented and these represent modifications of a framework established by Nijman and Nio (1975) in the light of our new data. Throughout middle to late Eocene time, shortening along the oblique ramp strongly modulated subsidence and depositional patterns in both basins. The structural culmination of the Mediano anticline successively localized and restricted the shelf break (Fig. 19A) during its early growth, and subsequently it acted as a boundary between contrasting depositional domains. Growth of the Mediano anticline caused syndepositional rotation of strata (Fig. 14B) along its western and eastern flanks (Holl and Anastasio, 1993) and slowing subsidence to the east (Fig. 19B). Despite rapid subsidence within the Ainsa Basin (40 cm/ky), the sediment supply was sufficient to fill the subsiding marine basin and ensuing deltaic deposition was focused along the axis of the basin (Fig. 19C). As uplift along the Mediano anticline diminished, the sandstones and conglomerates of the Escanilla system spread across the entire region (Fig. 19D). Because the amount of N–S translation along the basal thrust is poorly known in this area, the reconstructions do not account for the changing relative positions of the Ainsa and Tremp–Graus basins as the piggyback basin was carried southwards. By correlation with balanced sections to farther east in the SCU (see chapter

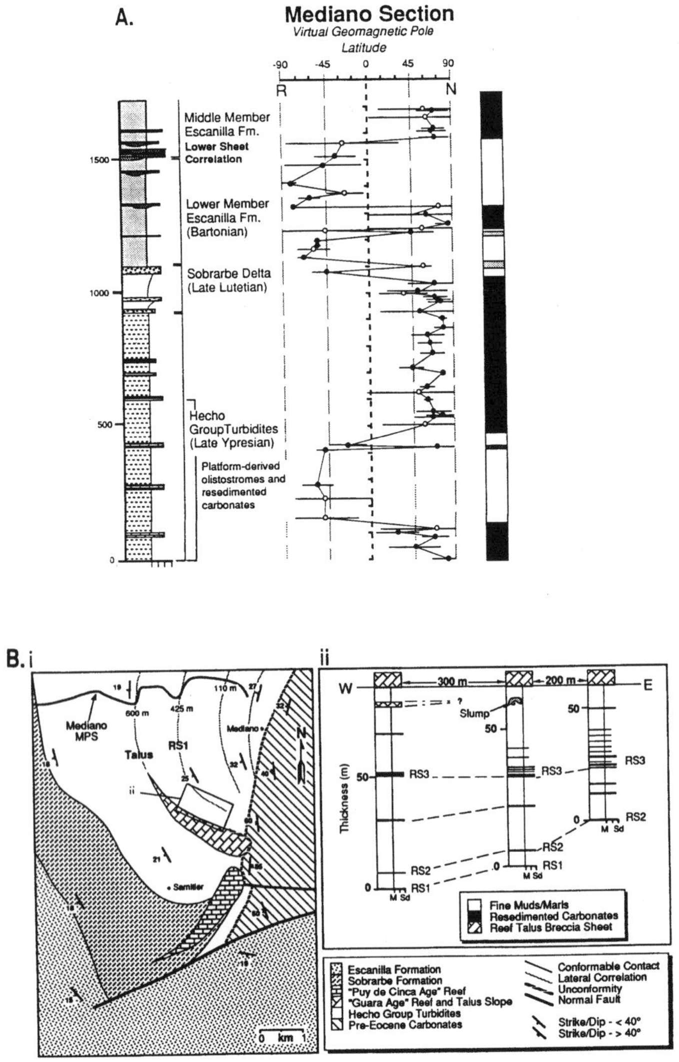

Fig. 14. A. Magnetic polarity stratigraphy of the Mediano section based on 92 sites. B. Syntectonic geometries and thickness variations along the western flank of the Mediano anticline in the region of Samitier. (i) Geologic sketch map. (ii) Panel diagram showing thickness variations beneath the lower reef-derived talus breccia. Map shows the correlation of these units to the appropriate stratigraphic height in the Mediano MPS.

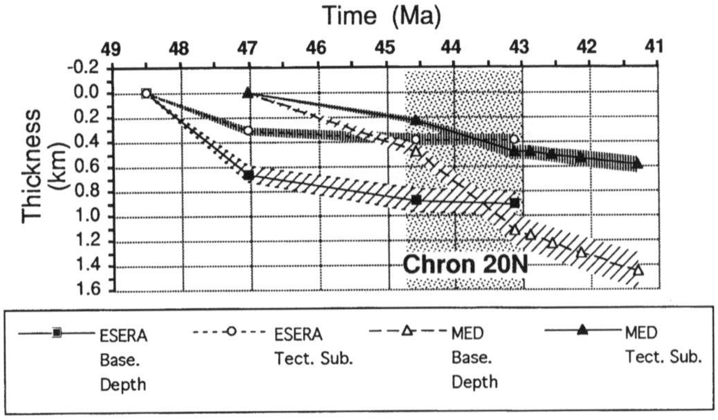

Fig. 15. Geohistory diagram showing sequential basement depth and tectonic subsidence for the Mediano and composite Esera magnetic sections. Opposing subsidence trends during chron 20 occur on opposite flanks of the Mediano anticline and are interpreted as resulting from relative uplift of the Tremp–Graus piggyback basin to the east that was contemporaneous with thrust-load induced subsidence in the Ainsa Basin to the west.

E11), ~20–25 km of translation may have occurred during the interval encompassed by these reconstructions. Relative motion of this magnitude would have only a modest impact on these paleogeographic reconstructions.

Acknowledgments

This research was supported by grants to DWB from the National Science Foundation (EAR-8517482, 8816181, and 9018951). Acknowledgment is gratefully made to the donors of the Petroleum Research Fund, administered by the American Chemical Society (grants 17625, 20591, and 23881). PAB would also like to thank AAPG Grants-in-Aid, USC Graduate School and the USC Department of Geological Sciences Graduate Student Research Fund for additional support during the completion of his dissertation.

References

Angevine, C.L., Heller, P.L. and Paola, C. (1990). *Quantitative sedimentary basin modelling*. Tulsa, American Association of Petroleum Geologists, 133 pp.

Bentham, P. (1992). The tectono-stratigraphic development of the western oblique ramp of the South-Central Pyrenean thrust system, northern Spain. PhD: University of Southern California, 253 pp.

Burbank, D.W., Puigdefábregas, C. and Muñoz, J.A. (1992). The chronology of the Eocene tectonic development of the eastern Pyrenees. *Geol. Soc. Am. Bull.*, **104**, 1101–1120.

Cuevas, M., Donselaar, M.E. and Nio, S.D. (1985). Eocene clastic tidal deposits in the Tremp–Graus Basin (provs. of Lérida and Huesca). In *Int. Assoc. Sedimentol., 6th European Regional Meeting, Lleida, Spain, Excursion Guidebook*, pp. 215–266.

Harland, W.B., Armstrong, R.L., Cox. A.V., Craig, L.E., Smith, A.G. and Smith, D.G. (1990). *A geologic time scale 1989*. Cambridge, Cambridge University Press, 263 pp.

Heller, P.L., Angevine, C.L., Winslow, N.S. and Paola, C. (1988). Two-phase stratigraphic model of foreland basin development. *Geology*, **16**, 501–504.

Holl, J.E. and Anastasio, D.J. (1993). Paleomagnetically derived folding rates, southern Pyrenees, Spain. *Geology*, **21**, 271–274.

Muñoz, J.A. (1991). Evolution of a continental collision belt: ECORS-Pyrenees crustal balanced cross-section. In McClay, K.R. (ed.), *Thrust tectonics*. Chapman & Hall, London, pp. 235–246.

Nijman, W.J. and Nio, S.D. (1975). The Eocene Montañana delta (Tremp–Graus Basin, provinces of Lérida and Huesca, southern Pyrenees, N. Spain). In *IX Congrés de Sedimentologie, Nice*, 18 pp.

Puigdefábregas, C., Collinson, J.D., Cuevas, J.L., Dreyer, T., Marzo, M., Mercade, L., Nijman, W., Vergés, J., Mellere, D. and Muñoz, J.A. (1989). Alluvial deposits of the successive foreland basin stages and their relation to the Pyrenean thrust sequences. In *4th Int. Conference of Fluvial Sedimentology, Excursion Guidebook*, 175 pp.

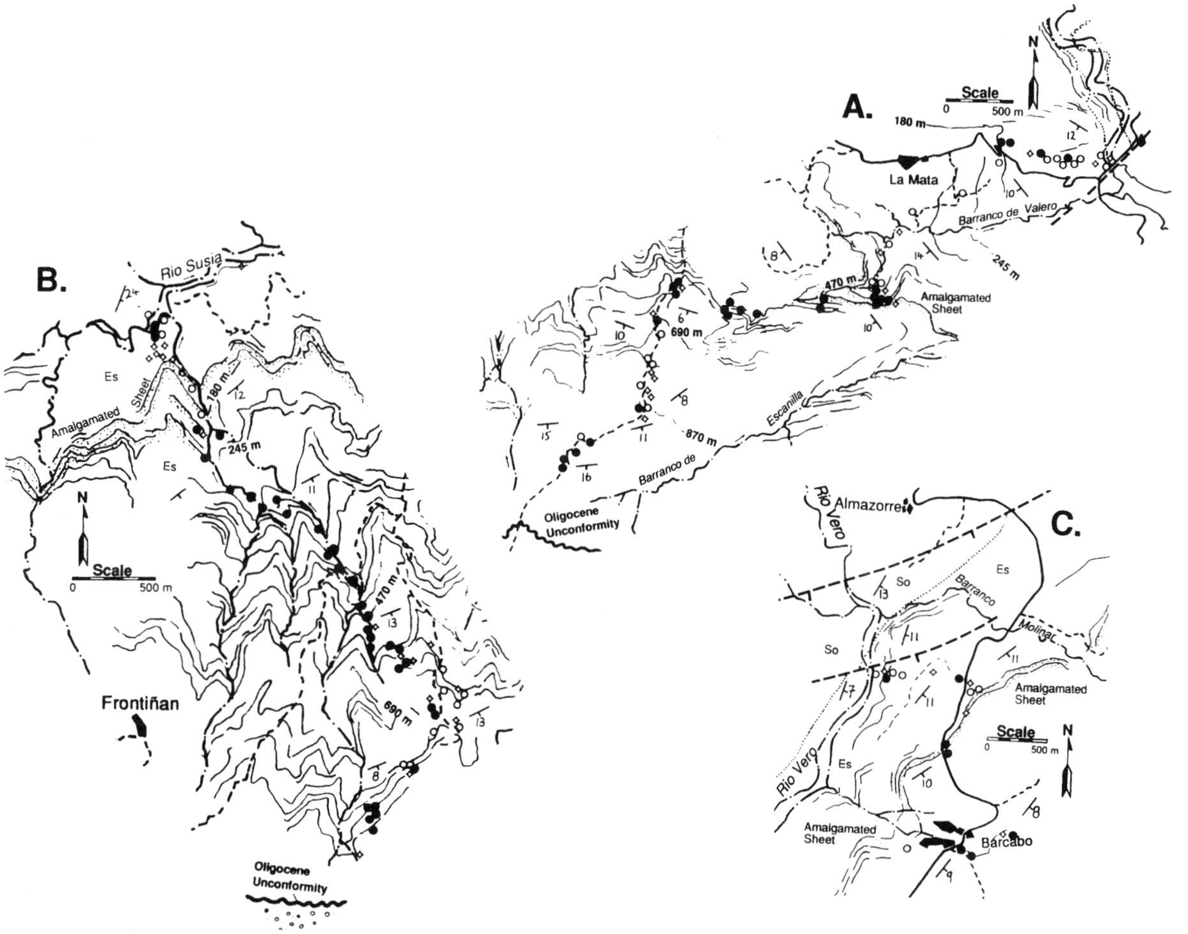

Fig. 16. A. Location map for the 58 sites of the Liguerre magnetic section in the southeastern Ainsa Basin. B. Location map for the 63 sites of the Eripol magnetic section in the west-central Ainsa Basin. C. Location map for the 18 sites of the Almazorre magnetic section in the southwestern Ainsa Basin.

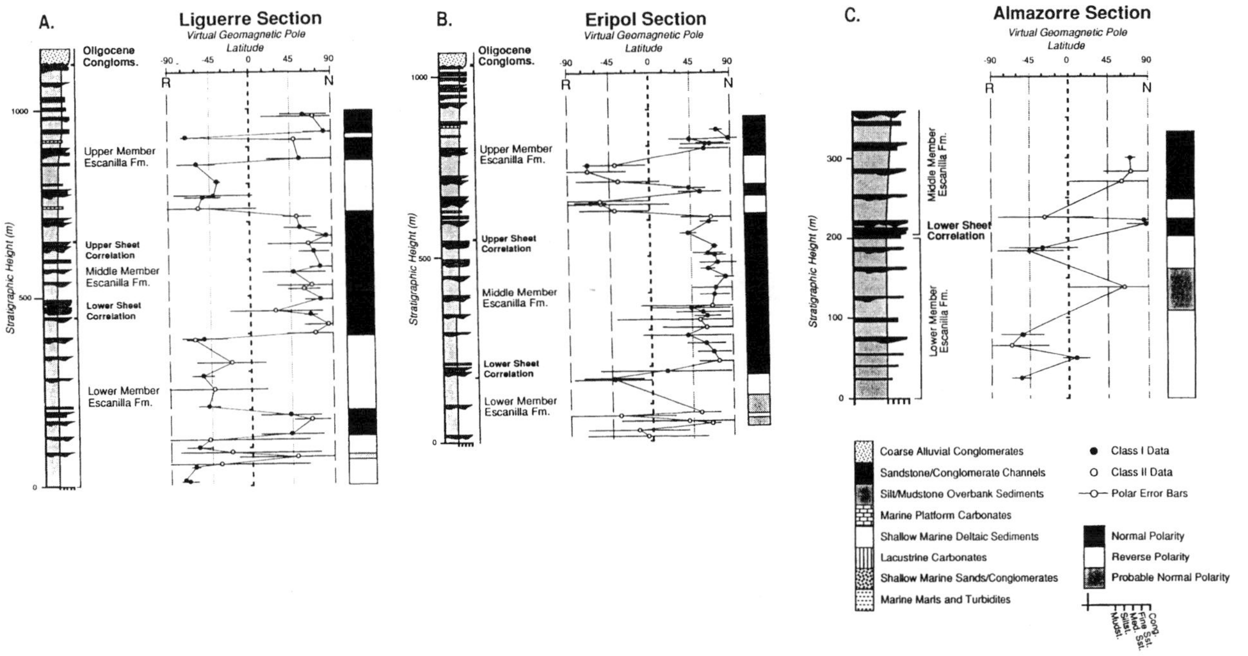

Fig. 17. A. Magnetic polarity stratigraphy (MPS) of the Liguerre section, ranging from the top of the Sobrabe delta to near the base of the Oligocene conglomerates. B. MPS of the Eripol section (Sobrarbe delta to the upper member of the Escanilla Formation). C. MPS of the Almazorre section (Sobrabe delta to the middle member of the Escanilla Formation).

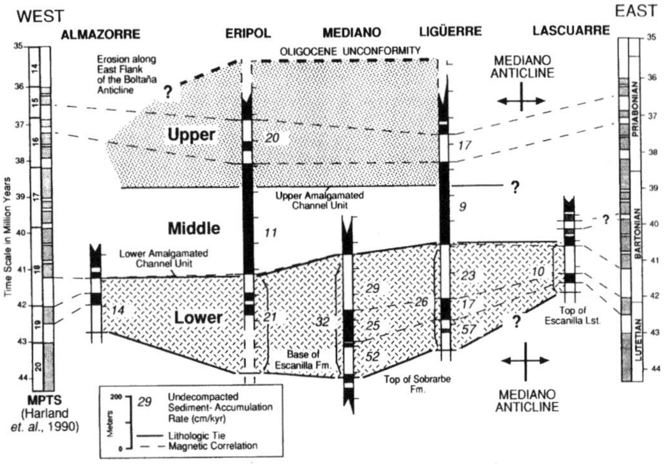

Fig. 18. Litho- and magneto-stratigraphic correlation from the Ainsa Basin to the western Tremp–Graus basin for the Escanilla Formation. Average undecompacted rates of sediment accumulation are shown adjacent to each section. These clearly depict the slow rates in the Tremp–Graus piggyback basin and show that subsidence and sediment accumulation within the Ainsa Basin were most rapid away from the deforming flanks of the bounding anticlines.

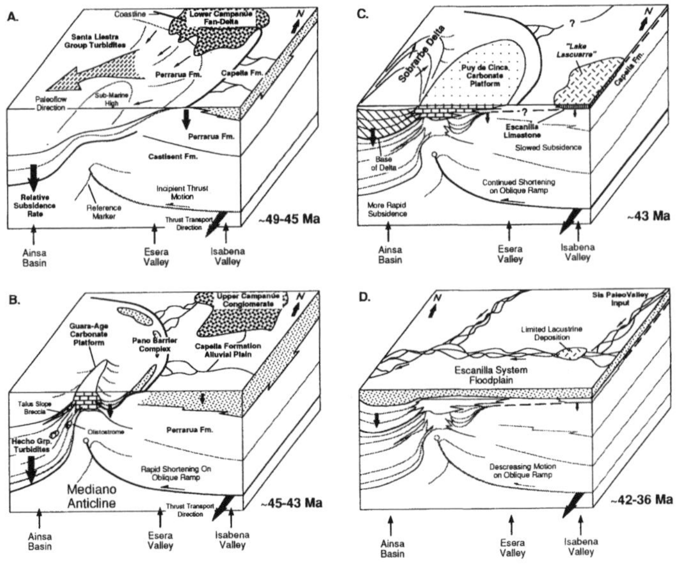

Fig. 19. Paleogeographic reconstructions: A. Latest Ypresian to early Lutetian reconstruction of depositional patterns in the vicinity of the western margin of the south-central unit. During early development of the Mediano anticline, the region of the oblique ramp represents a narrow shelf break between deep marine turbiditic deposition in the Ainsa Basin and shallow marine and continental deposition in the western Tremp–Graus Basin. The Campanue fan delta supplied abundant detritus to the Hecho Group turbidite basin across the shelf break. B. Middle Lutetian paleogeography during Capella and Pano Formation deposition. The main phase of uplift and rotation of the Mediano anticline occurs as a carbonate platform develops across the emergent structural culmination and breccias are shed from its unstable flanks into the Ainsa Basin. Capella alluvial deposits prograde towards the west in response to slowing subsidence and filling of the Tremp–Graus Basin. C. Late Lutetian deposition during which ponded conditions ('Lake Lascuarre') prevailed to the east (upstream) of the Mediano axis, the carbonate platform expanded eastwards from the anticlinal crest, and the Sobrarbe delta was channelized along the axis of the Ainsa Basin. D. Bartonian to Priabonian reconstruction during the slowing of motion along the oblique ramp. The Escanilla floodplain apparently spread across the crest of the Mediano anticline, although subsidence was still more rapid to the west of its culmination.

Puigdefábregas, C., Muñoz, J.A. and Vergés, J. (1991). Thrusting and foreland basin evolution in the Southern Pyrenees. In McClay, K.R. (ed.), *Thrust tectonics*. Chapman & Hall, London, pp. 247–254.

Séguret, M. (1972). Étude tectonique des nappes et séries décollées de la partie centrale du versant sud des Pyrénées. *Publication USTECA, Montpellier, Series Geologie et Structure*, **2**.

E14 Evolution of the Jaca piggyback basin and emergence of the External Sierra, southern Pyrenees

P.J. HOGAN AND D.W. BURBANK

Abstract

From Late Lutetian times through to the Oligocene, the Jaca Basin evolved as a piggyback basin transported to the south by the San Felices thrust sheet. During the Oligocene, emergence, imbrication, and erosion of the frontal parts of this thrust sheet created the External Sierras and re-organized the pre-existing drainage systems. We describe here new magnetostratigraphic dates for stratigraphic sections within both the Jaca and northern Ebro basins. The resultant chronologies provide a temporal framework for quantifying and synthesizing the mid-Tertiary depositional and structural evolution of the western Pyrenees.

Introduction

The Jaca Basin represents the westernmost structurally partitioned basin of the South Pyrenean foreland (Fig. 1). Previous studies (Puigdefàbregas, 1975) have delineated the detachment of the basin as a sole thrust propagated beneath it during late Eocene times, after which it survived as a long-lived piggyback basin accumulating detrital sediments at least until the mid Oligocene. During this interval, the External Sierra emerged as an important structural and topographic range defining the southern margin of the Jaca Basin. Because most of the piggyback strata, as well as those fronting the External Sierra in the northern Ebro Basin, were deposited subaerially and have only yielded a sparse faunal record, previous temporal controls on the evolution of the basin and bounding ranges have been imprecise. We describe here the results of recent magnetostratigraphic studies within the Jaca Basin and the northern Ebro Basin. Taken together, the chronologic control derived from these data permits a more reliable synthesis of the basin history and a clearer delineation of the rates and timing of both structural and stratigraphic events.

The procedures for magnetic analyses were the same as those described by Burbank *et al.* (1992) and Bentham and Burbank (see Chapter E13). Guided by the constraints derived from the available biochronologic data, the pattern of magnetic reversals was correlated with the new magnetic time scale of Cande and Kent (1992).

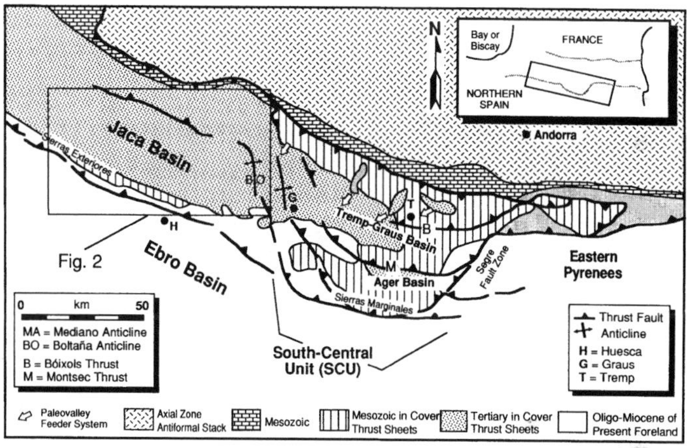

Fig. 1. Geologic map of the Pyrenees showing the study area of the Jaca Basin, northern Ebro Basin, and the External Sierra. Box shows location of figure 2.

Jaca Basin

The Jaca Basin (Fig. 2) contains at least 5 km of foreland-basin strata that were deposited during Eocene and Oligocene times (Puigdefàbregas, 1975). Early deposition in the northern basin was dominated by the Hecho Group turbidites and associated megaturbidite breccia sheets (Labaume and Seguret, 1985) that were coeval with the Guara carbonate platform along the southern basin margin. The Arguís marls were deposited in Bartonian times as the southern carbonate platform was drowned (Canudo *et al.*, 1988), and during the Priabonian the basin was filled from east to west by the Belsué–Atarés deltaic deposits which presaged the transition to terrestrial deposition. The ensuing fluvial strata of the Campodarbe Group are as much as 4 km thick (Puigdefàbregas, 1975) and largely accumulated in the piggyback Jaca Basin during its southward transport above a sole thrust localized in Triassic evaporites. Deformation related to propagation of the thrust tip and fault imbrication behind it gradually created the External Sierra, which delimits the present southern margin of the Jaca Basin and whose growth disrupted the pre-existing fluvial system flowing into and across the Jaca Basin.

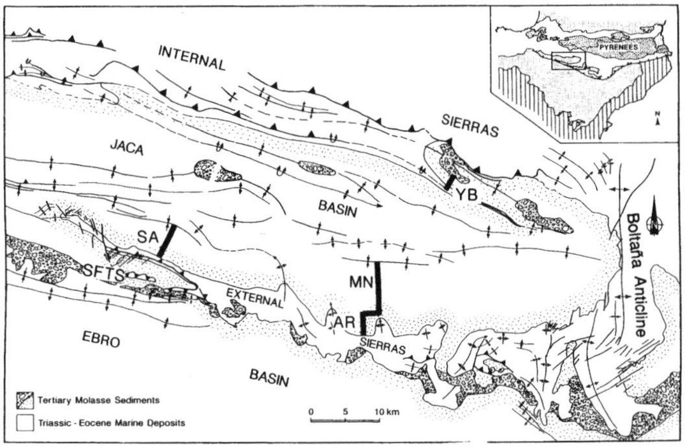

Fig. 2. Simplified geologic map of the Jaca Basin showing the location of the magnetic sections. SA: Salinas; AR–MN: Arguís–Monrepós; YB: Yebra de Basa; SFTS: San Felices thrust sheet.

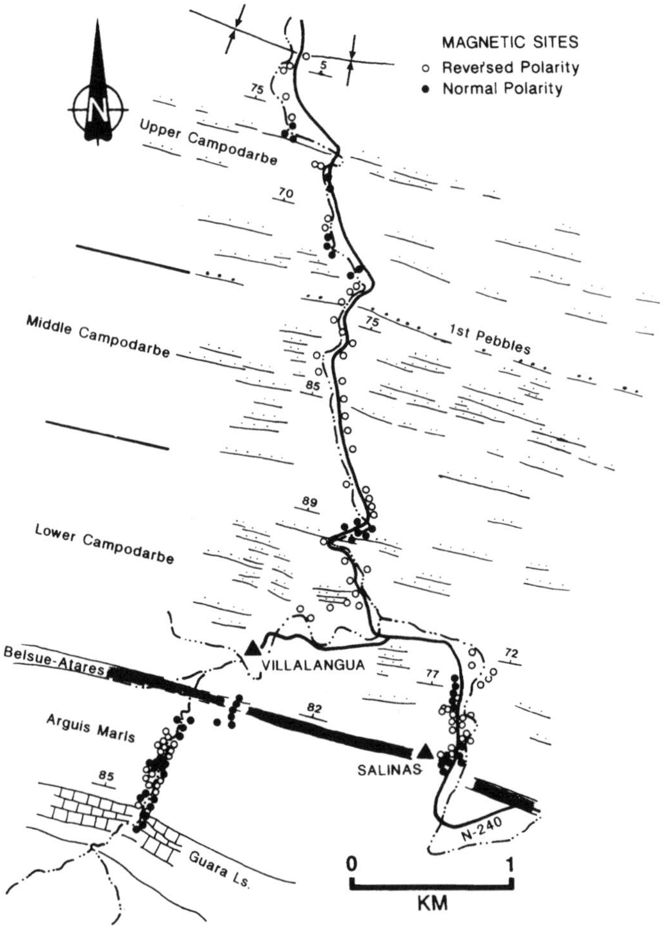

Fig. 3. Site location map for the 193 sites in the Salinas magnetic section, southern Jaca Basin.

Three major magnetic sections (Hogan, 1993) have been developed in the Jaca Basin (Fig. 2). These sections span most of the interval of foreland-basin deposition recorded in the southern Jaca Basin and part of the record in the northern basin. The Salinas section (Fig. 3) begins at the top of the Guara limestone, traverses the Arguís marls and Belsue–Atarés deltaic strata, and encompasses nearly 4 km of fluvial rocks in the Campodarbe Group. These strata are part of a nearly vertical limb of a post-depositional box fold developed above the Triassic evaporites (Hogan *et al.*, in review). Within the lower Campodarbe strata at ~1600 m (Fig. 4), there is a minor angular unconformity which is interpreted as resulting from a tilting of the distal basin margin due to early stages of deformation in the External Sierra (Hogan, 1993). The middle Campodarbe strata record a transition between the unconfined, laterally migrating, and easterly sourced fluvial system which characterizes the lower Campodarbe (Jolley and Hogan, 1989) and the entrenched, northerly sourced upper Campodarbe fluvial system. This change is interpreted as resulting from the topographic emergence of the External Sierra as a barrier to formerly northwestward flow across the northern Ebro–Jaca Basin foreland. The magnetostratigraphy from Salinas (Fig. 4) indicates that the beginning of the marl deposition occurred at ~42.5 Ma (Fig. 5), the Belsue–Atarés delta prograded across this region between 37.0 and 37.5 Ma, and fluvial deposition persisted until at least 29.5 Ma. The lower Campodarbe unconformity is dated at ~34.5 Ma and significant transitions in fluvial style separating lower, middle, and upper Campodarbe strata occur at ~33.5 and ~31.5 Ma. Despite the tectonism in the External Sierra, smooth trends in accumulation rates are visible as they accelerate during initial marl deposition, slow during deltaic deposition, and then accelerate again in the early stages of fluvial sedimentation (Fig. 6).

Situated about 30 km east of the Salinas section, the Arguís–Monrepós section (Fig. 2) is also located along the southern margin of the Jaca Basin and records the transition from marine to terrestrial deposition and subsequent changes in the style of fluvial sedimentation. The 4-km-thick section extends from the Guara limestone to the middle Campodarbe Group (Fig. 7). The magnetostratigraphy (Fig. 8) indicates that the section spans ~10 my (Fig. 5), ranging from ~42–32 Ma. Erosion has apparently removed the top of the section, such that, in comparison with Salinas, it does not encompass as full a record of piggyback deposition. Variations in sediment-accumulation rates (Fig. 6) at this locality are sensitive to the Arguís section's position on the flanks of a growing anticline (Puigdefàbregas, 1975) and to the position of the basal Monrepós section along the anticlinal crest (Fig. 7). Consequently, variations in accumulation rates reflect both the control exerted by local deformation, as well as changes in load-induced subsidence due to hinterland thrusting.

The Yebra de Basa section (Figs. 2 and 9) commences near the top of the Hecho Group turbidites and records the gradual upward transition from shallow marine and deltaic conditions to subaerial sedimentation in the coarse-grained conglomerates of the Santa Orosia fan. Correlation of the magnetic record from Yebra de Basa (Fig. 10) is ambiguous (Fig. 5), but suggests that turbiditic deposition terminated in this region by 40.0 Ma and that the Santa Orosia

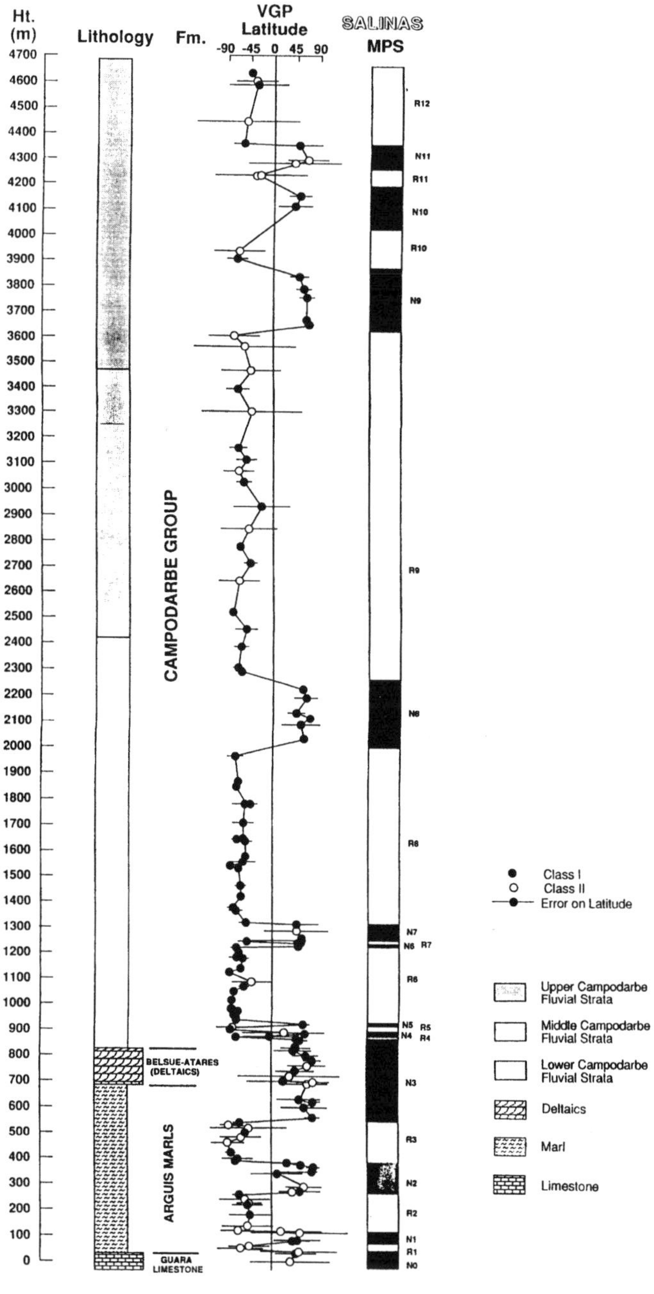

Fig. 4. Magnetic polarity stratigraphy from the Salinas section, ranging from the top of the Guara limestone to the top of the preserved Campodarbe Group strata in the central Jaca Basin.

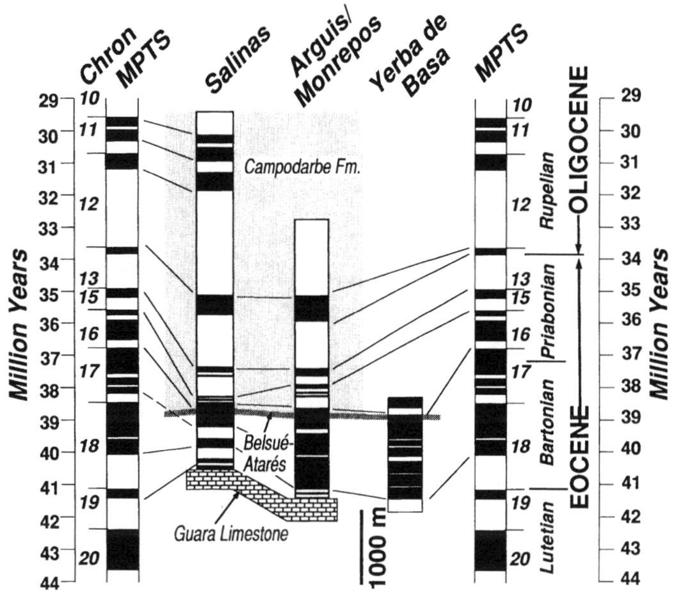

Fig. 5. Correlation of the local MPSs from the Jaca Basin with the magnetic polarity time scale (Cande and Kent, 1992).

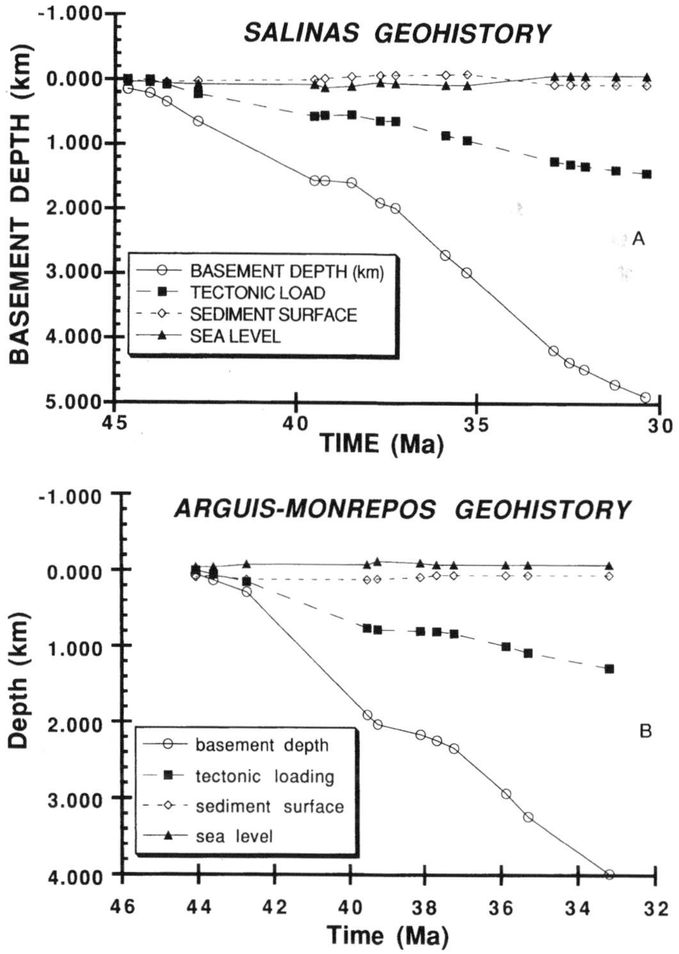

Fig. 6. Geohistory diagrams from Salinas (top) and Arguís–Monrepós (bottom) magnetic sections (time scale from Berggren and others, 1986).

fan delta prograded southward across the Belsue–Atarés delta at ~37.0 Ma.

Synthesis of the chronologic data with lithostratigraphic, structural, paleocurrent, and provenance data from the Jaca Basin permits a time-constrained reconstruction of the main phases of middle Eocene to early Oligocene deposition and deformation (Fig. 11). The drowning of the carbonate shelf during late Lutetian to Bartonian times occurs during a phase of rapid subsidence (Fig. 6).

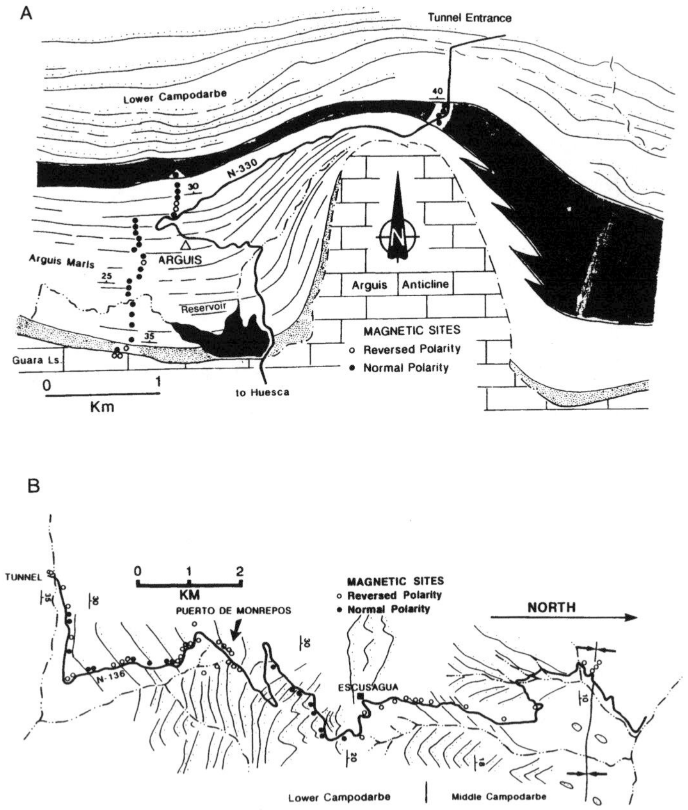

Fig. 7. Site location maps for the 121 sites in the Arguís (top)–Monrepós (bottom) magnetic section, southern Jaca Basin.

During this interval, early detachment occurred within the incipient External Sierra and a succession of north–south-oriented anticlines began to develop. Their growth continued during ensuing deltaic deposition and into the earliest stages of the terrestrial sedimentation in the piggyback Jaca Basin. This transition to terrestrial deposition is approximately coeval with one phase of hinterland thrusting (Fig. 11) that is recorded near the top of the Yebra de Basa section. Subsequently, a small interval of uplift on the San Felices thrust sheet (~34.5 Ma) was followed by large-scale southward translation of the basin from 33 Ma until the end of our record at 29 Ma.

External Sierras

The External Sierras (Fig. 2) mark the southern limit of major deformation in the west–central Pyrenean foreland (McElroy, 1990; Pocoví, *et al.*, 1990). During thrusting and imbrication of the San Felices thrust sheet, the External Sierra became emergent topographic features which acted as local source areas and exerted strong controls on the nearby depositional systems. Excellent exposures of the cross-cutting and overlapping relationships between the thrusts and related syntectonic strata (Fig. 12) permit the sequence of thrusting to be reconstructed unambiguously. In this study, four major conglomeratic depositional sequences (Con-

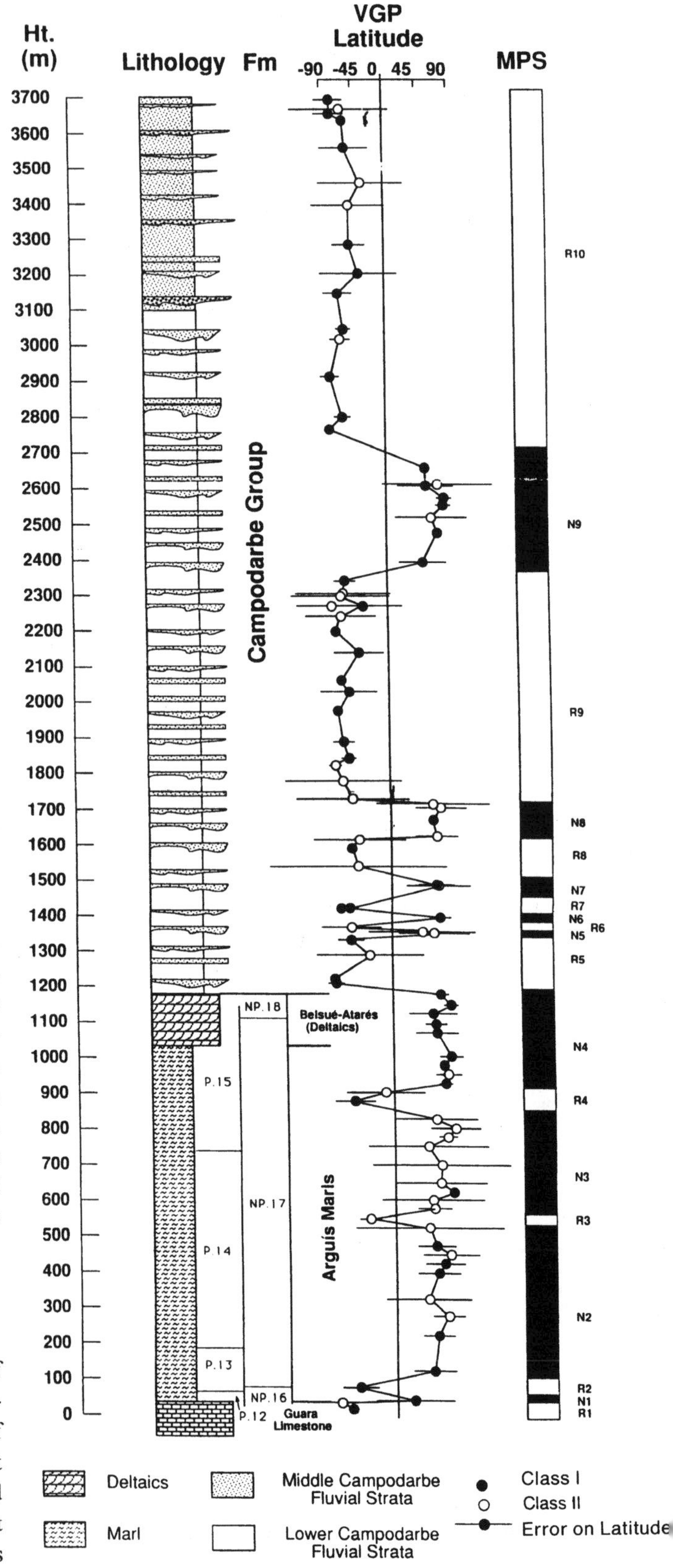

Fig. 8. Magnetic polarity stratigraphy for the Arguís–Monrepós section, ranging from the top of the Guara limestone to the middle of the Campodarbe Group strata.

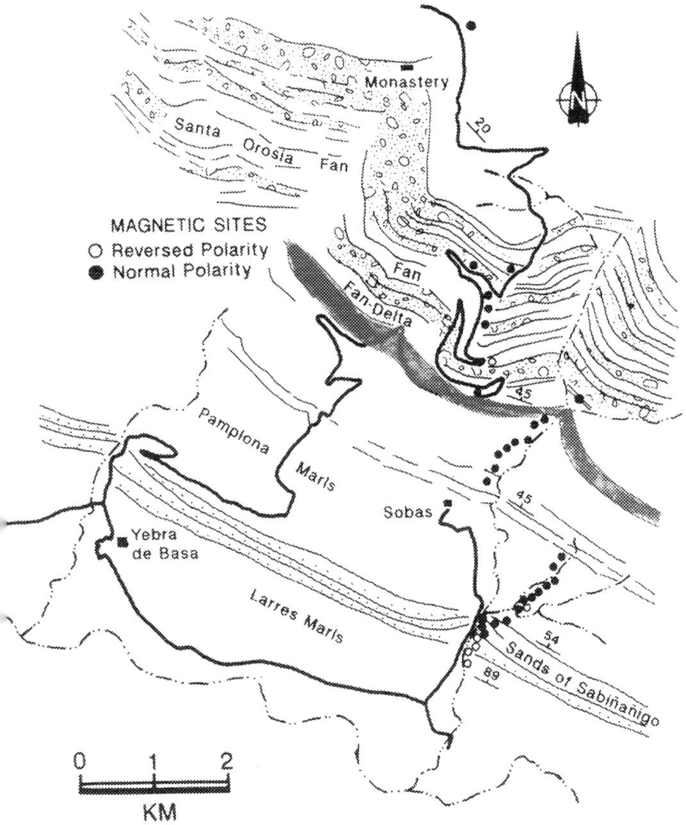

Fig. 9. Site location map for the 41 sites in the Yebra de Basa magnetic section, northern Jaca Basin.

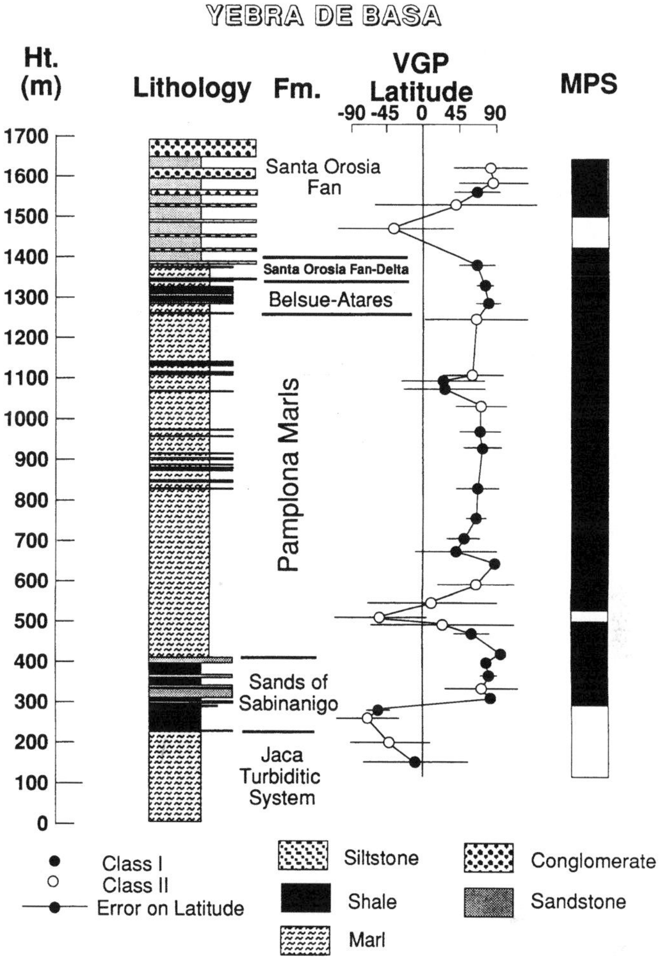

Fig. 10. Magnetic polarity stratigraphy for the Yebra de Basa section, ranging from the Hecho turbidites to the Santa Orosia fan.

glomerates 1 to 4) are further divided into separable subunits where these serve to delineate different tectonic events. Development of magnetic stratigraphies through the related syntectonic strata allows chronologic limits to be placed on the deformational history (Hogan, 1993). The initial major motion on the San Felices thrust sheet is recorded by the progressively rotated strata exposed at the western termination of the External Sierras. This thrust sheet is overlain by conglomerate 1A, which is cut by the Punta Cumún thrust (Fig. 13), which is in turn overlain by younger conglomerates. Farther to the east, similar relationships with the Riglos thrust, Linas thrust, and the Peña del Sol thrust also delineate a sequential breakback thrusting sequence (Fig. 14). Magnetic sections from

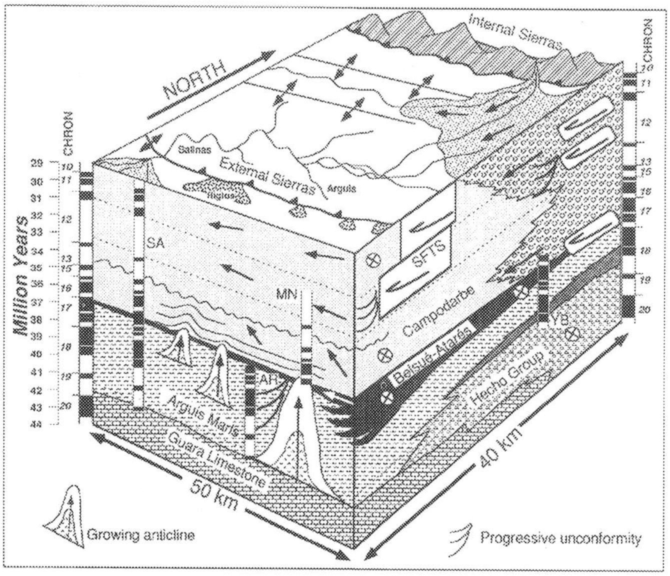

Fig. 11. Time–space reconstruction of lithofacies and structural relationships in the central Jaca Basin during middle Eocene to early Oligocene times. Hecho Group turbidites were coeval with both the Guara limestone and the overlying Arguís marls. Initial growth of the N–S anticlines in the southern Jaca Basin began in the Bartonian and indicates that at least the eastern part of the basin was detached from the basement at this time. Encroachment of proximal fan bodies along the northern basin margin occurred in response to hinterland thrusting in the early Priabonian and is synchronous with progradation of the Belsué–Atarés deltaic system across the southern parts of the basin. Broader detachment of the Jaca Basin occurred in the latest Priabonian as evidenced by unconformities in the Salinas section and changes in depositional style within the basin. Break-back thrusting within the External Sierras began shortly thereafter and continued throughout the Oligocene (Hogan, 1993).

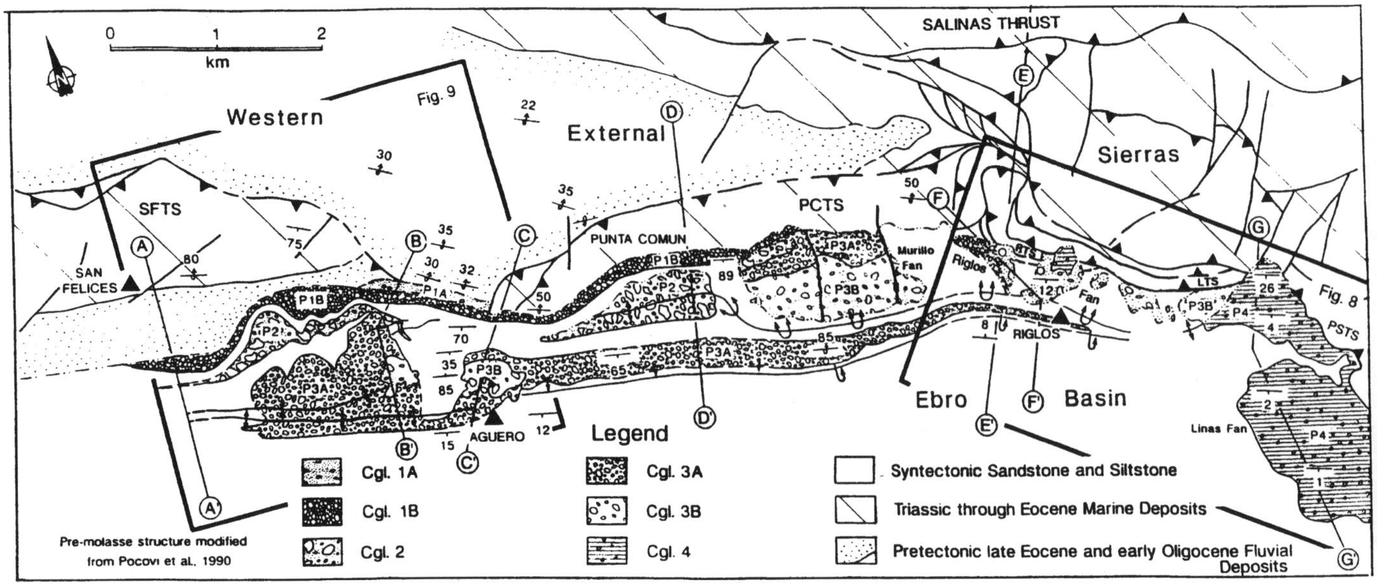

Fig. 12. Simplified geologic map of the western External Sierras from the Linas fan in the east to the San Felices fan in the west. Cross-cutting relationships between various conglomeratic bodies (labelled Cgl. 1A to Cgl. 4) and multiple thrusts provide clear evidence from hindwards imbricating thrusts.

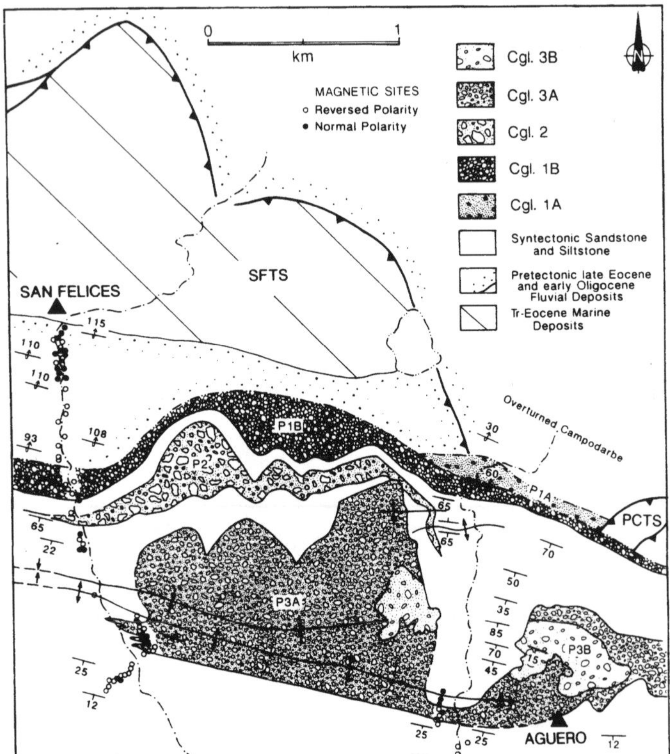

Fig. 13. Geologic map of the thrust sheets and syntectonic strata in the western External Sierras. The San Felices thrust sheet (SFTS) is overlain by conglomerate 1A which is cut by the Punta Común thrust sheet (PCTS). The PCTS is, in turn, overlain by conglomerate 1B. The locations of the 89 magnetic sites in the San Felices section and of 16 sites in the Agüero section are also shown.

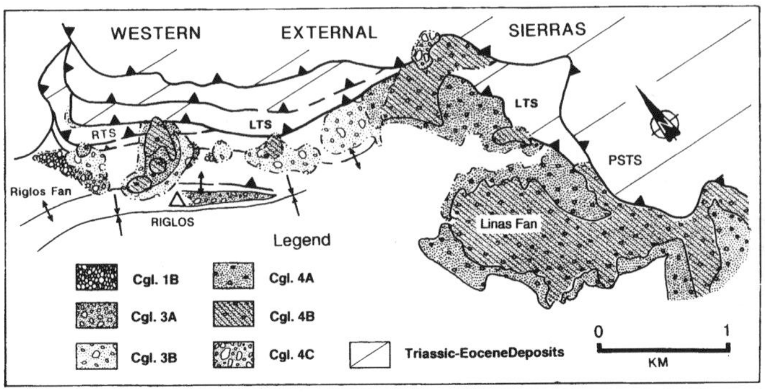

Fig. 14. Geologic map of the thrust sheets and syntectonic strata near Riglos, External Sierras. Clear breakback thrusting is shown by overlapping relationships with the Riglos thrust sheet (RTS), the Linas thrust sheet (LTS) and the Peña del Sol thrust sheet (PSTS).

San Felices (Fig. 15) and Agüero (Fig. 16) in the External Sierras and from Ayerbe (Fig. 17), which commences several kilometres south of Riglos, provide time control for the syntectonic strata (Fig. 18). In conjunction with temporal data from the Jaca Basin (Fig. 11), these show that major translation and imbrication of the San Felices thrust sheet persisted for at least 13 my and extended into the early Miocene. According to the magnetic dating, the emplacement of this thrust sheet occurred between 33.5 and 30.6 Ma, the Punta Común thrust sheet moved between 30.6 and 29.4 Ma, and major translation of the Riglos thrust sheet took place between ~24.7 and 24.0 Ma. The final, largely post-tectonic phase of conglomeratic deposition occurred during the late Chattian and the early Aquitanian (~23–24 Ma). Rates of shortening associated with translation of the growth of the External Sierra range from ~2.4 mm/y (33.5 to 30.6 Ma) to 1.5 mm/y (29.4 to 23 Ma) (Hogan *et al.*, in review).

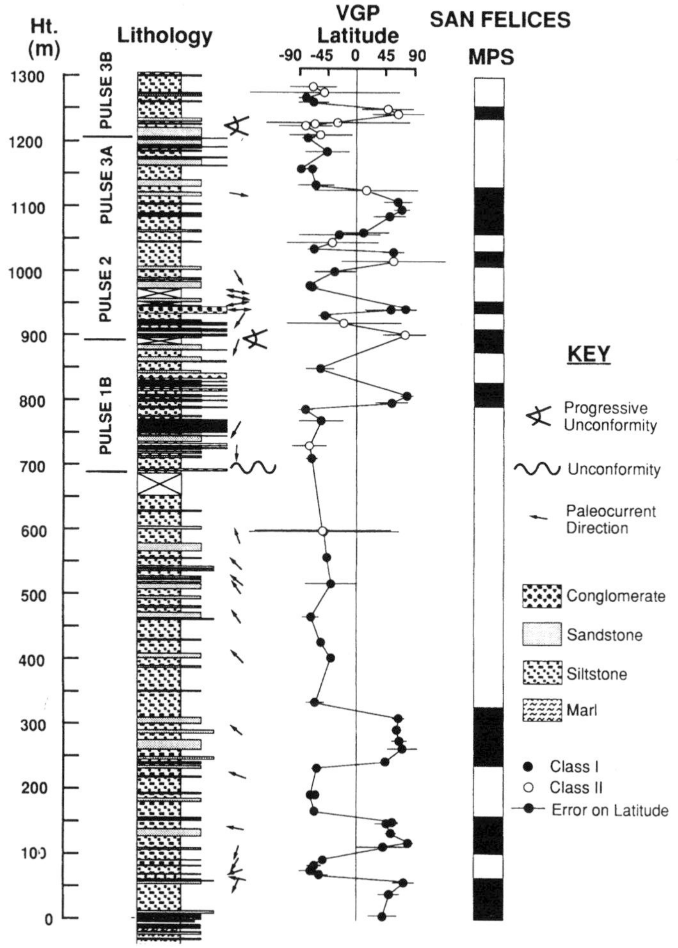

Fig. 15. Magnetic polarity stratigraphy for the San Felices section, ranging from the top of the Belsué–Atarés deltaic strata to conglomerate 3B.

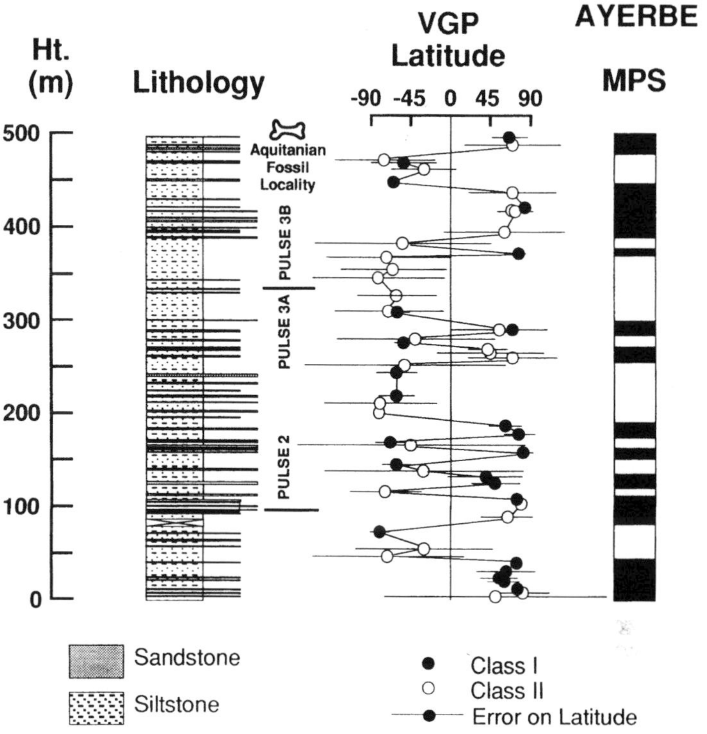

Fig. 17. Magnetic polarity stratigraphy for the Ayerbe section, ranging from below conglomerate 2 up to an 'Aquitanian' fossil locality.

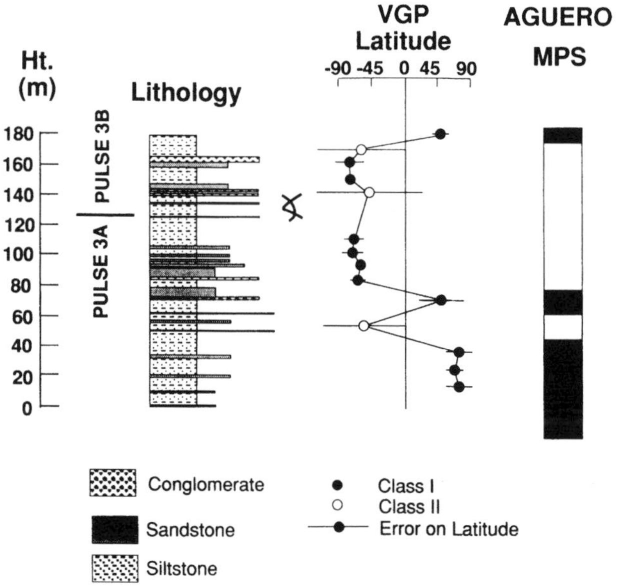

Fig. 16. Magnetic polarity stratigraphy for the Agüero section which encompasses conglomerates 3A and 3B.

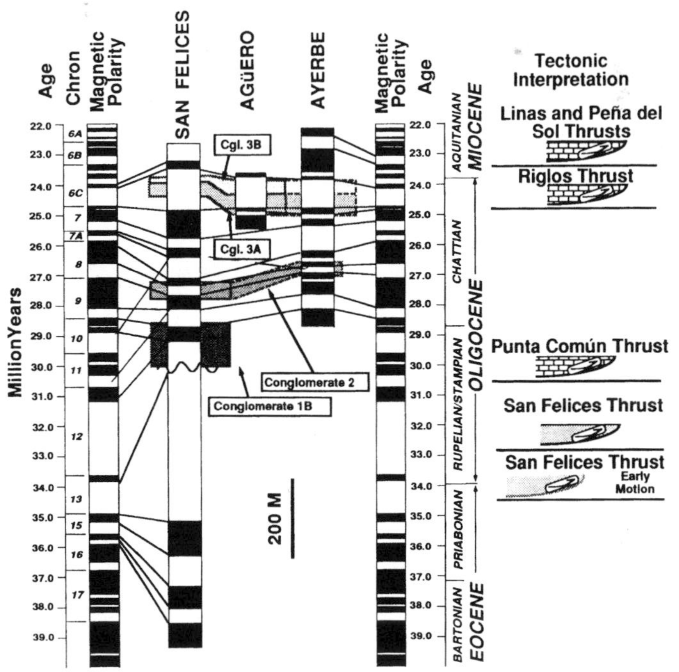

Fig. 18. Correlation of the local MPSs from the External Sierras and northern Ebro Basin with the MPTS (Cande and Kent, 1992). The temporal position of various dated conglomeratic units are shown, as well as the intervals of motion determined for the associated thrusts.

Although the rates of shortening decrease through time, the rates of bedrock uplift and erosion increase during the same interval. This is the latest well-dated, major thrusting in the southern Pyrenees.

Summary

These magnetic studies in the southern Pyrenees provide a more precise temporal framework than has previously been available for the analysis of the stratigraphic and structural record. Synthesis of these data with lithostratigraphic studies permits an improved reconstruction of the evolving foreland basin. These magnetic data are presented here in an effort to provide a detailed database which can be used to promote new studies that require higher resolution in order to correlate between spatially separated regions or to examine the interrelationships between the rates, timing, and style of depositional and deformational processes.

Acknowledgments

This research was supported by grants to DWB from the National Science Foundation (EAR–8517482, 8816181, and 9018951) and by Fulbright Fellowships to PJH. Acknowledgment is gratefully made to the donors of the Petroleum Research Fund, administered by the American Chemical Society (grants 17625, 20591, and 23881).

References

Burbank, D.W., Puigdefàbregas, C. and Muñoz, J.A. (1992). The chronology of the Eocene tectonic development of the eastern Pyrenees. *Geological Society of America Bulletin*, **104**, 1101–1120.

Cande, S.C. and Kent, D.V. (1992). A new geomagnetic polarity time scale for the Late Cretaceous and Cenozoic. *Journal of Geophysical Research*, **97**, (10), 13917–13953.

Canudo, J.I., Molina, E., Rivellene, J. Serra-Kiel, J. and Sucunza, M. (1988). Les événements biostratigraphiques de la zone prépyrénéene d'Aragon (Espagne), de l'Eocene moyen à l'Oligocène inférieure. *Revue de Micropaléontologie*, **31**, 15–29.

Hogan, P.J. (1993). Geochronologic, tectonic, and stratigraphic evolution of the southwest Pyrenean foreland basin, northern Spain. PhD, University of Southern California, 220 pp.

Hogan, P.J., Burbank, D.W. and Vergés, J. (in review). Emergence of the frontal thrust complex, External Sierra, SW Pyrenees: chronology and sedimentary response: *Tectonics*.

Jolley, E.J. and Hogan, P.J. (1989). The Campodarbe Group of the Jaca basin Pyrenean tectonic control of Oligo-Miocene river systems Huesca, Aragon, Spain. *Fourth International Fluvial Conference Excursion Guidebook, v. Servei Geològic de Catalunya*, **4**, 93–120.

Labaume, P., Seguret, M. and Seyve, P. (1985). A section across a turbiditic foreland basin and analogy with an accretionary prism. *Tectonics*, **4**, 661–685.

McElroy, R. (1990). Thrust kinematics and syntectonic sedimentation: the Pyrenean frontal ramp, Huesca, Spain. PhD, Cambridge University, 175 pp.

Pocoví, A., Millan, H., Navarro, J.J. and Martinez, M.B. (1990). Rasgos estructurales de la Sierra de Salinas y zona de los Mallos (Sierras Exteriores, Prepirineo, provincias de Huesca y Zaragoza). *Geogaceta*, **8**, 36–39.

Puigdefàbregas, C. (1975). La sedimentación molásica de la cuenca de Jaca. *Pirineos*, **104**, 118.

E15 Long-lived fluvial palaeovalleys sited on structural lineaments in the Tertiary of the Spanish Pyrenees

S.J. VINCENT AND T. ELLIOTT

Abstract

Major fluvial palaeovalleys located in transfer zones that link sediment-producing hinterlands with depositional basins are an important, but neglected, element of fluvial systems. Studies of present-day examples are limited and, to date, no examples have been described from the geological record. Tertiary syn-orogenic fluvial successions in the Spanish Pyrenees include regional-scale, unconformably-based, linear bodies of conglomerate interpreted as transfer zone palaeovalleys. These palaeovalleys are sited mainly in the external zones of the mountain belt, between the internal, Axial Zone on which the drainage basin was largely established and the depositional basins. One palaeovalley, here termed the Sis palaeovalley, is located in a growth syncline between two thrust-related, lateral structures. Subsidence within the syncline permitted a 1400 m plus succession of clast-supported cobble conglomerates to accumulate during an 11–21 Ma period (Middle Eocene to Oligo-Miocene). This palaeovalley served not one, but a series of evolving thrust-sheet-top and foreland basins during this period. The stability of the palaeovalley was governed by its structural siting which, using evidence from the stratigraphy immediately underlying the palaeovalley, is considered to be a long-lived lineament which was re-used during compressional deformation. An appreciation of the location and evolution of the palaeovalley has the potential to contribute significantly towards a better understanding of the development of the Pyrenean orogen.

Introduction

Fluvial systems in mountain belts comprise three main zones: (i) a drainage basin located primarily in the internal zones of the mountain belt; (ii) a sediment transfer zone comprising major, valley-bound trunk rivers sited in the external zones; and (iii) receiving basins which may either be thrust-sheet-top basins located behind the thrust front or a foreland basin beyond the thrust front (Schumm, 1977; Jones & Vincent, 1990; Fig. 1). Numerous studies have examined the deposits of fluvial systems in mountain belts and identified former drainage basins by integrating structural and provenance studies. By comparison, sediment transfer zones have been neglected despite their importance in linking sediment-producing hinterlands with receiving basins.

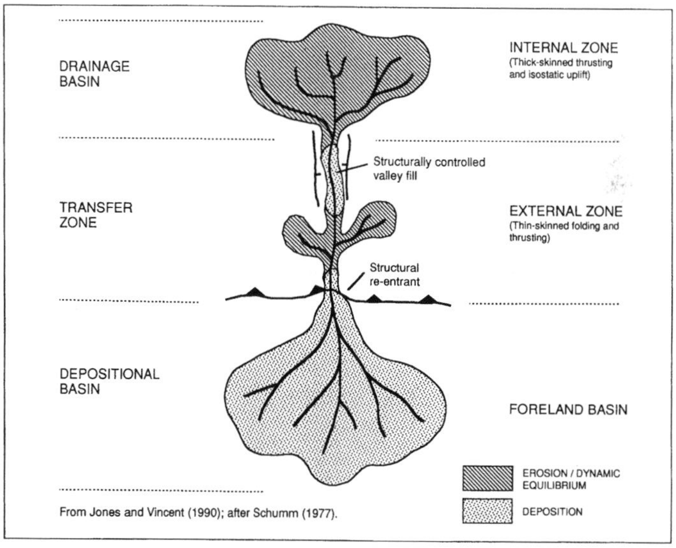

Fig. 1. Major elements of fluvial systems and their general relationships to mountain belts and foreland basins.

Exposures of Tertiary syn-orogenic fluvial deposits of the Spanish Pyrenees provide an opportunity to redress this balance. Included in these deposits are two regional-scale, unconformably-based, linear bodies of conglomerate that are aligned down the regional gradient of the mountain belt (Fig. 2). These bodies, known as the Sis and Gurp conglomerate bodies, are considered to represent well-preserved remnants of major palaeovalleys that acted as sediment transfer zones of the orogenic fluvial system during the Tertiary. This chapter is concerned with the western Sis conglomerate body and will document evidence that supports the interpretation of this body as a long-lived, structurally-controlled palaeovalley.

The Sis conglomerate body

The Sis conglomerate body is spectacularly exposed in the Isábena valley near the village of Serraduy, Huesca Province (Fig. 2). The body is approximately 20 km long, 7.5 km wide and more than 1400 m thick, and is located mainly on the preserved northern

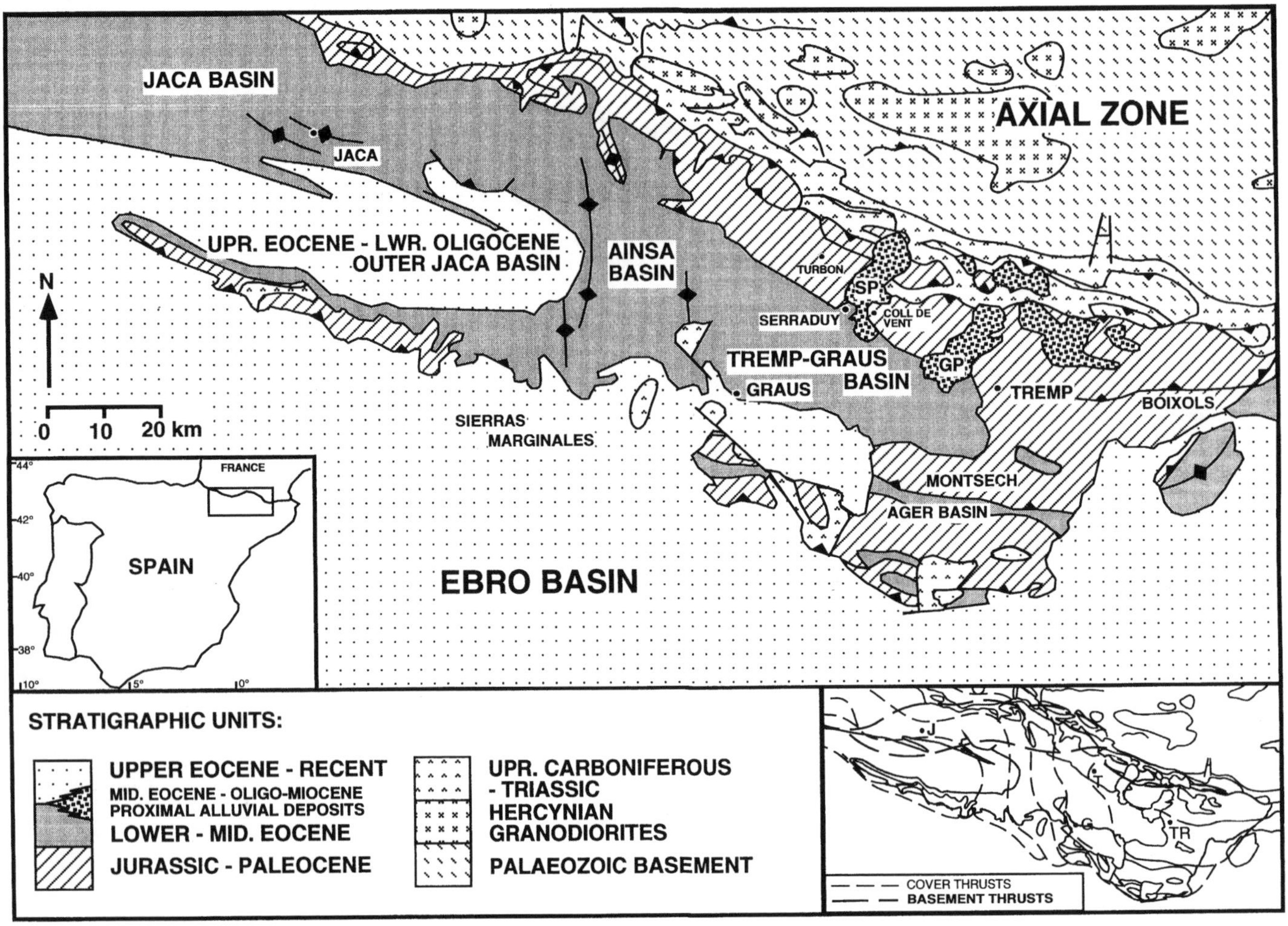

Fig. 2. Geological map of the south-central Pyrenees (modified after Choukroune & Séguret, 1973). Abbreviations: SP–Sis palaeovalley; GP–Gurp palaeovalley. Insets: left, Iberian sketch map with the study area highlighted, and right, an interpretation of the basement and cover thrust sheet geometries within the study region (after Cámara & Klimowitz, 1985). G=Graus; TR=Tremp; TA=Turbon.

margin of the upper cover rock thrust sheets and, to a lesser extent, on the internal zone antiformal stack. These structures acted as a passive roof duplex and triangle zone, respectively, during deformation (Muñoz, 1992; Fig. 2). In keeping with the definition of sediment transfer zones (Fig. 1) the body is sited in a structurally complex zone which lies largely between the internal zone antiformal stack upon which the drainage basin was established, and the depositional thrust-sheet-top and foreland basins to the south.

Key evidence for the interpretation of the Sis conglomerate body as a palaeovalley includes: (i) the NE–SW trend of the body, which is parallel to the southwesterly palaeocurrent, clast size and facies trends of the conglomerates; (ii) variations in thickness and facies across the conglomerate body, which support the notion of axial to marginal changes (e.g. thicker, coarser and more-highly-stacked axial successions, and small, locally sourced, fan systems shed laterally from its margins); (iii) relief of up to 120 m from axial to marginal positions on the basal unconformity of the conglomerate body with onlap onto its tectonically steepened margins; and (iv) syntectonic unconformities that are developed locally at the margins of the conglomerate body and die out towards its centre. These observations suggest that the conglomerate body is a well-preserved remnant of a palaeovalley, and discount the possibility of the body merely being the remains of a formerly more extensive, sheet- or fan-like deposit.

The fill of the palaeovalley is at least 1400 m thick and has an overall coarsening-upwards trend characteristic of the 'molasse' stage of foreland basin deposition. Coarse, cobble-grade conglomerates predominate, with subordinate facies including sandstones, siltstones–claystones, lacustrine limestones and coals. The conglomerates are clast-supported and occur in erosive-based units that are 2–5 m thick and laterally extensive. Units are either structureless or exhibit poorly defined flat to gently inclined bedding. Imbrication is sporadic and cross bedding rare. The deposits are interpreted to result from widespread, weakly channellised, sheet-like flows, possibly analogous to the upper parts of valley-confined braided river systems described by Boothroyd & Ashley (1975).

The palaeovalley succession is divisible into three major units, between 300–550 m thick, that are defined by significant changes in facies, alluvial architecture and provenance (Fig. 3). These units formed as a consequence of large-scale changes in the activity of the

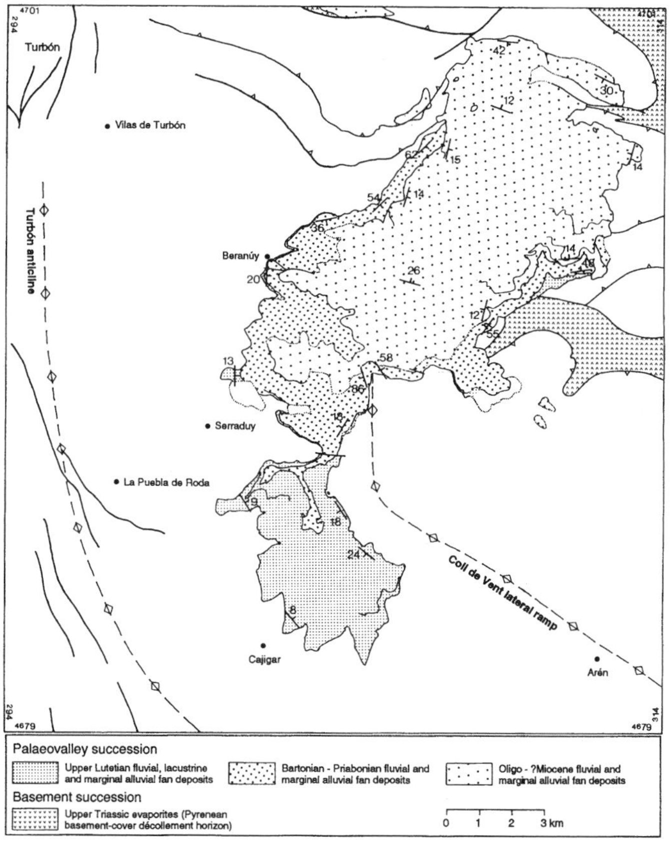

Fig. 3. Detailed geological map of the Sis palaeovalley showing: (i) the lateral structures which define its position, (ii) the synclinal nature of its three-stage fill which display progressive rotation, and (iii) the offset of the Triassic evaporite Axial Zone décollement across the palaeovalley trend. From Vincent (1993).

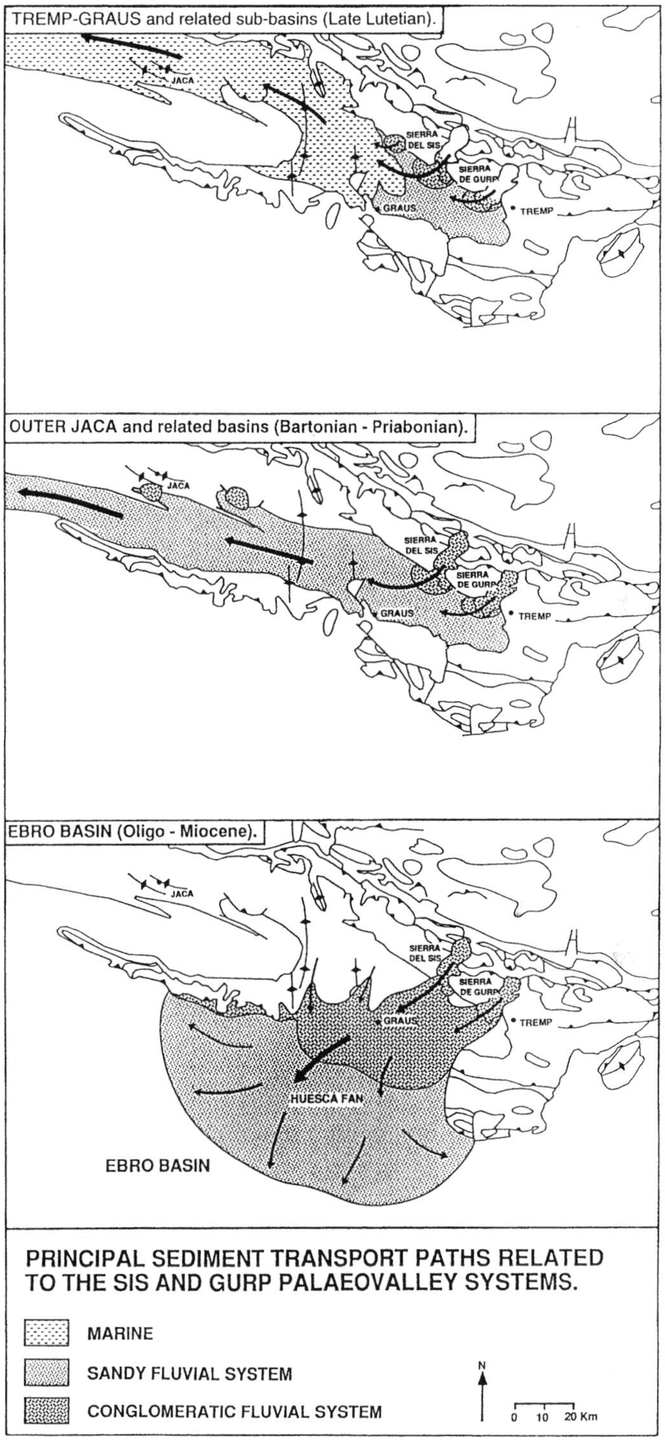

Fig. 4. Principal sediment dispersal routes, involving the Sis and Gurp palaeovalleys as fluvial transfer zone feeder systems to successive Eocene to Oligo-Miocene thrust-sheet-top and foreland basins. This three stage evolution is equivalent to the three stage fill of the Sis palaeovalley illustrated in Figure 3. Additional data are taken from key references cited in the text. The authors do not intend this to be an exhaustive palaeogeographic interpretation of the region (cf. Fig. 3 of Chapter E16).

Pyrenean Axial Zone, which also resulted in the reorganisation of the depositional basins down-dip of the palaeovalley (cf. Fig. 4). The stratigraphy of the palaeovalley can be further subdivided into 50–200 m thick, unconformity-bound, stratigraphic packages which may be traced across the exposure limits of the body. These smaller-scale sequences are interpreted to have formed as a result of local reactivation of structures that define the palaeovalley and phases of antiformal stack development which caused an increase in gradient of the fluvial system, and on occasions the disruption of the drainage network. Relative changes in base-level are not thought to have played a significant role in defining these sequences, due to the highly proximal nature of the Sis palaeovalley (Vincent, 1993).

Structural controls on the location of the Sis palaeovalley

The deposits of the Sis palaeovalley are preserved in a broad syncline which formed in response to uplift on a series of flanking structures developed at an oblique angle to the regional compression direction in both the cover and basement thrust sheets (Fig. 3). Within the cover thrust sheets these structures include the Coll de Vent and Turbón structures which form oblique/lateral ramp components of two cover thrust imbricates which segment the

Eocene thrust-sheet-top basin (Cámara & Klimowitz, 1985; Fig. 2). As a consequence of the westerly component of transpression on these cover thrust structures the easterly Coll de Vent lateral ramp forms the most important structural control on the position of the Sis palaeovalley. The lateral culmination of this ramp defines the eastern margin of the palaeovalley, with deposits accumulating in its footwall syncline to the west. The less-steeply-dipping hangingwall limb of the Turbón anticline forms the western limb of the Sis syncline. The palaeovalley is therefore located in an asymmetric syncline whose locus of subsidence is situated towards its eastern margin (Fig. 3). Those parts of the palaeovalley sited on the southern margin of the Pyrenean antiformal stack (the Nogueras zone) are also tectonically constrained within a synclinal form. This results from the development of a series of complex tectonic imbricate thrust systems, whose nature significantly varies on either side of the palaeovalley position, due to a basement discontinuity situated along its eastern margin (Fig. 3).

Evidence for the activity of the structures which define the palaeovalley position during sedimentation is provided by the progressive tightening of the palaeovalley syncline (Fig. 3). Syntectonic and angular unconformities developed at the margins of the body as the deposits were successively steepened. This activity also influenced alluvial stacking patterns of the depositional system with channel facies being concentrated and strongly amalgamated in the centre of the palaeovalley. Evidence for topographic relief at the margins of the palaeovalley are also indicated by the lateral supply of locally sourced alluvial fans and olistoliths up to 100 m in diameter. These deposits, derived from the surrounding limestone-dominated Mesozoic stratigraphy, were shed from the margins of the palaeovalley and interfinger with the more distant, internally sourced, fluvial systems which are directed axially down the palaeovalley and dominate its fill.

Longevity of the Sis palaeovalley system

The lower part of the Sis palaeovalley fill is of Middle Eocene age (Upper Lutetian; Cuevas Gozalo, 1989), whilst the uppermost part equates with the Collegats Group of Oligo-Miocene age (Bentham *et al.*, 1992; Vincent, 1993). Initially, sediments routed through the palaeovalley were deflected westwards to supply the axial, prograding depositional systems of the late stages of the Tremp–Graus basin and, more particularly, the Outer Jaca basin (Fig. 4). As the Ebro basin became the main site of deposition during the Oligo-Miocene, sediments supplied via the palaeovalley were directed SSW, entering the basin via a structural and topographic low in the developing thrust front which was coincident with a zone of lateral ramps. This region formed the apex of a large-scale Late Oligocene–Early Miocene alluvial fan, termed the Huesca fan system, which is one of the main elements of the Ebro basin fill (Hirst & Nichols, 1986; Chapter E16; Fig. 4). The Sis palaeovalley therefore supplied not one, but a series of Pyrenean basins which developed during orogenesis over an estimated 11–21 my period (Vincent, 1993).

Considered in their regional context, strata immediately underlying the palaeovalley are distinctive in two respects. Firstly, at several stratigraphic levels there is evidence of discrete, localised clastic input, as noted by previous workers (Puigdefábregas *et al.*, 1985; Cuevas Gozalo *et al.*, 1985; Eichenseer, 1988). Late Palaeocene alluvial facies of the Tremp Formation are thicker and more sandstone-dominated in the vicinity of the palaeovalley site than elsewhere along the northern margin of the Tremp–Graus Basin (Eichenseer, 1988). In the overlying Early Eocene Alveolina Limestone, prograding siliciclastic mouth bars interfinger, rather unusually, within transgressive carbonate facies (Eichenseer, *op. cit.*). Marine marls deposited following this transgression include coarse-grained fan delta systems (the Roda Sandstone and San Esteban Formations), which are uniquely developed in this region. Secondly, certain stratigraphic units display evidence of unusually strong tidal currents. For example, the Alveolina Limestone exhibits 10 m sets of northerly directed cross bedding that are interpreted as large carbonate sand-waves formed by flood-oriented tidal currents. Also, the Roda Sandstone contains prolific evidence of ebb- and flood-oriented tidal currents in small- to moderate-scale sets of cross bedding. These sets are associated with large-scale sets up to 20 m set height interpreted as sand-rich Gilbert-type deltas or ebb-tidal deltas that prograded into a tidally-influenced nearshore area (Puigdefàbregas *et al.*, 1985; Nio & Yang, 1991).

These features of the pre-palaeovalley stratigraphy, coupled with the evidence of the palaeovalley succession itself, demonstrate that highly distinctive patterns of sedimentation pertained in the Sis region for approximately 24–34 my (Thanetian to Oligo-Miocene). The distinctive character of the pre-palaeovalley stratigraphy is felt to be controlled by the Turbón, and in particular, Coll de Vent lateral structures which ultimately determined the position of the palaeovalley.

The Coll de Vent and Turbón structures are believed to be major, pre-existing lineaments which were re-used during compressional deformation (Cámara & Klimowitz, 1985; Puigdefàbregas & Souquet, 1986; Duguey, pers. commun.). In the Aptian–Albian the NNE–SSW trend of the structures was marked by significant thickness and facies variations during a transtensional phase of deformation and basin development. At this time the NNE–SSW structures were probably part of a suite of transfer faults at the terminations of north-dipping transtensional faults. Compression commenced in the Late Cretaceous, inverting the north-dipping master faults as frontal thrust structures and using the NNE–SSW structures as lateral or oblique ramps to these compressional structures (Simó *et al.* 1985; Souquet & Déramond, 1989). The Coll de Vent structure, which presently defines the western margin of the Bóixols thrust sheet (Fig. 2), underwent inversion commencing in the Thanetian (Late Palaeocene; Eichenseer, 1988), or possibly earlier (Souquet & Déramond, 1989). The basement-derived, pre-existing, control on this cover thrust feature is clearly seen in the continuation of the same trend within the basement-involved antiformal stack to its north where a significant offset in Triassic-cored thrust systems, and change in structural style, occurs across this feature (Fig. 3). Once the Coll de Vent structure became active, the Sis region became a major entry point for clastic sediment, initially acting as an embayment or re-entrant at the margin of the

Eocene thrust-sheet-top basin and subsequently becoming a palaeovalley to the later basins.

Conclusions

Unconformably-bound, linear bodies of conglomerate preserved in Tertiary fluvial deposits in the Spanish Pyrenees are interpreted as structurally sited, long-lived palaeovalleys which acted as feeder systems to a series of developing compressional basins. In the case of the Sis palaeovalley the preserved 1400 m plus thickness is almost certainly a small fraction of the total amount of sediment which was transported through the palaeovalley during its 11–21 My history. The upper part of the Eocene Tremp–Graus basin fill is 500 m thick, the Late Eocene–Oligocene Outer Jaca basin fill is 4000 m thick and the Oligo-Miocene Ebro basin also has a preserved thickness of 4000 m. Clearly, not all the sediment for these basins was routed through this palaeovalley, but it was a major conduit and is, therefore, regarded as a fluvial transfer zone in the sense of Schumm (1977). Long-lived transfer zone palaeovalleys serving a series of developing basins have not been reported previously from the geological record.

The Sis palaeovalley was preceded by a re-entrant at the margin of a basin which was an entry point for clastic sediment and, at selected times, a region of amplified tidal currents. The longevity and stable positioning of the re-entrant and palaeovalley are explained by their siting along long-lived, inherited structures which were re-used during compressional deformation. Growth on these structures during the history of the palaeovalley also accounts for the preservation of the palaeovalley deposits, with the deposits accumulating in a focused belt of net subsidence in the footwall of the structure.

In view of the longevity and structurally complex evolution of the palaeovalley, an analysis of the palaeovalley fill can contribute towards an improved understanding of Pyrenean internal zone uplift and exhumation (via provenance changes), external zone deformation (via active growth structures) and basin-fill sedimentation.

Acknowledgements

The authors wish to acknowledge C. Puigdefábregas, P. Heller and M. Kraus for their constructive comments on an earlier version of the manuscript, N.E.R.C. for provision of a studentship (GT4/89/GS/060) and additional support from Mobil North Sea Ltd., and an AAPG grant-in-aid award.

References

Bentham, P.A., Burbank, D.W. and Puigdefábregas, C. (1992). Temporal and spatial controls on the alluvial architecture of an axial drainage system: Late Eocene Escanilla Formation, southern Pyrenean foreland basin, Spain. *Basin Research*, **4**, 335–352.

Boothroyd, J.C. and Ashley, G.M. (1975). Processes, bar morphology, and sedimentary structures on braided outwash fans, Northeastern Gulf of Alaska. In Jopling, A.V. & McDonald, B.C. (eds.) *Glaciofluvial and glaciolacustrine sedimentation*. Special Publication of the Society of Economic Paleontologists and Mineralogists, Tulsa, **23**, 193–222.

Cámara, P. and Klimowitz, J. (1985). Interpretación geodinámica de la vertiente Centro-occidental surpirenaica (Cuencas de Jaca–Tremp). *Estudios Geológicos*, **41**, 391–404.

Choukroune, P. and Séguret, M. (1973). Tectonics of the Pyrenees; role of compression and gravity. In DeJong, K.H. & Schotten, R. (eds.) *Gravity and tectonics*. John Wiley & Sons, New York, 141–156.

Cuevas Gozalo, M.C. (1989). Sedimentary facies and sequential architecture of tide-influenced alluvial deposits. An example from the middle Eocene Capella Formation, South-Central Pyrenees, Spain. PhD Thesis, University of Utrecht, 152 pp.

Cuevas Gozalo, M., Donselaar, M.E. and Nio, S.D. (1985). Eocene clastic tidal deposits in the Tremp–Graus Basin. In Mila, M.D. & Rosell, J. (eds.) *Excursion Guidebook, VI European Regional Meeting, Lérida, Spain*. International Association of Sedimentologists, **Excursion 6**, 217–266.

Eichenseer, H. (1988). Facies geology of Late Maestrichtian to Early Eocene coastal and shallow marine sediments, Tremp–Graus Basin, Northeastern Spain. PhD Thesis, University of Tubingen, 237 pp.

Hirst, J.P.P. and Nichols, G.J. (1986). Thrust tectonic controls on Miocene alluvial distribution patterns, Southern Pyrenees. In Allen, P.A. & Homewood, P. (eds.) *Foreland basins*. Special Publication of the International Association of Sedimentologists, **8**, 247–258.

Jones, N.E. and Vincent, S.J. (1990). Late stage Fluvial Systems in Mountain Belts and Foreland Basins. *13th International Sedimentological Congress, Nottingham, England*, Poster Abstracts, 113.

Muñoz, J.A. (1992). Evolution of a continental collision belt: ECORS–Pyrenees crustal balanced cross-section. In McClay, K.R. (ed.) *Thrust tectonics*. Chapman and Hall, London, England, 235–246.

Nio, S.D. and Yang, C.S. (1991). Sea-level fluctuations and the geometric variability of tide-dominated sandbodies. *Sedimentary Geology*, **70**, 161–193.

Puigdefábregas, C. and Souquet, P. (1986). Tecto-sedimentary cycles and depositional sequences of the Mesozoic and Tertiary from the Pyrenees. *Tectonophysics*, **129**, 173–203.

Puigdefábregas, C., Samso, J.M., Serra-Kiel, J. and Tosquella, J. (1985). Facies analysis and faunal assemblages of the Roda Sandstone Formation, Eocene of the southern Pyrenees. *VI European Regional Meeting, Lérida, Spain*, Abstracts and poster abstracts, 639–642.

Schumm, S.A. (1977). *The fluvial system*. John Wiley & Sons, Chichester, England, 338 pp.

Simó, A., Puigdefábregas, C. and Gili, E. (1985). Transition from shelf to basin on an active slope, Upper Cretaceous Tremp area, Southern Pyrenees. *6th European Regional Meeting, Lérida, Spain*, Excursion Guidebook, 63–108.

Souquet, P. and Déramond, J. (1989). Séquence de chevauchements et séquences de dépôt dans un bassin d'avant-fosse. *Compte Rendus de l'Acadamie des Sciences, Paris*, **309**, 137–144.

Vincent, S.J. (1993). Fluvial palaeovalleys in mountain belts: an example from the south central Pyrenees. PhD thesis, University of Liverpool, 287 pp.

E16 Evolution of the central part of the northern Ebro basin margin, as indicated by its Tertiary fluvial sedimentary infill

P.F. FRIEND, M.J. LLOYD, R.McELROY, J. TURNER, A. VAN GELDER AND S.J. VINCENT

Abstract

Recent work on the evolution of the northern margin of the Ebro basin in its central sector, in Aragon, is reviewed, particularly in the light of new work on the mammal biostratigraphy, palaeomagnetism, tectonic structure and sedimentology of the Late Eocene to Early Miocene fluvial sediments. Three Sequences are distinguished in these sediments. During the deposition of Sequence A (Late Eocene to Early Oligocene) an 'eastern fluvial system' transported sediment axially westwards, a 'northern fluvial system' transported sediment from the Pyrenean axial zone towards the south, and there is no evidence of any feature, structural or topographic, along the line of the present Ebro basin margin. During the deposition of Sequence B (Early Oligocene to Late Oligocene) the 'northern fluvial system' replaced the 'eastern fluvial system', and the south Pyrenean piggy-back or thrust-sheet top basins became isolated from the Ebro basin by the first emergence of the Sierras Exteriores thrust ramp, which generated conglomerates to the south, and influenced sedimentation to the north. During the deposition of Sequence C (Late Oligocene to Early Miocene), sedimentation largely ceased north of the new thrust ramp, and the Ebro basin sedimentation became differentiated into the two major (Luna and Huesca) systems round the lateral edges of the ramp, and a series of small fan bodies in between.

Introduction

The central section of the northern margin of the Ebro basin is clearly defined structurally by the southern limit of Pyrenean deformation. This limit is marked by the emergence of the southernmost (Guarga–Gavarnie) thrust sheet (Fig. 1), which is now well understood from surface mapping (Puigdefàbregas, 1975; Nichols, 1987b; Millán and Pocoví, 1995), and from drilling and seismic work (Camara and Klimovitz, 1985). The topographic expression of this structure is known as the Sierras Exteriores (Fig. 2B), where the change in topography follows the line of emergence of the major frontal thrust, or an associated branch thrust. Outcrops of conglomerate fringe this structure to the south, and mark the line of uplift. To the east, the basin margin continues topographically as the Sierras Marginales (Fig. 2B), and has a broadly similar structural significance. To the west, in the west Jaca area (Fig. 2B), the major thrust is no longer emergent, and the structure known as the Peña flexure forms a rather less distinct topographic feature; this feature is also fringed by conglomerate outcrops. North of the Sierras Exteriores and Marginales lie the South Pyrenean thrust-top or 'piggy-back' basins (mainly the Jaca, Ainsa, and Tremp–Graus basins; see Fig. 2B; Ori and Friend, 1984); these developed progressively from east to west, along the southern edge of the Pyrenean Axial Zone of folded and thrusted Mesozoic and Palaeozoic rocks.

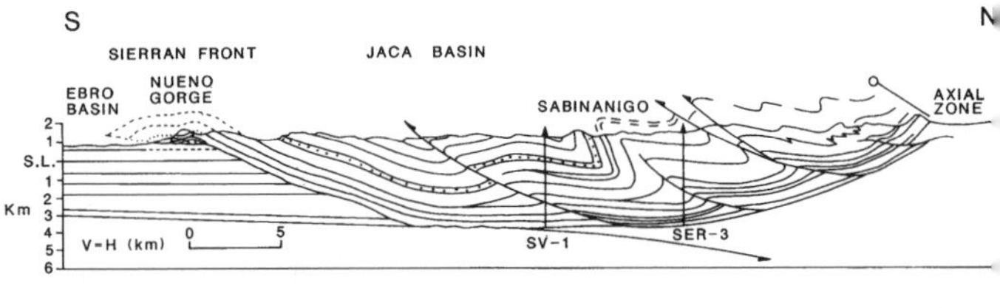

Fig. 1. True scale cross-section across the Jaca basin of the Southern Pyrenees, from the Axial Zone in the north to the Sierran front and Ebro basin in the south; for location see Fig. 2B.

This chapter reviews the evolution of this thrust-formed northern margin of the Ebro basin, as revealed by the distribution pattern of the Late Eocene to Early Miocene sediments that outcrop to its north and south.

The study area, shown in Figs. 2 and 3, lies roughly parallel to the trend of the Pyrenean Axial Zone, and also to the trend of the emergent Sierran frontal thrust ramp. In order to locate observations relative to the present outcrop geology, areas of Mesozoic and Palaeozoic outcrop are distinguished on our maps from those of Tertiary outcrop (IGME, 1980). However, it must be stressed that almost all of the older outcrops are of allochthonous rocks that have been considerably translated by thrusting (generally southwards) at different times during the Pyrenean orogeny.

Fig. 2B also shows isopachs for the Tertiary fill of the Ebro basin (Riba *et al.*, 1987); these demonstrate the pronounced, though rather oblique, thickening of sediments towards the Pyrenean margin. This thickening is the primary evidence that the Ebro basin,

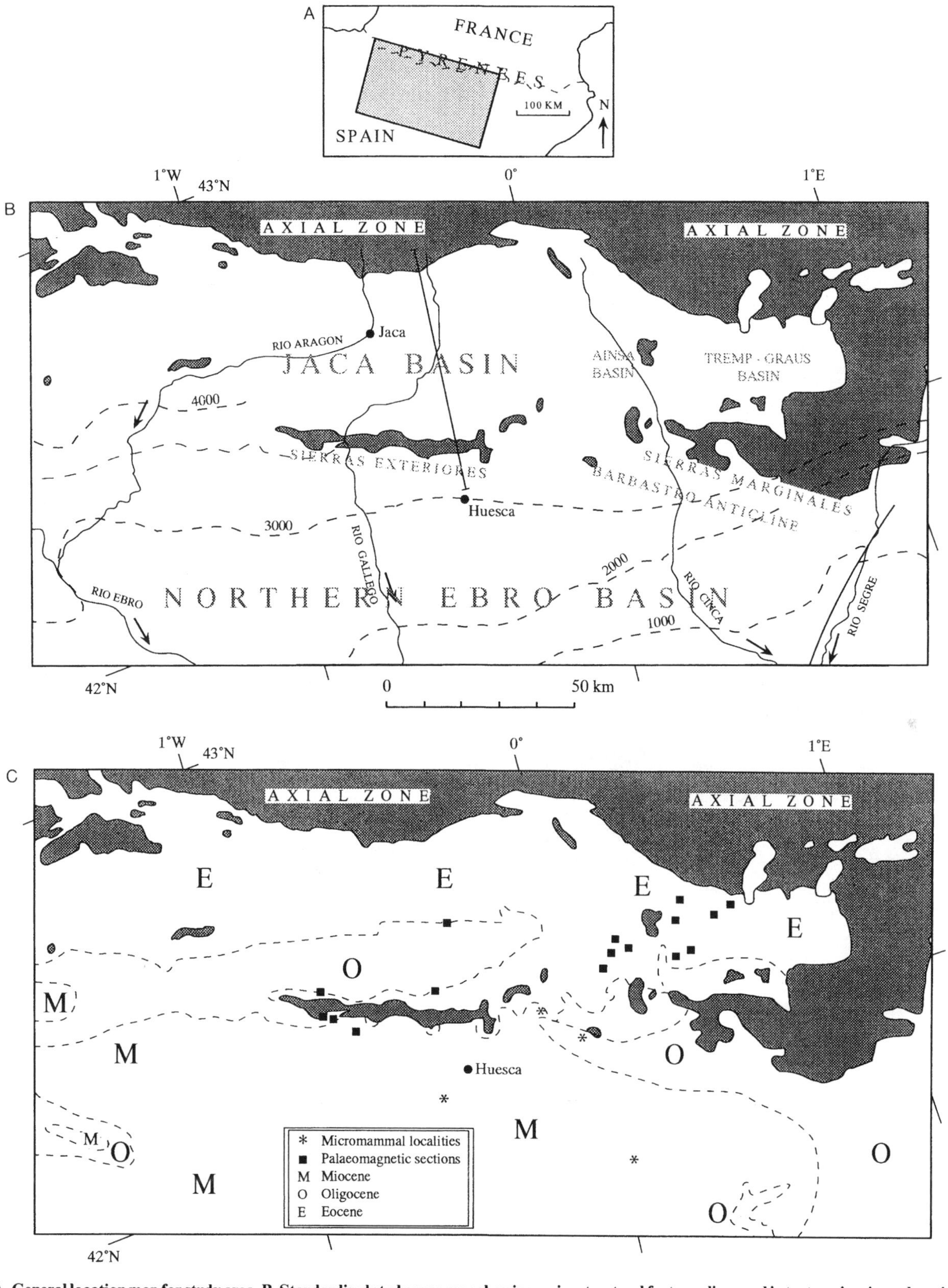

Fig. 2. A. General location map for study area. B. Standardised study area map showing major structural features discussed in text, major rivers, location of Fig. 1 section, and isopachs on the thickness of the Tertiary fill of the northern Ebro basin (Riba *et al.*, 1987). Mesozoic and Palaeozoic outcrop areas are shaded. C. Standardised study area, with Mesozoic and Palaeozoic outcrop area shaded, and boundaries of Palaeocene and Eocene, Oligocene and Miocene sedimentary outcrops marked from IGME 1/1M geological map (1980). Localities of the micromammal and palaeomagnetic work discussed in the text also indicated.

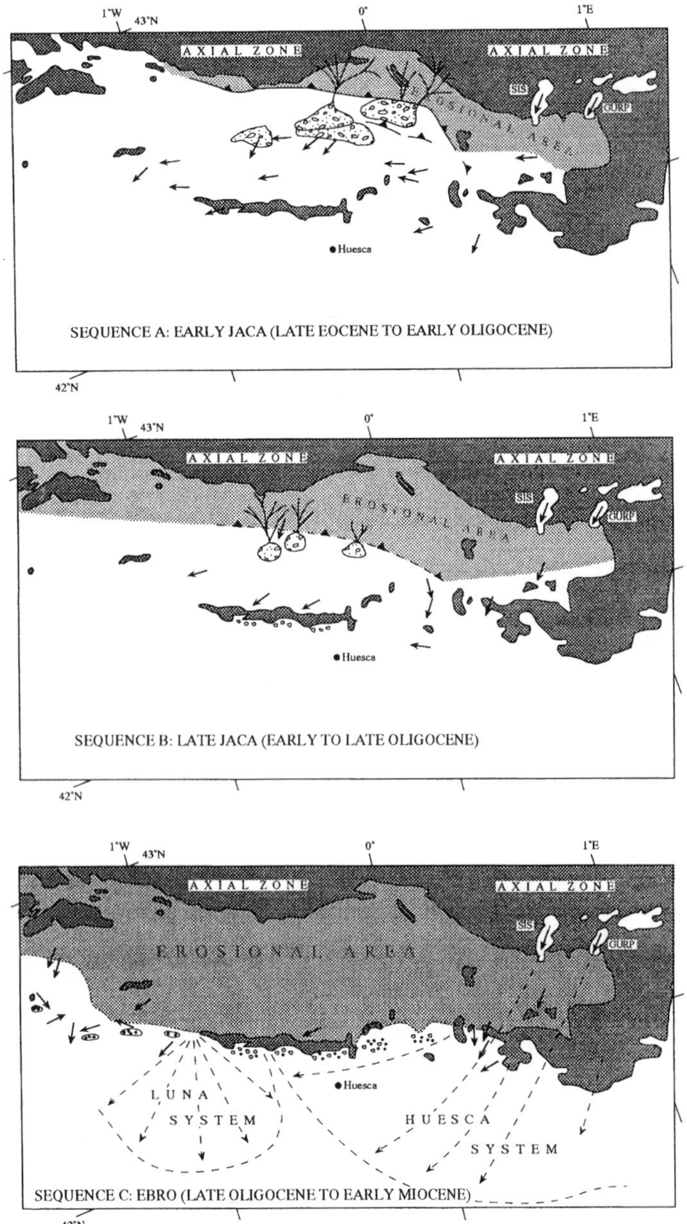

Fig. 3. Series of standardised study area maps constructed for the three sequences distinguished in this paper: A. Early Jaca, B. Late Jaca, C. Late Ebro. In each case, sources of the data, and general significance, are discussed in the text.

and its antecedent South Pyrenean piggy-back basins (Ori and Friend, 1984), were part of a foreland system generated by Pyrenean loading to the north.

Early investigations into the fluvial sedimentation of the northern Ebro Basin assumed that the spectacular conglomerate pinnacles that outcrop along the northern, Sierran, margin of the Ebro basin were proximal relicts of alluvial systems that had emerged from this Pyrenean thrust-front and then simply flowed south and basinwards, via an apron of sands, to the mud and evaporite basin centre. Later, more detailed study of some of the younger alluvial outcrops (Hirst and Nichols, 1986; Nichols, 1987b) showed that the pattern of river systems was more complex, and that the emergence of the south Pyrenean thrust sheet had, in fact, resulted in the deposition of a series of small, conglomeratic fan systems along the length of the frontal ramp. These small fan systems were bounded to the west, east and south by the much larger fluvial systems, the Luna and Huesca systems, whose location was controlled by major structures at the lateral ends of the main thrust ramp, as shown on Fig. 3C.

Dating the deposits

Puigdefàbregas and Souquet (1986) analysed the Mesozoic and Tertiary stratigraphic and structural history of the Pyrenees of France and Spain. Their analysis distinguished nine cycles, of which the last two are Tertiary in age:

Cycle 8: Transition to foreland basins (Palaeocene)
Cycle 9: Migrating foreland basins (Eocene to Early Oligocene)

For the purpose of this review, we distinguish three 'sequences' that correlate partly with Cycle 9 of the above analysis, and partly with later sediments described as post-Pyrenean by Puigdefàbregas and Souquet (1986), but clearly formed as part of the final episode of the Pyrenean orogeny itself. These three sequences are:

Sequence A – 'Early Jaca'; Late Eocene (Priabonian) to Early Oligocene (Early Rupelian)
Sequence B – 'Late Jaca'; Early Oliogene (Late Rupelian) to Late Oligocene (Early Chattian)
Sequence C – 'Ebro'; Late Oligocene (Late Chattian) to Early Miocene (Aquitanian to Burdigalian)

Aspects of the fluvial sediments of each of these sequences will be discussed in the sections that follow. However, a general statement about their dating and recognition must precede this.

Eocene, Oligocene and Miocene outcrops are distinguished across the whole of our study area on the IGME 1/1M geological map of the Iberian peninsula (1st edition, IGME, 1980). Two new approaches to dating the sediment outcrops covered by this map have since been developed: micromammal biostratigraphy and palaeomagnetic reversal stratigraphy.

Although fossils are very rare in the fluvial deposits, recent work using new micromammal (rodent tooth) biostratigraphy has met with success at some localities (Fig. 2C; Alvarez Sierra *et al.*, 1990; Cuenca *et al.*, 1989). This work has not significantly changed the age pattern suggested by earlier work.

Early palaeomagnetic studies in the Ebro basin (Turner *et al.*, 1984; Friend *et al.*, 1988) demonstrated the presence of a reversal stratigraphy in the fluvial and lacustrine sequences. The reversal sequences, however, were of too short a duration, and the sediments lacked independent dating, so they could not be correlated with the global magnetic polarity time scale. Recent work by D.W. Burbank and his group has shown that major studies of carefully chosen successions can yield correlatable reversal sequences in both the

Ebro basin and the thick sequences of the south Pyrenean basins. Fig. 2C shows the location in our study area of the successions studied by this group (Chapters E13 and E14). The work of Hogan and Burbank (Chapter E14) suggests that some of the Ebro basin fluvial sediment exposed immediately south of the thrust front is of Late Oligocene age, rather than its earlier interpretation as Early Miocene.

Sequence A, 'Early Jaca'

This sequence is defined by the Lower and Middle Campodarbe Group (Puigdefàbregas, 1975) of the Jaca basin. Dating of its lower boundary is based primarily on the age of the underlying marine deposits, but has also been independently confirmed by Hogan and Burbank (Chapter E14). The sequence appears to be Priabonian (Late Eocene) to early Rupelian (Early Oligocene) in age.

The Peraltilla Formation (Crusafont *et al.*, 1966), exposed to the south of the northern margin of the Ebro basin along the Barbastro anticline, and the Gaiardon and Ruesta members of the West Jaca area (Turner, 1990), also appear to be part of this sequence. Recent palaeomagnetic work on the southern margin of the Tremp–Graus basin and in the Ainsa basin (Chapter E13) suggests that the sequence is also represented in these areas.

Fig. 3A shows mean fluvial palaeocurrent directions, together with conglomerate fan bodies and their associated controlling faults (Jolley, in Friend *et al.*, 1989). Also shown are some palaeocurrent data from the west Jaca basin (Turner, pers. commun.; Turner, 1990), and from the Barbastro area (Van Gelder, pers. commun.).

In her analysis of the Jaca basin-fill sediments, Jolley (in Friend *et al.*, 1989), recognised that this sequence was comprised of two major river systems. The 'eastern fluvial system' (EFS), which flowed from the east to the WNW from a margin near the present location of the Boltaña anticline, deposited the major sandstones of the southern Jaca basin. In the northern region, the 'northern fluvial system' (NFS) flowed from the edge of the Axial Zone, and deposited a number of major conglomeratic fan bodies, which prograded from active fault scarps in the northernmost part of the Jaca basin.

In the southern Tremp–Graus and Ainsa basins (see Fig. 2B), sediment was generally transported towards the WNW parallel to the axial trend of the South Pyrenean basins and the Axial Zone. There is also evidence that some sediment was supplied to these basins via the Sis and Gurp palaeovalley systems, which were incised into the southern edge of the Axial Zone (see Chapter E15).

During the deposition of this sequence, there is no evidence of any uplift or supply of sediment from the area later uplifted to become the Sierras Exteriores.

Sequence B, 'Late Jaca'

This sequence is defined by the Upper Campodarbe Group (Puigdefàbregas, 1975) in the Jaca basin, and ranges in age from late Rupelian (Early Oligocene) to Chattian (Late Oligocene). Some of the sediments outcropping immediately north of the Barbastro anticline should also be included in this sequence, as should some of those outcropping in the deformed zone along the northern Ebro basin margin (see above, and Chapter E14).

Palaeocurrent averages from fluvial sediments, and the position of fault-related conglomerate bodies, are shown in Fig. 3B. This information comes mainly from the work of Jolley (in Friend *et al.*, 1989), but also from the work of Turner, in the West Jaca area (pers. commun.; see also Turner, 1990), and Van Gelder, from the area to the north of Barbastro (pers. commun.). Jolley used provenance and palaeocurrent data to imply that the whole of this sequence was deposited by the NFS and that the EFS was no longer active in the Jaca basin.

Further east, in the Tremp–Graus basin, major accumulation of conglomerates occurred in the northern palaeovalleys at this time (Chapter E15); the more southerly sediments of this sequence show clear evidence of a palaeoflow direction towards the south.

In this sequence, there is, for the first time, evidence from the deposition of syn-orogenic conglomerates (conglomerates 1 and 2, Chapter E14) of the emergence and consequent erosion of the Sierras Exteriores. In the Jaca basin to the north, palaeocurrent data do not record any deflection of fluvial distribution patterns. However, several other events record this thrust generated uplift: there is a syn-tectonic unconformity within the middle–upper Campodarbe Group strata immediately to the west of the western termination of the Sierras; pebbly sandstone units occur in the middle part of this unconformity, indicating increased palaeoslopes; alterations in the general fluvial depositional environment occur in the Jaca basin at, or shortly after, this time (for example entrenchment of the fluvial system, activation of growth folds, localised ponding of sediment (Jolley, 1987; Barbed *et al.*, 1988)). The emergence of the Sierras also implies the detachment and southward translation of the Jaca basin at this time. The uppermost sediments of this basin were cannibalised to supply conglomerate bodies to the south. It appears that conglomeratic alluvial fans were deposited only on the *southern* flank of the Sierras at this time because of the difference in relief between the higher Jaca basin and the lower Ebro basin on either side of the Sierran dividing line.

Sequence C, 'Ebro'

This sequence is defined by the Sariñena Formation, deposited to the south of the Sierras (Riba *et al.*, 1987). This formation is a rather collective term for the fluvial sediments deposited in this part of the Ebro basin in Aragon. There are a number of correlative units named for different facies across the Ebro basin, and much work remains to be done to analyse their stratigraphic and facies relationships. However, it is clear that this sequence should also include the Uncastillo Formation of the west Jaca area (Puigdefàbregas, 1975; Turner, 1990), some of the outcrops north of the Barbastro anticline, and the higher parts of the conglomerate bodies along the Sierran front (conglomerates 3 and 4 of Hogan and Burbank, Chapter E14).

Micromammal and palaeomagnetic work suggests that the sequence is probably Late Chattian (Late Oligocene) to Aquitanian (Early Miocene) in age, and may extend up into the Burdigalian.

Fig. 3C shows some palaeocurrent averages from Turner (pers. commun.; Turner, 1990) in the west Jaca basin, and Van Gelder (pers. commun.), north of the Barbastro anticline. The two major river systems, Luna and Huesca, defined and named by Hirst and Nichols (1986) on the basis of their sandstone-body architecture, palaeocurrent patterns and distinctive detritus, are also located on the figure. The Luna system has a very regular palaeocurrent pattern that defines an apex just west of the western end of the Sierras Exteriores (Jupp *et al.*, 1987; Nichols, 1987b, c). This radial system had a radius of about 40 km, with mud intervals separating the sandstone bodies (Friend *et al.*, 1986) indicating low gradients rather similar to those of the Himalayan megafans of eastern India (Singh *et al.*, 1993).

The depositional area of the Huesca system was larger than that of the Luna, and of similarly low gradient. Fig 3C involves some modification of the form of this system suggested by the earlier work of Hirst and Nichols (1986). Hirst (1991) presented palaeocurrent roses for the system, and also calculated the confidence region for the apex, using the approach of Jupp *et al.* (1987). These data indicated an apex further to the east than was suggested by earlier authors; the unpublished palaeocurrent averages of Van Gelder (Fig. 3C) tend to support the interpretations of Hirst. The Huesca depositional system, therefore, genuinely may not have been as symmetrical as the Luna. This may be because it was supplied from more than one palaeovalley system to the north. Work on the Sis and Gurp palaeovalley systems, some 40 km north of the Ebro margin, on the margin of the Pyrenean Axial Zone, suggests that they were still acting as sediment transfer or feeder systems during the deposition of this Sequence (Chapter E15). We suggest here that they may have acted as such for the Huesca system.

Sequence C also includes many of the smaller, conglomeratic alluvial fans which outcrop so spectacularly to the south of the Sierras Exteriores and provide very clear examples of discrete river systems. For each of these, tectonism, uplift, erosion and deposition can be investigated as parts of closed systems because significant mixing seems not to have occurred with the sediments of the larger Luna and Huesca systems. Published studies of these bodies include papers by Nichols (1987a), Friend *et al.* (1989), Pocovi *et al.* (1990), and Millan and Pocovi (1995). Hogan and Burbank (Chapter E14) present the results of work on the western part of the Sierras Exteriores, using a combination of magnetostratigraphy and cross-cutting relationships between conglomerates and thrust sheets to link phases of fan activity to distinct episodes of thrusting.

A typical example of these fans outcrops to the north of the village of Agüero, 5 km west of the Río Gallego (Fig. 2B). The internal features of the fan are nicely exposed by dissection of the present Agüero stream valley, the western side of which is sketched in Fig. 4. The fan exhibits a radial distribution of palaeocurrents and clast sizes, episodes of coarsening upwards (prograding early development), progressive unconformities and cumulative wedge structures (deposition during active tectonics), and episodes of fining upwards (stabilisation and retreat; Nichols, 1987a; Hogan, 1991). All conglomerate clasts deposited on the fan were sourced directly from thrust sheets uplifted immediately to the north within the Sierras. Clast-type counts in the conglomerates indicate several phases of unroofing of the thrust sheets, and there is now no doubt that tectonic uplift of the thrust sheets was the primary cause of initiation, positioning and rejuvenation of the fans, rather than geomorphic or climatic controls. It is also possible that the precise location of some of the branch or secondary thrusts was determined by thrust pinning caused either by the increased resistance to thrust motion incurred when attempting to uplift the durable and erosion-resistant limestones at the base of the thrust sheets, or by the blocking of thrust motion by the deposition of barriers of conglomeratic sediment.

References

Álvarez Sierra, M.A., Daams, R., Lacomba, J.I., Lopez Martínez, N., van der Meulen, A.J., Sese, C. and de Visser, J. (1990). Palaeontology and biostratigraphy (micromammals) of the continental Oligocene–Miocene deposits of the North–Central Ebro Basin (Huesca, Spain). *Scripta Geol.*, **94**, 77. Leiden.

Crusafont Pairo, M., Riba, O., Villena, J. (1966). Nota preliminar sobre un nuevo yacimiento de vertebratos Aquitanienses en Santa Cilia (Rio Formigá; Provincia de Huesca) y sus consequencias geológicas. *Notas y Communs. Inst. Geol. y Minero de España*, **83**, 7–14.

Cuenca, G., Aranza, B., Canudo, J.I. and Fuertes, V. (1989). Los micromamíferos del Miocene inferior de Panalba (Huesca). Implicaciones bioestratigráficas. *Geogaceta*, **6**, 75–77.

Friend, P.F., Hirst, J.P.P. and Nichols, G.J. (1986). Sandstone-body structure and river process in the Ebro Basin of Aragon, Spain. *Cuad. Geol. Ibérica*, **10**, *Fluvial sedimentation in Spain*, 9–30. Universidad Complutense, Madrid.

Friend, P.F., Brazier, S.A., Cabrera, L., Feistner, K.W.A. and Shaw, J. (1988). Magnetic reversal stratigraphy in the Late Oligocene successions of the Ebro Basin, near Fraga, Province of Huesca, Northern Spain. *Cuad. Geol. Ibérica*, **12**, *Paleomagnetismo*, 121–130. Universidad Complutense, Madrid.

Friend, P.F. (ed.), Hirst, J.P.P., Hogan, P.J., Jolley, E.J., McElroy, R., Nichols, G.J., Rodriguez Vidal, J. (1989). Pyrenean tectonic control of Oligo-Miocene river systems, Huesca, Aragon, Spain. Excursion Guidebook No. 4 (ed. M. Marzo and C. Puigdefábregas), *4th International Conference on Fluvial Sedimentology*, 132 pp.

Hirst, J.P.P. (1991). Variations in alluvial architecture across the Oligo-Miocene Huesca fluvial system, Ebro Basin, Spain. In Miall, A.D. and Tyler, N. (eds.), *The three-dimensional facies architecture of terrigenous clastic sediments and its implications for hydrocarbon discovery and recovery*. SEPM (Society for Sedimentary Geology). *Concepts in Sedimentology and Paleontology*, vol. 3. Tulsa, Oklahoma, pp. 111–121.

Hirst, J.P.P. and Nichols, G.J. (1986). Thrust tectonic controls on Miocene alluvial distribution patterns, southern Pyrenees. *Spec. Publ. Int. Assoc. Sedimentol.*, 247–258.

Hogan, P.J. (1991). Geochronologic, tectonic and stratigraphic evolution of the southwest Pyrenean foreland basin, northern Spain. Unpublished PhD thesis, Univ. of Southern California.

Jolley, E.J. (1987). Thrust tectonics and alluvial architecture of the Jaca Basin, southern Pyrenees. Unpublished PhD thesis, Univ. of Wales.

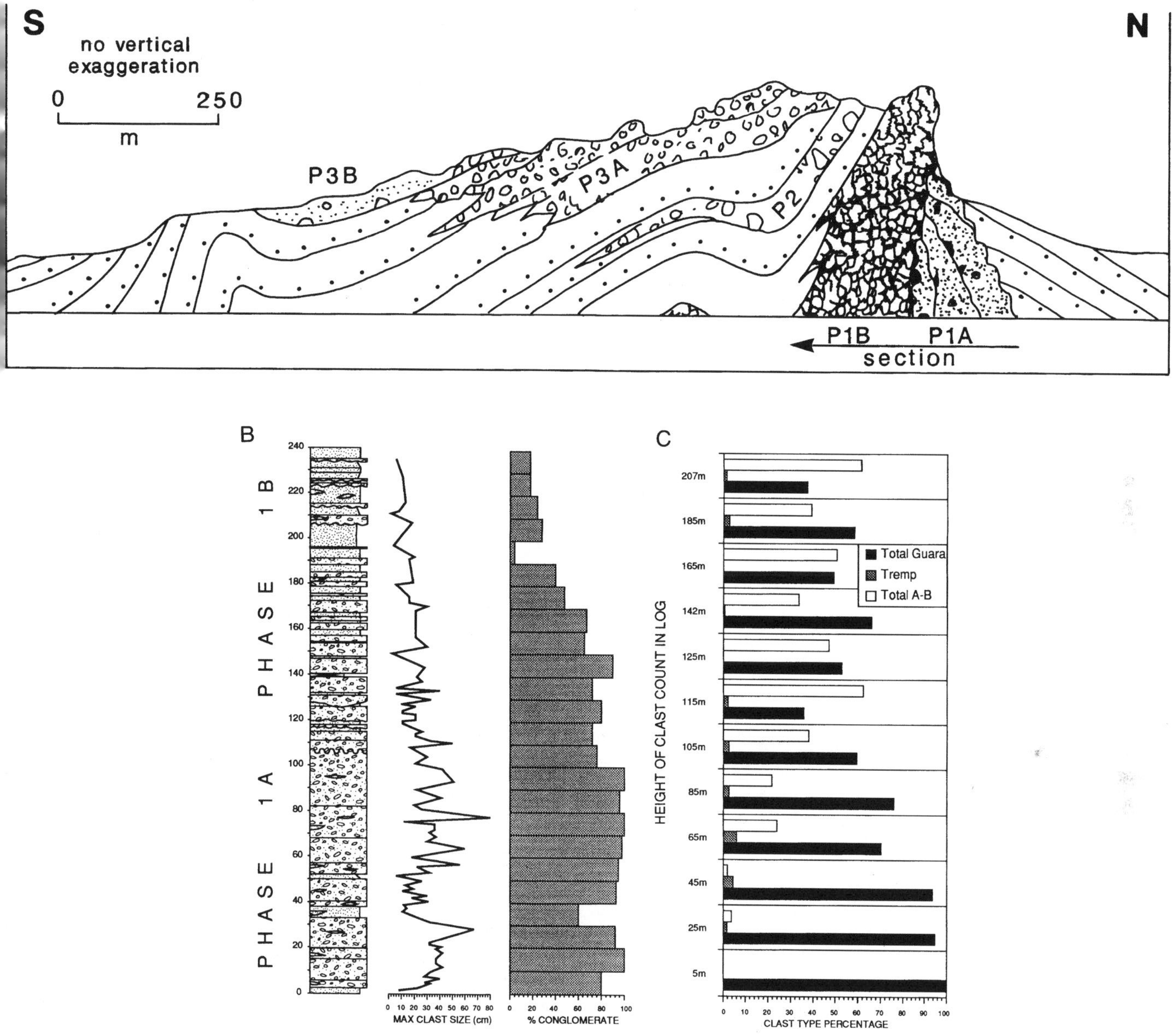

Fig. 4. Detailed study of evolution of the small alluvial fan body at Agüero (see Fig. 3C for location). A. Cross-section through the conglomerate-bearing sequence, where different conglomerate bodies are distinguished as P1A, P1B, P2, P3A and P3B (using the scheme of Hogan and Burbank, Chapter E14). B and C present the results of detailed study of the conglomerates through parts of the profile in A.

Jupp, P.E., Spurr, B.D., Nichols, G.J. and Hirst, J.P.P. (1987). Statistical estimation of the apex of a sediment distribution system from paleocurrent data. *Math. Geol.*, **19**, 319–333.

Millán, H.G. and Pocoví, A.J. (1995). Estructura y cinemática del sector occidental de las Sierras Exteriores surpirenaicas. *Rev. Soc. Geol. Esp.*, (in press.)

Nichols, G.J. (1987a). Syntectonic alluvial fan sedimentation, southern Pyrenees. *Geol. Mag.*, **124**, 121–133.

Nichols, G.J. (1987b). Structural controls on fluvial distributary systems: the Luna system, northern Spain. In Ethridge, F.G., Flores, R.M. and Harvey, M.D. (eds.). *Soc. Econ. Paleontol. Mineral., Spec. Publ.*, **39**, 269–277.

Nichols, G.J. (1987c). The structure and stratigraphy of the western External Sierras of the Pyrenees, northern Spain. *Geol. J.*, **22**, 245–259.

Ori, G.G. and Friend, P.F. (1984). Sedimentary basins formed and carried piggy-back on active thrust sheets. *Geology*, **12**, 475–478.

Pocoví, A., Millán, N., Navarro, J.J. and Martínez, M.B. (1990). Rasgos estructurales de la Sierra de Salinas y zona de los Mallos (Sierras exteriores, Prepirineo, provincias de Huesca y Zaragoza). *Geogaceta*, **8**, 36–39.

Puigdefábregas, C. (1975). La sedimentación molásica en la cuenca de Jaca. *Pirineos*, **104**, 1–188, with map.

Puigdefábregas, C. and Souquet, P. (1986). Tectonosedimentary cycles and depositional sequences of the Mesozoic and Tertiary from the Pyrenees. *Tectonophysics*, **129**, 173–203.

Riba, O., Reguant, S. and Villena, J. (1987). Ensayo de síntesis estratigráfica y evolutiva de la cuenca Terciaria del Ebro, Tema III, 3.8, vol. 2, pp. 131–159 in *Libro Jubilar J.M. Rios, 'Geologia de Espana'*, IGME, Madrid.

Singh, H., Parkash, B. and Gohain, K. (1993). Facies analysis of the Kosi megafan deposits. *Sed. Geol.*, **85**, 87–113.

Turner, J.P. (1988). Tectonic and stratigraphic evolution of the West Jaca thrust-top basin, Southwest Pyrenees. Unpublished PhD thesis, Univ. of Bristol, 225 pp.

Turner, J.P. (1990). Structural and stratigraphical evolution of the West Jaca thrust-top basin, Spanish Pyrenees. *J. Geol. Soc. Lond.*, **147**, 177–184.

Turner, P., Hirst, J.P.P. and Friend, P.F. (1984). A palaeomagnetic analysis of Miocene sediments at Pertusa, near Huesca, Ebro Basin, Spain. *Geol. Mag.*, **121**, 279–290.

E17 The Rioja Area (westernmost Ebro basin): a ramp valley with neighbouring piggybacks

M.J. JURADO AND O. RIBA

Abstract

This study deals with the evolution of the Rioja area in structural and stratigraphic terms. This area is located at the western end of the Ebro basin. Field data, particularliy subsurface data from hydrocarbon exploration activity, allow a new integrated approach. The Rioja area has experienced a particular evolution at least since the Mesozoic, and can be differentiated from neighbouring Ebro basin domains. Major thrusting along the northern (Cantabrian Chain) and southern (Iberian Chain) margins of the Rioja basin lead to a reduction in the extent of the original basin of approximately 70%. Syntectonic subsidence, coeval with the major thrusting on both margins, took place during Tertiary times. On the northern thrusted domain, piggy-back basins developed. Diapirism is also a significant feature of the Cantabrian region.

Geological setting

The geology of the Rioja area is dominated by a graben-like basin filled with continental sediments of Tertiary age (including the continental Garumnian facies). Thrust sheets form the northern and southern boundaries of the basin (see Riba and Jurado, 1993; and older papers, for example, Joly, 1922a, 1922b; Schriel, 1930; Saenz, 1942; Riba, 1955; Lotze, 1958; Colchen, 1966). The northern margin corresponds to the Montes Obarenes–Sierra de Cantabria range, a southwards-directed thrust unit which forms part of the westernmost Pyrenean thrust belt. The southern margin is bounded by the Demanda–Cameros ranges, with northwards-directed thrusts at the western limit of the Iberian Chain.

Allochthonous relationships in both the Pyrenees and the Iberian range have long been recognized (Fontbote *et al.*, 1986; Casas, 1990; Munoz, 1991; Riba, 1992b; Riba and Jurado, 1993), but the existence of large displacements, kilometres in extent, along the northern and southern margins of the Rioja basin, has only recently been documented using seismic and well data (Instituto Geológico y Minero de España, 1990a, 1990b; Lanaja *et al.*, 1987) (Fig. 2). Tertiary sediments have been recognised in exploration boreholes underlying the marginal thrust units (for example, Demanda-1, Fig. 1), and boreholes within the basin (Fig. 2 and Table 1: Rioja-1, Rioja-2, Rioja-3, Rioja-4, Rioja-5). The exploration boreholes also indicate that Tertiary sediments unconformably overly a basement consisting either of Mesozoic or Paleozoic rocks (Fig. 3). Seismic interpretation confirms that the magnitude of thrusting exceeds some 30 to 40 km on both margins (Casas, 1990; Riba *et al.*, 1993). Consequently, the original extent of the Rioja basin was much larger prior to thrusting.

Early stratigraphy and paleogeography

Stratigraphic and paleogeographic evolution of the Rioja area during the Mesozoic and Tertiary differed considerably from that of the neighbouring Ebro basin areas. The Rioja area had therefore become a differentiated domain within the Ebro basin at least since the Triassic.

The most striking stratigraphic and paleogeographic features recognized in the Rioja area are described below:

Paleozoic

Data are scarce on the Paleozoic rocks underlying the Tertiary in the Ebro basin. Two exploration boreholes (Rioja-1 and Rioja-3) reached Paleozoic slates, sandstones and quartzites.

Mesozoic

There is no evidence of **Permo-Triassic** sediments in any of the wells in the Rioja area. In contrast, Germanic facies Triassic sediments appear to have been deposited across the whole area of the present Ebro basin (Jurado, 1986, 1988, 1990). Bundsandstein (terrigenous), Muschelkalk (carbonates, evaporites and shales) and Keuper (evaporites and shales) are widely represented. A major area of subsidence in the central part of the basin has been documented. Paleogeographic maps (Lanaja, 1987; Jurado, 1989; ITGE, 1990) show that towards the Rioja area, Muschelkalk and Keuper facies wedge out. The absence of Triassic in this westernmost part of the Ebro basin could be due to either non-deposition, or to erosion. In either case, the Rioja area must have been a non-subsiding area, or a paleogeographic high, during Permo-Triassic times.

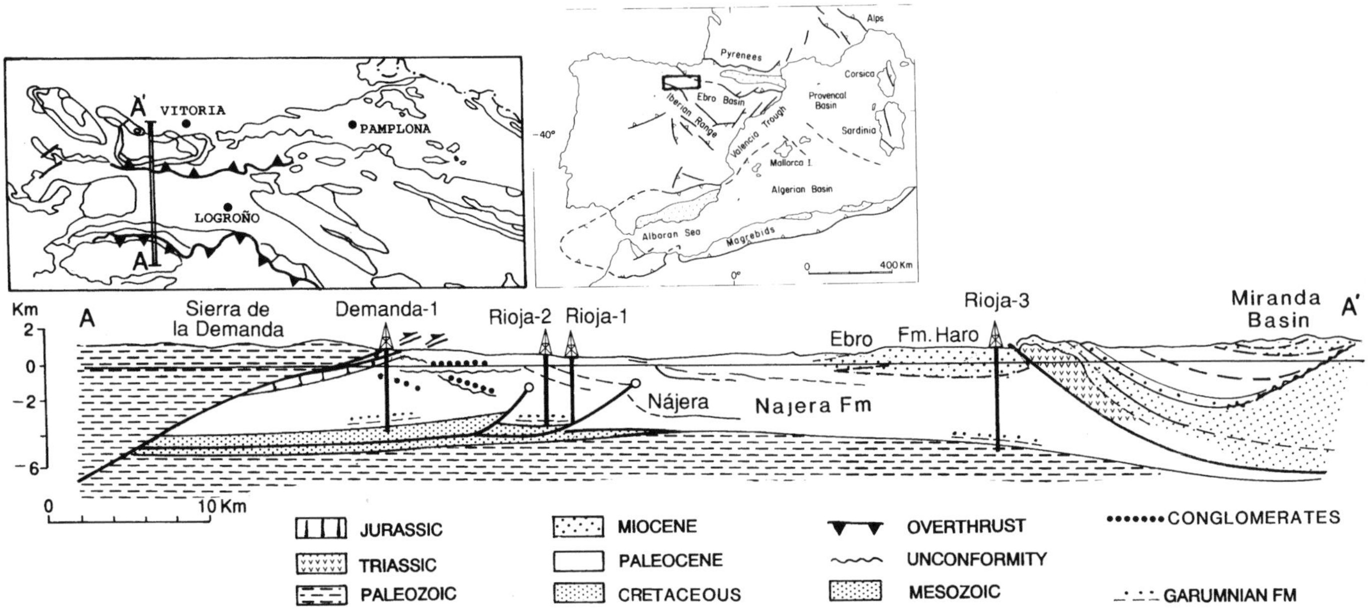

Fig. 1. Geological cross section of the Rioja area, modified from Guimerà and Riba (1992). The general structural location of well Demanda-1 is shown, though it is west of the section line.

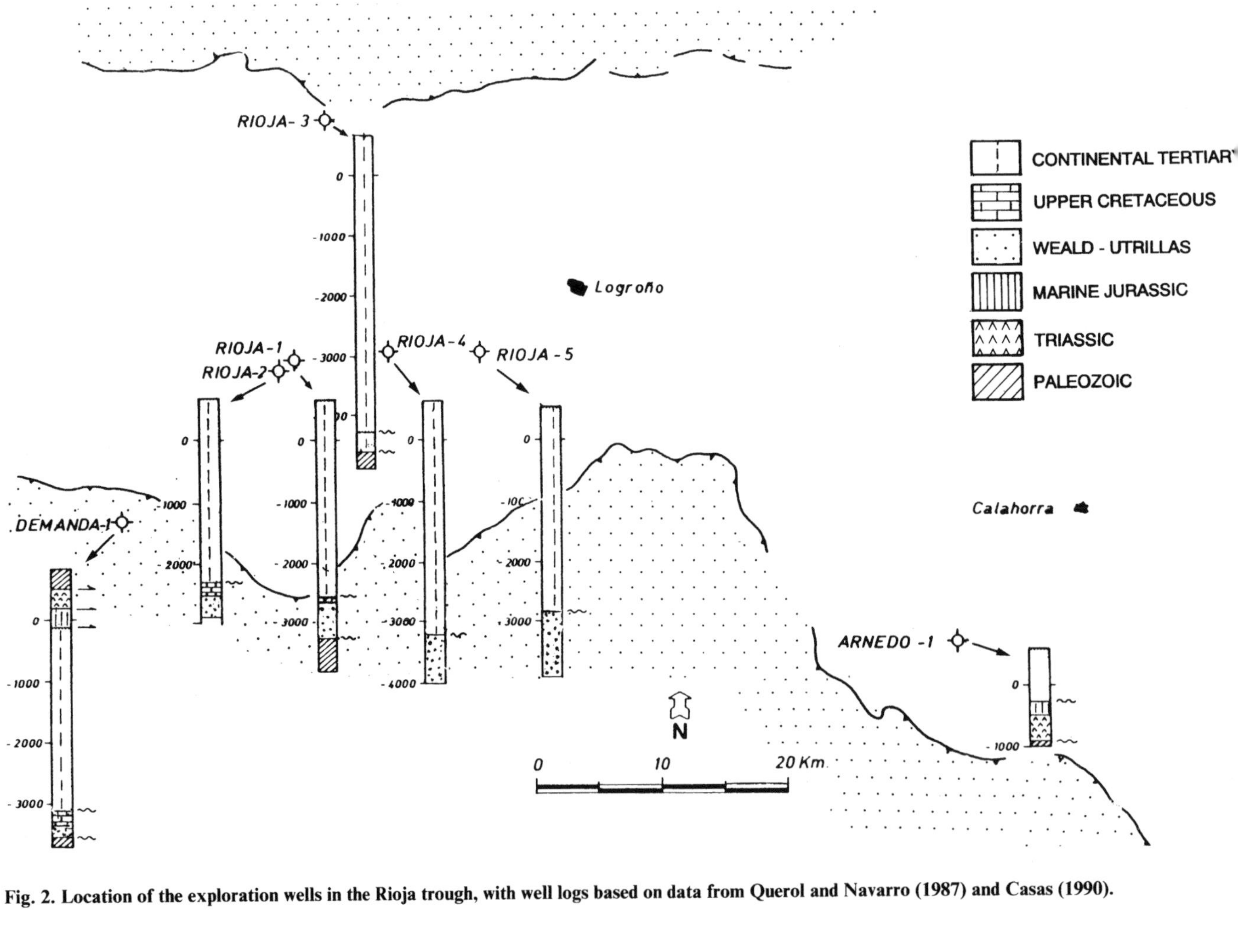

Fig. 2. Location of the exploration wells in the Rioja trough, with well logs based on data from Querol and Navarro (1987) and Casas (1990).

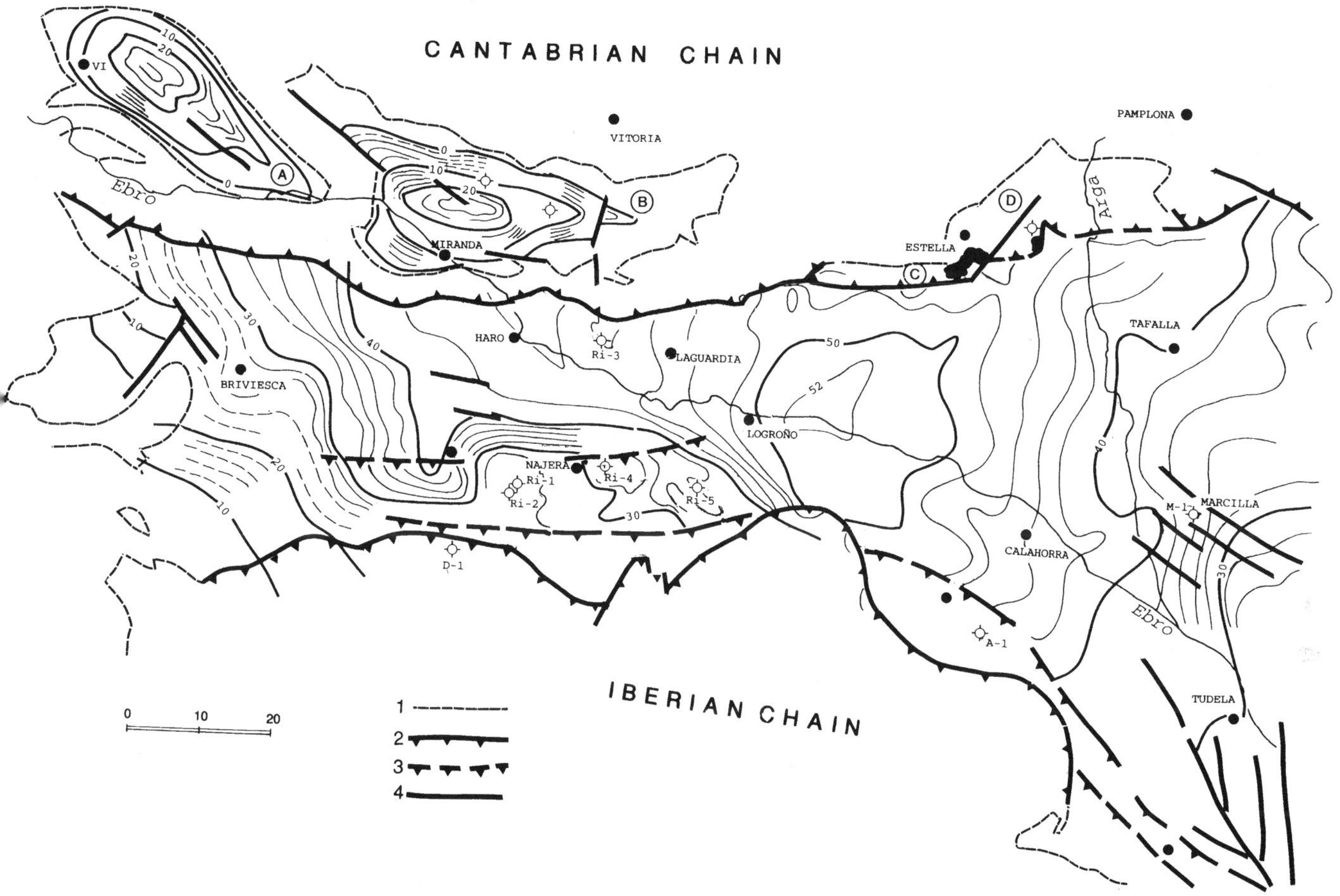

Fig. 3. Isobath map of the Tertiary–Garumnian base in western Ebro Basin. Contour interval 200 m (modified from Antón *et al.*, 1965, 1987; Lanaja *et al.*, 1987). 1: Continental Tertiary boundaries; 2: main thrusts; 3: blind thrusts; 4: basal faults. Tertiary piggyback basins: A: Villarcayo Basin. B: Miranda–Treviño Basin. C: Názar–Monjardín Basin. D: Sierra del Perdón-Estella Basin. See Table 1 for details of the wells.

Marine **Jurassic** sediments were deposited over a large part of the Ebro basin (Dahm, 1966; Ramirez-del-Pozo, 1971a, 1971b; Fontana *et al.*, 1993). The Jurassic succession appears to have undergone major erosion in the western and eastern parts of the Ebro basin, although in the central area thick marine carbonates and a basal Jurassic anhydrite unit are preserved in several wells (Jurado, 1991; Fig. 1). There is no evidence of deposition of Jurassic marine sediments in the Rioja basin.

The **Cretaceous** paleogeography of northern Spain (Basque Country, Navarre, Rioja and Old Castile) has been studied by several authors: Rios (1956); Saeftel (1959); Brinkmann *et al.* (1967, 1968); Aguilar (1970); Wiedmann *et al.*, (1976); Plaziat (1981); Floquet *et al.* (1982); Garcia-Mondejar *et al.* (1982); Salomon (1982); Lanaja *et al.* (1985); Alonso *et al.* (1991).

In the subsurface of the Rioja area, the Cretaceous is relatively thin compared with the successions in the northern and southern neighbouring basins, where very great thicknesses were deposited (Basque–Cantabrian and Iberian Chains). The Wealden and Utrillas detrital facies (Lower Cretaceous) of the Rioja area are variable in thickness, but do not exceed 1000 m in thickness (Rioja-1, Rioja-2, Rioja-4, Rioja-5 and Demanda-1 wells). Albian–Cenomanian to Turonian (and Maastrichtian?) marine facies are present but do not exceed 300 m in the Rioja-1, Rioja-2 and Demanda-1 boreholes.

Tertiary stratigraphy of the Rioja area

According to Tertiary subsurface maps (Fig. 3) of the Rioja area, based on well and seismic data (Anton-Plaza *et al.*, 1965; Julivert *et al.*, 1974; Lanaja *et al.*, 1987), the base of the Tertiary is approximately 5200 m below sea-level at its deepest. Underlying the Tertiary sediments, the basement is either Paleozoic (Rioja-3) or Cretaceous (Rioja-1, Rioja-2, Demanda-1).

Paleocene or Eocene marine transgressions from the northern Pyrenean–Basque basin did not reach the Rioja area, and there must have been a paleogeographic high in this area. Tertiary beds (including Garumnian facies) lie unconformably on an erosive surface that truncated Paleozoic, Upper and/or Lower Cretaceous (Wealden) sedimentary rocks.

Initially, the source area for the Garumnian and Paleogene of the Tertiary Rioja basin lay entirely to the south (as deduced from paleocurrent measurements). However, by the end of the Oligocene, sediments derived from both northern and southern areas are

Table 1. *Exploration wells located in the Rioja Trough*

Well	*Num.*	*Year*	*Operator*	*Locality*	*Tert. Bottom (m)*
Alloz-1	63	1959	Ciepsa	Alloz (Na)	1492
Arnedo-1	107	1962	Amospain	Arnedo (Ri)	834
Valdearnedo-1	278	1971	Ciepsa	Valdearnedo (Bu)	715
Rioja-1	357	1977	Campsa	(+) (Ri)	3378
Rioja-2	376	1977	Campsa	(+) (Ri)	3082
Rojas-NE-1	388	1979	Eniepsa	Rojas (Bu)	243
Rioja-3	408	1978	Campsa	Samaniego (Vi)	5120
Rioja-4	434	1980	Campsa	Alesón (Ri)	3830
Demanda-1	521	1983	Eniepsa	Ezcaray (Ri)	4172
Rioja-5	542	1984	Eniepsa	Navarrete (Ri)	3665

Notes:
(+) Torrecilla sobre Alesanco. (Ri = Rioja; Na = Navarra; Bu = Burgos prov. Vi = Alava prov.) Wells Alloz-1, Rojas-Ne-1 and Valdearnedo-1 are placed on allochthonous units and do not reach the Ebro basin itself.
Data from Lanaja *et al.* 1987, and ITGE 1990.

recognized in deposits overlying the Barbarin unconformity (dated as latest Oligocene or close to the Oligocene–Miocene boundary; Fig. 4; Riba, 1993). Conglomerates with a source area in the Pyrenees are recognized in the Estella area, in the Sierra del Perdón (S. Pamplona), and to the east of this latter mountain (for example, Peña Izaga).

More than 5000 m of continental, mainly terrigenous, sediments were deposited approximately at sea-level during Tertiary times in the Rioja area; these deposits indicate a steady subsidence which kept pace with sedimentation. Compressional events on both its northern and southern margin would have led to the emplacement by thrusting of allochthonous, hanging-wall sheets over the Tertiary autochthonous sediments of the Rioja basin. Evidence from progressive unconformities recognized in field studies indicates that thrust emplacement and sedimentation were broadly contemporaneous (Riba, 1976, 1977, 1992). In the Rioja area, subsidence was therefore coeval with both compression and thrust emplacement.

Conglomeratic units derived from a source area located in the southern margin (Iberian range) of the Rioja basin (Riba, 1955; Munoz, 1991) are bounded by unconformities and disconformities. Conglomeratic units with source areas in the Pyrenees and Basque Country are overthrust by the Sierra de Cantabria. The southward displacement of the northern margin of the Rioja basin appears to have been more important, and later than the northward displacement of the southern (Iberian chain) margin.

Upper Miocene, post-tectonic conglomerates have been recognized from the La Bureba area of Burgos Province, and unconformably overly the thrusts of both margins (Santurdejo, Santa Casilda, Poza de la Sal, Pancorbo conglomerates, etc.).

Accepting that the deposition of the continental Tertiary sediments (including the Garumnian facies) took place at elevations close to sea-level from Upper Maastrichtian to Lower Vallesian (Lower Tortonian, Upper Miocene) (Santafe *et al.*, 1982; Casas, 1990; Munoz, 1991), a period of 50 my, it is possible to estimate a

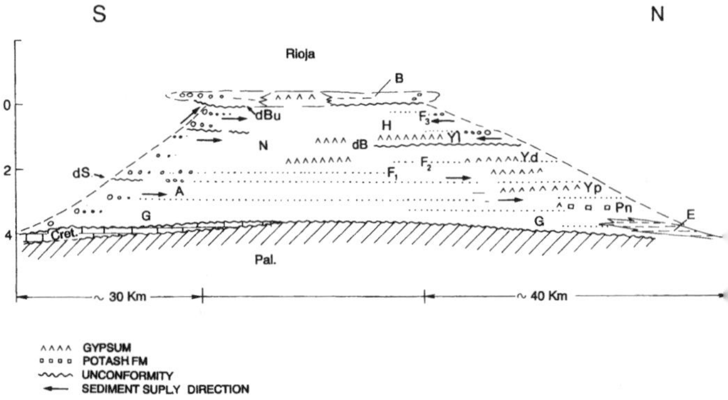

Fig. 4. Model of a tectonically restricted basin. N–S stratigraphic sketch showing the evolution of Garumnian facies and continental Tertiary in the Rioja Trough. Allochthonous units are not represented. Pal: undifferentiated Paleozoic. Cret: undifferentiated Cretaceous. G: Garumnian facies. E: marine Paleocene–Eocene. Pn: Navarra Potash Formation. Y: gypsum, different formations on the northern margin (Navarre). Yp: Puente la Reina Gypsum. Yd: Desojo Gypsum. Yl: Los Arcos Gypsum. A: Arnedo Formation. N: Nájera Formation. H: Haro Formation. B: Bureba formations belonging to Upper Miocene (Santurdejo Conglomerates, Cerezo de Río Tirón Gypsum, Altable Formation. d: angular unconformities, dB: Barbarín unconformity; dS: Serradero unconformity, dBu: Bureba unconformity. Fossiliferous levels, F_1: Tudela-1 (Burdigalian); F_2: Tudela-2 (Aquitanian); F_3: Cellórigo (Vallesian, Upper Miocene).

nett subsidence, or sediment accumulation rate. Disregarding sediment compaction, the value calculated is 104 mm per 1000 years.

Summary of structural history

The Mesozoic stratigraphy is characterized by erosional episodes, stratigraphic gaps and erosional unconformities (Muñoz, 1991), and the Rioja area was therefore a structural high or non-subsiding area from early to late Mesozoic times. The Upper

Table 2. *Paleontological Tertiary levels and localities in the Rioja trough*

Locality	*Age*
La Bureba (Burgos)	Turolian
Cellórigo (Rioja) (=Tortonian).	Lower Vallesian
Los Agudos (Rioja)	Agenian–Lower Aragonian to Upper Agenian
Fuenmayor-2 (Rioja)	Agenian
Carretil (Rioja)	Agenian
Autol (Rioja)	Lower Agenian
Bergasa (Rioja)	Upper Arvernian
Tudela II (Navarra)	Burdigalian
Tudela I (Navarra)	Aquitanian
Autol, Bergasa, Arnedo, Quel (Rioja)	Oligocene
Espronceda (Navarra)	Oligocene

Note:
A *Rhinoceros* was found in Cellórigo (Rioja), in steep dipping deformed Tertiary beds in the vicinity of the main thrust of the Sierra de los Obarenes (Crusafont *et al.*, 1966 and Santafé *et al.*, 1982); it has been dated as Vallesian (equivalent to marine L. Tortonian). This dating marks the most recent thrusting movement of the Sierra de los Obarenes – Cantabria. Consequently, the main compressive events reach very high stratigraphic levels in La Rioja.

Cretaceous marine regression then marked a change from a general extensional regime to a compressional regime. The Rioja area experienced steady subsidence during Tertiary times coeval with compression, and was not reached by any marine Tertiary transgressions. Marine deposits are recognized to the north of the Rioja area, and form part of the allochthonous units of the Pyrenees.

A major control of the subsidence of the Rioja trough is considered to be thrust loading on both northern and southern margins. The present-day configuration of the trough, with its reverse fault boundaries, is that of a *ramp valley* (Biddle *et al.*, 1985). The age of thrusting along the margins ranges from Uppermost Cretaceous (Maastrichtian) to Vallesian (Upper Miocene). Compressional events have reduced the original depositional area by more than 70%.

The Tertiary Rioja basin is thus defined as being structurally controlled and bounded, with offlap sedimentation and large-scale coarsening-upwards sequences at the basin margins due to the synsedimentary displacement of the source areas, and progradation of coarse terrigenous sediments towards the basin centre.

Piggyback basins of the northern, Cantabrian, margin

The term 'piggyback' was first applied by Ori & Friend (1984) to basins that were being filled while they were actually being transported as part of a thrust sheet. The original examples came from the Aragonese Pyrenees, and here we draw attention, for the first time, to some other examples from the area further to the west (see capital letters on Fig. 3):

(A) Villarcayo basin (Burgos) and
(B) Miranda–Treviño basin (Burgos and Alava). Both of these basins display strong asymmetry of sediment thickness and facies, and marginal progressive unconformities, which were controlled by active tectonic margins (Riba, 1977).
(C) Nazar–Monjardin basin (west of Estella, Navarra), formed close to the main overthrust of the northern margin (Riba, 1993).
(D) Sierra del Perdón-Estella (Navarra), a piggyback basin that also acted as a sedimentary bypass towards the main foreland Ebro basin, between the Estella and Alloz diapirs.
(E) Peña Izaga basin, 18 km ESE of Pamplona (Navarra), which clearly developed as an individual basin.

Diapirism

Diapirism of the Upper Triassic evaporites is strongly developed in the northern thrust units. It is a striking feature of the geological evolution of the area, particularly in the Basque country (see Kind, 1963, 1967; Brinkmann *et al.*, 1968; Serrano *et al.*, 1989, 1990). Halokinesis appears to have been active mainly from Upper Jurassic to Cretaceous times. During the compression from Upper Cretaceous to Upper Miocene times, diapirism appears to have been largely inactive, although earlier-formed salt structures were deformed during the later compressional episodes. The final stage of development of the diapirs took place from Upper Miocene times, producing the domes which correspond to the main Tertiary sedimentary depocentres (Salinas de Añana, Salinas de Rosío) or to the small depocentres in rim synclines (Maeztu, Salinas de Oro, etc). The Estella diapir (Pflug, 1967) is located on the intersection of the main northern Ebro basin margin overthrust and a sinistral NE–SW trending, transcurrent fault. Both these primary structures involved the local basement.

Elsewhere in the Rioja area, diapirism is also a result of the halokinesis of evaporites of Tertiary age, as in the Ebro basin (Ortí *et al.*, 1990). This is particularly a feature of a large area of the Ribera de Navarra, in the centre of the Ebro basin. The large anticlines of Marcilla and Tafalla have evaporitic cores underlain by flat-bottomed structures, and they appear to result from Pyrenean thrusting. The synformal basins of Miranda de Arga and Barasoain (Navarra) appear to have been piggyback basins, and have outstanding progressive unconformities (see illustrations in Riba, 1977 and 1992a). These kinds of structures are similar to those described for the potash basin of Catalonia (Vergés *et al.*, 1990).

Acknowledgements

We wish to thank D. Angel Rodríguez-Paradinas, (director of CIEPSA) for permission to use documentation for this study, and Repsol Exploración, for providing subsurface data. We acknowledge financial support from 'Proyecto DGICYT, núm PB91–0805'.

References

Aguilar Tomás, M.J. (1970). Sedimentología y paleogeografía del Albense de la Cuenca Cantábrica. Doctoral Thesis. Univ. Barcelona, 2 vols., 329 pp.

Alonso, A., Melendez, N. and Mas, R.J. (1991). Sedimentación lacustre durante el Cretácio de la Cordillera Ibérica, España. *Acta Geol. Hispanica*, **26/1**: 35–54.

Antón Plaza, J.A. and Riba, O. (1965). Mapa de isobatas de la base del Terciario de la zona Vasco-Cantàbrica (Burgos, Alava, Rioja, Navarra), E. 1:200 000. CIEPSA. Vitoria. (Unpubl.)

Babinot, J.F. and Freytet, P. *et al.* (1983). Le Sénonien supérieur continental de la France méridionale et de l'Espagne septentrionale. *Géol. Médit.*, **10/3–4**: 245–268.

Biddle, K.T. *et al.* (1985). Ramp valley: A basin bounded by reverse faults. *S.E.P.M. Spec. publ.*, **37**: 375–386.

Bomer, B. (1954). Trois aspects du contact entre Monts Celtibériques occidentaux et Bassin de l'Ebre. *Bull. Assoc. Géogr. Français*, **239–240**: 35–41.

Brinkmann, R. and Loegters, H. (1968). Diapirs in Western Pyrenees and Foreland, Spain. *AAPG. Me.*, **8**: 275–292.

Casas, A.M. (1988). La compresión alpina en un sector del borde norte de las sierras de Cameros (Depresión de Arnedo, La Rioja). *2nd. Congr. Geol. España, Granada, 1988. Com.*, **2**: 115–118.

Casas, A.M. (1990). El frente norte de las Sierras de Cameros: estructuras cabalgantes y campo de esfuerzos. Doctoral Thesis. Univ. de Zaragoza, 382 pp.

Castiella, J.J. and Del Valle, J. (1978). Mapa geológico de Navarra. A escala 1:200 000. *Diput. Navarra, Serv. Geol., Dir. de O. P.* Pamplona.

Colchen, M. (1966). Sur la tectonique tertiaire du massif paléozoïque de la Sierra de la Demanda et sa couverture mésozoïque et cénozoïque. *Bull. Soc. Géol. France*, **7** (8): 87–97.

Crusafont, M., Truyols, J. and Riba, O. (1966). Contribución al conocimiento de la estratigrafía del Terciario de Navarra y Rioja. *Not. y Com. IGME*, **90**: 53–76.

Dahm, H. (1966). Stratigraphie und Paläogeographie im Kantabrischen Jura (Spanien). *Beih. Geol. Jb.*, **44**: 13–54.

Desegaulx, P. and Moretti, I. (1988). Subsidence history of the Ebro Basin. *J. Geodynam.*, **10**: 9–24.

Floquet, M., Alonso, A. and Meléndez, A. (1982). Cameros–Castilla. El Cretácico superior. In: *El Cretácico España*. Ed. Univ. Complutense Madrid. 1 vol., pp. 387–455.

Fontana, B., Gallego, M.R., Jurado, M.J. and Meléndez, G. (in press). Correlation of subsurface and surface data for the Middle Jurassic between the Ebro Basin and central Iberian Chain (Eastern Spain). *Geobios*.

Fontboté, J.M., Muñoz, J.A. and Santanach, P. (1986). On the consistency of proposed models for the Pyrenees with the structure of the eastern parts of the belt. *Tectonophysics*, **129**: 291–301.

García-Mondéjar, J. and Pujalte, V. (1982). Región Vasco Cantábrica y Pirineo Navarro. In *El Cretácico de España*. Ed. Univ. Complutense Madrid. 1 vol., pp. 49–159.

Guiraud, M. and Seguret, M. (1985). A releasing solitary overstep model for the Late Jurassic–Early Cretaceous (Wealdien) Soria strike–slip basin (Northern Spain). *S.E.P.M. Spec. Publ.*, **37**: 159–176.

Instituto Geológico y Minero De España. (1971–1973). *Mapa geológico de España* E. 1:200 000. Sheets: 22 *Tudela*, 21 *Logroño*.

Instituto Geológico y Minero De España. (1976–1992). *Mapa geológico de España* E. 1:50 000, 2a Ed. Mem. expl. & sheets: 109: *Villarcayo*. 110: *Medina de Pomar*. 136: *Oña*. 137: *Miranda de Ebro*. 138: *La Puebla de Arganzón*. 139: *Eulate*. 140: *Estella*. 141: *Pamplona*. 168: *Briviesca*. 169: *Casalarreina*. 170: *Haro*. 171: *Viana*. 172: *Allo*. 173: *Tafalla*. 239: *Pradoluengo*. 240: *Ezcaray*.

Instituto Tecnológico Geominero De España. (1990a). *Documentos sobre la Geología del subsuelo de España*. No. 6: *Ebro-Pirineos*.

Instituto Tecnológico Geominero De España. (1990b). *Documentos sobre la Geología del subsuelo de España*. No. 7. *Cantábrica*.

Joly, H. (1922a). Note préliminaire sur l'allure générale et l'âge des plissements de la Chaîne Celtibérique (Espagne). *C. R. Acad. Sci. Paris*, **174**: 976–978.

Joly, H. (1922b). Sur la présence d'écailles ou de lambeaux de charriage dans la Chaîne Celtibérique (provinces de Saragosse, Logroño et Soria). *C.R. Acad. Sci. Paris*, **174**: 1185–1187.

Julivert, M. Fontboté, J.M. Ribeiro, A. and Conde, L. (1974). Mapa tectónico de la Península Ibérica y Baleares. E. 1:1 000 000. *IGME*. Mem. 113 pp.

Jurado, M.J. (1988). Rasgos litoestratigráficos y sedimentológicos de los depósitos evaporíticos triásicos en el subsuelo de la cuenca del Ebro. *Congr. Geol. de España, Comunicaciones*, vol. 1: 225–227.

Jurado, M.J. (1989). *El Triásico del subsuelo de la cuenca del Ebro*. Mem. 264 pp., 15 figs., 24 well logs, 23 maps, 13 seismic lines, 43 pl.

Jurado, M.J. (1990). El Triásico y el Liásico basal evaporíticos del subsuelo de la Cuenca del Ebro. In F. Ortí *et al.*, eds. *Form., evap. Cuenca Ebro*. ENRESA, GPPG. Univ. Barcelona. pp. 21–28, 9 figs. Barcelona.

Jurado, M.D. and Riba, O. (1986). Aspectos de la sedimentación triásica en facies Muschelkalk en la Cuenca del Ebro. *XI Congr. Esp. Sedimentol. Barcelona, 1986*, p. 98. Univ. de Barcelona; CSIC.

Kind, H.D. (1963–1967). Diapire und Altertiär im Sudöstlichen Baskenland (Nordspanien). *Beih. Geol. Jb.*, **66**: 127–174.

Lanaja, J.M., Querol, R. and Navarro, A. (1987). Contribución de la exploración petrolífera al conocimiento de la Geología de España. *I.G.M.E.*., 1 vol. 465 pp. 17 pl. Madrid.

Lotze, F. (1958). *Mapa Geológico de la zona Oeste de los Pirineos y del Este de la Cordillera Cantábrica. E. 1:200 000*. Ed. Comp. Petrolif. Ibérica. Madrid.

Lotze, F. (1973). *Geol. Karte Pyrenäisch – Kantabrischen Grenzgeb. E. 1:200 000*. Mainz, *Akad. Wiss. und Lit.* 22 pp. Abh. Math. – *Naturw. Kl*. Nr. 1.

Mensink, H. (1966). Stratigraphie und Paläogeographie des marinen Jura in den nordwestlichen Iberischen Ketten. *Beih. Geol. Jb.*, **44**: 55–102.

Muñoz, A. Pérez, A., González, A., Pardo, G. and Villena, J. (1991). Estratigrafía de los conglomerados miocenos adosados al frente cabalgante de la Sierra de Cameros. *Com. I Congr. Grupo Esp. Terciario, Vic*. pp. 227–232, 3 figs.

Muñoz-Jiménez, A. (1991). Análisis tectosedimentario del Terciario del sector occidental de la Cuenca del Ebro. Doctoral thesis. Univ. de Zaragoza, 496 pp.

Ori, G.G. and Friend, P.F. (1984). Sedimentary basins formed and carried piggyback on active thrust sheets. *Geology*, **12**: 475–478.

Ortega-Lozano, A. and Pérez-Lorente, F. (1984). El Terciario de la Depresión de Arnedo. *Berceo*, **2**: 99–113. Logroño.

Ortí, F. (1990). Introducción a las evaporitas de la Cuenca Terciaria del Ebro. In F. Ortí & J.M. Salvany, eds. *Form. evapor. Cuenca del Ebro*. ENRESA, GPPG. & Univ. Barcelona. pp. 62–66.

Pérez-Lorente, F. (1987). La estructura del borde norte de la Sierra de Cameros. *Bol. Geol. Min.*, **98**(4): 484–492.

Pflug, R. (1967). Der diapir von Estella (Nordspanien). *Beih. Geol. Jb.*, **66**: 21–62. Hannover. Idem. (1973).

Plaziat, J.C. (1981). Late Cretaceous to Late Eocene palaeogeographic evolution of Southwest Europe. *Palaeogeogr., Palaeoclimatol. Palaeoecol.*, **36**: 263–320.

Ramírez Del Pozo, J. (1971a). Bioestratigrafia y microfacies del Jurásico y Cretácico del Norte de España, región Cantábrica. 3 vols. *Edic. CIEPSA*, 357 pp. Madrid.

Ramírez Del Pozo, J. (1971b). Algunas observaciones sobre el Jurásico de Alava, Burgos y Santander. *Cuad. Geol. Ibérica*, **2**: 491–508.

Ramírez Del Pozo, J. (1973). Síntesis geológica de la Provincia de Alava. *Caja de Ahorros de Vitoria*, 66 pp. 1 geol. map. Vitoria.

Ramírez Del Pozo, J. (1987). Res. conf. Geologia del subsuelo en el sector meridional de la cuenca Vasco-Cantábrica. *Geogaceta*, **3**: 40–44, 5 figs. Madrid.

Rat, P. (1964). Problèmes du Crétacé inférieur dans les Pyrénées. *Geol. Rund.*, **53**: 205.

Riba, O. (1955a). Sur le type de sédimentation du Tertiaire de la partie ouest du Bassin de l'Ebre. *Geol. Rund.*, **43**: 363–371.

Riba, O. (1976). Syntectonic unconformities of the Alto Cardener, Spanish Pyrenees: a genetic interpretation. *Sedim. Geol.*, **15**: 213–233.

Riba, O. (1977). Tectogenèse et sédimentation: deux modèles de discordances syntectoniques pyrénéennes. *Bull. BRGM.* 2ème Sér., **1/4**: 383–401. Orléans.

Riba, O. (1992a). Las discordancias sintectónicas como elementos de análisis de cuencas. In Arche coord. *Sedimentologia* vol. 2, pp. 489–522, 23 figs. *C.S.I.C.* 2nd. Edn. Madrid.

Riba, O. (1992b). La Conca de l'Ebre. Consideracions generals. In *Història Natural dels Països Catalans*. vol. 2, pp. 135–148. *Ed. Enciclop. Catalana*. Barcelona.

Riba, O. (1992c). Las secuencias oblicuas en el borde norte de la Depresión del Ebro, Navarra y la discordancia de Barbarín. *Acta Geol. Hisp.*, **27**: 55–68.

Riba, O. and Jurado, M.J. (1993). Reflexiones sobre la geologia de la parte occidental de la depresión del Ebro. *Acta Geol. Hisp.*

Riba, O. and Pérez Mateos, J. (1961). Sobre una inversión de aportes sedimentarios en el brode norte de la cuenca terciaria del Ebro (Navarra). *2a. Reun. de Sedimentol Sevilla. Vol. com.pp. 201–221. Inst. Edafologia. CSIC.* Madrid.

Riba, O., Reguant, S. and Villena, J. (1987). Ensayo de síntesis estratigráfica y evolutiva de la Cuenca Terciaria del Ebro. In *Libro Jubilar J.M. Ríos*. vol. 2, pp. 131–159, 16 figs. *Inst. Geol. Min. Esp.*, Madrid.

Ríos, J.M. (1956). El sistema Cretáceo en los Pirineos de España. *Mem. IGME*, **57**: 1–128.

Saeftel, H. (1959). Paläogeographie des Albs in den Keltiberischen Ketten Spaniens. *Z. deutsch. Geol. Ges.* **111**(3): 684–711. Hannover.

Saeftel, H. (1961). Not. y Com. *IGME*, **26**: 109–134.

Saenz, C. (1942). Estructura general de la cuenca del Ebro. *Est. Geogr.*, **3**(7): 249–269.

Saez, A. and Salvany, J.M. (1990). Las formaciones evaporíticas de Barbastro y Puente la Reina (Eoceno superior – Oligoceno basal) de la Cuenca Surpirenaica. In Ortí *et al.*, eds. *Form. evap. Cuenca Ebro.* ENRESA, GPPG., Univ. Barcelona, p. 100.

Salomón, J. (1982). Cameros–Castilla. El Cretácico inferior. In *El Cretácico de España*. Ed. Univ. Complutense Madrid. 1 vol., pp. 345–378.

Salvany, J.M. and Duran, J.M. (1989). Las formaciones evaporíticas del terciario continental de la Cuenca del Ebro en Navarra y La Rioja, litoestratigrafía, petrologia y sedimentologia. Tesis, Univ. Barcelona. 397 pp.

Santafé-Llopis, J.V., Casanovas-Cladellas, M.L. and Alférez-Delgado, F. (1982). Presencia del Vallesiense en el Mioceno continental de la Depresión del Ebro. *Rev. R. Ac. C. Exact., Fis y Nat. Madrid*, **76**(2): 277–284. Madrid.

Schriel, W. (1930): Die Sierra de la Demanada und die Montes Obarenes. *Abh. Ges. Wiss. Göttingen, Math.–Phys.* Kl. N.F., **16/6**: 105. Berlin.

Serrano, A. and Martínez Del Olmo, W. (1990). Tectónica salina en el dominio Cántabro–Navarro: evolución, edad y origen de las estructuras salinas. In F. Ortí *et al.*, eds. *Form. evap. Cuenca Ebro. ENRESA, GPPG., Univ. Barcelona.* pp. 39–53.

Serrano-Oñate, A. Martínez del Olmo, W. and Cámara-Rupelo, P. (1989). Diapirismo del Trías salino en el dominio Cántabro-Navarro. In *Lib. Homenaje a R. Soler. Asoc. Geól. y Geofis. Esp. Petróleo (AGGEP)*, pp. 115–121, 6 figs. Madrid.

Vergés, J. and Muñoz, J.A. (1990). Thrust sequences in the southern Central Pyrenees. *Bull. Soc. Géol. Fr.*, **8** (6)/2: 265–271.

Wiedmann, J. (1962). Contribution à la paléogéographie du Cretacé Vascogothique et Celtibérique septentrional, Espagne. *Livre Mém. P. Fallot, Soc. Géol. Fr.*, **1**: 351–366.

Wiedmann, J. (1964). Le Cretacé supérieur de l'Espagne et du Portugal. *Est. Geol.*, **20**: 107–148.

Wiedmann, J. and Kauffmann, E.G. (1976). Mid-Cretaceous biostratigraphy of Northern Spain. *Ann. Mus. Hist. Nat. Nice*, **4**.

Wienands, A. (1966). Über den Muschelkalk in der Sierra de la Demanda. *N. Jahrb. Geol. Paläont. Mh.* (1966): 151–160.

Zoetmeijer, R., Desegaulx, P., Cloetingh, S., Roure, F. and Moretti, I. (1990). Lithospheric dynamics and tectonic–stratigraphic evolution of the Ebro Basin. *J. Geophys. Res.*, **95**, B3: 2701–2711.

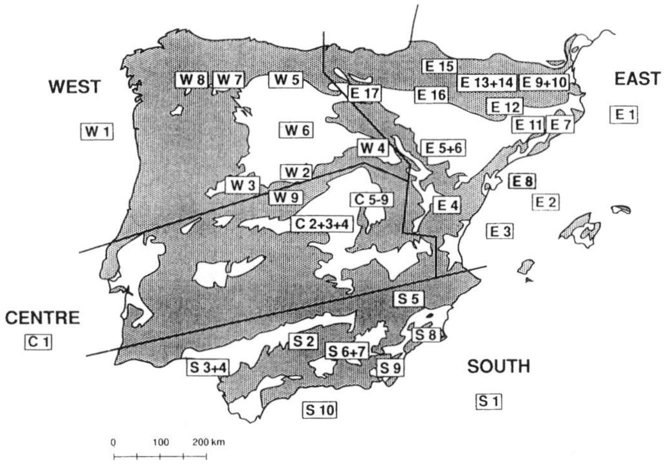

Part W
West

The castle at Peñafiel, 50 km east of Valladolid, has been built on outcrops of lacustrine and palustrine dolomitic marlstones of the 'Páramo facies' (see Chapter W6), of Miocene age.

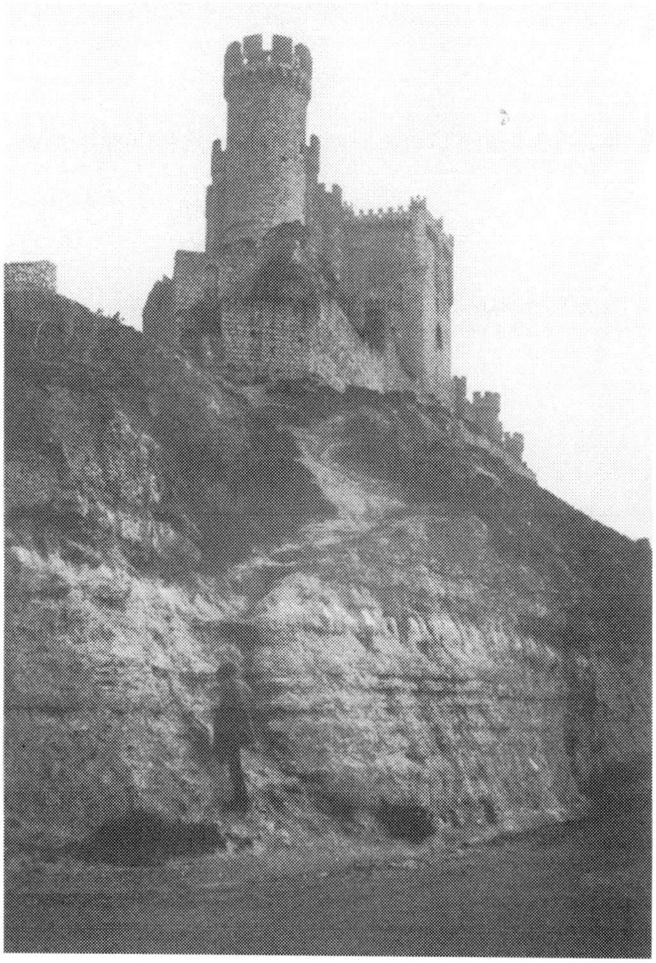

W1 The Duero Basin: a general overview

J.I. SANTISTEBAN, R. MEDIAVILLA, A. MARTÍN-SERRANO AND C.J. DABRIO

Abstract

The Duero basin occupies a large area in the north-west of the Iberian Peninsula. It has an approximately quadrangular shape, and three of its four corners are the sites of distinctive sub-basins that extend outwards from the main basin. The different margins of the sub-basins and the main basin tend to have distinctive histories of tectonic and sedimentary evolution.

Introduction

The Duero Basin is the largest Cenozoic basin in Spain with a surface area of almost 50 000 km^2. It occupies the major part of the north-west Iberian Peninsula. High-relief mountains composed of igneous and metamorphic rocks of Paleozoic age (mainly to the south and west) and siliciclastic and carbonate rocks of Mesozoic age (mainly to the east) bound the basin (Fig. 1). These borders formed during the Alpine Orogeny and played an important role in the geodynamic evolution of the basin.

The roughly quadrangular basin extends into three relatively narrow basins protruding near the corners (Fig. 1).

- The Ciudad Rodrigo Basin, in the south-western corner, is a half-graben oriented NE–SW that penetrates south-westerly into the Hercynic Massif. Its sedimentary record consists mainly of Paleogene deposits, although no Early Paleogene deposits have been found so far.
- The Almazán Basin is a complex area that extends to the east between the Iberian Range and the Central System. It was filled by siliciclastic and carbonate sediments, along with some rare evaporites, of Paleogene and Neogene age.
- La Bureba Corridor, in the north-eastern corner, is a narrow basin separating the Cantabrian Mountains from the Iberian Range. This Corridor acts as a linking area between the Ebro and Duero basins.

The Alpine structure of the Duero Basin

The Late-Hercynian structure of north-western Spain strongly influenced the structure of the Duero Basin. The main structural lineaments of the basement reacted under the new tectonic conditions imposed by the Alpine Orogeny, but new fault lines also appeared. However, the borders of the basin tended to evolve independently, and this fact is clear in the sedimentary record of the Duero Basin. It is necessary therefore to deal, at least briefly, with the structural features of the basin borders before describing the sedimentary record.

The present northern boundary consists of low-angle thrusts (a more detailed description can be seen in Chapter W5). These thrusts moved several kilometres towards the south, over the Duero Basin fill. They must be backthrust related to the subduction zone further north, where the Cantabrian Sea crust moves under the Iberian Plate (Boillot, 1984; Boillot & Malod, 1988). Sediments, ranging from Upper Cretaceous (Garumnian facies) to Oligocene, occur below these large thrusts. Neogene sediments onlap this structure.

The eastern border is a tectonic massif bounded by reverse faults with small horizontal displacements. Sediments affected by these faults range from Cretaceous to Oligocene in age. As in the northern border, Neogene sediments onlap the earlier structures.

Most of the southern boundary consists of reverse faults that affect Paleogene and Mesozoic sediments. Fault surfaces are relatively vertical in outcrop, but the fault dips decrease with depth. Neogene sediments usually onlap this border, but in some places they are affected by normal faults. There are also strike–slip faults along this border; their magnitude increases to the west.

The western border has mainly been passive, but some north to north-east trending faults acted during the Tertiary, modifying the disposition of the sediments.

The tectonic record of the internal parts of the Duero Basin is much poorer because Paleogene sediments do not crop out; thus, the record is limited to Neogene times. Here, the movements of basement faults decrease upwards and are hardly apparent at the surface. The only effect of the largest faults was to produce differential subsidence or to induce families of small faults.

Three major tectonic periods (Fig. 2) can be recognized in the

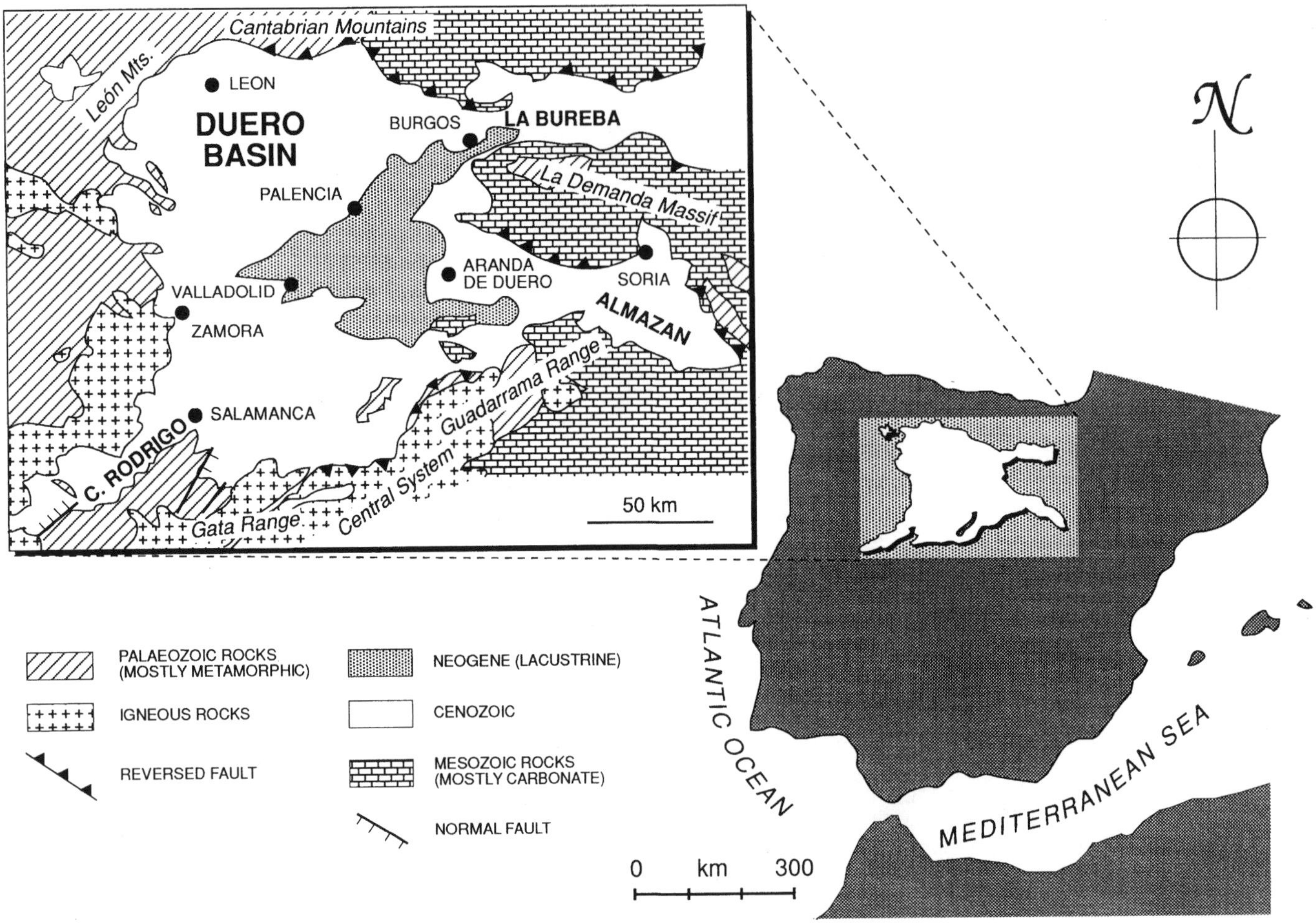

Fig. 1. Location map of the Duero Basin in the Iberian Peninsula.

Duero Basin. Activity during Upper Cretaceous to Paleocene times relates to the Mesozoic extensional regime. Activity during the Eocene and Oligocene reflects the compressional regime caused in Iberia by convergence of tectonic plates. The period from Miocene to Recent represents a new extensional stage with adjustments of the relief.

The sedimentary record of the Duero Basin

The stratigraphic framework of the Duero Basin has been studied and described since the nineteenth century (Botella, 1877, 1884; Dantín Cereceda, 1912; Ezquerra, 1837, 1845; among many others). The ideas of one of them, Eduardo Hernández-Pacheco, still remain the basis of present-day stratigraphical research. In the first third of the twentieth century, Hernández-Pacheco (1914, 1915, 1921, 1930) studied the stratigraphy, 'sedimentology' (of a rudimentary sort, but including the first description and interpretation of fossil point bar deposits published in the world), paleontology and tectonics of a large part of the basin.

Most of the sedimentary fill of the basin was deposited in terrestrial sedimentary environments. A great variety of lithofacies crop out.

Siliciclastic sediments are widespread. They range from gravels to muds, with a well-marked dependence on their source areas. The composition of sediments derived from the north and east is lithic, whereas those coming from the south and southwest are arkosic to lithic.

Carbonates are also well represented, but their largest volumes occur near the present basin centre.

Evaporites occur only in the eastern half of the basin, and mainly towards the centre and north-east.

One of the more remarkable results of previous works was the generation of an abundant nomenclature, rich in local names (for a general overview the reader is referred to Portero *et al.*, 1982, and Jiménez *et al.*, 1983). As a consequence, attempts to establish valid general stratigraphical frameworks for the whole basin were only partial, because it is very difficult to trace particular stratigraphical units across the basin. In general, the proposed frameworks rested upon lithostratigraphy, with little or no detailed tectonic analysis, and poor understanding of the importance of weathering processes. The resulting basin models showed a bull's-eye facies distribution,

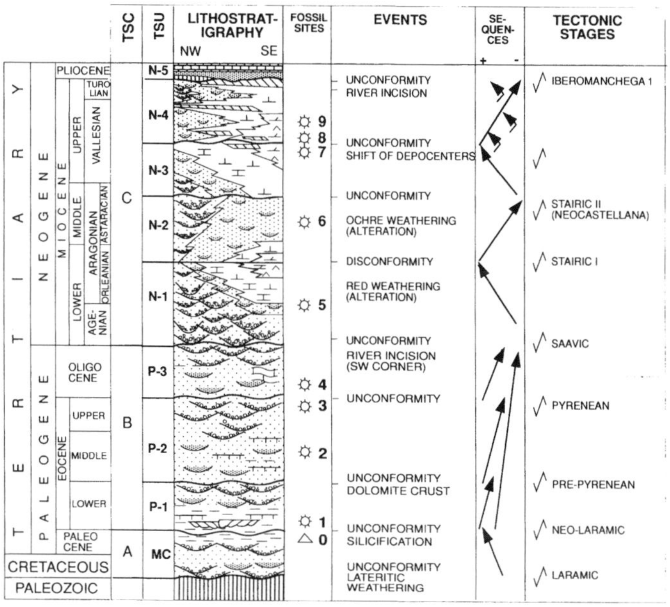

Fig. 2. Tertiary units of Duero Basin. *TSC*: tectonosedimentary complexes; *TSU*: tectonosedimentary units (*sensu* Megías, 1982). *Fossil sites*, 0: Absolute age (Kr/Ar) 58 Ma (Blanco *et al.*, 1982), 1: Sanzoles and Avedillo (Zamora), 2: Teso de la Flecha (Salamanca) and Corrales II (Zamora), 3: Molino del Pico and San Morales (Salamanca), 4: Camino Fuentes and El Molino (Ciudad Rodrigo Basin), 5: El Guijo (Salamanca), 6: Otero (Palencia), 7: Torremormojón (Palencia), 8: Los Valles de Fuentidueña (Segovia), 9: Torremormojón (Palencia). (Modified from Santisteban *et al.*, Chapter W2; Mediavilla *et al.*, Chapter W6).

with alluvial fans along the periphery and lacustrine environments in the centre. These models interpret the Duero Basin as an endorheic basin until Plio-Quaternary times.

A new era of research began, following systematic application of modern approaches to basin analyses. These studies include detailed work on tectonics, sedimentology, geomorphology and weathering profiles (García del Cura, 1974; Ordóñez *et al.*, 1980, 1981; Armenteros, 1986; Pozo, 1987; Martín-Serrano, 1988a, b, 1989, 1991; Mediavilla, 1985, 1986–87; Mediavilla & Dabrio, 1986, 1988; Mediavilla *et al.*, Chapter W6; Santisteban *et al.*, 1991a, 1991b, (Chapters W2 and W3; Bond, Chapter W4).

From a general point of view, we have divided the sedimentary record of the Duero Basin into three tectonosedimentary complexes (TSCs), composed, in their turn, of several tectonosedimentary units (TSUs *sensu* Megías, 1982). Each one of these complexes relates to a stage of basin evolution (Figs. 2 and 3):

TSC A, also referred to as the 'preorogenic complex', is of Upper Cretaceous to Paleocene age. This unit consists of siliciclastic, carbonate, and evaporitic deposits arranged in a fining-upwards (FU) sequence. Outcrops occur only in some areas along the margins of the basin, but they are not continuous laterally due to intense faulting. There are also outcrops in down-thrown fault blocks of the mountains surrounding the basin (Central System, Galicia basins; see Chapters W7 and W9). The deposits of TSC A usually occur in stratigraphical continuity with the Upper Cretaceous in the north, east and south-east borders. In other places (west and south-west borders) they rest unconformably upon a thick lateritic weathering profile that affects the Paleozoic basement. In these places, the rocks forming TSC A underwent a strong episode of silicification that increased towards the top. This TSC is interpreted as having been deposited in environments ranging from terrestrial (towards the west) to marine (towards the east): this pattern still reflects the Late Mesozoic palaeogeography.

TSC B, also referred to as the 'synorogenic complex', is Eocene to Oligocene in age. It consists of mainly siliciclastic sediments, along with scarce carbonates (except in the Almazán Basin where carbonates reach considerable thicknesses), which form a generally coarsening-upwards (CU) sequence. During this time preservation of weathering profiles was very poor. The deposits of TSC B form a fringe near the borders of the basin, where they rest unconformably upon both rocks of the TSC A and the pre-Tertiary basement. TSC B splits into several TSUs bounded by unconformities (an example is the S and W borders in Fig. 2). Some of these unconformities are of the progressive type (*sensu* Riba, 1976a, b), also referred to as cumulative wedging systems, related to movements of reverse faults or thrusts along the borders of the basin. Most of the deposits of TSC B were laid down in terrestrial environments (alluvial fan, fluvial); marine deposits occur only in the basin of Villarcayo, in the north-east corner (Montes *et al.*, 1989).

TSC C, also referred to as the 'postorogenic complex' (not implying lack of tectonic activity), is of Miocene to Recent age. It consists of siliciclastic, carbonate and evaporite deposits that form a fining-upwards (FU) sequence. It is best represented in the central and north-weastern parts of the basin, covering the previous deposits. TSC C consists of several TSUs (Fig. 2) which progressively onlap previous units and the borders. These rocks formed in terrestrial environments (alluvial fan, fluvial and lacustrine), which filled a basin with a shape roughly similar to the present Duero basin. Tectonic stability favoured the development of weathering profiles during the deposition of TSC C, both in the basin margins and the borders (red and ochre Mediterranean soils).

Gravel sheets (the so-called *raña*, plural *rañas*) covered large areas of the Hesperic Massif, the borders of the basin, and large areas of the basin. Many authors have considered these deposits as having time significance, but we must stress that they lack specific chronological meaning because 'raña' has formed in the basin at a number of different times since the Oligocene–Miocene, following episodes of fluvial incision (Martín Serrano, 1991).

Cenozoic evolution of the Duero Basin

During Mesozoic times, the area occupied at present by the Duero Basin was a marine and terrestrial area, open to the north and east, under an extensional regime. To the west and south, the neighbouring, emergent Hesperic Massif supplied sediments to these basins. The emergent Hesperic Massif underwent intense weathering under a tropical climate that generated lateritic profiles, tens of metres deep.

At the end of the Paleocene the compressional phase of the

TSC	AGE		NORTHERN BORDER 1	EASTERN BORDER 2	EASTERN BORDER 3	ALMAZAN BASIN 4		ALMAZAN BASIN 5	SOUTHERN BORDER 6	SOUTHERN BORDER 7	S.W. DUERO BASIN 8	BASIN CENTER 9
C	NEOGENE	Pliocene-Vallesian	Barrillos, Vidanes, Cegoñal, Guardo, Siliceous Aviñante, & Cantoral systems		Upper Nebreda, Retuerta, and Cuevaburgos systems	TSU T6	Jalón Group	TSU T5		Ochre Series	Ochre Series	TSU N5 TSU N4 TSU N3 TSU N2
		Astaracian	Modino, Quintana de la Peña, Puente Almuhey, Polygenic Aviñante, Upper Cuevas and Upper Candanedo systems	second cycle				TSU T4				
		Agenian-Orleanian				TSU T5				Red Series	Red Series	TSU N1
B	PALEOGENE	Oligocene	Lower Cuevas and Lower Candanedo systems			TSU T4	Henar Group	TSU T2			TSU P3	
		Upper Eocene		first cycle	Lower Nebreda, and Lower Covarrubias systems	TSU T3						
		Middle Eocene	Upper Vegaquemada complex		Rio Arlanza system	TSU T2		TSU T1	Polymictic sediments	Red Series	TSU P2	
		Lower Eocene						?			TSU P1	
A		Paleocene	Lower Vegaquemada complex			TSU T1			Siliciclastic sediments	Siliceous gravels, sands, and clays	TSU MC	
	Cretaceous											

Fig. 3. Tentative correlation of units in the Duero Basin resulting from the reinterpretation of data. 1: Colmenero *et al.*, 1982; García Ramos *et al.*, 1982; 2: Pol & Carballeira, 1982; 3: Pol & Carballeira, 1986; 4: Bond, W4; 5: Armenteros *et al.*, 1989; 6: Del Olmo & Martínez-Salanova, 1989; 7: Martín-Serrano & Del Olmo, 1990; Martín-Serrano, 1988b; 8: Santisteban *et al.*, Chapter W2; 9: Mediavilla *et al.*, Chapter W6.

Alpine Orogeny began causing uplift of the borders of the basin, and retreat of the marine environments towards the east and north-east. This phase induced progradation of alluvial systems towards the basin centre. Progressive uplifting along the margins of the basin caused the deformation of alluvial deposits. Major changes in paleogeography took place, changing from a smooth landscape, with small tectonic-induced highs, to a well-differentiated basin. Tectonic instability at this time prevented good development of weathering profiles. Tectonic movements resulted in the opening of several small basins in the Central System and Galicia, and the basin of Ciudad Rodrigo. All these basins share a similar record of sedimentation and weathering (see Chapters W3, W7 and W9).

The Early Neogene landscape of the basin was approximately similar to the present-day landscape. However, a long-lasting modification then began to take place, when the Atlantic fluvial network captured some of the endorheic fluvial systems in the south-eastern corner of the basin (Chapter W2). The resulting incision of these new exorheic rivers initiated the process of draining the basin and evacuating enormous volumes of sediment to the Atlantic Ocean, as the area connected to the exorheic drainage grew progressively larger.

In the meantime, subsidence related to faulting favoured the continuity of lacustrine deposition in the still endorheic central and north-eastern realms of the basin. Here, marginal alluvial fans fed fluvial systems connected to central lakes; tectonic stability in the basin (which does not imply absence or lack of tectonism or diastrophism) allowed the development and preservation of thick Mediterranean weathering profiles. The fill of the basin onlapped the eroded borders of the basin.

The coexistence of the ever-growing fluvial network in the south-western areas and the more-or-less restricted lacustrine realms on the opposite side, continued until the drainage of the whole basin was captured. At that point, the whole basin was connected to the Atlantic Ocean base level through the ancestral Duero River, and the last lacustrine environments had disappeared from the basin (see Chapter W6).

References

Armenteros, I. (1986). *Estratigrafía y sedimentología del Neógeno del sector suroriental de la Depresión del Duero.* Ediciones Diputación de Salamanca. Serie Castilla y León, **1**: 471 pp.

Armenteros, I., Dabrio, C.J., Guisado, R. and Sanchez de Vega, A. (1989). Megasecuencias sedimentarias del Terciario del borde oriental de la Cuenca de Almazán (Soria-Zaragoza). In C.J. Dabrio (ed.), *Paleogeografía de la Meseta Norte durante el Terciario. Stvd. Geol. Salmanticensia*, vol. **5**: 107–127.

Blanco, J.A., Corrochano, A., Montigny, R. and Thuizat, R. (1982). Sur l'age du debut de la sedimentation dans le bassin tertiaire du

Duero (Espagne). Attribution au Paléocène par datation isotopique des alunites de l'unité inferieure. *C. R. Acad. Sci. Paris*, **295** (II): 559–562.

Boillot, G. (1984). Some remarks on the continental margins in the Aquitaine and French Pyrenees. *Geol. Mag.*, **121** (5): 407–412.

Boillot, G. and Malod, J. (1988). The north and north-west Spanish continental margin: a review. *Rev. Soc. Geol. Esp.*, **1** (3–4): 295–316.

Botella, F. de (1877). España y sus antiguos mares. *Bol. Soc. Geogr.*, **II**. Madrid.

Botella, F. de (1884). Nota sobre la alimentación y desaparición de las grandes lagunas peninsulares. *Actas Soc. Esp. Hist. Nat.*, **XIII**: 79.

Colmenero, J.R., Garcia-Ramos, J.C., Manjon, M. and Vargas, I. (1982). Evolución de la sedimentación terciaria en el borde N. de la Cuenca del Duero entre los valles del Torio y Pisuerga (León–Palencia). *Temas Geol. Mineral.* IGME **VI**: 171–181.

Dantín Cereceda, J. (1912). Noticia del descubrimiento de restos de mastodonte y de otros mamíferos en el cerro del Cristo del Otero (Palencia). *Bol. Real Soc. Esp. Hist. Nat.*, **12**: 78–84.

Del Olmo, A. and Martínez-Salanova, J. (1989). El tránsito Cretácico–Terciario en la Sierra de Guadarrama y áreas próximas de las Cuencas del Duero y Tajo. In C.J. Dabrio (ed.), *Paleogeografía de la Meseta Norte durante el Terciario, Stvd. Geol. Salmanticensia*, vol. **5**: 55–69.

Ezquerra y del Bayo, J. (1837). Indicaciones geognósticas sobre las formaciones terciarias del centro de España. *An. de Minas*, **III**: 300–316. Madrid.

Ezquerra y del Bayo, J. (1845). Sobre los antiguos diques de la cuenca del Duero. *An. de Minas*. Madrid.

García del Cura, M.A. (1974). Estudio sedimentológico de los materiales terciarios de la zona centro-oriental de la Cuenca del Duero (Aranda de Duero). *Estud. Geol.*, **30**: 579–597.

García Ramos, J.C., Colmenero, J.R., Manjon, M. and Vargas, I. (1982). Modelo de sedimentación en los abanicos aluviales de clastos carbonatados del borde N. de la Cuenca del cuero. *Temas Geol. Mineral*, IGME. **VI**: 275–289.

Hernández-Pacheco, E. (1914). Los vertebrados terrestres del Mioceno de la Península Ibérica. *Mem. Real. Soc. Esp. Hist. Nat.*, **IX**: 443–488.

Hernández-Pacheco, E. (1915). *Geología y paleontología del Mioceno de Palencia.* Junta Ampl. Est. e Inv. Cientif. Comunicación de Inv. Paleont. y Prehist, no. 5. 295 pp.

Hernández-Pacheco, E. (1921). Descubrimientos paleontológicos en Palencia. Las tortugas fósiles gigantescas, *Ibérica*, 328–330. Tortosa.

Hernández-Pacheco, E. (1930). Sobre la extensión del Neógeno en el N. de la altiplanicie de Castilla la Vieja. *Bol. Real. Soc. Hist. Nat.*, **XXX**: 396–398.

Jiménez, E., Corrochano, A. and Alonso Gavilan, A. (1983). El Paleógeno de la Cuenca del Duero. In J.A. Comba (ed.), *Libro Jubilar J.M. Ríos, Vol. II. Geología de España*: 489–494. IGME, Madrid.

Martín-Serrano, A. (1988a). *El relieve del la región occidental zamorana. La evolución geomorfológica de un borde del Macizo Hespérico.* Instituto de Estudios Zamoranos Florián de Ocampo, Diputación de Zamora: 306 pp.

Martín-Serrano, A. (1988b). Sobre la transición Neógeno-Cuaternario en la Meseta. El papel morfodinámico de la raña. *II Congr. Geol. Esp.*, comun. **1**: 395–398.

Martín-Serrano, A. (1989). Características, rango, significado y correlación de las Series Ocres del borde occidental de la Cuenca del Duero. In C.J. Dabrio (ed.), *Paleogeografía de la Meseta Norte durante el Terciario, Stvd. Geol. Salmanticensia*, vol. **5**: 239–252.

Martín-Serrano, A. (1991). La definición y el encajamiento de la red fluvial actual sobre el Macizo Hespérico en el marco de su geodinámica alpina. *Rev. Soc. Geol. Esp.*, **4**: 337–351.

Martín-Serrano, A. and del Olmo, A. (1990). *Mapa Geológico de España a escala 1:50 000. Cartografía y Memoria del Mesozoico y Cenozoico del la Hoja no. 507, El Espinar.* ITGE-Servicio de Publicaciones del Ministerio de Industria y Energía, Madrid.

Mediavilla, R.M. (1985). *Estratigrafía y sedimentología del Neógeno de Palencia.* Tesis de Licenciatura. Dpto. Estratigrafía, Univ. Salamanca. 135 pp.

Mediavilla, R.M. (1986–87). Sedimentología de los yesos del Sector Central de la Depresión del Duero. *Acta Geol. Hisp.*, **21–22**: 35–44.

Mediavilla, R.M. and Dabrio, C.J. (1986). La sedimentación continental del Neógeno en el sector centro-septentrional de la depresion del Duero (provincia de Palencia). *Stvd. Geol. Salmanticensia*, **XXII**: 111–132.

Mediavilla, R.M. and Dabrio, C.J. (1988). Controles sedimentarios neógenos en la Depresión del Duero (sector central). *Rev. Soc. Esp.*, **1**: 187–196.

Megías, A.G. (1982). Introducción al análisis tectosedimentario: aplicación al estudio dinámico de cuencas. *Quinto Congreso Latinoamericano de Geología, Argentina. Actas*, **1**: 385–402.

Montes, J.J., Alonso Gavilan, A. and Dabrio, C.J. (1989). Estratigrafía y paleogeografía del Cretácico terminal-Paleógene del borde suroeste de la Cuenca de Villarcayo (Burgos). In C.J. Dabrio (ed.), *Paleogeografía de la Meseta Norte durante el Terciario, Stvd. Geol. Salmanticensia*, vol. **5**: 71–87.

Ordóñez, S., Garcia del Cura, M.A. and López Aguayo, F. (1980). Contribución al conocimiento de la Cuenca del Duero (Sector Roa–Baltanás). *Estud. Geol.*, **36**: 361–369.

Ordóñez, S., Garcia del Cura, M.A. and López Aguayo, F. (1981). Chemical carbonated sediments in continental basins: the Duero Basin. *Abstracts I.A.S. 2nd Eur. Mtg. Bologna*, 130–133.

Pol, C. and Carballeira, J. (1982). Las facies conglomeráticas terciarias de la región de Covarrubias (Burgos). *Temas Geol.Mineral. IGME*, **6** (2): 509–525.

Pol, C. and Carballeira, J. (1986). El sinclinal de Santo Domingo de Silos: Estratigrafía y paleogeografía de los sedimentos continentales (borde Este de la Cuenca del Duero). *Stvd. Geol. Salmanticensia*, **22**: 7–35.

Portero, J.M., Olmo, P., Ramírez, J. and Vargas, I. (1982). Síntesis del Terciario Continental de la Cuenca del Duero. *Temas Geol. Mineral. IGME*, **6** (1): 11–40.

Pozo, M. (1987). Mineralogía y sedimentología de la 'Facies de las Cuestas' en la zona central de la Cuenca del Duero: Génesis de sepiolita y paligorskita. Tesis Doctoral. Univ. Autón. Madrid. Fac. Ciencias. 536 pp.

Riba, O. (1976a). Syntectonic unconformities of the Alto Cardener, Spanish Pyrenees: a genetic interpretation. *Sed. Geol.*, **15**: 213–233.

Riba, O. (1976b). Tectogénese et sédimentation: deux modèles de discordances syntectoniques pyrénéenes. *Bull. Bur. Rech. Géol. Mineral.*, 2e sér., Sect. I, **4**: 383–401.

Santisteban, J.I., Martín-Serrano, A., Mediavilla, R. and Molina, E. (1991a). Introducción a la estratigrafía del Terciario del SO de la Cuenca del Duero. In J.A. Blanco; E. Molina & A. Martín-Serrano (eds.), *Alteraciones y paleoalteraciones en la morfología del oeste peninsular. Zócalo hercínico y cuencas terciarias.* Monogr. Soc. Esp. Geomorfol., **6**: 185–198.

Santisteban, J.I., Martín-Serrano, A. and Mediavilla, R. (1991b). El Paleógeno del sector suroccidental de la Cuenca del Duero: Nueva división estratigráfica y controles sobre su sedimentación. In F. Colombo (ed.), *Libro Homenaje a Oriol Riba, Acta Geol. Hisp* **26**(2): 133–148.

W2 Alpine tectonic framework of south-western Duero basin

J.I. SANTISTEBAN, R. MEDIAVILLA AND A. MARTÍN-SERRANO

Abstract

The tectonic activity in the south-western area of the Spanish Northern Meseta (Ciudad Rodrigo and Duero basins) during most of the Tertiary was determined by a transpressive regime that reactivated Hercynian to Late-Hercynian faults. The record of the Alpine Orogeny is complex because the sedimentary record indicates a compressive regime in the source areas coeval with the extensional to transpressive regime indicated by normal or strike–slip faults. This duality is due to the geotectonic position of this area between two compressive areas, the Cantabrian Range and the Central System, and the extensional Atlantic margin.

Introduction

The Duero basin is an intracontinental basin of cratonic type (*sensu* Sloss & Speed, 1974) bounded by mountain ranges that evolved relatively independently during the Tertiary (Fig. 1).

The northern border is the Cantabrian Mountains, made up of Mesozoic and Palaeozoic rocks affected by thrusts and low-angle reverse faults. Its history is related to the Alpine evolution of the Pyrenees.

The eastern border is the Iberian Range that extends between the Pyrenees and the Betics, the main Spanish compressive orogens.

The southern border is the Central System, bounded by high-angle reverse and strike–slip faults of Hercynian to Late Hercynian age, reactivated during Alpine Orogeny.

The western border is the Palaeozoic metasedimentary and igneous rocks of the western Spanish Meseta. It has a relatively passive tectonic history but was affected by the evolution of the Atlantic margin.

South-western border

The south-west corner of the Duero basin is at the junction of two tectonically different borders: one dominated by reverse and strike–slip faults (the southern edge), and the other dominated by vertical, low-magnitude movements (the western border). The morphological expression of the junction area is a half-graben oriented NE–SW, and filled with Paleogene and Neogene sediments: it is referred to as the *Ciudad Rodrigo Graben* ('Fosa de Ciudad Rodrigo').

The 'classic' relative chronology of alpine movements is based upon the assumption that the stratigraphic frameworks of the Duero and Ciudad Rodrigo basins are different. As a consequence, many authors consider that the palaeogeographic and tectonic evolution of these two basins was independent (Jiménez *et al.*, 1983; Corrochano & Carballeira, 1983).

However, detailed mapping by the present authors has revealed similar successions of Tertiary materials in the Duero and Ciudad Rodrigo basins (Fig. 2). This implies that they were connected during the Tertiary and underwent a common evolution (Santisteban *et al.*, 1991; see also Chapter W3).

The Alpine tectonics

Southern border

The southern border of the basin can be divided into two structural domains with different tectonic behaviour during the Alpine Orogeny: the Central System and a series of structures that will be referred to as the Border Massifs (Fig. 3).

The Central System

The evolution of the Central System has been explained in several ways: related to an intracontinental shear zone (Vegas *et al.*, 1986), as a rhombus-graben (Portero & Aznar, 1984), and related to thrust nappes or reverse faults (Warburton & Alvarez, 1989; Babín *et al.*, 1992; Vicente *et al.*, 1992). Diverse stages have been established for the Alpine Orogeny in the northern and southern margins of the Central System (Portero & Aznar, 1984; Vegas *et al.*, 1986; Capote *et al.*, 1990; Calvo *et al.*, 1991; Vicente *et al.*, 1992) (Fig. 3).

Capote *et al.* (1990) differentiated three faulting episodes or stages, and this is the most generally accepted division:

Iberian Stage: Mean horizontal compression N45–55E that ended with an almost radial distension with the same axis orientation. The age coincides with the

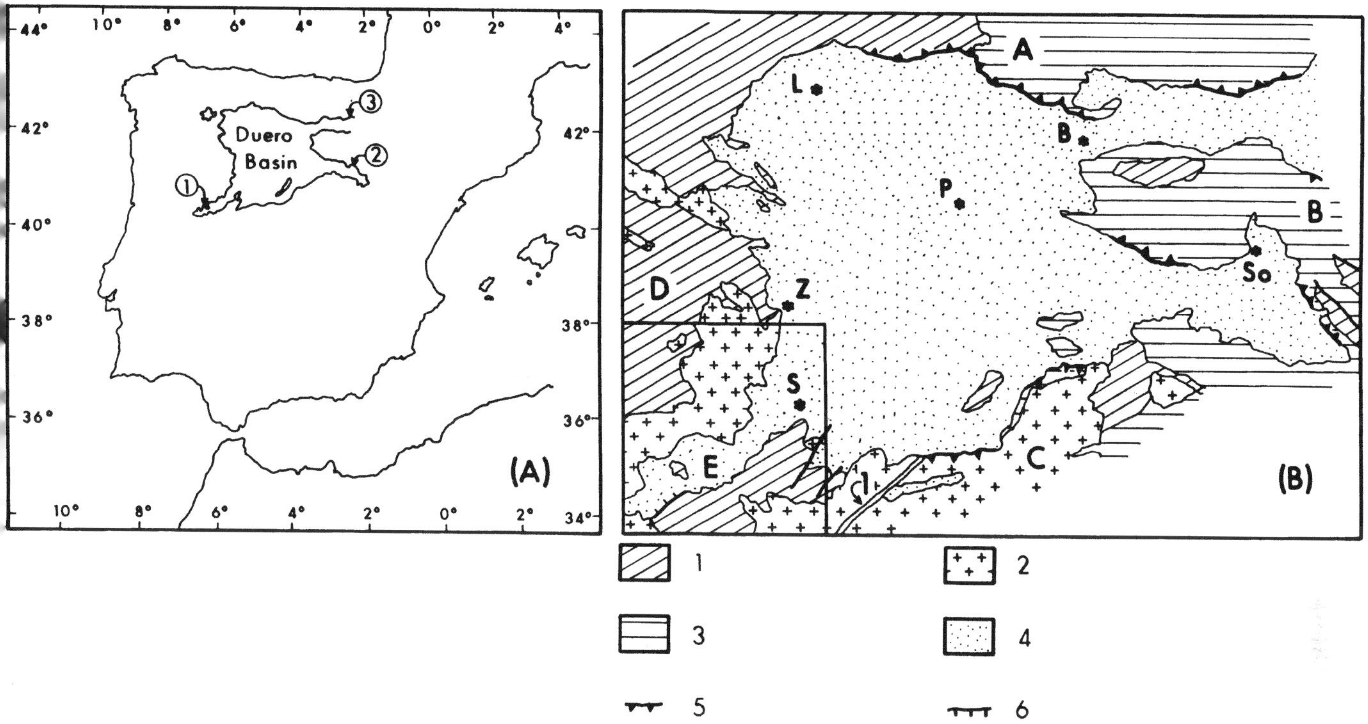

Fig. 1. A. Location map of Duero Basin in the Iberian Peninsula; 1: Ciudad Rodrigo Basin, 2: Almazán Basin, 3: La Bureba Corridor. B. Study area in Duero Basin; L: León, B: Burgos, P: Palencia, So: Soria, Z: Zamora, S: Salamanca; 1: Alentejo–Plasencia Fault, A: Cantabrian Range, B: Iberian Range, C: Central System, D: Western Border, E: Ciudad Rodrigo Basin. Key: Paleozoic, 1: metamorphic rocks; 2: igneous rocks; Mesozoic, 3: carbonates and siliciclastics; Cainozoic, 4: siliciclastics, carbonates and evaporites; Faults, 5: inverse fault; 6: normal fault.

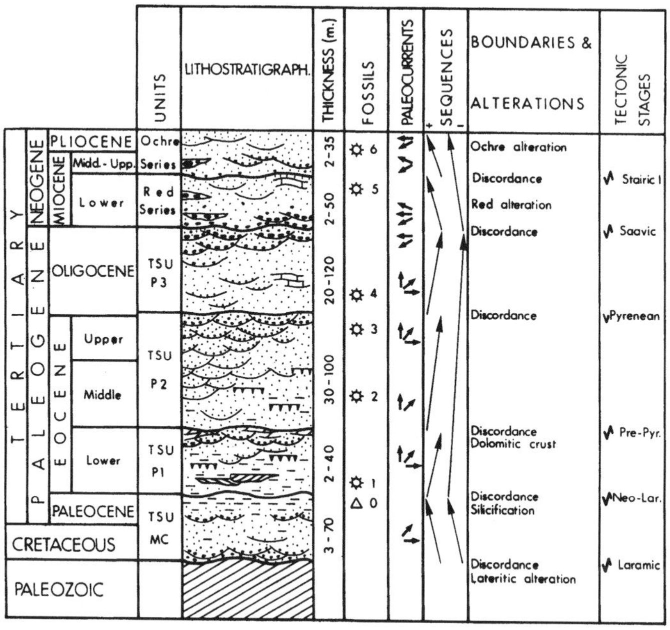

Fig. 2. Tertiary units of SW Duero Basin. Fossils, 0: Absolute age (Kr/Ar) 58 Ma (Blanco *et al.*, 1982), 1: Sanzoles and Avedillo (Zamora), 2: Teso de la Flecha (Salamanca) and Corrales II (Zamora), 3: Molino de Pico and San Morales (Salamanca), 4: Camino Fuentes and El Molino (Ciudad Rodrigo Basin), 5: El Guijo (Salamanca), 6: Benavente (Zamora). (Modified from Santisteban *et al.*, 1991.)

Oligocene–Early Miocene boundary, but movements affected Paleogene sedimentation more generally.

Guadarrama Stage: Maximum horizontal compression N140–155E that diminished with time. It took place in the Early–Late Miocene boundary (intra-Aragonian *sensu stricto*) and was responsible for the present reverse horst–graben structure.

Torrelaguna Stage: This was a minor phase with compression N160–200E, probably related to the previous one. Late Miocene to Quaternary.

The dates of these stages were deduced from the sediments of the closest basins affected by the faulting. This raises some doubts, particularly about the northern border of the Central System, because there is controversy concerning the age of sediments affected by the reverse faults of the Guadarrama Stage. Some authors (Corrales, 1982; Portero *et al.*, 1982; Corrochano *et al.*, 1983) consider these sediments as Early–Late Miocene, whereas others (Olmo & Martínez-Salanova, 1989; Santisteban *et al.*, 1991; see Chapter W3) consider them as Oligocene in age (Fig. 2). The last dating implies that the Iberian Stage was pre-Oligocene (possibly at the Eocene–Oligocene boundary, i.e. the Pyrenean phase of Brinkmann, 1931) and the Guadarrama Stage was Oligocene–Early Miocene (the Saavic phase of Brinkmann, 1931).

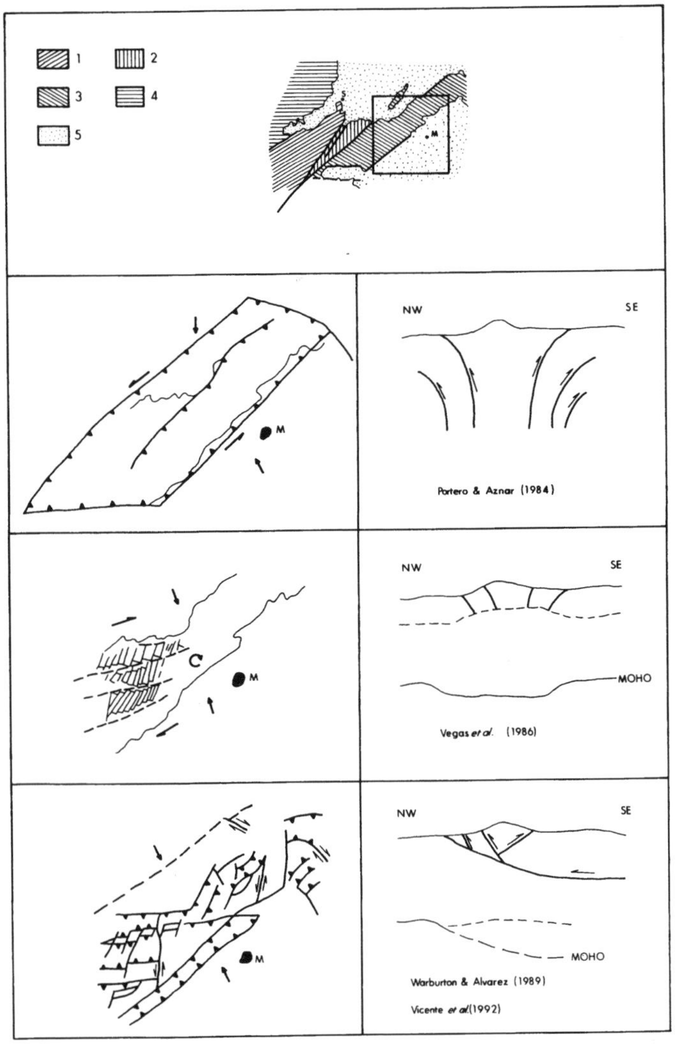

Fig. 3. Previous interpretations of the structural development of the Central System. The upper diagram shows the location of the main geologic zones: 1 – Border Massifs, 2 – transition zone, 3 – Central System (*sensu stricto*), 4 – Western Border, 5 – Tertiary deposits (Duero and Tajo basins). The lower three pairs of diagrams compare three different interpretations of the structural development of the area, with structural outline maps on the left, and cross-sections on the right (from Portero and Aznar, 1984; Vegas *et al.*, 1986; Warburton and Alvarez, 1989; and Vincente *et al.*, 1992).

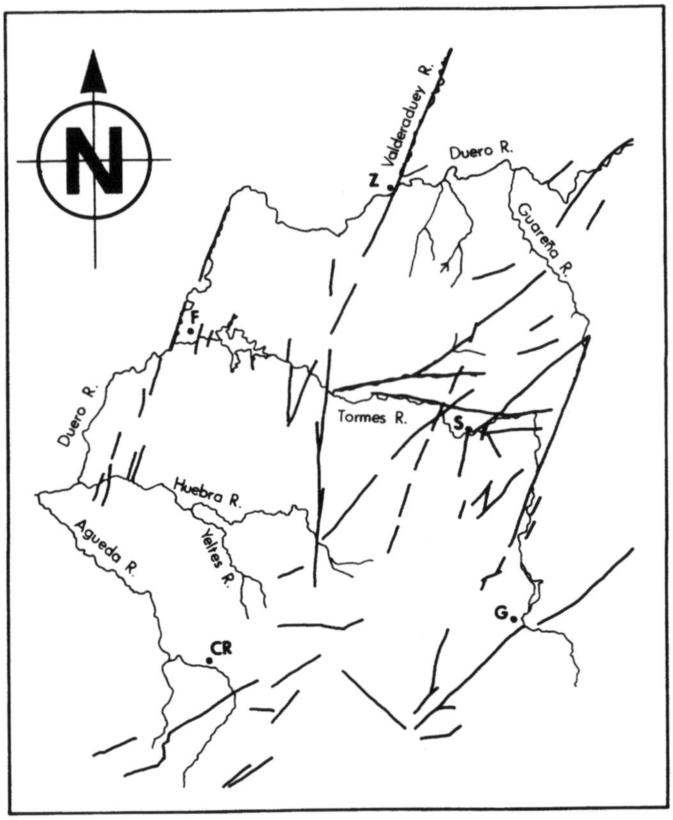

Fig. 4. Faulting sketch of Western Duero Basin deduced from field work and teledetection studies. Z: Zamora, F: Fermoselle, S: Salamanca, CR: Ciudad Rodrigo, G: Guijuelo. (From Santisteban *et al.*, in press.)

The Border Massifs

In the north-western side of the Central System a few Palaeozoic structural highs pertaining to the southern border of Duero basin have survived (Fig. 3). The structure of these massifs is quite different from the Central System *sensu stricto*.

They are bounded by normal or strike–slip faults with dominantly vertical movements and a configuration of horsts and grabens that extends NE–SW. These fault-blocks were horizontally displaced by faults trending NNE–SSW and they are also bounded by WNW–ESE faults (Fig. 4).

The border massifs preserve the best record of the alpine deformation of this area (Jiménez, 1972; Jiménez, 1973; Corrochano *et al.*, 1983), (Figs. 5 and 6A). Accordingly to Brinkmann's (1931) nomenclature the tectonic stages of this area are:

- Laramic phase (Late Cretaceous–Paleocene): faulting of basement affected by the Mesozoic lateritic weathering profile.
- Neo-Laramic phase (Paleocene–Eocene): high-angle faults (NNE–SSW, NE–SW and E–W) bring together Hercynian and Cretaceous–Paleocene rocks. Tilting of these sediments towards NE. There are normal, strike–slip and some, scarce, E–W reverse faults with small displacement.
- Pre-Pyrenean phase (Early–Middle Eocene): tilting and sinking of Lower Eocene sediments towards N and NE due to NE–SW and E–W normal and normal–strike–slip faults.
- Pyrenean phase (Upper Eocene–Oligocene): great reorganisation of the basin related to a stage of fault reactivation and major uplift of the borders of the basin. The horst and graben structure also affected the sedimentary basin. After this time these fracture areas are indicated by slight subsidence. The newly created structural highs were never covered (buried) by younger sediments.

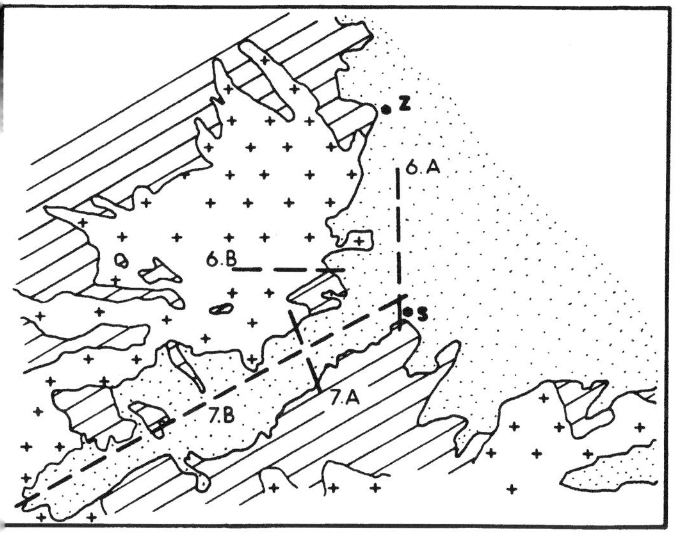

Fig. 5. Location of cross-sections in Figures 6 and 7.

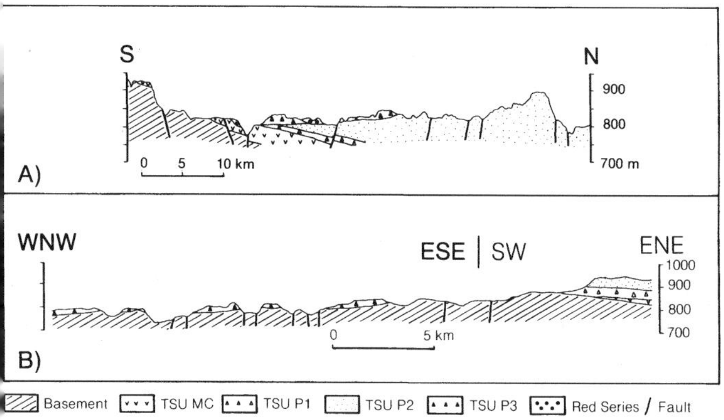

Fig. 6. A. S–N cross-section from Border Massifs towards Duero Basin. Note the horst located at N which serves as palaeo-threshold along the Oligocene and Neogene (modified from Santisteban *et al.*, 1991). B. W–E cross-section of Western Border; TSU P3 sediments are located in lower positions than previous units and are tilted towards the west in relation to NNE–SSW faults.

- Saavic phase (Oligocene–Early Miocene): movements of NE–SW and E–W normal faults that modified the basin extension and tilted previously defined blocks. NNE–SSW faults lowered blocks towards the west. Major uplift in the eastern and south-eastern areas generated a configuration very close to the present.
- Stairic phase (Early–Late Miocene, Pliocene): small extensional phenomena that lowered blocks towards the west.

The western border

This border has been considered inactive due to the scarcity of Tertiary deposits allowing the recognition of alpine movements, and the fragmentation of the old, plain landscape.

However, detailed study of small Tertiary sedimentary outcrops and weathering profiles has revealed at least three tectonic stages of post-Paleocene, pre-Oligocene and post-Oligocene ages. Related vertical displacements are about 100 m (Figs. 5 and 6B).

The first faulting stage affected igneous and metamorphic rocks with a superimposed lateritic weathering profile and silicification processes of Mesozoic age, related to the top of MC tectonosedimentary unit (MC TSU of Fig. 6), (Upper Cretaceous–Paleocene). Elsewhere, these faults are fossilised by the sediments of the P1 TSU (Lower Eocene); fault movements can be dated as Paleocene–Early Eocene. However, it may be argued that this is actually the result of a double faulting process (pre-Paleocene and Paleocene–Early Eocene).

Sediments of the younger P3 TSU (Oligocene) are located in topographically lower positions to the west of the previous units due to NNE–SSW and NE–SW fault systems. These structures extend to the Valderaduey faulting zone (Martín-Serrano, 1988). Igneous rocks often show S–C structures, related to these movements, that record normal displacements (Díez Montes, pers. commun., 1992). The distribution of sediments of the P3 TSU related to these faults, and the displacement of faults by other fault systems show that these movements are of pre-Oligocene age.

The last tectonic movements recorded here lowered blocks including Tertiary sediments towards the west, i.e. away from Valderaduey fracture zone. Two stages can be differentiated: a first subsidence of the sediments of P3 TSU towards the east, followed by rotation (tilting) of blocks and subsidence to the west. A minimum age cannot so far be given to these movements because of the absence of younger deposits. They are thought to be of post-Oligocene age.

The Ciudad Rodrigo Basin

This is a half-graben bounded to the south by a main NE–SW fault. In fact, this is not a single fault but a parallel system cut by a conjugate (secondary) NNE–SSW system that displaces the main system. There are also scarce NW–SE and WNW–ESE faults that displace the fault (Figs. 5, 7A and 7B). The basin border therefore has a complex structural history.

The high-angle dip of the fault planes makes it very difficult to determine the true components of movement. Gracia Plaza *et al.* (1981) and Jiménez & Martín-Izard (1987) described strike–slip components, whereas Alonso Gavilan & Polo (1986–87) found normal components. The accumulated vertical displacement amounts to 300 m (Jiménez & Martín-Izard, 1987). No reverse components have been found so far.

The first Alpine movements in the Ciudad Rodrigo Basin, supposed to be of the Laramic phase, caused domes trending NE–SW (Mingarro *et al.*, 1970). However, alpine faulting before the Eocene cannot be clearly identified due to the lack of previous sediments.

The Ciudad Rodrigo Basin was generated in the Early Eocene by the activity of the fault forming the southern boundary.

At the Eocene–Oligocene transition, new reactivation of faults

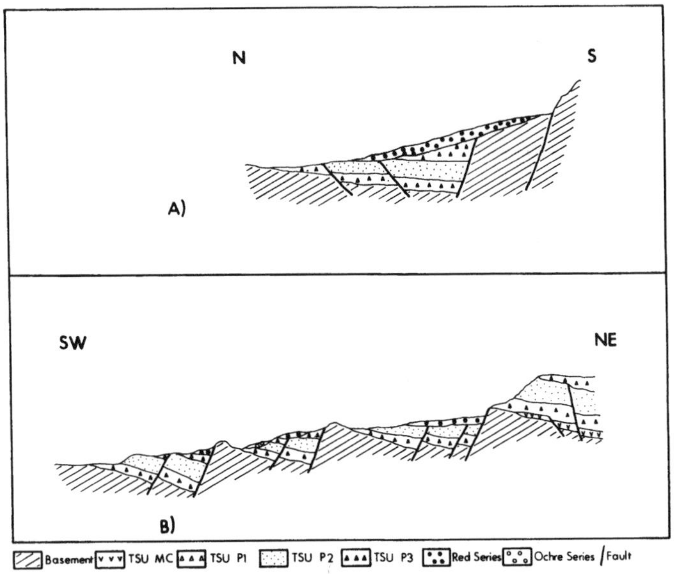

Fig. 7. Schematic cross-sections of the Ciudad Rodrigo Basin. A. N–S section showing the asymmetrical infill of the basin in relation to its southern border. B. SW–NE section showing the westward deepening of the basin. Fault dips are exaggerated.

lowered the northern blocks and the basin depocentres shifted towards the north.

A new reactivation at the Oligocene–Miocene boundary (the Saavic phase of Brinkmann, 1931) modified again the extent of the basin. Unlike other basins of the Hesperic Massif there was no sedimentary tectofacies development and sedimentation took place immediately after faulting. The basin deposits were not genetically related to the faults, which were buried rapidly after their movement.

Neogene alpine activity took place in the western and central areas of the basin. The main Neogene phase took place between the deposition of the Red (Lower Miocene) and Ochre Series (Middle Miocene–Pliocene). Then, a reactivation of NNE–SSW faults caused vertical movements of several tens of metres. Several faulting stages have been recorded in the eastern areas, where they break the top of the Red Series. The age of these movements cannot be established, but they show normal–dextral displacement (Gracia Plaza *et al.*, 1981).

Later movements into the basin acted during the Late Miocene–Pliocene (Moreno, 1991) and until the Quaternary. However, their importance and magnitude are less and they clearly indicate extension.

The Duero Basin

The Tertiary sediments of the Duero basin *sensu stricto* show features that indicate a close relationship between tectonics and sedimentation. These features are fractures and anomalous thicknesses of sediments related to buried fault systems (Figs. 5 and 6A).

Upper Cretaceous to Paleocene sedimentation took place in a low-relief landscape with irregular topography and a well-developed weathering profile.

A younger episode of faulting broke up this homogeneous pattern. This is observed only in the margins of the basin.

During Early–Middle Eocene times, the basin tilted towards the east and north-east. Although the surface expression of the faults was not strong, there were notable differences of subsidence related to deep faults.

In the transition from Late Eocene to Oligocene times a system of horst and graben was generated, and tilting to the east and north-east took place again.

A small fault episode, and the beginning of tilting to the west, were recorded at the Oligocene–Early Miocene boundary.

After that time, faulting took place that was related to almost radial extension. These were the intra-Miocene and Plio-Quaternary faulting episodes that were characterised by tectonic subsidence towards the west.

Tectonics and sedimentation

The evidence for all these tectonic movements was preserved in the stratigraphic framework of the basin. It is noteworthy that the climatic curve records increasing aridity, while sedimentary successions show a coarsening-upwards trend related to the progressive uplift of source areas. Similarly, changes of palaeogeography in successive units coincide with lines of possible tectonic origin. Also, areas of subsidence tend to be defined by lines parallel to the main faults.

Palaeodrainage patterns are most useful in interpreting the tectonic evolution. Channels tend to flow parallel to fault strikes but, when they flowed at right angles to fault lines, river deposits fossilised the faults. The geometry of the resulting units is almost tabular and this is considered as an indication of very limited tectonic activity. Thus, we consider that tectonic activity has been recorded as changes of palaeogeography related to subsidence or local faulting in the sedimentary basin. In contrast, the largest tectonic movements occurred near the source areas far away to the south.

Synthesis

Tectonic stages

From this work we can differentiate the following faulting episodes (according to Brinkmann's, 1931, nomenclature) (Fig. 2):

- Cretaceous–Paleocene: progressive uplift of the Hesperic Massif due to N–S compression (Laramic phase).
- Pre-Eocene: reactivation of NE–SW and NNE–SSW normal to strike–slip fault systems. ENE–WSW extension (Neolaramic phase).
- Early Eocene–Middle Eocene: fault favoured lowering towards the NE. Near radial extension (Pre-pyrenean phase).

- Late Eocene–Oligocene: generation of NE–SW horst–graben systems bounded by E–W faults and displaced by NNE–SSW ones. Major uplift of borders and source areas and reorganisation of the sedimentary basin. ENE–WSW extension (Pyrenean phase).
- Oligocene–Early Miocene: gentle sinking towards the north and west favoured by small slip normal faults. E–W extension (Saavic phase).
- Early Miocene–Middle Miocene: westward sinking favoured by N–S faults. E–W extension (Stairic I phase).
- Late Miocene–Recent: continued, pulsating, sinking towards the west. Near radial extension (Stairic II and following phases).

Geodynamic setting and evolutionary model

We propose an evolutionary model for this area of the Duero and Ciudad Rodrigo Basins as follows:

Uplift of the Hesperic Massif at the end of the Late Cretaceous, as a result of the convergence of Iberia and Eurasia. Since then until the Late Oligocene–Early Miocene the approach was recorded as compressive pulses (Paleocene–Early Eocene, Early Eocene–Middle Eocene, Late Eocene–Oligocene) directed NNE–SSW. These movements have been recorded in the sedimentary record as a coarsening-upwards macro sequence, composed of coarsening-upwards TSUs, reflecting the progressive uplift of the southern source areas.

The Iberian and Euroasiatic plates welded together in the Miocene and, since that time, behaved as a single plate. The compression due to the convergence of the African plate caused the last uplift and modification of the Central System and southern borders of the Duero Basin.

Extension has dominated in the basin since Miocene times, causing small changes such as little morphological scarps and modifications of the river drainage pattern.

This scheme covers all the main normal and strike–slip faults developed during the whole of the Tertiary. To understand this let us consider the position of the area in relation to the main Tertiary plate boundaries. Two areas of lithospheric convergence, the Pyrenees and the Betic Ranges, limited to the N and S the Iberian peninsula (Fig. 8). The western boundary was the divergence area of the Mid-Atlantic Ridge, whereas the eastern one was the compressive chain of the Iberian Range. This pattern generated an area of minimum compression to the west that underwent extension during most of Tertiary times (Proença Cunha & Pena dos Reis, 1992).

Another fact supports the different behaviour of the western and eastern areas: the eastern Central System was the locus of marine and coastal sedimentation during Late Mesozoic and Early Tertiary times whereas, at the same time, the Western Central System was an uplifted, terrestrial realm. According to this, the eastern areas experienced a more pronounced uplift during Alpine Orogeny than the western ones.

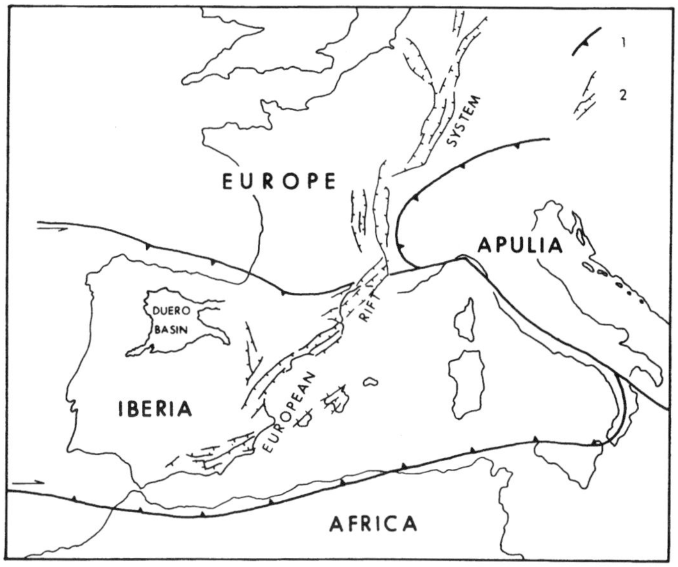

Fig. 8. Relative position of Iberian Peninsula and Duero Basin relative to main plate boundaries. N and S limits (Pyrenees and Betics) are compressive domains which acted during the Tertiary. This situation led to an W–E extensional regime in western Iberia.

According to the previous reasoning, we propose a hypothesis based on Simón Gómez (1984, 1990):

(a) Maximal compression occurs to the east of the basin (Pyrenees, Iberian and Betic Ranges). This might produce an arcuate deformation of the stress field so that western areas showed compression directions oblique to main faults. This could generate a transpressive regime.
(b) Changes in relation and/or direction of stress related to a crustal irregularity (like the Alentejo–Plascencia fault).
(c) Coeval compressive and extension fields. Extension prevailed in the study area. This hypothesis implies a change in stress relationships but does not require a change in stress direction.
(d) Thrust erosion simultaneous with its positioning (as proposed by Beaumont *et al.*, 1992) and passive behaviour of the Duero basin. Under these constraints, lithospheric overload produced a frontal furrow and a marginal ridge (dome) that favoured vertical instead of tangential movements. Forces acting on a faulted substratum reactivated older fault lines as 'normal' faults. In support of this hypothesis, geophysical data show a crustal thickening in the Central System and thinning towards the north-west (Martín Escorza, 1990; Babin *et al.*, 1992).

However, these are merely hypotheses and they now need to be tested by new studies.

Conclusion

The south-western area of the Spanish Northern Meseta (Ciudad Rodrigo and Duero basins) is characterised by tectonically active south and south-western boundaries and a relatively tectonically passive western border.

The tectonic activity in the area during most of the Tertiary was determined by a transpressive regime that reactivated Hercynian to Late-Hercynian faults; newly created faults are scarce. The main faulting stages have strike–slip to normal components. Brittle response of the crustal materials favoured faulting instead of folding. However, sedimentary units show a coarsening-upwards trend related to accelerated uplift of the source areas located to the south and south-east. This evidence indicates a compressive regime for areas located towards the east (Central System) during Paleogene times. Neogene deposition records extensional regimes.

The tectonic activity strongly changed the morphology and boundaries of this area generating and modifying systems of horsts and grabens.

There is a complex record of the Alpine Orogeny in the area, because the sedimentary record indicates a compressive regime in the source areas, coeval with an extensional to transpressive regime indicated by normal or strike–slip faults. This duality is due to the geotectonic position of this area between two compressive areas, the Cantabrian Range and the Central System, and the extensional Atlantic margin.

The south-western Duero Basin is considered to have been a moderately active area of cratonic type (*sensu* Sloss & Speed, 1974). It occupies an intermediate position between the largest areas of deformation of the Iberian Plate.

References

Alonso Gavilán, G. (1981). Estratigrafía y sedimentología del Paleógeno en el borde suroccidental de la Cuenca del Duero (Provincia de Salamanca). PhD Thesis, Salamanca Univ.: 435 pp.

Alonso Gavilán, G. and Cantano, M. (1987). La Formación Areniscas de Ciudad Rodrigo: ejemplo de sedimentación controlada por paleorrelieves (Eoceno, fosa de Ciudad Rodrigo). *Stvd. Geol. Salmanticensia*, **24**: 247–258.

Alonso Gavilán, G. & Polo, M.A. (1986–87). Evolución tecto-sedimentaria oligomiocénica del SO de la fosa de Ciudad Rodrigo. Salamanca. *Acta Geol. Hisp.*, **21–22**: 419–426.

Babín, R., Bergamín, J.F., Fernández Rodriguez, C., González Casado, J.M., Hernández Enrile, J.L., Rivas, A., Tejero, R. and Vicente, G. de (1992). Modelos gravimétricos para la corteza superior en el borde SE del Sistema Central Español. *Geogaceta*, **11**: 14–18.

Beaumont, C., Fullsack, P. and Hamilton, J. (1992). Erosional control on active compressional orogens. In K.R. McClay (ed.), *Thrust tectonics*: 1–18. London: Chapman & Hall.

Blanco, J.A., Corrochano, A., Montigny, R. and Thuizat, R. (1982). Sur l'age du debut de la sedimentation dans le bassin tertiaire du Duero (Espagne). Atribution au Paléocène par datation isotopique des alunites de l'unité inferieure. *C. R. Acad. Sci. Paris*, **295 (II)**: 559–562.

Brinkmann, R. (1931). Betikum und Keltiberikum im Sudostspanien. *Beitr. zur Geol. der West Mediterrangebiet*, **6**: 305–434. Berlín. Trad. J. Gómez de Llarena, Las Cadenas béticas y celtibéricas del Sureste de España. *Publ. Extr. Geol. Esp. C.S.I.C.*., **4**: 307–439.

Calvo, J.P., Vicente, G. de and Alonso Zarza, A.M. (1991). Correlación entre las deformaciones alpinas y la evolución del relleno sedimentario en la Cuenca de Madrid durante el Mioceno. *I Congr. Grupo Español del Terciario*, Comun.: 55–58.

Cantano, M. and Molina, E. (1987). Aproximación a la evolución morfológica de la 'Fosa de Ciudad Rodrigo'. Salamanca, España. *Bol. R. Soc. Hist. Nat. (Geol)*, **82** (1–4): 87–101.

Capote, R., Vicente, G. de and Gonzalez Cadado, J.M. (1990). Evolución de las deformaciones alpinas en el Sistema Central Español (S.C.E.). *Geogaceta*, **7**: 20–22.

Corrales, I. (1982). El Mioceno al sur del río Duero (sector occidental). *I Reunión sobre la Geología de la Cuenca del Duero, Salamanca 1979, Temas Geol. Mineral.*, **6**: 709–713.

Corrochano, A. (1977). Estratigrafía y sedimentología del Paleógeno de la provincia de Zamora. PhD. Thesis, Salamanca Univ., 336 pp. Unpublished.

Corrochano, A. and Carballeira, J. (1983). Las depresiones del borde suroccidental de la Cuenca del Duero. In J.A. Comba (ed.), *Libro Jubilar J.M. Ríos, Tomo II. Geología de España*: 513–521. IGME, Madrid.

Corrochano, A., Carballeira, J., Pol, C. and Corrales, I. (1983). Los sistemas deposicionales terciarios de la depresión de Peñaranda-Alba y sus relaciones con la fracturación. *Stvd. Geol. Salmanticensia*, **19**: 187–199.

Gracia Plaza, J.M., García Marcos, J.M. and Jiménez, E. (1981). Las fallas de 'El Cubito': Geometría, funcionamiento y sus implicaciones cronoestratigráficas en el Terciario de Salamanca. *Bol. Geol. Mineral.*, **92** (4): 267–273.

Jiménez, E. (1970). Estratigrafía y paleontología del borde sur-occidental de la Cuenca del Duero. PhD. Thesis, Salamanca Univ., 323 pp. Unpublished.

Jiménez, E. (1972). El Paleógeno del borde SW de la Cuenca del Duero. I: Los escarpes del Tormes. *Stvd. Geol. Salmanticensia*, **3**: 67–110.

Jiménez, E. (1973). El Paleógeno del borde SW de la Cuenca del Duero II: La falla de Alba-Villoria y sus implicaciones estratigráficas y geomorfológicas. *Stvd. Geol. Salmanticensia*, **5**: 107–136.

Jiménez, E. and Martin-Izard, A. (1987). Consideraciones sobre la edad del Paleógeno y la tectónica alpina del sector occidental de la cuenca de Ciudad Rodrigo. *Stvd. Geol. Salmanticensia*, **24**: 215–228.

Jiménez, E., Corrochano, A. and Alonso Gavilán, A. (1983). El Paleógeno de la Cuenca del Duero. In J.A. Comba (ed.), *Libro Jubilar J.M. Ríos, Tomo II. Geología de españa*: 489–494. IGME, Madrid.

Martín Escorza, C. (1990). Distensión–compresión en la cuenca de Campo Arañuelo. Implicación cortical. *Geogaceta*, **8**: 39–42.

Martín-Serrano, (1988). *El relieve de la región occidental zamorana. La evolución geomorfológica de un borde del Macizo Hespérico.* Instituto de Estudios Zamoranos Florián de Ocampo, Diputación de Zamora: 306 pp.

Mingarro, F., Mingarro, E. and López de Azcona, M.C. (1970). *Mapa Geológico de España E. 1:50 000, Hoja no. 500 (Villar de Ciervo).* Mapa y Memoria explicativa: 12 pp. IGME, Madrid.

Moreno, F. (1991). Superficies de erosión y tectónica neógena en el extremo occidental del Sistema Central Español. *Geogaceta*, **9**: 47–50.

Olmo, A. del and Martínez-Salanova, J. (1989). El tránsito Cretácico Terciario en la Sierra de Guadarrama y áreas próximas de las cuencas del Duero y Tajo. In C.J. Dabrio (ed.), *Paleogeografía de la Meseta Norte durante el Terciario, Stvd. Geol. Salmanticensia*, vol. **5**: 55–69.

Portero, J.M. and Aznar, J.M. (1984). Evolución morfotectónica y sedimentación en el Sistema Central y cuencas limítrofes (Duero y Tajo). *I Congr. Español Geol.*, **3**: 253–263.

Portero, J.M., Olmo, P. del, Ramírez del Pozo, J. and Vargas, I. (1982). Síntesis del Terciario continental de la Cuenca del Duero. *I Reunión sobre la Geología de la Cuenca del Duero, Salamanca 1979, Temas Geol. Mineral.*, **6**: 11–37.

Proença Cunha, P.M.R.R. and Pena dos Reis, R.P.B. (1992). Síntese da evoluçao geodinâmica e paleogeográfica do sector norte da Bacia Lusitânica, durante o Cretácico e Terciário. *III Congr. Geol. España y VIII Congr. Latinoamer. de Geol.*, Actas, **1**: 107–112.

Santisteban, J.I., Martín-Serrano, A. and Mediavilla, R. (1991). El Paleógeno del sector suroccidental de la Cuenca del Duero: Nueva división estratigráfica y controles sobre su sedimentación. In Colombo, F. (ed.), *Libro Homenaje a Oriol Riba, Acta Geol. Hisp.* **26** (2): 133–148

Simón Gómez, J.L. (1984). Compresión y distensión alpinas en la Cadena Iberica oriental. *Inst. Est. Turolenses*: 269 pp.

Simón Gómez, J.L. (1990). Algunas reflexiones sobre los modelos tectónicos aplicados a la Cordillera Ibérica. *Geogaceta*, **8**: 123–130.

Sloss, L.L. and Speed, R.C. (1974). Relationships of cratonic and continental-margin tectonic episodes. In Dickinson, W.R. (ed.), *Tectonics and sedimentation, S.E.P.M. Spec. Publ.*, **22**: 98–119.

Vegas, R., Vazquez, J.T. and Marcos, A. (1986). Tectónica alpina y morfogénesis en el Sistema Central Español: Modelo de deformación intracontinental distribuida. *Geogaceta*, **1**: 24–25.

Vicente, G. de, González Casado, J.M., Bergamín, J.F., Tejero, R., Babín, R., Rivas, A., H. Enrile, J.L., Giner, J., Sánchez Serrano, F., Muñoz, A. and Villamor, P. (1992). Alpine structure of the Spanish Central System. *III Congr. Geol. España y VIII Congr. Latinoamer. de Geol.*, Actas, **1**: 284–288.

Warburton, J. and Álvarez, C. (1989). A thrust tectonic interpretation of the Guadarrama Mountains, Spanish Central System. *Libro Homenaje a R. Soler. Mem. A.G.G.E.P.*: 147–155.

W3 South-western Duero and Ciudad Rodrigo basins: infill and dissection of a Tertiary basin

J.I. SANTISTEBAN, A. MARTÍN-SERRANO, R. MEDIAVILLA AND C.J. DABRIO

Abstract

In the south-western sector of the intracontinental Duero Basin, the post-Hercynian sedimentary record consists of Upper Cretaceous to Quaternary terrestrial sediments. Climates shifted from tropical, with poorly defined seasons (end of Cretaceous), to Mediterranean (Neogene). Tertiary deposits are divided into three tectonostratigraphic complexes. The Late Cretaceous–Paleocene, related to the end of the Mesozoic cycle, is characterised by a well-developed weathering profile that was eroded later. The Eocene–Oligocene, formed during the morpho-structural definition of the actual basin boundaries, consists of three unconformity-bounded units related to successive tectonic events of the Alpine Orogeny; by the end of this cycle, progressive incision of the Atlantic fluvial network led to capture of the fluvial systems of the southern Duero Basin and degradation (emptying) began. The Miocene–Pliocene, related to an extensional tectonic regime, represents the spreading of exorheic conditions to the whole basin that marked a complete hydrographic reorganisation. Deposition and aggradation continued in more central areas of the basin until the end of the Neogene, coeval with degradation of the south-western corner of the Duero Basin. The coexistence resulted from differential subsidence, hinge lines (uplift zones) separating sub-basins, and the dynamics of capture processes.

Introduction

The Duero Basin has been considered as an intracratonic basin (*sensu* Sloss & Speed, 1974). Its north, south and east margins are moderately tectonically active mountain ranges, whereas the western boundary is a relatively flat Hercynian border that remained essentially passive during Cainozoic times (Fig. 1). Outcrops of Paleogene sediments occur only at the edges of the basin, whereas Neogene deposits are best represented towards the inner parts of the basin. According to the classical ideas, the basin was filled by endorheic continental (alluvial and lacustrine) deposits of Tertiary age. Some areas underwent great subsidence as in the case of the eastern and southern edges where thicknesses reach up to 2000 m.

The south-western corner of the Duero basin can be described as the junction of the tectonically active southern margin and the passive western border. In this area there is a NE–SW-trending half-graben known as the *Ciudad Rodrigo Basin* (Fosa de Ciudad Rodrigo).

Previous studies (Alonso Gavilán & Cantano, 1987; Corrochano & Carballeira, 1983; Jiménez *et al.*, 1983) concluded that the stratigraphic frameworks and evolving palaeogeographies of the Duero and Ciudad Rodrigo Basins during the Paleogene were different (Fig. 2). The main support for this hypothesis was that outcrops of the same facies – and age – are at different topographic heights in the two basins. Also, the two basins are morphologically separated at present. The resulting model visualised two closed and isolated, independent, basins with endorheic, centripetal drainage patterns.

Recent investigation based upon detailed mapping reveals that the stratigraphic records in the Duero and Ciudad Rodrigo Basins are identical, and this strongly supports the conclusion that they were connected and followed a common evolution through the Tertiary (Santisteban *et al.*, 1991b; Martín-Serrano *et al.*, in press a, b, c; Martín-Serrano & Mediavilla, in press.; Martín-Serrano & Santisteban, in press, a, b). The new model consists of alluvial systems flowing through connected, tectonically configured areas, located some distance away from the areas under maximal deformation (see Chapter W2). In our interpretation, part of the topographic and geometric relationships between units are the result of the progressive capture of the fluvial network of the (south-western) Duero–Ciudad Rodrigo single basin by a fluvial pattern that drained towards the Atlantic. As a consequence, the former basin deposits were progressively eroded and evacuated to the west, with a new base-level.

According to Martín-Serrano *et al.*, (in press a, b, c), Martín-Serrano & Mediavilla (in press) and Martín-Serrano & Santisteban (in press, a, b) the stratigraphic succession common to the south-western Duero and Ciudad Rodrigo basins includes, in ascending order, several TSUs (tectonosedimentary units, *sensu* Megias, 1982), (Fig. 3):

– TSU **MC** = 'Siderolithic Unit' (*sensu* Millot, 1964) of Late Cretaceous–Paleocene age (Blanco *et al.*, 1982;

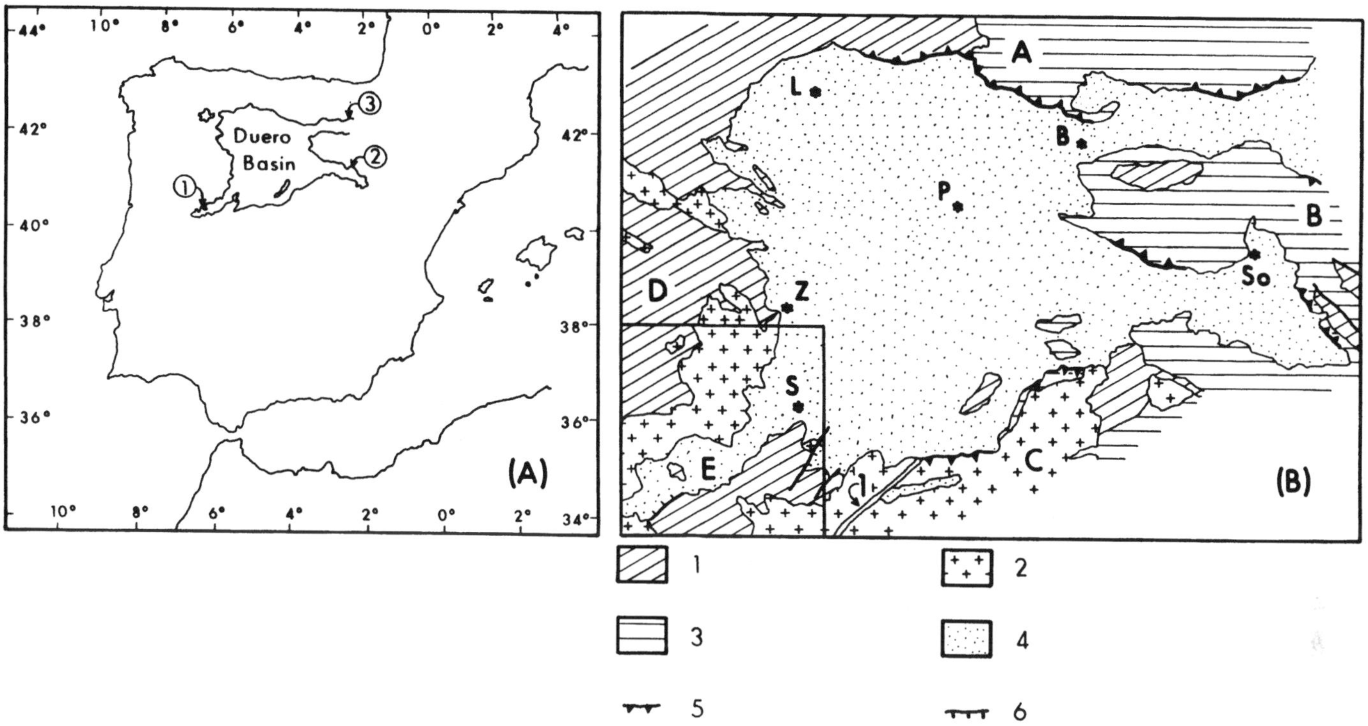

Fig. 1. A. Location map of the Duero Basin in the Iberian Peninsula; 1: Ciudad Rodrigo Basin, 2: Almazán Basin, 3: La Bureba Corridor. B. Study area; L. León, B: Burgos, P: Palencia, So: Soria, Z: Zamora, S: Salamanca; 1: Alentejo–Plasencia Fault, A: Cantabrian Range, B: Iberian Range, C: Central System, D: Western Border, E: Ciudad Rodrigo Basin. Key: Palaeozoic, 1: metamorphic rocks; 2: igneous rocks; Mesozoic, 3: carbonates and siliciclastics: Cainozoic, 4: siliciclastics, carbonates and evaporites; Faults, 5: reverse fault; 6: normal fault.

	SALAMANCA		ZAMORA		CIUDAD RODRIGO BASIN		
	1	2	3	4	5	6	7
PLIOCENE				Ochre			
MIOCENE Upper						Cabezuela	
MIOCENE Middle			Tierra de Campos Facies	Series		Conglomerates	
MIOCENE Lower	Cilloruelo Red Conglomerates	Armuña Conglomerates	Mirazamora Facies	Red Series			Variegated Conglomerates
OLIGOCENE	Molino del Pico Sandstones	Molino del Pico Sandstones	Upper Detritic Unit	Bellver Conglomerates & Sandstones (Upper Group)		Alamedilla Arkoses	Upper Arkosic Unit
	Mollorido Sandst.	Aldearrubia Sandstones					
	Aldearrubia Sandst.		Cubillos Limest.				
EOCENE	Cabrerizos Sandst.	Cabrerizos Sandstones	Clayey Unit	Yellow Silts (Lower Group)	Ciudad Rodrigo Series	Ciudad Rodrigo Formation	Lower Arkosic Unit
	Villamayor Sandst.				Tejoneras Series		
PALEOCENE	Río Almar Sandst.	Arapiles Conglom.	Zamora Facies	Zamora Facies			
	Salamanca Sandst.	Peña Celestina Mudst.					
CRETACEOUS	Amatos Sandst.	Terradillos Sandst.	Montamarta Facies	Montamarta Facies			
	Lower Conglomerate	Peña de Hierro Bed	Ferralitic Crust	Ferruginous Crust			

Fig. 2. Previous stratigraphic nomenclatures for Tertiary deposits of SW Duero Basin. 1: Jiménez (1970); 2: Alonso Gavilán (1981); 3: Corrochano (1977); 4: Martín-Serrano (1988); 5: Jiménez & Martín Izard (1987); 6: Alonso Gavilán & Polo (1986–87) and Alonso Gavilán & Cantano (1987); 7: Cantano & Molina (1987).

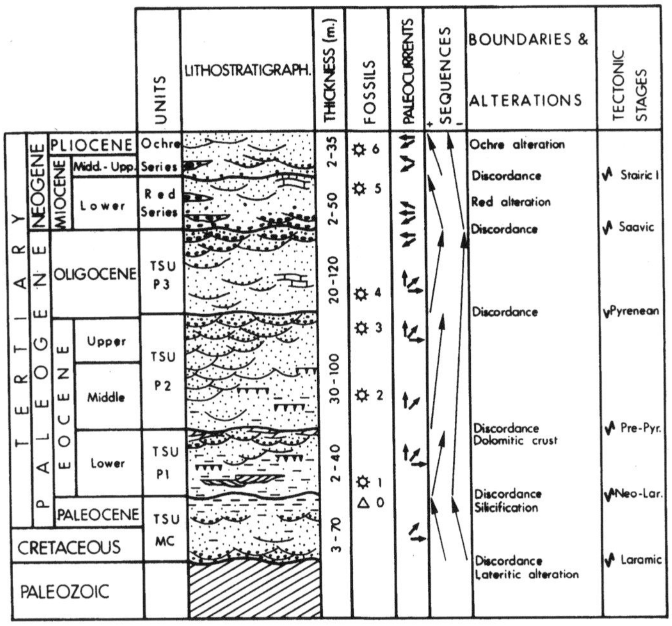

Fig. 3. Tertiary units of SW Duero Basin. Fossils, 0: Absolute age (Kr/Ar) 58 Ma (Blanco *et al.*, 1982), 1: Sanzoles and Avedillo (Zamora), 2: Teso de la Flecha (Salamanca) and Corrales II (Zamora), 3: Molino del Pico and San Morales (Salamanca), 4: Camino Fuentes and El Molino (Ciudad Rodrigo Basin), 5: El Guijo (Salamanca), 6: Benavente (Zamora). (Modified from Santisteban *et al.*, 1991b).

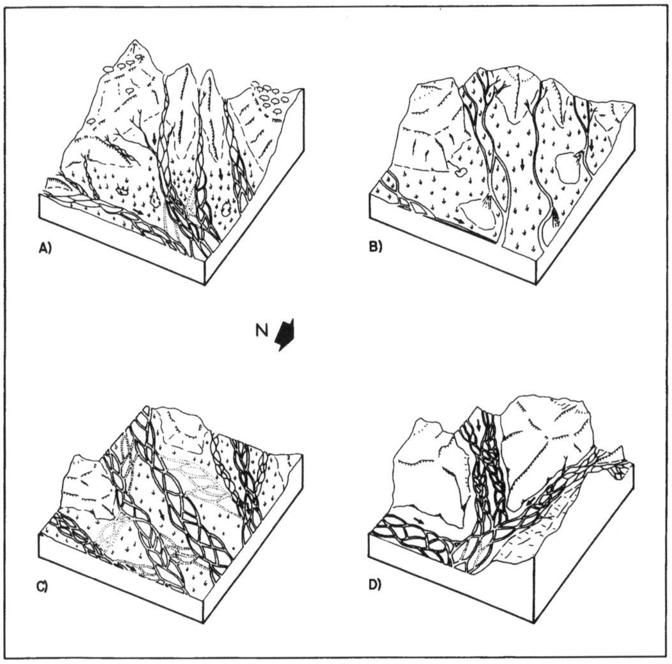

Fig. 4. Paleogeographical sketches for basin infill. A. Top of TSU MC; B. Middle of TSU P1; C. Middle of TSU P2; D. Bottom of TSU P3. Arrows: sense of flow.

Martín-Serrano, 1988; Molina *et al.*, 1989). This is not represented in the Ciudad Rodrigo Basin and is only recorded as a basal weathering profile in some outcrops.

- TSU **P1** (Santisteban *et al.*, 1991b) = 'Arkosic Unit' characterised by impregnations of a violet colour. Early Eocene.
- TSU **P2** (Santisteban *et al.*, 1991b) = 'Subarkosic–lithoarkosic Unit' characterised by clayey and/or carbonate cements. Middle–Late Eocene (Jiménez *et al.*, 1983).
- TSU **P3** (Santisteban *et al.*, 1991b) = 'Arkosic conglomerate Unit' with white–green colours. Oligocene (Polo *et al.*, 1987).
- Red Series (Martín-Serrano, 1988; Santisteban *et al.*, 1991a) = Conglomerate–clayey Unit, characterised by a reddish colour. Early Miocene (Mazo & Jiménez, 1982).
- Ochre Series (Martín-Serrano, 1988; Santisteban *et al.*, 1991a) = Conglomerate–sandy–clayey Unit, ochre in colour. It is poorly preserved in this area. Miocene–Pliocene.

Infill (aggradation)

There is only a poor record of Mesozoic sediments in the western border of the Duero Basin. This is very different from the eastern border. The Mesozoic is essentially represented by a well-developed lateritic profile located below the first post-Hercynian sediments (Bustillo & Martín-Serrano, 1980; Martín-Serrano, 1988; Molina *et al.*, 1989). The age of this profile is unknown but is thought to extend back to Mesozoic times.

The first post-Hercynian sediments of the south-western Duero Basin are terrestrial deposits laid down after the transgressive maximum of Latest Cretaceous age; they are related to the end-of-Cretaceous cycle. The ensuing regression, coupled with faulting to produce extensive terrestrial deposits in wide areas of the Hesperic Massif and high relief raised in the southern margin of the basin (Olmo & Martínez-Salanova, 1989). In general, the margins of the basin were quite different from the present, but in the western areas (the future Ciudad Rodrigo Basin) there have been no major changes. Here, sediments of braided systems (Fig. 4A) progressively buried the landscape of the late-Mesozoic weathering profile. Increased burial promoted a reduction of the braiding pattern and an increase in the flood plains. Low frequencies of avulsion and lateral shifting suggest continuous subsidence.

Later, in Early Eocene times, a faulting phase modified the basin margins creating the basin of Ciudad Rodrigo Basin as a half-graben connected to the larger Duero Basin. The uplift of the southern and western margins of the basin deeply changed the depocentres of the basin and the pathways of sediment movement. Rapid degradation of deeper horizons of the lateritic weathering profile initiated the deposition of arkoses by low braiding and moderate to high sinuosity fluvial systems favoured by the gentle topographic slopes (Fig. 4B). Low gradients and the development of swamps indicate high base levels.

The coarsening-upwards trend of the 'Arkosic Unit' records

greater uplift in the source areas; however, the succession is capped by a well-developed carbonate crust that records a long period of no sedimentation and erosion. At that time, faulting caused a tilting to the east, and depocentres shifted.

Middle–Late Eocene fluvial anastomosing to braided systems flowed to the east and north-east, down relatively steep slopes, but the base level was still high (Fig. 4C); this caused repeated sequences of channel fill and swamp deposits. Fluvial deposits buried the newly created reliefs.

In Latest Eocene times there was a 'major' tectonic phase in this border that meant the end of the former flat, open landscape in which the underlying unit was deposited. Faulting of basement and basin uplifted the basin margins and produced systems of horsts and grabens in the basin. These narrow valleys received an increasing volume of sediments as a consequence of the progressive uplift of the Central System at the south.

During Oligocene times, a tectonically delimited channel system flowed towards the east and north-east and joined the main drainage systems which flowed towards the N–NNE conditioned by a lower base level (Fig. 4D).

The coarsening-upwards trend observed in all these fluvial deposits (of both main and secondary systems of the network) records the increasing uplift of the source areas. The lack of proximal alluvial fan deposits in the area can be explained in terms of increased distance from maximum uplift. However, to the east (e.g. Ojos Albos Massif, Avila), coeval sediments are involved in reverse faulting and overlain by igneous rocks of the Central System.

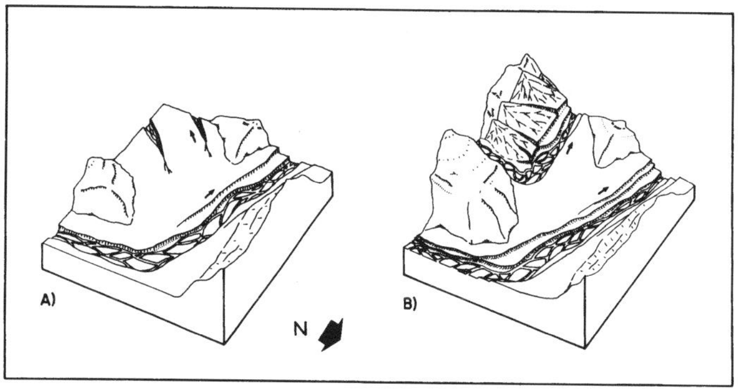

Fig. 5. Palaeogeographical sketches for basin emptying. A. Top of TSU P3; B. Red Series and Ochre Series. Arrows as in the previous figure.

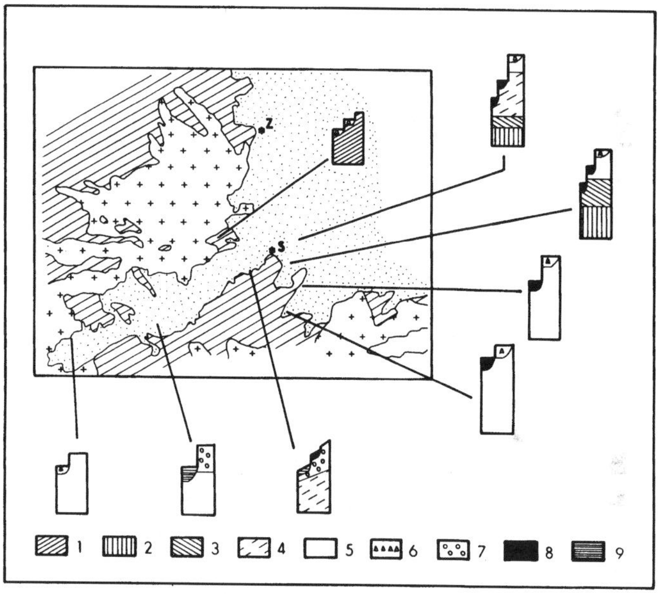

Fig. 6. Schematic logs of the Tertiary record showing relative position relationships between units. 1: Basement, 2: TSU MC, 3: TSU P1, 4: TSU P2, 5: TSU P3 (infilling), 6: TSU P3 (emptying), 7: Red Series (infilling), 8: Red Series (emptying), 9: Ochre Series (emptying).

Emptying (Degradation)

From Late Oligocene–Early Miocene onwards the stratigraphic record shows peculiar features. These and younger units consist of fluvial deposits, with palaeocurrents pointing to the west, which occur as almost tabular lithosomes. The bases of these bodies of sediment are highly erosive bases, but their upper surfaces are flat and determine the present geomorphology. The bodies occur at progressively lower topographic heights, and they are closely related to present river valleys or water divides. All these features indicate that these are old fluvial terraces (Fig. 5A). Alluvial fan deposits of Early Miocene to Pliocene age also occur with this topographic distribution, and in their distal parts they occur at progressively lower positions, forming alluvial terraces (Fig. 5B). In other cases it has been possible to correlate surfaces of erosion or weathering with these sedimentary surfaces by means of their topographic position or the mineralogy and petrology of the weathering profiles.

The age of these terraces is a most important discussion point. The lithology of these terraces and the Tertiary sediments of the Duero Basin are identical and a lithologic correlation is easily traced. This is supported by the parallel evolution of the mineralogy and petrology of these two areas: arkoses below the older surfaces, red sediments below the intermediate surfaces, and ochre sediments related to the youngest surfaces. There is also palaeontological evidence of the Early Miocene age (Mazo & Jimenez, 1982) of the Red Series of an intermediate terrace in the Ciudad Rodrigo Basin.

According to these arguments the definition of the fluvial network and the incision began, most probably, as early as Oligocene–Lower Miocene boundary times (older platforms with arkosic sediments of TSU **P3**). However, the incision process was not synchronous in the whole basin and the resulting terraces are diachronous in age (Martín-Serrano, 1988b, 1991). In some areas, the uppermost terrace is Oligocene–Early Miocene boundary in age (upper part of arkoses of **P3** TSU), whereas in others, the uppermost terrace consists of red sediments of the Red Series (Lower Miocene) or even the Upper Miocene to Pliocene Ochre Series (Mediavilla & Martín-Serrano, 1989) (Fig. 6). The distribution of terraces has been modified by a complex interaction of vertical displacements of base level (which strongly affected the lower

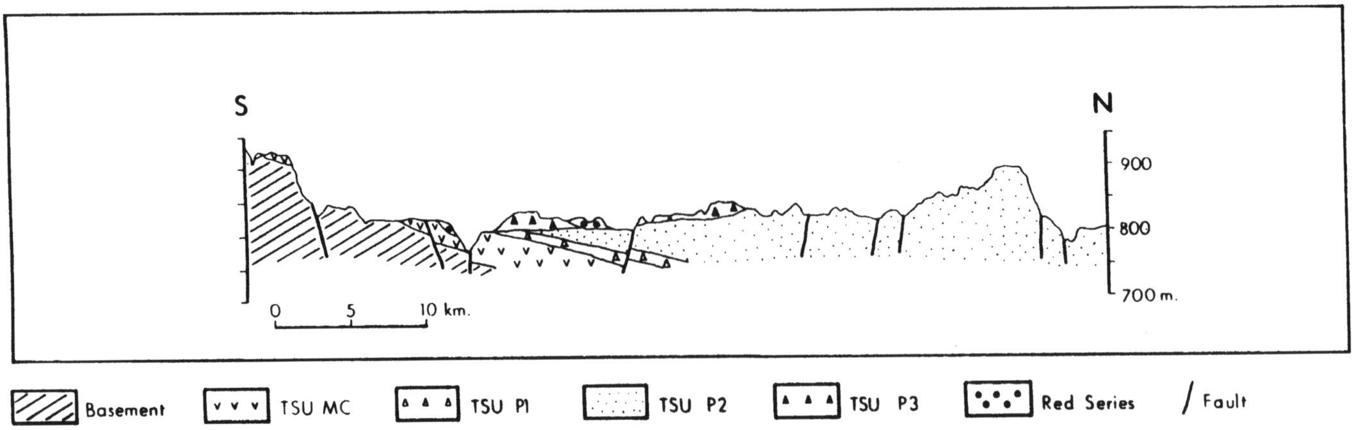

Fig. 7. South to north cross section showing relationship of units. Note the high relief located at north (near 900 m above sea level) that behaves as a threshold along Oligocene and Neogene. Moreover, TSU P3 and Red Series deposits are progressively at lower heights. (Modified from Santisteban *et al.*, 1991b.)

reaches of the streams), tectonic movements that usually modified the upper reaches of the streams, and processes of fluvial capture active in any area and age through Oligocene to Pliocene times.

Infill versus emptying

During the generation of the terraces in the Ciudad Rodrigo and south-western Duero Basins, a thick succession of lacustrine and fluvio-lacustrine sediments was deposited in the central parts of the Duero Basin (see Chapter W6). Coeval aggradation and degradation in one and the same basin can be explained by considering the palaeogeography of the basin at that time.

After the Late Eocene, the Pyrenean tectonic phase (Brinkmann, 1931) deeply changed the palaeogeography of the basin due to uplift of the northern, southern and eastern borders of the Duero Basin, and generation of horst-graben systems in the basin that divided some areas into smaller basins. Some of these new basins remained connected to the main basin but others evolved independently.

An example of these hinge lines or palaeogeographical thresholds is preserved as an east–west palaeo-hill, near the border between the provinces of Salamanca and Zamora (Fig. 7). At present, this uplift zone rises to 900 m acting as a divider of the hydrographic basins of the Tormes and Duero rivers. During the Oligocene and Miocene this palaeo-range separated two areas of different behaviour: the Duero and Tormes basins. The Oligocene arkoses of TSU **P3** are not represented north of these hills (in the Duero Basin, Province of Zamora), because this area was an uplifted massif from Eocene–Oligocene boundary times. Oligocene arkoses accumulated towards the south and east of this palaeo-hill but never buried it. After the Oligocene sedimentation, arkoses and palaeo-hills modified the boundaries of the sedimentary basin. During the Neogene, sedimentation took place towards the north of the palaeo-hills onlapping the arkoses, whereas to the south they formed terraces.

By the beginning of the Neogene, the fluvial systems of the southern area were captured by the Portuguese fluvial network draining to the west where the Atlantic Ocean offered a much lower base level. As rates of river incision largely surpassed the low rates of subsidence in this area there was an overall development of fluvial terraces. Larger subsidence rates in the more internal areas of the Duero Basin and hinge lines (uplift zones) prevented, but only temporarily, the capture of the whole Duero Basin. However, captures progressed to the east and north as more hinge lines were eroded and cut open. This, together with the progressive filling of the remaining Duero Basin, eventually resulted in the capture of the rest of the Duero fluvial network. The former endorheic Duero Basin turned into a generally exorheic basin by the end of the Neogene (Chapter W6).

The difference in the development of the exorheic regime is recorded in the present river profiles. Rivers of the south-western area (Huebra, Agueda and Tormes) join the main course of the Duero River between 200 and 400 m above sea level, whereas the younger rivers (Esla, Valderaduey, Guareña, Pisuerga, Adaja, Cega and Duratón) join it between 600 and 800 m above sea level. The differences in heights reflects the differences in age at which these rivers began their respective incisions. Moreover, the (older) southern rivers join the Duero downstream from a convex reach showing the present erosive position of the Duero River whereas the northern (younger) ones join the main course upstream of this reach. The history of the Duero is very complex and the present river courses are the result of several captures that took place during the Tertiary.

Conclusions

The Duero Basin is an intracontinental basin bounded by tectonically moderately active borders. In the south-western sector (including the Ciudad Rodrigo Basin), the post-Hercynian sedimentary record consists of terrestrial sediments of Late Cretaceous to Quaternary age.

The stratigraphic record of the Tertiary rocks is arranged in six TSUs (*sensu* Megías, 1982), which are grouped into three main tectonostratigraphic complexes:

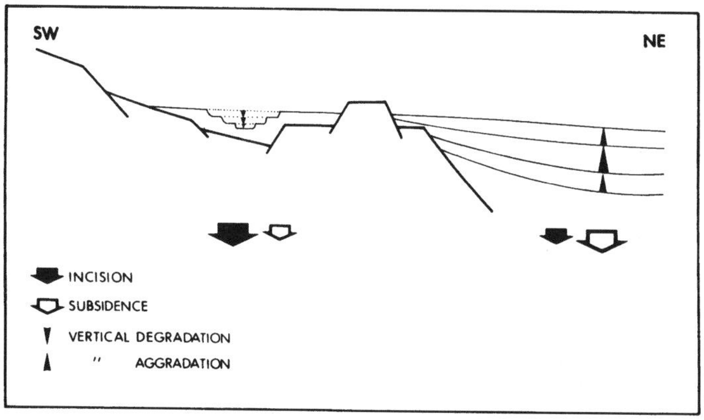

Fig. 8. Hypothetical sketch showing relationships between simultaneous basin infill and emptying. Differences in subsidence rates and paleohighs prevent complete basin drying.

- Late Cretaceous–Paleocene: related to the end of the Mesozoic cycle.
- Eocene–Oligocene: related to the morpho-structural definition of the actual basin boundaries.
- Miocene–Pliocene: tectonic and hydrographic reorganisation of the basin.

Climates shifted from tropical, with poorly defined seasons (end of Cretaceous), to Mediterranean (Neogene).

The base of the oldest TSU is characterised by an important weathering profile that was eroded afterwards. This profile records a long period of stability and subaerial exposure.

The second tectonosedimentary complex shows a coarsening-upwards trend that records the progressive uplift of the borders of the basin and the generation of marginal basins (connected to the Duero Basin). At that time, the instability of the basin was great enough to prevent the development of weathering profiles. The cycle is divided into three unconformity-bounded units (TSUs), related to the successive tectonic events of the Alpine Orogeny that were responsible for changes in basin palaeogeography. By the end of this cycle and the beginning of the next one the progressive incision of the fluvial network and degradation had begun.

The third complex records a new period of stability with development of weathering processes both in the basin and the neighbouring areas. The exorheic conditions extended to the whole basin. The tectonic regime was extensional.

Coeval with the degradational character of the south-western corner of the Duero Basin, deposition and aggradation continued in more central areas of the basin. The different behaviours reflect differences in subsidence, hinge lines (uplift zones) separating sub-basins, and the dynamics of the capture process that allowed the survival of an endorheic regime in the central areas of the Duero Basin until the end of the Neogene.

References

Alonso Gavilán, G. (1981). Estratigrafía y sedimentología del Paleógeno en el borde suroccidental de la Cuenca del Duero (Provincia de Salamanca). Doctoral Thesis, Salamanca Univ., 435 pp. Unpublished.

Alonso Gavilán, G. and Cantano, M. (1987). La Formación Areniscas de Ciudad Rodrigo: Ejemplo de sedimentación controlada por paleorrelieves (Eoceno, fosa de Ciudad Rodrigo). *Stvd. Geol. Salmanticensia*, **24**: 247–258.

Alonso Gavilán, G. and Polo, M.A. (1986–87). Evolución tecto-sedimentaria oligo-miocénica del SO de la fosa de Ciudad Rodrigo. Salamanca. *Acta Geol. Hispànica*, **21–22**: 419–426.

Blanco, J.A., Corrochano, A., Montigny, R. and Thuizat, R. (1982). Sur l'age du debut de la sedimentation dans le bassin tertiaire du Duero (Espagne). Attribution au Paléocène par datation isotopique des alunites de l'unité inferieure. *C. R. Acad. Sci. Paris*, **295** (II): 559–562.

Brinkmann, R. (1931). Betikum und Keltiberikum im Sudostspanien. *Beitr. zur Geol. der West Mediterrangebiet*, **6**: 305–434. Berlín. Trad. J. Gómez de Llarena, Las Cadenas béticas y celtibéricas del Sureste de España. *Publ. Extr. Geol. España C.S.I.C.*, **4**: 307–439.

Bustillo, M.A. and Martín-Serrano, A. (1980). Caracterización y significado de las rocas silíceas y ferruginosas del Paleoceno de Zamora. *Tecniterrae*, **36**: 1–16.

Cantano, M. and Molina, E. (1987). Aproximación a la evolución morfológica de la 'Fosa de Ciudad Rodrigo'. Salamanca, España. *Bol. R. Soc. Hist. Natl. (Geol)*, **82** (1–4): 87–101.

Corrochano, A. (1977). Estratigrafía y sedimentología del Paleógeno de la provincia de Zamora. Doctoral Thesis, Salamanca Univ., 336 pp. Unpublished.

Corrochano, A. and Carballeira, J. (1983). Las depresiones del borde suroccidental de la Cuenca del Duero. In J.A. Comba (coord.), *Libro Jubilar J.M. Ríos, Tomo II. Geología de España*: 513–521. IGME, Madrid.

Jiménez, E. (1970). Estratigrafía y paleontología del borde sur-occidental de la Cuenca del Duero. Doctoral Thesis, Salamanca Univ., 323 pp. Unpublished.

Jiménez, E. and Martín-Izard, A. (1987). Consideraciones sobre la edad del Paleógeno y la tectónica alpina del sector occidental de la Cuenca de Ciudad Rodrigo. *Stvd. Geol. Salmanticensia*, **24**: 215–228.

Jiménez, E., Corrochano, A. and Alonso Gavilán, A. (1983). El Paleógeno de la Cuenca del Duero. In J.A. Comba (coord.), *Libro Jubilar J.M. Ríos. Tomo II. Geología de España*: 489–494. IGME, Madrid.

Martín-Serrano, A. (1988a). *El relieve de la región occidental zamorana. La evolución geomorfológica de un borde del Macizo Hespérico.* Instituto de Estudios Zamoranos Florián de Ocampo, Diputación de Zamora: 306 pp.

Martín-Serrano, A. (1988b). Sobre la transición Neógeno-Cuaternario en la Meseta. El papel morfodinámico de la raña. *II Congr. Geol. España*, comun. **1**: 395–398.

Martín-Serrano, A. (1991). La definición y el encajamiento de la red fluvial actual sobre el Macizo Hespérico en el marco de su geodinámica alpina. *Rev. Soc. Geol. Esp*, **4**: 337–351.

Martín-Serrano, A. and Mediavilla, R. (in press). *Mapa y Memoria explicativa de la Hoja de Guijuelo*. ITGE-Servicio de Publicaciones Ministerio Industria y Energía, Madrid.

Martín-Serrano, A., Mediavilla, R. and Santisteban, J.I. (in press a). *La Fuente de San Esteban*. ITGE-Servicio de Publicaciones Ministerio Industria y Energía, Madrid.

Martín-Serrano, A., Santisteban, J.I. and Mediavilla, R. (in press b). *Mapa y Memoria explicativa de la Hoja de Barbadillo*. ITGE-Servicio de Publicaciones Ministerio Industria y Energía, Madrid.

Martín-Serrano, A., Santisteban, J.I. and Mediavilla, R. (in press c). *Mapa y Memoria explicativa de la Hoja de Matilla de los Caños*. ITGE-Servicio de Publicaciones Ministerio Industria y Energía, Madrid.

Martín-Serrano, A. and Santisteban, J.I. (in press a). *Mapa y Memoria explicativa de la Hoja de Salamanca.* ITGE-Servicio de Publicaciones Ministerio Industria y Energía, Madrid.

Martín-Serrano, A. and Santisteban, J.I. (in press b). *Mapa y Memoria explicativa de la Hoja de Las Veguillas.* ITGE-Servicio de Publicaciones Ministerio Industria y Energía, Madrid.

Mazo, A.V. and Jiménez, E. (1982). 'El Guijo', primer yacimiento de mamíferos miocénicos de la provincia de Salamanca. *Stvd. Geol. Salmanticensia*, **17**: 99–104.

Mediavilla, R. and Martín-Serrano, A. (1989). Sedimentación y tectónica en el sector oriental de la Fosa de Ciudad Rodrigo durante el Terciario. *XII Congr. Español Sedim.*, comun. **1**: 215–218.

Megías, A.G. (1982). Introducción al análisis tectosedimentario: aplicación al estudio dinámico de cuencas. *Quinto Congreso Latinoamericano de Geología, Argentina. Actas*, **1**: 385–402.

Millot, G. (1964). *Geologie des argiles.* Masson et Cie.: 499 pp. Paris.

Molina, E., Vicente, A., Cantano, M. and Martín-Serrano, A. (1989). Importancia e implicaciones de las paleoalteraciones y de los sedimentos siderolíticos del paso Mesozoico–Terciario en el borde suroeste de la Cuenca el Duero y Macizo Hercínico Ibérico. In C.J. Dabrio (ed.), *Paleogeografía de la Meseta Norte durante el Terciario. Stvd. Geol. Salmanticensia*, vol. **5**: 177–186.

Olmo, A. del and Martínez-Salanova, J. (1989). El tránsito Cretácico-Terciario en la Sierra de Guadarrama y áreas próximas de las cuencas del Duero y Tajo. In C.J. Dabrio (ed.), *Paleogeografía de la Meseta Norte durante el Terciario, Stvd. Geol. Salmanticensia*, vol. **5**: 55–69.

Polo, M.A., Alonso Gavilán, G. and Valle, M.F. (1987). Bioestratigrafía y paleogeografía del Oligoceno–Mioceno del borde SO de la Fosa de Ciudad Rodrigo (Salamanca). *Stvd. Geol.Salmanticensia*, **24**: 229–245.

Santisteban, J.I., Martín-Serrano, A., Mediavilla, R. and Molina, E. (1991a). Introducción a la estratigrafía del Terciario del SO de la Cuenca del Duero. In J.A. Blanco, E. Molina and A. Martín-Serrano (eds.), *Alteraciones y paleoalteraciones en la morfología del oeste peninsular. Zócalo hercínico y cuencas terciarias. Monogr. Soc. Española Geomorfol.*, **6**: 185–198.

Santisteban, J.I., Martín-Serrano, A. and Mediavilla, R. (1991b). El Paleógeno del sector suroccidental de la Cuenca del Duero: Nueva división estratigráfica y controles sobre su sedimentación. In F. Colombo (ed.), *Libro Homenaje a Oriol Riba, Acta Geol. Hispànica* **26** (2):133–148.

Sloss, L.L. and Speed, R.C. (1974). Relationships of cratonic and continental-margin tectonic episodes. In W.R. Dickinson (ed.), *Tectonics and sedimentation, S.E.P.M. Spec. Publ.*, **22**: 98–119.

W4 Tectono-sedimentary evolution of the Almazán Basin, NE Spain

J. BOND

Abstract

The Almazán Basin formed between the Iberian Range and the Central System, as a result of NE–SW and N–S Pyrenean compression causing reactivation of NE–SW- and NW–SE-oriented basement faults. The basin covers an area of *c.* 4200 km^2, is up to 55 km wide and 115 km long, and is elongated about a NW–SE axis. In the NE–SW profile, the basin is a half-graben, with up to 2.5 km of sediment deposited along the NE basin margin. In a NW–SE profile the basin deepens to the NW, with maximum basin subsidence (*c.* 3.5 km) occurring in the environs of Nolay. The NW basin margin is defined by a major dip–slip fault, downthrowing to the SE. Intrabasin faults, located between the towns of Almazán and Burgo de Osma, trend NE–SW, downthrow to the SE and have a wide range of vertical displacements, from *c.* 20 m up to *c.* 1.7 km. Sinistral strike–slip movement may have been accommodated on these intrabasin transfer faults. The NE basin margin comprises a SW-propagating thrust fault. The SE margin has low relief and is defined by listric normal faults in parts. In contrast the SW margin is steeper, and controlled by wrench faulting within the Palaeozoic basement and Mesozoic cover sequence.

An informal lithostratigraphy is devised for the Tertiary basin-fill, with folded Palaeogene sediments of the NE basin margin assigned to the Henar Group and flat-lying Miocene sediments of the central and southern basin assigned to the Jalon Group. A total of six tectonosedimentary units are defined, on the basis of unconformable contacts, lithofacies distributions and palaeontological data.

Non-marine sedimentation commenced in the Danian (T1) with deposition of palustrine limestones overlying the Late Cretaceous platform carbonates over the western margin of the Iberian Range. A major hiatus in sedimentation persisted from the Early Palaeocene to Mid Eocene. Alluvial sedimentation commenced at this time, following primary uplift of the Iberian Range, as a result of SW-directed regional compression (T2). In the following period of tectonic quiescence, sandstones and limestones of the Deza Fm. accumulated along the NE basin margin.

N–S compression in the Late Eocene (T3) caused reactivation of NE–SW-trending transfer faults, resulting in dip–slip faulting downthrowing to the SE, possibly accompanied by a minor component of left-lateral wrench movement. Rapid basin subsidence followed in the NE, with the development of a major progressive unconformity along the NE margin. The Gomara Fluvial System developed, draining the Iberian Range and Sierra de la Demanda to the NNE, and prograded south and westwards into the developing basin.

Repeated NNE–SSW compression in the Mid–Late Oligocene (T4) caused uplift of the Iberian Range and rejuvenation of alluvial sedimentation along the western flank of the Iberian Range, above an angular unconformity. This probably corresponds to the Castellian deformation event, as identified in the Tajo Basin, and is dated as Arvernian.

A later phase of Neocastellian deformation occurred in the Late Agenian. A pronounced angular unconformity is recognised around the basin margins, where alluvial facies onlap the Mesozoic cover sequence of the Iberian Range and Central System. Deposition of undeformed alluvial and lacustrine lithologies of the Jalón Group (T5) followed, from the latest Agenian to Late Aragonian.

During the Late Miocene and Pliocene small alluvial fans were sourced from the Iberian Range and shed sediment westwards into the maturing basin. Wide onlap of the Iberian Range occurred, with flat-lying lake sediments deposited unconformably on the basement sediments exposed in the Iberian Range.

Introduction

Relationships between basin evolution resulting from strike–slip fault displacement and consequent patterns of sedimentation in intraplate settings have been widely described from continent–continent collision zones (see review papers by Sylvester, 1988, and Woodcock, 1986). However, there are relatively few documented examples of basin formation due to reactivation of strike–slip faults within intracontinental compressional structural regimes. This chapter describes the formation of the Almazán Basin, a comparatively small, Spanish Tertiary basin, located at the junction of two mutually perpendicular wrench–fault systems. Basin subsidence was driven by the reactivation of deep-seated Palaeozoic basement faults following compression and internal

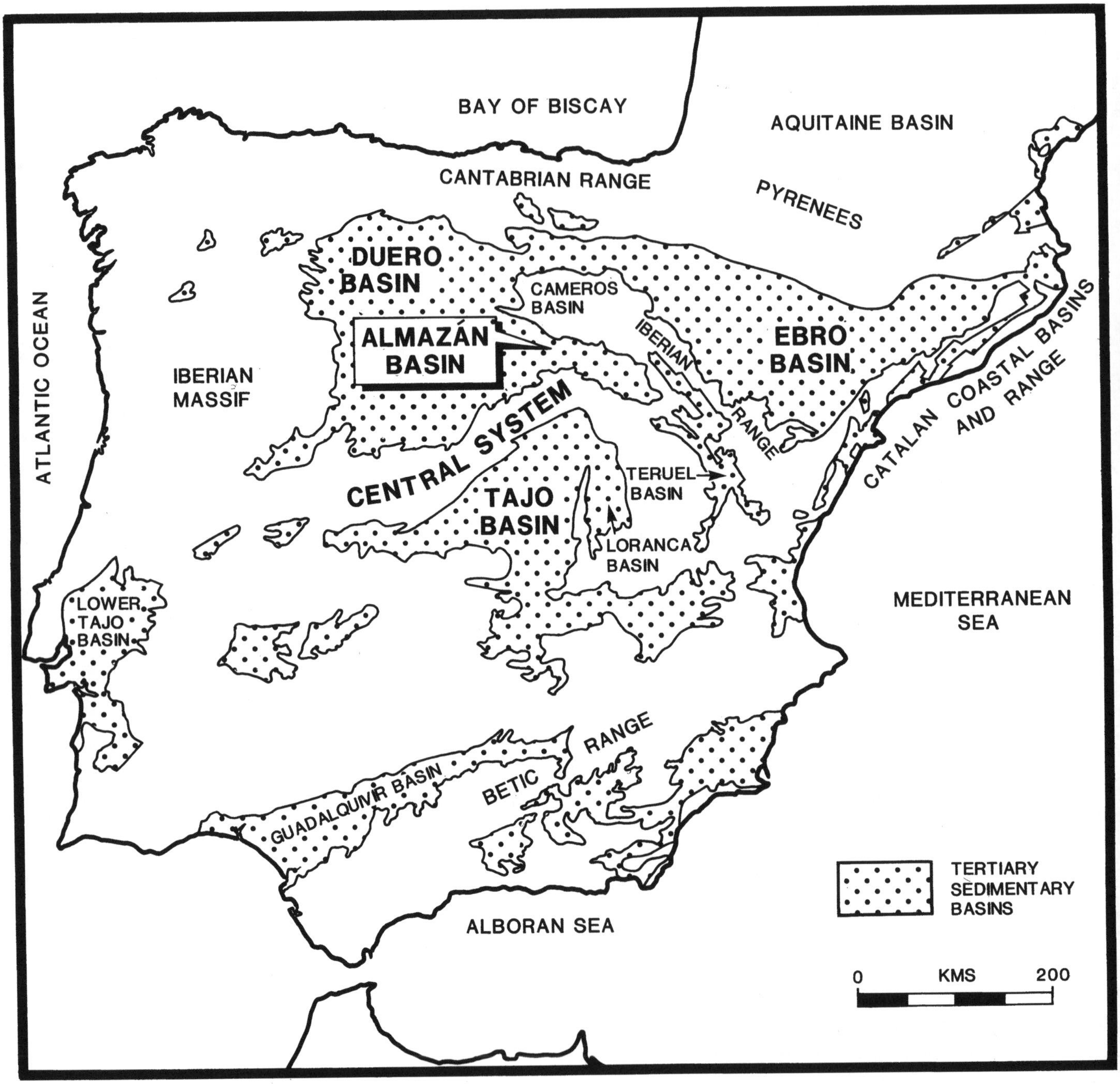

Fig. 1. Location map of the Almazán Basin.

deformation of the Iberian microplate during Pyrenean orogenesis. Good field exposures along the north-eastern and southern margins of the Almazán Basin provide the opportunity to document the spatial and temporal sedimentary response to reactivation of the basin boundary faults. A model is presented for the Tertiary tectonosedimentary evolution of the Almazán Basin, based on a detailed sedimentological field study by Bond (1989).

The Almazán Basin is located *c.* 200 km NE of Madrid, in the Provinces of Soria, Guadalajara and Zaragoza (Fig. 1). From Mid Eocene to Mid Miocene times the Almazán region comprised an intracontinental basin with internal drainage. The preservation of terrestrial vertebrate remains and widespread palaeosol profiles, together with extensive reddening of the sediments, provide evidence for non-marine deposition. The results of field mapping and sedimentary lithofacies analysis have shown that small alluvial fans (radius 0.5–8 km) fringed the fault-bounded basin margins. However, the main basin-fill was supplied by two large fluvial systems (radius 35–40 km). Sourced from the Iberian Range, the fluvial systems had apices located at structurally defined topographic lows, located at the NE and SE corners of the basin. Despite a significant volume of fluvial discharge being lost to surface evaporation and floodplain soak-away, sufficient runoff remained to supply contemporaneous lacustrine depositional systems. Excellent exposures of lacustrine and palustrine limestones and calcrete profiles provide evidence for periodic expansion and contraction of the lake margins, in response to seasonal climatic variations. The vertical evolution from freshwater lacustrine deposits to gypsiferous chalks in the Early to Mid Miocene is attributed to a long-term climatic evolution from seasonal, cool and humid conditions to a warmer, semi-arid climatic regime.

Previous work

Little previous geological work had been completed in the Almazán Basin, prior to the current study. Early workers (e.g., Aranzazu, 1877; Royo Gomez, 1922) focussed on the Miocene of the south-western Almazán Basin, establishing a tripartite lithostratigraphy, comprising basal conglomerates overlain by grey marls and capped by white limestones. The 1:50000 regional geological map was completed by Castells and de la Concha (1956a, 1956b, 1959), using a two-fold subdivision of the Miocene (Hahn, 1930). The first comprehensive sedimentological study of the south-western Almazán Basin was undertaken by Sánchez de la Torre (1963). More recently, sedimentary sections have been described by Capote *et al.* (1982) from the Miocene of the southern margin in the environs of Jaraba and Somaen.

During the 1980s the focus of attention shifted to the NE margin (Fig. 1), where Palaeogene deposits are in contact with folded Senonian carbonates of the Iberian Range. Lithofacies mapping of Tertiary sediments located between Alhama de Aragón and Almazul has been reported by Guisado *et al.* (1988) and Armenteros *et al.* (1989). Lithofacies distributions and sediment diagenesis have been described by Bond (1989).

Tectonic framework

Forming the south-eastern corner of the Tertiary Duero Basin, the Almazán Basin lies at the intersection of two mountain ranges: the Iberian Range and Central System (Fig. 1). Both comprise deformed Palaeozoic basement and Mesozoic cover sequences. The Almazán Basin originated from folding between the Iberian Range and Central System, arising from oblique crustal shortening about a predominantly NE–SW axis. The Cameros Basin, which crops out to the north, represents an inverted Mesozoic basin, uplifted in response to Alpine compression (Guiraud and Seguret, 1986; Platt, 1990). Detritus from the inverted Cameros Basin and Iberian Range was shed south and westwards, whereas clastics derived from the Central System were transported northwards into the subsiding Duero and Almazán Basins.

The Almazán Basin is from 30 km to 55 km wide, up to 115 km long, and is elongated about a NW–SE axis. The basin margins are defined by conjugate NE–SW- and NW–SE-trending faults which form a rectangle of Tertiary sediments, covering an area of *c.* 4200 km^2. Eo-Oligocene strata are exposed along the north-eastern basin margin, whereas Miocene sediments occur in central and southern areas. The basin has a relatively complex three-dimensional structure (Fig. 2a). A NE–SW-orientated cross-section reveals a half-graben with a deeper basement in the NE (*c.* 2.5 km), and shallowing westwards (*c.* 0.5 km, Fig. 2b). In a NW–SE profile the Tertiary thickens to the NW (Fig. 2c). Maximum basin subsidence occurred on a major NE–SW-striking, intra-basin fault zone (Fig. 2a). This high-angle fault downthrows to the SE. The depth to Top Cenomanian is *c* 3.9 km in the region of Nolay. Thus, a maximum of *c.* 3.5 km of Tertiary sediments accumulated in the basin depocentre, located on the hanging wall of this major fault zone.

Tertiary basin subsidence commenced in the Mid to Late Eocene when early Alpine compression reactivated NW–SE- and NE–SW-trending basement faults. This dominant fault pattern originates from late Hercynian times when much of western Europe was deformed in a right-lateral shear zone (Arthaud and Matte, 1977). Seismic and well data, made available for the purposes of this study by REPSOL, indicate that Late Jurassic and Early Cretaceous deposits are absent from the central and western zones of the Almazán Basin. This evidence indicates that the Almazán block underwent structural inversion at this time. Major Tertiary basin subsidence occurred along the NE and NW boundary faults.

The NW basin margin is defined by a NE–SW-striking, high-angle fault which was downthrown to the NW in Late Jurassic–Early Cretaceous times. This boundary fault was inverted in the Early Tertiary, with subsidence of the Almazán Basin to the SE. A series of NE–SW-striking basement faults occur between the towns of Almazán and Burgo de Osma, in the valley of the River Duero. Small lateral off-sets, of the order of 2–3 km, can be identified at Top Cenomanian (Fig. 2a). This probably represents a transfer zone, which has primarily accommodated dip–slip faulting, downthrowing to the SE, accompanied by subsidiary left-lateral displacement. Sinistral wrench movement appears to have been activated in the Oligocene as a result of N–S Pyrenean compression.

The NE basin margin is defined by SW-directed thrust faulting occurring within the western arm of the Iberian Range. Décollement has primarily taken place within the Triassic Keuper Marl, with subsidiary displacement exploiting evaporite horizons in the Jurassic 'Carniolas' facies. Tightly folded, asymmetric ramp-top anticlines are preserved in Late Cretaceous platform carbonates of the NE basin margin. Folds verge to the SW and strike 140–320°. Their western limbs are vertical or dip at a high angle towards an azimuth of 320°, whereas the eastern limbs have a shallower dip (*c.* 30–45°) towards the NE.

The effects of synsedimentary tectonism are pronounced along the NE basin margin. Both pre- and syntectonic sediments are folded into a progressive unconformity (Riba, 1976). To the N of Deza this deformed sedimentary wedge extends *c.* 8 km into the basin. However, along strike, to the SE, the width of the progressive unconformity diminishes and at Embid de Ariza it is confined within *c.* 1 km of the basin margin. The asymmetry of the progressive unconformity is attributed to increased basin subsidence along

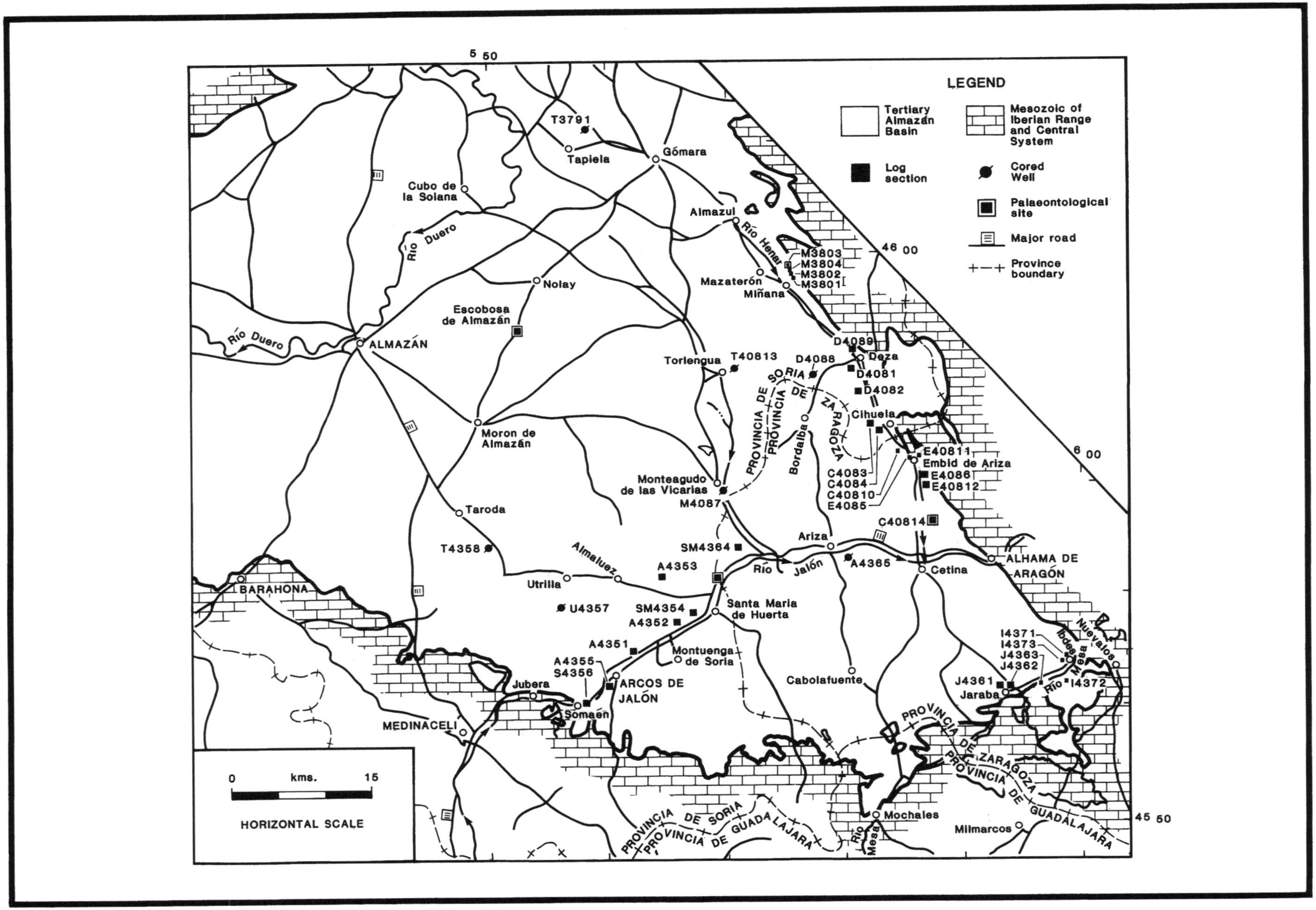

Fig. 2 (a). Almazán Basin with place names, major roads, logged sections, wells and palaeontological sites.

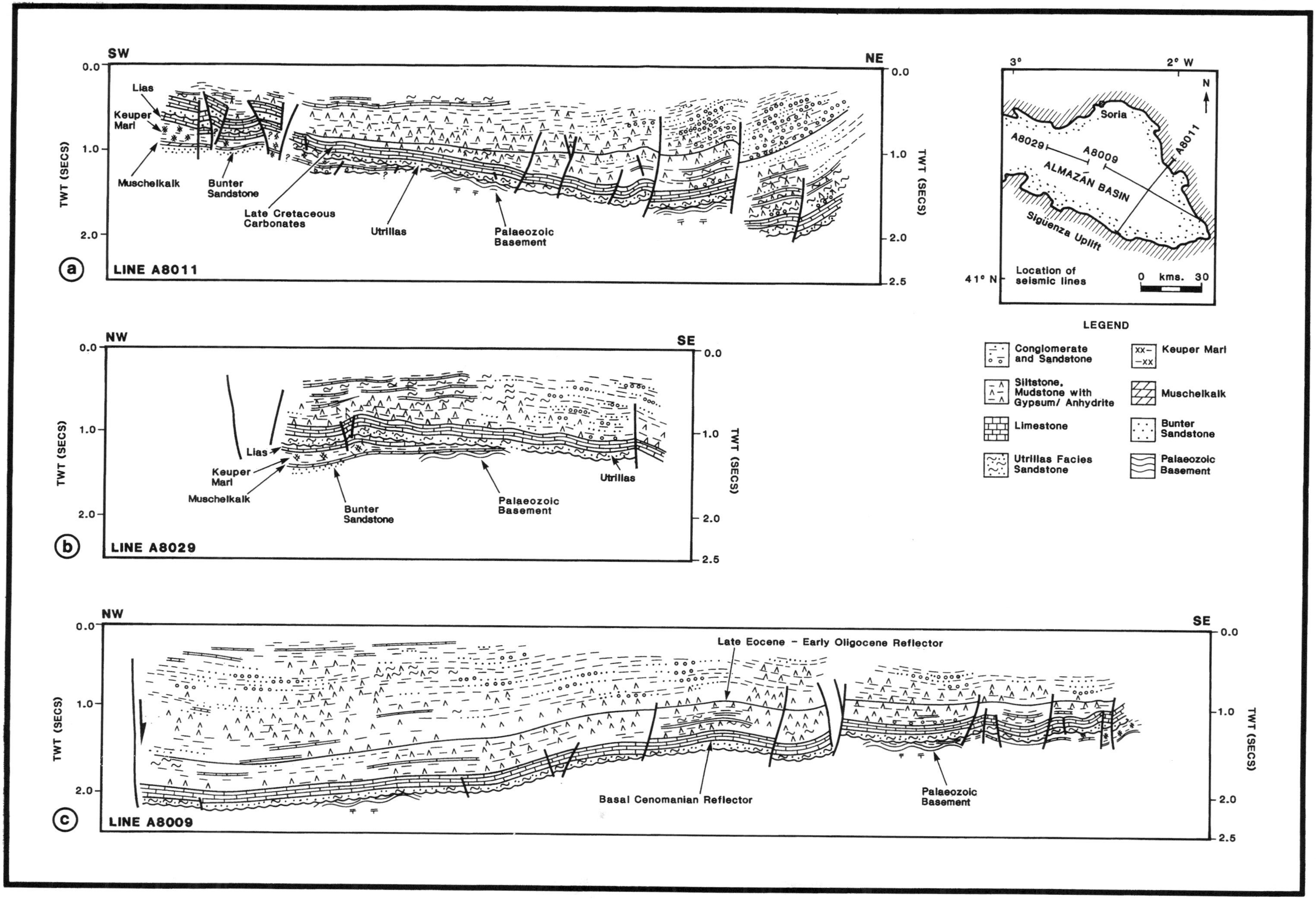

Fig. 2 (b). Interpreted seismic profiles acorss Almazán Basin.

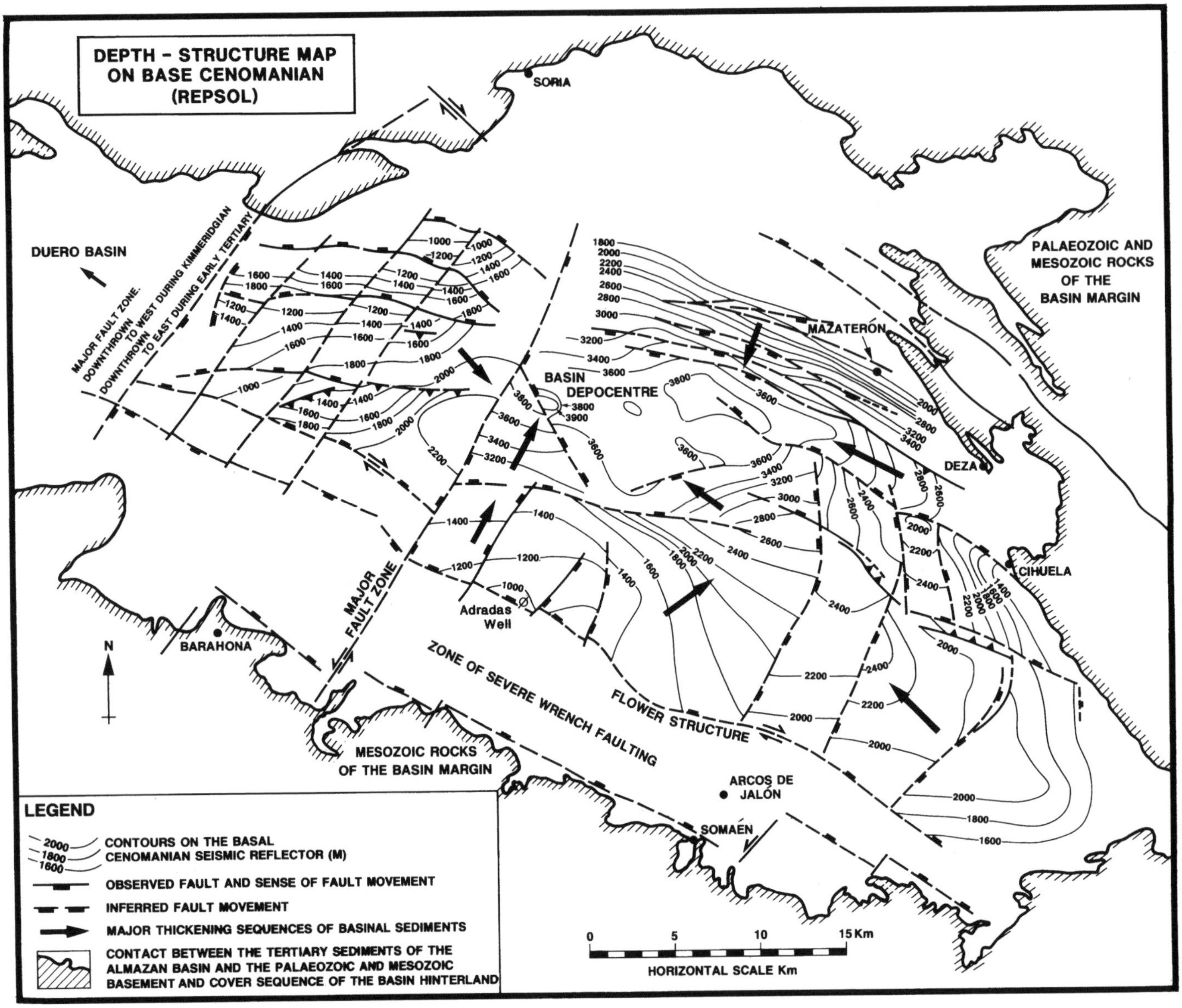

Fig. 2 (c). Structure of base of Cenomanian for the Almazán Basin.

strike to the NW, coincident with the intersection of the NE basin margin and a NE–SW-trending wrench fault zone. Alvaro *et al.* (1979) have suggested that right-lateral strike–slip movement was accommodated along the NW–SE-trending fault zone during the Tertiary.

The south-western border is bounded by a 5 km wide zone of wrench faulting within the Triassic and Jurassic cover sequence. It formed a prominent topographic feature in Miocene times. Debris-flow dominated alluvial fans accumulated at the break of slope, fed by the highlands of the Central System. Compressional structures within the Tertiary basin-fill, as identified from seismic data, are interpreted as the product of transpression, caused by right-lateral displacement along the SW margin. In contrast, the SE margin is defined by listric normal faults which appear to have generated low to moderate relief in Oligo–Miocene times.

Tectono-sedimentary evolution of the Almazán Basin

An informal lithostratigraphy has been devised for the Tertiary sediments of the Almazán Basin (Fig. 3). Six main tectono-stratigraphic units are recognised, comprising unconformity-bounded, heterogeneous sedimentary packages. Basin-fill sediments have been dated from five localities (Fig. 3), and where possible these data have been used to constrain the age of sedimentary sequences. Palaeogene sediments which crop out along the NE basin margin, and typically show evidence of syn- or post-depositional deformation, are assigned to the Henar Group. In contrast, undeformed Neogene lithofacies which post-date the Neocastillian deformation phase are considered to belong to the Jalón Group (Bond 1989). Each of the six tectonosedimentary units is considered below (Fig. 4).

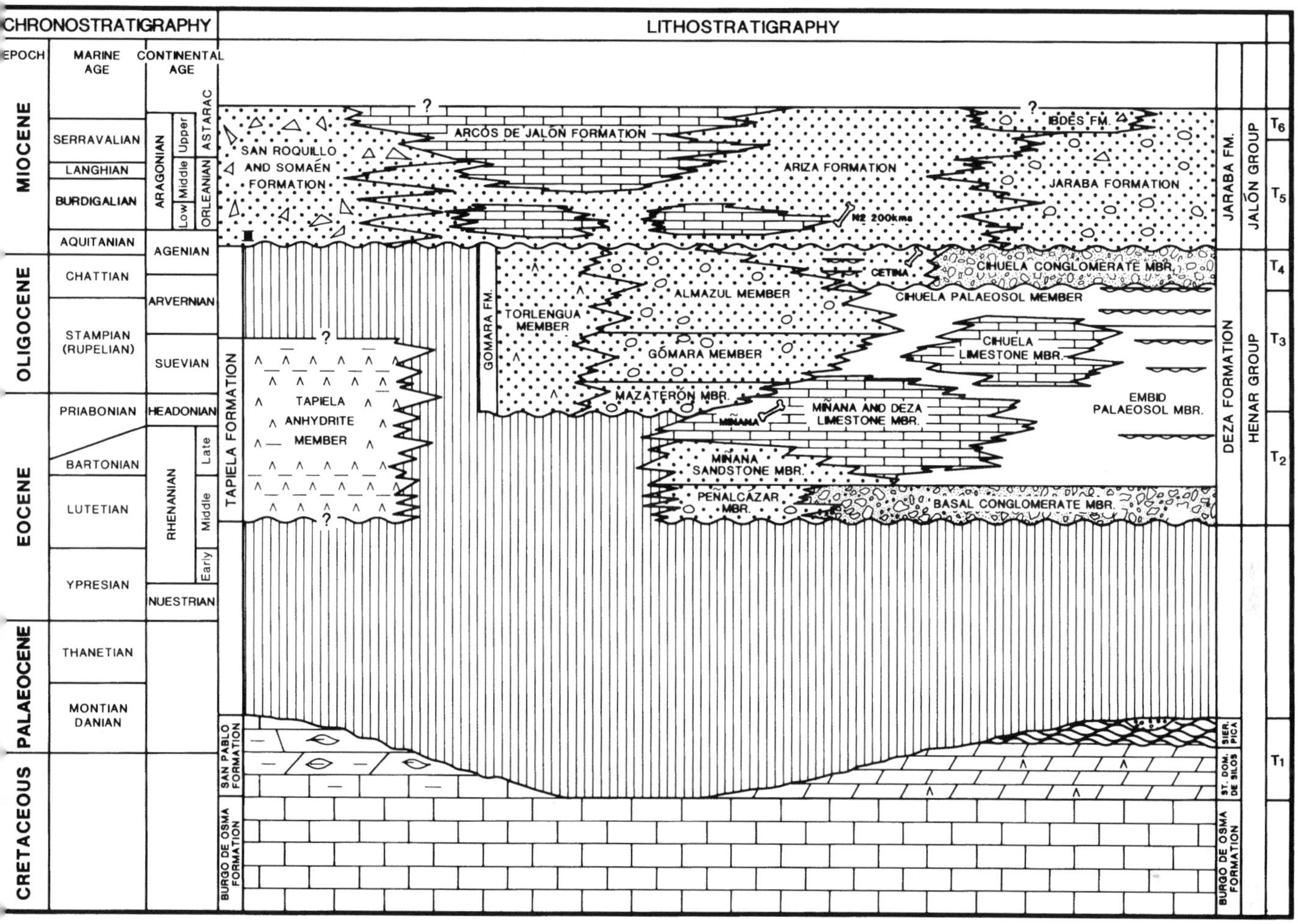

Fig. 3. Informal lithostratigraphy of the Almazán Basin.

T1 Late Santonian–Danian

A stable, shallow carbonate platform extended across the Iberian realm in Late Cretaceous times. Foraminiferal limestones of the Burgo de Osma Fm., described from Cihuela by Meléndez Hevia *et al.* (1985) were deposited in the Late Santonian. Representing a shallowing-upwards sequence, they were deposited within a restricted lagoonal environment. The Campanian–Maastrichtian marine regression culminated in the deposition of dolomicrites and anhydrites of the Santo Domingo de Silos Fm. (Floquet and Meléndez, 1982), which are exposed at Embid de Ariza. The dolomites accumulated within an inner carbonate platform and are overlain by sabkha evaporites. The anhydrites are unconformably overlain by nodular limestones, incised by small fluvial channels infilled by carbonate intraclasts and rounded black pebbles. The succession is capped by dolomitised calcretes featuring Microcodium. The nodular limestones and calcretes belong to the Sierra de la Pica Fm. (Floquet *et al.*, 1981) and are probably Danian in age. They are comparable to Garumnian lithologies described from the Sierra de Picofrentes by Saenz García. The nodular limestones accumulated in brackish to freshwater coastal swamps, thus marking the onset of Tertiary non-marine sedimentation in the Iberian realm.

Northwards in the Tapiela well (Fig. 1), drilled by ENUSA, 15 m of marls and dolomitic limestones with abundant plant debris, assigned to the San Pablo Fm., conformably overlie limestones of the Burgo de Osma Fm. The carbonaceous marls are closely comparable to Maastrichtian–Danian lithologies encountered in the Sierra de la Demanda. They probably accumulated in brackish lagoons, located on a lower coastal plain, and are broadly time equivalent to the Sierra de la Pica Fm.

Non-marine sedimentation commenced with deposition of marginal marine facies of the San Pablo Fm. In the Tapiela region these comprise carbonaceous marls and dolomitic limestones, which were deposited in brackish lagoons, and can be compared with the Garumnian facies. Further SE, in the environs of Embid de Ariza, the St. Domingo de Silos Fm. comprises dolomicrites and anhydrite breccias of sabkha origin. These are overlain by dolomitised calcretes and palustrine limestones (Floquet *et al.*, 1981). Both formations accumulated during the marine regression from the stable carbonate platform which extended over the Iberian realm in Late Cretaceous and Early Palaeocene times.

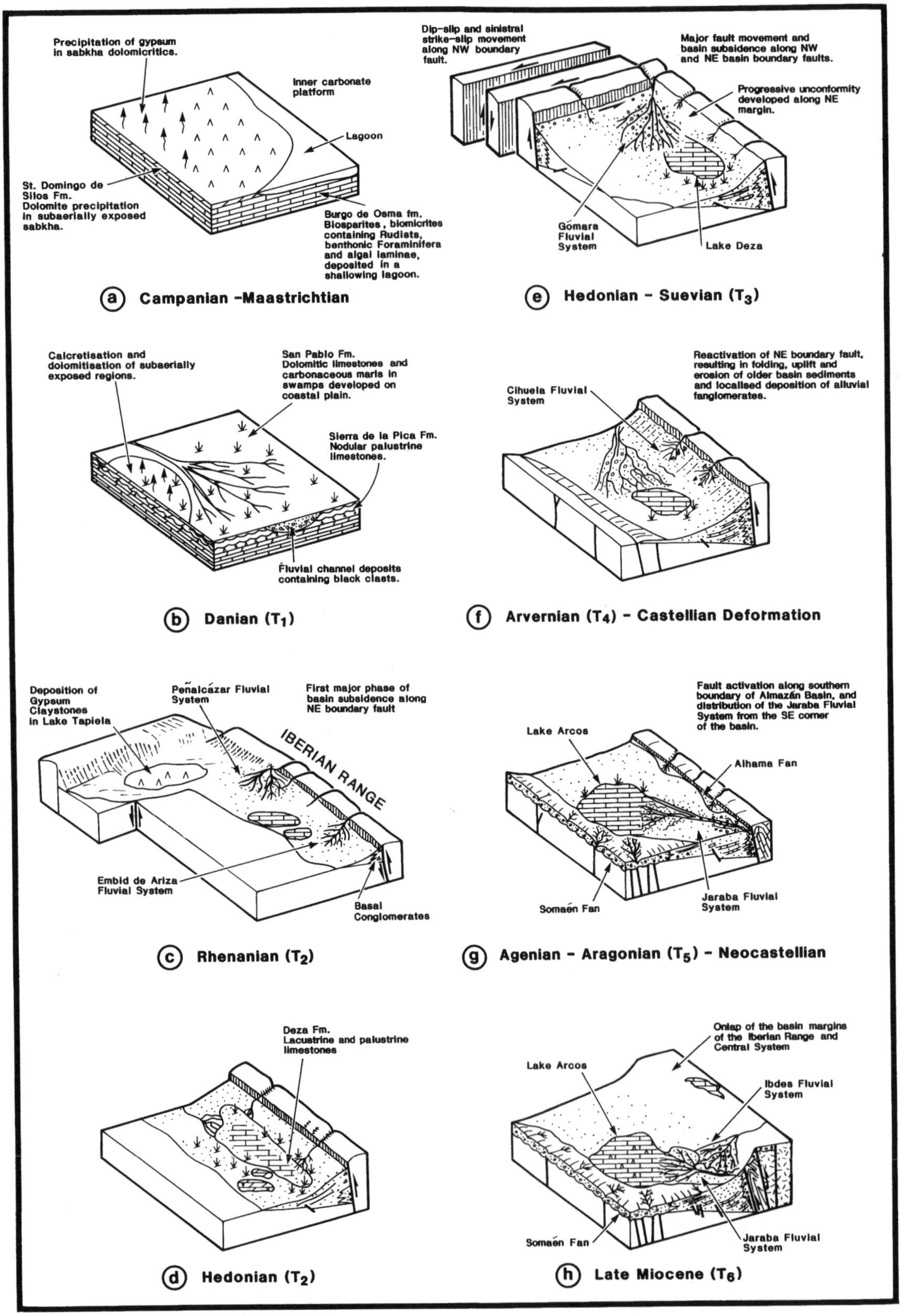

Fig. 4. Tectonosedimentary evolution of the Almazán Basin. (a) Campanian–Maastrichtian Burgo de Osma and St. Domingo de Silos Fm. (b) Danian (T1) Sierra de la Pica & San Pablo Fms. (c) Mid Eocene (T2) Basal Conglomerate. (d) Hedonian – Late Eocene (T2) Deza Fm. (e) Late Eocene – Early Oligocene (T3) Gomara Fluvial System Castellian Deformation. (f) Late Oligocene – Early Miocene Cihuela Cglm. Neocastellian (T4). (g) Late Agenian – Mid Aragonian Jaraba Fluvial System (T5). (h) Late Aragonian – Latest Miocene Ibdes Fluvial System (T6).

T2 Mid–Late Eocene

Following the period of subaerial exposure and karstification of the sabkha and coastal plain deposits during the Early Palaeocene, a hiatus in sedimentation persisted until Mid or Late Eocene times. Clastic sedimentation commenced in response to NE–SW compression of the Almazán block and uplift of the Iberian Range, to the east. The Peñalcázar and Basal Conglomerate Members formed an alluvial fringe along the NE basin margin, as seen between Deza and Almazul. At Embid de Ariza nodular limestones of the Sierra de la Pica Fm. have been eroded and unconformably overlain by conglomerates deposited by the Embid de Ariza fluvial system. The conglomerates form a large-scale fining-upwards sequence *c.* 200 m thick, which is folded into a progressive unconformity, providing evidence of syndepositional growth faulting along the western margin of the Iberian Range.

This primary phase of clastic deposition was followed in the Late Eocene by a period of tectonic quiescence. Along the NE basin margin, from between Embid de Ariza to Almazul, distal alluvial, fluvial and lacustrine sediments of the Deza Fm. are preserved. They accumulated in a variety of depositional settings, ranging from fluvio-deltaic environments, located on a distal floodplain, to shallow, carbonate lakes, with highly variable water levels and chemistries. The widespread development of nodular and mottled limestones, comparable to palustrine carbonates (Freytet and Plaziat, 1982), provides evidence that aquatic vegetation fringed the lakes.

To the north, in the Tapiela well, a thick (*c.* 200 m) sequence of interbedded anhydrite and grey marls overlies the 'Garumnian-like' facies of the San Pablo Fm. The thick evaporite sequence accumulated in a saline playa lake. It has not been possible to date these evaporites, and they may in fact be older than Mid Eocene. Similar basal sediments are encountered in the Tajo Basin, where unit T1 (Capote *et al.*, 1982) comprises 350 m of gypsiferous marls (Díaz-Molina *et al.*, 1985).

T3 Late Eocene to Early Oligocene

In Late Eocene–Early Oligocene times the basin was compressed about a N–S axis. This resulted in fault displacement at the intersection of the NE–SW-trending intra-basin fault zone and the NE basin margin. Accelerated basin subsidence occurred in the region of Gómara. Inversion of the Cameros Basin appears to have occurred around the same time. The Gómara Fluvial system drained the elevated hinterland of the southern Cameros Basin and Sierra Del Moncayo, and transported distinctive Liassic and Wealden lithoclasts south and southwestwards into the deepening Almazán Basin. Syndepositional faulting was accommodated along the NE basin margin as shown by the progressive unconformity developed in the region of Miñana and Mazaterón.

Southwards progradation of the Gómara Fluvial System lead to an influx of very-fine-grained clastic detritus in Lake Deza, and the focus of carbonate lake sedimentation appears to have shifted southwards, focussing on Cihuela.

T4 Mid to Late Oligocene

During the Mid Oligocene the Almazán Basin underwent N–S to NNE–SSW compression. This phase of crustal shortening probably corresponds to the Castellian deformation phase, identified in the Tajo Basin by Díaz-Molina and López-Martínez (1979) and dated as Arvernian. Tectonic rejuvenation of the NE basin margin resulted in increased rates of erosion of the basin hinterland in the Iberian Range. Alluvial fans built out from the basin margin in the region of Cihuela and Embid de Ariza (e.g., Cihuela Conglomerate Member). In contrast, fluvial sedimentation continued to the north. Sandstones and conglomerates of the Almazul Member were derived from the northern Iberian Range and transported in a southerly direction.

T5 Early to Mid Miocene

A Late Agenian or Early Aragonian phase of tectonic reactivation caused folding and faulting of the Cihuela Conglomerate Member, located to the south of Embid de Ariza. Subsidence of the southern central Almazán Basin progressed, probably driven by right-lateral wrench movement along the SW boundary fault and coupled with normal, listric fault displacement along the SE boundary fault. This phase of tectonism is correlated with the Neocastellian phase of folding, as seen in the Tajo Basin (Aguirre *et al.*, 1976), and marks the onset of the main pulse of Miocene sedimentation. A pronounced angular unconformity is developed around the basin margins where Miocene conglomerates onlap folded Palaeogene and Mesozoic strata. However, in general, the Miocene sediments are subhorizontal and undeformed.

The focus of Miocene sedimentation shifted to the southern–central basin. The Jaraba Fluvial System distributed clastics from the fan apex in the SE corner of the basin, in a radial dispersal pattern to the N and NW. A catena of soils matured in the Jaraba floodplain. Calcaric fluvent palaeosols evolved in the well-drained, strongly oxidised sediments of the proximal fan. In the medial fan, where subsidence rates were high and floodplain drainage moderate to good, fluvent soils and pseudogleys formed. In contrast, strongly gleyed calcaric aquents developed on the distal floodplain. During the same period small alluvial fans deposited fanglomerates in a fringe along the southern and south-eastern basin margins (e.g., Somaen and Alhama de Aragon Alluvial fans).

Early to Mid Miocene lacustrine sedimentation focused on Lake Arcos, located in the environs of Arcos de Jalón. The earliest lake sediments were gypsiferous marls. Lake chemistry and sedimentation patterns evolved in the Early Aragonian, with the change to a warmer, humid climate (López-Martínez *et al.*, 1987). Bioclastic, lacustrine and swamp limestones and marls were deposited at this time. Sedimentation in Lake Arcos culminated with the precipitation of gypsiferous dolomicrites, under the warm, dry climate of the Middle Aragonian.

T6 Late Miocene–Pliocene

During Late Miocene and possibly Pliocene times small alluvial fans continued to build out from the western margin of the Iberian Range. At Ibdes, located in the SE corner of the Almazán Basin, up to 20 m of conglomerates, sandstones and siltstones with root traces are preserved. They were deposited by a gravelly, braided fluvial system (*c.* 3 km radius). Palaeocurrent indicators demonstrate palaeoflow from the NNE. The varied clast composition is dominated by Cambro-Ordivician quartzites and pelites, with subordinate clasts of Triassic Bunter Sandstone, Cretaceous limestone and rare Utrillas sandstone. All clast types are exposed in the Iberian Range, within a 7 km radius.

Younger outliers of flat-lying algal limestones (Turolian–Ruscinian) unconformably overlie Triassic and Cambro-Ordovician lithologies of the Iberian Range. This evidence may indicate that onlap of the basin margins continued until the early Pliocene.

Conclusions

It is possible to define the geometry of the Almazán Basin from examination of key dip and strike lines from the two-dimensional seismic survey shot by REPSOL. The most active basin subsidence appears to have taken place along the NE and NW basin margins, with the development of a broad, syntectonic sedimentary wedge associated with overthrusting of the NE basin margin. A tectonostratigraphic framework has been devised for the Tertiary basin-fill, incorporating six main unconformity-bound sedimentary packages. Marine regression from the Iberian realm in the earliest Palaeocene was followed by a major sedimentary hiatus, lasting until Mid Eocene times. Alluvial sedimentation commenced in the Mid Eocene, triggered by uplift of the Iberian Range, resulting in fanglomerates being deposited along the NE basin margin. A period of relative tectonic stability followed during the Late Eocene, coincident with deposition of limestones of the Deza Fm. An increase in the clastic content of the Deza carbonates in the Latest Eocene–Early Oligocene coincides both with tectonic reactivation of the north-western margin of the Iberian Range (Castellian Deformation) and progradation of the Gómara Fluvial System towards the S–SW.

More recent uplift and erosion of the western arm of the Iberian Range took place in the Late Oligocene–Earliest Miocene (Neocastellian Deformation), with renewed clastic sedimentation along the basin margins. The focus of Miocene fluvial sedimentation shifted to the SE with deposition of the Jaraba Fluvial System. Fluvial channel and overbank facies were transported to the N–NW, derived from a fan apex located between Ibdes and Jaraba. Lacustrine sedimentation was centred in the SW, in the environs of Arcos de Jalón. Regional onlap of the basin margins continued throughout the Late Miocene and Earliest Pliocene, to produce outliers of alluvial conglomerates, lacustrine limestones and tuffas, unconformably overlying basement lithologies of the Iberian Range.

Acknowledgements

Firstly, I wish to thank REPSOL for granting permission to publish seismic data from the Almazán Basin. Secondly, thanks are extended to ENUSA (Soria), who made cores available for inclusion in this study. Thirdly, I would like to thank P.F. Friend and M. Díaz-Molina for their helpful technical advice and discussion throughout the study. The research was funded by a Natural Environment Research Council studentship (GT4/82/GS/108) held at the University of Cambridge.

References

Aguirre, E. *et al.* (1976). Datos paleontológicos y fases tectónicas en el Neógeno de la Meseta sur Española. *Trab. Neog. Cuat.*, **5**, 7–29.

Alvaro, M. *et al.* (1979). Un modelo de evolución geotectónia para la Cadena Celtiberica. *Acta Geol. Hisp.* **14**, 172–177.

Aranzazu, J.M. (1877). Apuntes para una descripción fisico-geológica de las provincias de Burgos, Logroño, Soria y Guadalajara. *Bol. Com. Mapa Geol. Esp.*, **IV**, 1–47, Madrid.

Armenteros, I. *et al.* (1989). Megasecuencias sedimentarias del Terciario del Borde Oriental de la Cuenca de Almazán (Soria–Zaragoza). *Stvd. Geol. Salmanticensia*, **5**, 107–127.

Arthaud, F. and Matte, P. (1977). Late Paleozoic strike–slip faulting in southern Europe and northern Africa: result of a right-lateral shear zone between the Appalachians and the Urals. *Bull. Geol. Ass. Am.*, **88**, 1305–1320.

Bond, J. (1989). Depositional environments and diagenesis of the Cenozoic sediments of the Almazán Basin, North-East Spain. Unpublished PhD Thesis, University of Cambridge, 241 pp.

Capote, R. *et al.* (1982). Evolución sedimentológica y tectónica del ciclo Alpino en el Tercio noroccidental de la rama Castellana de la Cordillera Ibérica. *Temas Geologico Mineros*, **5**, 390 pp., I.G.M.E., Madird.

Castells, J. and de la Concha, S. (1956a). *Mapa Geológico de España. Explicacion de la Hoja 462, Maranchón (Guadalajara).* Inst. Geol. Miner. España, Madrid.

Castells, J. and de la Concha, S. (1956b). *Mapa Geologica de Espana. Explicacion de la Hoja 434, Barahona (Guadalajara–Soria).* Inst. Geol. Mineral. Espana, Madrid.

Castells, J. and de la Concha, S. (1959). *Mapa Geológico de España. Explicación de la Hoja 435, Arcos de Jalón (Soma).* Inst. Geol. Miner. España, Madrid.

Daams, R. (1976). Miocene rodents (Mammalia) from Cetina de Aragón (Prov. Zaragoza) and Buñol (Prov. Valencia), Spain I. *Kon. Ned. Akad. v. Wet. Amsterdam, Proc. Series B*, **79**, 162–182.

Daams, R. and Meulen, A.J. van der (1984). Paleoenvironmental and paleoclimatic interpretation in the Upper Oligocene and Miocene of north central Spain. *Paléobiol. Cont. Montpellier*, **14**, 241–257.

Díaz-Molina, M. and López Martínez, N. (1979). El Terciario continental de la Depresión Intermedia (Cuenca). Bioestratigrafia y paleogeografia. *Estud. Geol.*, **35**, 149–167.

Díaz-Molina, M. *et al.* (1985). Wet fluvial fans of the Loranca Basin (central Spain). Channel models and distal bioturbated gypsum with chert. Excursion guide book, *IAS 6th European Regional Meeting, Lerida, Spain*, 149–185.

Floquet, M. and Meléndez, A. (1982). Características sedimentarias y paleogeográficas de la regresion finicretácica en el sector central de la Cordillera Ibérica. *Cuad. Geol. Iberica*, **8**, 237–257.

Floquet, M., Meléndez, A. and Pedauye, R. (1981). El Cretacico superior de la región de Alhama de Aragón (borde septentrional de la rama castellana, Cordillera Ibérica). In *El Cretácico de la Cordillera Ibérica (sector central)*. Libro guía Jornadas del Campo. Grupo español de Mesozoico. 23–26 Junio 1981, Univ. de Zaragoza, 166–207.

Freytet, P. and Plaziat, J.C. (1982). Continental carbonate sedimentation and pedodiagenesis: Late Cretaceous and Early Tertiary of Southern France. *Contrib. Sedimentol.*, **12**, 213 pp.

Guiraud, M. and Seguret, M. (1986). Releasing solitary overstep model for the late Jurassic–Early Cretaceous (Wealdian) Soria strike–slip basin (Northern Spain). In Biddle, K.T. and Christie-Block, N. (eds.), *Strike–slip deformation, basin formation and sedimentation. Spec. Publ. Soc. Econ. Palaeontol. Min.*, **37**, 149-175.

Guisado, R. *et al.* (1988). Sedimentación continental paleógena entre Almazul y Deza (cuenca de Almazán oriental, Soria). *Stvd. Geol. Salmanticensia*, **25**, 65–83.

Hahne, C. (1943). Investigaciones Estratigráficas y Tectónicas en las Provincias de Teruel, Castellón y Tarragona. *Publicaciones Extranjeras Sobre Geologia de Espana*, **2**, 51–98, Madrid.

Lopez-Martínez, N., *et al.* (1987). Approach to the Spanish continental Neogene synthesis and paleoclimatic interpretation. Proceedings of the 8th R.C.M.N.S. Congress, Budapest. Hungarian Geological Survey. *Ann. Inst. Geol. Publ. Hung.*, LXX, 383–391.

Melendez Hevia, A., *et al.* (1985). Stratigraphy, sedimentology and palaeogeography of Upper Cretaceous evaporitic-carbonate platform in the central part of the Sierra Iberica, Excursion 5. In *6th European regional meeting excursion guidebook of the IAS*, (Ed. M.D. Mila and J. Rosell), Lleida, 1985.

Riba, O. (1976). Syntectonic unconformities of the Alto Cardener, Spanish Pyrenees: a genetic interpretation. *Sed. Geol.*, **15**, 213–233.

Richter, G. (1959). Las Cadenas Ibéricas entre el Valle del Jalón y la Sierra de la Demanda. *Publicaciones Extranjeras Sobre Geologia de España*, 63–143.

Royo Gomez, J. (1922). El Mioceno Continental Iberico y su fauna malacológica. *Junta para Ampliación de Est. e Invest. Cient. Com. de Invest. Paleontol Prehist.*, **30**, Madrid.

Platt, N.H. (1990). Basin evolution and fault reactivation in the western Cameros Basin, Northern Spain, *J. Geol. Soc., London*, **147**, 165–175.

Sanchez de la Torre, L. (1963). El borde Mioceno en Arcos de Jalón. *Estud. Geol.*, **19**, 109–136.

Sylvester, A.G. (1988). Strike–slip faults. *Geol. Soc. Am. Bull.*, **100**, 1666–1703.

Woodcock, N.H. (1986). The role of strike–slip fault systems at plate boundaries. *Royal Soc. Lond. Phil. Trans., Series A*, **317**, 13–29.

W5 Tertiary basins and Alpine tectonics in the Cantabrian Mountains (NW Spain)

J.L. ALONSO, J.A. PULGAR, J.C. GARCÍA-RAMOS AND P. BARBA

Abstract

Two sub-aerial major Tertiary basins developed during the Alpine orogeny in the Cantabrian Mountains: the Oviedo Basin and the Duero Basin. The Oviedo Basin, now deeply eroded, is located in front of a thrust developed by the tectonic inversion of a Late Jurassic–Early Cretaceous rift-related normal fault. The northern margin of the Duero Basin, with a well-preserved succession up to 2500 m thick, occurs in front of a large basement-cored uplift developed as a result of thrusting over a long ramp connected to a midcrustal detachment. The movement along this thrust carried the Oviedo Basin southwards in a piggyback manner. Both basins show syntectonic unconformities ahead of the mountain front, associated with syndepositional thrust-related folds. Syntectonic alluvial fans developed along the mountain front. These are more extensive in the Duero Basin (with radii up to 25 km) than in the Oviedo Basin (with radii up to 5 km). In both basins, the succession was overall prograding and changes vertically from polymictic to siliceous conglomerates. This lithological variation is attributed to a climatic change towards more humid conditions. The later siliceous episodes were essentially post-tectonic in both basins. An erosion–deposition mass balance is calculated for the Duero Basin.

Introduction

The Cantabrian Mountains constitute the western extension of the Pyrenees (Fig. 1A); their westernmost part (Asturias and León provinces) is formed by Paleozoic basement, which represents a Hercynian, east-vergent, arcuate, thin-skinned foreland thrust and fold belt (Julivert, 1971; Perez-Estaún *et al.*, 1988), uplifted during the Alpine deformation over two synorogenic Tertiary basins: the Duero Basin in the South, and the Bay of Biscay in the North (Fig. 1A). Another smaller Tertiary basin, the Oviedo Basin, formed between them. Both the Oviedo and the Duero Basins are related to Alpine south-verging structures, whereas the Tertiary offshore basin, not described in this chapter, is related to north-verging Alpine structures (Boillot *et al.*, 1979; Derégnaucourt & Boillot 1982; Boillot & Malod, 1988). This offshore basin extends eastward into the northernmost structures of the Basque–Cantabrian Zone and the North Pyrenean Zone, with that same vergence.

The Alpine cycle began in this area with Permo-Triassic rifting and basin development (Lepvrier & Martínez-García, 1990). A later extensional phase, related to the opening of the Atlantic Ocean and the Bay of Biscay, followed during the Late Jurassic and Early Cretaceous times. The basin remained stable from Albian times until tectonic inversion during the Tertiary. The Alpine deformation and its relation to Tertiary basins in this part of the Iberian Peninsula is poorly understood relative to the central and western part of the Pyrenees, which have been the subject of many recent papers (Seguret, 1972; Cámara & Klimowitz, 1985; Puigdefàbregas *et al.*, 1986; Choukroune & ECORS Team, 1989; Roure *et al.*, 1989; Burbank *et al.*, 1992; Muñoz, 1992; Puigdefàbregas *et al.*, 1992; Teixell, 1992; Vergés *et al.*, 1992).

The main aim of this chapter is to explain the evolution of the Tertiary basins during the Alpine orogenic event. The application of a new structural model to the Alpine deformation in the Cantabrian Mountains (Pulgar & Alonso, in press) allows the main topographic features of the basement uplift that supplied the detritus that infilled the basins to be explained. Thus, the amount of eroded and deposited rock can be calculated. In addition, the depositional systems that filled each basin and the structure of their synorogenic deposits are described. Pre-existing data (in particular geological maps) combined with new field mapping and data on the structure and stratigraphy of the Tertiary synorogenic deposits have been used for the structural and stratigraphic analysis. Seismic reflection profiles supplied by *REPSOL Exploration* are used for the interpretation of the structures underlying the Tertiary sediments of the Duero Basin.

The Alpine structure of the Cantabrian Mountains

In the Cantabrian Mountains, the Alpine orogeny involved detachment and uplift of the Hercynian basement. However, in contrast to the southern Pyrenees, the Mesozoic cover in the Cantabrian Mountains remained undetached. The outcrop pattern of Paleozoic, Mesozoic and Tertiary rocks in the Cantabrian Mountains is shown in Fig. 1B, together with the Hercynian and

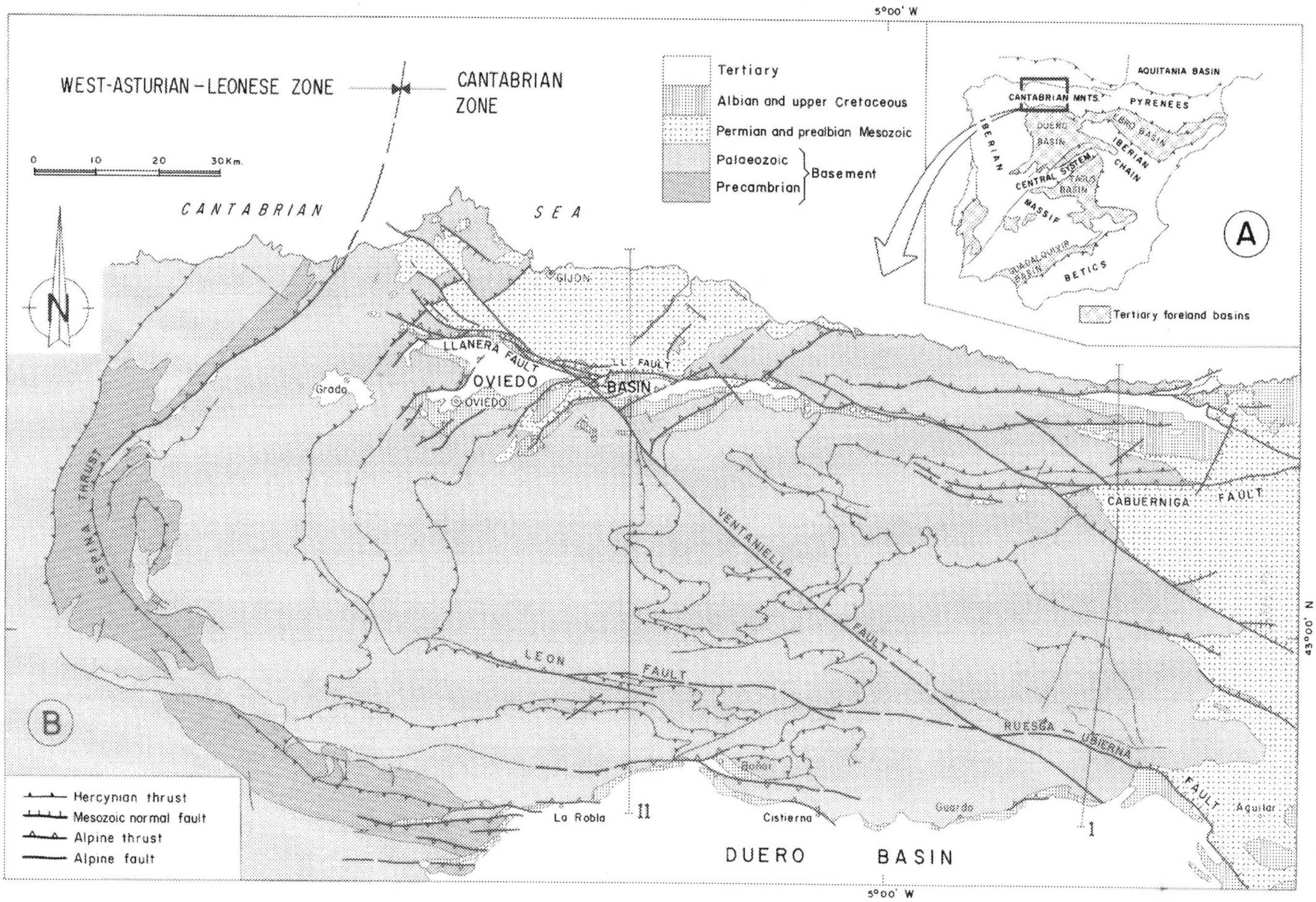

Fig. 1. A: Location map for the study area. B: Geological map of the Cantabrian zone, showing the main Hercynian thrusts, Mesozoic extensional faults and Alpine thrusts. The locations of cross-sections I and II in Fig. 2 are shown.

Alpine structural elements. The Alpine structure can be observed in the cross-sections of Fig. 2. Note that in both figures many Alpine faults reactivate earlier Hercynian thrusts or invert Mesozoic rift-related normal faults.

The overall structure of the Cantabrian Mountains consists of a regional monoclinal flexure leading to an extensive uplift of tens of kilometres. The structural style of this flexure may be explained in terms of fault–bend fold models in which the northern slope of the mountain range forms the 'dorsal culmination' (Butler, 1982) of a fault–bend fold whose displacement is lesser than the length of the ramp (Suppe, 1983, Fig. 3A). The 15° N dip of the Mesozoic cover in the dorsal culmination allows the angle of the ramp to be predicted, and the length of the culmination would represent a bed displacement of approximately 25 km (Fig. 2). Uplift of the Cantabrian Mountains can, therefore, be explained as a result of a basement thrust fault with a long ramp. A more detailed analysis of the relations between displacement, shortening and uplift in this mountain range is to be found in Pulgar & Alonso (in press).

The cross-sectons of Fig. 2 cut largely through Paleozoic rocks. However, the eroded structure of the Mesozoic cover has been reconstructed in cross-section I by projecting the structures of the Mesozoic rocks located to the east (Espina, 1992), where fold axes dip eastwards, onto the plane of the section (Fig. 1B). To the west of cross-section I, Alpine structures can be assumed to have a subhorizontal axial disposition since this area has a similar degree of erosion of the Tertiary rocks in the Oviedo Basin. Furthermore, small patches of Mesozoic rocks found on the west of the cross-section, in the extension of the Cabuérniga fault (Fig. 1B), have also been used to infer the height and disposition of the Mesozoic cover in cross-section II (Pulgar & Alonso, in press). The horizontal attitude of the top of the basement in this area can also be inferred from the presence of a level of summits that indicates uplift with no apparent tilting of the central massif of the mountain range.

The structural relief created by the monoclinal flexure is still evident in the present relief (Fig. 3, cf. Fig. 2); the present-day watershed is not far from the upper dorsal culmination wall hinge that must have represented the initial watershed in the beginning of the uplift of the mountain range. The winding trace of the present-day watershed results from the different erosion rates undergone by the lithological units of the basement, although the overall trace is

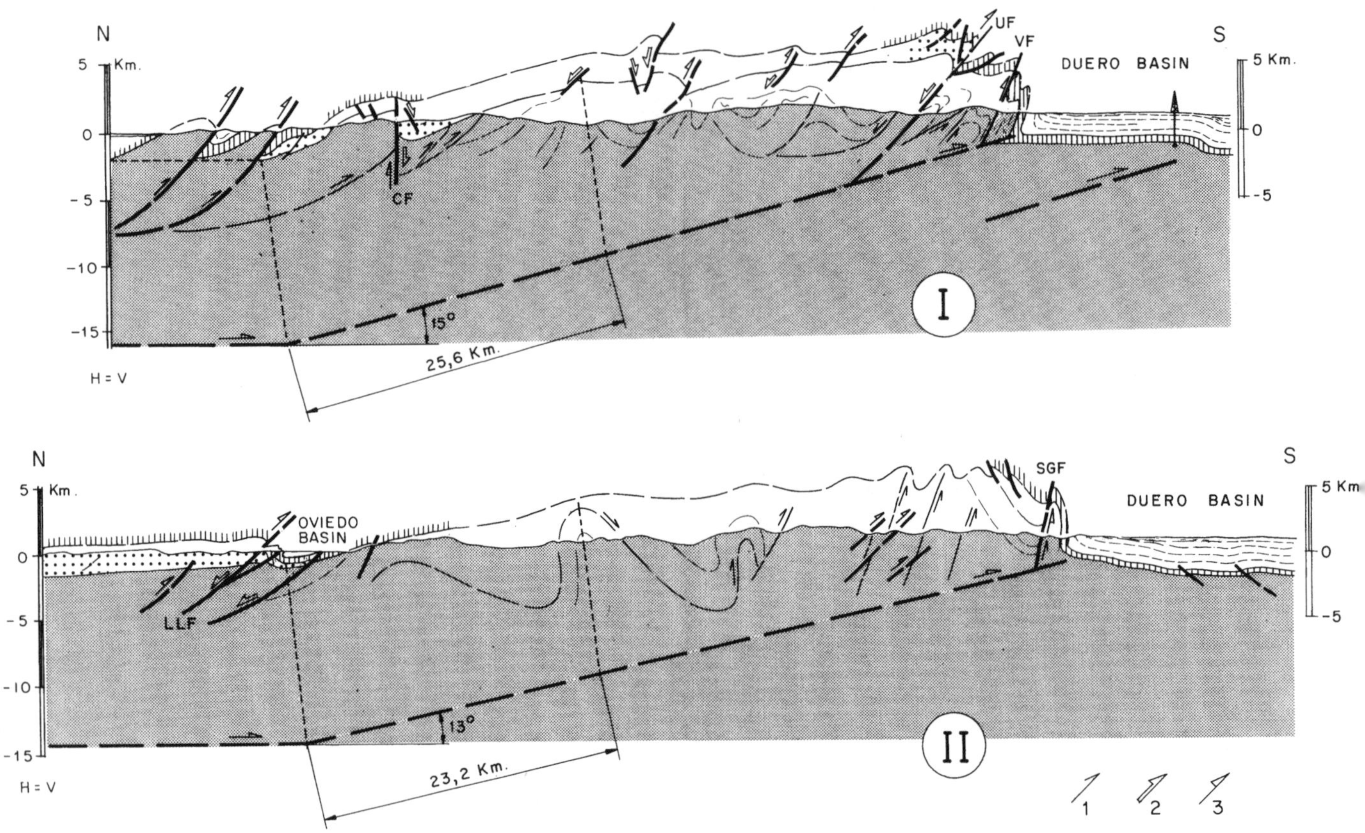

Fig. 2. Cross-sections through the Cantabrian Mountains. For location of cross-sections see Fig. 1. 1, Slip sense of the Hercynian thrust; 2, slip sense of the Mesozoic extensional faults; 3, slip sense of the Alpine thrusts. UF, Ubierna Fault; VF, Ventaniella Fault; SGF, Sabero-Gordón Fault; LLF, Llanera Fault; Cabuérniga Fault. H=V – horizontal=vertical (no vertical exaggeration).

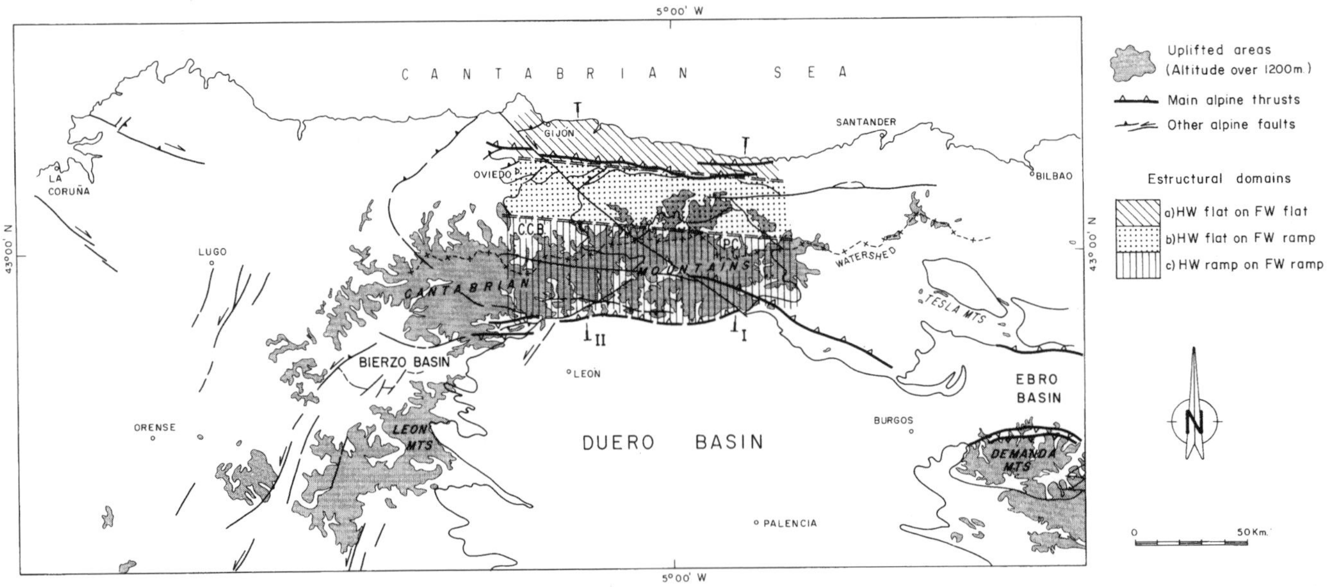

Fig. 3. Morpho-tectonic map showing structural interpretation of the Cantabrian relief. Compare with cross-sections in Fig. 2.

E–W. The area, whose height is over 1200 m, has many summits ranging between 1800 and 2000 m.

The basement thrust resulted in the folding and the overturning of the Mesozoic rocks (Fig. 2) ahead of the upper part of the ramp (see below) similar to uplifts found in other mountain ranges (Rodgers, 1987; DeCelles *et al.*, 1991; Schumaker, 1992) and by the experimental models (Chester *et al.*, 1988; Chester *et al.*, 1991). However, no major thrust can be seen on the surface. Several small reverse faults associated with the overturned fold limb do occur. In our viewpoint the displacement of the basement structure, estimated to be around 25 km, is accommodated by the frontal fold and the longitudinal shortening of the hanging wall, evidenced by the cover–basement structural relationships (Fig. 2).

The Mesozoic cover rocks are overturned along most of the mountain front (Fig. 4A). In the easternmost area, the Mesozoic cover has a second-order asymmetric fold whose axis shows a variable E and W plunge (Fig. 4A and cross-section III–III′ in Fig. 5). To the east, this fold involves Stephanian beds. East of La Robla (Fig. 4A), another fold with the same asymmetry also affects the Paleozoic basement. The folds in the Boñar area are associated with a fault that can be interpreted as a backthrust rotated by simple shear related to the larger thrust responsible for the uplift of the mountain range (Fig. 4A and cross-section IV–IV′ in Fig. 5). In this area, inversion is less remarkable, perhaps because the displacement was largely accommodated by the backthrust. The geometry of this structure is similar to those from some experimental works (Serra, 1977; Chester *et al.*, 1991; Liu Huiqui *et al.*, 1992) or natural examples (Teixell, 1992).

Minor thrusts with a WNW trend occur all along the front zone. These are NE-dipping with an angle less than bedding which results in stretching of the overturned limb and omission of strata. These thrusts affect all the Mesozoic, Palaeozoic and Tertiary rocks (Fig. 4A and cross-sections III–III′ and V–V′ in Fig. 5).

As has already been said, the behaviour of the basement was not rigid. Instead, the pre-Alpine structures, Hercynian folds and thrusts were reactivated to varying extents (Fig. 1B). The Alpine deformation of the basement is particularly important along the southern border of the mountain range, where the Stephanian formations have undergone a rotation and overturning similar to the Cretaceous rocks (Figs. 4 and 5). This implies that most of the post-Stephanian deformation in this sector of the Cantabrian Zone, which caused the overturning to the south of the Hercynian thrusts (Alonso, 1989), must be Alpine. In the NE corner of the Duero Basin, reactivation and steepening of the SSW-dipping Hercynian thrusts occurred in the Precambrian rocks of the Narcea Antiform (Figs. 4 and cross-section V–V′ in Fig. 5). The antiform tightening presumably caused the tilting of Mesozoic rocks and reactivation of Hercynian thrusts by flexural-slip mechanisms. This explains the present steep attitude of these thrusts. The seismic profiles show that these faults extend eastwards as buried thrusts, although with only minor displacements (cross-section II in Fig. 2).

The relationships between the structure of the Cantabrian Mountain Range and the Duero and Oviedo synorogenic Tertiary basins can be seen in Figs. 1B and 2; the northern margin of the Duero Basin can be considered as a true foreland basin ahead of the mountain front. The uplift of the range must have also caused a crustal flexure giving rise to a foredeep on the margin of the Duero Basin (1.5 km subsidence). The Oviedo Basin can be considered as a piggyback basin developed as a result of tectonic inversion of the Llanera extensional fault, which causes a minor uplift. The Tertiary succession shows syntectonic unconformities in basin-margin sequences in the Oviedo and Duero Basins, implying contemporaneous rotation of beds and sedimentation (Figs. 4A and B, 5 and 7).

Stratigraphy and depositional environments

Northern margin of the Duero Basin

The Duero Basin is a wide endorheic Tertiary basin whose northern margin is formed by an assemblage of alluvial fans developed at the foot of the southern slope of the Cantabrian Mountains. Fig. 6 shows four synthetic stratigraphic columns of the Tertiary succession in this area. In the Duero Basin, the Tertiary rocks lie disconformably on the Mesozoic. The lower stratigraphic levels form a coarsening-upward sequence known as Vegaquemada Fm. (Evers, 1967). The upper part is mostly conglomeratic and consists of several alluvial fans of different size and lithological composition (Fig. 4C and D). On the basis of variation in fan morphology towards the inner part of the basin, two groups can be differentiated:

(a) Highly efficient alluvial systems whose drainage basin was deeply sourced in the inner part of the northern mountain range. These were major fans that deposited the polymictic conglomerates in the lower part (Candanedo, Modino, Puente Almuhey and lower Aviñante systems) and the siliceous conglomerates in the upper part (Barrillos, Vidanes, Cegoñal, upper Aviñante, Guardo and Cantoral systems). However, in the western margin of the basin, a siliceous fan, with a higher proportion of sand than the upper clay-rich siliceous fans, is found in the lower part of the succession (La Robla alluvial system). This later relationship represents an anomaly in the above-mentioned arrangement (Fig. 4C and D). The depositional slope of the siliceous systems (perhaps with the exception of La Robla) was smaller than that of the polymictic systems.

(b) Poorly efficient alluvial systems whose drainage basin lay just at the mountain front. These were small, usually coalescent fans, consisting of calcareous clasts derived from steep calcareous relief (Cuevas system, lower and middle part of the succession, and Quintana de la Peña system, upper part of the succession) (Fig. 4C and D). Their source rocks are Cretaceous limestones in the Cuevas system and mainly Carboniferous limestones in the Quintana de la Peña system. Their depositional slope was higher than those in group (a).

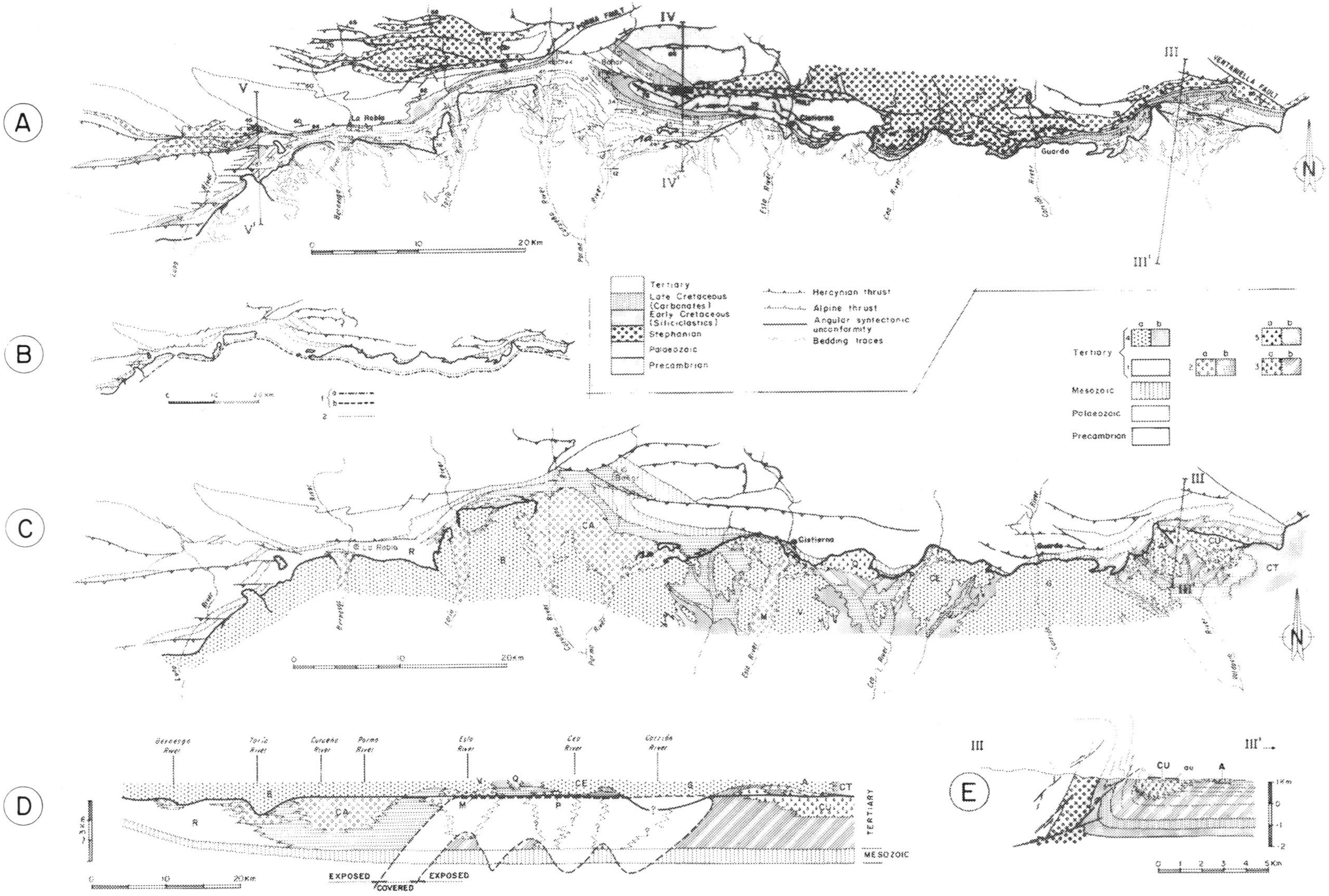

Fig. 4. A, Structural map of the northern border of the Duero basin. The locations of cross-sections in Fig. 5 are shown. B, Structural sketch showing the inferred tip of the angular syntectonic unconformity. 1a: buried, 1b: eroded; 2: trace location of the longitudinal section of Fig. 4D. C, Geological map showing the composition and names of different alluvial fans. Lower systems: 1. Quartzitic system with sandy matrix (R–La Robla); 2. Polymictic systems consisting of Paleozoic and Mesozoic clasts (CA–Candanedo, M–Modino, P–Puente Almuhey); 3. Polymictic systems consisting of Mesozoic clasts (CU–Cuevas). Upper systems: 4. Quartzitic system with argillaceous–sandy matrix (B–Barrillos, CE–Cegoñal, G–Guardo, CT–Cantoral, V–Vidames); 5. Polymictic systems mainly consisting of Paleozoic clasts (Q–Quintana de la Peña). a: proximal deposits, predominantly conglomerates. b: distal deposits. D, Longitudinal section along the northern border of the Duero Basin showing the lateral relationships between different depositional systems. Key as in C. E, Cross-section showing the disposition of proximal and distal facies. Geological map after Colmenero *et al.* (1978), García-Ramos *et al.* (1978), Manjón *et al.* (1978) and incorporating our own data.

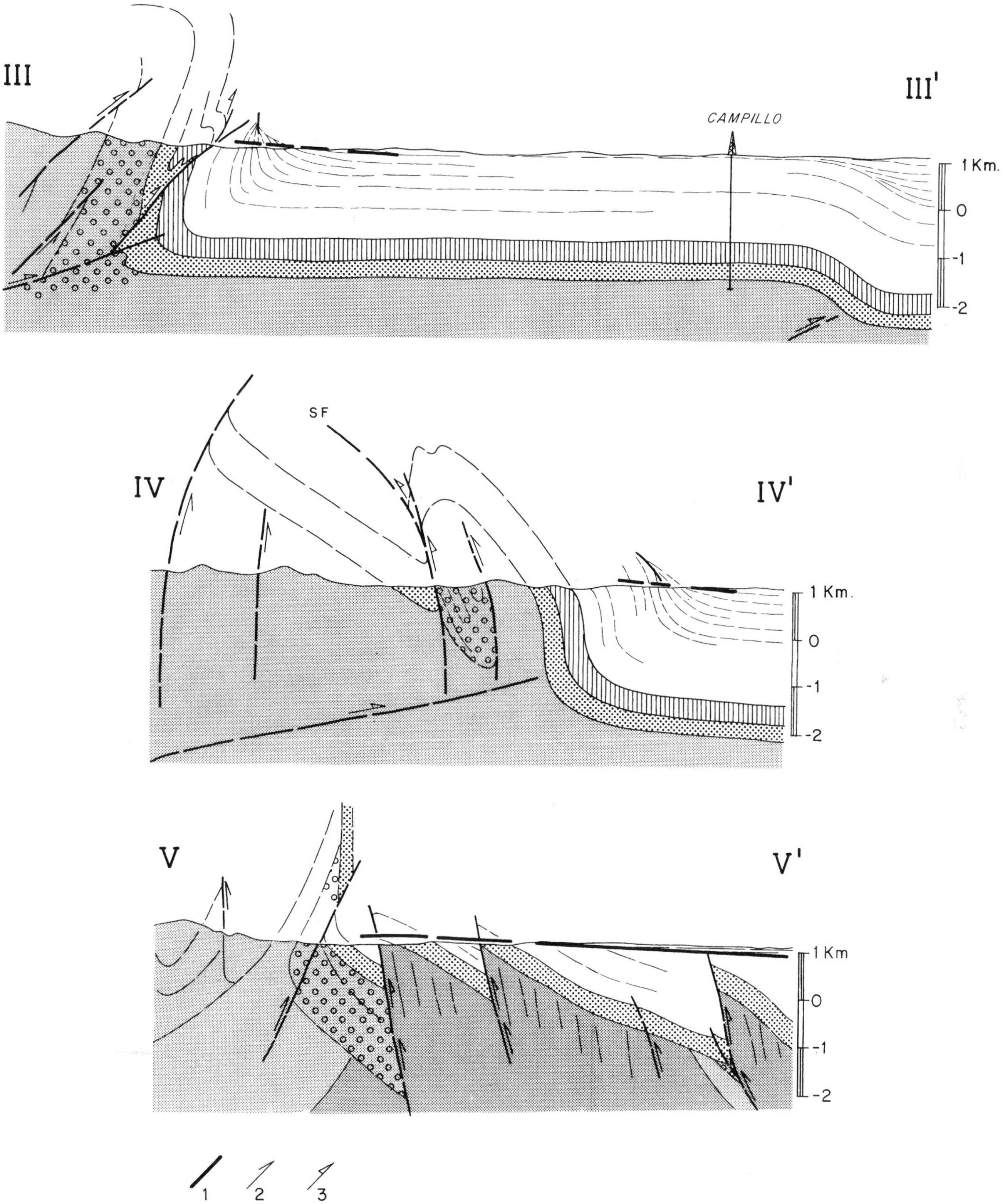

Fig. 5. Cross-sections through the northern border of the Duero Basin. Location and key in Fig. 4 A. 1, intraformational angular unconformity; 2, slip sense of Hercynian thrusts; 3, slip sense of the Alpine thrusts. SF, Sabero Gordón Fault.

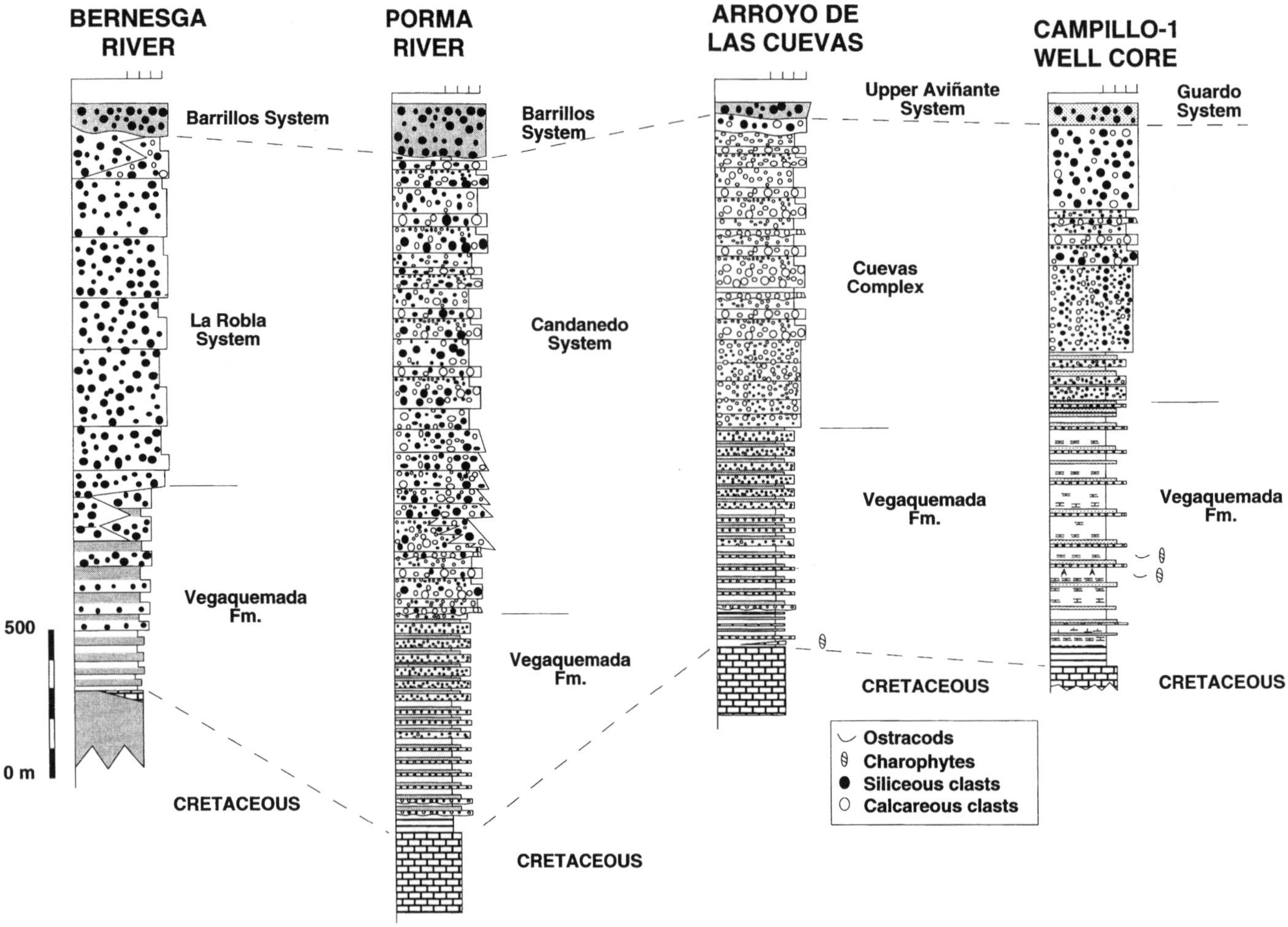

Fig. 6. Stratigraphic sections of the northern border of the Duero Basin. For location see Figs. 4 and 5.

The clastics were mainly supplied from the north, although in the westernmost part of the basin the Paleozoic siliceous clasts have, at least partially, a northwestern origin (La Robla alluvial system, see below). However, part of the sandy materials was probably derived from the erosion of Cretaceous sands. La Robla fan characteristics suggest the longitudinal infill of a subsidising trough parallel to the mountain front. On the other hand, the Cretaceous–Tertiary transition beds are better preserved to the east of Guardo, which suggests a provenance change eastwards to more lateral or to more distal from the Cuevas and La Robla alluvial systems, respectively.

For further details of the sedimentation in the margin of the Duero Basin in the area of study, see Ciry (1939), Mabesoone (1959), Evers (1967), Colmenero *et al.* (1982a, b), García-Ramos *et al* (1982a, b), Majón *et al.* (1982a, b), Sánchez de la Torre (1982), García-Ramos *et al.* (1986), Corrochano (1989) and Remondo & Corrochano (1992).

Oviedo Basin

The sedimentary record of the Oviedo Basin is incomplete, with a maximum preserved thickness of approximately 400 m. The only data available on its age are from the base of the series, which, in Oviedo, is Upper Eocene according to data from charophytes and mammalian remains (González Regueral & Gómez de Llarena, 1926; Truyols & García-Ramos, 1992). The base onlaps southwards over a karstified Cretaceous paleorelief, and thus in the north there are older Tertiary deposits overlying the unconformity. A geological map and cross-section of the Oviedo Basin are shown in Fig. 7. A synthetic column of the basin in the Oviedo area is illustrated in Fig. 8.

The northern margin of the basin consists of a fringe of small coalescent alluvial fans of poorly efficient transport power with radii of up to 4 or 5 km. Debris cones can also be recognized. Conglomerate clasts within the fans consist mainly of Cretaceous limestones with only a small proportion of siliceous clasts (Fig. 7). The clastics were supplied from the north by erosion of the Mesozoic and Paleozoic rocks situated in the hanging wall of the Llanera Fault.

The alluvial plain areas further from the northern margin consist predominantly of pinkish and reddish-orange mudstones that, near the periphery of the lacustrine areas, are grey and greenish. These mudstones are interbedded with sandstone channel bodies de-

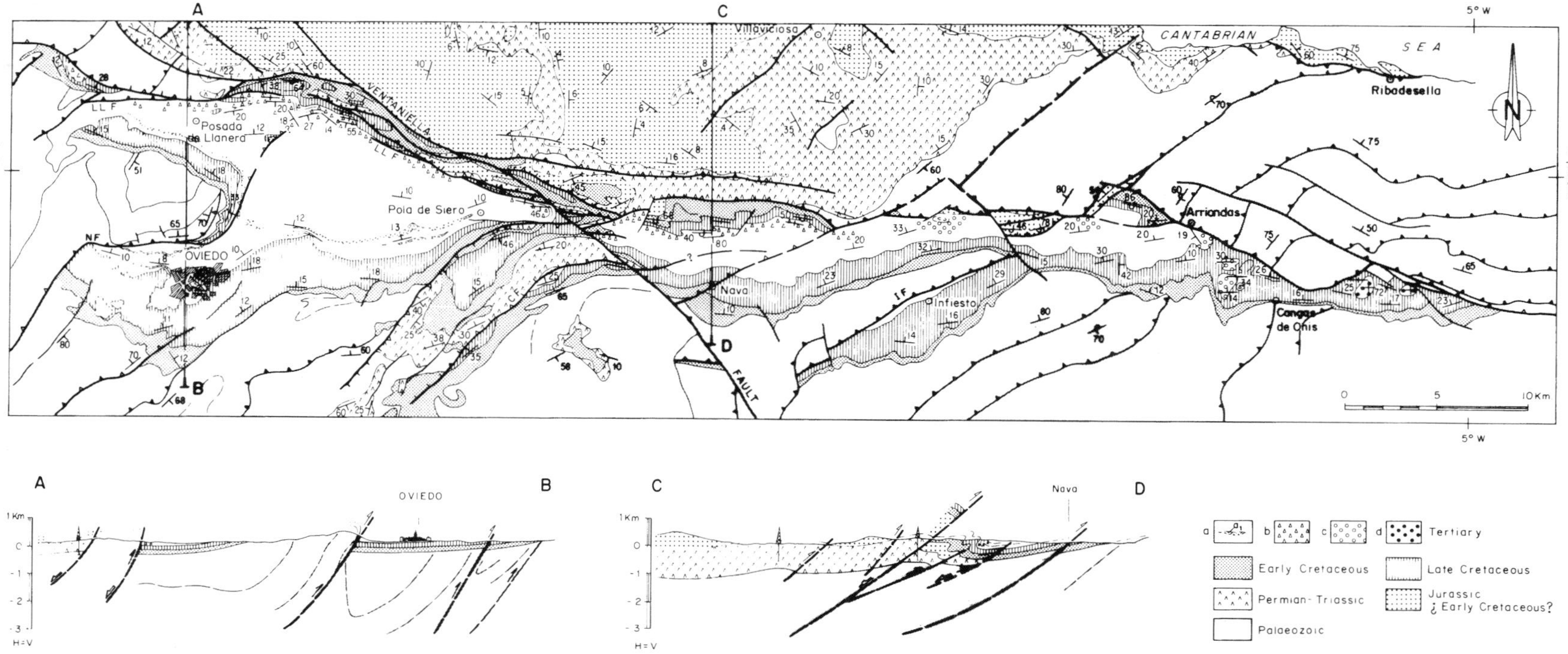

Fig. 7. Geological map and cross-sections of the Oviedo Basin. a: Predominantely mudstone with lacustrine limestones (a_1). b: Polymictic systems consisting of Mesozoic clasts. c: Polymictic systems consisting of Paleozoic and Mesozoic clasts. d: Quartzitic system with argillaceous–sandy matrix. Geological map mainly after Almela & Ríos (1962) incorporating our own data, also including data from Llopis Lladó (1950) and Gutierrez-Claverol (1972).

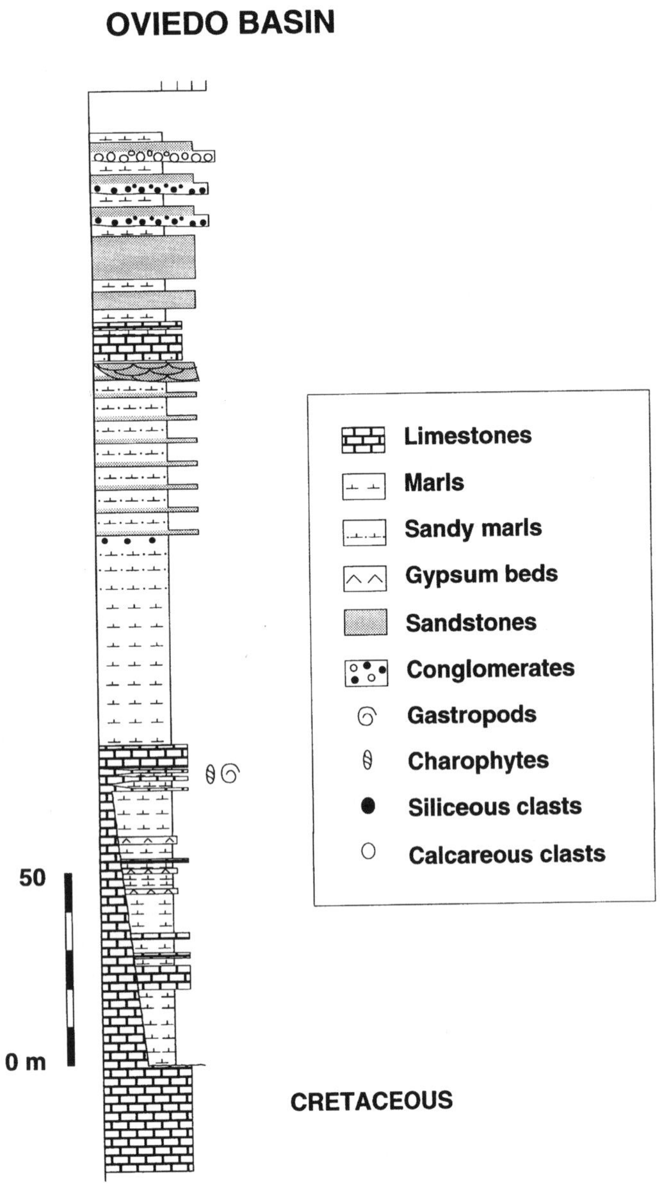

Fig. 8. Stratigraphic section of the Oviedo Basin.

posited from ephemeral currents. In the central and southern parts of the basin, several beds (3 to 4 m thick) of white micritic, marly limestones occur. These are of lacustrine origin and contain gastropods, ostracods and algae (mainly charophytes). Gypsiferous evaporitic horizons are also found. The small thickness of these limestone beds suggests deposition in ephemeral and very shallow lakes with muddy waters that underwent periodic drying. The occurrence of caliches and evaporites together with a scarcity of vegetal remains suggests the climate was semiarid during this period.

The uppermost part of the Tertiary succession outcrops in two areas; to the west of Oviedo (around the village of Grado, Fig. 1B) and to the east of Cangas de Onís (Fig. 7). In these areas it consists of relatively thin alluvial beds composed of siliceous conglomerates and sandstones. The drainage sense of this alluvial system is opposite to that of the lower polymictic conglomerates.

The types of facies, their distribution and the paleocurrents imply that the basin must have been endorheic, at least during the Paleogene.

Tectonics–sedimentation relationships

Structural and sedimentary evolution of basin margins

In both the Duero and Oviedo basins a wedge of non-marine sediments prograded from the basin margin during Alpine deformation. Syntectonic unconformities in the active margin sedimentary series in both basins show contemporaneous rotation of beds and sedimentation during deformation. More proximal sections were eroded during rotation, while a more complete record is preserved towards the foreland where deposits remain undeformed (Figs. 4E and 5).

In the Oviedo basin, syntectonic unconformities occur at the base of the Tertiary succession (cross-section C–D in Fig. 7), whereas in the Duero Basin, syntectonic unconformities are in the late Tertiary succession (Fig. 5). Nevertheless, uplift of the mountain range must have taken place earlier since proximal alluvial fan deposits occur in the lower part of the succession in some sectors (e.g., 5 km to the west of Guardo) of the margin of the Duero Basin (Cuevas system, Fig. 4D). In the rest of the margin, the apex of the fans was never too far away from the present mountain front.

Duero Basin

Syntectonic unconformities are better preserved in the Duero Basin than in the Oviedo Basin because the upper part of the Tertiary succession is present (Fig. 5). In the Duero Basin, closer to the mountain front, the Tertiary–Mesozoic disconformity is subvertical or overturned, with an isopachic geometry in the first 1000 m of the succession. In the upper part of the succession, bedding dips become progressively more gentle and younger beds eventually become horizontal. This pattern corresponds to a syntectonic unconformity in Riba's sense (Riba, 1976). The bedding traces in Tertiary rocks (Fig. 4A) show the progressive decrease in dip, as well as the local angular unconformity dying out southwards and the onlap of the younger beds (Figs. 4 and 5). In some cases (e.g., to the east of Guardo, Fig. 4A), the syntectonic unconformity has a complex character with two angular unconformities with progressive dip changes. However, the bedding dip changes laterally and, therefore, the geometry of the syntectonic deposits also changes.

The syntectonic unconformity outcrops in the eastern and western parts of the northern margin of the Duero Basin (Fig. 4A). In the central sector (Guardo–Cistierna sector) the progressive unconformity becomes almost completely covered in angular unconformity by the younger deposits onlapping even over the preorogenic Mesozoic and Paleozoic sediments (Fig. 4A and B). The complete trajectory of the unconformity tip line has been inferred in Fig. 4B, and the deformed sector of the Tertiary basin seems to form a fringe of similar thickness along the mountain front.

In the sector to the southeast of Guardo, the reflection seismic profiles show another later progressive unconformity associated with a minor uplift approximately 14 km wide, located ahead of today's mountain front and buried under Tertiary sediments (cross-section I in Fig. 2 and III–III′ in Fig. 5). This unconformity implies the southwards propagation of the deformation, in the eastern sector, while to the west the main mountain front remained active.

The angular unconformity at the base of the upper fans is associated with onlap onto a paleorelief as the fans backfilled up the valleys (locally with up to 200 m of incision preserved) from which they were sourced. These relationships are better observed in a section longitudinal to the mountain front (Fig. 4D). Obviously, the steeper paleorelief occurs when the unconformity is on the Paleozoic basement (Quintana de la Peña, Puente Almuhey and Cegoñal fans). The bedding traces in the Tertiary (Fig. 4A) allow both the paleorelief and the correlation of the fans to be observed.

The basal conglomeratic deposits on this angular unconformity still have a polymictic or calcareous character (upper part of Candanedo, Modino, Cuevas, Puente Almuhey and lower Aviñante systems), similar to the fan sediments unconformably beneath them. Subsequently, the other highly efficient fan systems, with gentler slope, developed. They are formed by siliceous conglomerates with abundant sandy–clayey matrix (Barrillos, Vidanes, upper Aviñante, Cantoral, Guardo and Cegoñal systems). The disappearance of the carbonate clasts in the uppermost fans may be related to a climatic change to more humid conditions or to the progressive development of a hierachy and entrenchment of the feeding fluvial networks by preferential erosion in the siliciclastic formations. The rare calcareous clasts that reached the basin would have eventually disappeared due to intrastratal dissolution within the highly permeable gravelly bodies. The climatic change is also evidenced in the basin by the sudden vertical decrease in Paleozoic carbonate levels.

These siliceous conglomeratic deposits higher in the sequence were essentially formed after the deformation. In the area closer to the mountain front, to the east of Cistierna, they show dips of about 15° (Fig. 4A). However, in the eastern zone, closer to the mountain range, these deposits lie subhorizontal and are progressively unconformable on the frontal fold of the El Campillo uplift south of Guardo (Fig. 5).

Concerning the sedimentary evolution, the Tertiary succession shows a coarsening-upward general trend, although the uppermost part of the polymictic conglomerates usually fines upwards. The overlapping of adjacent fans represents a probable explanation for superimposing more distal and more proximal facies creating vertical fluctuations in the dominant coarsening-upward sequential trend.

From about the Pliocene–Pleistocene boundary to the present, the foreland has been an erosional instead of a depositional area. Many of the rivers that flowed transverse to the mountain range have gradually entrenched their own fans (except for the Aviñante and the Cantoral system E of Guardo). In the beginning, there must have been a minor and short entrenchment of the feeding channel on the upper fan apexes and the sediments would have settled down the river as a pediment mantle. They form thin and widespread bodies with fan morphologies and very smooth slopes known as 'rañas'. In some areas (e.g., to the E of Guardo), small alluvial cones, with radii of around 1 km, deposited at the foot of the mountain front, formed by calcareous breccias sourced from the Carboniferous limestones. Evidence of catastrophic flooding caused by glacial melting (García-Ramos *et al.*, 1986) suggests these deposits probably formed at the beginning of the Pleistocene interglacial periods. The entrenchment of the present fluvial system is probably related to the drainage change of the Duero Basin from endorheic to exorheic into the Atlantic Ocean.

Most of the Tertiary drainage network excavated into the Paleozoic basement is likely to have been controlled by the same factors as that of today. The apexes of major fans are located at Alpine structural lows (Candanelo alluvial system whose entry is located in the Boñar area) or in areas where the rocks are easily eroded (La Robla alluvial system, probably entering the basin through the Precambrian rocks of the Narcea Antiform).

Oviedo Basin

The sediments of the Oviedo Basin are generally tilted slightly northward. However, along the northern margin they are cut across by the Llanera thrust, or display progressive unconformities associated with the thrust-related folds (Fig. 7). The reactivation of the Hercynian thrusts with NE–SW trend that affect the Oviedo Basin (Naranco, Carbayín, Infiesto Faults, etc., Fig. 7), seems to have taken place after the movement of the Llanera thrust. This is supported by the fact that no clastic wedges occur in front of those faults and because there are outcrops of Palaeozoic rocks in their hangingwall whose fragments do not occur in the Tertiary series of this basin.

From the viewpoint of sedimentary evolution, the Tertiary succession presents an overall progradational trend. This can be seen in the sectors of the basin where a thicker succession is preserved (e.g., sectors north of Oviedo and east of Posada de Llanera) (Fig. 8).

The siliceous conglomerates in the upper part of the succession, outcropping to the west of Oviedo (Grado) and to the east of Cangas de Onís, are horizontal, essentially postdating the Alpine deformation. They probably represent a part of the succession equivalent to the uppermost siliceous fans of the Duero Basin.

Erosion–deposition mass balance in the Duero Basin

The Duero Tertiary Basin is a foreland basin whose geometry and infill are determined by three Alpine mountain ranges that surround it: the Cantabrian Mountains on the north, the Iberian Chain on the east and the Central System on the south (Fig. 1A). Crustal thickening during Alpine deformation in these mountain ranges caused bending of the lithosphere with progressive deepening at the base of the basin, developing troughs or foredeeps ahead of the mountain fronts (Fig. 9). In a cross-section from the

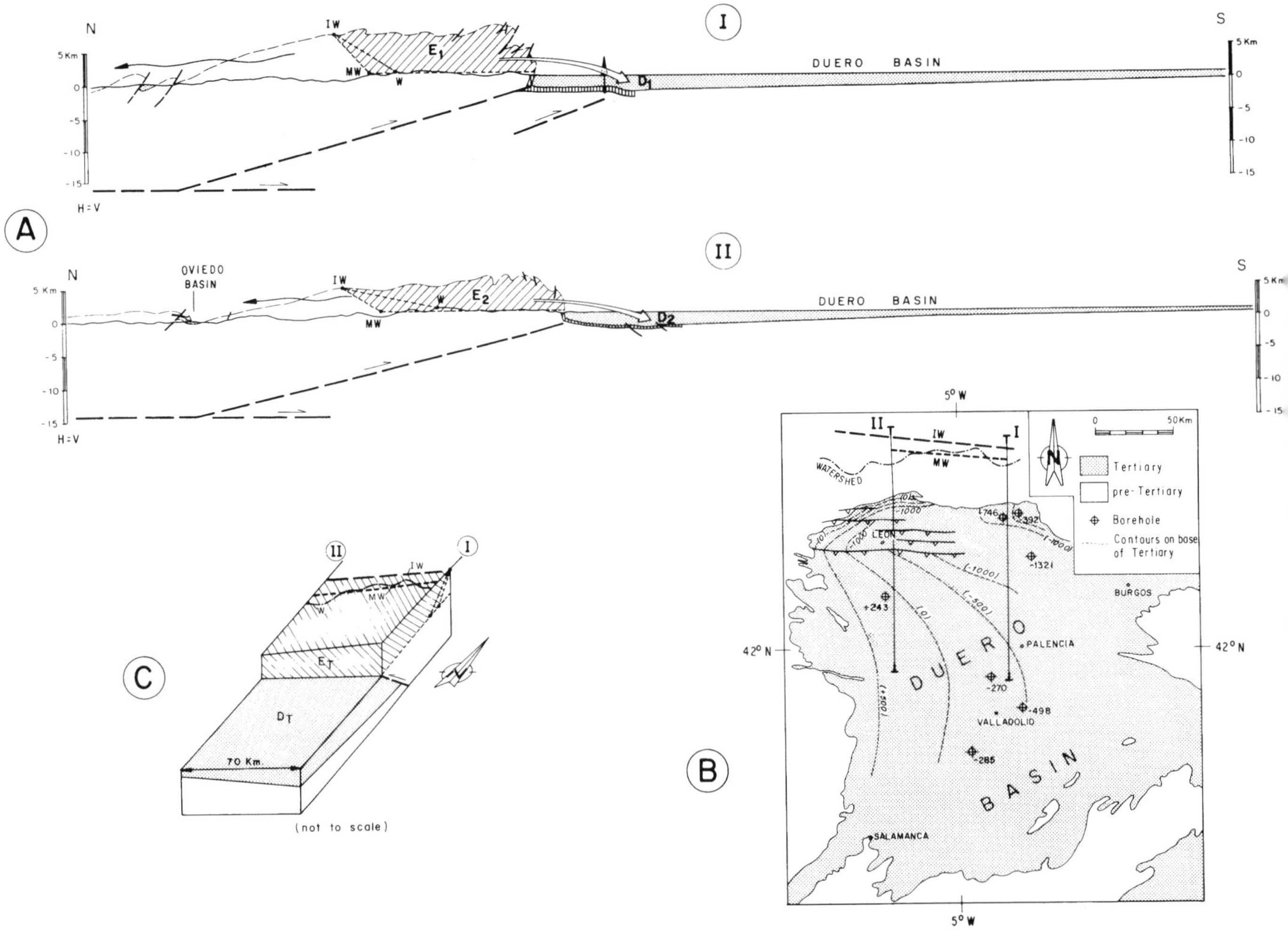

Fig. 9. A, Cross-sections through the Cantabrian Mountains and the northern part of the Duero Basin, showing the erosion–deposition mass balance. B, Sketch map of the Duero Basin showing contour of base of Tertiary and location of cross-sections in A. C, Three-dimensional sketch of erosion–deposition mass balance. See text for explanation. MW=mean watershed; IW=initial watershed. H=V – horizontal=vertical (no vertical exaggeration).

Cantabrian range to the basin centre, the geometry of the basin fill is wedge-shaped with the floor of the basin at depths greater than 1.5 km below sea level in the north (Fig. 9). The basin plunges eastwards, probably due to the larger tectonic load caused by the Iberian Range.

The reconstruction of the Alpine structure of the Cantabrian Mountains, illustrated in Fig. 2, is consistent with a mass balance (eroded and deposited) between the range and the basin (Fig. 9). Because of the winding trace of the present watershed, a mean watershed has been used to work out the amount of eroded mass, assuming that it has shifted from the initial watershed at the upper hinge of the regional monoclinal flexure. The watershed shifting is supported by the facts that: (a) the denudation of the mountain range has always been, and is still today, much more remarkable on the northern slope, where the Tertiary basins are completely eroded, in contrast with the well-preserved Duero Basin; (b) the regional slope and the fluvial entrenchment are much bigger on the northern slope than on the southern slope, due to the closer proximity of their base level to the watershed; and (c) capture phenomena are also recorded. The southwards shifting of the watershed also seems to be controlled by the lithology; hence, in the area where quartzites and limestones are more abundant (Ponga Region), the watershed is located more to the north than in the zones where more-easily-eroded shales are dominant, like the Central Coal Basin and the Pisuerga–Carrión province (CCB and PC in Fig. 3). To calculate the amount of deposited mass, it has been assumed that detritus from the Cantabrian Zone filled the Duero Basin, to a parallel line situated between the Palencia and Valladolid localities (Fig. 9B), since the Tierra de Campos Facies and Serna Facies, which are distal facies of the alluvial fans from the north (Portero *et al.*, 1982), reach that far.

Fig. 9 illustrates the mass balance both in cross-sectional area (Fig. 9A) and in volume (Fig. 9C). The values obtained for the eroded and deposited area respectively are $E_1 = 162.5$ km^2 and $D_1 = 190.2$ km^2 for section I, $E_2 = 130.7$ km^2 and $D_2 = 145.8$ km^2 for section II (see Fig. 9). The balance in volume has been calculated using the mean area of both sections multiplied by the width of the basin that separates them (70 km). The resulting values are: eroded volume $E_T = 10\,262$ km^3 and deposited volume $D_T = 11\,742.5$ km^3. The results for eroded and deposited volumes are reasonably consistent, considering that the sediment dispersal pattern is unknown. The difference between the volumes, and the higher volume for the Duero Basin, may be due to the contribution of the detritus from the basement located to the W of the basin, in which some fans originated (Corrales *et al.*, 1986) or to the different degree of compaction undergone by the eroded and the deposited rocks.

Conclusions

In the Cantabrian Mountains, the Alpine orogeny involved the Hercynian basement and the Mesozoic cover remained undetached, in contrast to the southernmost Pyrenees. The overall structure of the Cantabrian Mountains consists of a large monoclinal flexure that caused an extensive uplift of tens of kilometres. This can be explained in terms of fault–bend fold models. In this model, the northern slope of the mountain range is the dorsal culmination of a fault–bend fault. The 15° N dip of the cover in the dorsal culmination allows the ramp angle to be predicted, and the length of the culmination would represent the displacement of the bed, which is about 25 km. The uplift of the mountain chain is thus explained by means of a basement-involved thrust with a long ramp. The northern margin of the Duero Basin is the foreland basin situated ahead of the mountain front. The Oviedo Basin, behind the dorsal culmination wall, can be interpreted as a piggyback basin developed from the tectonic inversion of the Llanera extensional fault and minor uplift.

In the mountain fronts related to the aforementioned basement thrusts, the deformation in the upper part of the Mesozoic succession was accommodated by means of thrust-related folds. Hence the Tertiary rocks in these basin margins display syntectonic unconformities showing contemporaneous rotation of beds and sedimentation. The Duero Basin is better preserved from erosion than the Oviedo Basin and the syntectonic unconformities are therefore better preserved too. The syntectonic unconformities outcrop in the eastern and western sectors of the Duero Basin's northern margin; in the central sector the deposits involved in the progressive unconformity are almost completely unconformably overlain by younger deposits that onlap over the Mesozoic and Paleozoic succesion.

From the sedimentary viewpoint, the northern margin of the Duero Basin was formed by a set of alluvial fans developed at the foot of the southern slope of the Cantabrian Mountains. Highly efficient alluvial systems had their drainage area deeply sourced inside the northern mountain range. These coexisted with alluvial systems of poorly efficient transport power whose drainage area was close to the mountain front. Older alluvial fans consisted of polymictic conglomerates, whereas younger ones consisted of siliceous conglomerates, except in the western part of the basin where the whole composition is siliceous. The lateral lithological changes were related to different source areas, and the vertical ones to climatic changes and erosion depth in the drainage area.

The reconstruction of the Alpine structure on the Cantabrian Range is consistent with a reasonable mass balance (eroded and deposited) between the range and the Duero Basin.

The northern margin of the Oviedo Basin consisted of a fringe of small coalescent alluvial fans of polymictic conglomerates. The alluvial plain areas, further away from the northern margin, were mainly formed by mudstones. In the central and southern zones of the basin, several levels of white limestones of lacustrine origin occur, as well as different evaporitic horizons. The uppermost part of the Oviedo Basin fill contains siliceous conglomerates and sandstones from alluvial systems whose drainage sense was opposite to the lower polymictic conglomerates.

As a whole the Tertiary succession is prograding in both basins. In the Duero Basin the maximum progradation of conglomerates towards the inner part of the basin took place at the end of the Tertiary. A decrease in subsidence and a climatic change to more humid conditions could have contributed to this progradation.

Acknowledgements

This work received financial support from C.I.C.Y.T. Projects GE091–1086–C02–02 and PB92–1013. We wish to thank A. Marcos, L.P. Fernández and D.L. Brown for comments and help with the English version of the manuscript. We also thank REPSOL Exploración for releasing seismic data for the northern margin of Duero Basin.

References

Almela, A. and Ríos, J.M. (1962). *Investigación del Hullero bajo los terrenos Mesozoicos de la costa Cantábrica (zona de Oviedo–Gijón–Villaviciosa–Infiesto)*, Empresa Nacional Adaro de Investigaciones Mineras, Madrid, 159 pp.

Alonso, J.L. (1989). Fold reactivation involving angular unconformable sequences: theoretical analysis and natural examples from the Cantabrian Zone (Northwest Spain), *Tectonophysics*, **170**, 57–77.

Boillot, G., Dupeuble, P.A. and Malod, J. (1979). Subduction and tectonics on the continental margin off Northern Spain, *Mar. Geol.*, **32**, 53–70.

Boillot, G. and Malod, J. (1988). The north and north-west Spanish continental margin: a review, *Rev. Soc. Geol. Esp.*, **1**, 295–316.

Burbank, D.W., Vergés, J., Muñoz, J.A. and Bentham, P. (1992). Coeval hindward- and forward-inbricating thrusting in the south-central Pyrenees, Spain: timing and rates of shortening and deposition, *Geol. Soc. Am. Bull.*, **104**, 3–17.

Butler, R.W.H. (1982). The terminology in structures in thrust belts, *J. Struct. Geol.*, **4**, 239–245.

Cámara, P. and Klimowitz, J. (1985). Interpretación geodinámica de la vertiente centro-occidental surpirenaica, *Est. Geol.*, **41**, 391–404.

Chester, J.S., Logan, J.M. and Spang, J.H. (1991). Influence of layering and boundary conditions on fault-bend and fault-propagation folding, *Geol. Soc. Am. Bull.*, **103**, 1059–1072.

Chester, J.S., Spang, J.H. and Logan, J.M. (1988). Comparison of thrust rock models to basement-cored folds in Rocky Mountain foreland, *Geol. Soc. Am. Mem.*, **171**, 65–74.

Choukroune, P. and ECORS Team (1989). The Ecors Pyrenean deep seismic profile reflection data and the overall structure of an orogenic belt, *Tectonics*, **8**, 23–39.

Ciry, R. (1939). Étude géologique d'une partie des provinces de Burgos, Palencia, León et Santander., *Bull. Soc. Hist. Nat. Toulouse*, **74**, 1–519.

Colmenero, J.R., García-Ramos, J.C., Manjón, M. and Vargas, I. (1982a). Evolución de la sedimentación terciaria en el borde N de la Cuenca del Duero entre los valles del Torío y Pisuerga (León-Palencia), *Temas Geológico–Mineros, IGME*, **6**, 171–181.

Colmenero, J.R., Manjón, M., García-Ramos, J.C. and Vargas, I. (1982b). Depósitos aluviales cíclicos en el Paleógeno del borde N de la Cuenca del Duero (León-Palencia), *Temas Geológico–Mineros, IGME*, **6**, 185–196.

Colmenero, J.R., Vargas, I., García-Ramos, J.C., Manjón, M., Crespo, A. and Matas, J. (1978). *Mapa Geológico de España E. 1:50 000, Hoja No. 132 (Guardo)*, IGME, Madrid.

Corrales, I., Carballeira, J., Flor, G., Pol, C. and Corrochano, A. (1986). Alluvial systems in the north-western part of the Duero Basin (Spain), *Sediment. Geol*, **47**, 149–166.

Corrochano, A. (1989). Facies del Cretácico terminal y arquitectura secuencial de los abanicos aluviales terciarios del borde N de la Depresión del Duero (valle de Las Arrimadas, León), *Stud. Geol. Salmant.*, **Esp. 5**, 89–105.

DeCelles, P.G., Gray, M.B., Ridgway, K.D., Cole, R.B., Srivastava, P., Pequera, N. and Pivnik, D.A. (1991). Kinematic history of a foreland uplift from Paleocene synorogenic conglomerate, Beartooth Range, Wyoming and Montana, *Geol. Soc. Am. Bull.*, **103**, 1458–1475.

Derégnaucourt, D. and Boillot, G. (1982). Structure géologique du Golfe de Gascogne, *Bull. Bur. Rech. Geol. Mineral. France*, **2–I**, 149–178.

Espina, R.G. (1992). La estructura del borde occidental de la Cuenca Vasco-Cantábrica en el área de Campóo (Cantabria–Palencia, Norte de España), *Actas del III Congreso Geológico de España y VIII Congreso Latinoamericano de Geología*, **1**, 294–298.

Evers, H.J. (1967). Geology of the Leonides between the Bernesga and Porma rivers, Cantabrian Mountains, NW Spain, *Leidse Geol. Meded.*, **41**, 83–151.

García-Ramos, J.C., Colmenero, J.R. and Manjón, M. (1986). Un modelo muy peculiar de abanicos aluviales en el límite meridional de la Sierra del Brezo (N de Palencia), *Acta Salmaticensia, Ciencias*, **50**, 93–112.

García-Ramos, J.C., Colmenero, J.R., Manjón, M. and Vargas, I. (1982a). Modelo de sedimentación en los abanicos aluviales de clastos carbonatados del borde N de la Cuenca del Duero, *Temas Geológico–Mineros, IGME*, **6**, 275–289.

García-Ramos, J.C., Manjón, M. and Colmenero, J.R. (1982b). Utilización de minerales pesados y de espectros litológicos como ayuda en la identificación del área madre y en la separación de los diferentes sistemas de abanicos aluviales. Terciario del borde N de la Cuenca del Duero, *Temas Geológico–Mineros, IGME*, **6**, 293–301.

García-Ramos, J.C., Vargas, I., Manjón, M., Colmenero, J.R., Crespo, A. and Mitas, J. (1978). *Mapa Geológico de España E. 1:50 000, Hoja No. 131 (Cistierna)*, IGME, Madrid.

González Regueral, J. and Gómez de Llarena, J. (1926). Hallazgo de restos fósiles de un mamífero terciario en Oviedo, *Bol. R. Soc. Esp. Hist. Nat.*, **23**, 399–406.

Gutierrez-Claverol, M. (1972). Estudio geológico de la Depresión Mesoterciaria central de Asturias, PhD Thesis, Oviedo.

Julivert, M. (1971). Décollement tectonics in the Hercynian Cordillera of Northwest Spain, *Am. J. Sci.*, **270**, 1–29.

Lepvrier, C. and Martínez-García, E. (1990). Fault development and stress evolution of the post-Hercynian Asturian Basin (Asturias and Cantabria, northwest Spain), *Tectonophysics*, **184**, 345–356.

Liu Huiqui, McClay, K.R. and Powell, D. (1992). Physical models of thrust wedges, in *Thrust tectonics*, pp. 71–81, ed. McClay, K.R., Chapman & Hall, London.

Llopis Lladó, N. (1950). *Mapa Geológico de los alrededores de Oviedo*, Serv. Geol. Inst. Est. Asturianos, Oviedo.

Mabesoone, J.M. (1959). Tertiary and Quaternary sedimentation in a part of Duero Basin, Palencia (Spain), *Leidse Geol. Meded.*, **24**, 31–180.

Manjón, M., Colmenero, J.R., García-Ramos, J.C. and Vargas, I. (1982a). Génesis y distribución espacial de los abanicos aluviales siliciclásticos de Terciario superior en el borde N de la Cuenca del Duero (León–Palencia), *Temas Geológico–Mineros, IGME*, **6**, 357–370.

Manjón, M., García-Ramos, J.C., Colmenero, J.R. and Vargas, I. (1982b). Procedencia, significado y distribución de diversos sistemas de abanicos aluviales con clastos poligénicos en el Neógen del borde N de la Cuenca del Duero, *Temas Geológico–Mineros, IGME*, **6**, 373–388.

Manjón, M., Vargas, I., Colmenero, J.R., García-Ramos, J.C., Crespo, A. and Matas, J. (1978). *Mapa Geológico de España E. 1:50 000, Hoja No. 130 (Vegas del Condado)*, IGME, Madrid.

Muñoz, J.A. (1992). Evolution of a continental collision belt: ECORS–Pyrenees crustal balanced cross-section, in *Thrust tectonics*, pp. 235–246, ed. McClay, K.R., Chapman & Hall, London.

Perez-Estaún, A., Bastida, F., Alonso, J.L., Marquínez, J., Aller, J., Alvarez-Marrón, J., Marcos, A. and Pulgar, J.A. (1988). A thin-skinned tectonics model for an arcuate fold and thrust belt: the Cantabrian zone (Variscan Ibero-Armorican arc), *Tectonics*, **7**, 517–537.

Portero, J.M., del Olmo, P., Ramírez del Pozo, J. and Vargas, I. (1982). Síntesis del Terciario continental de la Cuenca del Duero, *Temas Geológico–Mineros, IGME*, **6**, 11–37.

Puigdefábregas, C., Muñoz, J.A. and Marzo, M. (1986). Thrust belt development in the Eastern Pyrenees and related depositional sequences in the southern foreland basin, in *Foreland basins*, pp. 229–246, ed. Allen, P.A. & Homewood, P., vol. 8, Special Publication of the International Association of Sedimentologists.

Puigdefábregas, C., Muñoz, J.A. and Vergés, J. (1992). Thrusting and foreland basin evolution in the Southern Pyrenees, in *Thrust tectonics*, pp. 247–254, ed. McClay, K.R., Chapman & Hall, London.

Pulgar, J.A. and Alonso, J.L. (in press). Inverted extensional basin and basement uplift in the Basque–Cantabrian Mountains, Western Pyrenees, N Spain, *Tectonics*.

Remondo, J. and Corrochano, A. (1992). *Estudio del Terciario del valle del rio Cea entre Almanza y Puente Almuhey: estratigrafía y sedimentología*, III Congreso Geológico de España, Salamanca. *Actas*, **1**: 191—195.

Riba, O. (1976). Syntectonic unconformities of the Alto Cardener, Spanish Pyrenees: a genetic interpretation, *Sediment. Geol.*, **15**, 213–233.

Rodgers, J. (1987). Chains of basement uplifts within cratons marginal to orogenic belts, *Am. J. Sci.*, **287**, 661–692.

Roure, F., Choukroune, P., Berastegui, X., Muñoz, J.A., Villien, A., Matheron, P., Bareyt, M., Seguret, M., Cámara, P. and Deramond, J. (1989). ECORS deep seismic data and balanced cross sections: geometric constraints on the evolution of the Pyrenees, *Tectonics*, **8**, 41–50.

Sánchez de la Torre, L. (1982). Características de la sedimentación miocena en la zona norte de la Cuenca del Duero, *Temas Geológico–Mineros, IGME*, **6**, 701–705.

Schumaker, (1992). Paleozoic structure of the central basin uplift and adjacent Delaware Basin, West Texas, *AAPG Bull.*, **76**, 1804–1824.

Seguret, M. (1972). *Étude tectonique des nappes et séries décollées de la partie centrale du versant sud des Pyrénées*, Montpellier, 155 pp.

Serra, S. (1977). Styles of deformation in the ramp regions of overthrust faults, *Wyoming Geological Association, Annual Field Conference, 29th, Guidebook*, 487–498.

Suppe, J. (1983). Geometry and kinematics of fault-bend folding, *Am. J. Sci.*, **283**, 684–721.

Teixell, A. (1992). Estructura alpina de la transversal de la zona axial pirenaica, PhD Thesis, Universitat de Barcelona.

Truyols, J. and García-Ramos, J.C. (1992). El Terciario de la cuenca de Oviedo y el yacimiento de vertebrados de Llamaquique, *Bol. Ciencias Natural, R.I.D.E.A.*, **41**, 77–99.

Vergés, J., Muñoz, J.A. and Martínez, A. (1992). South Pyrenean fold and thrust belt: the role of foreland evaporitic levels in thrust geometry, in *Thrust tectonics*, pp. 255–264, ed. McClay, K.R., Chapman & Hall, London.

W6 Lacustrine Neogene systems of the Duero Basin: evolution and controls

R. MEDIAVILLA, C.J. DABRIO, A. MARTÍN-SERRANO AND J.I. SANTISTEBAN

Abstract

Vertical aggradation of Neogene fluvial and lacustrine deposits occurred until the Late Neogene in central and northern areas of the Duero Basin, coeval with river incision in the south-western corner of the basin. The whole basin became exorheic in the Latest Neogene. We have differentiated five tectonosedimentary units (TSUs) of basinal extent, bounded by unconformities or breaks in the sedimentary record. Deposits in each TSU consist of alluvial-fan deposits in areas close to the active northern and eastern margins, and fluvial deposits along the western margin. These systems converged in the lower, subsiding areas of the basin occupied by carbonate–evaporite lacustrine systems.

Tectonics and climate controlled sedimentation. The main faults active from the Neogene to the Present reflect Late Hercynian basement fractures that were re-activated during the Alpine Orogeny, both fracturing blocks and modifying landscapes, and creating or modifying the areas of subsidence. Analysis of climatic variations during the Miocene shows that deposition of saline materials occurred in dry TSUs (1, 2) and, particularly, in humid TSUs (3, 4). Climate does not seem to have been a determining factor for the formation of evaporites. However, it was a very important factor in determining both the amount of water that reached the basin and, eventually, also the extent of the lacustrine systems.

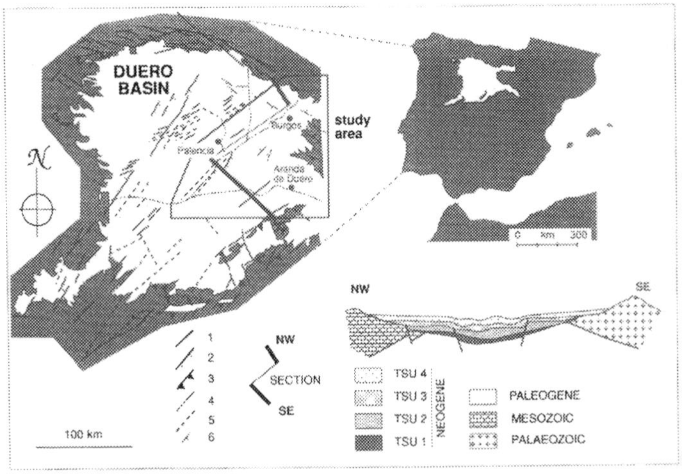

Fig. 1. Tectonic map of the Duero Basin (from Baena *et al.*, in press) and composite cross section of the central zone. 1: Fractures; 2: faults with indication of the sinking block; 3: overthrust; 4: fracture (supposed); 5: river; 6: strike and dip.

Introduction

By the end of Oligocene times, the northern and southern margins of the Duero Basin experienced final compressional stresses and were uplifted by reverse faults above the Paleogene deposits of the basin (Barba, pers. commun., 1992; see also Chapter W2). During the Neogene, sedimentation took place in an extensional regime, with major vertical movements, which continue at present.

Gentle subsidence in the south-western margin during the Neogene favoured the headward erosion of the Atlantic fluvial network that began to be defined in Uppermost Oligocene–Early Neogene times (Martín-Serrano, 1991; see also Chapter W3). Coeval to river incision at this side of the basin, high subsidence rates in the central and northern areas of the basin (Fig. 1) caused vertical aggradation of Neogene fluvial and lacustrine deposits until the end of the Neogene when the whole basin became exorheic. We have differentiated five tectonosedimentary units (TSU, *sensu* Megias, 1982 and Pardo *et al.*, 1989) of basinal extent separated by unconformities or breaks in the sedimentary record (Figs. 2 and 3).

In general, each of these TSUs consists of alluvial-fan deposits in areas close to the active north and east margins, and fluvial deposits along the west margin. At that time, Paleogene arkoses formed the southern margin along the Honrubia–Pradales Range, which constituted a relatively passive area as suggested by the scarce development of alluvial deposits.

All these systems converge in the lower, subsiding areas of the basin (Fig. 2) occupied by the carbonate–evaporite lacustrine systems studied in this paper. There is an abundant literature concerning these deposits (e.g. Hernández-Pacheco, 1915; Royo Gómez, 1926; San Miguel de la Cámara, 1946; García del Cura, 1974; Ordóñez *et al.*, 1980; Ordóñez *et al.*, 1981; Portero *et al.*, 1982;

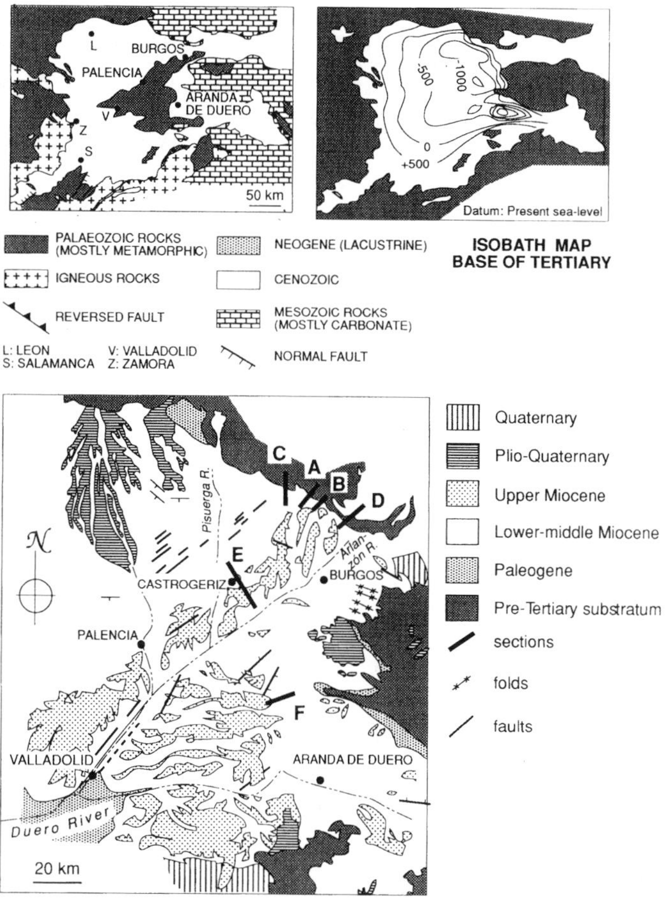

Fig. 2. Lacustrine Neogene deposits in the Duero Basin. A to F refer to cross-sections shown on Fig. 4.

Armenteros, 1986; Mediavilla & Dabrio, 1986, 1988, 1989a, 1989b; Corrochano & Armenteros, 1989; Pineda, in press; Pineda & Arce, in press; Mediavilla & Picart, in press; Piles & Picart, in press; López Olmedo & Enrile, in press; Nestares & Wouters, in press).

TSU 1

This reaches a maximum visible thickness of 50 m to the north of the study area (Fig. 3). Here it dips to the south-west. Near the Cantabrian and Iberian ranges it consists of alluvial gravels, sands and alluvial muds that rest unconformably upon Mesozoic and Paleogene deposits (Pineda & Arce, in press), (Fig. 4A). These coarse-grained sediments change to lacustrine carbonates and evaporites towards the south-west and north-east.

Lacustrine sediments crop out discontinuously in a NE–SW fringe parallel to the Pisuerga and Arlanzón river valleys. Maximum thicknesses are measured east of these rivers where evaporitic deposits (dolostones, microlenticular gypsum and gypsarenites) are best represented. There are also paludal fringes with carbonate sedimentation in positions laterally equivalent to these saline lakes (Fig. 8A).

As a whole, lacustrine sediments onlap Paleogene sediments in the southern margin (Fig. 5) and Mesozoic deposits of the Cantabrian Range in the north (Fig. 4B). Deposits of TSU 1 form an expansive and shallowing-upwards sequence topped by a major karstification profile (Pineda & Arce, in press). The deposition of TSU 1 records an important tectonic reactivation (basal unconformity) and the progressive infill of the basin under a delayed diastrophic regime.

TSU 2

This rests disconformably upon TSU 1 and covers the greatest area (Figs. 3, 4B and 4C). It crops out in the central Duero Basin with a maximum thickness of 60 m.

Near the margins of the basin, TSU 2 consists of coarse-grained siliciclastic sediments (conglomerate, sand, mud) of alluvial-fan and proximal fluvial systems flowing from source areas located to the north, north-east and east. These deposits fine out to more distal fluvial systems in the south. In the Valladolid–Palencia area, the siliciclastic sediments change to lacustrine limestones, dolostones and marls. Probably, these were formed in shallow lakes surrounded by paludal or swampy fringes (Fig. 8B). In the pattern suggested, fluvial systems flowed to topographically lower areas supposed to lie somewhere to the west of the central areas. However, there are no lacustrine Neogene deposits in these areas and we suggest that these fluvial systems drained outside the basin. The vertical stacking of the siliciclastic sediments is explained by a high rate of subsidence in the basin (Fig. 6).

Deposits of TSU 2 represent the progradation of fluvial over lacustrine facies with a coarsening-upwards trend that is thought to record tectonic movements in both the source areas and the basin. Mediavilla & Dabrio (1989) and Pineda & Arce, (in press) showed that NNW–SSE and WNW–ESE faults affected the sediments of TSU 2 along the northern margin (Fig. 4D). In the central areas there are fractures with block tilting, and also layers with palaeoseismites (Baena *et al.*, in press). The fractures observed at the surface are not really important, but they reflect the reactivation of the older, deeper Pisuerga scissors fault that lowered the NW block to the north-east and the SE block to the south-west, as demonstrated by careful thickness measurements.

In the central areas of the basin the unit is topped by dark layers of clay and limestone, interpreted as fluvio-paludal deposits. In other places these layers are paleosoils (Portero & del Olmo, 1982a, b, c; Mediavilla & Dabrio, 1986). These sediments form a continuous layer, of regional extent, considered as a marker level that indicates a sedimentary hiatus.

TSU 3

This unit rests unconformably upon TSU 2, and it onlaps the Mesozoic and Paleogene deposits of the northern and southern margins respectively (Figs. 4D and 5). The thickness is 25–45 m; the maximum thicknesses occur NE and SW of the Pisuerga River (Fig. 7). To the north-east there is a zone of subsidence with two maxima

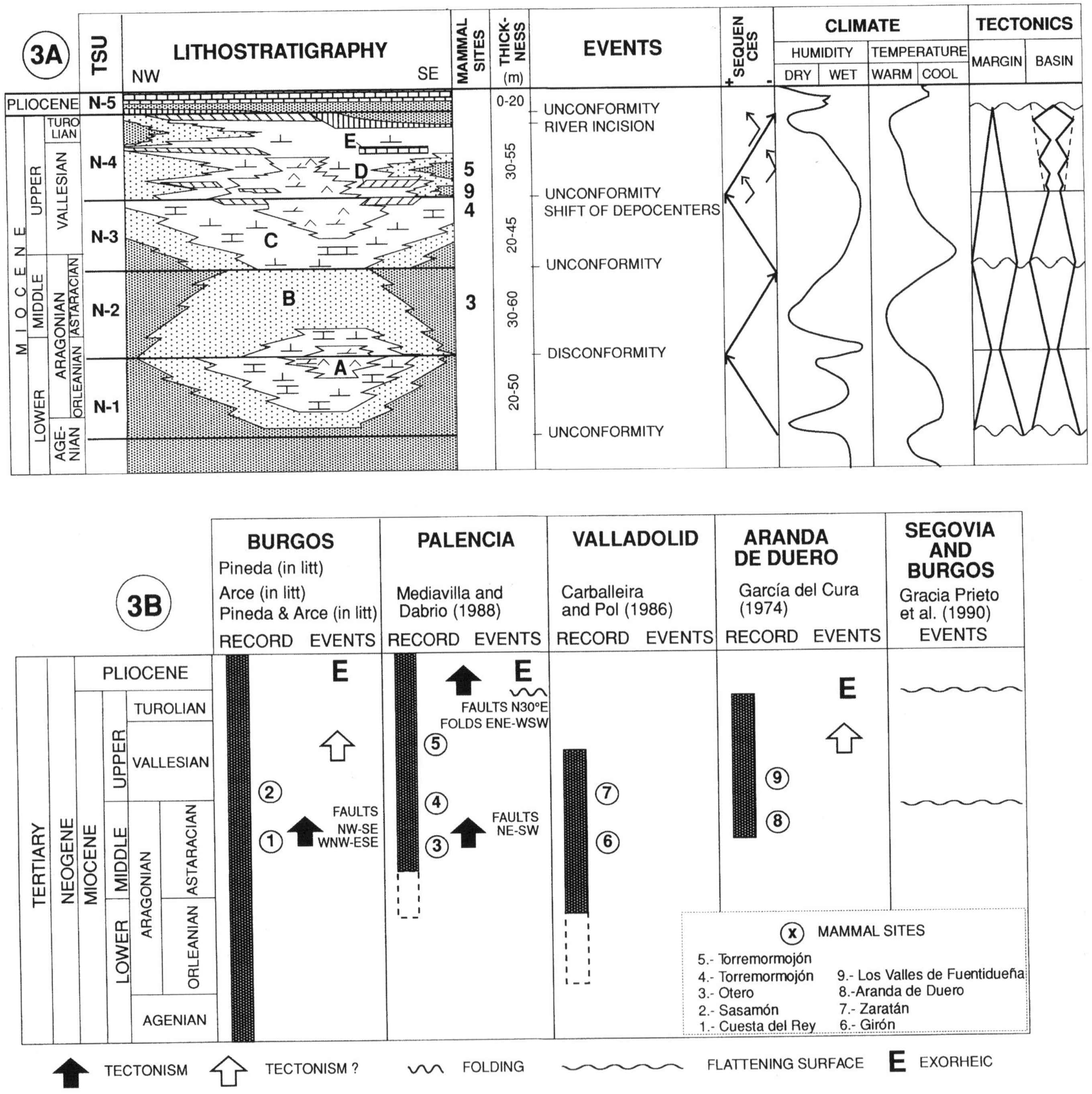

Fig. 3. Stratigraphic framework of the Neogene deposits of the Duero Basin, climatic and tectonic variations. Climatic curve after Lopez Martinez *et al.* (1985). Classic denominations: A, Facies Dueñas (Portero & Del Olmo, 1982); B, Tierra de Campos; C, Cuestas (Hernández Pacheco, 1915); D, Páramo 1; E, Páramo 2 (San Miguel de la Cámara, 1946). B, Sedimentary record, tectonic events and mammal sites in the Duero Basin. TSU –tectonosedimentary units.

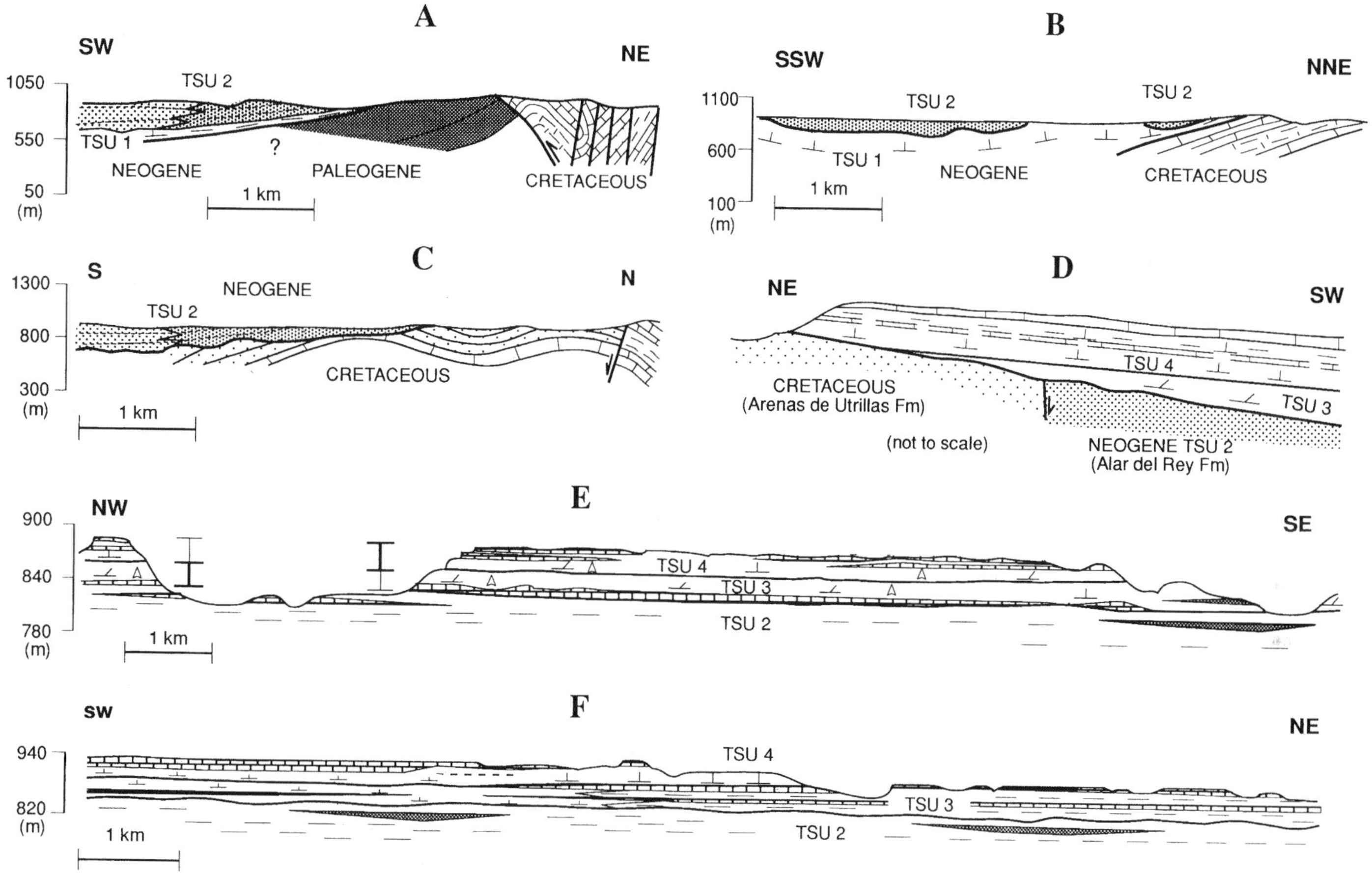

Fig. 4. Geologic cross-sections to show lateral relationships and changes of facies and thickness in the Neogene TSU of the Duero Basin. Location of sections indicated in Fig. 2. A, B, C and D: after Pineda (in press); E: after Pavón *et al.*, (in press); F: after López Olmedo *et al.*, (in press).

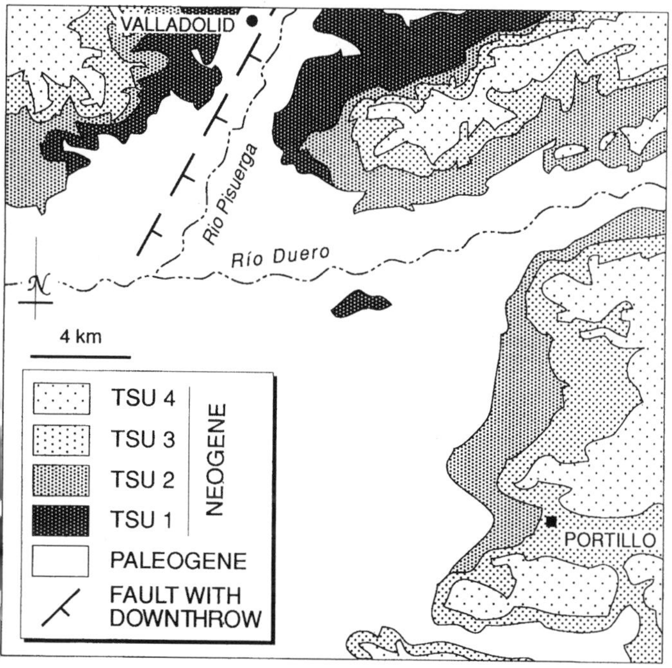

Fig. 5. Onlap of Neogene TSUs upon the Paleogene substratum in the area of Valladolid and Portillo. Modified from Portero and Del Olmo (1982), and Del Olmo and Portero (1982).

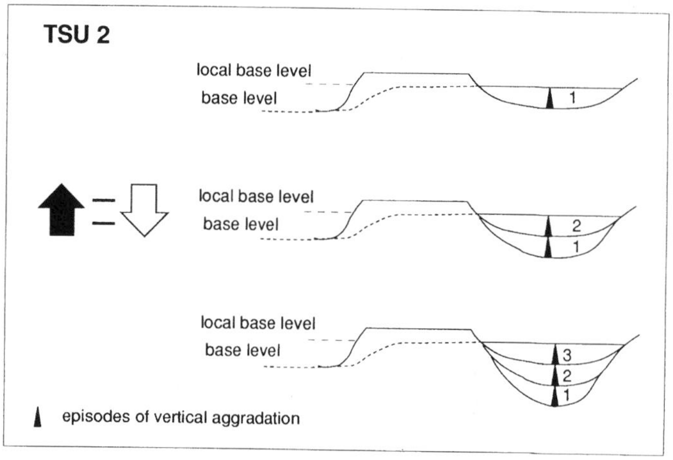

Fig. 6. Coeval dissection and vertical aggradation during the deposition of TSU 2 in central and south-western Duero Basin.

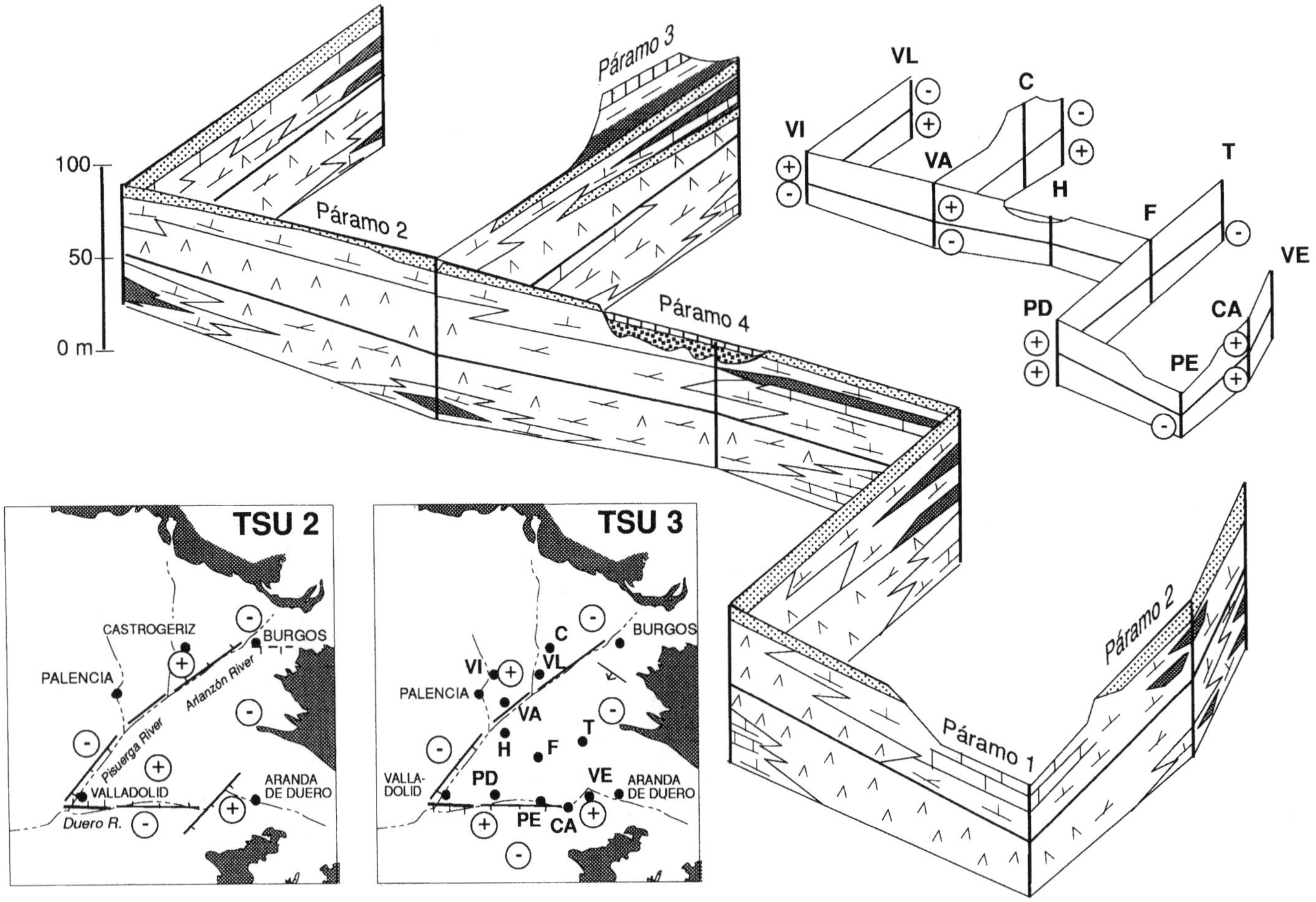

Fig. 7. Correlation panel for Neogene deposits in central Duero Basin. Same key as Figure 8 with the addition of: VL: Valbuena; VI: Villajimena; VA: Valdeolmillos; VE: Valdezate; H: Hontoria; F: Fombellido; T: Tórtoles de Esgueva; PD: Peñalba de Duero; PE: Peñafiel; CA: Castrillo de Duero. (+) maximum thickness (subsidence) and (−) minimum thickness (subsidence) for lacustrine deposits. Location of sections indicated in palaeogeographic maps (compare with Fig. 1 for scale).

(Fig. 4E). To the south-east, the areas of maximum subsidence coincide with the E–W direction of the Duero River.

TSU 3 records one of the events of major lacustrine expansion and displacement of the fluvial systems towards the north-east border and towards the west.

The lacustrine systems include several subenvironments. In the shallow marginal areas (Figs. 4F and 7) sand to mud units related to the development of alluvial systems alternate with episodes of organic clays or paludal fossiliferous limestones.

Towards the central areas, far away from alluvial influence, chemical sedimentation increased. Facies associations are mainly of the carbonate–evaporite type (limestones, dolostones, primary microlenticular gypsum or gypsarenites with ripple cross-lamination). In some places there are also lenticular sand and mud bodies of deltaic origin.

The areal distribution of subenvironments (Fig. 8C, D) is governed by: 1. the existence of topographically lower, more strongly subsiding, areas controlled by fractures, which became depocentres; and 2. the supply of fresh water and siliciclastic sediments from source areas located to the north-west, north-east and south-east.

The deposits of TSU 3 generally form a shallowing-upwards sequence that reflects the expanding character of this unit (carbonate–evaporite deposits increasingly dominate upwards). This sequence records basin infill with a decreasing diastrophic regime.

The top of the unit is marked by a change in the sedimentary polarity, a displacement of the areas of maximum subsidence and depocentres (Fig. 7), and progradation of carbonate marginal lacustrine facies towards the central areas of lakes and widespread pedogenesis.

The paleosol formation does not necessarily imply an anomalous event. Periodic desiccation could have been a common feature. We assume that it was a basin-wide feature because of two facts: its occurrence in all the measured stratigraphic sections (despite the diversity of sedimentary environments involved), and the significant change in physical and chemical conditions of the basin, environmental distribution and sedimentary processes detected just above this layer. Accordingly, we consider that this pedogenic episode indicates an extensive sedimentary discontinuity.

TSU 4

The thickness of TSU 4 is 35 to 55 m (Fig. 3) with the maximum located some distance to the south-east of the areas with thickest TSU 3 (Fig. 7). TSU 4 is a complex unit composed of three sedimentary cycles each recording a progradation of alluvial upon lacustrine deposits, later expansion of lacustrine facies and final progradation of carbonate marginal facies to the central parts of the lacustrine areas.

The facies architecture in the first two cycles is similar to TSU 3, although there was a shift of depocentres between these units, producing variations in the areal distribution of sedimentary environments (Fig. 8D). As in TSU 3, the facies pattern was controlled by: 1. location of fresh water and sediment supply that, by the end of the second, upper cycle, prevented the deposition of evaporitic facies; and 2. the existence of fault-related, subsiding areas that became depocentres.

The third cycle shows a similar pattern but there was no deposition of evaporites due to dilution of the lacustrine waters. At the top of this cycle, and in stratigraphic positions equivalent to the development of the lacustrine systems, there are fluvial systems, that drain to the west. The proximity of these systems to the south-western zones of the basin – which were the first to be captured by the exorheic Atlantic fluvial network – suggests that perhaps they were the first Neogene fluvial systems to be captured (Figs. 8E and 9). However, it is difficult to verify the equivalence of these isolated (encased) deposits or to confirm that they belong to the third cycle, because of the lack of connection of outcrops due to erosion.

The prograding character of the alluvial deposits inside each cycle, particularly in the two younger cycles, has been explained by successive tectonic reactivations (tectonic uplifts) of the basin margins (Portero *et al.*, 1982, in press; García del Cura, 1974; Pineda & Arce, in press). However, Pineda & Arce (in press) indicate that in the margins and neighbouring areas there is no evidence of deformation due to these tectonic movements. On the other hand, the sedimentary characteristics of fluvial deposits, even in locations close to the margins, indicate deposition by high-sinuosity fluvial systems in low-gradient zones (Pineda & Arce, in press; Armenteros, 1986). Moreover, the transition from alluvial to lacustrine facies is gradual (Fig. 3).

All these features lead us to conclude that the progradations of the fluvial systems resulted from oscillations of base level, perhaps related to 'pulses' of subsidence in the centre of the basin due to fluctuations of the rate of subsidence.

The unconformable nature of the lower boundary (unconformity recorded as a shift of depocentres with respect to those of the underlying TSU 3) and the upper boundary (unconformity due to a tectonic phase with faulting and folding related to adaptation to subsurface fractures) of this rock unit indicates that these three cycles form a single TSU.

TSU 5

This unit rests discordantly upon TSU 4 but in positions topographically lower than it. It is represented by alluvial-fan deposits (gravels) in the northern and southern borders, whereas in central areas, where it is located E of the Pisuerga River, it consists of 5 to 15 m of fluvial conglomerates changing vertically to floodplain muds, pedogenic crusts and paludal carbonates.

This TSU was previously interpreted as part of the basin infill (Mediavilla & Dabrio, 1989), but recent research (Mediavilla *et al.*, in press) has related it to basin emptying as the first terrace of this area.

Controls of sedimentation

Tectonics

The main faults active from the Neogene to the Present reflect Late Hercynian basement fractures that were reactivated, faulting the surface, modifying landscapes, and creating and modifying the previously established areas of subsidence.

The Neogene sedimentary record of the Duero Basin illustrates the relationships of tectonics and sedimentation. Tectonics controlled the sedimentation of all TSUs in different ways in the various areas, as recorded in the units described. Increase of activity in the margins (TSU 2) produced retraction of lacustrine environments, whereas decreasing activity favoured expansion of lacustrine deposits (TSUs 1, 2, 4). During deposition of unit TSU 4 there was a hybrid pattern with progradation of fluvial systems and expansion of lacustrine deposits. This was a result of accelerated diastrophic activity in the central areas (subsidence) that promoted progradation of alluvial systems but decreasing activity in the margins that allowed expansion units and covering of marginal areas.

Tectonics also influenced sedimentation by modifying paleogeographies, creating and moving the low areas that focused the drainage pattern. These modifications of the geometry of the basin, and of the areas of sediment and fresh-water supply produced a diversity of facies patterns in the sedimentary record of the basin.

Climate

Evaporites are one of the most popular criteria used in interior basins as climatic indicators.

Analysis of climatic variations during the Miocene (Fig. 3) shows that deposition of saline materials occurred both in dry periods (TSUs 1, 2) and humid periods (TSUs 3, 4). It is noteworthy that maximum development occurred particularly in humid periods. According to this, it seems clear that climate was not a determining factor for the generation of evaporites. In our opinion, the disappearance of saline lakes cannot be related to a particular climatic evolution because, in the Duero Basin, the end of evaporite deposition coincided with a climatic trend to more arid conditions (TSU 4). However, climate was a very important factor determining the amount of water that reached the basin and, eventually, also the

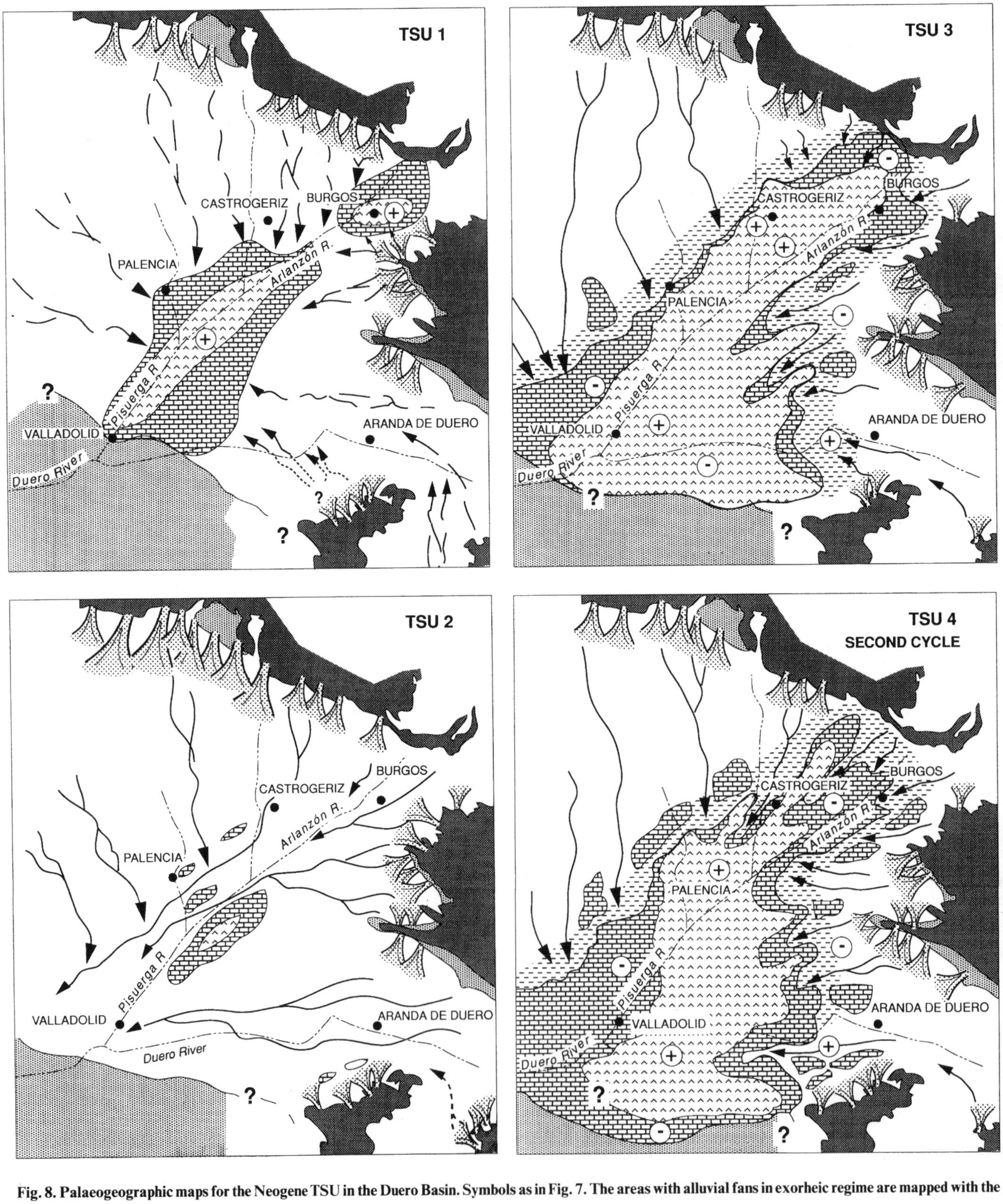

Fig. 8. Palaeogeographic maps for the Neogene TSU in the Duero Basin. Symbols as in Fig. 7. The areas with alluvial fans in exorheic regime are mapped with the present morphology in plan.

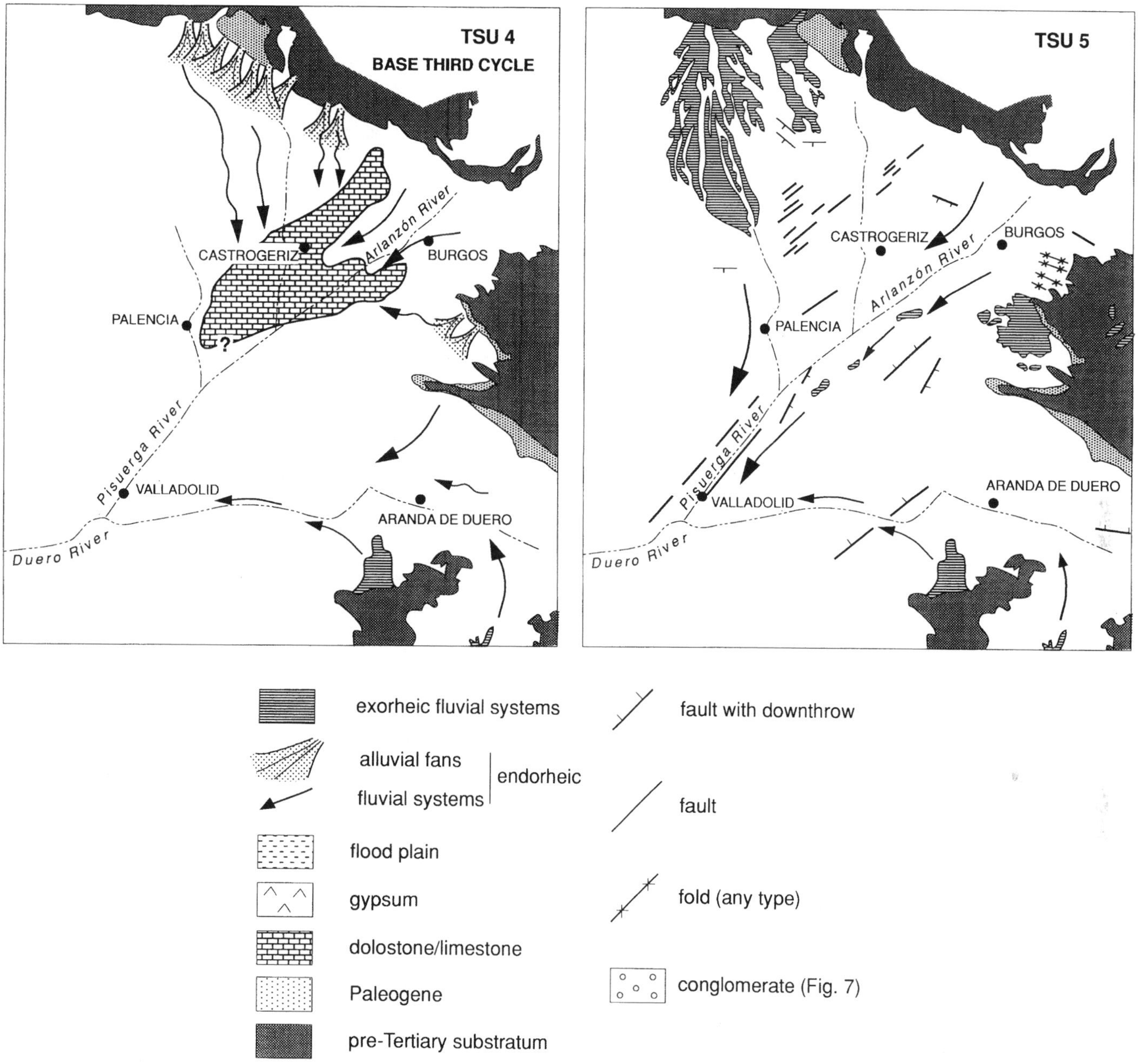

Fig. 8. *cont.*

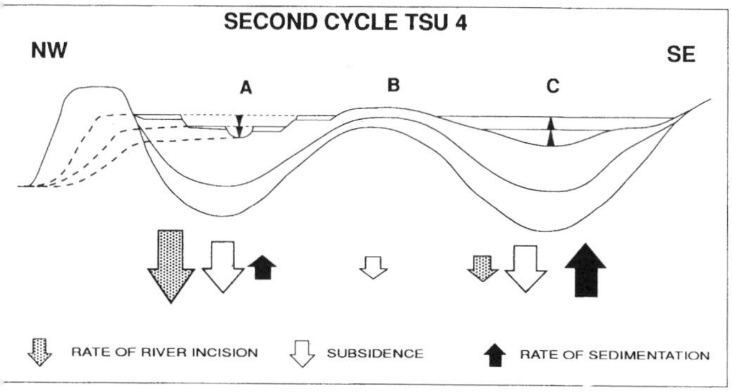

Fig. 9. Schematic ideal cross-section in the south-eastern Duero Basin during the sedimentation of the second cycle of TSU 4, with coeval endorheic filling and exorheic regime. The variables subsidence, rate of sedimentation, and rate of river incision are mutually related.

extent of lacustrine systems. The extent of lacustrine deposits, particularly evaporites, during TSU 3 (Fig. 8A, C) is greater than during TSU 1. Although TSU 1 and 3 were influenced by greater diastrophism, TSU 3 coincided with a period of humid climate.

References

Armenteros, I. (1986). *Estratigrafía y sedimentología del Neógeno del sector suroriental de la depresión del Duero*. Ediciones Diputación de Salamanca. Serie Castilla y León, 1: 471 pp.

Baena, J., Moreno, F., Nozal, F., Alfaro, J.A. and Barranco, L.M. (in press). *Mapa Neotectónico de España 1:1 000 000*. ITGE, Madrid.

Corrochano, A. and Armenteros, I. (1989). Los sistemas lacustres neógenos de la Cuenca terciaria del Duero. *Acta Geol. Hisp.*, **24** (3–4): 259–279.

García del Cura, M.A. (1974). Estudios sedimentológicos de los materiales terciarios de la zona centro-oriental de la Cuenca del Duero (Aranda de Duero). *Estud. Geol.*, **30**: 579–597.

Hernández-Pacheco, E. (1915). Geología y paleontología del Mioceno de Palencia. *Trab. Com. Inv. Paleontol. Prehist.*, **5**: 1–75.

López Martínez, N., Agustí, J., Cabrera, L., Calvo, J.P., Civis, J., Corrochano, A., Daams, R., Díaz, M., Elízaga, E., Hoyos, M., Morales, J., Portero, J.M., Robles, F., Santisteban, C. and Torres, T. (1985). Approach to the Spanish continental Neogene synthesis and paleoclimatic interpretation. *VIIIth Congress of the Regional Committee on Mediterranean Neogene Stratigraphy. Symposium on Late Cenozoic Mineral Resources. Hungarian Geological Survey, Abstracts.*

López Olmedo, F. and Enrile, A. (in press). *Mapa Geológico de España, E. 1:50 000, 2ª serie (MAGNA). Memoria explicativa de la Hoja 374 (Peñafiel)*. Madrid. ITGE, Servicio de Publicaciones del Ministerio de Industria y Energía.

López Olmedo, F., Enrile, A. and Cabra, P. (in press). *Mapa Geológico de España, E. 1:50 000, 2ª serie (MAGNA). Memoria explicativa de la Hoja 313 (Antigüedad)*. Madrid. ITGE Servicio de Publicaciones del Ministerio de Industria y Energía.

Martín-Serrano, A. (1988a). *El relieve de la región occidental zamorana. La evolución geomorfológica de un borde del Macizo Hespérico.* Instituto de Estudios Zamoranos Florián de Ocampo, Diputación de Zamora: 306 pp.

Martín-Serrano, A. (1988b). Sobre la transición Neógeno–Cuaternario en la Meseta. El papel morfodinámico de la raña. *II Congr. Geol. España, Comun.* **1**: 395–398.

Martín-Serrano, A. (1988c). Sobre la posición de la raña en el contexto morfodinámico de la Meseta. Planteamientos antiguos y tendencias actuales. *Bol. Geol. Mineral.*, **99** (6): 855–870.

Martín-Serrano, A. (1989). Características, rango, significado y correlación de las series ocres del borde occidental de la Cuenca del Duero. *Stvd. Geol. Salmanticensia*, **5**: 239–252.

Martín-Serrano, A. (1991). La definición y el encajamiento de la red fluvial actual sobre el Macizo Hespérico en el marco de su geodinámica alpina. *Rev. Soc. Geol. Esp.*, **4**: 337–351.

Martín-Serrano, A. and Olmo, A. (1990). *Mapa Geológico de España, E. 1:50 000, 2ª serie (MAGNA). Cuaternario, Terciario y Mesozoico de la Hoja 507 (El Espinar)*. Madrid. ITGE, Servicio de Publicaciones del Ministerio de Industria y Energía.

Mediavilla, R. and Dabrio, C.J. (1986). La sedimentación continental del Neógeno en el sector centro-septentrional de la depresión del Duero (Provincia de Palencia). *Stvd. Geol. Salmanticensia*, **22**: 111–132.

Mediavilla, R. and Dabrio, C.J. (1988). Controles sedimentarios neógenos en la Depresión del Duero (sector central). *Rev. Soc. Geol. Esp.*, **1**: 187–196.

Mediavilla, R. and Dabrio, C.J. (1989a). Las calizas del Páramo en el sur de la provincia de Palencia. *Stvd. Geol. Salmanticensia*, **5**: 273–291.

Mediavilla, R. and Dabrio, C.J. (1989b). Análisis sedimentologico de los conglomerados de Tariego (Unidad 4. Neógeno de la Depresión del Duero). *Stvd. Geol. Salmanticensia*, **5**: 293–310.

Mediavilla, R. and Picart, J. (in press). *Mapa Geológico de España, E. 1:50 000, 2ª serie (MAGNA). Memoria explicativa de la Hoja 312 (Baltanás)*. Madrid. ITGE, Servicio de Publicaciones del Ministerio de Industria y Energía.

Megias, A.G. (1982). Introducción al análisis tectosedimentario: aplicación al estudio dinámico de cuencas. *Quinto Congreso Latinoamericano de Geología, Argentina. Actas*, **1**: 385–402.

Nestares, E. and Wouters, P. (in press). *Mapa Geológico de España, E. 1:50 000, 2ª serie (MAGNA). Memoria explicativa de la Hoja 344 (Esguevillas de Esgueva)*. Madrid. ITGE, Servicio de Publicaciones del Ministerio de Industria y Energía.

Ordóñez, S., García del Cura, M.A. and López Anguayo, F. (1980). Contribución al conocimiento de la Cuenca del Duero (Sector Roa–Baltanás). *Estud. Geol.*, **36**: 361–369.

Ordóñez, S., García del Cura, M.A. and López Aguayo, F. (1981). Chemical carbonated sediments in continental basins: the Duero Basin. *Abstracts I.A.S. 2nd. Eur. Mtg. Bologna*, 130–133.

Pardo, G., Villena, J. and González, A. (1989). Contribución a los conceptos y a la aplicación del análisis tectosedimentario. Rupturas y unidades tectosedimentarias como fundamento de correlaciones estratigráficas. *Rev. Soc. Geol. Esp.*, **2**: 199–219.

Pavón, J., García, J.M. and Manjón, M. (1973). *Mapa Geológico de España, E. 1:50 000, 2ª serie (MAGNA). Memoria explicativa de la Hoja 236 (Astudillo)*. Madrid. ITGE, Servicio de Publicaciones del Ministerio de Industria y Energía.

Piles, E. and Picart, J. (in press). *Mapa Geológico de España, E. 1:50 000, 2ª serie (MAGNA). Memoria explicativa de la Hoja 373 (Quintanilla de Onésimo)*. Madrid. ITGE, Servicio de Publicaciones del Ministerio de Industria y Energía.

Pineda, A. (in press). *Mapa Geológico de España, E. 1:50 000, 2ª serie (MAGNA) Memoria explicativa de la Hoja 166 (Villadiego)*. Madrid. ITGE, Servicio de Publicaciones del Ministerio de Industria y Energía.

Pineda, A. and Arce, M. (in press). *Mapa Geológico de España, E. 1:50 000, 2ª serie (MAGNA). Memoria explicativa de la Hoja 200 (Burgos)*. Madrid. ITGE, Servicio de Publicaciones del Ministerio de Industria y Energía.

Portero, J.M. and del Olmo, P. (1982a). *Mapa Geológico de España, E. 1:50 000, 2ª serie (MAGNA). Memoria explicativa de la Hoja 273 (Palencia)*. Madrid. IGME, Servicio de Publicaciones del Ministerio de Industria y Energía.

Portero, J.M. and del Olmo, P. (1982b). *Mapa Geológico de España, E. 1:50 000, 2ª serie (MAGNA). Memoria explicativa de la Hoja 372 (Valladolid)*. Madrid. IGME, Servicio de Publicaciones del Ministerio de Industria y Energía.

Portero, J.M. and del Olmo, P. (1982c). *Mapa Geológico de España, E. 1:50 000, 2ª serie (MAGNA). Memoria explicativa de la Hoja 400 (Portillo)*. Madrid. IGME, Servicio de Publicaciones del Ministerio de Industria y Energía.

Portero, J.M., del Olmo, P., Ramírez, J. and Vargas, I. (1982). Síntesis del Terciario continental de la Cuenca del Duero. *Temas Geol. Min., IGME*, **6** (1): 11–40.

Royo Gómez, J. (1926). *Terciario continental de Burgos*. XIV Congr. Geol. Intern. Guía Excursiones, A-6: 69 pp.

San Miguel de la Cámara, M. (1946). *Mapa Geológico de España, E. 1:50 000 (1ª Serie). Hoja 346 (Aranda de Duero)*. Madrid. IGME.

Santisteban, J.I., Martín-Serrano, A., Mediavilla, R. and Molina, E. (1991). Introducción a la estratigrafía del Terciario del SO de la Cuenca del Duero. In J.A. Blanco, E. Molina and A. Martín-Serrano (eds.), *Alteraciones y paleoalteraciones en la morfología del oeste peninsular. Zócalo hercínico y cuencas terciarias. Monogr. Soc. Española Geomorfol.*, **6**: 185–198.

W7 North-western Cainozoic record: present knowledge and the correlation problem

A. MARTÍN-SERRANO, R. MEDIAVILLA AND J.I. SANTISTEBAN

Abstract

Tertiary deposits of the north-western Iberian Peninsula are heterogeneous because they occur in several morpho-structural positions as isolated and dispersed basins and outcrops. The quality of the palaeontological record is usually very poor and there are scarce data. Correlation depends on a wide range of criteria which are not always equivalent: palaeontology, mineralogy and petrology, geomorphology, tectonics and comparison with better-known and better-dated facies in regions nearby. The results lack homogeneity and there are notable discrepancies.

Introduction

Tertiary sediments of Galicia, Bierzo and the Cantabrian Range (north-western Spain) occur in small basins and isolated outcrops.

Galicia Basins

Many geographical features and the Tertiary basins of Galicia are related to fault-systems trending N–S, NNE–SSW, NE–SW, E–W and WNW NW–ESE SE (Fig. 1).

Lugo Basins

The most important outcrops of Tertiary sediments in Galicia are located in Lugo, between the central peneplain and the eastern mountains. The main basins (Terra Cha, Sarria and Monforte) are asymmetrical: Tertiary sediments rest unconformably upon Palaeozoic rocks in the west and are separated from crystalline basement (igneous rocks) by N20–30 and N50–60 faults (Fig. 2).

The sedimentary fills of the Sarria and Monforte basins, up to 200 m thick, have been divided in various stratigraphical schemes (Birot & Solé, 1954; Brell & Doval, 1974, 1979; Virgili & Brell, 1975; Martín-Serrano, 1979, 1980, 1982; del Olmo, 1986; Vergnolle, 1984, 1985, 1987, 1990) with several lithostratigraphic units which were grouped into the Monteforte Formation (Vergnolle, 1988). The sedimentary record of the larger basins includes two superposed megasequences of opposed trend (Figs. 3, 4), but it is difficult to differentiate these trends in the smaller, moderately subsiding basins. The lower, fining-upwards megasequence comprises:

- Green arkoses with minor clay matrix (smectite) arranged in channel-form bodies separated by mud layers. These are interpreted as the deposits of rivers that flowed from the granite area of Villalba–Chantada.
- Red-green alluvial-fan conglomerates. These are deposits of small, but thick, alluvial cones rooted in the metamorphic and granitic source areas of the eastern edges of the basins.
- Banded, red-green indurated clay series in the areas where the arkoses and conglomerates coalesce. This facies is more extensive upwards and there is also a progressive increase in the content of carbonate and the generation of dolomitic crusts associated with palygorskite and sepiolite (Sarria Basin).

The upper, coarsening-upwards megasequence contains green arkosic gravel as the last deposits of the basins infill.

SE Galicia

Three distinct basins occur in the small tilted (towards the S and SE) mountain blocks of Orense: Maceda, Ginzo de Limia and Verín.

Maceda is a complex, 160 m deep, half-graben bounded to the NNE by a large fault scarp. The lithofacies are different from those of the Lugo basins: the largest part of the basin fill is a fining-upwards fluvial unit of fine-grained sediments (Del Corno and De la Vega Members of Brell, 1975; De los Milagros Member of del Olmo, 1986). They are white, grey and ochre ferruginous sand and clay (55–65% kaolinite) with organic matter, deposited in fluviatile environments. The marginal sediments are red muddy sands and gravel, comparable with one of the lithofacies of the Monforte Formation; they are interpreted as alluvial-fan deposits related to the tectonic activity of the border.

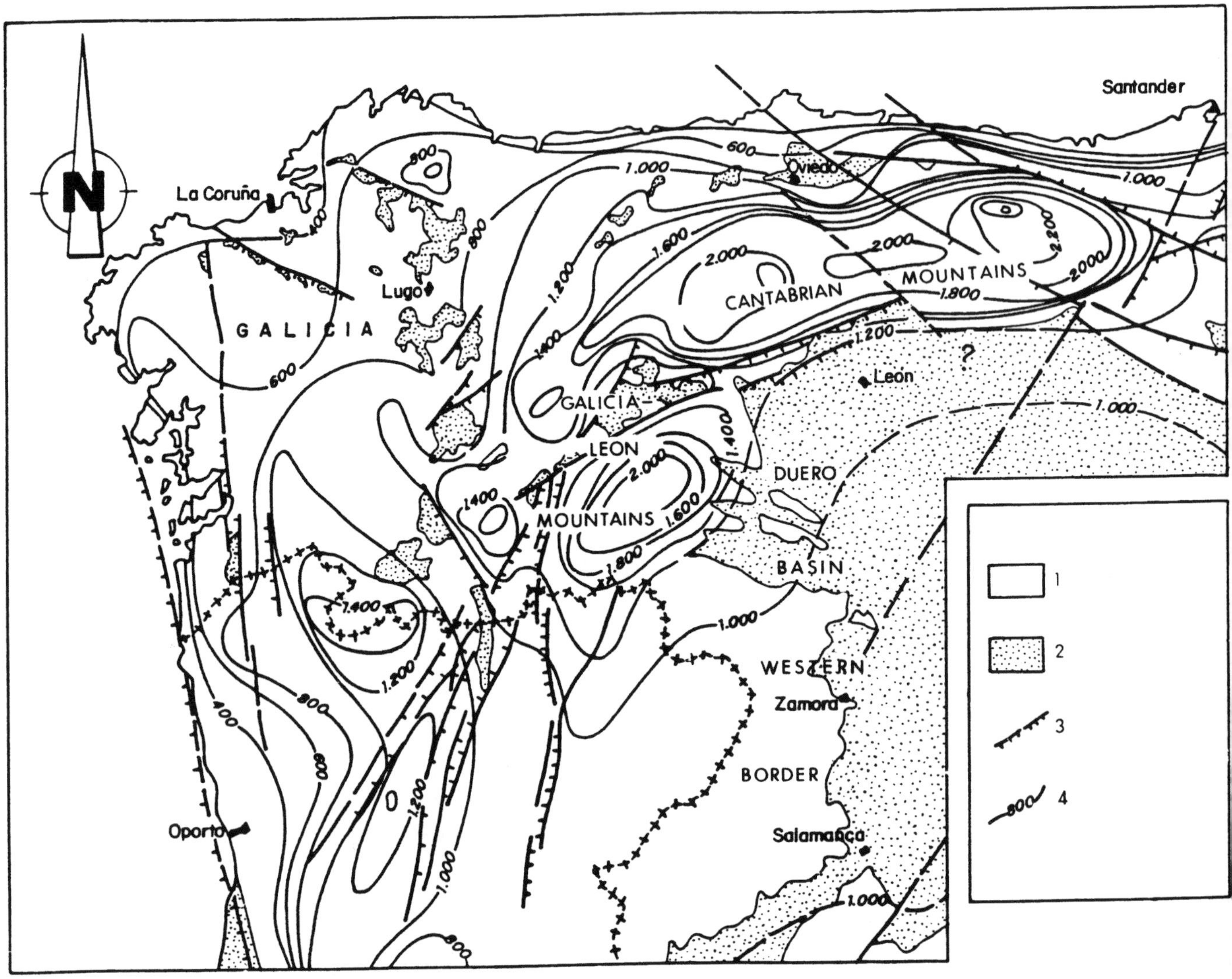

Fig. 1. Morphostructural situation of the north-western Tertiary. 1: Palaeozoic basement and Mesozoic; 2: Cainozoic; 3: Morphotectonic lines; 4: Isohypses of relief contours.

The Ginzo de Limia Basin is a sub-Recent tectonic trough bounded by N70, N–S and N120 faults and filled with alluvial sediments. The fill consists of 250 m of Tertiary sediments, including a lignite and clay layer, located between 80–130 m in depth, and interbedded between two thick sandy series (IGME, 1979–1984).

The narrow Verín Basin, in the morphostructural region of Tras-Os-Montes, reaches a maximum depth of 100 m and is filled with red and grey clay, sand and gravel.

Littoral Basins

These are small, isolated outcrops. Some of them are related to a structural corridor (the so-called Depresión Meridiana, Nonn, 1966): Porriño, Deva and Salvatierra. Other outcrops (Fazaouro Formation) are located on the Cantabrian side (Moucide, Burela and Lorenzana), and they are interpreted as a piedmont (Vergnolle, 1988) made up of fine siliciclastic sediments related to a regressive episode. These are *siderolithic* facies, with mature constituents from a mineralogical and petrological point of view: kaolinite and unalterable minerals. They are comparable with Maceda, Ginzo de Limia, Puentes de García Rodríguez and Meirama basin fills. The Meirama Basin contains the best developed Cainozoic record of these facies.

The basins of San Saturnino–Pedroso–Moeche–Puentes–Roupar–Moiñonovo and Lendo–Meirama–Visantña–Juanceda–Lanza–Orros–Boimil are related to dextral faults (Fig. 5) that strongly influenced the sedimentation (Maldonado, 1977; García Aguilar, 1987; Monge, 1987; Bacelar *et al.*, 1988, 1991; Santanach *et al.*, 1988). They were small, complex, compartmentalised, asymmetrical and strongly subsiding basins (Fig. 6) filled with green, grey and blue kaolinitic muds and sands. In some basins (Pedroso, Puentes de García Rodríguez, Roupar, Juanceda and Meirama) there is lignite as well. They were filled by small alluvial and fluvial systems associated with lacustrine and paludal environments (lig-

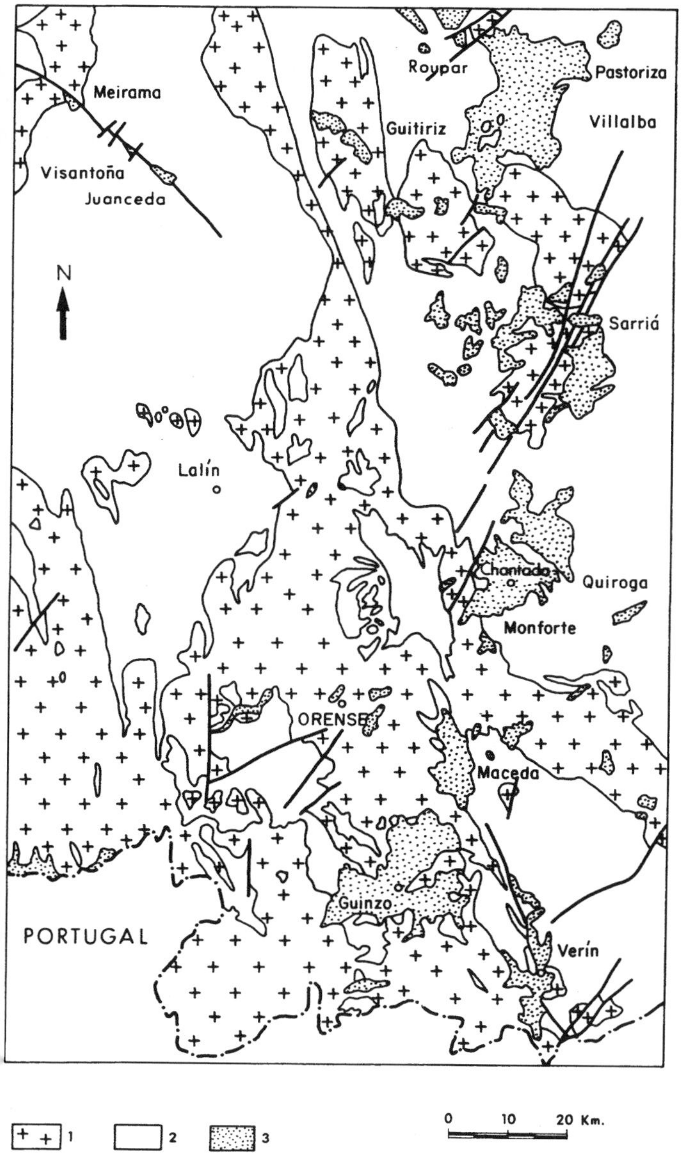

Fig. 2. Tertiary basins in eastern Galicia. 1. Igneous rocks; 2. Palaeozoic and pre-Palaeozoic basement; 3. Cainozoic.

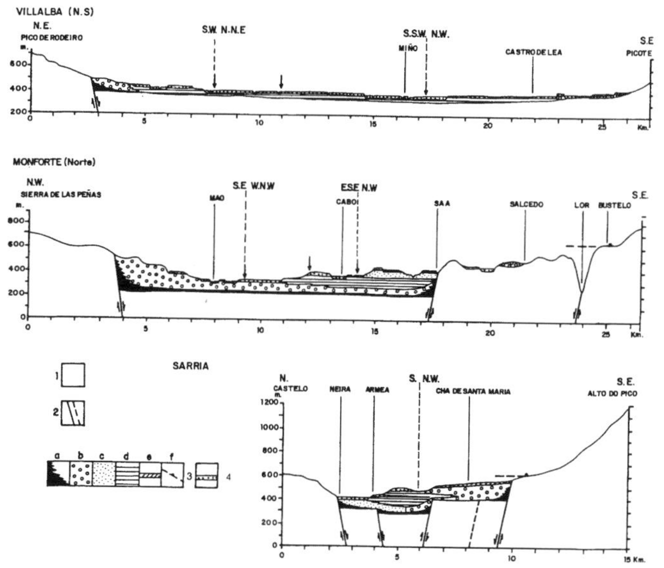

Fig. 3. Lithostratigraphy of the main basins of Lugo Province (from Vergnolle, 1988). 1. Basement; 2. faults; 3. Cainozoic deposits; a: breccias; b: conglomerates; c: arkoses; d: clays; e: dolomitic crust; f: blocks of infill; 4: Quaternary alluvial sheet.

nite and carbonaceous clays) and were controlled by their structural evolution. Successive extensional or compressional phases in the basin of Puentes were recorded as expansions and retractions of the lignite beds. Equilibrium between subsidence and sedimentation favoured the development of thick coal deposits in the Miramar Basin (Santanach *et al.*, 1988).

Sil corridor

This is the depressed axial area of a NE–SW-trending mountain system. The complex Bierzo Basin (Fig. 7), is the most important Tertiary development. It is a depression with two large basins (Ponferrada–Villafranca and Bembibre) surrounded by secondary subbasins to the north (Parradaseca, Finolledo, Fabero and Noceda) and south (Las Médulas and Carucedo). The other southern depressions (El Barco de Valdeorras, La Rua, Quiroga and Quintela) are small, narrow, fault-bounded deep basins.

The up to 700 m thick sedimentary fill of the Bierzo Basin can be divided into several units (Fig. 8) (Vidal Boix, 1941, 1954; Birot & Solé, 1954; Pannekoek & Sluiter, 1964; Delamire-Bray, 1977; Herail, 1979, 1981, 1984) which serve as a model for the remaining Sil basins:

- Toral Formation: fluviatile gravel and sands and paludal limestone and dolostones (crusts). These are green to pink polymictic smectitic-rich sediments with feldspar, micas and polymineral grains. Source areas are far away to the west with a few local sources (Fig. 9).
- Santalla Formation: beige-ochre or grey-red siliciclastic conglomerates and muddy sands deposited in continuously flowing, high-competence fluvio-torrential alluvial fans.
- Las Médulas Formation: matrix-rich siliciclastic gravel with boulders, deposited in high-competence, intermittently flowing, alluvial fans.
- These deposits are topped by a weathered *raña*-like sheet of gravel.

All Sil basins underwent a similar evolution as witnessed by the similarity of their sedimentary records. The Quiroga Formation forms the lower infill of the Sil basin, and is laterally equivalent to the Toral Formation; the overlying Monforte Formation has fluvial facies comparable with the Santalla Formation (Figs. 10 and 12).

Morphotectonic features are evident in the Bierzo Basin (NE–SW and E–W borders) and all along the Sil Valley (with narrow and

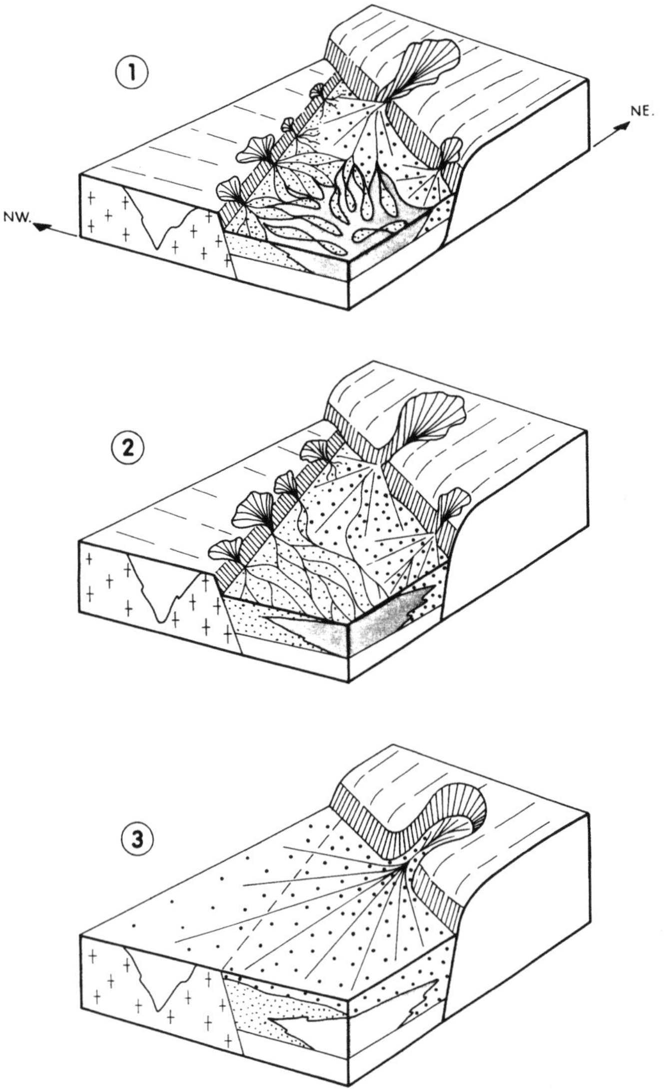

Fig. 4. Paleogeographical reconstruction of the basins filled by the Monforte Formation (simplified from Vergnolle, 1988).

deep N50–60 and N140 basins) (Birot & Sole, 1954; de Groot, 1974; del Olmo, 1986; Vergnolle, 1988). The infill of all these small basins shows a coarsening-upwards sequence. Strong deformation can also be seen affecting the Toral or Quiroga Formations. This deformation has not affected younger sediments such as the Santalla or Las Médulas Formations. The former are syntectonic deposits synchronous with the opening of the Bierzo Basin and Lugo basins.

Asturias outcrops

The Cantabrian Range, an old massif uplifted and deformed from Lutetian to Neogene times, is divided into mountain massifs, narrow and deep fluvial valleys and basins (Fig. 11). The *Surco Prelitoral* (pre-littoral furrow) is a tectonic scar with fragments of Permian and Mesozoic rocks that contains almost all the regional Tertiary record in discontinuous outcrops (La Espina, Grado, Oviedo, Infiesto, . . .) with a relative concordance of facies (Llopis Lladó & Martínez Alvarez, 1958, 1959, 1960; IGME, 1979–84, 1986; Truyols *et al.*, 1991).

The continental Tertiary deposits, up to 250 m thick, between Oviedo and Infiesto rest upon the karstified top of Upper Cretaceous carbonate rocks. They can be divided into two units (Truyols *et al.*, 1991): a lower mud unit with yellow-red or grey-green sands is overlain by red-yellow clays and marls with calcareous sands and brecciated limestone rich in gastropods and ostracods of Oligocene age. To the north, these two units change laterally into calcareous gravels and breccias (*Pudinga de Posada*).

Three units have been differentiated in Grado:

- A lower fining-upwards sequence of conglomerates and sandstone with carbonate cement and laminated, burrowed muds of braided river origin.
- A thick intermediate coarsening-upwards unit of clay and laminated mud passing upwards into conglomerates and sandstone with intercalations of laminated, burrowed muds. This is interpreted as sediments of ephemeral river systems and flood-plain deposits including limestone and marls.
- The top of the section consists of alluvial-fan, massive gravel, supplied from the south.

The westernmost fragments of this Tertiary Asturias Basin occur isolated and perched high near Pola de Alande. They are alluvial-fan deposits that change upwards to fluvial facies; they are related to a structural line. The proximal fan facies are massive gravels with sands and burrowed muds; distal fan and fluvial facies consist of gravels, sands and green burrowed muds (IGME, 1986).

All these oucrops are the remains of a unique basin that extended from east to west with an active northern boundary (*Franja Móvil Intermedia*) which governed sedimentation and strongly influenced the terminal tectofacies (*Pudinga de Posada*). Faults broke up the outline of the basin and parts of it were eroded some time afterwards.

Stratigraphic correlation, fragmentation, Cainozoic variety and dispersion

As previously indicated, the Tertiary of the north-west is varied because it is located in several morphostructural positions, which are isolated from each other. Furthermore, outcrops are usually of very poor quality and the palaeontological record is scarce. As a consequence, correlation depends on a wide range of criteria which are not always equivalent: palaeontology, mineralogy and petrology, geomorphology, tectonics and comparison with better known and dated facies in regions nearby. As expected, the results lack homogeneity and contain notable discrepancies (Fig. 12).

Palaeontological data are scarce, frequently confused and in disagreement with other data; in fact these data play a negative role because they impose forced correlation. The reason is that the

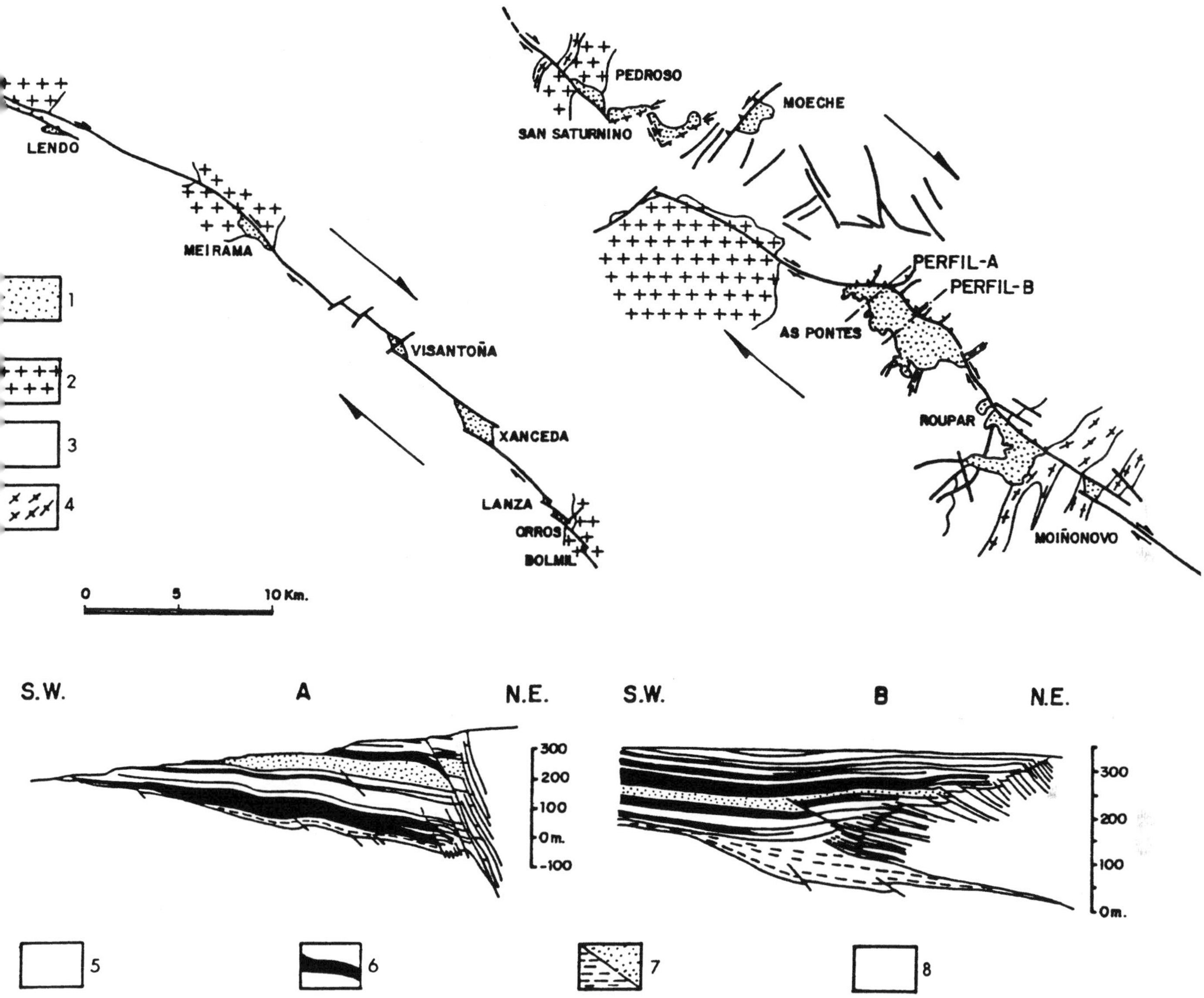

Fig. 5. Tectonic basins of Meirama and Puentes de García Rodríguez with two transverse cross-sections (Perfil-A and Perfil-B) to illustrate the deformation style of sediments in Puentes de García Rodríguez basin (from Bacelar *et al.*, 1988 and Santanach *et al.*, 1988). 1: Cainozoic; Palaeozoic, 2: Igneous rocks; 3: Quartzites and slates; 4: Schists and other metasediments; 5: Palaeozoic basement; 6: Coal dominated interbedding (siliciclastic beds subordinated); 7: Siliciclastic alternances (without coal beds); 8: Coal–siliciclastics alternation.

palaeontological assemblages are poor: only some layers of vertebrates and charophytes near Oviedo, dated as Rhenanian–Headonian in age (Truyols *et al.*, 1991), and palynomorphs in lignite beds in the Coruña basins, attributed to the Middle–Upper Miocene to Pleistocene (Medus, 1963; Nonn & Medus, 1963; Menéndez Amor, 1975; Maldonado, 1977; Araujo *et al.*, 1988). Recently, these sites also yielded vertebrates of Oligocene–Lower Miocene age (Esteban *et al.*, 1989).

The problem is that Cainozoic outcrops show a great variability of thickness, facies, sequential arrangement and petrologic and mineralogical composition. Such a diversity seems incompatible with the idea of a unique (and coeval) palaeogeography and palaeoclimate as especially suggested by the basins in Galicia. In some areas, there are siderolithic facies and lignite (Atlantic basins), whereas in others, arkosic sediments dominate (Lugo basins). Is there a single geological background for all these basins? An affirmative answer implies that the age ranges between Upper Oligocene and Neogene and also requires an explanation for the diversity of facies in terms of local palaeogeographical and palaeogeomorphological features (Birot & Solé, 1954; Nonn, 1966; Brell & Doval, 1974). If the answer is negative, the diversity has to be explained as the result of stratigraphic superposition (Fig. 13), which implies discontinuous sedimentation in steps and a prolongation of the time-duration of the stratigraphic record (Martín-

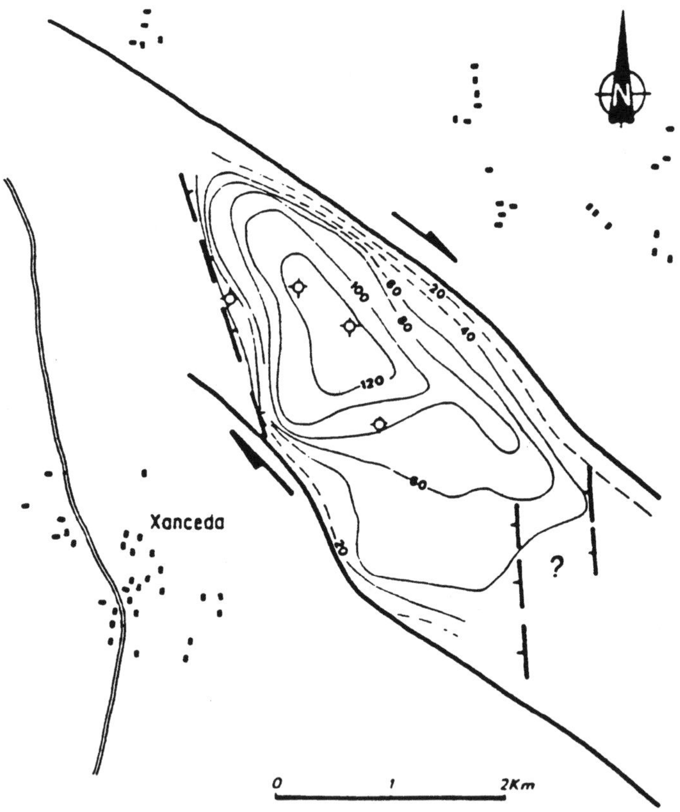

Fig. 6. Structural scheme of Xanceda Basin. Basement isobaths deduced from electric and mechanical vertical drillings (from Santanach *et al.*, 1988).

Serrano, 1982; Vergnolle, 1988). In this second mode, it is possible to correlate the basins of Galicia with the Duero Basin via the Bierzo Basins.

As with other major geological belts, the north-western Cainozoic basins were controlled by several fault systems leading to the analysis of the different basins on the basis of particular features. These features can be correlated to other lithostratigraphic or geomorphologic attributes, but these comparisons lead straight back to the basic problem: poor knowledge of the age and succession of geotectonic and geomorphologic events due to the poor chronostratigraphy of the associated sediments. On the other hand, erosion surfaces have been largely obliterated by faulting. For this reason, a confusing variety of geotectonic hypotheses have been proposed; for example: that the Pyrenean, Saavic or former Alpine phases were responsible for initial tilting and that more recent topographic changes and instability (evident through the Neogene) extends until Plio-Quaternary times.

This problem can only be overcome by the combination of geomorphologic, tectonic and stratigraphic studies. Systematic analysis should focus on geomorphologic investigation because, in these Tertiary studies, relief is not only the palaeogeomorphological context of the basins but also the result of the Alpine geotectonic framework. This approach makes it clear how important the study of weathering profiles, as 'alterites' is likely to be. These occur both still resting upon the basement and also incorporated into the Tertiary sediments, because through time the main bulk of alterites must have been eroded and incorporated into the basin fills. Then it should be possible to compare sediments and weathering profiles and in consequence correlate them with planation surfaces. With a special link like the alterites, we can plan to correlate the lithostratigraphy of basins with the evolution of relief and vice-versa.

References

Araujo, P., Hernández-Enrile, J.L. and Seara, J.R. (1988). Tectónica extensional y estructura de la Cuenca de Maceda (Galicia Meridional). *II Congr. Geol. España*, **2**: 107–110.

Bacelar, J., Alonso, M., Kaiser, C., Sánchez, M., Cabrera, L., Sáez, A. and Santanach, P. (1988). La cuenca terciaria de As Pontes (Galica): su desarrollo asociado a inflexiones contractivas de una falla direccional. *II Congr. Geol. España*, Simp.: 113–121.

Bacelar, J., Alonso, M., Kaiser, C., Sánchez, M., Cabrera, L., Ferras, B., Sáez, A. and Santanach, P. (1991). Relaciones tectónica-sedimentación de la cuenca cenozoica de As Pontes (A Coruña). *I Congr. Grupo Español de Terciario*, Comun.: 25–27.

Birot, P. and Solé Sabaris, L. (1954). Recherches morphologiques dans le Nord-Ouest de la Péninsule Ibérique. *Mem. et Doc. Centre Doc. cart. et Géograph.* C.N.R.S., **4**: 11–61.

Brell, J. and Doval, M. (1974). Un ejemplo de correlación litoestratigráfica aplicado a las cuencas terciarias del NO de la Península. *Estud. Geol.*, **30**; 631–638.

Brell, J. and Doval, M. (1979). Relaciones entre los sedimentos neógenos de Galicia y las alteraciones de su substrato. Interpretación paleoclimática. *Acta Geol. Hispànica*, **14**: 190–194.

Delmaire-Bray, M.M. (1977). Les grandes étapes de l'individualisation du bassin du Bierzo (Leon, Espagne) àpartir du Neogène. *Méditerranée*, 19–34.

Esteban, J., Lopez, N. and De La Peña, A. (1989). Prospección de micromamíferos fósiles en la zona de Galicia. *Informe interno, Proyecto Mapa Neotectónico*, ITGE.

García Aguilar, J.M. (1987). Caracterización estratigráfica y tectonosedimentaria de la cuenca lignitífera de Meirama (A Coruña). *Cuad. Lab. Xeol. Laxe*, **11**: 37–49.

Groot, R. De (1974). *Quantitative analyses of pediments and fluvial terraces applied to the basin of Monforte de Lemos, Galicia, NW Spain.* Amsterdam, 127 pp.

Herail, G. (1979). La sedimentación terciaria en la parte occidental del Bierzo (León, España) y sus implicaciones geomorfológicas. *I Reunión sobre la Geología de la Cuenca del Duero, Salamanca 1979. Temas Geol. Min*, **6**: 323–337.

Herail, G. (1981). El Bierzo: géomorphogénese fini-tertiaire d'un bassin intra-montagneux (Espagne). *R.G.P.S.O.*, **52** (2): 217–232.

Herail, G. (1984). Geomorphologie et litologie de l'or detritique. Piemonts et bassins intramontagneux du Nord-Ouest de l'Espagne. C.N.R.S.: 1–456.

IGME (1979–84). Proyecto para la investigación de lignito en la región de Galacia. Fases I.III. Unpublished.

IGME (1986). Exploración lignitífera en la región Astur-Galaica y experiencia piloto en la cuenca del Duero (Borde zamorano-leonés). Internal report. 226 pp.

Llopis Lladó, N. and Martínez Alvarez, J.A. (1958). Contribución al conocimiento del Terciario de los alrededores de Oviedo. *Monograf. Geol.*, **9**: 287–304.

Llopis Lladó, N. and Martínez Alvarez, J.A. (1959). Estudio hidrogeológico del Terciario de los alrededores de Grado. *Estud. Geol.*, **24**: 287–304.

Llopis Lladó, N. and Martínez Alvarez, J.A. (1960). Sobre el Terciario

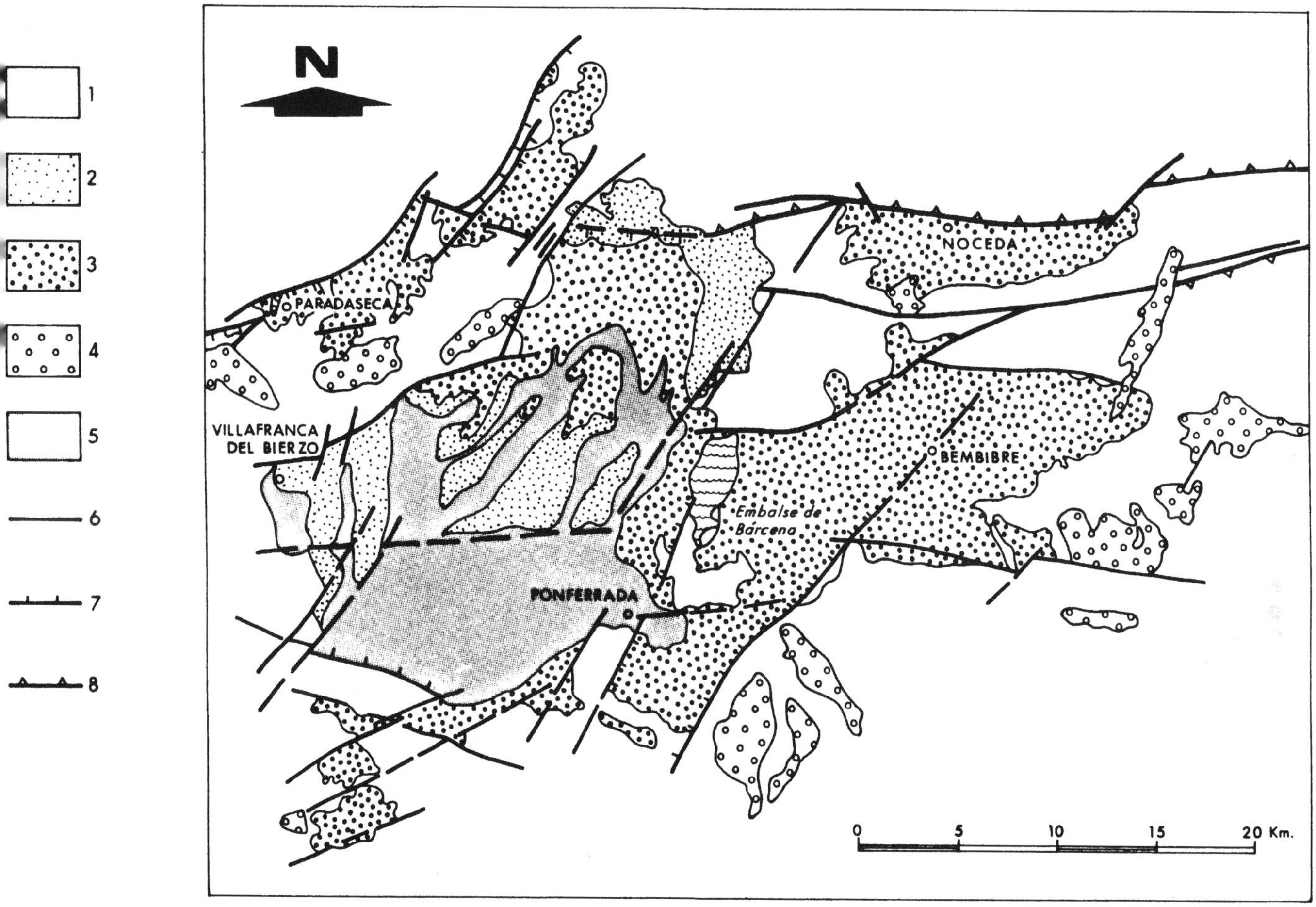

Fig. 7. Structure and sedimentation of Bierzo Basin (from Herail, 1984 and IGME, 1979–1984 and 1986). 1. Basement; 2. Toral Formation; 3. Santalla and Las Médulas Formation; 4. uppermost gravels; 5. terraces (Quaternary?); 6. faults; 7. normal faults; 8. reverse faults.

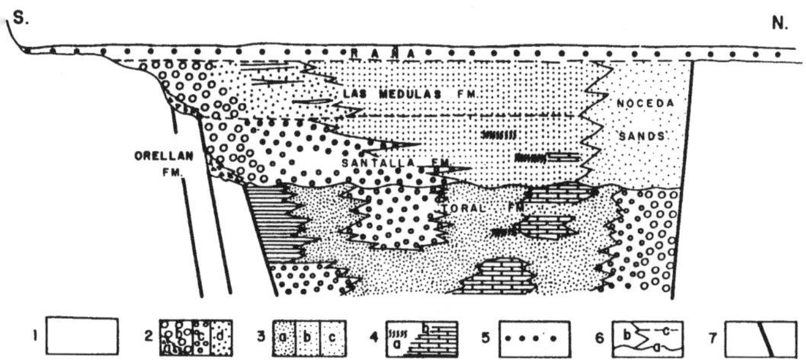

Fig. 8. Lithostratigraphy of the Bierzo Basin fill (from Herail, 1984). 1. Basement; 2. gravels: a. monogenic breccias; b. quartzite and sandstone blocks; c. rounded gravels; d. matrix-rich poor rounded gravels; 3. sandy, silty and muddy facies: a. smectite-rich, feldspar and quartzose sands; b. illite-rich, schist and quartzose sands; c. illite-rich, quartzose sands; 4. carbonates (limestones and dolostones): a. isolated concretions; b. massive and tabular levels; 5. *raña*, highly altered quartzite pebbles and blocks; 6. sedimentary discontinuity: a. discordance; b. lateral change of facies; c. vertical facies transition; 7. faults.

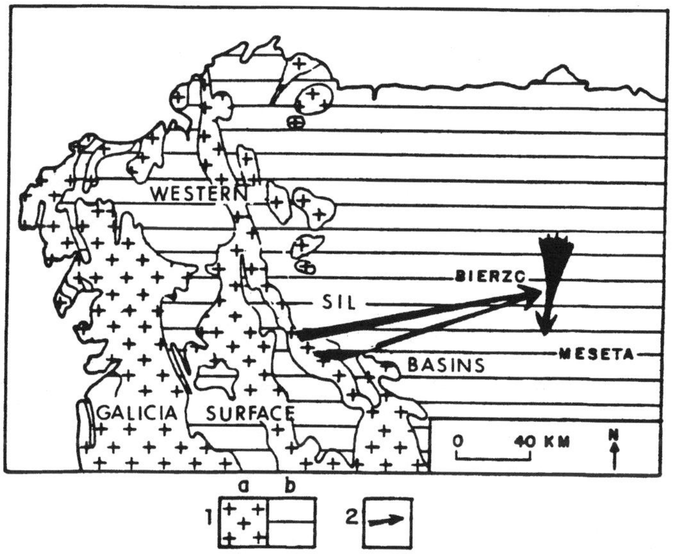

Fig. 9. Palaeocurrent directions measured in Lower Miocene deposits of north-western Iberian Peninsula (Vergnolle, 1988). 1. Basement: a. granitoids; b. metamorphic rocks; 2. palaeocurrents.

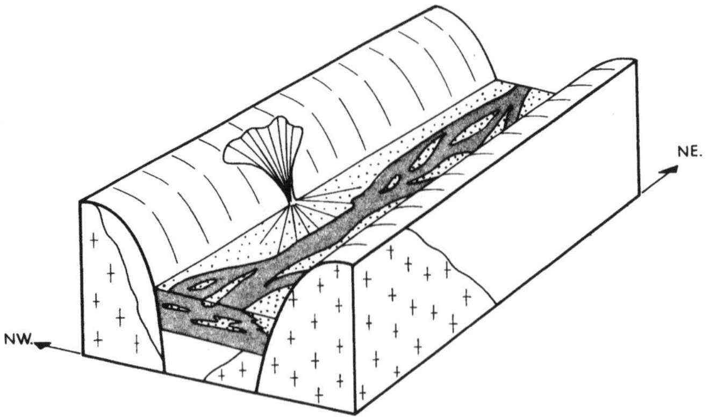

Fig. 10. Palaeogeographical scheme for Quiroga Formation (from Vergnolle, 1988).

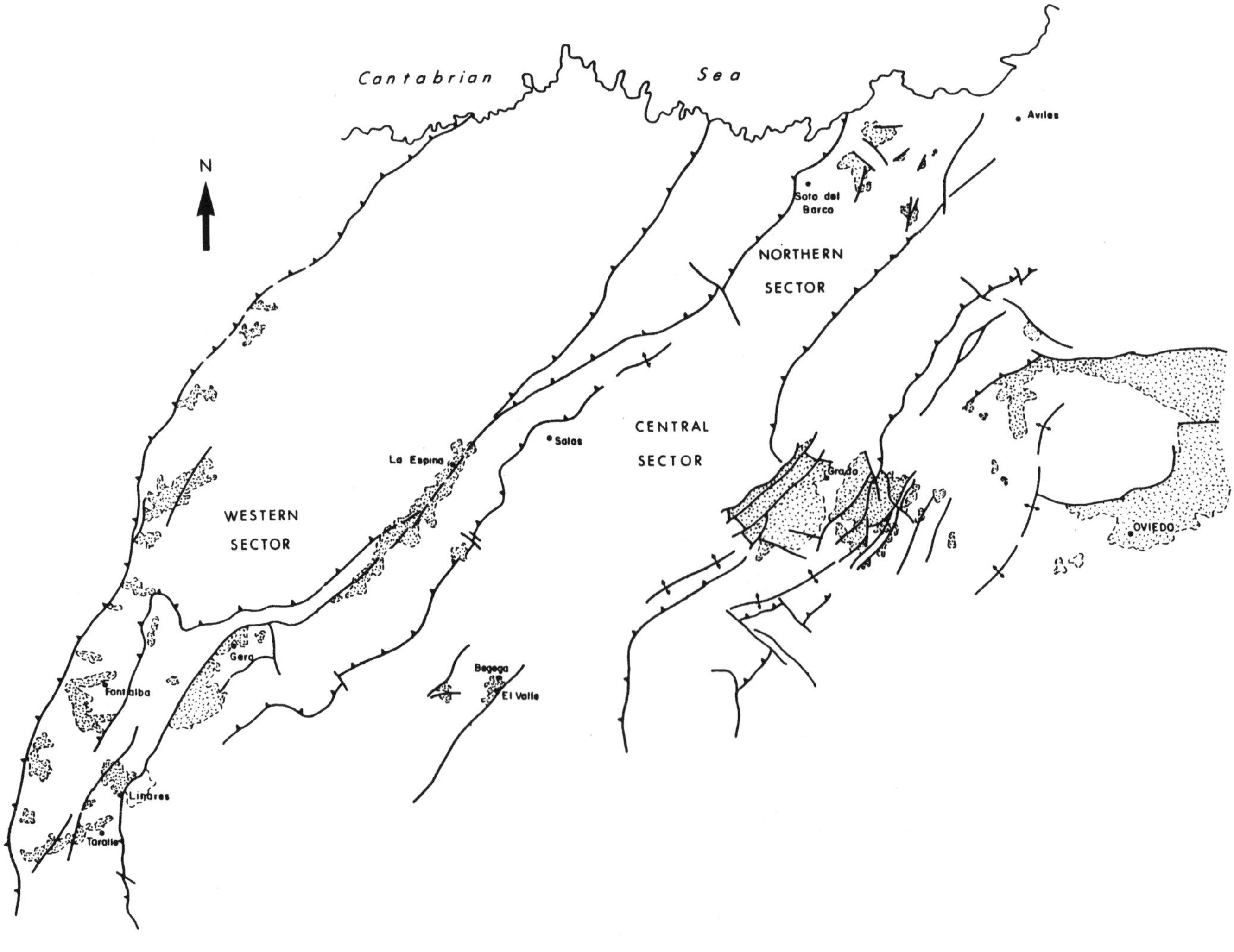

Fig. 11. Tertiary outcrops in western Asturias (IGME, 1979–1984).

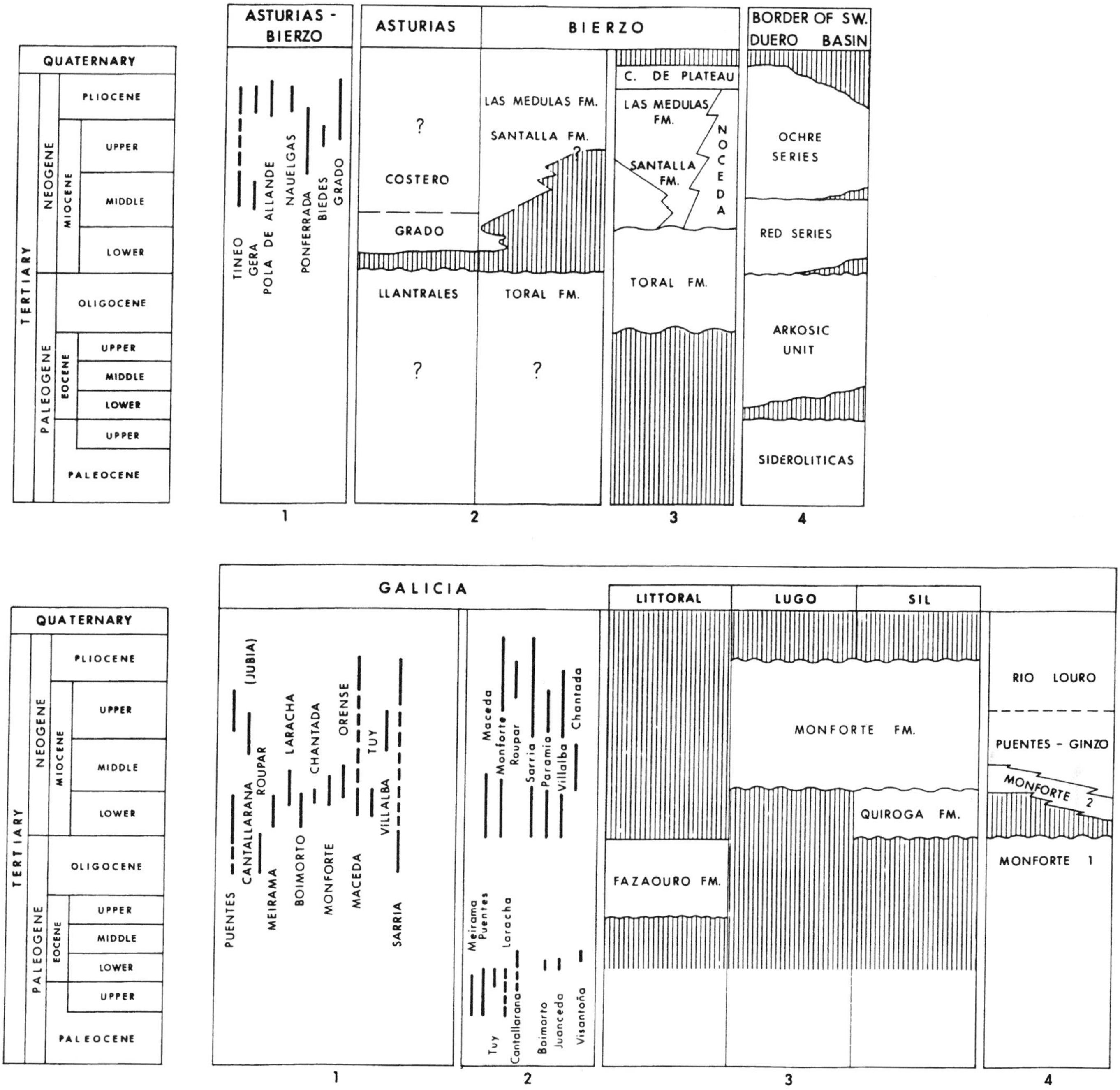

Fig. 12. Chronostratigraphy of Tertiary deposits in north-western Iberian Peninsula. Upper table: 1. Brell & Doval (1974); 2. IGME (1986); 3. Herail (1984); 4. Santisteban *et al.* (1991). Lower Table: 1. Brell & Doval (1974); 2. Martín-Serrano (1982); 3. Vergnolle (1988); 4. IGME (1986).

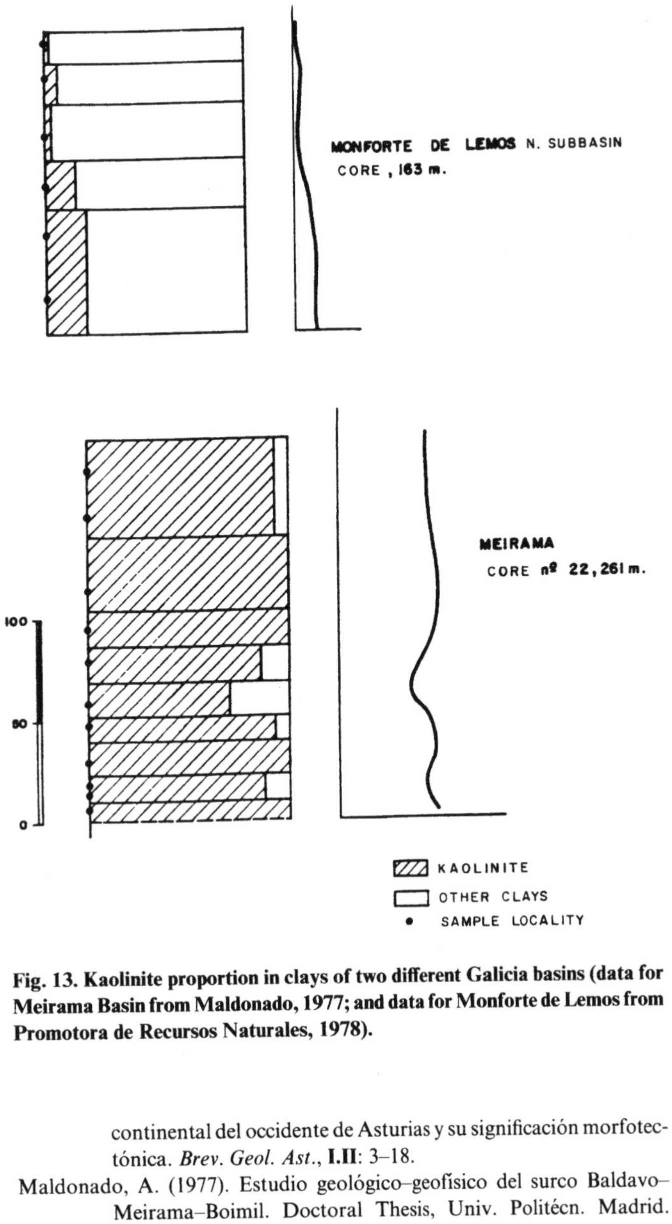

Fig. 13. Kaolinite proportion in clays of two different Galicia basins (data for Meirama Basin from Maldonado, 1977; and data for Monforte de Lemos from Promotora de Recursos Naturales, 1978).

continental del occidente de Asturias y su significación morfotectónica. *Brev. Geol. Ast.*, **I.II**: 3–18.

Maldonado, A. (1977). Estudio geológico–geofísico del surco Baldavo–Meirama–Boimil. Doctoral Thesis, Univ. Politécn. Madrid. ETSIM. Madrid.

Martín-Serrano, A. (1979). El conocimiento del lignito y del Terciario en Galicia. Exposición y crítica. *Tecniterrae*, **31**: 46–55.

Martín-Serrano, A. (1982). El Terciario de Galicia. Significado y posición cronoestratigráfica de sus yacimientos de lignito. *Tecniterrae*, **48**: 1–23.

Martín-Serrano, A. (1990). Nouvelles hypothèses concernant la signification géologique du lignite de Galice (Nord-Ouest de l'Espagne). *Ind. Min., les Techn.*, Juin: 249–258.

Menendez Amor, J. (1975). *Hoja Geológica E. 1:50000 MAGNA no. 22 (Puentedeume)*. IGME.

Medus, J. (1963). Contribution Palynologique a la connaisance de la flore et de la vegetation neogene de l'ouest de l'Espagne; étude des sediments recents de Galice. Thèse 3ème. cycle, Univ. Montpellier. 91 pp.

Monge, C. (1987). Estudio sedimentológico de la cuenca terciaria de Meirama. Un ejemplo de cuenca sobre una falla de salto en dirección. *Cuad. Lab. Xeol. Laxe*, **11**: 51–67.

Nonn, H. (1966). Les regions cotiéres de la Galice (Espagne). *Publ. Fac. des Lettres Univ. Strasbourg*, **3**: 1–591.

Nonn, H. and Medus, J. (1963). Primeros resultados geomorfológicos y palinológicos referentes a la cuenca de Puentes de García Rodríguez (Galicia). *Not. Com. Inst. Geol. Mineral. Esp.*, **71**: 87–94.

Olmo Sanz, A. Del (1986). Estudio sedimentario de las cuencas terciarias y cuaternarias de Monforte de Lemos, Maceda y Quiroga. *Cuad. Lab. Xeol. Laxe*, **10**.

Santanach, P., Baltuille, J.M., Cabrera, L., Monge, C., Sáez, A. and Vidal Romani, J.R. (1988). Cuencas terciarias gallegas relacionadas con corredores de fallas direccionales. *II Congr. Geol. España*, simp.: 123–133.

Sluiter, W.J. and Pannekoek, A.J. (1964). 'El Bierzo'. Etude sédimentologique et géomorphologique d'un bassin intra-montagneux dans le Nord-Ouest de l'Espagne. *Leidse Geol. Meded.*, **30**: 141–181.

Truyols, J., García-Ramos, J.C., Casanovas-Cladellas, M.J. and Santafé Llopis, J.V. (1991). El Terciario de los alrededores de Oviedo. *I Congr. Grupo Español de Terciario*, comun.: 334–336.

Vergnolle, C. (1984). Lithostratigraphie des bassins tertiaires du Nord-Est de la Galice (Espagne). *Mélanges de la Casa de Velázquez*, **20**: 372–392.

Vergnolle, C. (1985). Géométrie du remplissage des bassins de Sarria et de Monforte (Galice, Espagne) et évolution géomorphologique régionale. *Mélanges de la Casa de Velázquez*, **21**: 331–346.

Vergnolle, C. (1987). Tertiary geomorphological evolution of the marginal bulge of the north-west of the Iberian Peninsula, and lithostratigraphy of the grabens of the north-east of Galicia (Spain). In Gardiner, V. (ed.), *International geomorphology, 1986*, Part II: 1063–1072. John Wiley & Sons.

Vergnolle, C. (1988). Morphogenese des reliefs cotiers associes a la marge continentale nord-espagnole. L'Exemple du nord-est de la Galice. Thèse de Doctorat 'noveau régime', Univ. Toulouse Le Mirail: 217 pp.

Vidal Boix, C. (1941). Contribucíon al conocimiento morfológico de las cuencas de los rios Sil y Miño. *Bol. R. Soc. Esp. Hist. Nat.*, **39**: (3–4): 121–161.

Vidal Boix, C. (1954). Geología de los Montes Aquilianos y borde meridional de la depresión del Bierzo (León). *Homenaje a E. Hernández Pacheco, Bol. R. Soc. Esp. Hist. Natl.*, spec. publ.: 677–695.

Virgili, C. and Brell, J.M. (1975). Algunas características de la sedimentación durante el Terciario en Galicia. *I Cent. R. Soc. Esp. Hist. Natl.*, vol. extr.: 515–523.

W8 Onshore Cenozoic strike–slip basins in NW Spain

L. CABRERA, B. FERRÚS, A. SÁEZ, P.F. SANTANACH AND J. BACELAR

Abstract

There are many Cenozoic (Late Oligocene(?)–Early Miocene) non-marine onshore basins in NW Spain. Among these onshore basins two NW–SE alignments of small strike–slip basins occur. These strike–slip faults, which can be related to similar faults recorded in the offshore continental margin zones, have affected the NW Galician region and have given rise to a number of small basins which evolved independently, under the influence of local tectonic structures. Tectonic events resulted in quick changes in local drainage conditions which gave rise to poorly drained depressions. Development of palustrine, marsh–swamp and lacustrine systems, which were sometimes of a very small size, was enhanced by these geomorphic conditions. Workable coal deposits resulted from this tectonic and sedimentary evolution. Understanding of the structure and stratigraphy of these basins should contribute to a better understanding of the recent evolution of the westernmost end of the Northern Iberian continental margin.

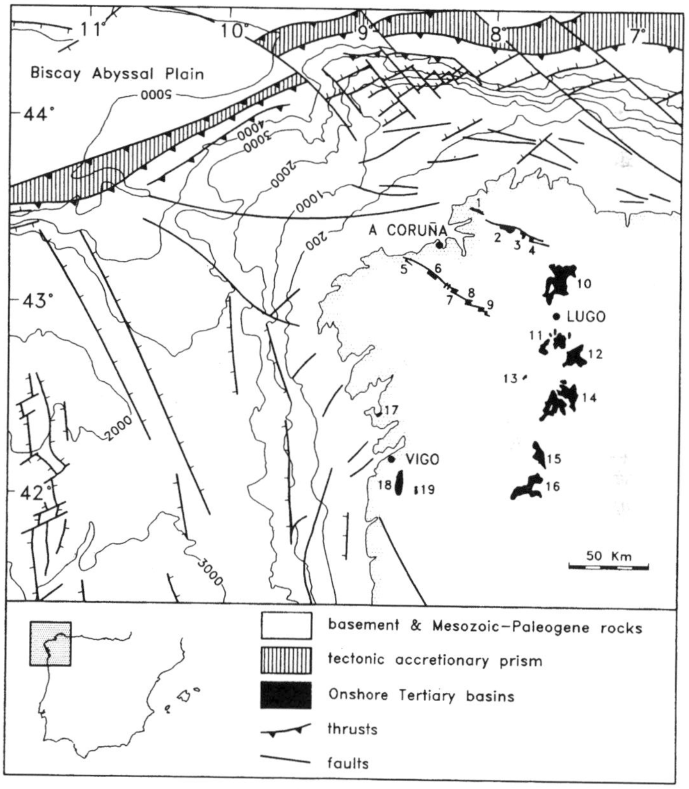

Fig. 1. General structural setting of the onshore non marine Galician basins in NW Spain (adapted and modified from Boillot *et al.*, 1989). The orientation of the onshore strike–slip systems and their related basins (1 to 9) corresponds well to that recorded for some of the offshore faults. Note that some of these offshore faults displaced the Early Paleogene (Paleocene–Eocene) tectonic accretionary prism. In Eastern Galicia there are other basins with a noticeable areal extent and oriented in a NNE–SSW direction (10 to 16). A few more basins (17 to 19) are located near the present western shore zone. *Strike–slip onshore basins. Northern system*: 1. Pedroso, 2. As Pontes, 3. Roupar, 4. Moiñonovo. *Southern system:* 5. Lendo, 6. Meirama, 7. Visantoña, 8. Xanceda, 9. Lanza-Orros and Boimil. *Other onshore basins*. 10. Villalba, 11. Paramo, 12. Sarria, 13. Chantada, 14. Monforte, 15. Maceda, 16. Xinzo da Limia, 17. Dena, 18. Tui, 19. Monçao.

Introduction

There are many small Tertiary non-marine basins in Galicia, NW Spain (Birot and Solé Sabarís, 1954; Nonn and Medus, 1963; Virgili & Brell, 1975; Martín Serrano, 1979; 1982; Santanach *et al.*, 1988). Some of these basins (1 to 9 in Fig. 1) occur in two NW–SE alignments, related to two strike–slip faults systems located in NW Galicia (Monge, 1987; Santanach *et al.*, 1988).

This chapter deals with the description of the general features of these strike–slip-related Cenozoic basins. Most of the sedimentary infills in these NW–SE-oriented basins consist of terrigenous dominated alluvial and lacustrine–paludine deposits, but some of them include thick, workable brown coal seams generated in swamp and marsh environments. In some basins (As Pontes and Meirama) the seams have been intensively mined from the 1960s to the present. Subsurface data which resulted from coal exploration surveys (IGME, 1979–1984) and from mining have improved knowledge of the structure and stratigraphy of these basins.

The interest of these basins, apart from their economic importance, derives from the possibility of detailed analysis of the tectonic influence on their evolution. Moreover, comprehension of their structure and stratigraphic record should contribute to better understanding of the recent evolution of the westernmost end of the Northern Iberian continental margin.

Geological setting

The strike–slip Cenozoic basins in Galicia (NW Spain) are located in the westernmost onshore zone of the Northern Iberian continental margin. This margin evolved as an extensional feature affected by some major lateral slip faults until the latest Cretaceous (Boillot and Malod, 1988). Nevertheless, during the Paleocene–Eocene interval, the margin became convergent as a result of the Eurasian–Iberian plate approach (Dewey *et al.*, 1989). In the eastern zones of the continental margin this process led to a continental collision and to the building of the Pyrenees. In the western segment of the margin the convergence led to the southward subduction of the European plate beneath the Iberian one (Mauffret *et al.*, 1978; Boillot *et al.*, 1979; Boillot, 1986; Boillot and Malod, 1988). The final structural development of these western zones of the continental margin started with a Paleogene subduction of the Bay of Biscay oceanic lithosphere under the continental Iberian lithosphere. This process resulted in, among other features, development of a marginal trench, upbuilding along the present northern Spanish continental margin of a tectonic accretionary prism (Boillot and Malod, 1988) and the generation in the offshore and onshore zones of folds, reverse and strike–slip faults and some minor related basins (Santanach *et al.*, 1988; see also Chapter W5).

The strike–slip fault systems recognized in the offshore zones of the Northern Galician margin (Fig. 1), and which affected and displaced the Paleocene–Eocene accretionary prism (Boillot *et al.*, 1988), can be traced into the onshore zones where they have given rise to several strike–slip-related basins. Among other onshore strike–slip faults which can be recognized along the northern Iberian margin (i.e. Urbenia, Vidio, Ventaniella, see Chapter W5), the Galician strike–slip systems are located in a westernmost position and display a variety of structural and sedimentary features. The activity of these strike–slip systems, considered in the framework of the continental margin evolution, was a late process.

The strike–slip systems and their related basins

Two major strike–slip zones with related Cenozoic basins can be distinguished in the NW Galician region: the Pedroso–As Pontes–Moiñonovo and the Lendo–Meirama–Boimil zones (Figs. 1 and 2). In these minor but complex strike–slip systems there are, apart from the major strike–slip faults, other faults whose slip depends on their orientation in relation to the resulting regional to local stress field. However, most of the tectonic structures recorded in these fault systems can be attributed to a N–S compression (Santanach *et al.*, 1988).

Cenozoic basins in both fault zones are related to major NW–SE fault systems and share some structural and stratigraphic features.

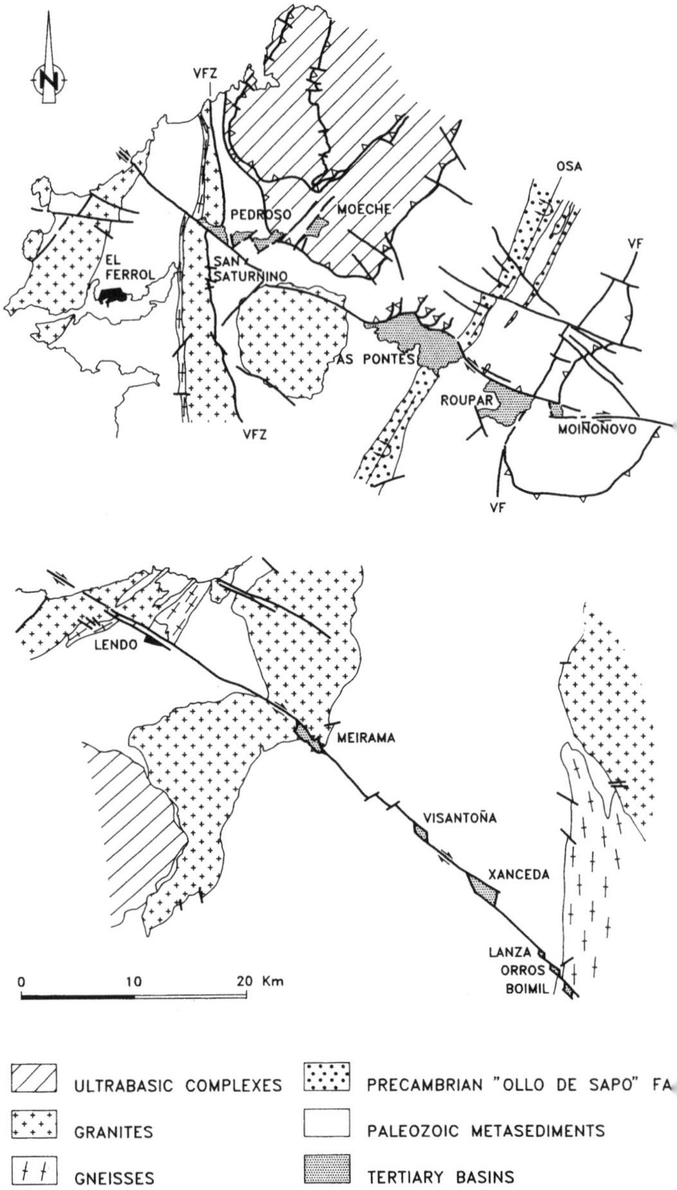

Fig. 2. Main structural features of the Pedroso–As Pontes–Moiñonovo (top) and Lendo–Meirama–Boimil (bottom) strike–slip systems and their associated basins. Note the relation of these basins to the existence of restraining fault junctions (top, Pedroso area), restraining bends (top, As Pontes and Roupar basins), restraining oversteps (bottom, Lendo, Meirama) and releasing oversteps related to right-slip right-overstepping faults (top, Moiñonovo; bottom, Visantoña, Xanceda, Lanza, Orros and Boimil). Lateral motion along the master faults of both systems resulted in the displacement of earlier (Hercynian) structures which affected the Precambrian–Early Paleozoic basement. Relationships between tectonic structures and the basin infills point to an essentially Cenozoic age (Late Oligocene? – Miocene) for these movements.

Most of these main faults, and others in Galicia which display similar orientations, were considered as Late Hercynian strike–slip faults (Parga, 1969; Arthaud and Matte, 1975). Nevertheless, the relationships between the tectonic structures and the associated Cenozoic basin infills point to the syntectonic development of the latter (García Aguilar, 1987; Monge, 1987; Santanach *et al.*, 1988; Bacelar *et al.*, 1988, 1992). As a consequence, a Cenozoic age for the activity of these faults is more likely.

The precise dating of the sedimentary infill of these basins has not yet been possible, due to the scarcity of well-characterized fossil assemblages. The scarce mammal remains (Esteban *et al.*, pers. commun.) found in the As Pontes basin have suggested a Late Oligocene–Early Miocene age. Some authors (Baltuille *et al.*, 1990, 1992) suggested, on the basis of the palynological data, a Late Oligocene(?)–Early Miocene age for some of the basins. Others (Nonn and Medus, 1963; Medus, 1965) suggested on the basis of similar data a Middle to Late Miocene age for the same basins. Reptilian remains found in the As Pontes basin have not contributed to more accurate dating (Cabrera *et al.*, in press).

The northern strike–slip fault zone (Pedroso–As Pontes–Moiñonovo) and its related basins

This strike–slip fault zone is up to 55 km long and 10 km wide, being noticeably wider and more complex than the southern strike–slip zone (Figs. 1 and 2). Two minor fault alignments occur, showing a slightly oblique orientation in relation to the fault-zone boundaries. One of them stretches from the coastline as far as the Pedroso–Moeche segment. The other corresponds to the As Pontes–Moiñonovo zone. Both fault alignments are separated by a zone of positive relief related to a compressive fault relay. Along the faults included in the strike–slip zone, the Precambrian and Paleozoic basement rocks are affected by horizontal right-lateral offsets of up to one kilometre.

The Pedroso, San Saturnino and Moeche basins are located at the SE end of a right-lateral strike–slip NW–SE-oriented master fault which stretches from the coastline (Fig. 2). At this SE termination the basin margins were formed by the activity of minor NE–SE-oriented strike–slip faults with a certain reverse component. Thus, the origin of these basins is related to the right-lateral slip activity of the major fault as well as to that of the associated minor left-lateral slip faults. The basins show a complex geometry and can be divided into minor sub-basins. Subsurface data show the noticeably deep sinking of the basement rocks (more than 225 m in the Pedroso basin) and the clear basin assymmetry. In the Moeche basin the basement has been tilted towards a bounding NW fault, where maximum depths are attained (Santanach *et al.*, 1988).

The As Pontes and Roupar basins are located along a NW–SE dextral strike–slip major fault. Both basins show a compressive character and developed in relation to the restraining bends of the right-lateral strike–slip major fault (Figs. 2 and 3). The As Pontes basin (Figs. 2, 3, 4 and 5) is the best known basin in the zone, thanks to the geological data which have resulted from brown coal mining (Bacelar *et al.*, 1988, 1992).

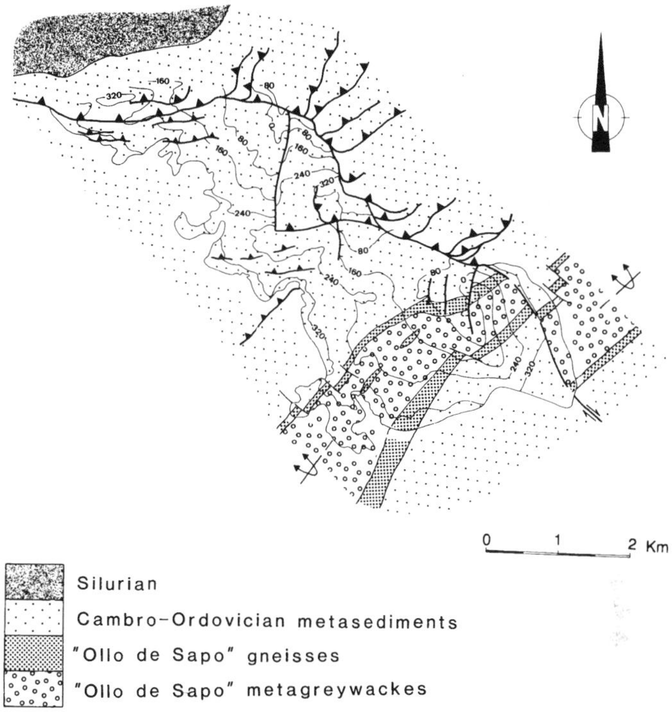

Fig. 3. Main structural features, isobath (sea level datum) and subcrop basement map of the As Pontes Basin (see Figs. 1 and 2 for location). The Late Oligocene (?)–Miocene As Pontes Basin is one of the better known in Galicia thanks to extensive well exploration and to open pit coal mining. The basin, 7 km long and 1.5 to 2.5 km wide, shows a NW–SE orientation, parallel to the strike–slip fault system orientation. The basin basement and the surrounding basin areas consist of Precambrian and Early Paleozoic rocks deformed during the Hercynian orogeny (Manera Bassa *et al.*, 1979). The basin is assymmetrical and it is divided into two sub-basins. The basement dips gently towards the northern basin margins, attaining the maximum depth (80 m below sea level). The northern basin margins are bounded by tectonic structures and the southern ones are defined by the unconformity between the Precambrian–Early Paleozoic basement and the bottom of the Cenozoic basin infill. The maximum basin infill thickness (400 m) is attained near the northern margins and decreases clearly towards the southern margin, where a clear southward onlap occurs. The basin was generated in a complex dextral-slip fault zone. The northern margin is bounded by a NW–SE dextral strike–slip fault with a total slip nearest to 1 km (note the displacement of the Hercynian anticline structures). The master fault has a double restraining bend geometry (*sensu* Christie-Blick and Biddle, 1985). There are thrust-sheet systems developed in these zones with a maximum southwards displacement of 800 m in the Hercynian basement rocks. These thrusts propagated in 'break-back' sequence and, laterally, take a NE–SW orientation. Interaction between the northern contractive structure and a N–S normal fault system resulted in the generation of two assymmetrical sub-basins which acted as main depocentres. Along the northern margin the basin infill is deformed noticeably and involved in thrust-sheet systems. In the inner basin zones, the infill is affected only by N–S normal faults and small E–W thrusts which involve the basement. There are minor folds related to the above mentioned structures (thrusts and normal faults) and probably also to differential compaction. All the major and minor structures described in the basin are coherent with a N–S Cenozoic (Late Oligocene (?)–Miocene) shortening.

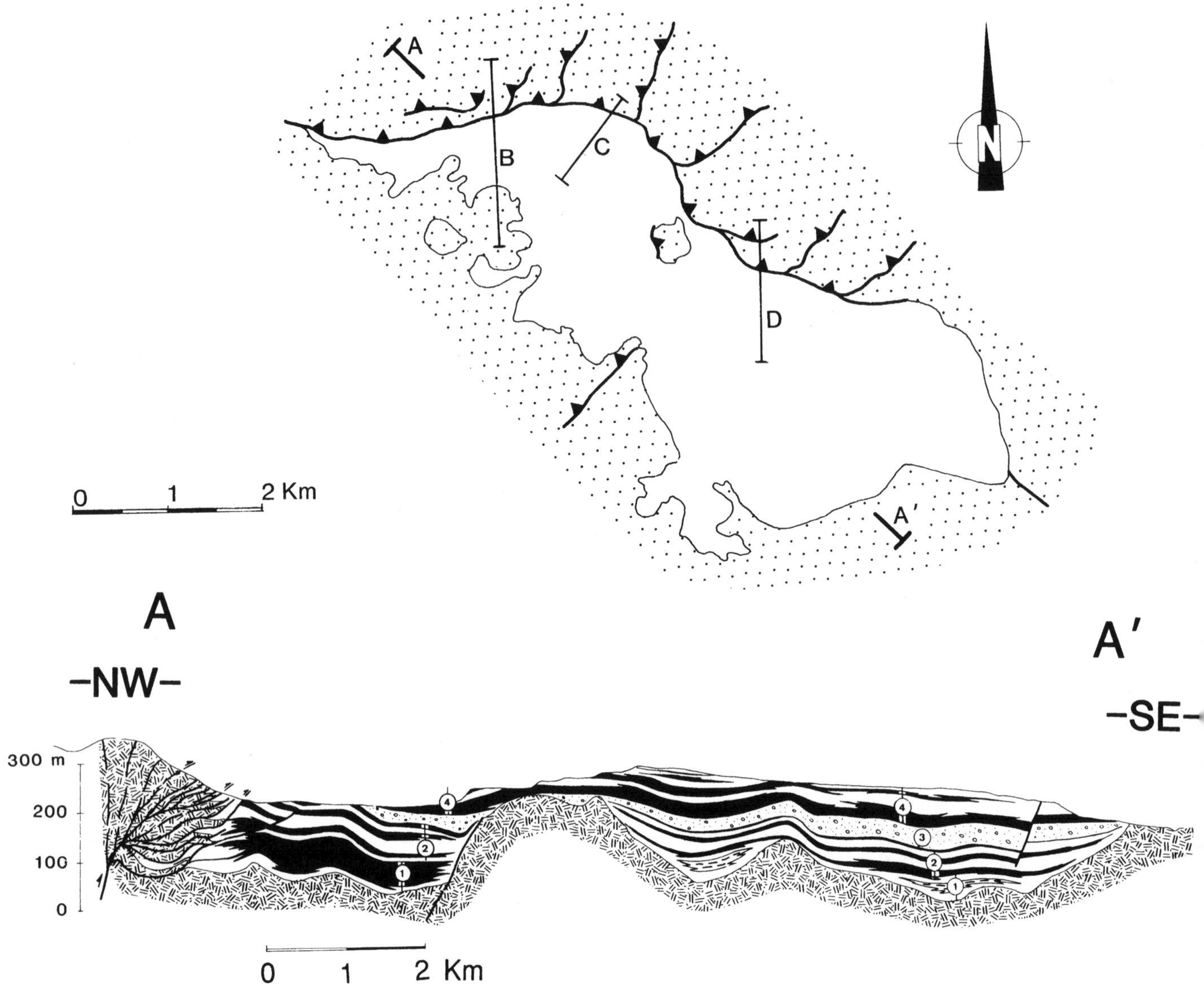

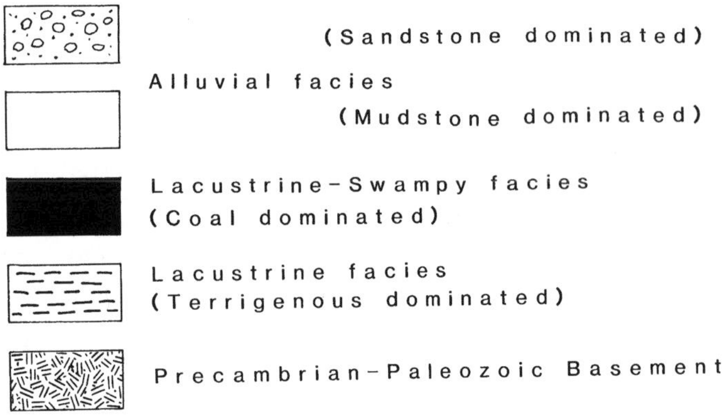

Fig. 4. Longitudinal cross-section of the As Pontes basin showing its general structural and stratigraphic features in the central basin zones. Note the break-back thrust sequence related to the northern basin margin in the western sub-basin and the tectonic threshold linked to the activity of a normal N–S fault and which was finally overlapped by the basin infill. Numbers 1 to 4 indicate a four-fold subdivision of this basin infill, based on the relative development of the terrigenous dominated (alluvial and lacustrine) and the coal dominated (lacustrine-swamp) facies assemblages. This sedimentary infill records: 1: Early evolutionary stage dominated by alluvial and lacustrine–swampy deposits. Note the diverse facies development in both sub-basins with a thick coal packet in the western sub-basin. 2: Stage with more equal alluvial and lacustrine–swampy sedimentation. Coal deposition extended to the whole basin. 3: Generalized spreading and dominance of alluvial–fluvial sedimentation. 4: Late-final stage characterized by a renewal of coal accumulation alternating with alluvial deposits which became dominant and finally exclusive at the upper stratigraphic levels.

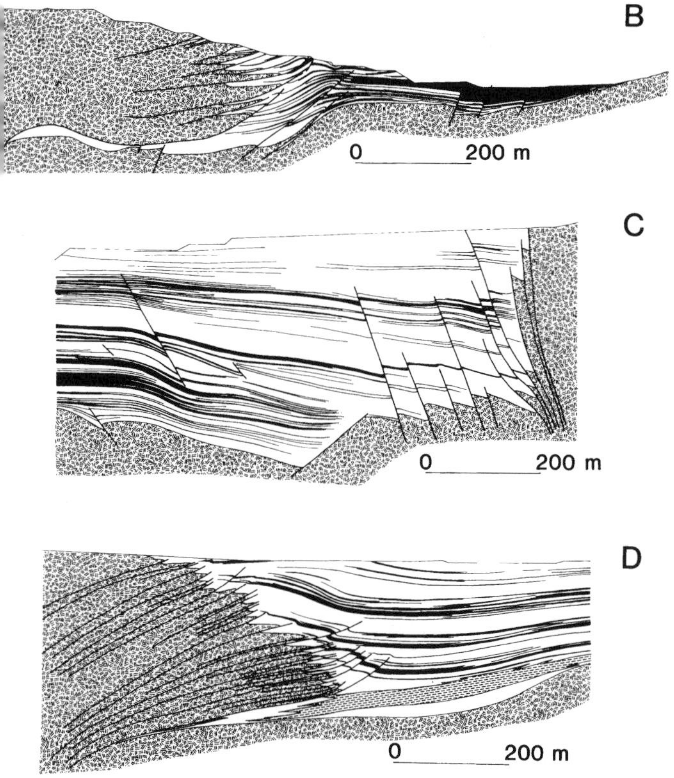

Fig. 5. Cross-sections of the As Pontes basin (see Fig. 4 for locations). Note the break-back sequence of thrust sheet emplacement in sections B and D and that the displacement and final emplacement of the major earlier sheets took place during and slightly after deposition of the lower part of the basin infill. Section C shows the development of more vertical reverse faults as well as that of a normal fault which bounded the western margin of the structural threshold in the central part of the basin. This fault only affected the lower part of the basin infill. The facies assemblage relationships show the successive spreadings and retreats which affected the alluvial and swampy lacustrine facies assemblages and which make up the main sequential arrangement observed in the basin.

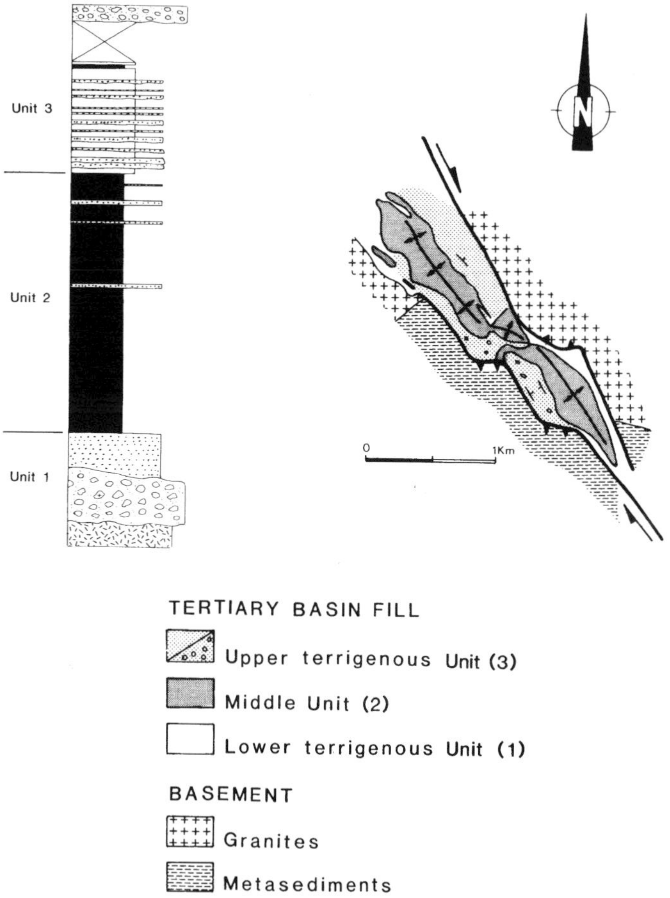

Fig. 6. The main structural and stratigraphic features of the Meirama basin (based on Monge, 1987, and Santanach *et al.*, 1988). This basin is located in the central part of the southern strike–slip zone and has an areal extent of about 2 km². It is elongated with a NW–SE oriented long axis 2.6 km in length. The basin is up to 600 m width and its margins, which display a stepped pattern, are related to NW–SE and E–W oriented faults which display right and left lateral slip, respectively. Normal faults which are perpendicular to the major basin axis also occur. The maximum basement depths (up to 250 m) are located along the NE basin margin and in its SE end while they decrease towards the NW and the SW margin. The basin infill is strongly affected by folds whose axes are parallel to the longitudinal basin axis and adapted to its stepped margin pattern. This basin infill deformation, together with the distribution, orientation and types of motion along the faults, point to the fact that the evolution of the Meirama basin was related to a restraining overstep of right strike–slip faults. The basin infill records the development of an early stage of deep weathering of the Paleozoic basement rocks (granite and slates). Early alluvial–colluvial sandy deposits are overlain by a thick brown coal accumulation with minor mudstone beds which record the persistent development of marshy–swampy paleoenvironments in a poorly drained setting. An upper alluvial progradational sequence, which consists of mudstones, sandstones and conglomerates, overlies the brown coal unit.

The Roupar basin structure is complex and subdivided into several minor sub-basins. The basin was formed along the same major fault which generated the As Pontes basin. This fault displays here a more reverse component due to its orientation. The maximum basement depth recorded in the basin is 60 m.

The Moiñonovo basin has a very restricted areal extent (1.5 to 2 km²) and a large basement sinking (recorded basement depth up to 98 m). This basin was formed in a pull-apart extensional zone related to a releasing overstep that developed between the major fault and another overstepped minor fault located more to the SW.

The southern strike–slip fault zone (Lendo–Meirama–Boimil) and its related basins

This strike–slip fault zone is up to 55 km long and only 3.5 km wide, being noticeably narrower than the northern strike–slip zone. Only one well-defined NW–SE-oriented structural alignment exists. Along the faults included in the strike–slip zone the Precambrian and Paleozoic basement rocks are affected by horizontal right-lateral offsets of up to 1.7 km slip (Monge, 1987). Several basins are included in this fault zone, all of them being related to NW–SE-oriented faults. Most of these basins are very small, with areal extents ranging between a few hundred square metres and half a square kilometre (Lendo, Visantoña, Lanza, Orros and Boimil basins). Only two basins (Meirama and Xanceda) have larger areal

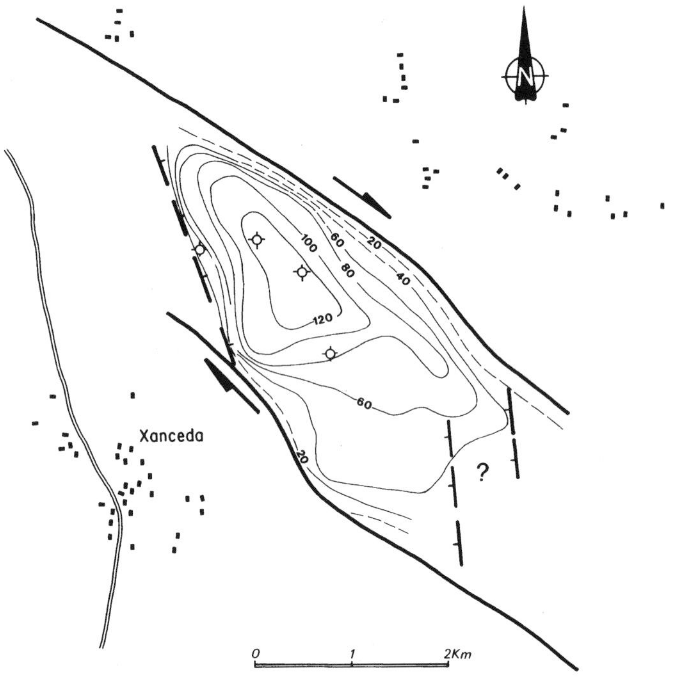

Fig. 7. The Xanceda basin is a rhomb-like graben with around 4 km^2 of areal extent. Subsurface data (VES – vertical electrical sounding – and some wells) resulting from coal exploration show that the Xanceda basin is clearly asymmetrical, with the maximum basement depth slightly displaced towards the NE margin. The basement depth decreases towards the NW and SE margins. The NE and SW basin margins are related to faults whose existence is deduced from the isobath lines of the basement established on the basis of subsurface data. The basin infill is tilted towards the NE but it is not folded. Basin infill in the central basin zone is characterized by fine-grained deposits (green and greyish green mudstones and siltstones) which are interbedded with organic matter-rich mudstones and with lignites. This facies assemblage records the persistent development of poorly drained conditions, with development of ponded alluvial and marshy–swamp paleoenvironments. The positions of wells are marked on the map.

extents (between 2 and 4 km^2). While the Lendo and Meirama basins are located in zones of restraining overstep, the remaining basins are essentially small-scale pull-apart basins that developed in releasing overstep zones.

In the Lendo basin, the fault which bound the northern basin margin has an orientation closer to the E–W direction. This fact would have resulted in its more reverse slip, which differs from that of the other minor basins in the strike–slip fault zone.

Coal mining in the Meirama basin (Figs. 2 and 6) has provided a better knowledge of the basin structure and stratigraphic features (Maldonado, 1977; García Aguilar, 1987; Monge, 1987).

The Xanceda basin is the largest one in this southern strike–slip system (Figs. 2 and 7). This basin formed in a releasing overstep zone which resulted in pull-apart extension.

The remaining basins in the strike–slip system do not attain the areal extent of Meirama and Xanceda. Most of them (Visantoña, Lanza, Orros, Boimil) are located in releasing overstep zones, this fact resulting in relatively large local extension and basement sinking in relation to the areal extent of the basins. Thus, in Visantoña, with a basin areal extent of 0.25 km^2, a basement depth of 156 m has been recorded. Moreover, this basin displays a very characteristic subrhomboidal shape.

Sedimentary infill in the strike–slip-related basins

Paleogeographic and paleoclimatic constraints

The sedimentary record in the strike–slip Galician Cenozoic basins was developed under conditions of complete isolation from the sea. The strike–slip evolution which affected the NW Galician region gave rise to the differentiation of a number of small basins which evolved independently, under the influence of local tectonic structures. As in similar strike–slip settings (Christie-Blick and Biddle, 1985) tectonic movements resulted in quick changes in the local drainage conditions (i.e. diverted or beheaded drainage). Moreover, they gave rise to poorly-drained depressions (undrained depressions, sag ponds, endorheic swamps or lakes) or, in contrast, to the uplift of structural high zones. This fact profoundly influenced the generation and evolution of the sedimentary systems, which were sometimes very small in size.

Palynological assemblages (Medus, 1965; Nonn and Medus, 1963; Baltuille *et al.*, 1990, 1992), fossil plant leaves and crocodile remains (Cabrera *et al.*, in press) found in some of the Galician basins (As Pontes and Meirama) point to the fact that warm and humid tropical–subtropical paleoclimatic conditions were dominant during most of the early basin infill sedimentation. These conditions gave rise to deep basement weathering (Brell and Doval, 1979). A gentle trend to slightly more temperate and drier conditions could have developed during sedimentation of the upper basin infills (Medus, 1965).

Main features of the sedimentary systems and basin infill organization

The basins related to the NW–SE strike–slip systems in Galicia were mostly infilled by terrigenous contributions fed by alluvial systems of varying importance. Development of palustrine, marsh–swamp and lacustrine systems was enhanced by the above-mentioned geomorphic conditions.

The smallest basins, usually related to releasing oversteps (Fig. 2), were essentially 'sinkholes' where water and sediment contributions of local, small-sized, alluvial–fluvial systems were ponded. These deposits display rather poor facies differentiation due to the small basin size. Mudstones, sandstones and minor gravel deposits, which show poorly developed sequential trends, are the most widespread facies in these basins.

In the larger basins (i.e. As Pontes, Meirama, Xanceda; see Figs. 2 to 7) a more varied array of alluvial–fluvial paleoenvironments developed, and lacustrine to marsh–swamp systems related to distal alluvial zones were frequent and even persistent elements in the sedimentary basin evolution. Non-channelized matrix supported conglomerates and muddy sandstones, deposited by mud flow and debris flow processes, were clearly dominant in the proximal alluvial zones, which, in some cases, evolved into middle and terminal channelized zones where mudstones, sandy mudstones and sandstones became dominant. Noticeably well-developed lacustrine episodes have also been recorded in the early evolutionary stages of some of the basins (i.e. As Pontes, Roupar; see Fig. 4). Some of these lakes were shallow and probably developed under internal drainage conditions, since athalassic saline lacustrine gastropods have been recorded. Organic-rich laminites and clay–carbonate rhythmites have been recorded in some cases (i.e. As Pontes Basin) when the lake environments were deeper and persistent. Marsh–swamp paleoenvironments developed more frequently than well-developed lacustrine conditions and minor to major coal deposition took place in a number of basins (i.e. As Pontes, Meirama, Xanceda).

Sequential organization in the basins with thicker infill was rather diverse as a consequence of the varied structural situations which arose in the diverse zones of the strike–slip systems. The wide basin infills in those basins with exclusively alluvial sedimentation include a succession of progradational and retrogradational terrigenous sequences, which record the successive spreadings and retreats of proximal and middle alluvial system zones. In those basins where lacustrine and marsh swamp environments developed frequently and/or persistently, alluvial spreadings and retreats were often punctuated by correlative changes in areal distribution of these environments. The higher-rank megasequences recorded in the larger basins (i.e. in As Pontes and Meirama basins; Figs. 5 and 6) were probably related to changing subsidence–sedimentation rates linked to tectonic changes, although this process was also coincident with paleoclimatic change. On the other hand the minor sequences could be related to either tectonic or paleoclimatic pulses.

Concluding remarks

The onshore Galician (NW Spain) strike–slip fault zones and their associated Cenozoic basins record the late stages of Cenozoic (Oligocene–Miocene) deformation along this segment of the Northern Spanish continental margin. The NW–SE strike–slip Galician basins resulted from a N–S compression which affected the westernmost end of the convergent plate boundary which stretched from the Pyrenees to the end of the Northern Iberian margin.

The study of the structure of the As Pontes and Meirama basins shows the development during Cenozoic time of a N–S shortening in the NW Galician region. This N–S shortening gave rise to the dextral movement recorded in the NW–SE-oriented strike–slip fault zones. These dextral movements were probably Late Oligocene(?) and/or Miocene in age and resulted in syntectonic basin infills. It must be stressed that the right slips observed in the Precambrian and Paleozoic basement rocks could be generated simply by this Cenozoic fault activity. Thus, the Late Hercynian origin and activity of the strike–slip faults recorded in the Galician continental margin (Arthaud and Matte, 1975) has to be re-examined.

Basin generation in the strike–slip fault systems was related to:

1. contractive terminations (Pedroso, San Saturnino, Moeche);
2. restraining bends linked to the major faults (As Pontes, Roupar);
3. restraining oversteps (Meirama);
4. releasing oversteps with resulting pull-apart extensional processes (Moiñonovo, Visantoña, Xanceda, Lanza, Orros, Boimil).

Examination of the relationship between tectonics and sedimentation in the best-known basins shows the areally restrictive evolution of the basins linked to restraining bends (As Pontes and Roupar Basins) and restraining oversteps (Meirama). Major, workable brown coal deposits and lacustrine facies occur with noticeable development in these contracting basins. This fact can be related to the peculiar subsidence evolution in these tectonically restricted settings.

The relationships of the diverse basins with single structures and the absence of reliable and refined paleontological data prevent direct correlation between the diverse basins. Basin-infill sequential analysis is possible but can not be used for correlation.

Acknowledgements

The authors thank the scientific editors of this book, Cristino J. Dabrio and Peter F. Friend for their useful suggestions to improve and clarify the text. This paper has been financed by the CICYT Project AMB92–0311 of the Spanish Ministry of Education and Science. B.F. received financial support from ENDESA through the University of Barcelona–ENDESA agreement managed by the Fundació Bosch i Gimpera (Project 907). The authors greatly appreciate the cooperation of the Mina de Puentes geological team for the development of this work.

References

Arthaud, F. and Matte, P. (1975). Les décrochements tardihercyniens du sud-ouest de l'Europe. Géometrie et essai de reconstruction des conditions de la déformation. *Tectonophysics*, **25**: 139–171.

Bacelar, J., Alonso, M., Kaiser, C., Sánchez, M., Cabrera, L., Sáez, A. and Santanach, P. (1988). La cuenca terciaria de As Pontes (Galicia): su desarrollo asociado a inflexiones contractivas de una falla direccional. *II Congr. Geol. España, Granada 1988, Simposios*: 113–121.

Bacelar, J., Cabrera, L., Ferrús, S.B., Sáez, A. and Santanach, P. (1992). Control tectónico sobre la acumulación de lignitos de la cuenca terciaria de As Pontes (A Coruña, NW de España). *III Congr. Geol. España y VIII Congr. Latinoamericano de Geología, Salamanca 1992, Simposios* (2): 227–238.

Baltuille, J.M., Becker-Platen, J.D., Benda, L. and Ivanovic Calzaga, Y. (1990). A contribution to the division of the Neogene in Spain using palynological investigations. *Abstracts IX RCMNS Congress*, pp. 34–40. Barcelona.

Baltuille, J.M., Becker-Platen, J.D., Benda, L. and Ivanovic Calzaga, Y. (1992). A contribution to the subdivision of the Neogene in Spain using palynology. *Newsl. Stratigr.*, **27** (1–2): 41–57.

Birot, P. and Solé Sabarís, L. (1954). Récherches geomorphologiques dans le NW de la Peninsule Ibérique. *Mém. Doc.*, **4**: 9–61.

Boillot, G. (1986). Comparison between the Galicia and Aquitaine margins. *Tectonophysics*, **129**: 243–255.

Boillot, G. and Malod, J. (1988). The north and north-west Spanish continental margin: a review. *Rev. Soc. Geol. Esp.*, **1** (3–4): 295–316.

Boillot, G., Dupeuble, P-A. and Malod, J. (1979). Subduction and tectonics on the continental margin off northern Spain. *Mar. Geol.*, **32**: 53–70.

Brell, J. and Doval, M. (1979). Relaciones entre los sedimentos de Galicia y las alteraciones de su substrato. Interpretación paleoclimática. *Acta Geol. Hisp.*, **14**: 190–194.

Cabrera, L., Jung, W., Kirchner, M., Sáez, A. and Schleich, H.H. (in press). Crocodilian and Palaeobotanical findings from the tertiary lignites of the As Pontes Basin (Galicia, NW Spain). (Crocodylia, Plantae). *Cour. Forsch.–Inst. Senckenberg.*

Christie-Blick, N. and Biddle, K.T. (1985). Deformation and basin formation along strike–slip faults. In K.T. Biddle & N. Christie-Blick (eds.), *Strike–slip deformation, basin formation and sedimentation, Soc. Econ. Paleontol. Mineral., Spec. Publ.* **37**: 1–34.

Dewey, J.F., Helman, M., Turco, E., Hutton, D.H.W. and Knott, S.D. (1989). Kinematics of the Western Mediterranean. In M.P. Coward, D. Dietrich & R.G. Parks (eds.), *Alpine tectonics. Geol. Soc. London Spec. Publ.*, **45**: 265–283.

García Aguilar, J.M. (1987). Caracterización estratigráfica y tectosedimentaria de la cuenca lignitífera de Meirama (A Coruña). *Cuad. Lab. Xeol. Laxe*, **11**: 37–49.

IGME (1979–1984). Proyecto para la investigación de lignito en la región de Galicia. Fases I–III. Unpublished.

Maldonado, A. (1977). Estudio geológico-geofísico del surco Baldayo-Meirama-Boimil. Unpublished Thesis, Univ. Politécnica Madrid ETSIM.

Maldonado, A. (1979). Nuevos datos sobre la génesis del yacimiento de lignitos límnicos de Meirama (La Coruña). *Bol. Geol. Min.*, **90**(5): 468–474.

Manera Bassa, A., Barrera Morate, J.L., Cabal García, J.M. and Bacelar, J. (1979). Aspectos geológicos de la cuenca terciaria de Puentes de García Rodríguez (provincia de La Coruña). *Bol. Geol. Min.*, **95**: 452–461.

Mauffret, A., Boillot, G., Auxietre, J. and Dunand, J. (1978). Évolution structurale de la marge continentale au Nord-Ouest de la peninsule Ibérique. *Bull. Soc. Géol. Fr.*, **7** (20–4): 375–388.

Martín-Serrano, A. (1979). El conocimiento del lignito y del Terciario de Galicia: Exposición crítica. *Tecniterrae*, **S–203**: 46–54.

Martín-Serrano, A. (1982): El Terciario gallego. Significado y posición cronoestratigráfica de sus yacimientos de lignito. *Tecniterrae*, **S-255**: 19–41.

Medus, J. (1965). Contribution palynologique à la connaissance de la flore et de la vegétation néogene de l'ouest de l'Espagne: Étude des sediments récents de Galice. Thèse 3ème Cycle, Univ. Montpellier.

Monge, C. (1987). Estudio sedimentológico de la cuenca Terciaria de Meirama. Un ejemplo de una cuenca sedimentaria sobre una falla de salto en dirección. *Cuad. Lab. Xeol. Laxe*, **11**: 51–67.

Nonn, H. and Medus, J. (1963). Primeros resultados geomorfológicos y palinológicos referentes a la cuenca de Puentes de Garcia Rodríguez (Galicia). *Not. y Com. I.G.M.E.*: **71**: 87–94.

Parga, J.R. (1969). Sistemas de fracturas tardihercínicas del Macizo Hespérico. *Trab. Lab. Geol. Laxe*, **37**: 17 pp.

Santanach, P., Baltuille, J.M., Cabrera, L., Monge, C., Sáez, A. and Vidal-Romaní, J.R. (1988). Cuencas terciarias gallegas relacionadas con corredores de fallas direccionales. *II Congr. Geol. España, Granada 1988, Simposios*: 123–133.

Virgili, C. and Brell, J.M. (1975). Algunas características de la sedimentación durante el Terciario en Galicia. *I Cent. R. Soc. Esp. Hist. Natl., vol. ext.*: 515–523. Madrid.

W9 Tertiary of Central System basins

A. MARTÍN-SERRANO, J.I. SANTISTEBAN AND R. MEDIAVILLA

Abstract

The rise of the Central System due to reactivation of Late Hercynian fault systems during the Alpine Orogeny directly affected the structure and stratigraphic framework of the basins nearby that were being filled at the same time. The sedimentary record is the essential key to understanding the tectonic and palaeo-morphological history of the Central Range, and vice-versa. Relating the filling of the basins with the definition of the mountain range, pre-arkosic, arkosic and post-arkosic stages have been proposed. However, it is difficult to support the previous idea that the arkosic stage continued throughout the Late Tertiary to finish in Middle Pliocene times with the deposition of the 'Páramos (limestone)'. The arkoses of the Central System are of Eocene–Oligocene age and the highest alluvial-fan deposits may be of Aragonian age. There is only a poor record of the remaining Tertiary and Quaternary sediments, because of active river incision during this time in the basins, the ranges and elsewhere in the Spanish Meseta.

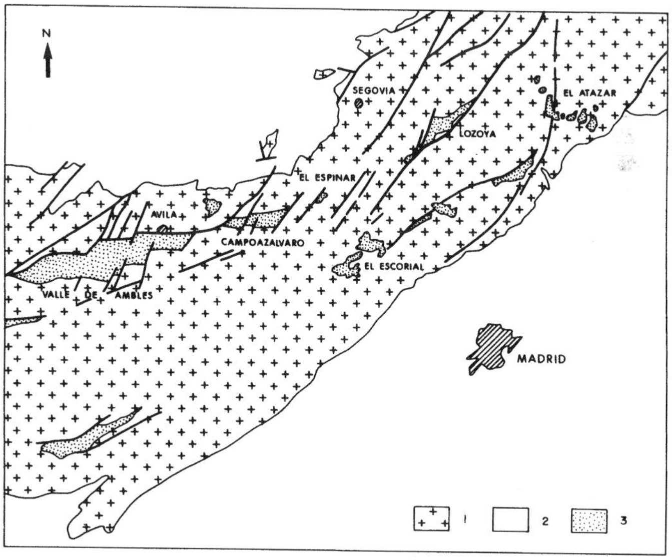

Fig. 1. Tertiary outcrops in western Central System. 1, Pre-Tertiary substratum; 2, Tertiary deposits of Duero and Tajo basins; 3, Tertiary deposits of the inner basins.

Introduction

The Central System is a complex inverse horst–graben system that developed during the Tertiary. The Plasencia fault separates two morphostructural domains: a western domain with large mountain blocks and basins oblique to the range, and an eastern one with large highs and minor basins parallel to the general structure (Fig. 1).

Western basins

The western Spanish Central System is defined by two fault families: ENE–WSW to NE–SW and NW–SE to WNW–ESE (Moreno, 1990). The most important sediment accumulations are preserved in its lowermost areas: Ciudad Rodrigo, Moraleja (Castelo Branco), Coria and Zarza de Granadilla.

The *Ciudad Rodrigo Basin* is a half-graben bounded to the south by a complex NE–SW-trending fault. Its infill is controlled by:

(a) asymmetry, with thicknesses up to 600 m S of Ciudad Rodrigo (Fernandez Amigot, 1981);

(b) compartmentalisation by transverse contemporaneous uplifts; and

(c) progressive abandonment of the sedimentary relationship with the Duero Basin, due to capture by the Portuguese fluvial network.

The oldest sedimentary rocks found in the western Duero Basin are the Cretaceous to Palaeocene *siderolithic* series (Jiménez, 1970, 1977; Blanco *et al.*, 1982; Molina *et al.*, 1989), which crop out only in the easternmost Ciudad Rodrigo Basin. These sediments are petrologically and mineralogically mature (quartz and kaolinite). Silcretes and ferricretes are abundant (Corrochano, 1977; Bustillo & Martín-Serrano, 1980; Blanco & Cantano, 1983).

The largest part of the sedimentary record is of Middle–Late Palaeogene age and it can be divided into three units of arkosic composition:

– The lower one (Early Eocene?) crops out to the north of

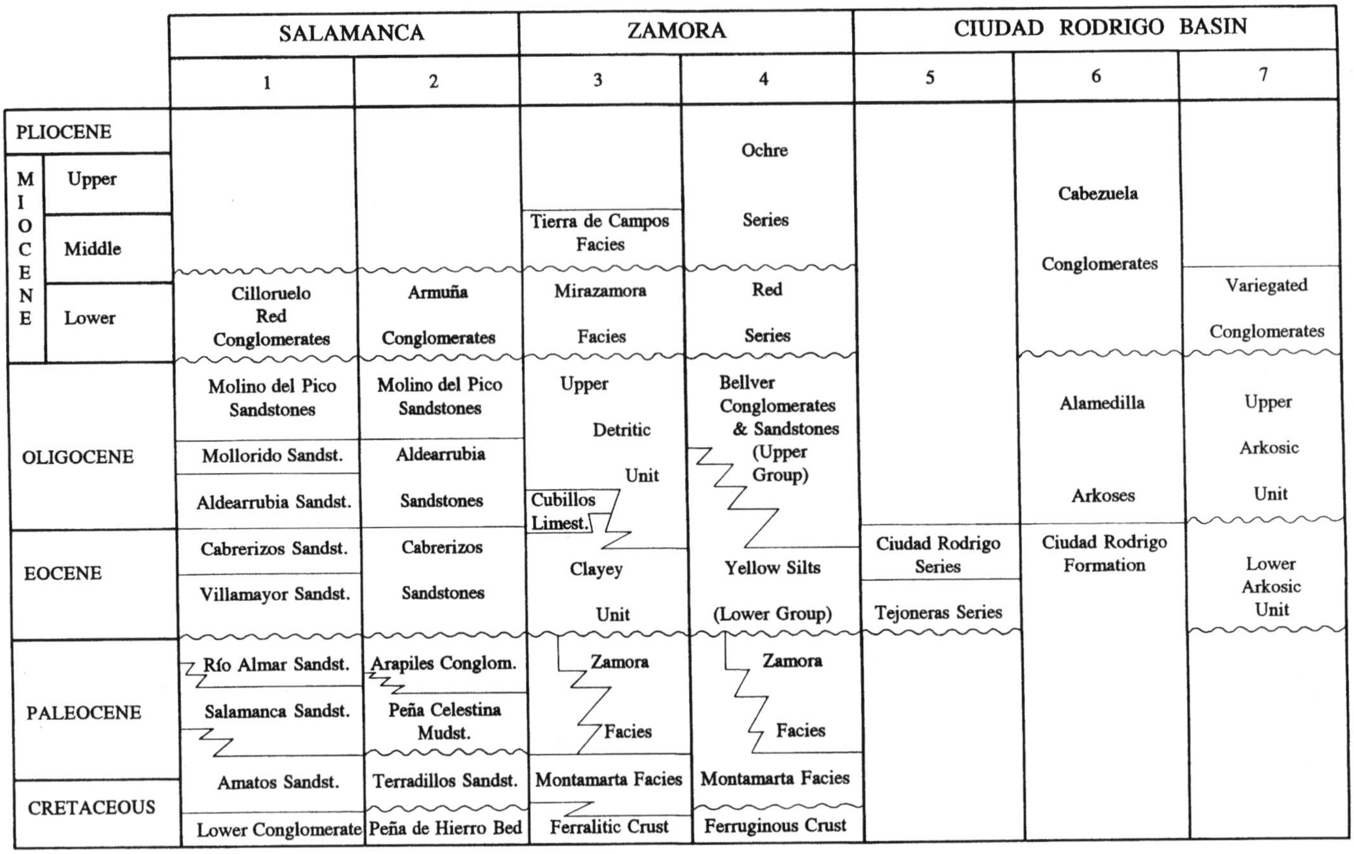
SALAMANCA
ZAMORA
CIUDAD RODRIGO BASIN
1
2
3
4
5
6
7
PLIOCENE
MIOCENE
Upper
Middle
Lower
OLIGOCENE
EOCENE
PALEOCENE
CRETACEOUS
Cilloruelo Red Conglomerates
Molino del Pico Sandstones
Mollorido Sandst.
Aldearrubia Sandst.
Cabrerizos Sandst.
Villamayor Sandst.
Río Almar Sandst.
Salamanca Sandst.
Amatos Sandst.
Lower Conglomerate
Armuña Conglomerates
Molino del Pico Sandstones
Aldearrubia Sandstones
Cabrerizos Sandstones
Arapiles Conglom.
Peña Celestina Mudst.
Terradillos Sandst.
Peña de Hierro Bed
Tierra de Campos Facies
Mirazamora Facies
Upper Detritic Unit
Cubillos Limest.
Clayey Unit
Zamora Facies
Montamarta Facies
Ferralitic Crust
Ochre Series
Red Series
Bellver Conglomerates & Sandstones (Upper Group)
Yellow Silts (Lower Group)
Zamora Facies
Montamarta Facies
Ferruginous Crust
Ciudad Rodrigo Series
Tejoneras Series
Cabezuela Conglomerates
Alamedilla Arkoses
Ciudad Rodrigo Formation
Variegated Conglomerates
Upper Arkosic Unit
Lower Arkosic Unit

Fig. 2. Various stratigraphic sections proposed for the Palaeogene deposits of Salamanca and Zamora provinces by previous workers (after Santisteban *et al.*, 1991.). 1: Jiménez (1970); 2: Alonso Gavilán (1981); 3: Corrochano (1977); 4: Martín-Serrano (1988); 5: Jiménez & Martín-Izard (1987); 6: Alonso Gavilan & Polo (1986–87) and Alonso Gavilán & Cantano (1987); 7: Cantano & Molina (1987).

Ciudad Rodrigo (Tejoneras Series; Jimenez & Martín-Izard, 1987), resting unconformably upon the *sidero-lithic* series of Salamanca (Santisteban *et al.*, 1991). It consists of white arkoses with variegated spots; grain size varies widely. This unit was deposited in proximal braided river systems.

- The Middle Eocene (Jiménez, 1977, 1982) intermediate unit crops out in many places with a thickness of up to 100 m both in Ciudad Rodrigo and in the SW Duero Basins (Figs. 2 and 3). It is composed of beige-greenish or reddish arkosic to lithoarkosic sandstone and mud, forming a coarsening-upwards megasequence. The muds are burrowed and hardened (calcretes and silcretes). They are interpreted as deposits of braided to sinuous fluvial systems that flowed towards the east and north-east and had well-developed flood plains.
- The third, Oligocene (Cantano & Molina, 1987; Polo *et al.*, 1987), unit is composed of coarse-grained arkoses with idiomorphic large-sized feldspars. They are rich in smectitic clays and show a little cementation. This unit reaches more than 200 m in thickness and is extensive.

Neogene sediments of presumed Early–Middle Miocene age are represented in the eastern Duero Basin, forming a gently dipping piedmont that erodes – and buries – the Palaeogene deposits. The age is based on detailed stratigraphical correlation. They are polymictic, heterometric red conglomerates deposited in alluvial cones passing distally into red sands and muds with paludal and pedogenic (edaphic) carbonates. The youngest deposits are Upper Miocene–Pliocene siliciclastic, ochre sediments deposited in alluvial plains of similar appearance to those of *rañas* related to the initiation of fluvial dissection (Mediavilla & Martín-Serrano, 1989).

The Alagón basins are morphostructural and stratigraphic copies of the Ciudad Rodrigo Basin. Thickness reaches 900 m in the Coria Basin. For instance, the Moraleja–Castelo Branco Basin is a half-graben lowered towards the NW by the Ponsul Fault (Dias & Cabral, 1989). All these small basins are the remains of a single, larger basin that underwent faulting and erosion (Fig. 4). As a consequence they all show the same lithofacies (Bascones & Martín Herrero, 1982a, b; Bascones *et al.*, 1982a, b, 1984a, 1984b; Ugidos *et al.*, 1985). The most prominent features are:

- The major part of their sedimentary fill consists of fluvio-lacustrine arkoses, subdivided into two litho-

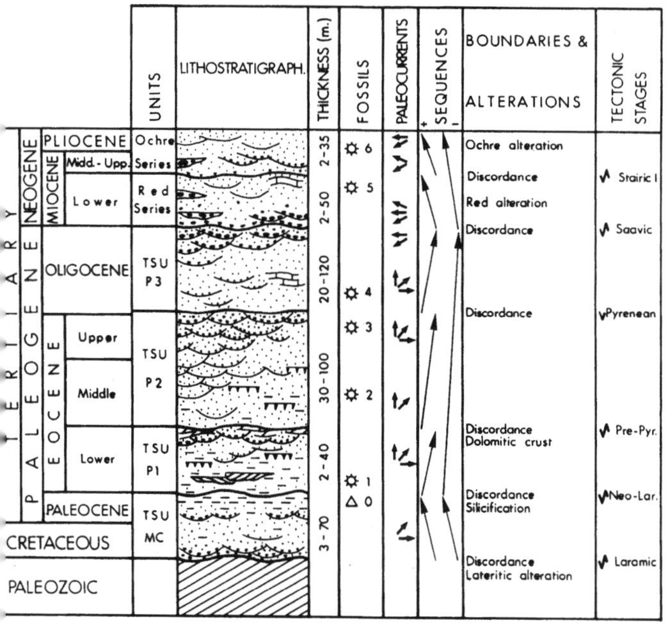

Fig. 3. Stratigraphic section of Paleogene deposits in the south-western Duero Basin (after Santisteban *et al.*, 1991). O: Absolute age (K/Ar) 58 Ma (Blanco *et al.*, 1982), 1: Sanzoles and Avedillo (Zamora), 2: Teso de la Flecha (Salamanca) and Corrales (Zamora), 3: Molino del Pico and San Morales (Salamanca), 4: Camino Fuentes and El Molino (Ciudad Rodrigo Basin), 5: El Guijo (Salamanca), 6: Benavente (Zamora).

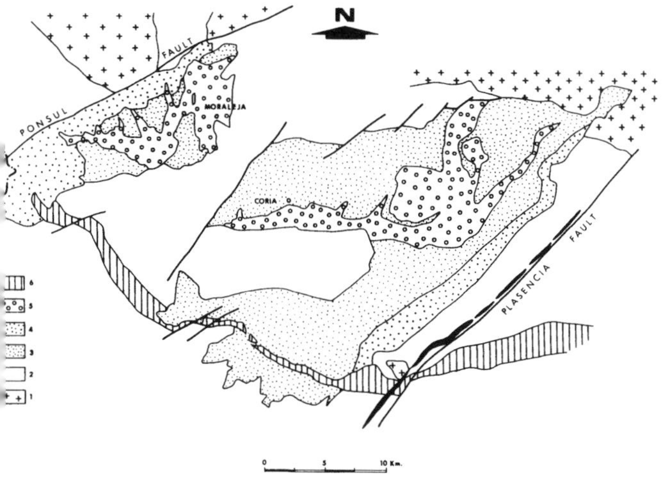

Fig. 4. Sedimentary infill of Alagón, Coria and Moraleja basins (adapted from Bascones *et al.*, 1982 b, 1984 a, b, c). 1, Igneous rocks; 2, schists; 3, quartzites; 4 and 5, Tertiary (4 fluvio-lacustrine arkoses; 5, coarse grained sediments of the marginal fringes); 6, recent deposits related to fluvial network.

facies: 1. white-grey and ochre gravely sands and polymictic micro conglomerates indurated by carbonates and clays; this lithofacies dominates the sedimentary fill; 2. grey, green and brown muddy sands, burrowed smectitic muds and sands in channel-form bodies with carbonate concretions and fish bones (Clupeidae).

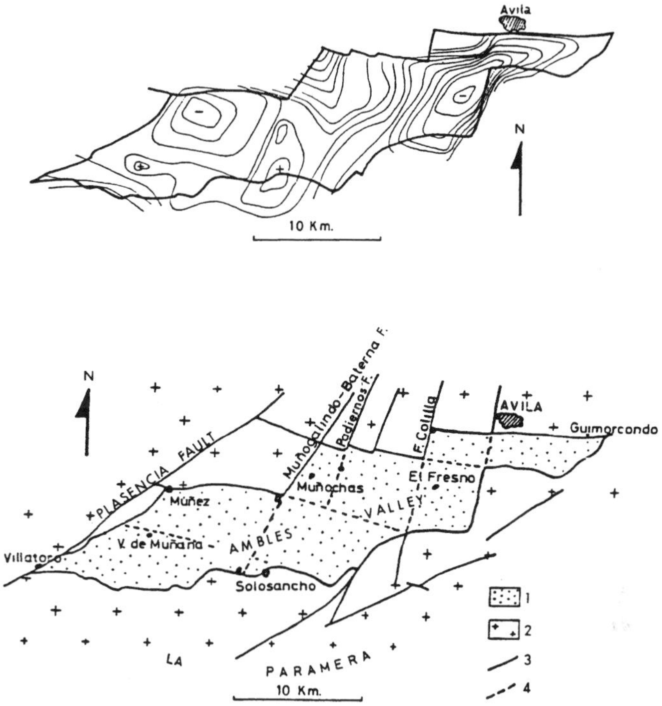

Fig. 5. Maps of structural and Bouguer anomalies of Amblés Basin (Garzon *et al.*, 1981). 1: Tertiary sediments; 2: Hercynian basement; 3: fault; 4: buried fault.

– Along the faults bounding the basins there are marginal ribbons of red muds and reddened polymictic conglomerates.

The largest part of the sedimentary fill of the Ciudad Rodrigo Basin is attributed to an Eocene–Oligocene age on the basis of detailed stratigraphic correlation with the area of Salamanca and Zamora, coupled with palynological dating. Stratigraphic and morphostructural similarities between the Ciudad Rodrigo Basin and the basins placed towards the south, support the extension of this correlation to their arkosic sediments. Tertiary sediments, mainly of Palaeogene age, should be present at both margins of the Central mountain range in the two tectonically controlled, subsiding troughs filled with fluvial sediments.

Internal basins of the eastern area

Small, discontinuous basins occur in an ENE–WSW direction for more than 150 km in the Gredos–Guadarrama–Somosierra massif. Such basins are affected and displaced by E–W and NE–SW faults. The most important of these small basins are the Amblés and Campo Azálvaro basins, where sedimentation reached thicknesses of 1000 and 400 m of sediments respectively (Fig. 5).

Eastward from Avila, basins contain siliciclastic and carbonate sediments of Late Cretaceous age. *Siderolithic facies sensu stricto* are restricted to Campo Azálvaro, Ambles and Alto Alberche basins. However, the sedimentary infill of these basins began with

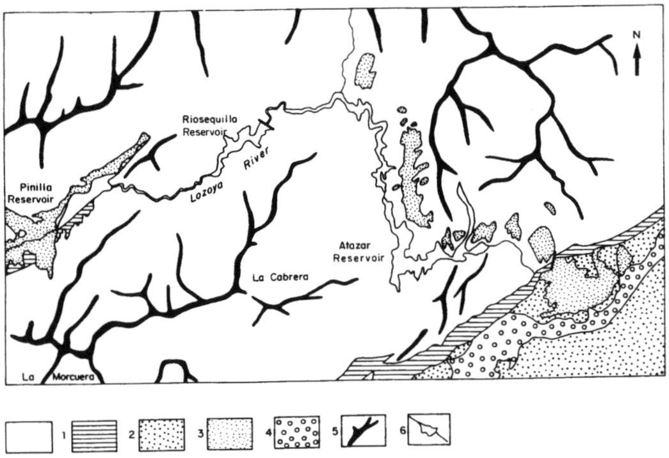

Fig. 6. Tertiary deposits related to the Lozoya River. Unshaded, substratum. 1, Mesozoic deposits of the Central System. 2, Tertiary deposits of the Madrid Basin. 3, Tertiary deposits related to the Lozoya River. 4, Jarama River's alluvium. 5, Main mountain divides. 6, Reservoirs.

arkoses containing Cretaceous rock fragments in the lower part. Stratigraphic correlation and palaeontological data from Los Barros (Amblés Valley) (Garzón & López, 1978) indicate a Middle–Late Palaeogene age for these sediments. This arkosic unit forms a fining-upwards megasequence composed of conglomerates with carbonate cement, arkosic sands and muds with caliches (carbonate–palygorskite–sepiolite) deposited in alluvial fan and braided-river environments. The palaeocurrent direction is opposite to the present regional morpho-structure (Martín-Serrano & del Olmo, 1990b; del Olmo, 1990a; Martínez Salanova & del Olmo, 1990).

Sediments of possible Neogene age are coarse-grained arkoses forming two megasequences with a restricted and incomplete areal distribution. Those of Early–Middle Miocene age are very coarse-grained sediments (conglomerates with boulders and sands with hydromorphic and edaphic features) with grain-size decreasing rapidly distally. They form a fining-upwards megasequence composed of alluvial-fan braided river sequences in concordance with the present morpho-structure. They form the morphological top of the basin infill (Martín-Serrano & del Olmo, 1990b; del Olmo, 1990a, b, 1991). Upper Neogene sediments are siliciclastic, of fluvial origin, and only represented in a few outcrops (Martín-Serrano & del Olmo, 1990b; de Olmo, 1990b, 1991), apparently related to the present fluvial network (Fig. 6).

The generation of the basins is related to post-Intra-Oligocene movements of N10–30° E and N60–100° E fault systems (*Iberian Stage*), and their Palaeogene arkosic infill (and locally also the Neogene) is affected by reverse or strike–slip faults N20–40° E and N75° E that define their borders (Capote *et al.*, 1987, 1990a, b, c, d, 1991; de Vicente, 1988).

Morphostructural evolution and sedimentary infill

The rise of the Central System is related to Alpine reactivation of Late Hercynian fault systems. The evolution of these movements directly affected the structure and sedimentary infill of the basins, and the sedimentary record is the essential key to understanding the tectonic and palaeomorphological history of the Central Range (and vice versa). The evolution of the Central System is deduced from the sediments found in nearby areas (pre-arkosic, arkosic and postarkosic stages; Garzón *et al.*, 1982), because the division links the filling of the basins with the definition of the mountain range.

Until the Upper Palaeogene (Middle–Late Eocene in Salamanca; Eocene–Oligocene in the Gredos–Guadarrama) a mature relief was developed (*superficie poligénica fundamental*) with smooth residual lineaments of Hercynian strike as in Peña de Francia, Tamames . . . (Pedraza, 1978; Garzón, 1980; Martín-Serrano, 1988; Fernández, 1988). The *siderolithic* or *siliceous sandstones* of Zamora and Salamanca (Jiménez, 1970; Corrochano, 1977; Bustillo & Martín-Serrano, 1980; Alonso Gavilán, 1981; Martín-Serrano, 1988), *Lower Tertiary Unit* (Garzón, 1980; Garzón *et al.*, 1981) or *pre-arkosic cycle* (Pedraza, 1978; Garzón *et al.*, 1982) of the Avila basins, are related to the Cretaceous sediments found west of the Campo Azálvaro meridian (Garzón *et al.*, 1982; Martín-Serrano & del Olmo, 1990a; Molina *et al.*, 1989), which, in the easternmost areas of the range, are interbedded with marine Cretaceous carbonates (Alonso, 1981).

All the Mesozoic siliciclastic lithofacies (*Utrillas*, *Weald* . . .) include alterite debris. The lack of carbonate fragments in the Palaeocene or pre-Lutetian and/or Cretaceous pre-arkosic sediments implies Cretaceous shelf preservation. Intensely weathered areas without Mesozoic sediments remained as source areas. Central System uplift did not take place during this episode (Martín-Serrano & del Olmo, 1990b).

The dispersion, isolation and deformation suffered by the outcrops of *siderolithic* rocks located close to mechanically deformed basin borders and their absence in the western basins support the previous arguments. Their sedimentation area is bounded by a line oblique to the range but parallel to the maximum extent of marine Cretaceous sediments (Fig. 7).

The uplift of the mountain ranges started from a Mesozoic inheritance: a smooth relief acting as source area of the pre-Lutetian continental deposits.

Morphostructural reorganisation of the area started with the *Arkosic cycle*. Its deposits rest unconformably upon Cretaceous sediments in the eastern basins. The lowermost arkosic sediments, near the bottom of the basins, include abundant rock fragments derived from Mesozoic outcrops. The rapid transition from these sediments (rich in fragments of granite and carbonate rocks) to those of exclusively arkosic nature suggests limited erosion of Cretaceous sediments and a slow areal uplift. The source area must have been a non-altered substratum.

No Cretaceous sediments have been found in the western basins; however, it is not difficult to deduce the age of uplift of the mountains. Near Salamanca, arkoses *sensu lato* fossilise an irregular, faulted substratum of Palaeozoic and *siderolithic* rocks. Fault directions are N30, N120 and E–W. As in the Amblés and Campo Azálvaro basins, a rapid, conspicuous change of petrology and

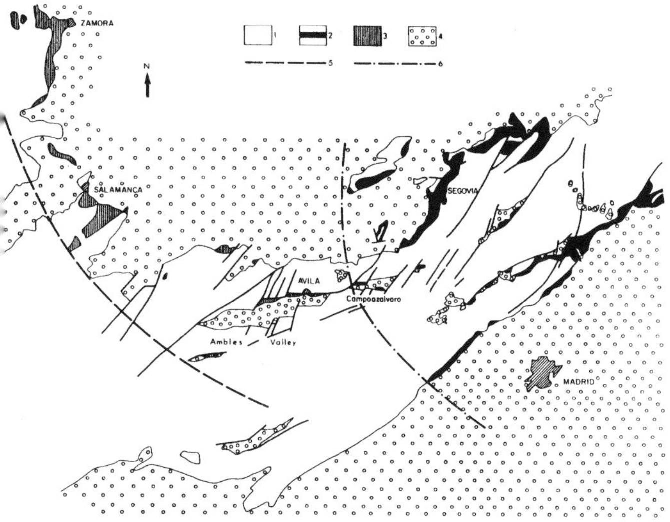

Fig. 7. Main outcrops of *siderolithic* facies and carbonate marine sediments of Cretaceous age of Central System. Probable limit of sedimentation. 1, basement; 2, Cretaceous marine sediments; 3, *siderolithic* facies; 4, Tertiary continental sediments; 5, marine sedimentation edge; 6, *siderolithic* sedimentation edge.

mineralogy between the *siderolithic* facies (of mature composition: quartz, quartzite and kaolinite) and the arkosic unit (rich in feldspar, neoformed smectites and labile minerals and rock fragments derived from metamorphic source areas) is easily observed. Non-granite fragments characterise the deposits of the western basins made up of interbedded more-or-less arkosic lithofacies.

The extent of the sedimentary basin during the Middle–Late Palaeogene in the western sector was much wider than the area of the present basins. There are several arguments:

- The lithofacies of the Ciudad Rodrigo Basin and the Alagón basins are very similar.
- As most of the basin fills are fine grained it can be assumed that environments were geographically extensive and surrounded by topographically smooth areas (plains).
- All the basins exhibit the same sequential evolution.

Moreover, the palaeocurrent patterns do not conform to the present morphology of the basins. All this supports the conclusion that the arkosic stage did not result from the morphological evolution of the Central System into basins and ranges (Garzón *et al.*, 1982). The lithologies of the Palaeogene sediments of the Ciudad Rodrigo or Alagón basins (burrowed smectitic muds of fluvio-lacustrine origin with fish remains) point to slow subsidence and open, flat landscapes instead of rapid orographic reorganisation. Such an environmental context is very different from the present; this is supported by the fact that tectofacies related to faults bounding the basins are present only in the latest stages of basin infill. At that stage, there was concordance between the sedimentary record and its morphostructural context. How was the *Iberian phase* (Oligocene–Early Miocene) recorded in the marginal tectofacies? Moreover, was it related to the coarsening-upwards trend of the arkosic unit *sensu lato* of these western basins? Tectofacies of the Gredos, Guadarrama and Somosierra are attributed to the *Guadarrama stage* (Intra-Aragonian), a tectonic phase connected to compression of the Betic Cordillera that caused the present pattern of the Central System and deposition of relatively thick alluvial-fan sediments by large alluvial cones at the top of most of the basin fills. These may be of the same age as the red alluvial-fan deposits of eastern Ciudad Rodrigo Basin, but their relationship with the coarser lithofacies topping the sedimentary infill of the Alagón basins is unclear. Both tectofacies may be coeval (synchronous) but they may also be two well-differentiated stages recorded along the whole Central System.

It is difficult to maintain that the arkosic stage continued through the Late Tertiary to end in Middle Pliocene times with the deposition of the 'Páramos (limestone)' unit as suggested by Garzon *et al.* (1982). The arkoses of the Central System are of Eocene–Oligocene age and the highest alluvial-fan deposits may be of Aragonian age (Martín-Serrano & del Olmo, 1990b; del Olmo, 1990a, b, 1991). There is only a poor record of the remaining Tertiary and Quaternary sediments because of the active river incision at that time in the basins, the ranges and elsewhere in the Spanish Meseta (Martín-Serrano, 1991).

References

Alonso, A. (1981). *El Cretácico de la Provincia de Segovia (Borde N. del Sistema Central).* Seminarios de Estratigrafía, Universidad Complutense de Madrid, Serie Monografías, **7**: 271 pp.

Alonso Gavilán, G. (1981). Estratigrafía y sedimentología del Paleógeno en el borde suroccidental de la Cuenca del Duero (Provincia de Salamanca). PhD Thesis, Universidad de Salamanca, 435 pp.

Bascones, L. and Martín, D. (1984). *Mapa Geológico de España a escala 1:50 000. Cartografía y Memoria de la Hoja no. 622, Torrejoncillo.* ITGE, Servicio de Publicaciones del Ministerio de Industria y Energía, Madrid.

Bascones, L., Martín, D. and Corretge, L.G. (1982a). *Mapa Geológico de España a escala 1:50 000. Cartografia y Memoria de la Hoja no. 575, Zarza la Mayor.* ITGE, Servicio de Publicaciones del Ministerio de Industria y Energía, Madrid.

Bascones, K., Martín, D. and Corretge, L.G. (1982b). *Mapa Geológico de España a escala 1:50 000. Cartografía y Memoria de la Hoja no. 621, Coria.* ITGE, Servicio de Publicaciones del Ministerio de Industria y Energía, Madrid.

Bascones, L., Martín, D. and Garcia de Figuerola, L.C. (1984a). *Mapa Geológico de España a escala 1:50 000. Cartografía y Memoria de la Hoja no. 596, Moraleja.* ITGE, Servicio de Publicaciones del Ministerio de Industria y Energía, Madrid.

Bascones, L., Martín, D. and Ugidos, J.M. (1984b). *Mapa Geológico de España a escala 1:50 000. Cartografía y Memoria de la Hoja no. 597, Montehermoso.* ITGE, Servicio de Publicaciones del Ministerio de Industria y Energía, Madrid.

Blanco, J.A. and Cantano, M. (1983). Silicification contemporaire a la sedimentation dans l'unité basale du Paleogene du bassin du Duero (Espagne). *Sci. Geol. Mem.*, **72**: 7–18.

Blanco, J.A., Corrochano, A., Montigny, R. and Thuizat, R. (1982). Sur l'age du debut de la sedimentation dans le bassin tertiaire du Duero (Espagne). Atribution au Paléocène par datation isotopique des alunites de l'unité inferieure. *C. R. Acad. Sci. Paris*, **295** (II): 559–562.

Bustillo, M.A. and Martín-Serrano, A. (1980). Caracterización y significado de las rocas silíceas y ferruginosas del Paleoceno de Zamora. *Tecniterrae*, **36**: 1–16.
Cantano, M. and Molina, E. (1987). Aproximación a la evolución morfológica de la 'Fosa de Ciudad Rodrigo'. Salamanca. España. *Bol. R. Soc. Hist. Nat. (Geol.)*, **82**: 87–101.
Capote, R., González Casado, J.M. and Vicente, G. de (1987). Análisis poblacional de la fracturación tardihercínica en el sector central del Sistema Central Ibérico. *Cuad. Lab. Xeol. Laxe.*, **11**: 305–314.
Capote, R., González Casado, J.M. and Vicente, G. de (1990a). *Mapa Geológico de España a escala 1:50 000. Cartografía y Memoria de la Hoja no. 508, Cercedilla.* ITGE, Servicio de Publicaciones del Ministerio de Industria y Energía, Madrid.
Capote, R., González Casado, J.M. and Vicente, G. de (1990b). *Mapa Geológico de España a escala 1:50 000. Cartografía y Memoria de la Hoja no. 509, Torrelaguna.* ITGE, Servicio de Publicaciones del Ministerio de Industria y Energía, Madrid.
Capote, R., González Casado, J.M. and Vicente, G. de (1990c). *Mapa Geológico de España a escala 1:50 000. Cartografía y Memoria de la Hoja no. 533, San Lorenzo del Escorial.* ITGE, Servicio de Publicaciones del Ministerio de Industria y Energía, Madrid.
Capote, R., González Casado, J.M. and Vicente, G. de (1990d). *Mapa Geológico de España a escala 1:50 000. Cartografía y Memoria de la Hoja no. 557, San Martín de Valdeiglesias.* ITGE, Servicio de Publicaciones del Ministerio de Industria y Energía, Madrid.
Capote, R., González Casado, J.M. and Vicente, G. de (1991). *Mapa Geológico de España a escala 1:50 000. Cartografía y Memoria de la Hoja no. 558, Prádena.* ITGE, Servicio de Publicaciones del Ministerio de Industria y Energía, Madrid.
Corrochano, A. (1977). Estratigrafía y sedimentología del Paleógeno de la Provincia de Zamora. PhD Thesis, Universidad de Salamanca, 336 pp.
Dias, R.P. and Cabral, J. (1989). Neogene and Quaternary Reactivation of the Ponsul Fault in Portugal. *Comun. Serv. Geol. Portugal*, **75**: 3–28.
Fernández, M.P. (1988). Geomorfología del sector comprendido entre el Sistema Central y el Macizo de Santa María de Nieva (Segovia). PhD Thesis, Universidad Complutense de Madrid (unpublished).
Fernández Amigot, J.A. (1981). Prospección e investigación de yacimientos uraniníferos en la Provincia de Salamanca. *Tecniterrae*, **43**: 45–47.
Garzón, M.G. (1980). Estudio geomorfológico de uns transversal en la Sierra de Gredos oriental (Sistema Central Español). PhD Thesis, Universidad Complutense de Madrid (unpublished).
Garzón, M.G. and Lopez, N. (1978). Los roedores fósiles de Los Barcos (Avila). Datación del Paleógeno continental del Sistema Central. *Est. Geol.*, **34**: 574–578.
Garzón, M.G., Ubanell, A.G. and Rosales, F. (1981). Morfoestructura del Valle de Amblés (Sistema Central Español). *Cuad. Geol. Ibérica*, **7**: 655–665.
Garzón, M.G., Pedraza, J. and Ubanell, A.G. (1982). Los modelos evolutivos del relieve del Sistema Central Ibérico (sectores de Gredos y Guadarrama). *Rev. R. Acad. Cienc. Exac. Fis. Nat. de Madrid*, **76**: 475–496.
Jiménez, E. (1970). Estratigrafía y paleontología del borde sur-occidental de la Cuenca del Duero. PhD Thesis, Universidad de Salamanca, 323 pp.
Jiménez, E. (1977). Sinopsis sobre los yacimientos fosilíferos paleógenos de la Provincia de Zamora. *Bol. Geol. Mineral.*, **85** (5): 357–346.
Jiménez, E. (1982). Quelonios y cococrilos fósiles de la Cuenca del Duero. Ensayo de biozonación del Paleógene de la Cuenca del Duero. *Stvd. Geol. Salmanticensia*, **17**: 125–127.
Jiménez, E. and Martín-Izard, A. (1987). Consideraciones sobre le edad del Paleógeno y la tectónica alpina del sector occidental de la cuenca de Ciudad Rodrigo. *Stvd. Geol. Salmanticensia*, **24**: 215–228.
Martín-Serrano, (1988). *El relieve de la región occidental zamorana. La evolución geomorfológica de un borde del Macizo Hespérico.* Instituto de Estudios Zamoranos Florián de Ocampo, Diputación de Zamora: 306 pp.
Martín-Serrano, A. (1991). La definición y el encajamiento de la red fluvial actual sobre el Macizo Hespérico en el marco de la geodinámica alpina. *Rev. Soc. Geol. España*, **4**: 337–351.
Martín-Serrano, A. and Olmo, A. del (1990a). *Mapa Geológico de España a escala 1:50 000. Cartografía y Memoria de la Hoja no. 507, El Espinar.* ITGE, Servicio de Publicaciones del Ministerio de Industria y Energía, Madrid.
Martín-Serrano, A. and Olmo, A. del (1990b). *Mapa Geológico de España a escala 1:50 000. Cartografía y Memoria del Mesozoico y Cenozoico de la Hoja no. 508, Cercedilla.* ITGE, Servicio de Publicaciones del Ministerio de Industria y Energía, Madrid.
Martínez-Salanova, J. and Olmo, A. del (1990c). *Mapa Geológico de España a escala 1:50 000. Cartografía y Memoria del Mesozoico y Cenozoico de la Hoja no. 509, Torrelaguna.* ITGE, Servicio de Publicaciones del Ministerio de Industria y Energía, Madrid.
Mediavilla, R. and Martín-Serrano, A. (1989). Sedimentación y tectónica en el sector oriental de la Fosa de Ciudad Rodrigo durante el Terciario. *XII Congr. Español Sedimentol.*, comun. **1**: 215–218.
Molina, E., Vicente, A., Cantano, M. and Martín-Serrano, A. (1989). Importancia e implicaciones de las paleoalteraciones y de los sedimentos siderolíticos del paso Mesozoico–Terciario en el borde suroeste de la Cuenca el Duero y Macizo Hercínico Ibérico. In C.J. Dabrio (ed.), *Paleogeografía de la Meseta Norte durante el Terciario, Stvd. Geol. Salmanticensia*, vol. esp. 5: 177–186.
Moreno, F. (1990). Superficies de erosión y fracturas en el enlace entre la Meseta Norte y la Illanura extremeña (Salamanca–Cáceres). *I Reunión Nacional de Geomorfología, Teruel*: 39–49.
Olmo, A. del (1990a). *Mapa Geológico de España a escala 1:50 000. Cartografía y Memoria del Mesozoico y Cenozoico de la Hoja no. 533, San Lorenzo del Escorial.* ITGE, Servicio de Publicaciones del Ministerio de Industria y Energía, Madrid.
Olmo, A. del (1990b). *Mapa Geológico de España a escala 1:50 000. Cartografía y Memoria del Mesozoico y Cenozoico de la Hoja no. 557, San Martín de Valdeiglesias.* ITGE, Servicio de Publicaciones del Ministerio de Industria y Energía, Madrid.
Olmo, A. del (1991). *Mapa Geológico de España a escala 1:50 000. Cartografía y Memoria del Mesozoico y Cenozoico·de la Hoja no. 458, Prádena.* ITGE, Servicio de Publicaciones del Ministerio de Industria y Energía, Madrid.
Pedraza, J. (1978). Estudio geomorfológico de la zona de enlace entre las Sierras de Gredos y Guadarrama (Sistema Central Español). PhD Thesis, Universidad Complutense de Madrid, 540 pp.
Polo, M.A., Alonso Gavilán, G. and Valle, M.F. (1987). Bioestratigrafía y paleogeografía del Oligoceno-Mioceno del borde SO de la Fosa de Ciudad Rodrigo (Salamanca). *Stvd. Geol. Salmanticensia*, **24**: 229–245.
Santisteban, J.I., Martín-Serrano, A. and Mediavilla, R. (1991). El Paleógeno del sector suroccidental de la Cuenca del Duero: Nueva división estratigráfica y controles sobre su sedimentación. In F. Colombo (ed.), *Libro Homenaje a Oriol Riba Arderiú, Acta Geol. Hispànica*, **26**: 133–148.
Ugidos, J.M., Rodríguez, M.D., Martin, D. and Bascones, L. (1985). *Mapa Geológico de España a escala 1:50 000. Cartografía y Memoria de la Hoja no. 576 Hervás.* ITGE, Servicio de Publicaciones del Ministerio de Industria y Energía, Madrid.
Vicente, G. de (1988). Análisis poblacional de fallas. El sector de enlace Sistema Central-Cordillera Ibérica. Zona central (Zona de Tamajón). *Rev. Mat. Proc. Geol.*, **2**: 213–228.

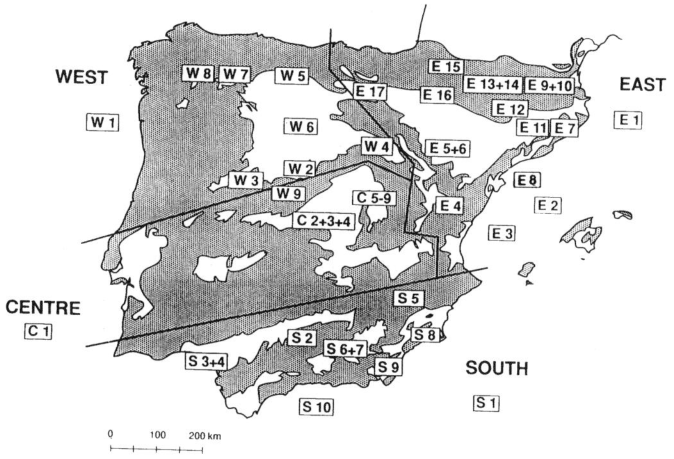

Part C
Centre

Central Loranca Basin, near Avía de la Obispatía. Outcrops of the Tórtola fluvial fan show details of a river channel bar that formed where the flow expanded at a local channel confluence (photo: M. Díaz-Molina).

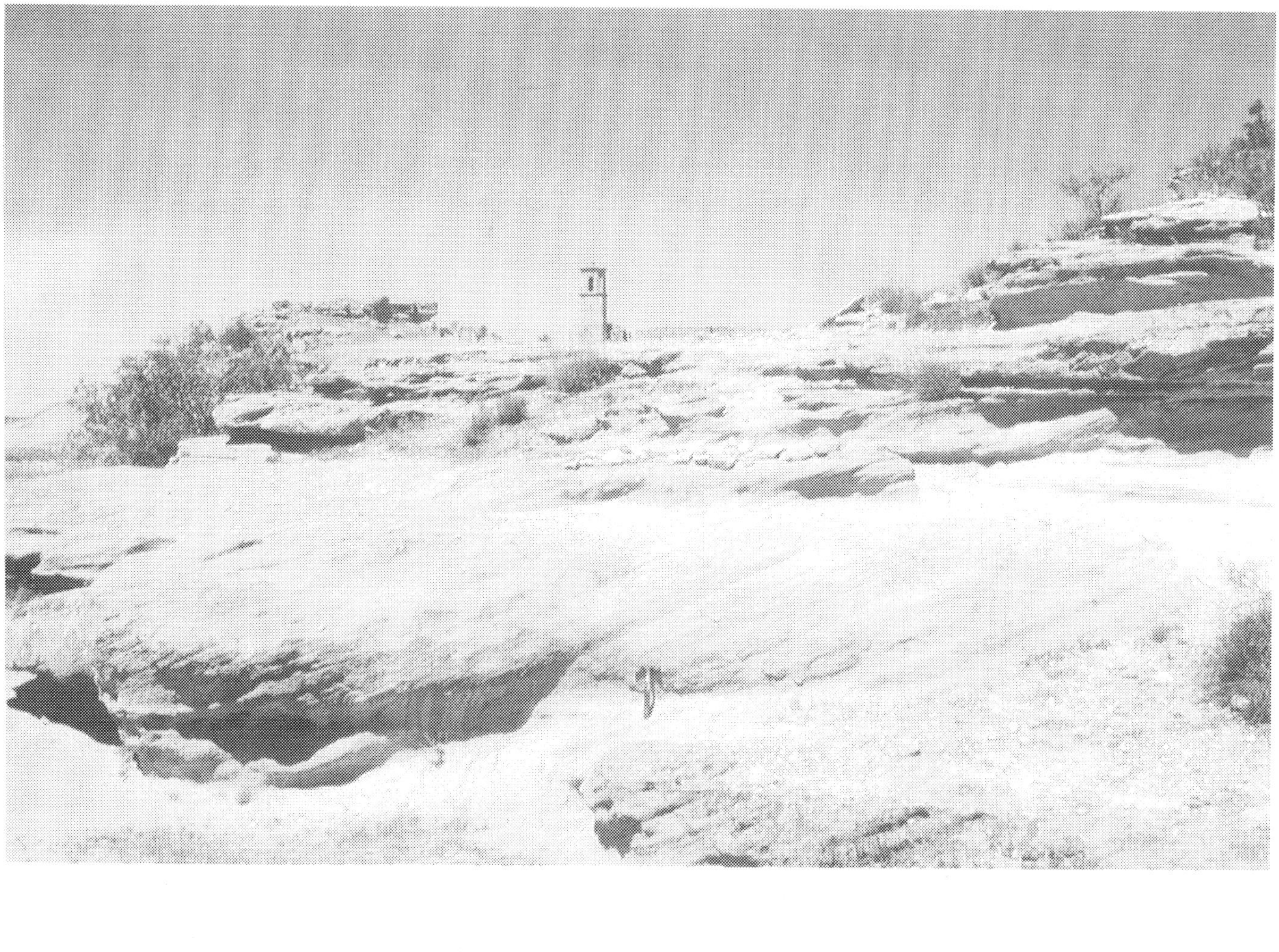

C1 Structure and Tertiary evolution of the Madrid basin

G. DE VICENTE, J.M. GONZÁLEZ-CASADO, A. MUÑOZ-MARTÍN, J. GINER AND M.A. RODRÍGUEZ-PASCUA

Abstract

The Madrid basin is intracratonic and triangular in form, being bounded on its three sides by Tertiary mountain ranges: the Spanish Central System in the north, the Toledo mountains in the south, and the Iberian and Altomira ranges in the east. The Altomira range separates the Loranca basin from the Madrid basin (Fig. 1). Each of these mountain ranges has a different structure and Tertiary geological evolution. The kinematic history of each of these borders of the Madrid basin reflects differences in the transmission of stresses from the active Iberian plate boundaries where the Betic and Pyrenean chains themselves had distinctive kinematic histories.

Geological setting

The crust of the Madrid basin and its surroundings is characterised by: 1. a Variscan basement (Palaeozoic metamorphic rocks with granitoids); 2. a pre-tectonic cover of Mesozoic and Early Cenozoic age (the former thinning gradually and then disappearing westwards); and 3. a syn- to post-tectonic cover of alluvial, fluvial and lacustrine sediments of Late Oligocene to Late Miocene age.

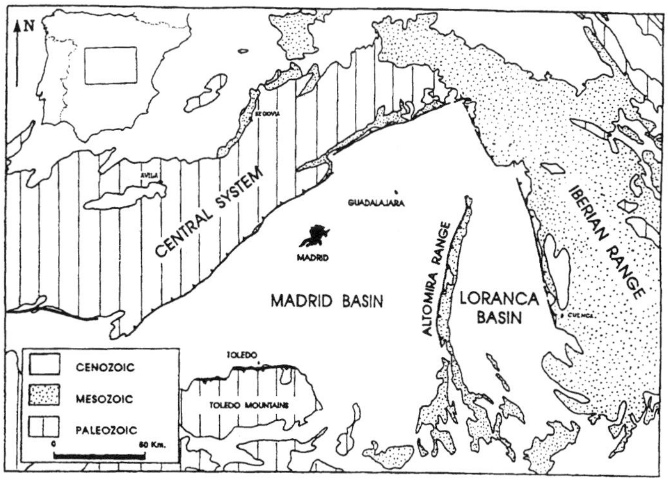

Fig. 1. Simplified geological map of the central Iberian Peninsula.

The structure of the borders of the Madrid Basin

Spanish Central System (SCS)

This is a linear mountain range that trends ENE–WSW, forming the Guardarrama Mountains, and then changes to trend E–W in the west, where it forms the Gredos Mountains. The altitude of the range is commonly 1500 m higher than the surface of the Madrid basin. The basement forms a core, flanked by the pre-tectonic cover. Typically, box folds and basement-cored anticlines are developed, and interpreted as the results of forced folding and reverse faulting. The contact between this mountain structure and the syn- and post-tectonic basin is generally a reverse fault or thrust (trending N60° E) with a throw of more than 2000 m. The shortening represented by this thrusting is balanced by transfer faults (N140° E dextral and N10° E sinistral). In general, the Central System is a pop-up structure rooted, at a depth of 11 ± 2 km, in a basal detachment that dips gently southward. Restored sections suggest shortening during this phase of compression of between 19% and 22%. The stress tensor associated with this uplift has a sigma one (σ_1) direction orthogonal to the range (N155° E across the Guadarrama, and N–S across the Gredos), and $R=0.7$ ($R=(\sigma_2-\sigma_3)/(\sigma_1-\sigma_3)$) from De Vicente and González-Casado, 1991). Some of the Tertiary structures result largely from reactivation of main Variscan and late Variscan structures (Vegas *et al.*, 1990) (Fig. 2).

Toledo Mountains

These are very similar to the SCS in that they have a Variscan basement and a thin sedimentary pre-tectonic cover. These rocks overthrust the sedimentary rocks of the Madrid basin. The thrust has an E–W trend, dips southwards 40–50° and is associated with N20° E, sinistral, and N150° E, dextral, transfer faults. The stress tensor has a sigma one (σ_1) direction of N–S, as in the Gredos Mountains.

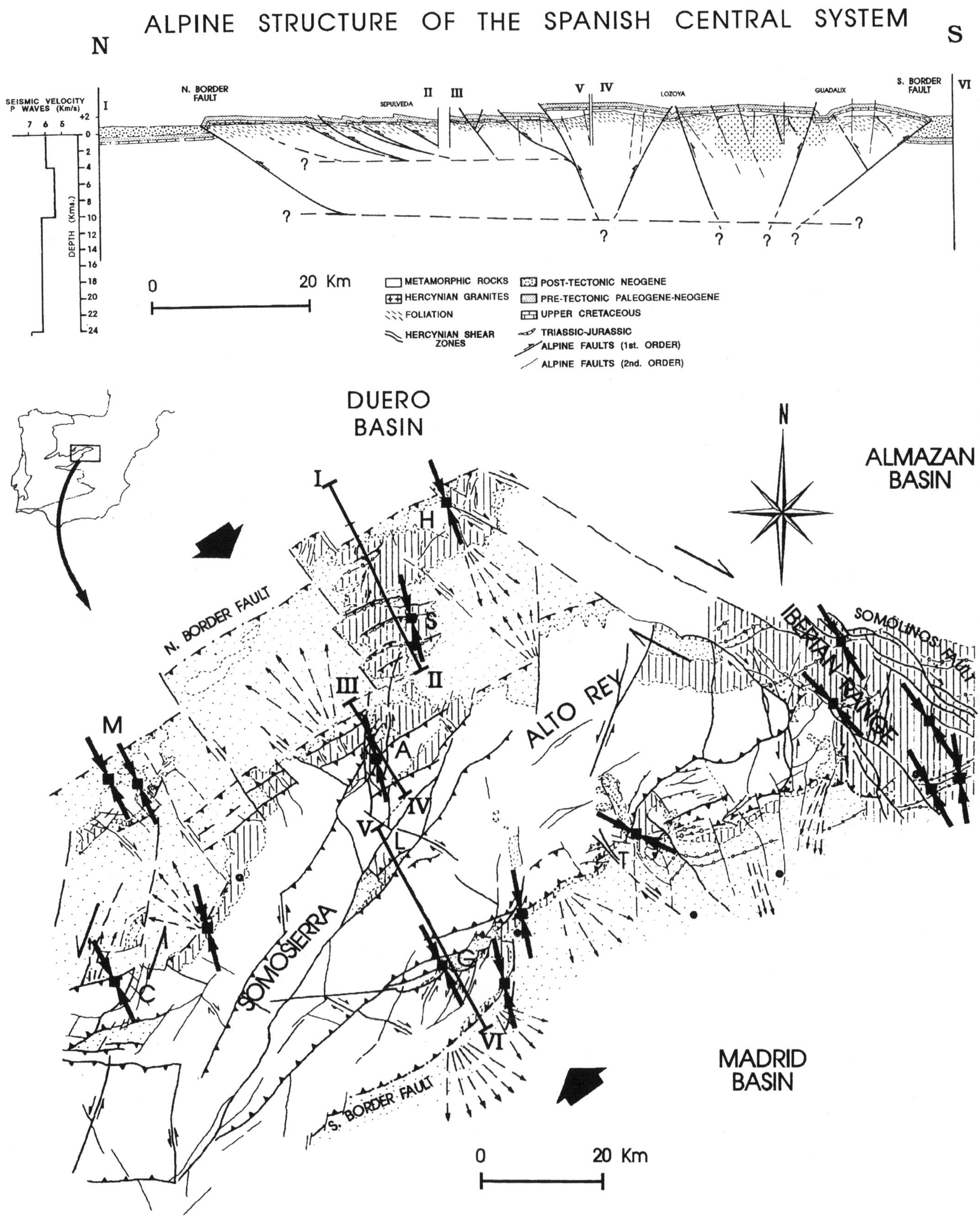

Fig. 2. Sketch section representing the structure of the Spanish Central System, with seismic data from Suriñach (1988).

Iberian and Altomira Ranges

In these ranges, the sedimentary pre-tectonic cover reaches a thickness of 1500 m. There is a well-defined décollement surface between a lower unit, consisting of Variscan basement and Early Triassic sediments, and an upper unit consisting of Late Triassic, Jurassic, Cretaceous and Early Cenozoic sediments. The Varisan basement only outcrops in the Iberian Ranges, in small rhomb-shaped corridors elongated parallel to the trend of the Range (N140° E), and is absent in the Altomira Range.

In the Iberian Range there are two main directions of faults and folds, N140° E and N60° E. The NW–SE set of folds trends subparallel to the direction of the range and is characterised by straight axial traces and box geometries. The NE–SW folds trend transverse to the direction of the range, and verge to the SE.

The whole of the western part of the range (the Castilian branch) appears to consist largely of compressional and extensional duplexes, associated with a group of anastomosing, N140° E-trending, strike–slip faults. These are interpreted as positive and negative flower structures with small associated basins (Rodríguez Pascua, 1993).

The compressive directions deduced from paleostress analysis of faults and stylolites show two successive sigma one (σ_1) orientations N150° E and N10° E. The strike–slip N140° E faults moved dextrally during the first shortening phase, and then like dextral reverse-slip faults during the second phase.

The Altomira Range is a narrow fold and thrust belt, trending N–S, dipping to the east, and separating the Loranca and Madrid basins (Fig. 1). To the south, this structure changes to a NW–SE (Iberian) orientation, and to the north, it disappears below the basinal area where the Madrid and Loranca basins join. Paleostress analyses again show two successive sigma one (σ_1) orientations. In the first episode, the sigma one orientation varied from area to area, being N70° E in the northern part of the Loranca basin and N110° E in the middle of the Altomira Range. In the second episode, the sigma one orientation of N155° E was general throughout the region. Recently, Muñoz-Martín (1993) has demonstrated that the first shortening episode resulted from the addition of two regional stress tensors with N60° E (the Pyrenean push) and N140° E (the Betic push). It appears that, during this episode, the Altomira Range was extruded towards the west (Fig. 3) during the general N–S compression. The western boundary of the extrusion appears to have been located at the termination of the Triassic décollement surface.

The structure of the Madrid Basin

The Tertiary sedimentary fill of the Madrid basin is thousands of metres thick, reaching a maximum close to the northern margin in the Alcobendas depression, where it is 3500 m thick. Geological studies (Calvo *et al.*, 1990, 1991; Sánchez Serrano *et al.*, 1993), along with seismic (Querol, 1989) and gravity (Babin *et al.*, 1993) surveys, have shown that the Cenozoic evolution of the Madrid basin involved the activity of numerous sets of faults that bounded small sub-basins, and resulted in several depocentres. Faults with the same trends (N60° E, N140° E and N–S) occur in the borders of the Madrid basin.

Within the Madrid basin, two rather different regions can be distinguished to the NE and SW, separated by an important fault, trending N140° E (see Chapter C2). This fault moved sinistrally in the Late Oligocene–Early Miocene, due to E–W compression, and dextrally in the Middle Miocene (N160° E). This important structure may have been localised by a contrast in the abundance of granitoids in the Variscan basement.

Structural evolution of the mountain ranges of Central Spain during the Tertiary

During the convergence and slip between the European, Iberian and African plates, stresses applied from the Pyrenean and Betic borders were transmitted to the central part of the Iberian Peninsula in several diachronous episodes. In some ways, Central Spain acted as a foreland for the Betics, or for the Pyrenees, or for both together. On the one hand, the Iberian Range and the Demanda Sierra appear to be associated with Pyrenean compression, with a maximum horizontal shortening along N10° E, whereas the Central System, the Toledo Mountains and the Portugese chains (Arrábida, Sintra, Ponsul) (Ribeiro *et al.*, 1990) are associated with Betic compression, with a maximum horizontal shortening along N150° E.

The episodes of stress transmission towards the central Iberian Peninsula were Eocene–Early Miocene for the Pyrenean compression, and Late Oligocene–Early Miocene for the Betic compression. During the Late Oligocene to Early Miocene, both stress regimes acted, and were resolved in the Madrid basin into an E–W shortening (Fig. 3A). Subsequently, a later episode, related to the opening of the Valencia trough in response to crustal extension (Fonbote *et al.*, 1990, Banda and Santanach, 1992) migrated from east to west deforming the Maestrazgo region in the Early to Middle Miocene, and the Madrid basin in the Pliocene to Quaternary.

The tectonic evolution of the borders of the Madrid basin can be divided into three episodes:

1. Eocene to Early Miocene (Fig. 3A)

The progressive westward movement of the Altomira Range occurred after the deposition of the late Paleogene formations and before the Early Miocene Lower Unit. The mean shortening direction was N110° E, and resulted from the addition of the Betic and Pyrenean compression components. The Pyrenean compression (N10° E) caused oblique movements (with reverse slip) on the N60° E trending faults of the Central System. In the Iberian Range, the movements were complicated: the N140° E faults moved mainly dextrally, but also formed folds and thrusts. The Toledo Mountain border moved either in a reverse sense or dextrally.

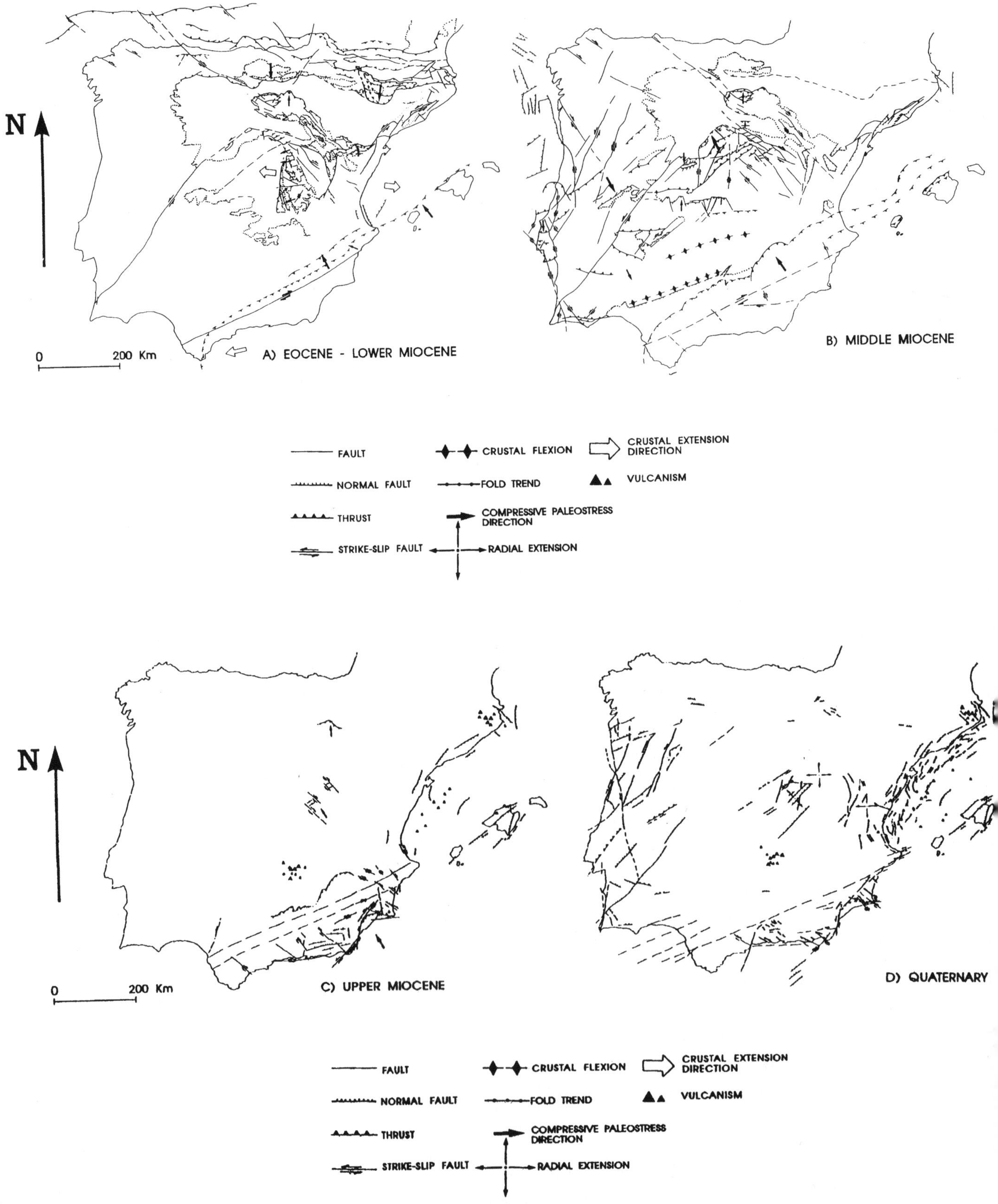

Fig. 3. Kinematic evolution of the crust of central Spain, with closed and open arrows indicating directions of compression and extension respectively, and triangles indicating vulcanism. A, Eocene to Early Miocene; B, Middle Miocene; C, Late Miocene; D, Quaternary.

2. Middle Miocene (Fig. 3B)

During the Aragonian (late Early Miocene to early Late Miocene), the Central System was uplifted. This is shown by the close relationship between the Intermediate Unit (Middle Miocene) and the Central System (see Chapter C2). The shortening direction was N150° E (the Guadarrama shortening of Capote *et al.*, 1990), and the Central System faults trending N60° E moved as reverse faults. The Altomira Range remained largely passive, as shown by the onlap of the Intermediate Unit against this range, although some minor folds (trending N60°) were superimposed on earlier structures. In the Iberian Range, the N140° E faults moved dextrally, and extensional movements began to the east (e.g. the Teruel Basin).

3. Late Miocene to Quaternary (Fig. 3C, D)

During the Late Miocene, the fluvial pattern of the Madrid basin changed from endorheic (internal) to exorheic, probably in response to the end of regional compression. The activity of N–S-trending faults was normal, whereas NW–SE- and NE–SW-trending faults moved dextrally. During this interval radial extension of the Iberian Range began, and spread progressively westwards to produce the NE–SW graben in the Madrid basin (Giner, 1993) in which the Tajo river flows. At present, activity is occurring on faults that trend N140° E.

Acknowledgements

This work has been financed by the CICYT project PB91/0397.

References

Babín, R., Bergamín, J.F., Fernández-Rodríguez, C., González-Casado, J.M., Hernandez-Enrile, J.L., Rivas, A., Tejero, R. and De Vicente, G. (1993). Modelisation gravimetrique de la structure alpine du Systeme central Spagnol (secteur noreste). *Bull. Sóc. Géol. Fr.*, **164**(3): 385–393.

Banda, E. and Santanach, P. (1992). The Valencia trough (Western Mediterranean): an overview. *Tectonophysics*, **208**: 183–202.

Calvo, J.P., Hoyos, M., Morales, J. and Ordóñez, S. (1990). Neogene stratigraphy, sedimentology and raw materials of the Madrid Basin. *Paleontol. Evol.*, mem. esp. no. 2: *Iberian Neogene Basins*: 62–95.

Calvo, J.P., De Vicente, G. and Alonso Zarza, A.M. (1991). Correlación entre las deformaciones alpinas y la evolución del relleno sedimentario de la Cuenca de Madrid durante el Mioceno. *I Cong. Grup. Esp. Terciario*: 55–58.

Capote, R., De Vicente, G. and González Casado, J.M. (1990). Evolución de las deformaciones alpinas en el Sistema Central Español (SCE). *Geogaceta*, **7**: 20–22.

De Vicente, G. González-Casado, J.M. (eds.) (1991). Las deformaciones alpinas del Sistema Central Español. *III Reunión de la Comisión de Tectónica. Soc. Geol. Esp.*: 140 pp.

Fontboté, J.M., Guimerá, J., Roca, E., Sabat, F., Santanach, P. and Fernández, F. (1990). The cenozoic geodynamic evolution of the Valencia Trough (Western Mediterranean). *Rev. Soc. Geol. Esp.*, **3**(4): 249–259.

Giner-Robles, J.L. (1993). *Neotectónica del borde oriental de la Cuenca de Madrid.* Tesis de Licenciatura. Universidad Complutense de Madrid: 300 pp.

Muñoz-Martín, A. (1993). Campos de esfuerzos alpinos y reactivación de fallas en el sector de enlace entre la Sierra de Altomira y la Cordillera Ibérica. Tesis de Licenciatura. Universidad Complutense de Madrid: 250 pp.

Querol, R. (1989). *Geología del subsuelo de la Cuenca del Tajo.* E.T.S.I. Minas de Madrid (Dpto. de Ingeniería Geológica): 48 pp.

Ribeiro, A., Kullberg, M.C., Manupella, G. and Phipps, S. (1990). A review of Alpine tectonics in Portugal: foreland detachment in basement and cover rocks. *Tectonophysics*, **184**: 357–366.

Rodríguez-Pascua, M. (1993). *Cinemática y dinámica de las deformaciones alpinas en la zona del Alto Tajo (Guadalajara).* Tesis de Licenciatura. Universidad Complutense de Madrid: 250 pp.

Sanchez Serrano, F., González Casado, J.M. and De Vicente, G. (1993). Evolución de las deformaciones alpinas en el borde suroriental del Sistema Central español (zona de Tamajón, Guadalajara). *Bol. Geol. Min.*, **104**(1): 3–12.

Suriñach, E. (1988). Crustal structure in Central Spain. *European Geotraverse Workshop*: 187–197.

Vegas, R., Vázquez, J.T., Suriñach, E. and Marcos, A. (1990). Model of distributed deformation, block rotation and crustal thickening for the formation of the Spanish Central System. *Tectonophysics*, **184**: 367–378.

C2 Neogene tectono-sedimentary review of the Madrid Basin

G. DE VICENTE, J.P. CALVO AND A. MUÑOZ-MARTÍN

Abstract

The Neogene sedimentary record of the Madrid Basin, central Spain, comprises three major stratigraphic units (the Lower, Intermediate and Upper Miocene Units), separated by sedimentary discontinuities which were linked with variations in the tectonic activity of the basin margins. The distribution of the depositional systems that formed the Lower Unit was, to a large extent, controlled by the emplacement of the Altomira Range during the Late Oligocene and Early Miocene. The sedimentation that formed the Intermediate Unit during most of the Middle Miocene and part of the Late Miocene was mainly controlled by uplift of the Central System. Finally, the sedimentary features of the Upper Unit reveal clearly a drastic change in regional stress tensors affecting the area. The stress fields became extensional in the Late Miocene, in contrast to the compressional fields that existed during the deposition of the earlier units.

Introduction

The general evolution of the sedimentary basins and ranges of the central Iberian peninsula (see Chapter C1) provides the context within which the sedimentary filling of the Madrid Basin may be easily related to the deformation of its borders (the Central System, Toledo Mountains, Iberian and Altomira Ranges). In this chapter, we describe briefly the main features of the Neogene sedimentary record of the Madrid Basin, placing emphasis on the distribution of major depositional systems and their control by the various macrostructures that were active during each period of regional deformation.

Neogene sedimentary record of the Madrid Basin

The Madrid Basin was filled by Tertiary sediments between 2000 and 3500 m thick. These Neogene sediments disconformably overlie Paleogene and/or Cretaceous formations in the marginal parts of the basin, though seismic profiles suggest that the contact is conformable in much of the subsurface more centrally.

During most of the Miocene, the basin was occupied by lakes and peripheral alluvial systems, forming a concentric facies pattern. This pattern is characteristic of the stratigraphic units distinguished as the Lower and Intermediate Units of the Miocene (Alberdi *et al.*, 1984; see also Chapter C3). Differences in both the extent and nature of the alluvial fans or fluvial distributary systems (Alonso Zarza *et al.*, 1993) indicate tectonic reactivation throughout the Early and Middle Miocene. The situation is clear along the edges of the Central System and the Altomira Range. In the latter area, the alluvial fans that developed during the Early Miocene changed with time into progressively finer and more thinly bedded deposits. During the Middle Miocene, lake systems extended into the vicinity of the Altomira Range, suggesting that this border was tectonically inactive at this time. The opposite situation is recognised along the northern margin of the basin, beside the Central System, where large, prograding alluvial systems characterise the Middle Miocene (Calvo *et al.*, 1989). In further contrast, the Toledo Mountains appear to have been mainly active during the Early Miocene.

The Upper Unit of the Miocene in the Madrid Basin consists mainly of fluvial terrigenous deposits and shallow lacustrine carbonates which occupied the central and eastern parts of the Basin. The facies pattern is markedly different from the earlier patterns, and is interpreted to result from a change in stress regime during the Late Vallesian (mid Late Miocene, see Chapter G3). Finally, small amounts of Pliocene sediments present in the basin show a similar pattern to that deduced for the Upper Unit.

Relationships between tectonics and sedimentation

Analysis of the stress-field tensors that caused the Tertiary deformation of the centre of the Iberian peninsula shows the following pattern of relationships between tectonics and sedimentation during the Neogene:

(A) During the Late Oligocene and Early Miocene (Fig. 1A), the Central System was rather inactive tectonically. Along the NE border of the Madrid Basin, the emplacement of the Altomira Range produced a maximum horizontal shortening varying within the quadrant from N70° E to N120° E. In the central parts of the Madrid Basin, the facies distribution of the Lower Unit of the Miocene was controlled by strike–slip faults trending N140° E (Rodríguez-Aranda, 1990).

(B) During the Middle Miocene (Figs. 1B and 2), the distribution

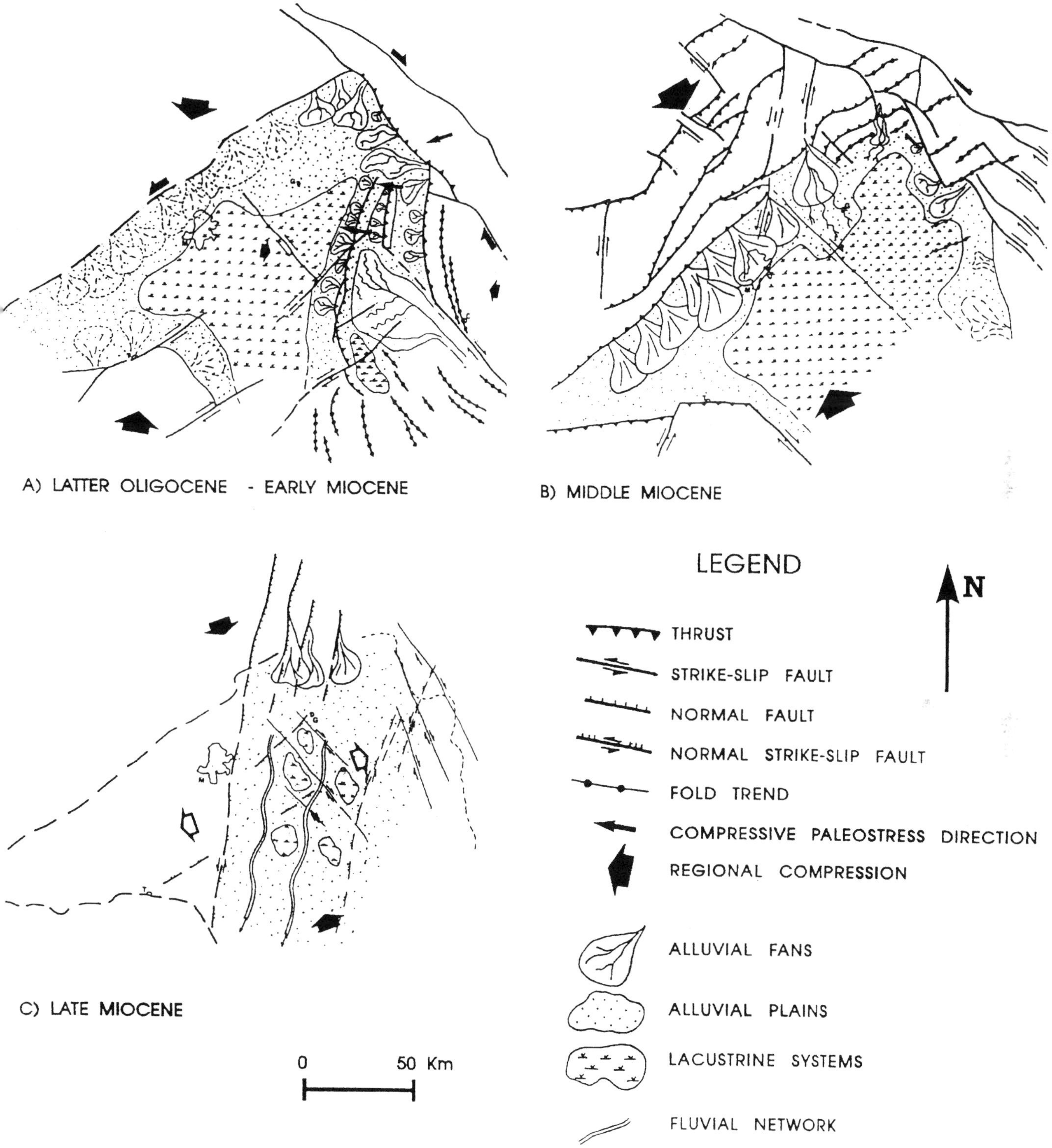

Fig. 1. Structural and environmental evolution of the Madrid basin: A, Late Oligocene–Early Miocene; B, Middle Miocene (paleogeographic data for the eastern border modified from Díaz-Molina *et al.*, 1989); C, Late Miocene.

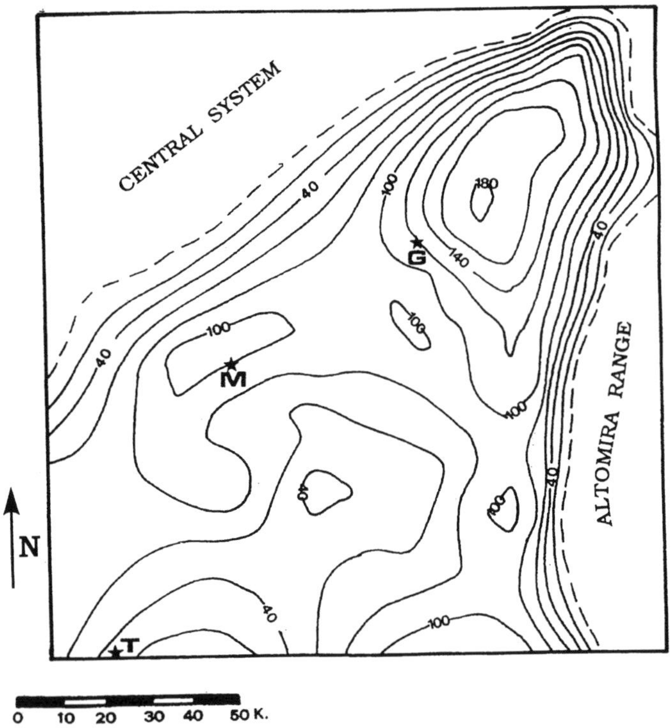

Fig. 2. Isopach map (metres) for the Intermediate Unit of the Madrid basin, based on more than twenty thicknesses evenly distributed across the basin. G, Guadalajara; M, Madrid; T, Toledo.

of facies in the Intermediate Unit was closely related to the uplift of the Central System thought to result from compression in the direction N150° E. The sediments of the Intermediate Unit onlap the relief of the Altomira Range, showing that this border was tectonically passive during this period. Study of the mechanical relationships of the faults in the Iberian Range makes it clear that there is a population that formed at this stage, distinctly younger than those related to the emplacement of the Altomira Range (Muñoz-Martín, 1993). The isopach map for the Intermediate Unit shows a linear feature of minimum sediment accumulation, trending N140° E, between depocentres. This feature is interpreted as a strike–slip fault zone that is consistent in direction with the main direction of compressive stress deduced from the contemporaneous kinematics of the Iberian Range.

(C) During the Late Miocene, the general compressional regime that existed during the two earlier stages changed to an extensional regime. The sedimentation pattern of the basin changed from endorheic (internally drained) to exorheic (Fig. 1C). The extensional kinematics became more pronounced through time, so that, through the Late Miocene and Pliocene, faults commonly had strike–slip and normal components of movements. Faults during the Quaternary sediments are usually normal.

Conclusions

The Neogene sedimentary record of the Madrid Basin comprises three main stratigraphic units which are separated by major sedimentary discontinuities. The facies distribution within each sedimentary unit, and the breaks in sedimentation, are closely related to changes from a compressional to an extensional regime, and variations in the principal shortening direction. These changes are summarised in Fig. 3.

References

Alberdi, M.T., Hoyos, M., Junco, F., López-Martínez, N., Morales, J., Sesé, C. and Soria, M.D. (1984). Biostratigraphy and sedimentary evolution of continental Neogene in the Madrid area. *Paléobiologie continentale*, Montpellier, **14**: 47–68.

Alonso-Zarza, A.M., Calvo, J.P. and García del Cura, M.A. (1993). Paleomorphological controls on the distribution and sedimentary styles of alluvial systems, Neogene of the NE of the Madrid Basin (central Spain). In *Alluvial sedimentation* (M. Marzo & C. Puigdefábregas, eds.), *Spec. Publ. Int. Ass. Sed.*, **17**: 227–292.

Calvo, J.P., Alonso-Zarza, A.M. and García del Cura, M.A. (1989). Models of Miocene lacustrine sedimentation in response to varied source areas and depositional regimes in the Madrid Basin, central Spain. *Palaeogeogr., Palaeoclimatol., Palaeoecol.*, **90**: 199–214.

Díaz-Molina, M., Arribas, J. and Bustillo, A. (1989). The Tórtola and Villalba de la Sierra fluvial fans: Late Oligocene – Early Miocene, Loranca Basin, Central Spain. *Fourth International Conference on fluvial sedimentology, Barcelona–Sitges, Spain.* Field Trip 7, p. 74.

Muñoz-Martín, A. (1993). *Campos de esfuerzos Alpinos y reactivación de fallas en el sector de enlace entre la Sierra de Altomira y la Cordillera Ibérica.* Tesis de Licenciatura, Universidad Complutense de Madrid: 250 pp.

Rodríguez-Aranda, J.P. (1990). *La sedimentación neógena en el margen oriental de la Cuenca de Madrid (Barajas de Melo-Tarancón): transición de abanicos aluviales a evaporitas.* Tesis de Licenciatura. Universidad. Complutense de Madrid: 143 pp.

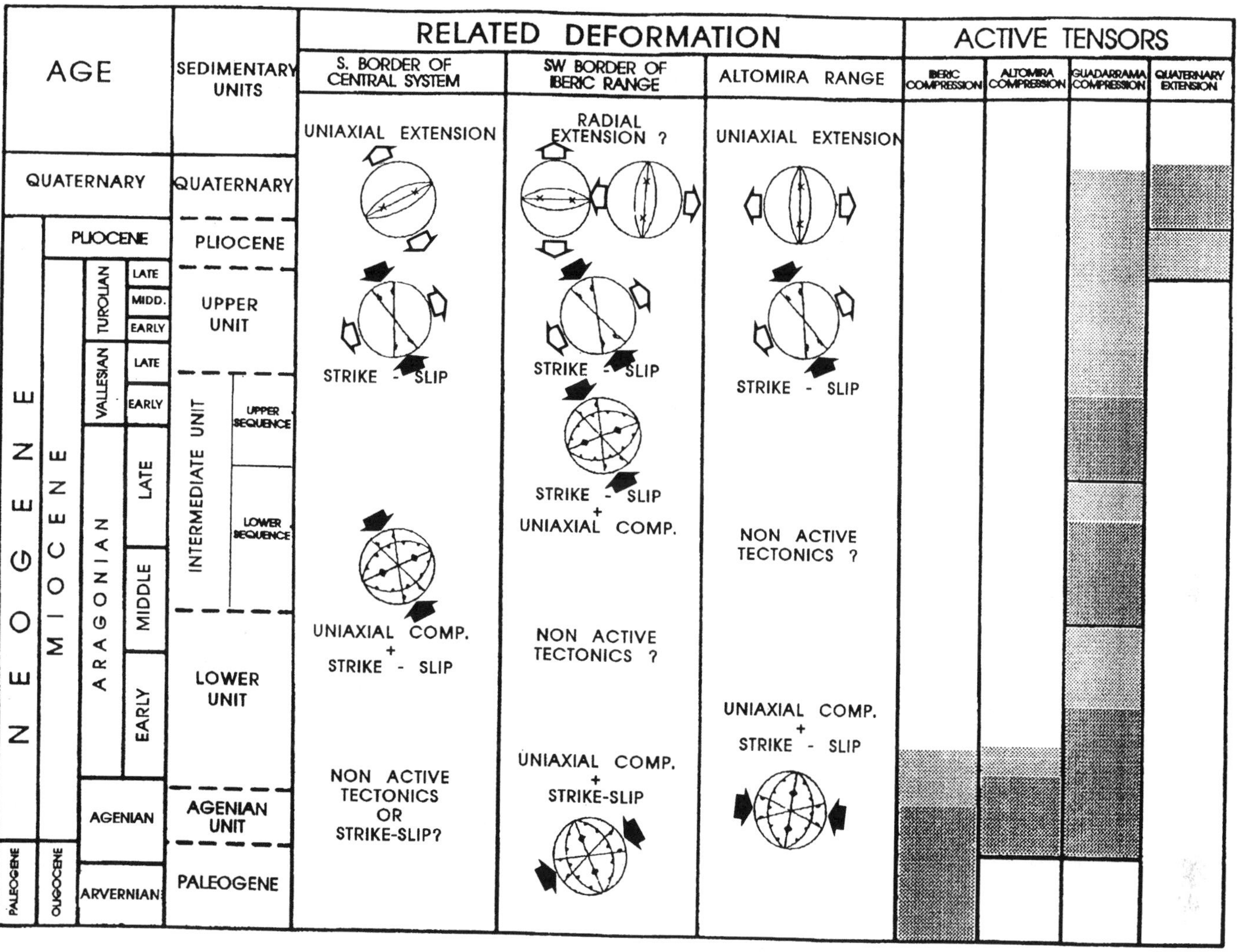

Fig. 3. Table indicating the evolution through the Neogene and Quaternary of stress patterns and related deformation.

C3 Sedimentary evolution of lake systems through the Miocene of the Madrid Basin: paleoclimatic and paleohydrological constraints

J.P. CALVO, A.M. ALONSO ZARZA, M.A. GARCÍA DEL CURA, S. ORDÓÑEZ, J.P. RODRÍGUEZ-ARANDA AND M.E. SANZ MONTERO

Abstract

Miocene sedimentary successions of the Madrid Basin, central Spain, contain well-developed lake complexes that show distinctive facies associations through time. Lake systems formed during the Ramblian and Early Aragonian (Lower Unit) were predominantly evaporite lakes in which highly soluble salts precipitated. During the Middle to Late Aragonian and Vallesian (Intermediate Unit), sedimentation in the lakes was dominated by carbonate and gypsum. An expansion of fresh-water carbonate lakes is recorded towards the top of the unit. During the Turolian (Upper Unit), lacustrine sedimentation took place in a network of interconnected shallow carbonate lakes and fluvial subenvironments. In this paper we discuss the correlation or agreement between the paleoclimatic interpretation that can be inferred from lithofacies analysis and that furnished by mammal assemblages evolving throughout the Miocene in the Iberian Peninsula. Our results indicate that other factors besides climate (e.g., tectonics, source areas, and paleohydrological regimes) influenced decisively the nature of sediments in the lake complexes.

Introduction

Lacustrine sedimentary sequences can potentially provide a very powerful archive of paleoenvironmental evolution in continental basins. The sedimentary evolution of lake complexes through time can provide meaningful information about the record of past climate changes on at least local or regional scales (Talbot & Kelts, 1989). Climatic influence is usually expressed in terms of temperature (cool versus warm) and humidity (dry versus moist), both parameters controlling the input/output balance in lake systems. Recognition of climate-influenced sedimentation patterns appears to be easier in closed-lake basins than open-lake basins because signatures derived from changing evaporation rates should be particularly well developed in the sediments.

There is abundant literature on very thick lacustrine successions that represent permanence of lake complexes through long time intervals (see Smith, 1990, for a review). Some examples are those furnished by the Devonian–Orcadian basin (Parnell *et al.*, 1990), the Triassic–Jurassic Newark Supergroup (Gore, 1989; Olsen, 1990) or by many Chinese basins that contain Cretaceous and Tertiary lacustrine successions (Tian *et al.*, 1983). A main conclusion to be extracted from all these examples is that thick lacustrine sequences always record a complex history of paleoenvironmental changes that can be related either to periodic (sometimes erratic) climatic fluctuations or to variation of deformational tectonics through time.

The Madrid Basin, central Spain, provides an additional case study for recognizing and interpreting tectonic and climatic constraints on the sedimentary evolution of a large continental basin in which lake systems developed widely through a long time interval spanning most of the Tertiary. This chapter deals exclusively with the Miocene record, which accounts for one third (about 800–1000 m in thickness) of the total sedimentary filling of the basin. By comparing charts of relative temperature and humidity curves proposed for the Spanish Miocene (Morales, 1992; Van der Muelen & Daams, 1992; Calvo *et al.*, 1993) with the evolutionary trend of lacustrine sedimentation in the Madrid Basin, a discussion arises about the agreement between both sets of data. Emphasis is also placed on other factors (source areas, hydrologic pattern, tectonic regime) besides climate that were decisive in controlling changes in lacustrine sedimentation.

Geological context of the Madrid Lake systems

The geology of the Madrid Basin has been described by Martín Escorza (1976), Megías *et al.* (1983), Junco & Calvo (1983), and Calvo *et al.* (1990), among others. The basin, located in the centre of Spain (Fig. 1), covers more than 10000 km^2, and lacustrine systems may have reached 6000 or 7000 km^2 at their maximum extent. Tectonically, this was an intracratonic basin with a rather complex deformational history due to diachronism of tectonic movements affecting the basin margins throughout the Late Oligocene and Neogene (see Chapter C2). A summary of the geodynamic context of the Iberian Peninsula and subsequent deformational pattern of the Madrid Basin during this period can be found in Anadón *et al.* (1989). Paleolatitude of the region during the Tertiary was not much different from that of the present day (Smith *et al.*, 1981; Anadón *et al.*, 1989).

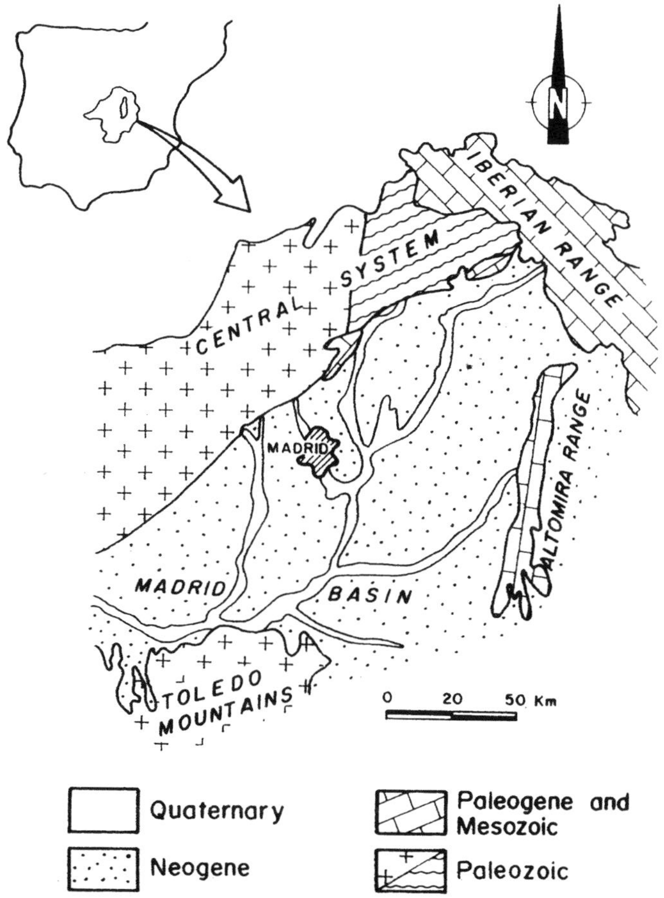

Fig. 1. Schematic map of the Madrid Basin.

Fig. 2 sketches the development of lake complexes in the Madrid Basin during the Miocene. The general stratigraphic scheme for the Miocene is represented by three main units (Lower, Intermediate, Upper) that are separated by major sedimentary disconformities recognizable across the whole basin (see Junco & Calvo, 1983; Alberdi *et al.*, 1984; Calvo *et al.*, 1989b, for detailed lithostratigraphic description of these units). Lacustrine facies associations show great variation from Lower to Upper Miocene, which is indicative of significant change in the hydrologic and depositional patterns of the lake systems through time. In the Lower Unit, lake systems are evaporite-dominated and show a complex association of saline minerals (Ordóñez *et al.*, 1991). This association is thought to have been deposited in closed perennial saline lakes that were fringed by arkosic or litharenitic alluvium (Fig. 3). In most of the Intermediate Unit, lacustrine facies consist predominantly of carbonate and gypsum. These deposits interfinger with mudstones (locally sepiolite and bentonite) in marginal areas of the lake system (Calvo *et al.*, 1989a; Ordóñez *et al.*, 1991). Lakes that developed during the Middle and Upper Aragonian are interpreted as shallow, slightly saline, perennial lakes. Towards the upper part of the Intermediate Unit (Vallesian), lake facies became predominantly carbonate and episodes of deeper lacustrine conditions have been locally recognized (Bellanca *et al.*, 1992).

The Upper Unit contains fresh-water carbonate lacustrine facies that occur interbedded with, and are spaced out laterally by, fluvial deposits. These facies are interpreted to result from deposition in a mosaic of hard-water, shallow, relatively small lakes whose spatial distribution was controlled by approximately N–S lineations.

Fig. 3 shows several idealized sketches representing the paleogeographical evolution of the Madrid Basin throughout the Miocene. The geometry of the lake systems is outlined in the sketches. During both the Lower and the Middle to Upper Aragonian, the depositional pattern corresponds to that of a closed hydrologic and geomorphic basin (Eugster & Kelts, 1983) displaying a characteristic concentric facies zonation. A similar pattern is recognized for the Vallesian, although by this period the lake system reached a maximum extent, the lacustrine carbonate facies onlapping locally the eastern basin margin. This pattern changed abruptly during the Upper Unit due to different deformational strains affecting the area throughout the uppermost Vallesian and Turolian. A change from compressive to extensional strains has been invoked to explain this variation in paleogeography of the Miocene units (Calvo *et al.*, 1991; see Chapters C1, C2).

The interplay between basin geological evolution and climate in explaining the sedimentary trend of the Madrid lake systems is discussed below.

Discussion

On the basis of facies associations and stratigraphic characteristics of the continental succession, the evolution of lacustrine systems in the Madrid Basin can be divided into four main stages (Fig. 2).

The first stage (I in Fig. 2) corresponds to the development of saline lake systems during the Ramblian and lowermost Aragonian. Lithostratigraphically, the lacustrine facies belong to the Lower Unit and are characterized by relatively to highly soluble saline deposits (anhydrite, gypsum, halite, thenardite, glauberite, polyhalite, and others) (García del Cura, 1979). This saline mineral association is characteristic of strongly evaporating conditions promoting the formation of highly concentrated brines. However, this is in contrast with the general trend shown by Neogene paleoclimatic curves (Fig. 2) proposed for the Iberian Peninsula (Calvo *et al.*, 1993) as well as by those suggested for North Central Spain (Van der Meulen & Daams, 1992). Data furnished by mammal associations indicate that rather moist and cold conditions prevailed in most of Spain during this period.

Two explanations may be suggested for this paradox: 1. in contrast with the mentioned general climatic pattern, the Madrid Basin constituted a local 'rain shadow' region in central Iberia; and 2. extensive recycling of Mesozoic and Paleogene evaporite formations that were tectonically uplifted in eastern parts of the basin (Altomira and Iberian Ranges) produced a supply of highly concentrated brines in spite of unfavourable climatic conditions for thermal evaporation. The importance of evaporite recycling in Spain during the Alpine cycle has been outlined by Ortí *et al.* (1988), and the real effectiveness of this mechanism in the eastern part of the

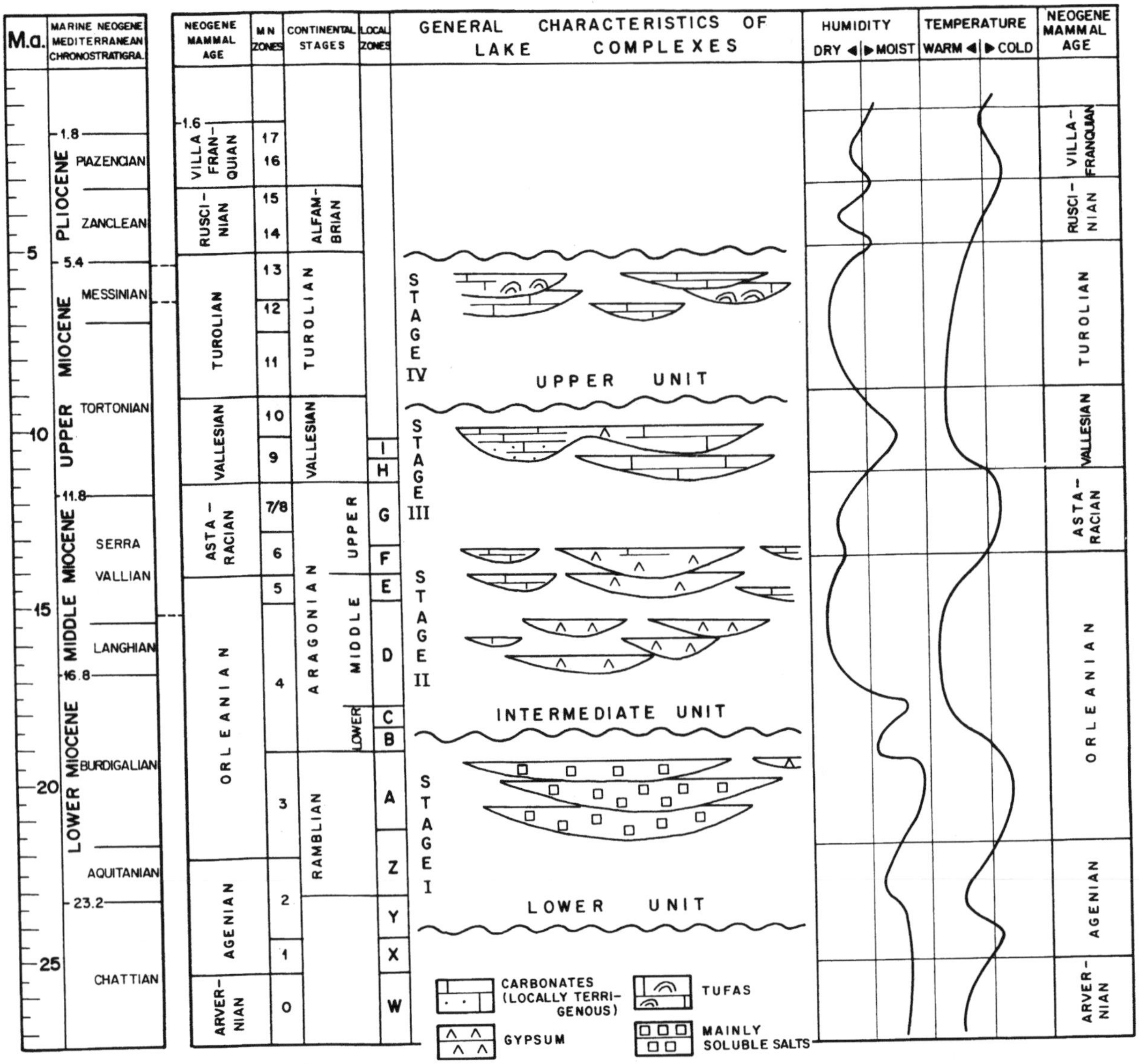

Fig. 2. Integrated sketch for lithostratigraphy of the Miocene sedimentary record of the Madrid Basin showing main evolutionary stages of the lake complexes. Paleoclimatic curves at right are those proposed in Calvo *et al.* (1993).

Madrid Basin has been demonstrated by Rodríguez-Aranda *et al.* (1991). On the other hand, a probable modern analogue for the formation of evaporites under non-extreme climatic conditions may be found in the Northern Great Plains of Canada (Last, 1989).

Two stages have been distinguished within the Intermediate Unit (Fig. 2). Stage II covers most of the sedimentary record of this unit. The lake facies are predominantly carbonate and gypsum. The deposition of the unit began with marked progradation of alluvial fan systems, which was related to the uplift of the Central System in western and northern parts of the basin. A climatic shift towards more humid conditions (Fig. 2) also probably contributed to this progradation. Drier and warmer conditions prevailed during most of the Middle Aragonian, which is in agreement with the common occurrence of sepiolite and smectite deposits in the basin (Calvo *et al.*, 1989a; Ordóñez *et al.*, 1991). In spite of the presence of favourable climatic conditions for the formation of highly soluble saline facies, no deposits like those recognized in the Lower Unit formed in this period. The main reason invoked for such an absence is that recycling of older evaporite formations was not so effective as it was in Stage I. During the sedimentation of the Intermediate Unit (Middle Aragonian to Vallesian), the Altomira Range behaved as a tectonically non-active area and its potential as a source of solutes was drastically reduced (Calvo *et al.*, 1989b). In contrast, the main tectonic activity during this period influenced the Central System (Calvo *et al.*, 1990), an area in which older evaporite formations are scarce or absent.

There is agreement among the several authors (Morales, 1992; Van der Meulen & Daams, 1992) that humid conditions prevailed in the Iberian Peninsula during the Vallesian. This assessment fits well with the expansion of fresh-water carbonate lakes in the

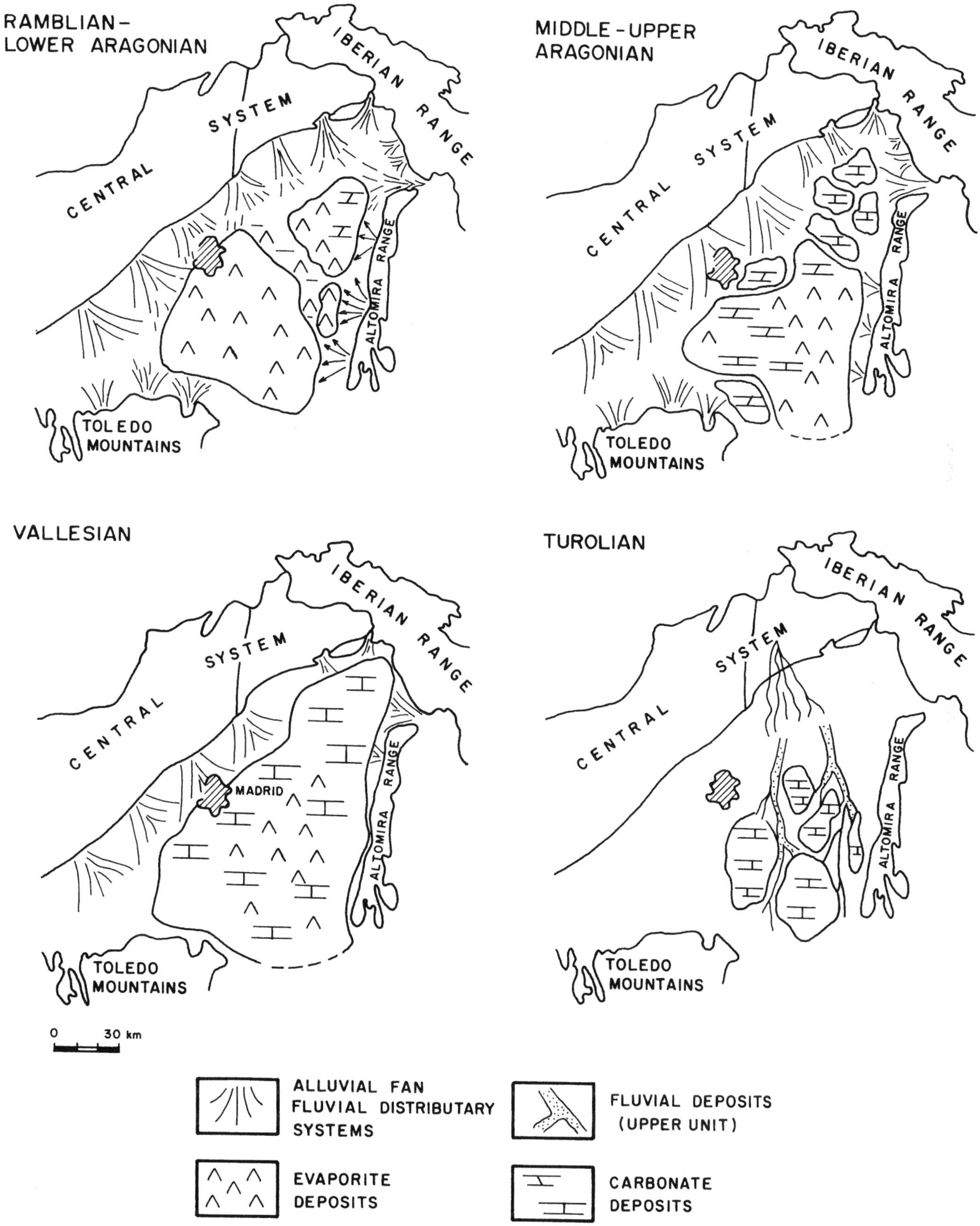

Fig. 3. Paleogeographic evolution of alluvial and lacustrine systems in the Madrid Basin through the Miocene.

Madrid Basin during this period (Figs. 2 and 3). Thus, Stage III records the growth of shallow to moderately deep carbonate lakes (Bellanca *et al.*, 1992; Alonso-Zarza *et al.*, 1992) and these characterized a period of relative tectonic quiescence in the basin.

The Upper Unit of the Miocene in the Madrid Basin is formed of a lower subdivision of fluvial deposits and an upper subdivision of lacustrine beds characteristic of very shallow, irregularly distributed, fresh-water carbonate lakes. Carbonate-rich fluvial deposits (stromatolites, oncoids, and tufa) are recognized as well (Ordóñez and García del Cura, 1983). This wide occurrence of fresh-water carbonates is apparently in contrast with the general climatic pattern proposed for this period (Morales, 1992; Alcalá, 1992). The Turolian was characterized by semi-arid (dry and warm) climatic conditions leading to a savannah-like paleoenvironment. This pattern changed slightly towards the end of the Turolian.

The absence of sedimentary deposits indicative of dry and warm conditions during the Turolian may be related to both tectonic and hydrological factors. As mentioned earlier, there was a distinct change from a concentric depositional pattern, typical of closed lake basins, during the Intermediate Unit to an interconnected network of fluvial streams and carbonate ponds in the Upper Unit. This new scenario may have destroyed the environmental conditions favourable for the development of concentrated brines in the lakes in spite of favourable climatic conditions. In addition, extensive precipitation of carbonate in the lakes was probably favoured by active leaching of older Miocene carbonate formations in the basin (Calvo *et al.*, 1989b; see also Chapter C4). Similar conditions for extensive accumulation of carbonate, especially tufas, in a semi-arid region, are found in recent interconnected carbonate lakes of central Spain (Ordóñez *et al.*, 1986).

Conclusions

Stacked Miocene complexes of the Madrid Basin provide a useful case study for testing agreement between paleoclimatic data furnished by mammal assemblages and those inferred from lithofacies associations and depositional trends. Both sets of data seem to fit well for the Middle and Late Aragonian as well as the Vallesian time intervals. However, there is a marked disagreement in the Ramblian–Early Aragonian and Turolian. The study suggests that other factors besides climate could have determined the nature of lake facies in the basin. The tectonic factor seems to have been critical, controlling the uplift of the source areas and the paleohydrologic regime of the basin.

Acknowledgments

We thank A. Blanco for his help in preparing drawings included in this chapter. The work was financed by the Project DGICYT PB89–0032.

References

Alberdi, M.T., Hoyos, M., Junco, F., López-Martínez, N., Morales, J., Sesé, C. and Soria, M.D. (1984). Biostratigraphy and sedimentary evolution of continental Neogene in the Madrid area. *Paléobiol. Cont.*, Montpellier, **14**, 47–68.

Alcalá, L. (1992). *Micromamíferos Neógenos de la Fosa de Teruel–Alfambra.* Tesis Doctoral, Univ. Complutense, Madrid, 519 pp.

Alonso-Zarza, A.M., Calvo, J.P. and García del Cura, M.A. (1992). Palustrine sedimentation and associated features – grainification and pseudo-microkarst – in the Middle Miocene (Intermediate Unit) of the Madrid Basin, Spain. *Sed. Geol.*, **76**, 43–61.

Anadón, P., Cabrera, L., Juliá, R. and Roca, E. (1989). Lacustrine oil–shale basins in Tertiary grabens from NE Spain (Western European Rift system). *Palaeogeogr., Palaeoclimatol., Palaeoecol.*, **70**, 7–28.

Bellanca, A., Calvo, J.P., Censi, P., Neri, R. and Pozo, M. (1992). Recognition of lake level changes in Miocene lacustrine units, Madrid Basin, Spain. Evidence from facies analysis, isotope geochemistry and clay mineralogy. *Sed. Geol.*, **76**, 135–153.

Calvo, J.P., Alonso Zarza, A.M. and García del Cura. M.A. (1989a). Models of Miocene marginal lacustrine sedimentation in response to varied source areas and depositional regimes in the Madrid Basin, central Spain. *Palaeogeogr., Palaeoclimatol., Palaeoecol.*, **90**, 199–214.

Calvo, J.P., Ordóñez, S., García del Cura, M.A., Hoyos, M., Alonso Zarza, A.M., Sanz, E. and Rodríguez-Aranda, J.P. (1989b). Sedimentología de los complejos lacustres miocenos de la Cuenca de Madrid. *Acta Geol. Hisp.*, **24**, 281–294.

Calvo, J.P., Hoyos, M., Morales, J. and Ordóñez, S. (1990). Neogene stratigraphy, sedimentology and raw materials of the Madrid Basin. *Iberian Neogene Basins. Paleont. Evol.*, mem. esp. **2**, 63–95.

Calvo, J.P., Daams, R., Morales, J., López-Martínez, N., Agustí, J., Anadón, P., Armenteros, I., Cabrera, L., Civis, J., Corrochano, A., Díaz-Molina, M., Elizaga, E., Hoyos, M., Martín-Suarez, E., Martínez, J., Moissenet, E., Muñoz, A., Pérez-García, A., Pérez-González, A., Portero, J.M., Robles, F., Ruiz Bustos, A., Santisteban, C., Torres, T., Van der Meulen, A. and Vera, J.A. (1993). Up-to-date of the Spanish continental Neogene synthesis and paleoclimatic interpretation. *Rev. Soc. Geol. Esp.*, **6**, 29–40.

Eugster, H. and Kelts, K. (1983). Lacustrine chemical sediments. In *Chemical sediments and geomorphology* (Goudie, A.S. & Pye, K., eds). Academic Press, London, 321–368.

García del Cura, M.A. (1979). *Las sales sódicas, calcosódicas y magnésicas de la Cuenca del Tajo.* Fund. Juan March, Ser. Universitaria, 39 pp.

Gore, P.J.W. (1989). Toward a model for open- and closed-basin deposition in ancient lacustrine sequences: the Newark Supergroup (Triassic–Jurassic), eastern North America. *Palaeogeogr., Palaeoclimatol., Palaeoecol.*, **70**, 29–52.

Junco, F. and Calvo, J.P. (1983). Cuenca de Madrid. *Geología de España*, IGME, Madrid, II, 534–542.

Last, W.M. (1989). Continental brines and evaporites of the northern Great Plains of Canada. *Sediment. Geol.*, **64**, 207–222.

Martín Escorza, C. (1976). Actividad tectónica durante el Mioceno de las fracturas del basamento de la Fosa del Tajo. *Estud. Geol.*, **32**, 509–522.

Megías, A.G., Ordóñez, S. and Calvo, J.P. (1983). Nuevas aportaciones al conocimiento geológico de la Cuenca de Madrid. *Rev. Mat. Proc. Geol.*, **1**, 163–192.

Morales, J. (1992). Las faunas de mamíferos del Neógeno de Europa. In *Paleontología de vertebrados: faunas y filogenia, aplicación y sociedad*, (U. Aldibia, ed.). Serv. Edit. Univ. País Vasco, 235–256.

Olsen, F. (1990). Tectonic, climatic, and biotic modulation of lacustrine ecosystems – Examples from Newark Supergroup of Eastern North America. In *Lacustrine basin exploration: case studies and modern analogs.* (Katz, B.J., ed.). A.A.P.G. Mem. **50**, 209–224.

Ordóñez, S. and García del Cura, M.A. (1983). Recent and Tertiary fluvial carbonate in Central Spain. *Spec. Publ. Int. Assoc. Sedimentol.*, **6**, 485–497.

Ordóñez, S., González Martín, J.A. and García del Cura, M.A. (1986). Sedimentación carbonática actual y paractual en las lagunas de Ruidera. *Rev. Mat. Proc. Geol.*, **4**, 229–255.

Ordóñez, S., Calvo, J.P., García del Cura, M.A., Alonso Zarza, A.M. and Hoyos, M. (1991). Sedimentology of sodium sulphate deposits and special clays from the Tertiary Madrid Basin (Spain). *Spec. Publ. Int. Assoc. Sedimentol.*, **13**, 39–55.

Ortí, F., Rosell, L., Utrilla, R., Inglés, M., Pueyo, J.J. and Pierre, C. (1988). Reciclado de evaporitas en la Península Ibérica durante el ciclo Alpino. *II Congr. Geol. España, Granada, Comunicaciones*, **1**, 421–424.

Parnell, J., Marshall, J. and Astin, T. (1990). *Field guide to lacustrine deposits of the Orcadian Basin, Scotland.* 13th Int. Sedimentol. Congress, Nottingham, 47 pp.

Rodríguez-Aranda, J.P., Calvo, J.P. and Ordóñez, S. (1991). Transición de abanicos aluviales a evaporitas en el Mioceno del borde oriental de la Cuenca de Madrid (Sector Barajas de Melo-Illana). *Rev. Soc. Geol. Esp.*, **4**, 33–50.

Smith, A.G., Hurley, A.M. and Briden, J.C. (1981). *Phanerozoic paleocontinental maps.* Cambridge Univ. Press, 102 pp.

Smith, M.A. (1990). Lacustrine Oil Shale in the geologic record. In *Lacustrine basin exploration: case studies and modern analogs.* (Katz, B.J., ed.). A.A.P.G. Mem., **50**, 43–60.

Talbot, M.R. and Kelts, K. (eds.) (1989). The Phanerozoic record of lacustrine basins and their environmental signals. *Palaeogeogr., Palaeoclimatol., Palaeoecol.*, **70**, 1–304.

Tian, Z., Chang, C., Huang, D. and Wu, C. (1983). Sedimentary facies, oil generation in Meso-Cenozoic continental basins in China. *Oil Gas J.*, **May 16**, 120–126.

Van der Meulen, A. and Daams, R. (1992). Evolution of Early–Middle Miocene rodent faunas in relation to long-term palaeoenvironmental changes. *Palaeogeogr., Palaeoclimatol., Palaeoecol.*, **93**, 227–253.

C4 Paleomorphologic features of an intra-Vallesian paleokarst, Tertiary Madrid Basin: significance of paleokarstic surfaces in continental basin analysis

J.C. CAÑAVERAS, J.P. CALVO, M. HOYOS AND S. ORDÓÑEZ

Abstract

Several paleokarst surfaces are recognized within the Neogene sedimentary record of the Madrid Basin, central Spain. During the Late Vallesian, one of these paleokarsts was related to a major break in the Miocene sedimentary sequence. The paleokarst developed diachronically over fresh-water carbonates in central and eastern parts of the basin. A variety of both exokarstic and endokarstic morphologies and associated sediments (breccias, siliciclastic infill deposits, speleothems) are distinguished. The paleokarst records a complex history of dissolution, collapse and infilling processes that were related to long-term subaerial exposure of carbonate. In this chapter, emphasis is placed on paleotectonic constraints leading to the formation of paleokarst. Paleokarst development was related to a drastic change from compressive to extensional strains in the area during the Late Vallesian.

Introduction

As stated by Wright (1991a), 'paleokarst refers to karstic features formed in the past, related to an earlier hydrological system or landscape'. Most specific karstic features deal with small- to large-scale dissolution processes that affect highly soluble rocks, mainly carbonates and evaporites, and lead to well-developed secondary porosity (Ford & Williams, 1989; Smart & Whitaker, 1991). The importance of paleokarst as a relevant aspect in stratigraphic analysis or in exploration strategies is reflected in a rather abundant recent literature on this topic (Wright, 1982; Esteban & Klappa, 1983; James & Choquette, 1988; Bosak *et al.*, 1989; Wright, 1991b).

Most paleokarsts recognized in the sedimentary record can be attributed to meteoric waters, as most of them show morphological patterns similar to recent meteoric karst (Ford, 1988; Ford & Williams, 1989). However, other origins for paleokarsts have been suggested as well (Wright, 1991b). Meteoric diagenetic changes related to paleokarst could develop in continental or oceanic domains (Smart & Whitaker, 1991) so that karstic products may be widely different. The differentiation is probably relevant for paleokarst surfaces that resulted from eustasy-related exposure of large carbonate platforms. Many recent studies on paleokarst have been focused on this geological context, leaving aside some other, probably quite specific, features that occur in paleokarsts developed in non-marine successions.

In this chapter, we describe a paleokarst system formed in the continental Madrid Basin, central Spain (Fig. 1), during the late Miocene. The paleokarst is clearly associated with and represents a break in the Miocene sedimentary sequences. Besides the morphological characteristics and diagenetic facies related to this paleokarst, we place emphasis on the paleoclimatic and paleotectonic constraints leading to its formation. A brief discussion follows later on the validity of the case study and the proposed model for paleokarsts developed in continental series.

Geologic context of the intra-Vallesian paleokarst

The paleokarst studied is located towards the upper part of the Tertiary succession of the Madrid Basin. The complete succession ranges between 2000 and 3500 metres in thickness, depending on the part of the basin (Calvo *et al.*, 1990), and is formed of both Paleogene and Neogene terrestrial strata. The Neogene deposits have been divided stratigraphically into three main Miocene units (Fig. 1) and two Pliocene sedimentary cycles, each separated by unconformities from the underlying units. In many cases paleokarsts occur in relation to those unconformities. More-or-less detailed previous studies of the paleokarsts between the different units include those of Calvo *et al.* (1984) (for paleokarst between the Miocene Lower and Intermediate Units), Calvo *et al.* (1980), Ordoñez *et al.* (1985) and Cañaveras (1991) (for paleokarst between the Intermediate and Upper Units), and Sanz *et al.* (1991, 1992) (for paleokarsts developed at the top of the Miocene Upper Unit and through the Pliocene).

The intra-Vallesian paleokarst outlines the sedimentary discontinuity (unconformity) between the Miocene Intermediate and Upper Units in the central part of the Madrid Basin (Fig. 2). In this area, the Intermediate Unit, ranging in age from the Middle Aragonian to the lower part of the Late Vallesian (Calvo *et al.*, 1990), is formed by a 50–200 m thick succession of terrigenous and/or chemical deposits passing upwards into fresh-water lacustrine

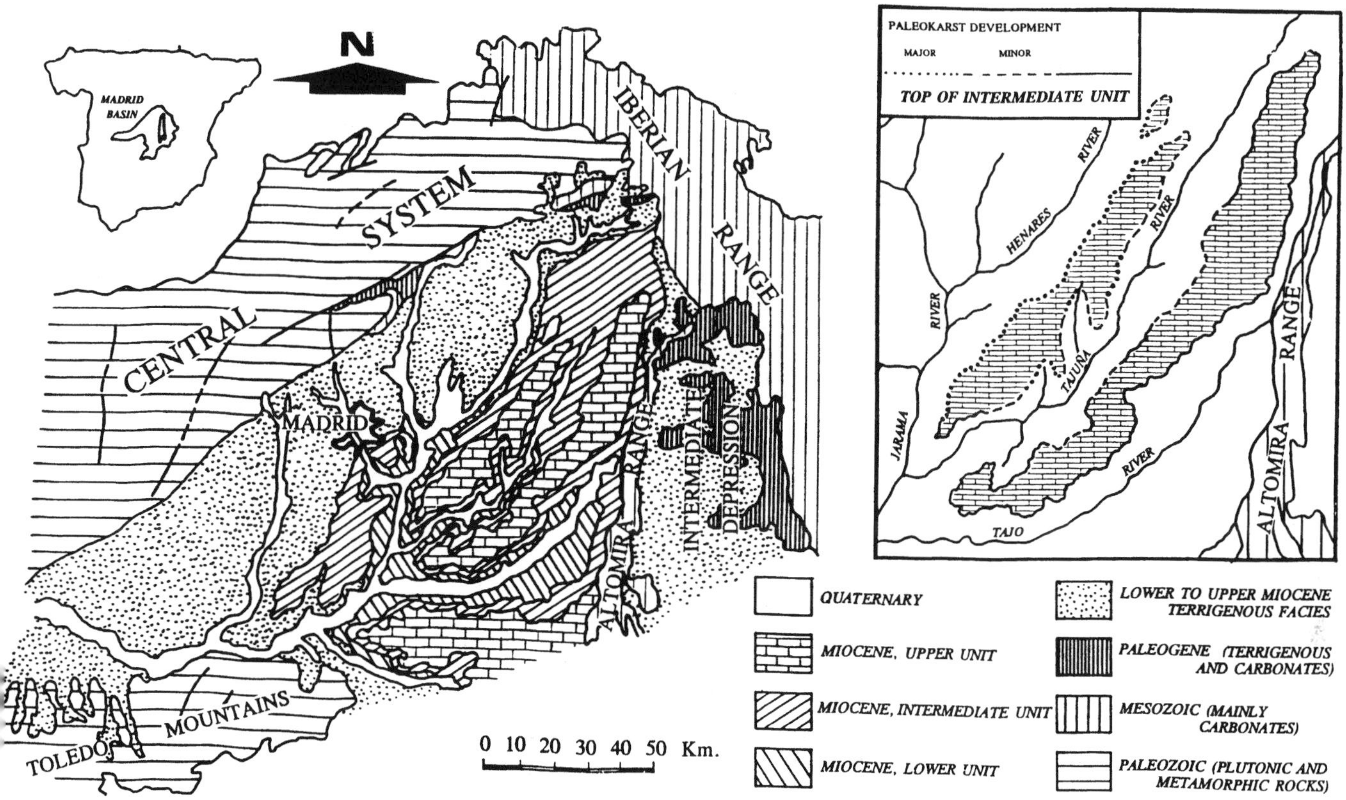

Fig. 1. Regional geologic setting of the intra-Vallesian paleokarst. On the left, schematic map of the Madrid basin (modified from Calvo *et al.*, 1989). On the right, locality map of the top of the Intermediate Unit and the intra-Vallesian paleokarst.

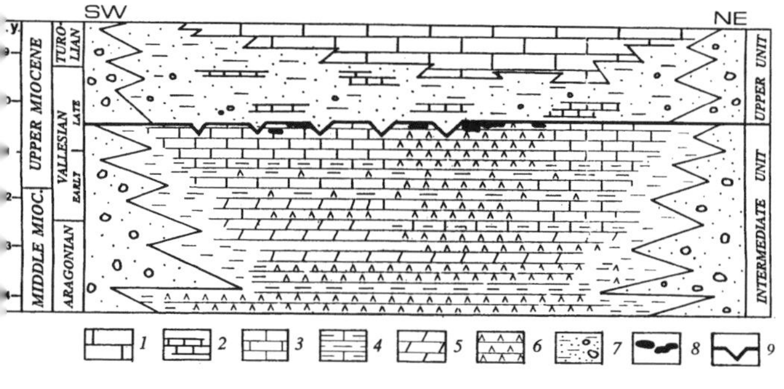

Fig. 2. Integrated lithostratigraphic scheme for the sedimentary filling of Madrid basin during the middle to late Miocene. 1, fluvial–lacustrine limestones; 2, palustrine–pedogenic limestones; 3, lacustrine limestones; 4, marls and marly carbonates; 5, dolostones; 6, gypsum; 7, lutites, sands, sandstones and gravels; 8, chert; 9, paleokarstic surface.

carbonate deposits. These carbonate deposits, largely affected by karstification, lie horizontally and can be traced out through most of the basin.

The intra-Vallesian paleokarst surface is overlain by terrigenous and carbonate deposits belonging to the Upper Unit. The age of the lower deposits of this unit is not precisely known, because determinative mammal faunas have not yet been found; thus, uncertainty remains on the Late Vallesian or Early Turolian age of the deposits and with the subsequent time interval represented by paleokarst.

There was a change in the tectonic evolution of the basin between the deposition of the Intermediate and Upper Units, as pointed out by De Vicente *et al.* (1990) and De Vicente *et al.* (Chapter C2). According to their work, the sedimentation of the Intermediate Unit took place under a compressive regime resulting in a depositional pattern characteristic of closed hydrologic and geomorphic lacustrine basins (Calvo *et al.*, 1989). This pattern changed dramatically during the Upper Unit when, as a result of extension, the sedimentation took place with a pattern of fluvial–lacustrine systems distributed from north to south in the basin, probably reflecting exorheic conditions (Fig. 3). As discussed below, this drastic change in basin deformational styles and concomitant depositional pattern is considered important in the development of the intra-Vallesian paleokarst.

Finally, paleoclimatic conditions, inferred mainly from mammal assemblages, were characterized by relatively cool and wet periods in the centre and northern part of the Iberian Peninsula, at least during the Early Vallesian (Van der Meulen & Daams, 1992) and probably during some of the Late Vallesian (López-Martínez *et al.*, 1987). In the Madrid Basin, these paleoclimatic conditions prevailed during the deposition of the uppermost part of the Intermediate Unit (Sesé *et al.*, 1990). No conclusive data have been furnished either for the uppermost part of the Vallesian or for the beginning of the Turolian in the Madrid Basin, although it has been

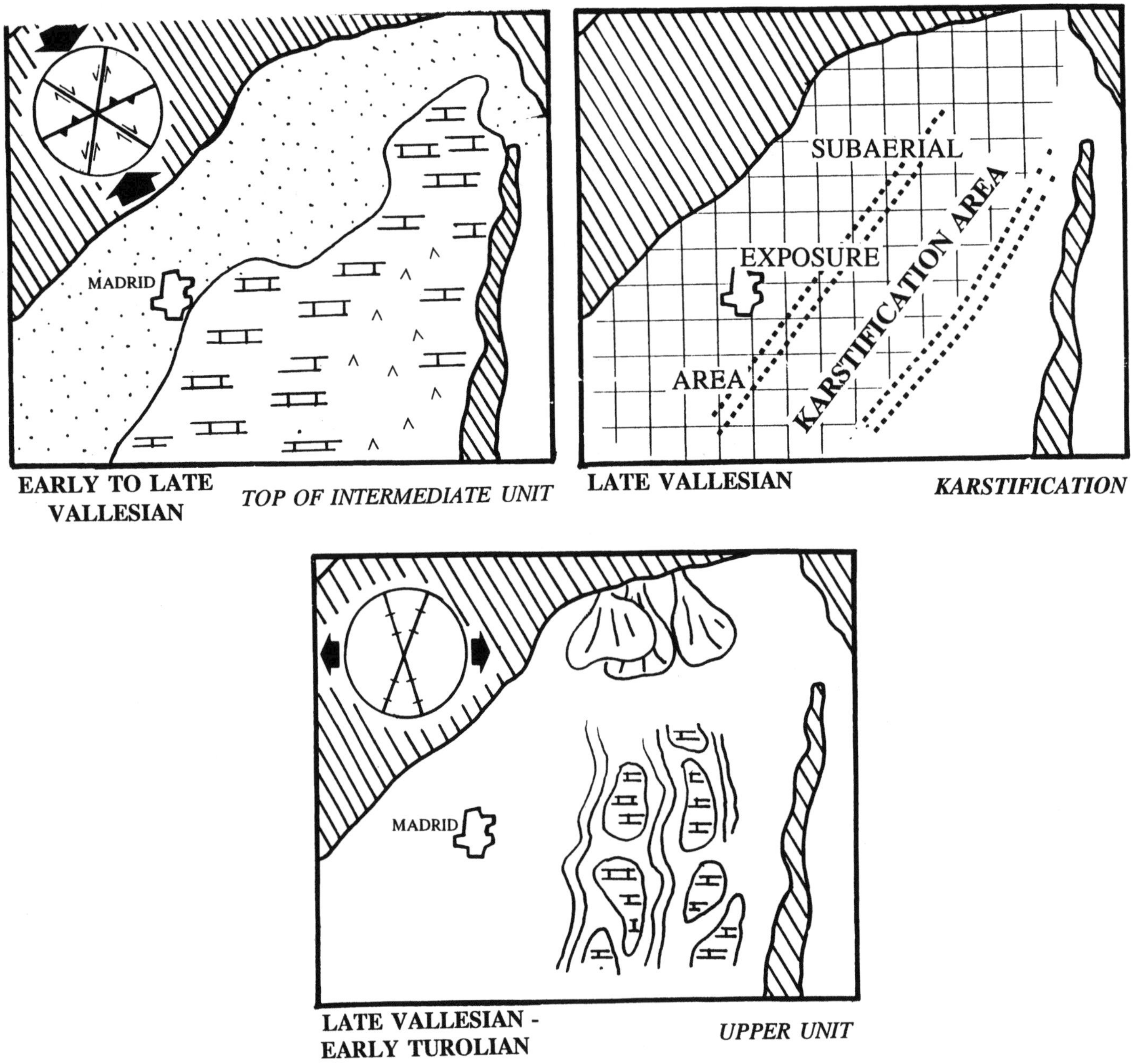

Fig. 3. Idealized paleogeographical sketches of the Madrid basin through the late Miocene. Symbols distinguish areas of bed rock (diagonal lines), sub-aerial exposure of sediment (grid), siliciclastic sediments (stipple), alluvial fans, carbonate sediments (bricks), evaporite sediments (inverted Vs), channel-fills (ribbons).

suggested (López-Martínez *et al.*, 1987) that warmer and drier climatic conditions prevailed through most of the Turolian in the Iberian Peninsula.

Areal distribution and general features of the paleokarst

Outcrops allowing good observation of the paleokarst are located along topographic highs (mesetas) between the valleys of the Henares, Jarama, Tajuña and Tajo rivers (Fig. 1). In these areas the paleokarst profiles or zones of karstic diagenetic alteration extend downwards for up to 10–30 metres. Both upper and lower limits of the paleokarst are highly irregular at outcrop scale. The top surface displays morphologies indicative of erosion and syn- or post-karstic collapse phenomena that result in differences of 2–15 metres in topography (Figs. 4 and 5). The lower limit is transitional from strongly recrystallized carbonates into lacustrine carbonate strata displaying well-preserved depositional fabrics (micrites and biomicrites).

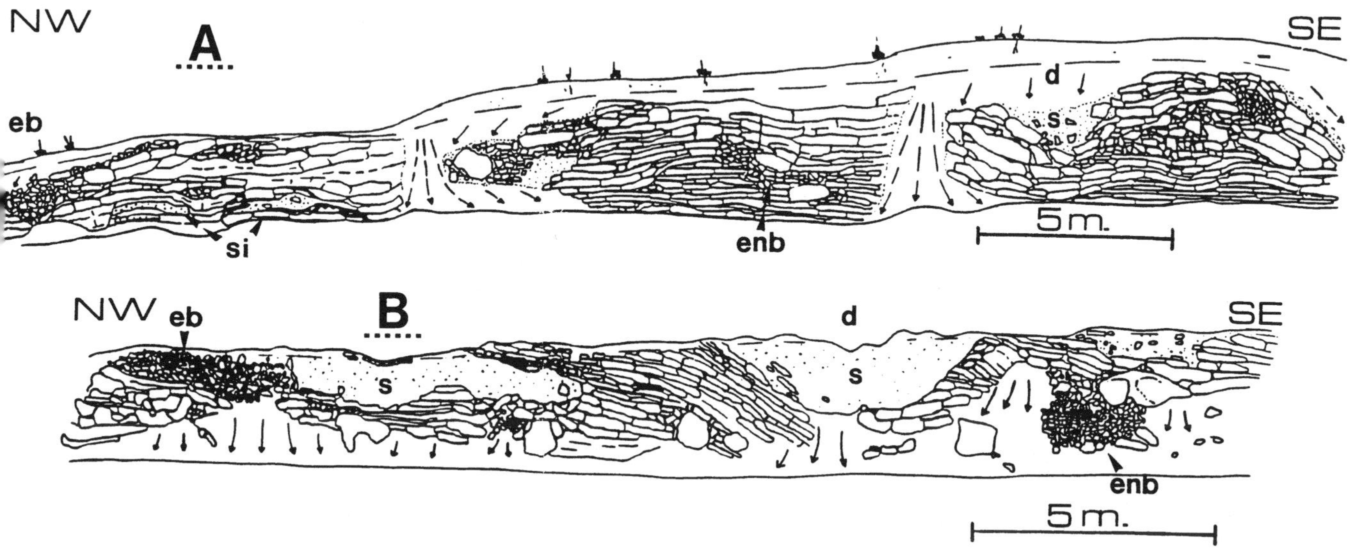

Fig. 4. Sections of paleokarst outcrops in the central area of the Madrid Basin. Top: Campo Real section. Bottom: Torres de la Alameda section. d: doline, s: exokarstic siliciclastic breccia, eb: exokarstic breccia, enb: endokarstic breccia, si: siliciclastic infill deposits.

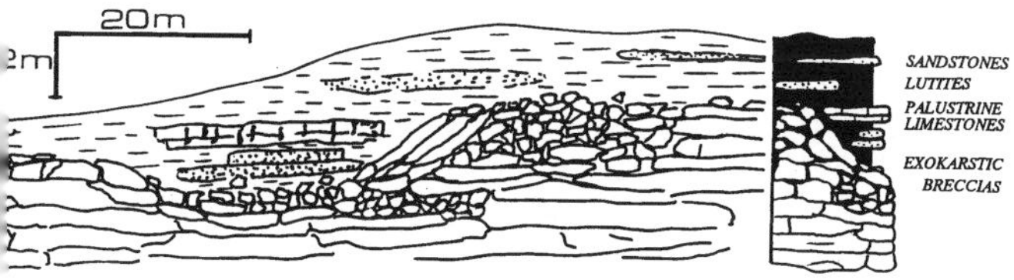

Fig. 5. Schematic section showing exokarstic features of intra-Vallesian paleokarst.

Exokarstic morphology and sediments

Exokarstic morphological features, mainly eroded depressions and dolines, are widely distributed, though not abundant, and show very irregular sections of variable size (metres to a few tens of metres). They are interpreted as resulting from gravitational collapses due to the erosion and dissolution of near-surface caves. The exokarstic infill-facies are varied and consist of: 1. chaotic clast-supported carbonate breccias; 2. siliciclastic deposits (lutites, sands and gravels) related to downwards infill from the overlying stratigraphic unit; and 3. mixed facies. These facies are described below.

Chaotic clast-supported breccias

This facies comprises poorly sorted breccias of the host carbonate rock as well as fragments of karst-generated deposits (speleothems and internal sediment) in a marly, occasionally sandy matrix. Clast size ranges from a few centimetres up to more than one metre, the clast shapes varying from very angular to subangular. Locally, these clasts display more-or-less clearly imbricated patterns towards the centre of the depressions. The breccia passes gradationally into the substrate carbonate rock. The geometry of these breccias allows them to be classified as collapse and/or mantling breccia (James & Choquette, 1988) (Fig. 4).

Siliciclastic deposits

This facies comprises usually lutites and sands, although gravels are locally prominent. Compositionally, the sandstones consist of subarkose whereas the lutite mineralogy is dominated by illite and smectite, both compositions being characteristic of the detrital beds of the overlying Upper Unit. The siliciclastic deposits cover discontinuously the paleokarst surface, forming channels and their onlap relationships with the carbonate substrate providing clear evidence that their formation was conditioned by the existence of the paleokarstic surface. Poorly developed carbonate soil profiles veneer the paleokarst, and thin oncolitic carbonate beds have been locally observed (Fig. 5).

Mixed facies

This facies consists of matrix-supported carbonate breccias in a marly and/or sandy (occasionally gravelly) matrix. These mixed facies are transitional materials between entirely exokarstic deposits and fluvial deposits belonging to the lower part of the Upper Unit (Fig. 6A).

Endokarstic morphology and sediments

The intra-Vallesian karst is characterized by a rather complex network of horizontally elongated caves connected by vertical to oblique irregular conduits. Cave dimensions are variable, ranging from 10–60 cm in height and extending laterally several

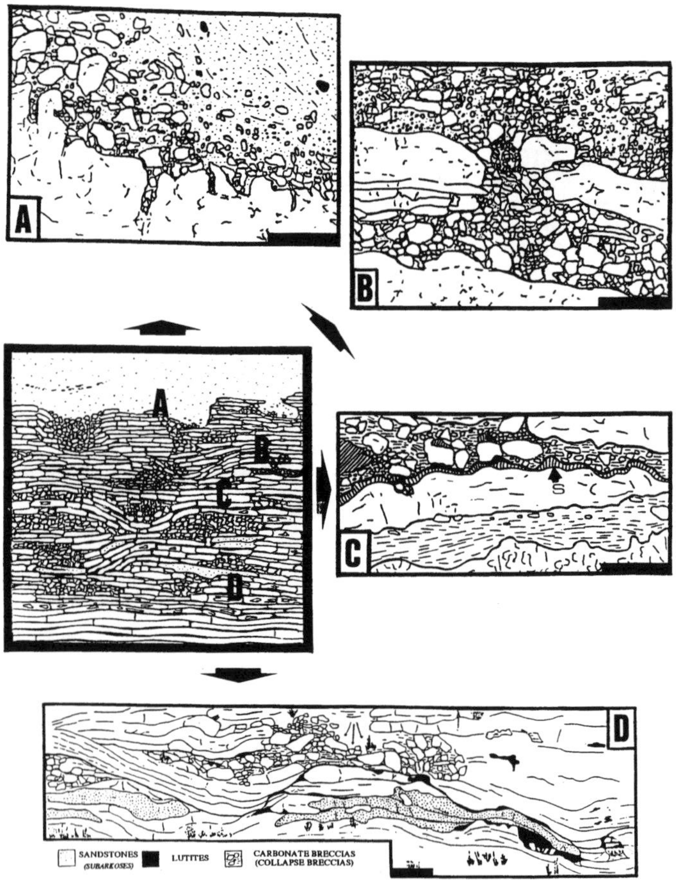

Fig. 6. Summary sketch of the intra-Vallesian paleokarst features. The sketch located at the central-left part of the figure indicates the common occurrence of features described in A, B, C and D through the paleokarst profile. A, Detail of the bottom of a doline showing host rock (bottom and left) and exokarstic deposits (mixed facies and siliciclastic sediments) (scale bar: 0.5 m). B, Tabular-shape bodies of endokarstic collapse breccias (scale bar: 0.5 m). C, Speleothems (s), fibrous calcite coating on cave floors and as endokarstic breccia fragments (scale bar: 0.2 m). D, Siliciclastic infill deposits (scale bar: 1 m).

metres, rarely more than 10 metres. The shape of the main cave system was generally controlled by stratification of the carbonate substrate and is thought to have been mainly developed by solution enlargement processes in a phreatic environment. The solution process may have been effective through discontinuities related to bedding planes and joints, though an additional control by other host-rock heterogeneities is not impossible.

Three types of deposits accumulated in cave interiors and conduits and are considered to be the most distinctive features of the endokarst. These are: 1. carbonate breccias; 2. siliciclastic infill deposits; and 3. speleothems. The description and interpretation of these facies is given below.

Carbonate breccias

This facies consists of clast-supported breccias of variable size that form tabular to irregular-shaped sedimentary bodies whose size ranges from 10–30 centimetres (in tabular-shape bodies) up to more than 1 metre (in some irregular-shape bodies) (Fig. 6B). Clast size ranges from a few centimetres up to several decimetres, the shape varying from very angular to subangular. Commonly, these materials are associated with other endokarstic deposits (especially siliciclastic infill deposits) and also with exokarstic deposits; in upper parts of the karst profile the endokarstic carbonate breccias are mixed frequently with the exokarstic ones and it is then difficult to differentiate between them.

The origin of these breccias is related to internal gravitational collapses and crumbling of the caves, the process being triggered by erosion, dissolution and subsequent mechanical instability of cave walls, ceilings, floors and pillars.

Siliciclastic infill deposits

This facies comprises both fine- to medium-grained sandstones and lutites that very often occur intercalated in the cave interiors. Thin layers exclusively formed of lutites coat locally walls and floors. Sandstones and intercalated mudstones occur as 20–60 cm thick tabular bodies that can be followed for up to 1–5 m in length (Fig. 6D). Lutite intercalations range from 2 to 6 cm in thickness and are commonly eroded by the overlying sandstone layers. Locally, these materials are associated and/or intercalated with chemical deposits (cements and speleothems). These siliciclastic infill facies are restricted to the southern parts of the study area (Fig. 4A).

The petrological characteristics of the sandstones are quite similar to those of the exokarstic siliciclastic deposits and the clastic deposits that form the Upper Unit covering the karst surface. On this basis, we propose an allochthonous origin for these deposits, which is supported by the recognition of sandstone-filled conduits connecting internal parts of the karst system with the overlying stratigraphic unit (petrographic and mineralogical studies confirm this proposal). However, some lutite occurrences have not so clear an origin.

Speleothems

Although quantitatively not very important, speleothems are widely distributed along the area of paleokarst occurrence, especially to the N and NW.

Two speleothem types, fibrous and banded, are distinguished. Both types occur as more-or-less continuous coatings on cave floors, walls and ceilings. Besides these occurrences, speleothems are found as fragments in endo- and exokarstic breccias; surprisingly, the thickest (up to 30 cm) specimens of speleothems have been found within the breccia facies (Fig. 6C).

Fibrous speleothems show a palisade fabric formed by anastomosing bundles of fibrous and columnar calcite crystals with rhomb terminations (similar to those described by Folk & Assereto, 1976; Kendall & Broughton, 1978; Dixon & Wright, 1983). The thickness of this speleothem type ranges from 1 to 15 cm. It is especially abundant in upper levels of the paleokarst section, where it occurs either on host-rock carbonate or even on siliciclastic infill deposits.

Banded speleothems display millimetre to centimetre thick lamination and usually are paler coloured than the fibrous speleothems. They are typically composed of columnar (palisade) and dendritic crystals of calcite (probably after aragonite). Banded speleothems are less abundant than the fibrous ones and form coatings or flowstones on different substrates (host-rock carbonates, breccias, other speleothems).

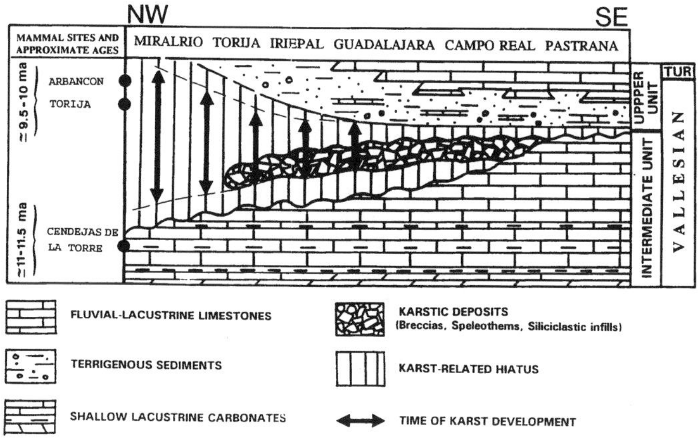

Fig. 7. Summary schematic diagram of karst development and related Miocene units. Paleogeographic distribution of the paleokarst is represented by different localities in the uppermost part of the diagram. Approximate time intervals are indicated by the stratigraphic position of mammal sites.

Discussion

Both endokarstic and exokarstic morphologies and deposits in the intra-Vallesian paleokarst reflect a rather complex history of solution and further collapse processes. These processes resulted in partial destruction of the primary structure of the carbonate strata in which karst developed as well as in the transformation of the initial composition (low magnesian calcite, high magnesian calcite, dolomite, gypsum) and texture of the carbonates. Surficial erosion and sedimentary infill caused porosity reduction (dekarstification in the sense of Smart & Whitaker, 1991) although many of the products created by karst processes (carbonate breccias and, to some extent, siliciclastic deposits) retain considerable porosity.

The intra-Vallesian paleokarst is a shallow, tabular-shaped and depositionally controlled karst and is considered to be a buried (exhumed in northern parts) karst that formed soon after the deposition of carbonates of the Intermediate Unit. These assessments are supported by the geometric relationships between the paleokarst surface and the overlying (Upper Unit) deposits. The diagenetic facies represented by this paleokarst fits better those features characteristic of karstification under wet ('humid') climatic conditions (Wright, 1991a) as no clear evidence of deposits reflecting arid conditions (for instance, calcrete crusts) has been found associated with the paleokarst. Paleoclimate curves proposed for the Iberian Peninsula during the Miocene (López-Martínez *et al.*, 1987) indicate that relatively warm and moist conditions prevailed in the area until the upper Late Vallesian.

The paleotectonic context of the Madrid Basin during the Vallesian (De Vicente *et al.*, Chapter C2) provides some clues for understanding constraints on paleokarst formation in continental domains. The generation of paleokarst involves emergence of previously deposited materials and further modification of the depositional sedimentary pattern in the basin. In our case, the development of paleokarst was related to a change from a compressive to an extensional deformation regime (Fig. 3) in the basin.

The paleokarst was associated with an unconformity that separates two depositional sequences (Intermediate and Upper Units). This stratigraphic discontinuity is interpreted as a third-order regional unconformity according to the hierarchy proposed by Esteban (1991). The paleokarst surface varies in space and time across the basin. The stratigraphic relationships between the paleokarst and the underlying unit seem to indicate that both development and fossilization of the paleokarst were diachronous (Fig. 7). As mentioned above, the time interval represented by the paleokarst is not precisely fixed due to lack of chronologic information within the deposits that cover the unconformity. The longest karstification periods, probably spanning 1–2 million years, occurred in the northern and central parts of the basin (Fig. 7).

As a result of subaerial exposure of the carbonates of the top of the Intermediate Unit, a paleorelief formed and induced rearrangement of depositional systems. A main paleogeographical development was the formation of valleys with a network of approximately N–S fluvial streams, probably corresponding to an exorheic pattern. The distribution of the paleokarst was restricted to areas in which carbonates of the top of the Intermediate Unit were present (Fig. 3). Some areas with these carbonates lack paleokarst because they did not undergo emergence and record continuity in sedimentation.

Conclusions

The case study provides a good example of paleokarst formation in continental formations and its relationship with stratigraphic discontinuities. In our opinion, the most relevant aspect is that paleokarst formation was mainly controlled by tectonic readjustments in the basin. Thus the paleokarst development is thought to have resulted from a drastic change from compressive to extensional strains leading to new depositional patterns and rearrangement of the base level in the basin. This mechanism is suggested as a model for other basins filled by terrestrial successions in which eustatic changes of base level cannot be invoked.

Acknowledgments

This work was supported by DGICYT projects PB89–0032 and PB89–0047.

References

Bosak, P., Ford, D.C., Glazek, J. and Horacek, I. (1989). *Paleokarst – a systematic and regional review*. Academia Praha/Elsevier, Prague and Amsterdam, 725 pp.

Calvo, J.P., García del Cura, M.A. and Ordoñez, S. (1980). Fábricas diagenéticas, retrodiagenéticas y karstificación en calizas continentales (sector NE de la Cuenca de Madrid. *Rev. Inst. Inv. Geol.*, **34**: 135–148.

Calvo, J.P., Ordoñez, S. and García del Cura, M.A. (1984). Caracterización sedimentológica de la Unidad Intermedia del Mioceno de la zona Sur de Madrid. *Rev. Mat. y Proc. Geol. Madrid*, **2**: 145–176.

Calvo, J.P., Alonso-Zarza, A.M. and García del Cura, M.A. (1989). Models of Miocene marginal lacustrine sedimentation in the Madrid Basin (Central Spain). *Palaeogeogr., Palaeoclimatol., Palaeoecol.*, **70**: 199–214.

Calvo, J.P., Hoyos, M., Morales, J. and Ordoñez, S. (1990). Neogene stratigraphy, sedimentology and raw materials of the Madrid Basin. *Paleontol. Evol.*, **2**: 63–95.

Cañaveras, J.C. (1991). Caracterización petrológica y geoquímica del karst del techo de la Unidad Intermedia de la Cuenca de Madrid. Tesis de Licenciatura. 184 pp.

De Vicente, G., Calvo, J.P. and Alonso-Zarza, A.M. (1990). Main sedimentary units and related strain fields of the Madrid Basin (central Spain) during the Neogene. *IX R.C.M.N.S. Congress, Barcelona*: 121–122.

Dixon, J. and Wright, V.P. (1983). Burial diagenesis and crystal diminution. The origin of crystal diminution in some limestones from South Wales. *Sedimentology*, **30**: 537–546.

Esteban, M. (1991). Paleokarst: Practical applications. In Wright, V.P. (ed.), *Paleokarst and paleokarstic reservoirs. P.R.I.S. Occ. Publ. Series, 02,* Univ. of Reading. 89–119.

Esteban, M. and Klapa, C. (1983). Subaerial exposure environment. *Am. Assoc. Petrol. Geol. Mem.*, **33**: 1–54.

Folk, R.L. and Assereto, R. (1976). Comparative fabrics of length-slow and length-fast calcite and calcitized aragonite in a Holocene speleothem, Carlsbad Caverns, New Mexico. *J. Sed. Petrol.*, **46**: 486–496.

Ford, D.C. and Williams, P.W. (1989). *Karst geomorphology and hydrology*. Unwin Hyman, MA, 601 pp.

James, N.P. and Choquette, P.W. (eds.) (1988). *Paleokarst*. Springer Verlag, New York, 416 pp.

Kendall, A.C. and Broughton, P.L. (1978). Origin of fabrics in speleothems composed of columnar calcite crystals. *J. Sed. Petrol.*, **48**: 519–538.

López-Martínez, N., Agusti, J., Cabrera, L., Calvo, J.P., Civis, J., Corrochano, A., Daams, R., Diaz, M., Elizaga, E., Hoyos, M., Martínez, J., Morales, J., Portero, M., Robles, F., Santiesteban, C. and Torres, T. (1987). Approach to the Spanish continental Neogene synthesis and paleoclimatic interpretation. *Ann. Inst. Geol. Publ. Hungarici*, **70**: 383–391.

Ordoñez, S., García del Cura, M.A., Hoyos, M. and Calvo, J.P. (1985). Middle Miocene paleokarst in the Madrid Basin (Spain). A complex karstic system. *6th European Reg. Meet. IAS. Lleida, Abstract*: 624–627.

Sanz, E., Calvo, J.P., García del Cura, M.A. and Ordoñez, S. (1991). Origin and diagenesis of calcretes in Upper Miocene limestones, Southern Madrid Basin, Spain. *Rev. Soc. Geol. Esp.*, **4**: 127–142.

Sanz, E., Calvo, J.P. and Ordoñez, S. (1992). Litoestratigrafía y sedimentología del Neógeno en el sector S de la Cuenca de Madrid (Mesa de Ocaña). *III Cong. Geol. de España y VIII Cong. Latinoamericano de Geología, Salamanca. Actas*, **1**: 212–216.

Sese, C., Alonso-Zarza, A.M. and Calvo, J.P. (1990). Nuevas faunas de micromamíferos del Terciario continental del NE de la Cuenca de Madrid (Prov. de Guadalajara). *Estud. Geol.*, **46**: 433–451.

Smart, P.L. and Whitaker, F.F. (1991). Karst processes, hydrology and porosity evolution. In Wright, V.P. (ed.), Paleokarst and paleokarstic reservoirs. P.R.I.S. Occ. Publ. Series, 02, Univ. of Reading. 1–55.

Van Der Meulen, A.J. and Daams, R. (1992). Evolution of Early–Middle Miocene rodent faunas in relation to long-term palaeoenvironmental changes. *Palaeogeogr., Palaeoclimatol., Palaeoecol.*, **93**: 227–253.

Wright, V.P. (1982). The recognition and interpretation of paleokarst: two examples from the Lower Carboniferous of South Wales. *J. Sed. Petrol.*, **52**: 83–94.

Wright, V.P. (1991a). Paleokarst: types, recognition, controls and associations. In Wright, V.P. (ed.), *Paleokarst and paleokarstic reservoirs. P.R.I.S. Occ. Publ. Series. 02,* Univ. of Reading. 56–88.

Wright, V.P. (ed.) (1991b). *Paleokarst and paleokarstic reservoirs. P.R.I.S. Occ. Publ. Series, 02,* Univ. of Reading. 158 pp.

C5 Tectono-sedimentary analysis of the Loranca Basin (Upper Oligocene–Miocene, Central Spain): a 'non-sequenced' foreland basin

J.J. GÓMEZ, M. DÍAZ-MOLINA AND A. LENDÍNEZ

Abstract

The western margin of the Loranca Basin is the Sierra Altomira, which separates it from the Madrid Basin. The eastern margin is formed by the Iberian Range, which supplied most of the Tertiary continental sediment that fills the Loranca Basin. This basin fill is divided into five units (I to V), ranging from Middle Eocene to Late Miocene in age. The tectonic and sedimentary evolution of the Basin has been analysed using outcrop and seismic mapping of the units and their folding, faulting and unconformities.

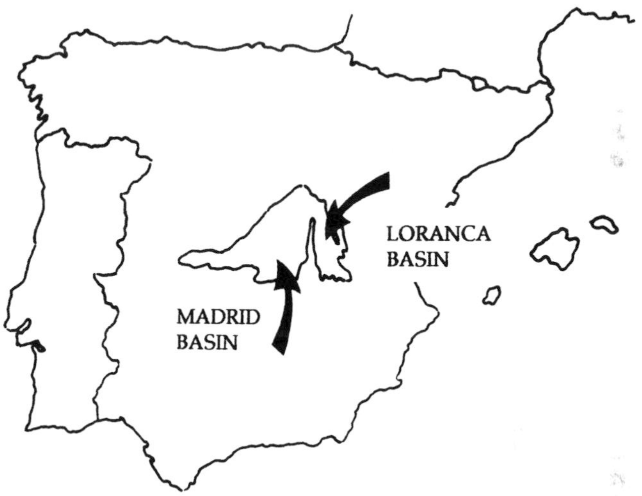

Fig. 1. Location map of the Loranca Basin which, together with the Madrid Basin, forms the Tajo Basin.

Introduction

The Loranca Basin, located in Central Spain, is separated from the Madrid Basin by the Sierra Altomira and bounded to the east by the Iberian Range (Fig. 1). The basin was filled by pre- and synorogenic Tertiary continental sediments, mainly sourced from the Iberian Range fold and thrust belt, which acted as the main active foreland.

The general stratigraphy of the Loranca Basin has been described by several authors (Vilas-Minondo & Pérez González, 1971; Meléndez Hevia, 1971; Viallard, 1973; Díaz-Molina, 1974; García-Abbad, 1975; Díaz-Molina & López-Martínez, 1979; Díaz-Molina *et al.*, 1985, 1989; Torres & Zapata, 1986–1987a, b; Torres *et al.*, 1992).

Knowledge of the surface geology of the basin has been improved in the last few years by extensive regional mapping by ITGE. The results of this work have been reviewed and compiled in Fig. 2. The subsurface geology of the whole Tajo Basin has been outlined by Querol (1989), and the area has been included in wider tectonic interpretations related to the Tertiary orogenic belts of eastern Spain (Guimerá & Alvaro, 1990; Hernaiz *et al.*, 1990; Banks & Warburton, 1991).

General structure of the basin

The northern half of the Loranca Basin was considered in the 1970s as an area of interest for oil exploration. Consequently, data, provided by a grid of seismic lines and a well (Torralba-1) which intersected the sediments of the basin down to the Paleozoic basement, are now available.

Interpretation of the seismic lines has allowed detailed reconstruction of the structure of the bottom of the Tertiary basin. A mapped reflector is located near the top of the Villalba de la Sierra Fm., regionally considered to contain the Cretaceous–Tertiary boundary (Fig. 3).

Subsurface mapping and structural trends

Subsurface contouring reveals the presence of a fold and thrust belt at the bottom of the Tertiary basin, where three main structural orientations can be observed. The western portion of the basin is dominated by N–S- to NNE–SSW-trending structures, parallel to the outcropping Altomira thrust belt. In the central portion of the basin, NE–SE trending structures sub-parallel to the Iberian Range fold and thrust belt are the prevailing trends, and in the eastern part structural orientations are mainly NNW–SSE, comparable with the Bascuñana thrust belt. As a result of the

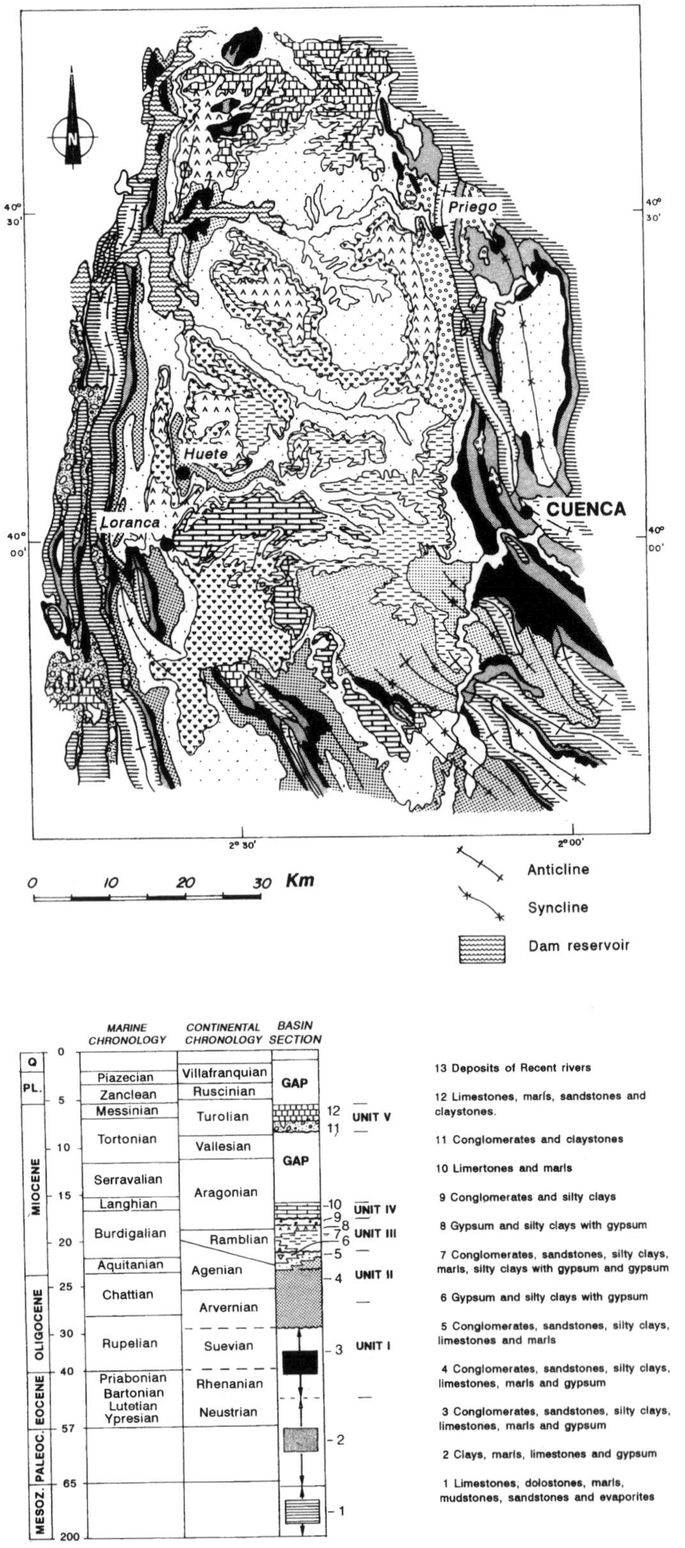

Fig. 2. Geological compilation map of the Loranca Basin and key representing the stratigraphic basin section.

convergent structural trends of its boundaries, the basin has a roughly triangular shape (Figs. 1 and 3).

Thrust and fold system

Basement and Permo-Triassic rocks constitute the relatively autochthonous basement and cover. These tectonic units are affected by late Hercynian normal faulting, which generated local half-graben basins, filled by Permo-Triassic deposits (Fig. 4). There was only occasional inversion tectonics when some of these normal faults affecting the basement, or Alpine normal listric faults active during part of the Mesozoic, were rejuvenated as thrust faults. In general, the basement tends to become shallower to the west, and only rarely has it been folded by Alpine deformation (Fig. 4, A–A′).

Triassic Keuper evaporites contain the sole thrust constituting the main décollement level, which extends at least over all those parts of the basin covered by the seismic surveys. The presence of another deeper basal detachment level has been suggested by many authors (Banks & Warburton, 1991), on the basis of a low-velocity zone located at 7–11 km depth (Banda *et al.*, 1981). As can be mapped at the surface, some subsidiary flats and thrusts have been occasionally developed in other incompetent lithologies higher in the section, such as the Upper Cretaceous marly units. The thrust system developed at the bottom of the basin should be generally considered as blind, since no thrust faults have been reported in surface mapping. However, there are indications of late minor activity of some of the thrust faults mapped at depth, and this late faulting reaches the surface.

Most of the thrust ramps are facing to the W in the area close to the Sierra Altomira, to the SW in the central portion of the basin and to the W–SW in the area near the Bascuñana thrust belt. In most cases, displacements along thrust planes are very small. However, the convergence of the basin borders towards the N caused more important displacement in thrust faults, and consequently higher values in shortening for the northern area (section A–A′ in Fig. 4) than in the central and southern areas covered by seismic survey (sections B–B′ and C–C′, Fig. 4). Associated with the thrust system there are back-thrusts on which displacement values are also quite small, but enough to form pop-up structures.

Folds in the central portion of the basin are frequently gently dipping. However, more steeply dipping fault-propagation folds are common in the front and back of the belt, coinciding with the western and eastern margins of the basin. Partly because of salt migration, some of the ramping anticlines are typically box shaped. Folds nearly perpendicular to the general structural trends can be related to lateral ramps or to wrench faults, visible in the surface to the east of the basin.

Salt tectonics

Salt tectonics also plays an important role in synsedimentary deformation and consequently in basin configuration. Halokinetic behaviour of Triassic salt located close to the décollement surface conditioned migration of evaporites towards the anticlinal

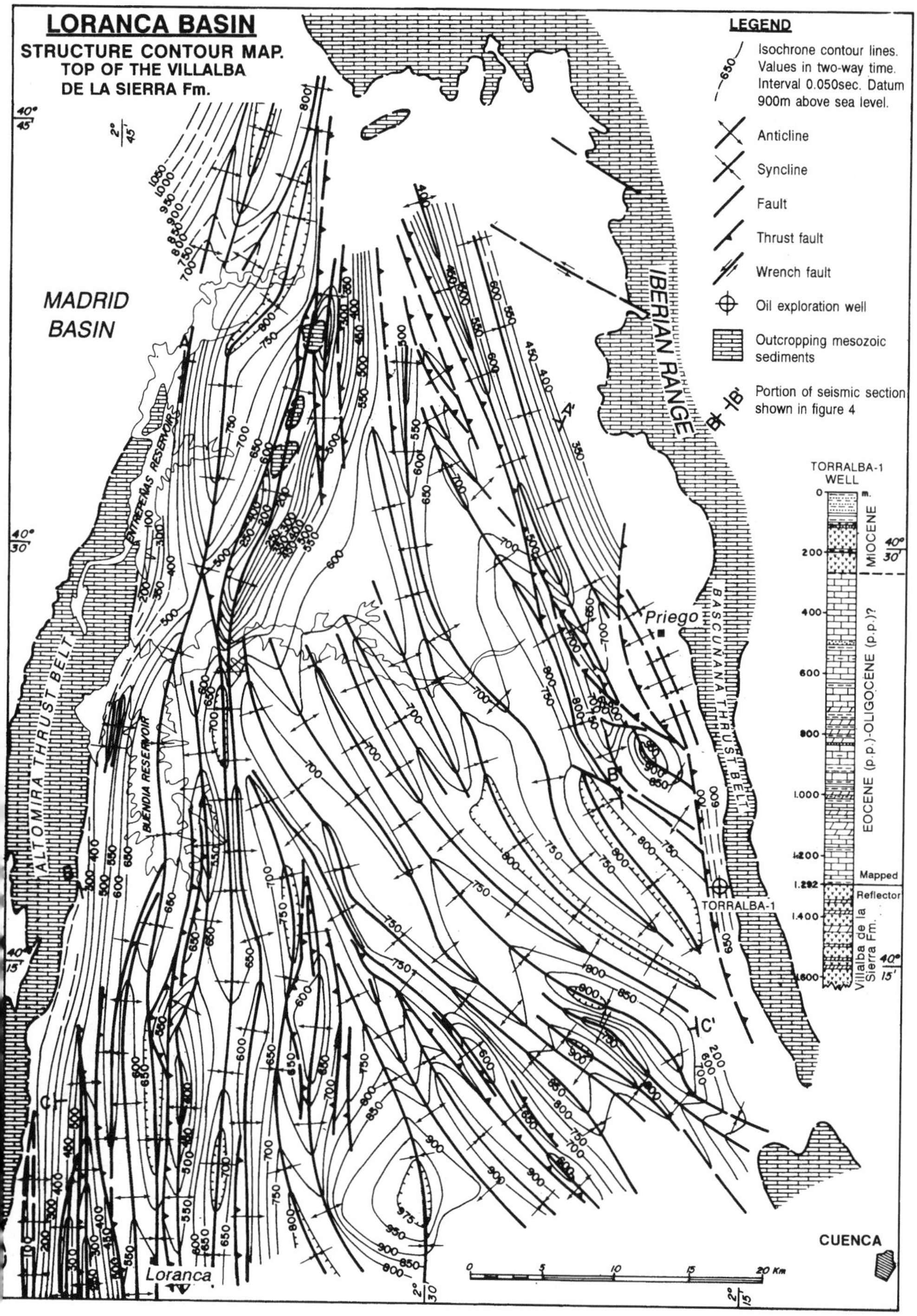

Fig. 3. Structure contour map of a reflector near the top of the Villalba de la Sierra Fm. Log corresponds to the upper portion of the Torralba-1 well. Transects A–A′, B–B′ and C–C′ are shown in Fig. 4.

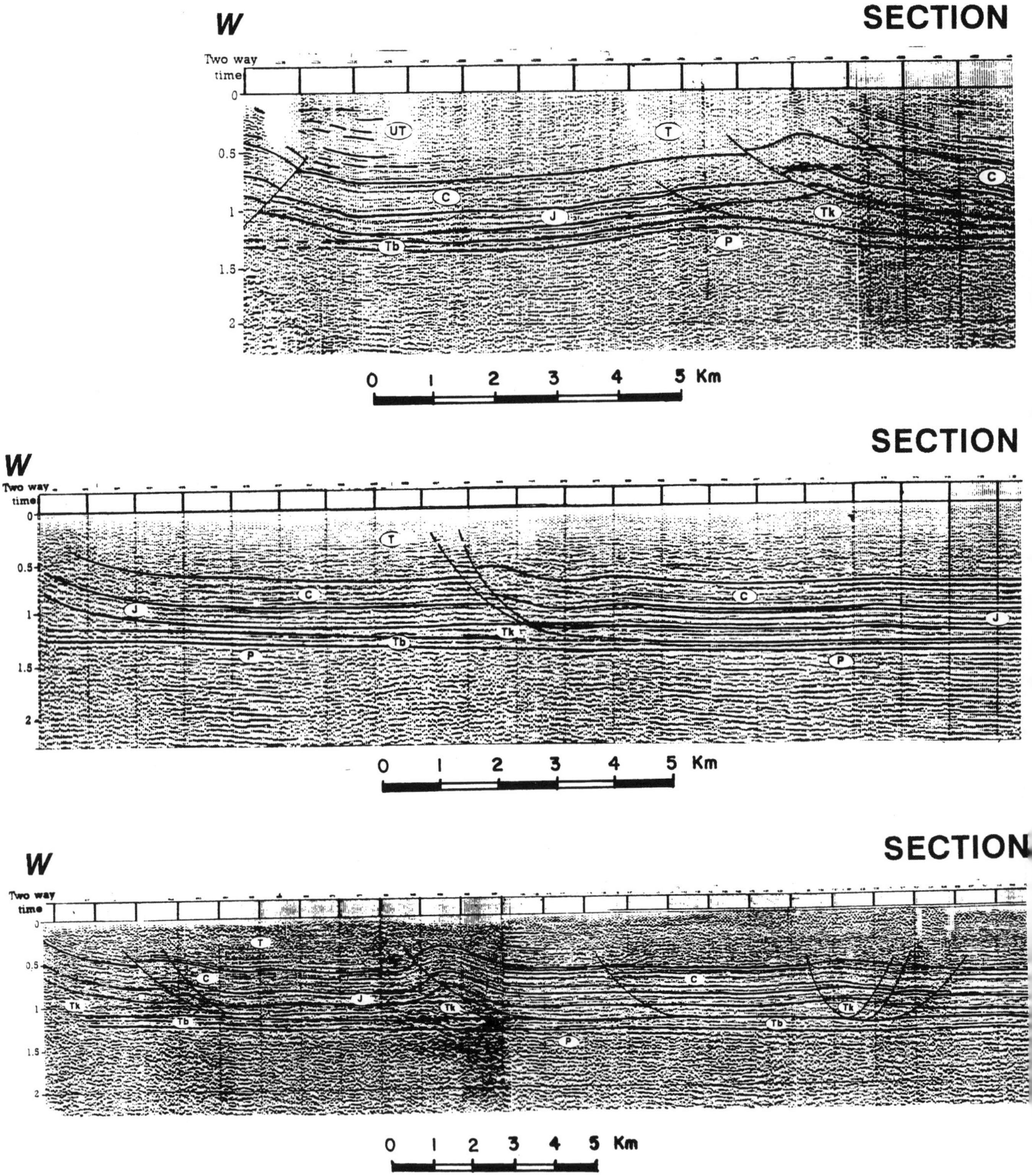

LEGEND

(T) *Tertiary sediments (continental facies)*

(UT) *Unconformity within Tertiary sediments*

(C) *Cretaceous (carbonates, siliciclastics & evaporites)*

Fig. 4. Seismic sections representative of the northern (A–A′), central (B–B′) and southern (C–C′) parts of the structure contour map shown in Figure 3.

A-A'

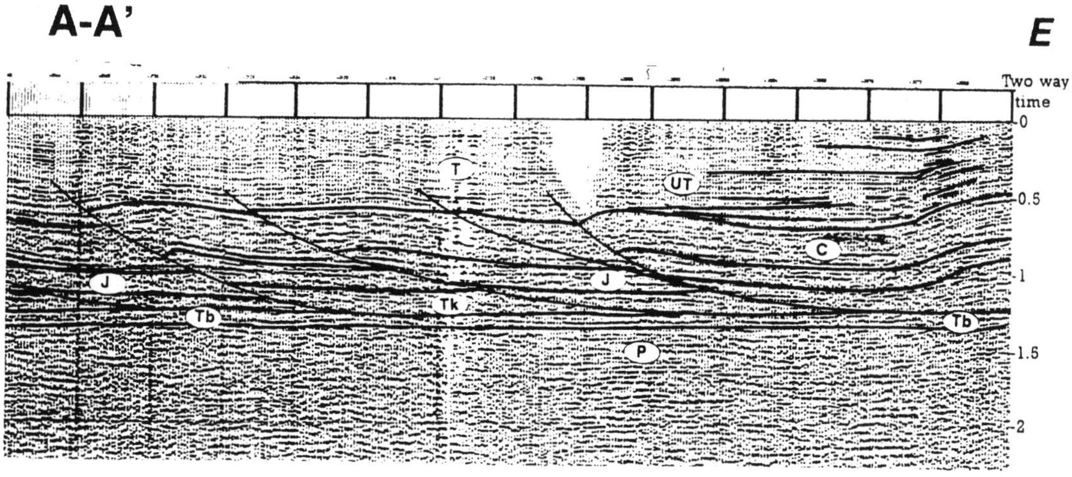

B-B'

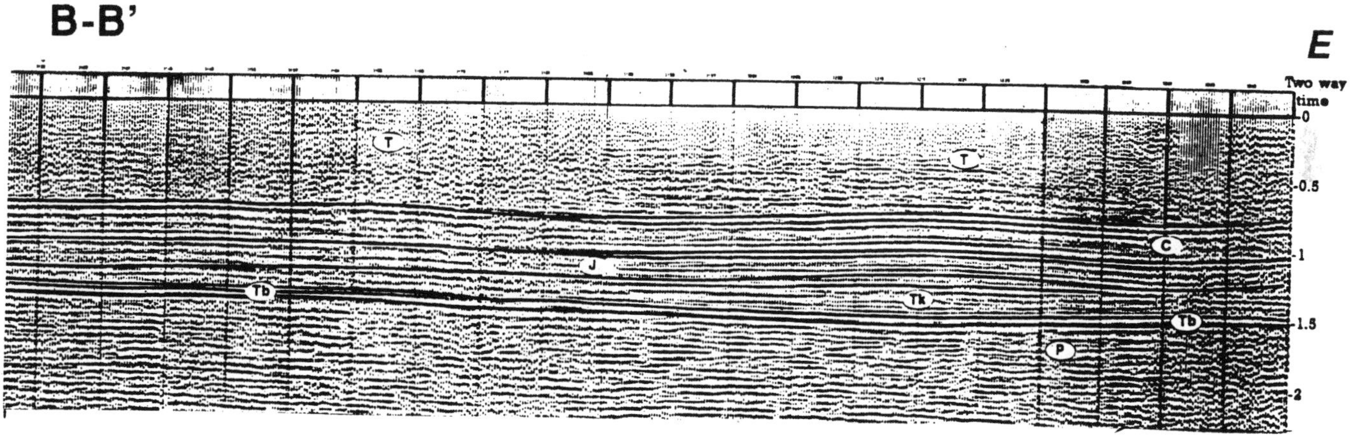

C-C'

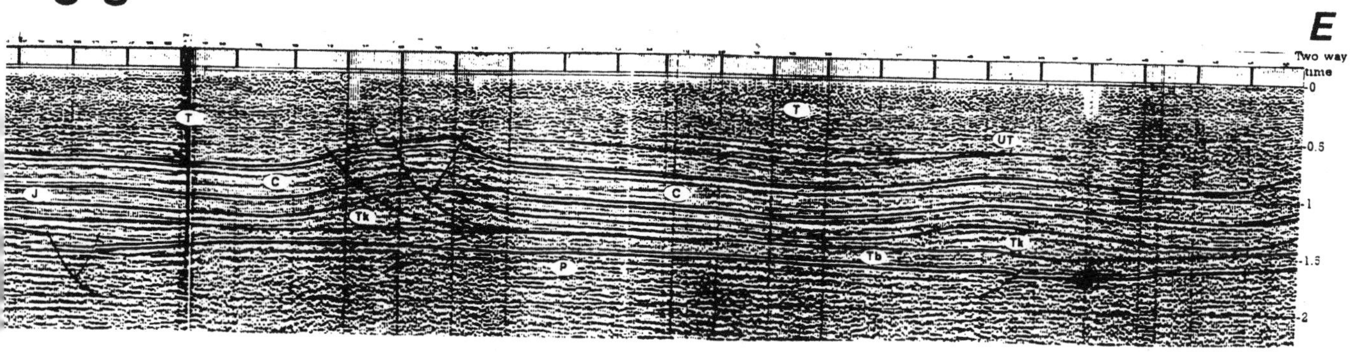

(J) *Jurassic (carbonates & marls)*

(Tk) *Triassic Keuper facies (anhydrite with interbedded salt & lutites)*

(Tb) *Triassic & Permian (Buntsandst. siliciclastics & Muschelkalk carbonates)*

(P) *Paleozoic (metamorphic & sedimentary rocks)*

structures. Salt redistribution generated primary and secondary rim synclines (Hopkins *et al.*, 1966), which, coupled with folding and thrusting, strongly controlled the location of depocentres, facies distribution, geometry of internal unconformities and general basin evolution.

The western limit of the fold–thrust belt, interpreted by some authors as the front of the Pyrenean System (Banks & Warburton, 1991) could have been influenced by the conjuction of two main factors: a shallower position of the basement towards the west (Fig. 4), and a probable decrease in the proportion of evaporites in the Keuper section, which were replaced by siliciclastics, which hindered the extension of the décollement processes.

Tectono-sedimentary analysis of the basin

Four unconformity-bounded stratigraphic units have been distinguished in the Tertiary succession of this basin. The unconformities seem to be correlative with some of the sedimentary breaks identified in many of the Spanish Tertiary basins (López-Martínez *et al.*, 1987).

Unit I

This unit, equivalent to the Lower Detrital Unit (Díaz-Molina, 1974) accumulated in a sedimentation area that was more extensive than the present Loranca Basin. Deposition of this unit took place from the Eocene to Late Oligocene and was partly simultaneous with the generation of the early structures causing the uplift of the Sierra Altomira. As a result, the Loranca Basin was separated from the Madrid Basin during deposition of this unit.

Unit I is dominated by fluvial deposits showing paleochannel trends to the W, N and SE. The stratigraphic succession is composed of conglomerates, sands, sandstones, silty clays, limestones, marls and gypsum. Thickness values reach 270 m, reflecting relatively low sedimentation rates. The unit is bounded by unconformities and probably contains internal breaks. However, angles of internal unconformity are too low to be clearly observed in field outcrops and neither internal unconformities nor sedimentary breaks have been detected at a basinal scale.

Unit II

Unit II, equivalent to the Upper Detrital Unit (Díaz-Molina, 1974) is composed of two depositional systems named the Tórtola and Villalba de la Sierra. This unit, up to 900 m thick, was deposited from Late Oligocene to Early Miocene, onlapping previously deposited units.

Syn-sedimentary tectonic activity generated progressive unconformities on the limbs of some anticlinal structures as well as along the basin margins (Fig. 4). A corridor connecting the Loranca and Madrid basins, located to the north of the Sierra Altomira persisted during part of the deposition of this unit. Three main stages of evolution have been distinguished, and are represented in Fig. 5 as A, B and C.

During stage A the Tórtola and Villalba de la Sierra fans occupied most of the basin. Due to the growth of early NW-trending folds in the southern area, the Tórtola fan did not reach part of the southwestern margin of the basin, where the multiple channel system was laterally replaced by floodbasin deposits, including lacustrine carbonates and alluvial fan deposits sourced from the Altomira thrust belt (Fig. 5A).

Stage B was characterized by the development of NW-trending structures in the central portion of the basin. It represents a progressive abandonment of the fluvial systems and an increase of supply to local alluvial fans sourced from the Altomira and Iberian Range fold and thrust belts. River systems were controlled by primary synclines, while floodbasin deposits were mainly associated with anticlinal structures. In the distal portion of the alluvial fans, wet areas, where deposition of evaporites and lacustrine sediments took place, were developed.

Stage C represents a slow down in tectonic activity, which conditioned a gradual expansion of lacustrine and playa-lake sediments. Gypsum deposits were dominant, with minor fluvial deposits (Fig. 5C), overlapping previously deposited units.

Unit III

Units III and IV form the so-called Terminal Unit (García-Abbad, 1975). Unit III corresponds to tectonic rejuvenation of the Iberian Range, which sourced the Valdeganga depositional system, of similar facies to the ones present in Unit II. The Valdeganga system accumulated a stratigraphic succession up to 100 m thick, which occupied most of the Loranca Basin during Early Miocene times (Fig. 6A), although the facies distribution was not so strongly influenced by basin internal folds as in the second stage of Unit II (Fig. 5B).

Decrease in tectonic activity within the basin resulted in expansion of gypsum facies over most of the basin (Fig. 6B). The Altomira western margin was virtually inactive at this time (Fig. 6A, B) whilst the growth of the N–NW-trending Bascuñana thrust belt in the eastern margin started. The uplift of the Bascuñana structure favoured migration of Triassic evaporites towards the core of this structure (824 m intersected in Torralba-1 well) generating, for the first time, a well-defined eastern margin for the Loranca Basin, as well as an active sub-basin elongated along the western limb of the Bascuñana belt (Fig. 6C). This sub-basin was filled up with a 350 m thick succession of local coalescing alluvial fan deposits showing internal progressive unconformities. The lateral stratigraphic equivalents are uncertain due to correlation difficulties. They could be represented by a thin succession of gypsiferous rocks or by a sedimentary break affecting most of the basin as indicated, in some areas, by processes of gypsum karstification, which are synchronous with this interval of deformation.

Unit IV

This sedimentary unit consists of lacustrine carbonates (up to 60 m thick) deposited during part of the Aragonian, and

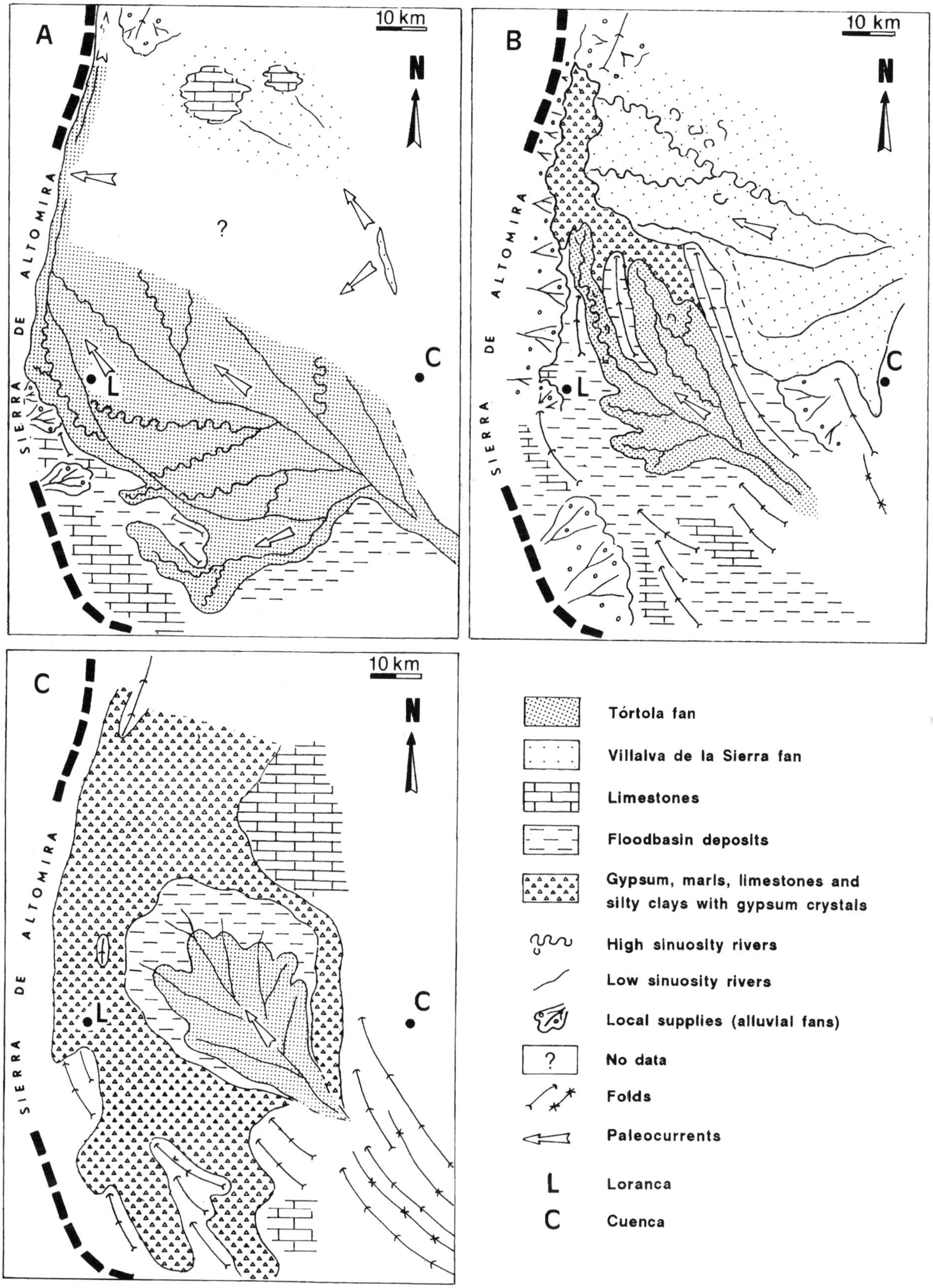

Fig. 5. Paleogeographic maps representing the three stages of the evolution of Unit II. Villalba de la Sierra facies reconstruction in the northern portion of map A is based on drilling data published by Torres *et al.* (1992). See text for explanation.

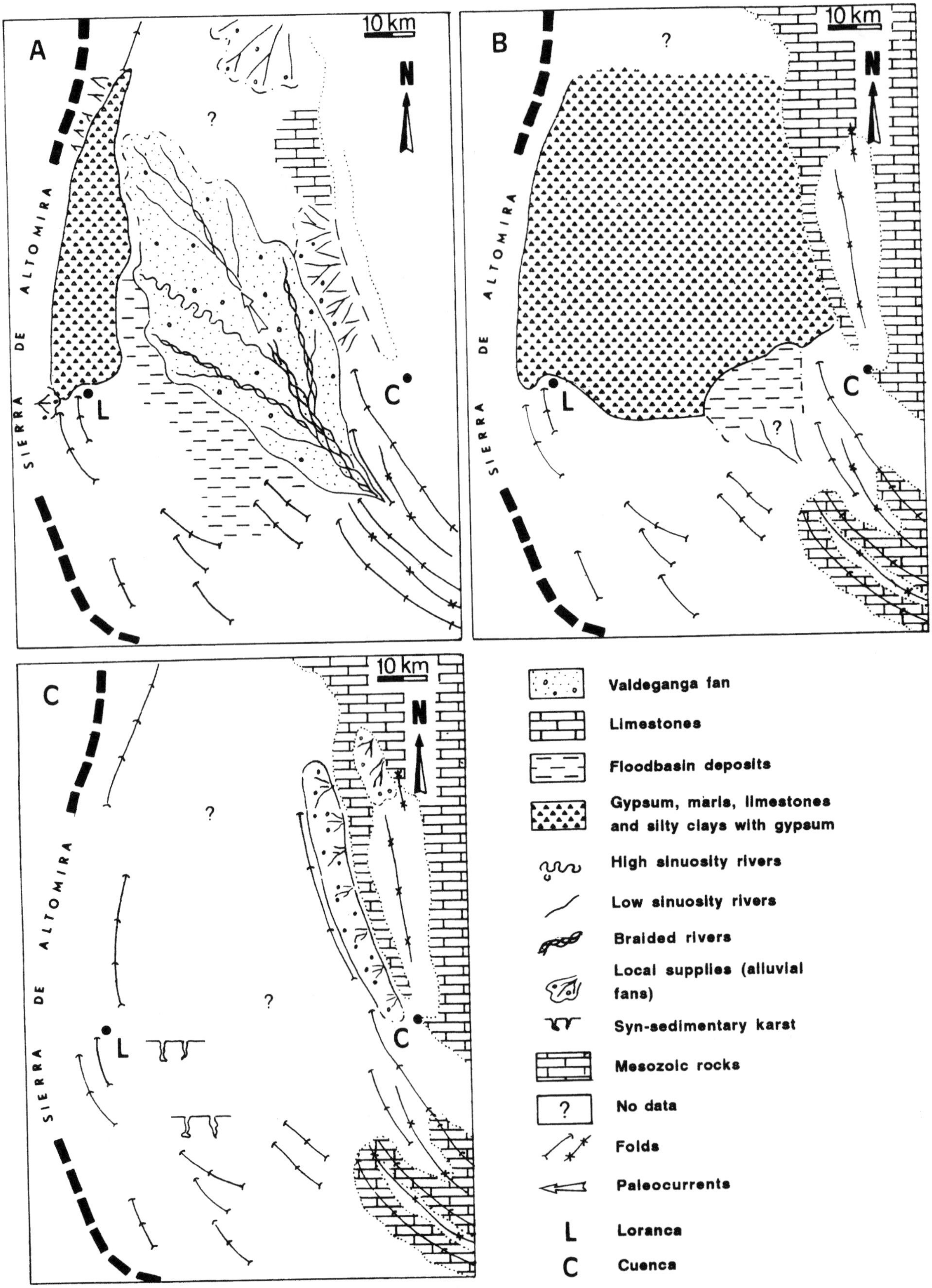

Fig. 6. Paleogeographic maps corresponding to tectonic and facies evolution during deposition of Unit III. See text for explanation.

extending over the entire basin. Although the unit has been folded along the eastern Bascuñana margin of the basin, where it onlaps the exposed Mesozoic rocks (Arribas-Mocoroa *et al.*, 1990), it generally corresponds to a relative pause in tectonic activity. The unit preceeds an important gap in sedimentation during part of the Aragonian and the Vallesian.

Unit V

Exposures of this unit, Turolian in age, are only found in the Sierra Altomira and surrounding areas as well as in the southern and northern areas of the basin (Fig. 2). In the northern area the stratigraphic sequence is formed by fluvial and lacustrine sediments, while only lacustrine facies occur in the southern outcrops, where there is a total thickness of 70 m.

Reactivation of the Sierra Altomira caused the development of alluvial fans, which were deposited at the edge of the Madrid and Loranca Basins as well as in the elongated depressions located between the growing anticlines of the Altomira belt. Fluvial, alluvial fan and lacustrine deposits constituting Unit V are always involved, whether in thrusting within the western margin, or affected by late folding in the interior of the basin. Even the youngest Tertiary sediments of the basin are deformed, indicating that deposition was synorogenic.

Conclusions

The Loranca Basin can be considered as a Foreland or A-Subduction basin (Bally & Snelson, 1980) floored by continental crust, filled with shallow marine to continental deposits, mainly derived from the Iberian Range fold–thrust belt, and having a structural style dominated by compressional folds with local intraformational unconformities attesting syndepositional tectonism (cf. Miall, 1984).

Tectono-stratigraphic analysis performed in the Loranca Basin allows us also to draw some conclusions concerning general questions about the sequence of thrusting, continuity/discontinuity of deformation processes and the changes of vector orientations with time.

It is generally accepted by many authors that deformation in most fold–thrust belts migrates following a specific sequence of movement. The sequence can be either towards the foreland, giving a piggyback, forward or 'cratonward' sequence of thrusting, or towards the fold–thrust belt, giving a backwards or hindward thrust sequence (Boyer & Elliot, 1982; Bally & Oldow, 1985). Most of the described examples accept the piggyback model as the most widely developed in foreland basins related to fold–thrust belts. However, paleogeographical reconstruction of the Loranca Basin shows a sequence of folding and thrusting which does not fit with these generally accepted models.

The first movements in the Loranca Basin (Eocene (part) to Late Oligocene (part)) initiated the development of the N–NE-trending Altomira belt, which is located in the most forward position of the system. The direction of maximum compression is estimated to have been nearly W–E, with a cratonwards main displacement. The growth of Altomira defined the early western boundary of the basin.

The movements continued during deposition of Unit II (part of Late Oligocene to part of Early Miocene), generating mainly NW-trending forward structures (vergent to the SW) which developed within the basin area. These structures are in a relatively hindward position with respect to the previously formed structures. Decrease of tectonic activity towards the upper portion of Unit II resulted in a progressive abandonment of the fluvial systems, an increase of supply associated with local alluvial fans, and a gradual expansion of lacustrine and playa-lake sediments. During this interval the maximum compressional direction is supposed to have been oriented mainly NE–SW, representing a significant change in the orientation of the shortening axis with respect to the previous unit. Discontinuity in the deformational process is indicated by expansion of gypsum deposits over most of the basin during stage C of Unit II. Cessation or at least attenuation of deformation probably contributed to the production of the widespread unconformity separating units II and III.

Unit III (Early Miocene) represents a rejuvenation of the Iberian Range belt, east of the basin, and later on, a relatively quiet period, with the exception of the rising of the Bascuñana structure (Fig. 3). The vergence is to the west, indicating a forward displacement as in previous units, and a further change in the direction of maximum compression direction to a W–NW orientation is also recorded. The growth of the Bascuñana anticlinal structure was partially implemented by the migration of salt from adjacent structures.

Discontinuity in tectonic deformation is again indicated by the relative quiescence recorded during Unit IV (Aragonian), only interrupted by the growing of the Bascuñana structure. Unconformities bounding the top and bottom of this unit are also present all over the basin and a hiatus affecting part of the Aragonian and the Vallesian has been demonstrated by biostratigraphical data.

Finally, tectonic activity re-started during deposition of Unit V (part of Turolian), which was involved in folding and thrusting. The structures constituting the Altomira and the bottom of the Loranca Basin fold–thrust belts were rejuvenated. This renovation of tectonic activity distinctly disrupted the general backwards sequence of thrusting within the basin in previous units.

As reflected in the literature, many foreland basins are interpreted as piggyback or backward basins on the basis of the sequence of migration of thrusting. In terms of kinematics, these are the most simple patterns of development. But, there is a group of foreland basins, characterized by the Loranca Basin, where the pattern of evolution is not as simple as in the general models. Each basin must be analysed independently, in order to reconstruct its own evolutionary pattern, and we propose the Loranca Basin as a reference example of a 'non-sequenced' foreland basin.

Acknowledgments

Research funded in part by the Commission of the European Communities in the framework of the Joule Programme (1989–1992), Subprogramme: Hydrocarbons.

References

Arribas-Mocoroa, M.E., Martínez-Salanova, J. and Díaz-Molina, M. (1990). Sedimentología de una unidad carbonatada lacustre del Mioceno Inferior. Sector nororiental de la cuenca de Loranca (Provincia de Cuenca, España). *Bol. Geol. Min.* **101**(6): 858–871.

Bally, A.W. and Oldow, J.S. (1985). Structural styles and the evolution of sedimentary basins. *AAPG short course, held in conjunction with the SEPM mid-year meeting*. 238 pp.

Bally, A.W. and Snelson, S. (1980). Realms of subsidence. In A.D. Miall, ed., *Facts and principles of world petroleum occurence, Can. Soc. Petrol. Geol. Mem.*, **6**: 9–94.

Banda, E., Suriñach, E., Aparicio, A., Sierra, J. and Ruiz de la Parte, E. (1981). Crust and upper mantle structure of the Central Iberian Meseta. *Geophys. J.R. Astron. Soc.* Oxford, **67**: 779–789.

Banks, C.J. and Warburton, J. (1991). Mid-crustal detachment in the Betic system of southern Spain. *Tectonophysics*, **191**: 275–289.

Boyer, S.E. and Elliot, D. (1982). Thrust systems. *Am. Assoc. Petrol. Geol. Bull.* **66**(9): 1196–1230.

Díaz-Molina, M. (1974). Síntesis estratigráfica preliminar de la serie terciaria de los alrededores de Carrascosa del Campo. *Estud. Geol.*, **35**: 241–251.

Díaz-Molina, M. and López-Martínez, N. (1979). El Terciario continental de la Depresión Intermedia (Cuenca). Bioestratigrafía y paleogeografía. *Estud. Geol.* **35**: 149–167.

Díaz-Molina, M., Bustillo-Revuelta, M.A., Capote, R. & López Martínez, N. (1985). Wet fluvial fans of the Loranca Basin (central Spain). Channel models and distal bioturbated gypsum with chert. *I.A.S. 6th European Regional Meeting, Lérida, Spain. Exc. Guide-book*: 149–185.

Díaz-Molina, M., Arribas-Mocoroa, J. and Bustillo-Revuelta, M.A. (1989). The Tórtola and Villalba de la Sierra fluvial fans: Late Oligocene–Early Miocene, Loranca Basin, central Spain. *4th International Conference on Fluvial Sedimentology, Barcelona–Sitges, Spain. Field Trip* 7, 74 pp.

García-Abbad, F. (1975). *Estudio geológico de la región del pantano de Alarcón*. Publ. Fac. Ciencias de la Universidad Complutense de Madrid, 175 pp.

Guimerá, J. and Alvaro, M. (1990). Structure et évolution de la compression alpine dans la Chaîne ibérique et la Chaîne côtière catalane (Espagne). *Bull. Soc. Géol. France.* (**8**), **VI**, **2**: 339–348.

Hernaiz, P.P., Galan, G., Diaz de Neira, A., Enrile, A., López-Olmedo, F., Rey, J., Delgado, G. and Cabra, P. (1990). *Thrust tectonics in the Maestrazgo Region (Eastern Spain)*. INYPSA. Internal report. 5 pp.

Hopkins, H.R., Shumaker, R.C. and Larsen, W.N. (1966). *Salt tectonics*. Esso Production Research Company. Internal report. 47 pp. 65 figs.

Instituto Tecnológico y Geominero de España. *Surface geological maps at the scales of 1:50000 and 1:200000*.

López-Martínez, N., Agustí, J., Cabrera, L., Calvo, J.P., Civis, J., Corrochano, A., Daams, R., Díaz, M., Elízaga, E., Hoyos, M., Martínez, J., Morales, J., Portero, J.M., Robles, F., Santisteban, C. and Torres, T. de (1987). Approach to the Spanish continental synthesis and palaeoclimatic interpretation. *Ann. Ins. Geol. Publ. Hung.*, **LXX**: 383–391.

Meléndez Hevia, F. (1971). *Estudio geológico de la Serranía de Cuenca en relación a sus posibilidades petrolíferas*. Publ. Fac. Ciencias de la Universidad Complutense de Madrid, serie A, 175 pp.

Miall, A.D. (1984). *Principles of sedimentary basin analysis*. Springer-Verlag. 490 pp.

Querol, R. (1989). *Geología del subsuelo de la Cuenca del Tajo*. E.T.S. de Ingenieros de Minas de Madrid. 48 pp.

Torres, T. de and Zapata, J.L. (1986–1987a). Caracterización de dos sistemas de abanicos aluviales húmedos en el Terciario de la Depresión Intermedia (Cuenca–Guadalajara). *Act. Geol. Hisp.*, **21–22**: 45–53.

Torres, T. de and Zapata, J.L. (1986–1987b). Paleotopografía y distribución de paleocorrientes de abanicos aluviales de la Depresión Intermedia (Cuenca–Guadalajara). *Act. Geol. Hisp.*, **21–22**: 56–61.

Torres, T. de, García Cortés, A., Mansilla, H. and Quintero, I. (1992). Upper Oligocene palustrine deposits in the 'Depresion Intermedia' basin (prov. of Cuenca and Guadalajara, central Spain): borehole cores interpretation. *III Congreso Geológico de España y VIII Congreso Latinoamericano de Geología, Salamanca*. Simposios **1**: 149–157.

Viallard, P. (1973). Recherches sur le cycle alpin dans la Chaîne Ibérique Sudoccidentale. Thèse Université Paul Sabatier, Toulouse, 445 pp.

Vilas-Minondo, L. and Pérez-Gonzalez, A. (1971). Contribución al conocimiento de las series continentales de la mesa manchega (Cuenca) *Bol. R. Soc. Esp. Hist. Nat. (Geol.)*, **69**: 103–114.

C6 Paleoecology and paleoclimatology of micromammal faunas from Upper Oligocene – Lower Miocene sediments in the Loranca Basin, Province of Cuenca, Spain

R. DAAMS, M.A. ÁLVAREZ SIERRA, A.J. VAN DER MEULEN AND P. PELÁEZ-CAMPOMANES

Abstract

In the Loranca Basin, sediments are exposed ranging in age from the Middle Eocene to the Turolian. The fossil micromammal record of this considerable time interval is, however, not continuous. For paleoecological study we have chosen a composite section in the Río Mayor valley where stratigraphical control allows us to superpose six faunas, dating from the Late Oligocene to the Early Miocene. In this interval, which covers some 4–5 million years, two relatively dry, and two relatively humid, periods are recognized.

Introduction

A first attempt at a paleoecological approach to Spanish micromammal faunas was made by Van de Weerd & Daams (1978), who recognized an alternation of relatively humid and dry periods during the Neogene of the Calatayud–Teruel Basin. Daams & Van der Meulen (1984) further developed the model, and constructed relative humidity and temperature curves for part of the Neogene of North Central Spain. Additional information and the application of principal component analysis subsequently resulted in the construction of more-detailed relative-humidity and temperature curves (Van der Meulen & Daams, 1992) for part of the Neogene of the Calatayud–Teruel Basin.

The Upper Oligocene and Lower Miocene of the Loranca Basin also yielded numerous micromammal faunas, but our taxonomical study is unfortunately not advanced enough for principal component analysis or the construction of detailed climatic curves. For this study, we have chosen a succession of six localities, which are all located in the Río Mayor valley and whose stratigraphic position can be controlled lithostratigraphically (Fig. 1). All localities are situated in the Tórtola Fluvial Fan System, as characterized by Díaz and Tortosa (Chapter C7). The lowermost locality of this succession (Canales) is of Late Oligocene age (MP28) and the uppermost one (Cabeza Rubia) of Early Miocene (MN2). The time interval covered by this succession is approximately 4–5 million years. The biochronological frameworks used in this paper are those of Schmidt-Kittler (1987) for the Upper Oligocene and De Bruijn *et al.* (1992) for the Lower Miocene.

Biostratigraphy (Table 1)

One of the main criteria for biostratigraphic subdivision of Upper Oligocene sediments is the *Issiodoromys minor – I. terminus* (Theridomyidae) lineage. For the uppermost Oligocene and Lower Miocene, the *Rhodanomys hugueneyae – Ritteneria manca* lineage (Eomyidae) is one of the most important criteria for biostratigraphic subdivision. Various lineages of *Eucricetodon* may also serve for local biostratigraphic subdivision, but this material is presently under study. Representatives of these lineages are present in our composite section in the valley of the Río Mayor. The six studied faunas cover five biozones.

The locality of Canales is correlated to Zone V of the Upper Oligocene (Alvarez *et al.*, 1987) because of the presence of *Issiodoromys limognensis*. Another taxon that points at this level is *Eomys* cf. *zitteli* (Eomyidae). This level may be correlated to MP28 as defined at the Mainz Symposium (Schmidt-Kittler, 1987).

The locality of Parrales contains *Issiodoromys pseudanaema*, representing the next evolutionary stage of the *I. minor–I. terminus* lineage, and *Rhodanomys transiens* as the most characteristic elements of this fauna. It is correlated to level W of the Upper Oligocene (Alvarez *et al.*, 1987) and it may be correlated to MP29.

The fauna of Moncalvillo lacks *Issiodoromys*, but it contains *Rhodanomys schlosseri*, which represents the next evolutionary stage of the *R. hugueneyae – R. manca* lineage. This fauna is correlated to Zone X (Daams & Van der Meulen, 1984) of the Oligocene–Early Miocene and to MN1 (De Bruijn *et al.*, 1992).

The faunas of Pozo 1 and Moheda contain *Ritteneria molinae* as the direct descendant of *Rhodanomys schlosseri* and are consequently placed in Zone Y1 of the Early Miocene boundary interval. The presence of two species of *Eucricetodon* (Cricetidae) is also characteristic of this zone. This zone may be correlated to MN2.

The fauna of Cabeza Rubia contains *Ritteneria manca* as the direct descendant of *R. molinae*, and is correlated to Zone Y2 of the Lower Miocene.

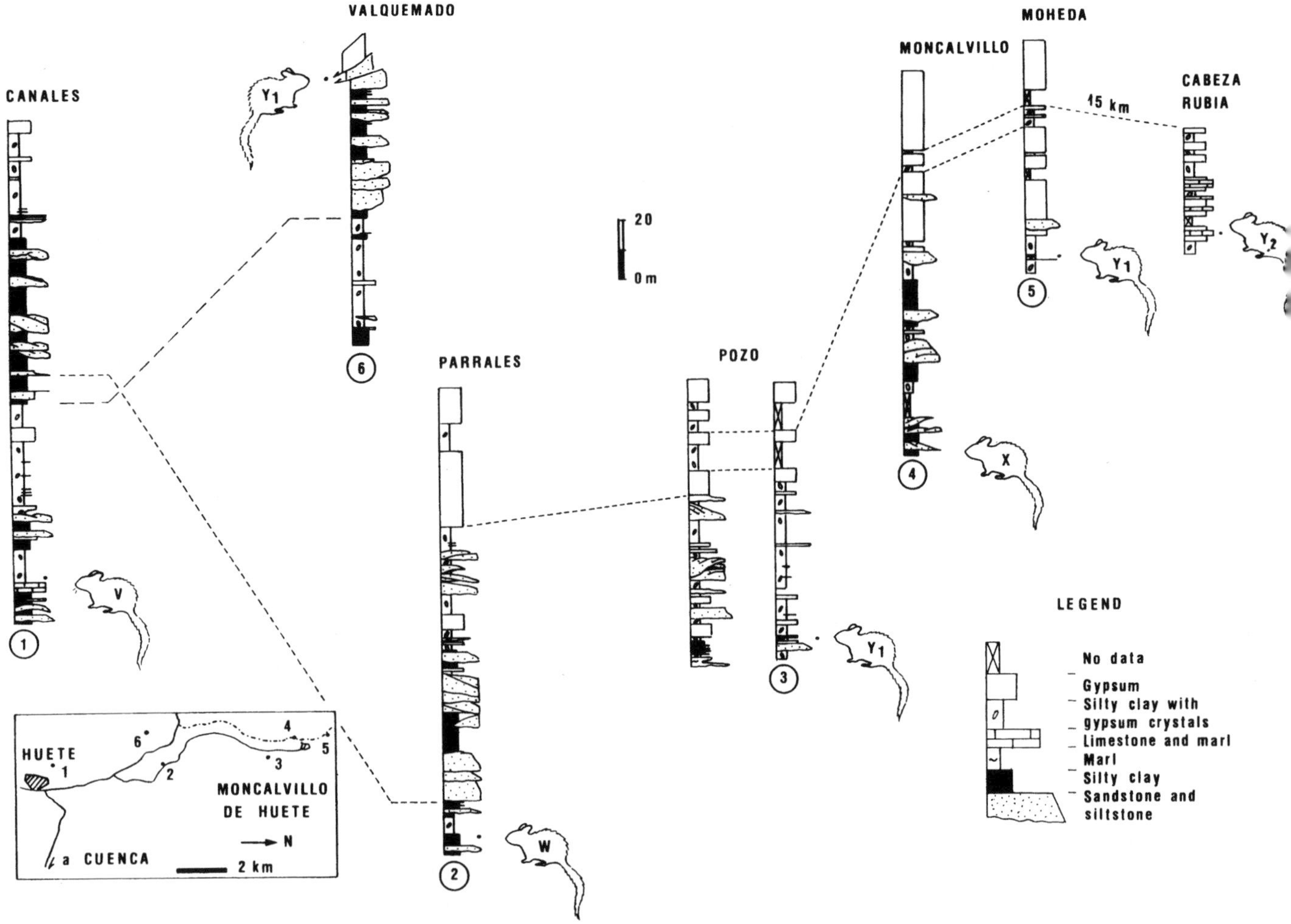

Fig. 1. Lithological logs of the Upper Detrital unit along the Mayor river valley. Fossiliferous levels are indicated by the position of the snouts of the rodents.

Paleoecological assumptions

The prerequisites for a reliable quantitative approach are detailed qualitative studies of the micromammal faunas and a dense documentation through time. In the Loranca Basin micromammal faunas are relatively scarce and most of the faunas are still under study. Therefore, only a tentative subdivision into relatively wet and dry periods can be given for the Late Oligocene and Early Miocene.

Before starting a paleoecological interpretation, we made a number of basic assumptions on habitat preferences of various rodent groups. The ecology of living representatives was extrapolated to the fossil ones, whenever possible.

- **Sciurinae** in our faunas are essentially ground squirrels of open country.
- **Zapodidae** (hopping mice) live in relatively variable environments. In North America they are found both on the Great Plains and in forests with dense undergrowth. In China representatives of this family live near rivers in cool forests at great altitudes (Grzimek, 1967). They all require relatively cool conditions.
- **Gliridae** (dormice). Van der Meulen & De Bruijn (1982) grouped living and fossil glirid species on the basis of characteristic features of the upper first and second molars and extrapolated the ecology of the living representatives to the fossil species. Their grouping, slightly changed by Daams & Van der Meulen (1984), will be used in this chapter.
- **Eomyidae** (an extinct family) are supposed to have been forest dwellers (Van de Weerd & Daams, 1978).
- *Eucricetodon* belongs to an extinct subfamily of the Cricetidae (hamsters). This genus is supposed to have preferred open country (Daams & Freudenthal, 1990).
- *Issiodoromys* belongs to the extinct Theridomyidae family. This taxon has large bullae tympanicae and extremely hypsodont teeth. The combination of these features is characteristic of open-country dwellers (Lavocat, 1951).

Table 1. *The distribution of rodent genera and species in six locations from the Upper Oligocene and Lower Miocene of the valley of the Río Mayor in the Loranca Basin*

CAN[1]	PAR[2]	MON[3]	PZ1[4]	MOH[5]	CAB[6]	Locality	
21						*Issiodoromys limognensis*	Theridomorpha
	6					*Issiodoromys pseudanaema*	
11	2					*Pseudocricetodon* sp.	CRI
4						*Adelomyarion* sp.	
5	4	10	33	19	20	*Eucricetodon* spp.	
7						*Plesiosminthus* sp.	ZAP
	61					*Plesiosminthus schaubi*	
		17				*Plesiosminthus myarion*	
38						*Eomys* cf. *zitteli*	Eomyidae
	16					*Eomys huerzeleri*	
	7					*Rhodanomys transiens*	
		42				*Rhodanomys schlosseri*	
			20			*Ligerimys* sp.	
			6	4		*Ritteneria molinae*	
					23	*Ritteneria manca*	
7						*Peridyromys* sp.	Gliridae
8						Gliridae indet.	
		2				*Vasseuromys autolensis*	
	2	0	X	X		*Armantomys bijmai*	
	3	17	X	X	21	*Peridyromys murinus*	
		10	X	X		*Pseudodyromys* sp.	
			X	X	9	*Pseud. simplicidens*	
					18	*Praearm. crusafonti*	
					4	*'Perid.' brailloni*	
1						Sciuridae indet.	SCI
		3	9	5		*Heteroxerus* spp.	
					5	*Heteroxerus rubricati*	
76	110	582	155	335	187	Number of M12	
V	W	X	Y1		Y2	Local zones	
28	29	1	2			MP/MN zones	
Upper Oligocene		?	Lower Miocene			Series	

Notes:
Numbers represent percentages. In Pozo 1 and Moheda the four Gliridae species are indicated with an X, as they are still under study and not satisfactorily separated. Together they form 32% and 72% respectively of the rodent fauna. CRI = Cricetidae; ZAP = Zapodidae; SCI = Sciuridae.
[1]Canales, [2]Parrales, [3]Moncalvillo, [4]Pozo 1, [5]Moheda, [6]Cabeza Rubira.

Increase and decrease in the relative numbers of representatives of forested areas are interpreted as expansion and shrinking respectively of the forest area. These changes are in turn controlled by changes in the humidity of the climate.

Paleoecology (Table 2)

The fauna of Canales contains a mixture of elements typical of dry and wet biotopes respectively. The most abundant representative of a dry biotope is *Issiodoromys* (21%), but Eomyidae (38%), typical of wet conditions, forms the most abundant group of this fauna. The Gliridae fauna has a low diversity and the few dormice present probably preferred dry conditions. The ecological preference of *Pseudocricetodon* (21%), belonging to an extinct subfamily of the Cricetidae (hamsters), is still unknown.

The fauna of Parrales is remarkable for its high content of Zapodidae (61%), accompanied by 23% Eomyidae. The Gliridae fauna still has a low diversity and the two species present are typical

Table 2. *Relative abundance of rodents grouped according to their ecological preference*

	Dry				Wet					
Locality	*Issiodor*	*Eucricet*	*Gliridae* dry	*Sciuridae*	*Plesiosm*	*Eomyidae*	*Gliridae* wet	Unknown	NM1M2	Sample Size (kg)
CAB		20	48	5		23	4		187	5375
NOH		19	72	5		4			335	6525
PZ1		33	32	9		26			162	2300
MON		10	26	3	17	42	2		582	5725
PAR	6	4	5		61	23		2	110	3725
CAN	21	5	7	1	7	38		21	76	6200

Notes:
Numbers are percentages.
NM1M2 – Total number of first and second molars. Generally the first and second molars from both upper and lower dentition of rodents are the largest elements. Premolars and third molars are not considered in the counting as not all rodents have premolars, and several rodent taxa have very small third molars that become easily lost during processing of the sample.

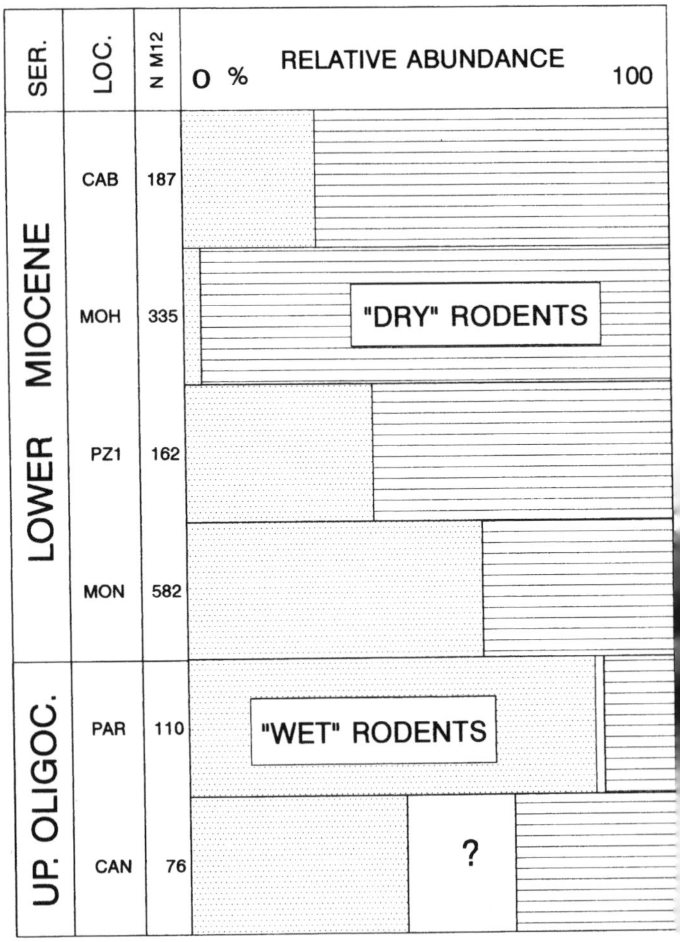

Fig. 2. Succession of relative humidity in the Late Oligocene/Early Miocene of the Loranca Basin.

of dry environments. Humid conditions prevailed more than at the time of the previous fauna of Canales.

In the fauna of Moncalvillo the Zapodidae and Eomyidae exchanged their relative abundance compared with that of Parrales. Eomyidae (42%) predominate over the Zapodidae (17%). The Gliridae are more diverse (four species), although one species (*Vasseuromys autolensis*) is a typical forest dweller. The other three species are characteristic of dry conditions. Although wet conditions prevailed, a tendency towards drier conditions compared with the previous fauna of Parrales is observed because of an increase of relative frequency of *Eucricetodon* and a slightly more diverse Gliridae fauna.

In the fauna of Pozo 1 representatives of dry environments predominate; *Eucricetodon* (33%) is represented by two species and the Gliridae (32%) by four which are all characteristic of dry conditions. The systematic study of these glirid species is still under way and therefore the relative abundance of the separate species cannot be given yet. Two species of ground squirrels are also present, but in low percentages (9%), and Eomyidae form 26% of the fauna. Dry conditions prevailed at the time of the fauna of Pozo 1.

The fauna of Moheda represents an even drier phase than that of Pozo 1. *Eucricetodon* (19%) is represented by two species and the diverse Gliridae, characteristic of dry environments, form the majority of the fauna (72%). The ground squirrel *Heteroxerus* (5%) is also represented by two species. The fauna of Moheda contains only 4% of Eomyidae. In a fauna of the same age (Valquemado) in the valley of the Río Mayor, an incomplete lower jaw of a tapir was found (Cerdeño, 1988). Present-day representatives of this taxon live in tropical zones in forested environment. The most abundant animal in this large-mammal fauna is a running rhinoceros, however, typical of savannah-like environments (Morales, 1989).

In the fauna of Cabeza Rubia the representatives of dry conditions are somewhat less frequent; *Eucricetodon* and *Heteroxerus* are represented by one species each; instead of four 'dry' Gliridae species there are three, and the number of Eomyidae increased compared to Moheda.

In summary we may recognize the following periods (Fig. 2):

- a moderately humid one represented by the fauna of Canales;
- a more humid one represented by the fauna of Parrales;
- a relatively dry one represented by the fauna of Moheda;
- a return to more humid conditions represented by the fauna of Cabeza Rubia.

As far as the temperature is concerned, our interpretations are more doubtful. The predominance of *Plesiosminthus* (Zapodidae) in Parrales may point to relatively cool conditions, and the successive decrease of this taxon may point to temperature increase.

Daams & Van der Meulen (1984) observed a complementary variation in the frequency of the Gliridae *Peridyromys murinus* and *Microdyromys* in the Lower Miocene of the Calatayud–Teruel Basin, which implies temperature change. These trends are most clear when the Gliridae family is set at 100%. *Peridyromys murinus* would tolerate lower temperatures and *Microdyromys* higher ones. In our composite section in the Loranca Basin *Peridyromys murinus* is present in variable percentages, but *Microdyromys* is absent, so that this complementary signal cannot be observed. But if we assume that low percentages of *P. murinus* indicate relatively high temperatures, then the time interval covered by the faunas of Pozo 1 and Moheda would have been relatively warm compared with earlier and later ones on the basis of the scarcity of *P. murinus*. This tendency is supported by the presence of a tapir, characteristic of tropical conditions, in Valquemado at the same stratigraphic level as Moheda (Morales, 1989).

Acknowledgements

We gratefully acknowledge financial support by the CICYT (project PB85–0022, 1986–1988) and in part by the Commission of the European Communities, JOULE program (1989–1992).

References

Alvarez Sierra, M.A., Daams, R., Lacomba, J.I., López Martínez N. and Sacristán, M.A. (1987). Succession of micromammal faunas in the Oligocene of Spain. *Münchner Geowiss. Abh.*, A, **10**: 43–47.

Cerdeño, E. (1988). Primeros datos sobre el esqueleto postcraneal de *Protapirus cetinensis* (Tapiridae). *Geogaceta*, **5**: 21–24.

Daams, R. and Freudenthal, M. (1990). The Ramblian and Aragonian: limits, subdivision, geographical and temporal extension. In *European neogene Mammal Chronology*. NATO ASI Series, A **180**: 51–59.

Daams, R. and van der Meulen, A.J. (1984). Paleoenvironmental and paleoclimatic interpretation of micromammal faunal successions in the Upper Oligocene and Miocene of north central Spain. *Paléobiol. Cont.*, **14**: 241–257.

De Bruijn, H., Daams, R., Daxner-Höck, G., Fahlbusch, V., Ginsburg, L. Mein, P. and Morales, J. (1992). Report of the RCMNS working group on fossil mammals, Reisensburg 1990. *Newsl. Stratigr.*, **26** (2/3): 65–118.

Grzimek, B. (1967). *Grzimeks Tierleben. Enzyklopädie des Tierreiches.* Kindler Verlag. Band 11.

Lavocat, R. (1951). Révision de la Faune des mammifères oligocènes d'Auvergne et du Velay. *Fac. Sci. Univ. Paris, Série A*, No. 2445.

Morales, J. (1989). Los yacimientos de grandes mamíferos del Terciario de Cuenca. In *La fauna del pasado en Cuenca.* Actas del I Curso de Paleontología, Instutio 'Juan de Valdés'. Serie '*Actas Académicas*' **1**: 167–188.

Schmidt-Kittler, N. (ed.) (1987). European reference levels and correlation tables. *Münchner Geowiss. Abh.*, A, **10**: 13–31.

Van der Meulen, A.J. and de Bruijn, H.(1982). The mammals from the Lower Miocene of Aliveri (Island of Evia, Greece). Part 2. The Gliridae. *Proc. Kon. Ned. Akad. Wet.*, B, **85**(4): 485–524.

Van der Meulen, A.J. and Daams, R. (1992). Evolution of Early–Middle Miocene rodent faunas in relation to long-term paleoenvironmental changes. *Palaeogeogr., Palaeoclimatol., Palaeoecol.*, **93**: 227–253.

Van de Weerd, A. and Daams, R. (1978). Quantitative composition of rodent faunas in the Spanish Neogene and paleoecological implications. *Proc. Kon, Ned. Akad. Wet.*, B, **81**: 448–473.

C7 Fluvial fans of the Loranca Basin, Late Oligocene – Early Miocene, central Spain

M. DÍAZ-MOLINA AND A. TORTOSA

Abstract

During Late Oligocene to Early Miocene times the Loranca Basin was filled by a thick sequence of continental deposits comprising three large depositional systems, Tórtola, Villalba de la Sierra and Valdeganga. Stratigraphic and sedimentological methods enable palaeoenvironmental reconstructions and the tectonic deformation history to be related to long-term palaeogeographic changes. Each depositional system consists of a fluvial fan and associated floodbasin and distal deposits, which were dominated by saline lakes. Syn-sedimentary alluvial fans fringed the basin margins, reflecting their differential tectonic movements. Topographic changes induced by fold generation controlled facies distribution evolution. The large fluvial fans contained many channel systems, and all the channel types identified in recent rivers are found. In proximal fan areas, braided channels predominated, but meander loops, and braided and other low-sinuosity channel types are found extensively across the fan surfaces. The stratigraphic successions are characterized by a high content of finer-grained deposits. Their sandstone composition reveals source areas composed of sedimentary rocks for all the depositional systems. Differences in composition of the main sediment supplies reflect diversity in the relative proportion of the different sedimentary rocks in the catchment areas. In contrast, variation in composition down the fans reflects local tributary mixing situations, both between fans or between more local alluvial systems.

Introduction

The Loranca Basin is located in the central part of Spain, within the Iberian Range (Fig. 1). This basin was formed during Eocene to Late Oligocene times and contains continental deposits locally folded along with the underlying Mesozoic rocks.

From the Late Oligocene to the Early Miocene the Loranca Basin was filled by two coalescing depositional systems (Díaz-Molina *et al.*, 1989), the Tórtola and the Villalba de la Sierra fluvial fans, and their associated environments. In the Early Miocene, the Valdeganga depositional system also contributed to the filling of the Loranca Basin (Díaz-Molina *et al.*, 1985).

The deposits of the Tórtola and Villalba de la Sierra depositional systems make up the Upper Unit and the overlying Valdeganga depositional system formed part of the Terminal Unit of the basin. Three stages are recognized during sedimentation of the Upper Unit. Stage 1 corresponds to the most active time interval of the fans (Fig. 1A). During stage 2 (Fig. 1B), the Tórtola fan receded while the Villalba de la Sierra fan was first reactivated and then later gradually abandoned. Stage 3 (Fig. 1C) is characterized by the dominance of gypsum deposits. Climatic variation during the Late Oligocene–Early Miocene (Daams & Van der Meulen, 1984; see also Chapter C6) consisted of progressively drier and warmer conditions that coincided with the gradual retraction of the Tórtola and Villalba de la Sierra fans (see also chapter C8).

The apexes of the fans were located along the eastern margin of the basin (Fig. 1A, B, C, D) and they have been eroded. The catchment basins of the fans are not preserved at all, because tectonic deformation in Neogene times strongly modified the Late Oligocene to Early Miocene landscape, mainly consisting of Cretaceous and Jurassic limestones, dolostones, sandstones and gypsum, and Triassic sandstones, mudstones and gypsums.

On the fan surfaces, floodplain elements were low-sinuosity channel fills consisting of gravels and sands, meander loops, abandoned meandering channels, crevasse splays, levees, and floodbasin deposits of silty clays. To establish the pattern of fluvial sedimentation, palaeocurrent data were collected throughout the area (Díaz-Molina, 1978; Díaz-Molina *et al.*, 1989). In the distal parts of the fluvial systems, thin siltstone and sandstone sheets are more abundant and fluvial sediments interdigitate with saline lacustrine deposits.

The Tórtola fan

The Tórtola fluvial fan is up to 94 km long, with a maximum width of 40 km, covering an area of 2500 km^2 (Fig. 1A) and its exposed stratigraphic succession is 700 m thick (Figs. 2 and 3). Mapping reveals that it does not have the form of a cone segment. In spite of its lacking a cone-like form it radiated downslope from the apex area developing a multiple channel system. The apex of the Tórtola fan has been partially eroded, and it

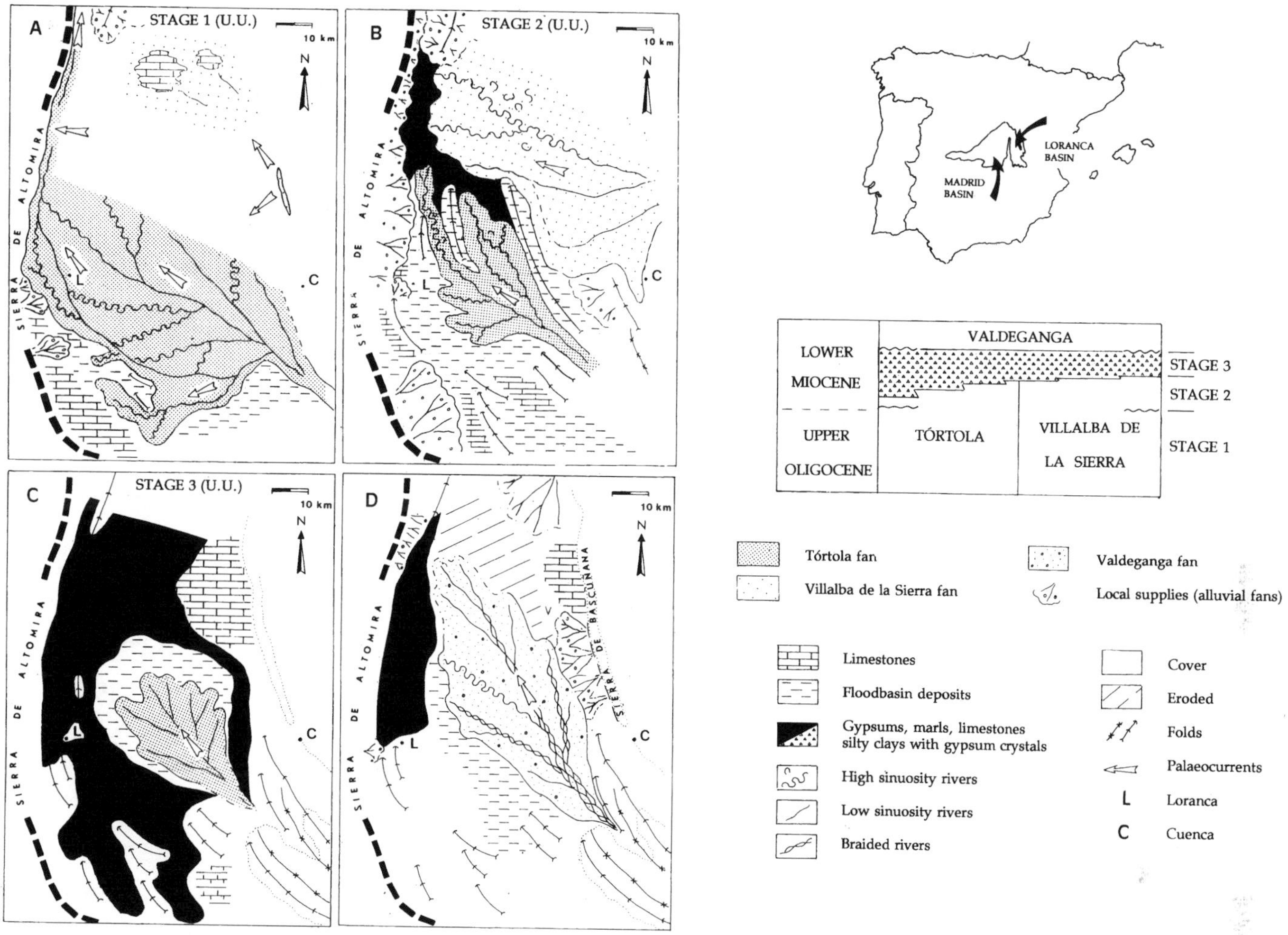

Fig. 1. Location map of the Loranca Basin, general stratigraphy in the central part of the basin and reconstructions of the Tórtola, Villalba de la Sierra and Valdeganga depositional systems.

was also covered in part by the Valdeganga fan apex (Díaz-Molina *et al.*, 1989). The fluvial network became restricted with time (Fig. 1B), and as the fluvial system retreated it was replaced by deposits of wet-land environments (Figs. 1B and 3, see also Chapter C8).

Distributary and tributary areas are distinguished using the analysis of palaeocurrents and variations in river type and depth. In the Tórtola fan there is a general downstream change from low- (braided) to high-sinuosity river types in the areas with a distributary pattern, though in most of the sections low- and high-sinuosity channels are present (Figs. 2 and 3). Other distributary areas have been recognized fringing the south-west lateral areas of the Tórtola fan where internal anticlines were present.

Towards the north-west margin of the basin, a tributary situation has been discovered, and at the Buendia locality to the north, low-sinuosity channels dominated (Díaz-Molina *et al.*, 1985). The tributary zone to the west of the Loranca Basin is also characterized by a noticeable growth of channel density. The channels of this area flowed to the north and passed to the Madrid Basin (Fig. 1A). Along the Sierra de Altomira margin, discharge in the channels could have been higher because of the tributary pattern, and this increase could have also caused an increase in both meander wavelength and channel width (Schumm, 1977). Another control to be considered is the generation of a local base level by the compressional shortening of the northern area (see Chapter C5).

Bed-load, braided-river palaeochannels (Schumm, 1981) are restricted to the proximal fan where diffuse gravel sheet deposits (Hein & Walker, 1977) are predominant. Gravity flow deposits, formed by debris and mud flows, are also found in the proximal area. Mixed-load braided and meandering channels occurred on the whole fan surface (Figs. 4 and 5), though meandering channels predominated downfan (Figs. 2 and 3). High-sinuosity rivers left meander-loop deposits, which frequently display laterally stacked sequences deposited by adjacent point bars, separated by reactivation surfaces (Díaz-Molina, 1978, 1984).

Sandstones located close to the Tórtola apex (VL, Fig. 6) have a litharenite composition ($Q_{54}F_7R_{39}$ – Fig. 7). Rock fragments are micritic and sparitic limestones and occasional dolostones, while intrabasinal components are scarce because of proximity to the

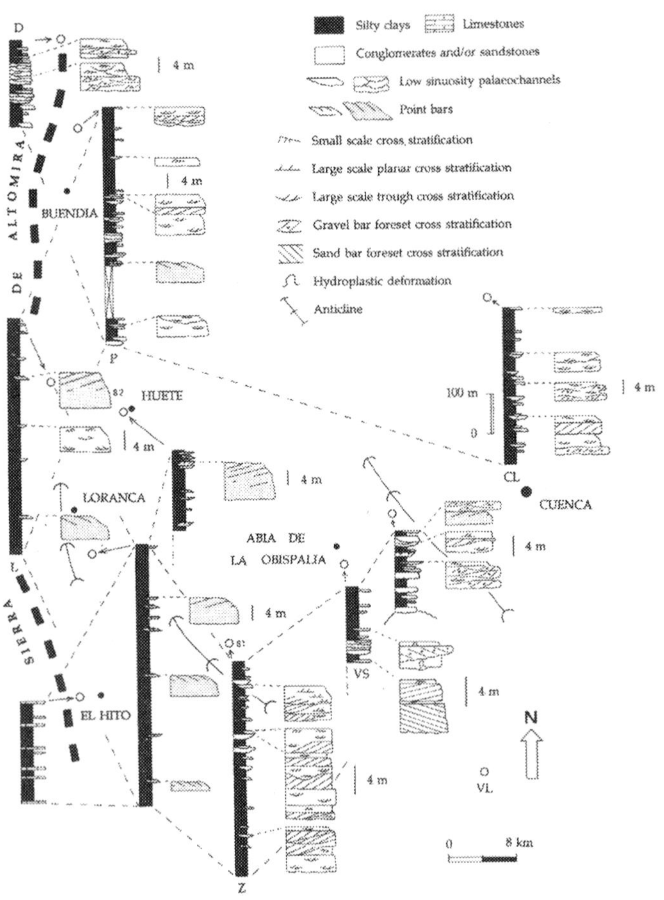

Fig. 2. Sedimentary logs through the Tórtola and Villalba de la Sierra fan deposits, corresponding to stage 1 in the evolution of the Upper Unit of the basin. Each log is marked by a letter code and located on the map by an open circle. Filled circles locate villages and towns.

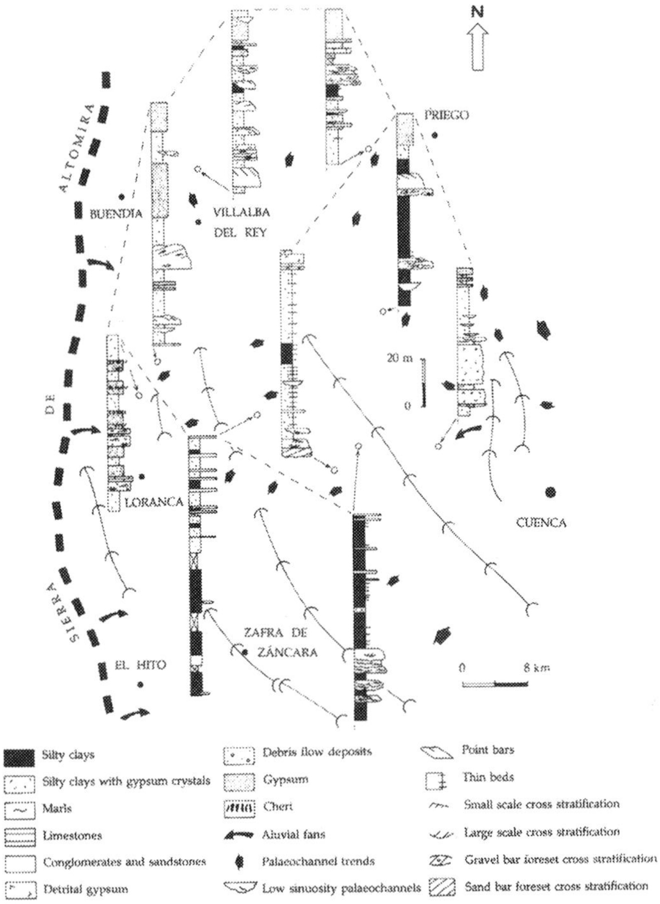

Fig. 3. Sedimentary logs corresponding to stage 2. Tórtola and Villalba de la Sierra fans interdigitate with lacustrine saline deposits. Internal anticlines controlled channelized areas distribution and alluvial fan sediments fringed the west margin of the basin as well as part of the east margin.

apex. Sandstone composition evolves downfan to a more mature framework content, as carbonate rock fragments (CE) decrease. In contrast, intrabasinal material became more notable (Fig. 7), particularly limeclasts derived from pedogenetic carbonates and palustrine–lacustrine limestones (Díaz-Molina *et al.*, 1989).

The internal anticlines of the basin, located in the south-western marginal areas of the Tórtola fan (Fig. 1A), provided local supplies of higher feldspar content to the basin (Fig. 7). These deposits derived from erosion of the underlying and more arkosic stratigraphic unit (Lower Unit, Díaz Molina *et al.*, 1989). This contribution is marked by an increase in the F/Q + F index (see L in Fig. 6).

In the same way, the presence of local alluvial fans along the Sierra de Altomira margin is reflected in the compositional evolution of these sandstones as shown in Fig. 7. These local systems supplied sediments with high CE content along the tributary area parallel to the Sierra de Altomira (see increment in the CE/E index in P, Fig. 6).

The mean modal sandstone composition is similar throughout the entire stratigraphic sequence, so it can be inferred that the lithology of the source areas remained constant during its deposition.

The Villalba de la Sierra fan

Lateral stratigraphic equivalents of the Tórtola fan are found in what nowadays is the eastern flank of the Bascuñana anticline (stage 1, Fig. 1A), where palaeocurrents and palaeochannel trends indicate that another fan was being built from the east. These deposits, mainly sandstones and silty clays, are considered to have been part of the Villalba de la Sierra fan, which was tectonically reactivated (stage 2, Fig. 1B), supplying conglomerates in the northern area of the Loranca Basin. The Villalba de la Sierra fan probably covered an area of a similar extent to that occupied by the Tórtola fan, though in the central area of the basin only the upper part of its stratigraphic succession, corresponding to stage 2, can be observed (Fig. 3). A partial reconstruction for stage 1 of the Villalba de la Sierra fan, in the north-central area of the basin (Fig. 1A), is based on drilling data (Torres *et al.*, 1992).

On the eastern flank of the Bascuñana anticline (stage 1, Fig. 1A), palaeochannels are filled by sandstones and conglomerates. The palaeochannels are of low-sinuosity and mixed-load types. The sandstones are subarkoses ($Q_{76}F_{15}R_9$ – Fig. 7), indicating larger

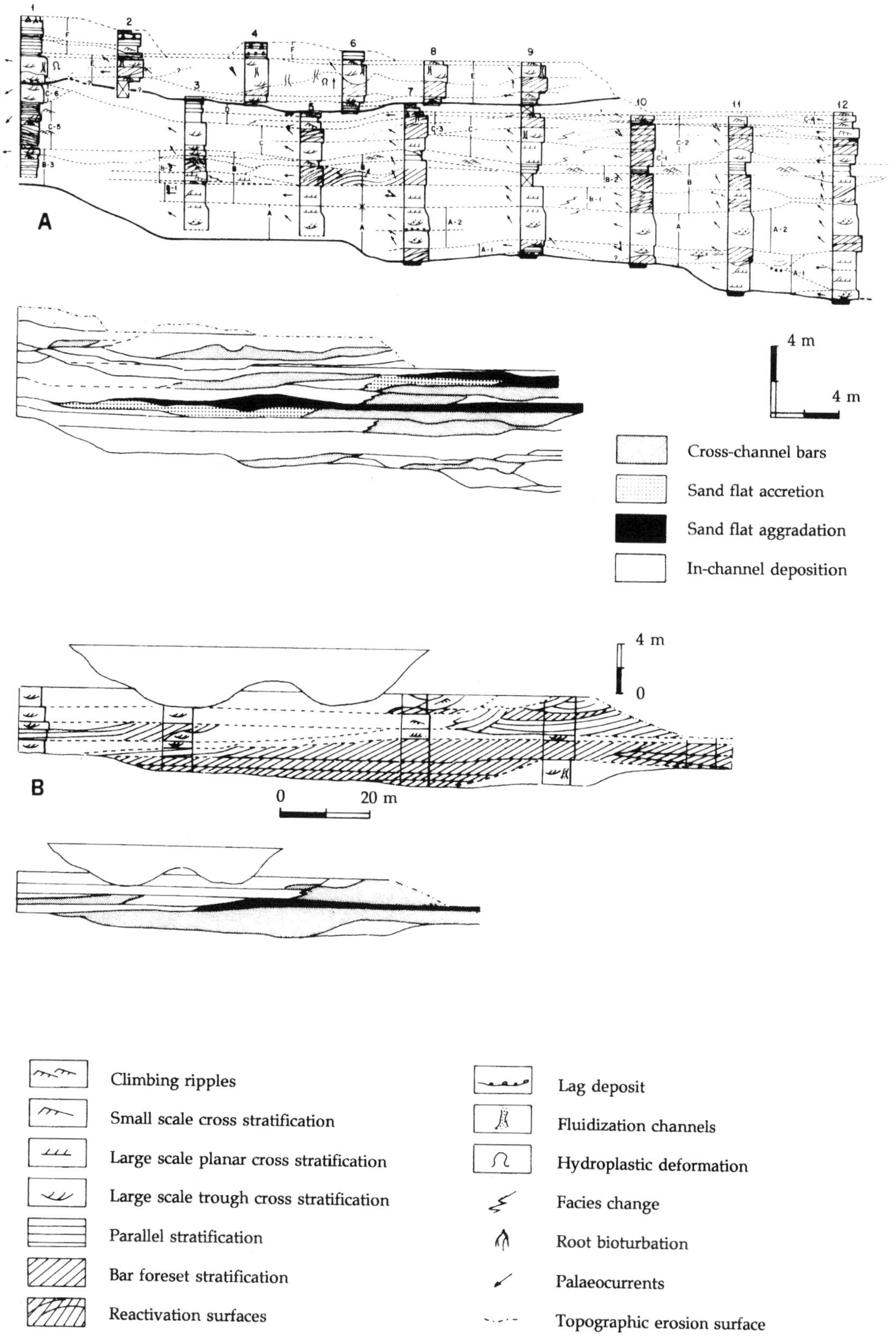

Fig. 4. Examples of braided channels; their location is indicated in Figure 2. Braiding is identified on the basis of: the presence of minor imbricated channels (Allen, 1965; William & Rust, 1969); channel incision on minor bars (Smith, 1971; Blodget & Stanley, 1981); and preservation of bed forms that are characteristic of the building of alluvial islands (Cant & Walker, 1978).

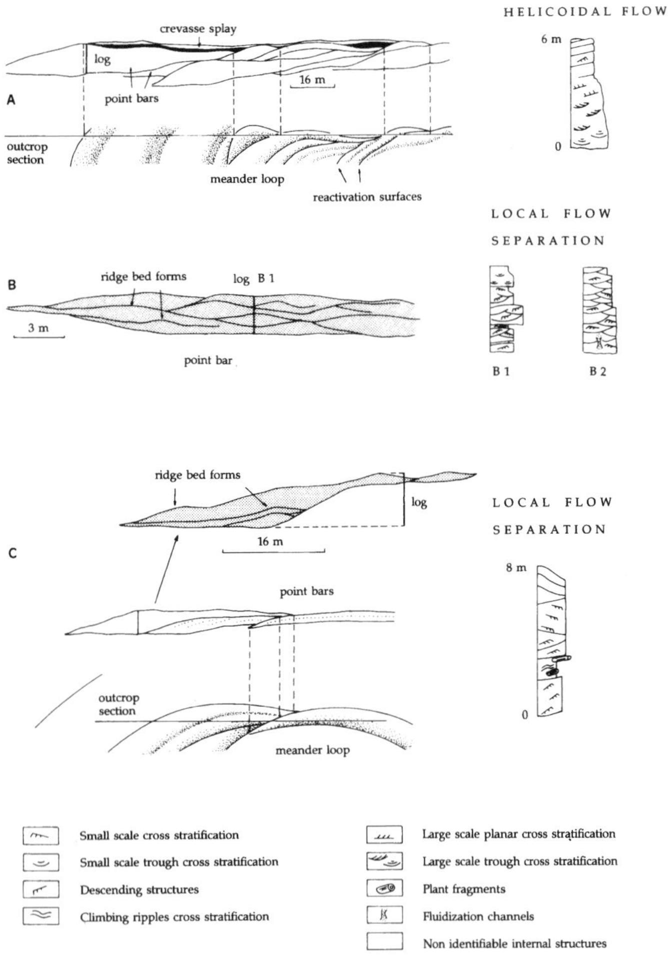

Fig. 5. Meander loops show a great variability of facies in the study area. Two main types are recognized, one shows sequences related to helicoidal flow (A), while others built up by ridge bed forms (B1 and C) are explained by local flow separation (Bagnold, 1960; Leopold *et al.*, 1960; Nanson, 1980). When ridge bed forms are not preserved, point bars show sets of ripples migrating upstream or up and down the point bar (B2). A and B belong to the Tórtola fan, C to the Villalba de la Sierra fan.

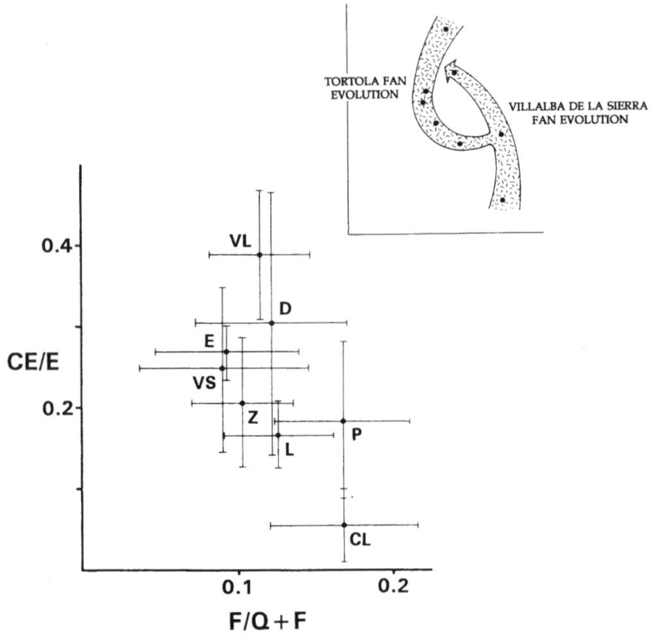

Fig. 6. Compositional variation in Tórtola and Villalba de la Sierra fans as a function of CE/E index (extrabasinal carbonates to total extrabasinal grains) versus F/Q+F index (K-feldspar grains to quartz plus K-feldspar grains). Locations VL, E, VS, Z, L, P, D and CL are shown in Figure 2.

exposures of sandstones in the source area compared with the Tórtola fan (CL, in Fig. 6). Rock fragments are similar to those described in samples from the Tórtola fan apex. The highest content of finer sediments in the stratigraphic succession, and the presence of intrabasinal micritic components in the sandstone framework, reveal that the exposed sediments were distant from an eroded apex area. The contribution of the Villalba de la Sierra fan to the western tributary area is only manifested from the Buendía locality to the north (Fig. 7), where an important increase in the F/Q+F index appears (see P, in Fig. 6). On passing to the Madrid Basin the channel system incorporated new sediments from a north-eastern lateral fan (D in Fig. 6, and Fig. 7).

In the second-stage sedimentary (Figs. 1B and 3) palaeochannels are mainly of the braided type and were filled up by conglomerates and sandstones, where gravel bars and amalgamated channels were the dominant features. Laterally, meander loops built by ridge bed forms are also present in the exposures of the central part of the basin (Fig. 5). These meander loops contain sandstones, though occasionally conglomerates also appear. In this sedimentation stage, sandstone composition is litharenitic, reflecting a change in the source area lithology related to the tectonic reactivation, which exposed more calcareous Cretaceous rocks in the eastern margin of the basin.

The Valdeganga fan

This fan covers 19 300 km², with a maximum width of 35 km (Fig. 1D). In the Valdeganga fan apex coarse conglomerates occur, forming a thick distinct body composed of stacked and imbricated palaeochannels. The conglomeratic body passes radially into individual channels (Fig. 8). In spite of the down-fan evolution to finer-grained sediments, most of the fan surface was dominated by braided channels. However, in the lateral parts of the fan, deposits left by high-sinuosity rivers also occur. Although the Tórtola and Valdeganga fans belong to different stratigraphic units, the Valdeganga fan can be considered a reactivation of the Tórtola fan, whose apex is in part covered and eroded by the Valdeganga fan apex.

Sandstones filling the channel system show a very similar composition to that of the Tórtola fan ($NCE_{56}CE_{36}CI_8$–$Q_{53}F_7R_{40}$). The percentage of the different quartz types reflects an increase in compositional maturity between the Tórtola and Valdeganga fans, with an enrichment in the most stable types underlining the

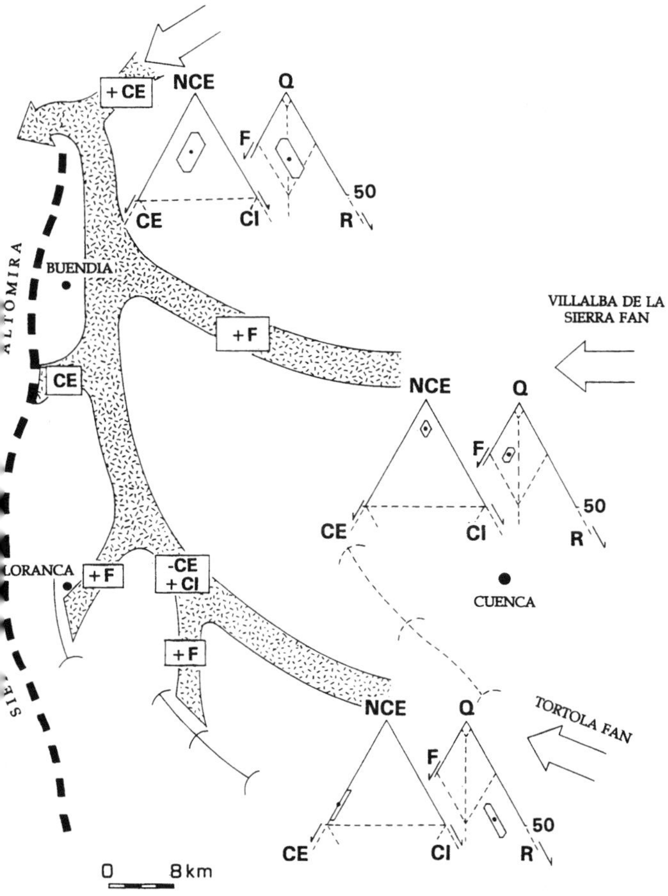

Fig. 7. Compositional evolution along the main distributary area of the Tórtola and Villalba de la Sierra fans during the first stage of sedimentation. The mean composition is represented in the diagrams NCECECI (Zuffa, 1980) and QFR (Pettijohn *et al.*, 1973). CE: Extrabasinal carbonates. CI: Intrabasinal carbonates. F: K-feldspar; Q: Quartz; R: Rock fragments; NCE: Non-carbonate extrabasinal.

importance of recycling processes. The compositional variation in the central areas is of the percentages of quartz and of rock fragments, which increase and decrease, respectively, the longer the transport. The area sampled lacks detritus with a local source, because of the absence of connection between the Valdeganga channel system and the peripheral alluvial fans.

Conclusions

The Tórtola, Villalba de la Sierra and Valdeganga fluvial fans were dominated by individual channels which were part of multiple channel systems, and they are considered as 'wet' fans (Schumm, 1977) formed by perennial stream flow. Recent examples are the Riverine Plain and the Kosi River fan (Schumm, 1977; Gole & Chitale, 1966; Wells & Dorr, 1987).

The palaeogeographic reconstructions shown in the figures represent a long stratigraphic interval. Detailed stratigraphic correlations are difficult because of the nature of fluvial deposits and the fact that channel types cannot be precisely correlated as stratigraphic indicators. Nevertheless, lateral and vertical distributions of channel types show the same kind of complexity observed in the drainage system of the Kosi River fan (Wells & Dorr, 1987), with a high abundance of meandering river deposits. In these examples, braided palaeochannels show facies characteristics similar to those observed in perennial recent rivers (Fig. 4). Besides, meander belts show complex meander loops formed by adjacent point bars (Fig. 5), revealing steady-state equilibrium during the time between channel incision and channel abandonment.

Continuing tectonic deformation initiated a shifting pattern of palaeoenvironments and consequently there is considerable lateral variability of facies in the basin sections. The basin structure and the differential tectonic activity along the basin margins strongly controlled the orientation, shape and evolution of the depositional systems (Fig. 1). Within the basin, fold generation deflected channelized areas, which were located in the topographic minima of the primary synclines. The loading of alluvial fans along the basin margins also controlled the lateral extent of the actively agrading fluvial systems.

In most of the basin, the fans show single or weakly amalgamated palaeochannels encased by great quantities of overbank sediments. Only in the preserved apex deposits of the Tórtola and Valdeganga fans is the content of finer deposits lower, and palaeochannels formed thicker units by amalgamation. Fan head trenches (Schumm, 1977) and downfan pattern distribution explain the downstream evolution from amalgamated channels to isolated channels with a progressive growth in the thickness of fine-grained sediments. The opposite evolution is exhibited where the main tributary area of the Tórtola and Villalba de la Sierra fans passes to the Madrid basin, and this may be explained by local channel concentration in a tectonically formed valley (Fig. 1A).

Successive palaeoenvironmental reconstructions can be contrasted with information from sandstone compositional analysis. Tórtola fan sandstones only vary laterally, showing radial evolution, local supplies from syn-sedimentary folds and the occurrence of channel tributary areas. Later folding in the south-east margin of the basin generated the Tórtola head trenching and reworking of its deposits in the Valdeganga fan. A conspicuous compositional difference between the deposits of stages 1 and 2 of the Villalba de la Sierra fan is explained by syn-sedimentary tectonic deformation in the north-east margin of the basin, which rejuvenated the drainage basin and generated new denudation levels formed by calcareous Cretaceous rocks.

Acknowledgments

This research was funded in part by the CICYT (PB85–0022, 1986–1988) and in part by the Commission of the European Communities, JOULE programme (1989–1992).

References

Allen, J.R.L. (1965). A review of the origin and characteristics of recent alluvial sediments. *Sedimentology*, **5**: 89–191.

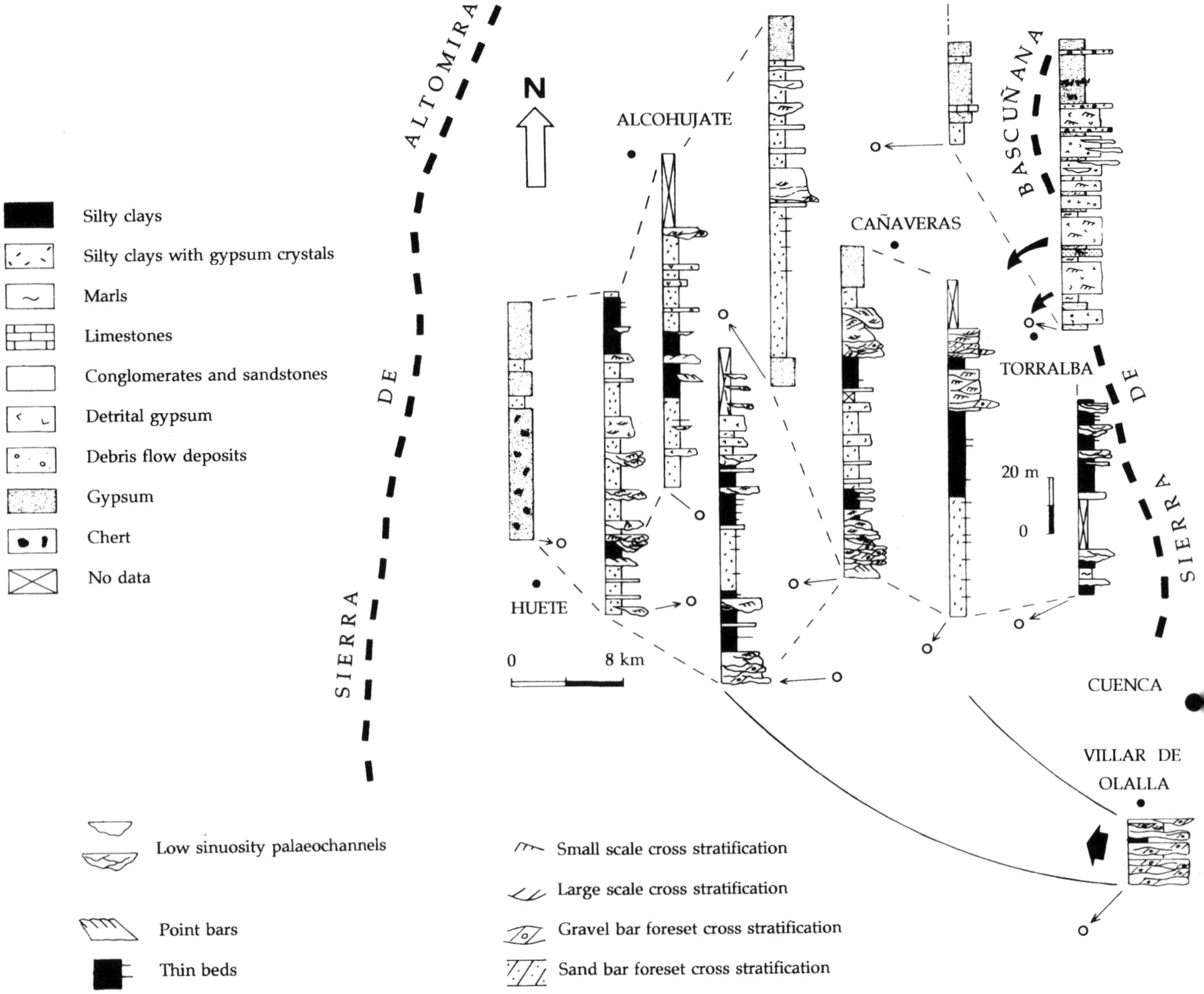

Fig. 8. Sedimentary logs through the Valdeganga depositional system. The Sierra de Bascuñana is the new east margin of the basin.

Bagnold, R.A. (1960). Some aspects of the shape of river meanders. *U.S. Geol. Surv. Prof. Pap.*, **282–E**: 132–144.

Blodgett, R.H. and Stanley, K.O. (1980). Stratification, bedforms and discharge relations of the Platte braided River system, Nebraska. *J. Sediment. Petrol.*, **50**: 139–148.

Cant, D.J. and Walker, R.G. (1978). Fluvial processes and facies sequences in the sandy braided South Saskatchewan River, Canada. *Sedimentology*, **25**: 625–648.

Daams, R. and Meulen, A.J. van der, (1984). Palaeoenvironmental and palaeoclimatic interpretation in the Upper Oligocene and Miocene of north central Spain. *Paléobiol. Cont.*, Montpellier, **14**: 241–257.

Díaz-Molina, M. (1978). Bioestratigrafía y paleogeografía del Terciario al este de la Sierra de Altomira. Tesis doctoral. Facultad de Ciencias Geológicas, UCM. 370 pp.

Díaz-Molina, M. (1984). Geometry of sandy point bar deposits, examples of the Lower Miocene, Tajo Basin, Spain. *I.A.S. 5th European Regional Meeting, Marseille, France. Abstracts*, 140–141.

Díaz-Molina, M., Bustillo-Revuelta, M.A., Capote, R. and López-Martínez, N. (1985). Wet fluvial fans of the Loranca Basin (central Spain). Channel models and distal bioturbated gypsum with chert. *I.A.S. 6th European Regional Meeting, Lérida, Spain Exc. Guide-book*, 149–167.

Díaz-Molina, M., Arribas-Mocoroa, J. and Bustillo-Revuelta, A. (1989). The Tórtola and Villalba de la Sierra fluvial fans: Late Oligocene–Early Miocene, Loranca Basin, Central Spain. *4th International Conference on Fluvial Sedimentology, Barcelona–Sitges Spain. Field Trip* **7**, 74.

Gole, C.V. and Chitale, S.V. (1966). Inland delta building activity of Kosi River. *Am. Soc. Civil Eng. J. Hydraul. Div.*, **HY–2**: 111–126.

Hein, F.J. and Walker, R.G. (1977). Bar evolution and development of stratification in the gravelly, braided Kicking Horse River British Columbia. *Can. J. Earth Sci.*, **14**: 562–570.

Leopold, L.B., Bagnold, R.A., Wolman, M.G. and Brusch, Jr, M.B. (1960). Flow resistance in sinuous or irregular channels. In *Physiographic and hydraulic studies of rivers. U.S. Geol. Surv. Prof. Pap.* **282–D**: 109–134.

Nanson, G.C. (1980). Point bar and flood plain formation of the meander ing Beatton River, northeastern British Columbia, Canada. *Sedimentology*, **27**: 3–30.

Pettijohn, F.P., Potter, P.E. and Siever, R. (1973). *Sand and sandstones*. Springer-Verlag, New York, Heidelberg, Berlin: 618 pp.

Schumm, S.A. (1977). *The fluvial system*. John Wiley & Sons. New York, London, 338 pp.

Schumm, S.A. (1981). Evolution and response of the fluvial systems, sedimentologic implications. *SEPM, Spec. Publ.*, **31**: 19–29.

Smith, N.S. (1971). Transverse bars and braiding in the Lower Platte River, Nebraska. *Geol. Soc. Am. Bull.*, **82**: 3407–3420.

Torres, T. de, García Cortés, A., Mansilla, H. and Quintero, I. (1992). Upper Oligocene palustrine deposits in the 'Depresión Intermedia' Basin (provs. of Cuenca and Guadalajara, central Spain): borehole cores interpretation. *III Congreso Geológico de España y VIII Congreso Latinoamericano de Geología, Salamanca. Simposios*, **1**: 149–157.

Wells, N.A. and Dorr, Jr, J.A. (1987). A reconnaissance of sedimentation of the Kosi alluvial fan of India. *SEPM, Spec. Publ.*, **39**: 51–56.

Williams, P.F. and Rust, B.R. (1969). The sedimentology of a braided river. *J. Sediment. Petrol.*, **39**: 649–679.

Zuffa, G.G. (1980). Hybrid arenites: their composition and classification. *J. Sediment. Petrol.*, **50**, 21–29.

C8 Saline deposits associated with fluvial fans, Late Oligocene – Early Miocene, Loranca Basin, Central Spain

J. ARRIBAS AND M. DÍAZ-MOLINA

Abstract

During sedimentation of stage 2 of the 'Upper Unit', four saline units developed in the central part of the Loranca Basin. These deposits occur associated with the palaeogeographic limits of the Tórtola and Villalba de la Sierra fluvial fans. Analysis of facies, sequences and their distribution through time permit examination of the evolution of saline environments during fan evolution. Saline facies distribution is asymmetric and controlled by fan activity. Primary gypsum is the dominant lithology of the saline deposits. The absence of anhydrite or more-soluble minerals reflects low salinity of the brines. Triassic and Cretaceous gypsiferous source rocks, basin configuration, and progressive aridity were the controls on the development of these saline environments.

Introduction

The Loranca Basin is located in the central part of Spain and it was filled by continental deposits from Late Oligocene to Miocene (Turolian) times. During 'Upper Unit' deposition, from Late Oligocene to Early Miocene (Agenian), three different 'stages' can be distinguished in the evolution of the Tórtola and Villalba de la Sierra depositional systems (Díaz Molina *et al.*, 1989; see also Chapters C5 and C7). The first stage corresponds to the most active time interval of the Tórtola and Villalba de la Sierra humid fans, when the Loranca Basin was connected to the Madrid Basin. During the following stage, syn-sedimentary folding started to plug the area connecting the Loranca Basin with the Madrid Basin, and local base levels were being established with development of wet areas of around 600 km² (Fig. 1). Gypsum precipitates represent the principal facies of these wet areas, still related to the humid fans. The last stage is characterized by the dominance of saline deposits in the basin. The top of stage 3 (22–22.5 Ma, according to Daams *et al.*, Chapter C6) coincides with the upper limit of the TB1.4 third-order cycle defined by Haq *et al.* (1988). Triassic and Late Cretaceous gypsiferous formations in the source area have been invoked as the source of the gypsum in the Loranca Basin (Díaz Molina *et al.*, 1989). The progressive aridity indicated by the facies characteristics of the second stage deposits is also well supported by

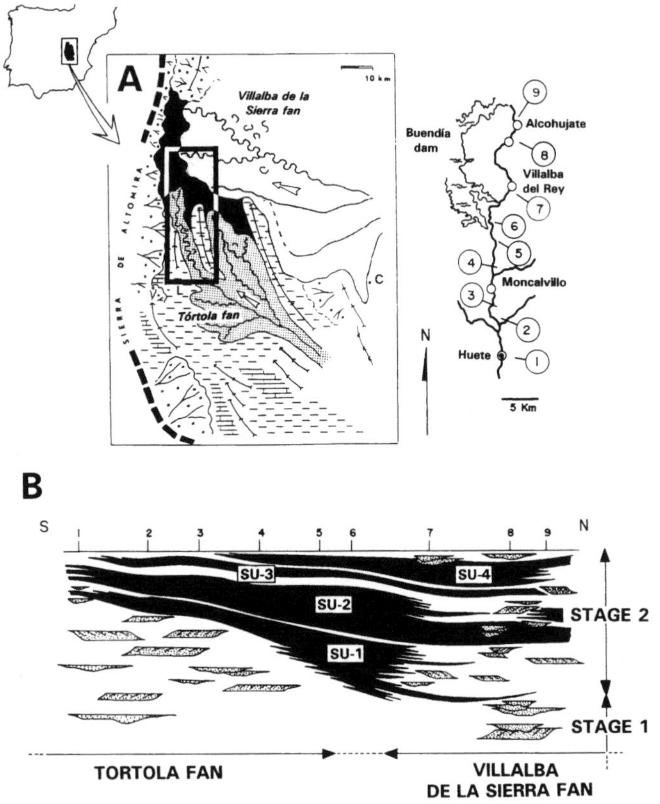

Fig. 1. A, Location of the study area on a reconstructed paleoenvironmental map during deposition of Upper Unit (stage 2) in the Loranca Basin. B, Cross-section reconstruction of stage 2 showing the relationships between saline units (SU in black) and detrital supplies from fluvial fans (in white). Numbers refer to localities on location map.

palaeoclimatic interpretation of the micromammal faunal succession (Daams and Van der Meulen, 1984).

Detailed reconstructions of stage 2 along a north–south cross-section in the central part of the basin allow us to analyze relationships between carbonate–gypsiferous episodes and fluvial deposits (Fig. 1). Four principal carbonate–gypsiferous episodes, here named saline units, appear alternating with continuous silty

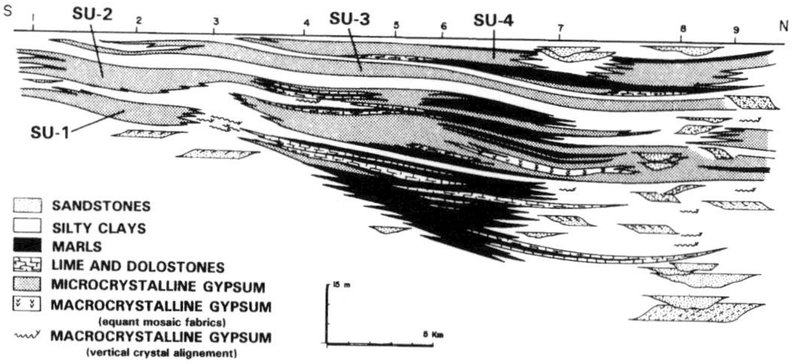

Fig. 2. Facies distribution of stage 2 in the studied cross-section. Numbers refer to localities on Fig. 1A. SU: saline unit.

clay levels (Fig. 1). Laterally, the saline units pass into the siliciclastic deposits of the fans, representing the depocentre of stage 2. Facies analysis, sequences and their distribution along the cross-section permit us to interpret the saline deposits and to document their lateral association with the humid fluvial fan environments.

Facies

The heterolithic composition of the stage 2 deposits provides a great variety of facies, divided in this chapter into (a) siliciclastic facies, (b) carbonate facies and (c) gypsum facies. Fig. 2 shows the facies distribution throughout the studied cross-section.

Siliciclastic facies

Sandstones with gypsum cement

These are channelized deposits usually showing well-developed lateral accretion with a variable thickness (0.5–7 m). Large- and small-scale cross-stratification are the dominant sedimentary structures. Though detrital components are mainly extrabasinal and siliciclastic, carbonate intrabasinal components are also present (see Chapter C7), demonstrating sediment reworking in the basin itself. The characteristic feature of these sandstones is the presence of poikilitic gypsum cement, interpreted as an early diagenetic event related to the saline environments of this stage (Díaz-Molina *et al.*, 1989).

Silty clays with or without gypsum crystals

These form massive reddish levels. Thicker levels developed to the north, directly related to sandstone deposits of the Villalba de la Sierra fan. Frequently scattered displacive gypsum occurs as single lenticular crystals (up to 5 mm in size) or as aggregates (rosettes). The concentration of gypsum crystals increases upward when layers evolve to other saline facies. These deposits represent distal floodbasin environments, in turn forming the outer fringes of lacustrine systems ('dry mud flat' of Hardie *et al.*, 1978).

Carbonate facies

These facies are well developed in the first saline unit (Fig. 2), mainly associated with marls. Most of the carbonate facies are white thin limestones (less than 0.5 m thick). Dolostones show similar textures and structures to limestones, and thus both facies are discussed here in the same section. Mudstones and wackestones with peloids, intraclasts and scarce charophyte debris are the most abundant microfacies. Bioturbation is intense, in the form of vertical and subvertical tubules with meniscus structures ranging from 0.5 cm to 1 cm in diameter. Lenticular gypsum crystals are also present to a very variable degree (5 to 50%). These crystals are less than 0.25 mm in size, exhibiting a random orientation in the micritic matrix. When associated with bioturbation tubules, the gypsum crystals follow the meniscus orientation. These deposits are related to shallow carbonate lacustrine environments (Arribas, 1986). The presence of displacive gypsum crystals is evidence for salinity increase of the interstitial waters during early diagenesis. In addition, a convergence between these facies and facies of higher salinity (microcrystalline gypsum facies) exists, providing a continuous suite of facies in terms of gypsum crystal content.

Occasionally, limestones and dolostones show algal lamination in thin levels (less than 0.2 m) in association with tepees, mud-cracks and bird's-eye structures. These deposits are related to marginal lacustrine environments with episodic subaerial exposures.

Gypsum facies

Primary gypsum facies show a wide textural variety related to distinct saline subenvironments. Facies can be grouped in terms of gypsum crystal size as microcrystalline (< 1 mm) and macrocrystalline (> 1 mm) facies.

Microcrystalline gypsum facies

These appear as white tabular, indurated, bioturbated layers of 0.5 m thick. A succession of layers can produce composite beds up to 7 m thickness. The principal elements are lenticular gypsum crystals with random orientation, embedded in a small amount of micritic matrix. Crystal overgrowth of gypsum, outlined by matrix inclusions, is frequent. Bioturbation (chironomidae?) is intense and equivalent to that observed in the carbonate facies. Similar facies have been described in marginal areas of ephemeral saline lakes (Ortí, 1987), and are generated below a shallow water table of low salinity, where interstitial growth of lenticular gypsum takes place (Ortí, 1987). At the sides of the studied area, this facies is associated with cherts (Bustillo and Díaz Molina, 1980; Arribas *et al.*, 1991). Occasionally, at the top of this facies an equivalent porous and crumbly facies appears, indicating dissolution processes (Lowenstein and Hardie, 1985) by meteoric waters (lake retraction), or brine dilution (flooding).

Pure microcrystalline gypsum with parallel and ripple lamination also occurs as beds less than 0.2 m thick, showing a marked orientation of lenticular crystals. Crystal overgrowth is abundant, and in this case bioturbation is not intense. This resedimented gypsum facies is frequent in lacustrine environments (Schreiber, 1978; Ortí, 1979–82; Ortí, 1987; Arakel, 1980). Mechanical remobilization also took place in a lake-margin subenvironment.

Powdery and friable microcrystalline gypsum occurs as thin

(<0.2 m) homogeneous beds associated with silty clays with gypsum crystals. Usually lenticular crystals are smaller than 0.25 mm, appearing corroded and embedded in a cryptocrystalline matrix of gypsum and micrite. These features are related to pedogenic processes (Stoops and Ilaiwi, 1981; Warren, 1982; Watson, 1985), occurring around lacustrine borders.

Macrocrystalline gypsum facies

Fabrics consisting of coarse gypsum crystals (generally more than 2 mm) are the principal feature of this facies. Based on crystal morphology and orientation, two main subfacies can be distinguished.

Macrocrystalline gypsum with vertical alignment forms discontinuous beds, 0.3 m thick associated mainly with silty clays or microcrystalline gypsum facies. They consist of twinned lenticular gypsum crystals between 10 and 25 mm in size, showing a vertical growth arrangement. A transitional contact with the overlying deposits results from a decrease in gypsum crystal content. Crystals show abundant matrix inclusions tracing the successive stages of crystal growth. This facies results from displacive interstitial growth of gypsum in a precursor sediment. The vertical alignment of gypsum crystals is related to capillary evaporation of interstitial waters in subaerial environments (Rosen and Warren, 1990). Channelized macrocrystalline gypsum facies are also associated with silty clays with gypsum crystals. In this case, the channels are thin (<0.5 m), filled by abraded twinned and untwinned lenticular gypsum crystals embedded in a silty-clay matrix. These channels represent ephemeral streams dissecting saline and mud flats.

Macrocrystalline gypsum with equant mosaic fabrics consists of tabular beds of prismatic gypsum crystals (4–20 mm in size) of variable thickness (0.3–1.5 m), showing fining- or coarsening-upwards sequences. Monocrystalline gypsum units exhibit a dense fabric with minor amounts of micritic matrix. Frequently, the gypsum crystals are mechanically abraded, showing crystal overgrowth or subsequent carbonate cement. This facies represents primary saline deposits generated by free precipitation from brines on the saline pond floor, with or without mechanical remobilization by currents (Schreiber, 1978; Arakel, 1980; Ortí, 1979–82; Warren, 1982). The maximum development of this facies occurs in the second saline unit near Villalba del Rey.

Facies distribution and sequences

The above facies are associated with ephemeral gypsiferous lacustrine environments. The prevalence of saline facies with abundant displacive growth of gypsum over primary saline facies denotes continental sebkha environments with calcium sulphate precipitation. The absence of more soluble minerals reveals the low concentration of sulphate brines (Ortí *et al.*, 1989). Based on the classic models of facies distribution in ephemeral saline lakes (Hardie *et al.*, 1978), subenvironments can be characterized as follows.

Mud flat. This comprises the distal deposits of humid-area fluvial fans with a dominance of siliciclastic fine-grained deposits without gypsum crystals. Sandstone deposits are scarce, showing poikilitic gypsum cement when present.

Saline mud flat. Fine-grained sediments persist, appearing in association with displacive lenticular gypsum crystals, corresponding to marginal deposits of a playa-lake environment (Reeves, 1978). Sequences with powdery microcrystalline gypsum facies and with macrocrystalline gypsum with vertical alignment facies (Fig. 3, A-1 and A-2) are related to illuviation and evaporation processes respectively (Warren, 1982; Watson, 1985; Rosen and Warren, 1990). The growth strategy of macrocrystalline gypsum facies, due to upward capillary movement of sulphate-saturated groundwater, requires a near-surface freatic level. Thus, this facies indicates close proximity to the marginal lacustrine areas. Reworking of gypsum from these facies by ephemeral currents produces channelized deposits of macrocrystalline gypsum.

Lake margin. This subenvironment is characterized by the dominance of bioturbated microcrystalline gypsum facies, in association with other saline facies. Sequence B-1 in Fig. 3 represents a marginal retraction of a playa-lake, showing an upward facies change related to subaerial exposure ('crumbly' facies and macrocrystalline vertically aligned gypsum facies). The inner marginal areas are represented by a succession of microcrystalline gypsum beds (Fig. 3, B-2), exhibiting in each bed a sequence related to an increase in salinity caused by lacustrine retraction or desiccation. This sequence was initiated with a rich micritic layer with intraclasts and abraded gypsum crystals that evolved to a bioturbated microcrystalline gypsum facies. In turn, this facies evolved into pure gypsum with scarce alabastrine gypsum nodules and minor amounts of micrite matrix. The top is characterized by bird's-eye structures and vertical fractures filled by micrite and gypsum cement. Throughout the sequence, parallel and ripple laminations are also present. The absence of nodular anhydritic facies, or their equivalent in secondary gypsum, outlines the lack of extreme aridity and the low concentration of sulphate brines in this sabkha environment.

Lake centre. The presence of macrocrystalline gypsum facies showing equant mosaic fabrics typifies this poorly represented subenvironment. Micro- and macrocrystalline facies with resedimented gypsum crystals are also frequent. These deposits are the only ones that can be related to free precipitation from brines on the lake floor, generated from a more stable water body.

Carbonate environments. As mentioned earlier, a close relationship between microcrystalline gypsum facies and limestone-dolostone facies exists. However, in the first saline unit, limestones and dolostones only are associated with marls, and represent a lacustrine environment that differed from the above-described gypsum environments. In this case a more diluted lacustrine sedimentation is inferred, with retraction–desiccation sequences similar to those described by Arribas (1986).

Significance of saline deposits

Sedimentary sequences appearing throughout the four saline units correspond to evaporitic cycles due to lacustrine retraction–expansion events induced by climatic changes. In addi

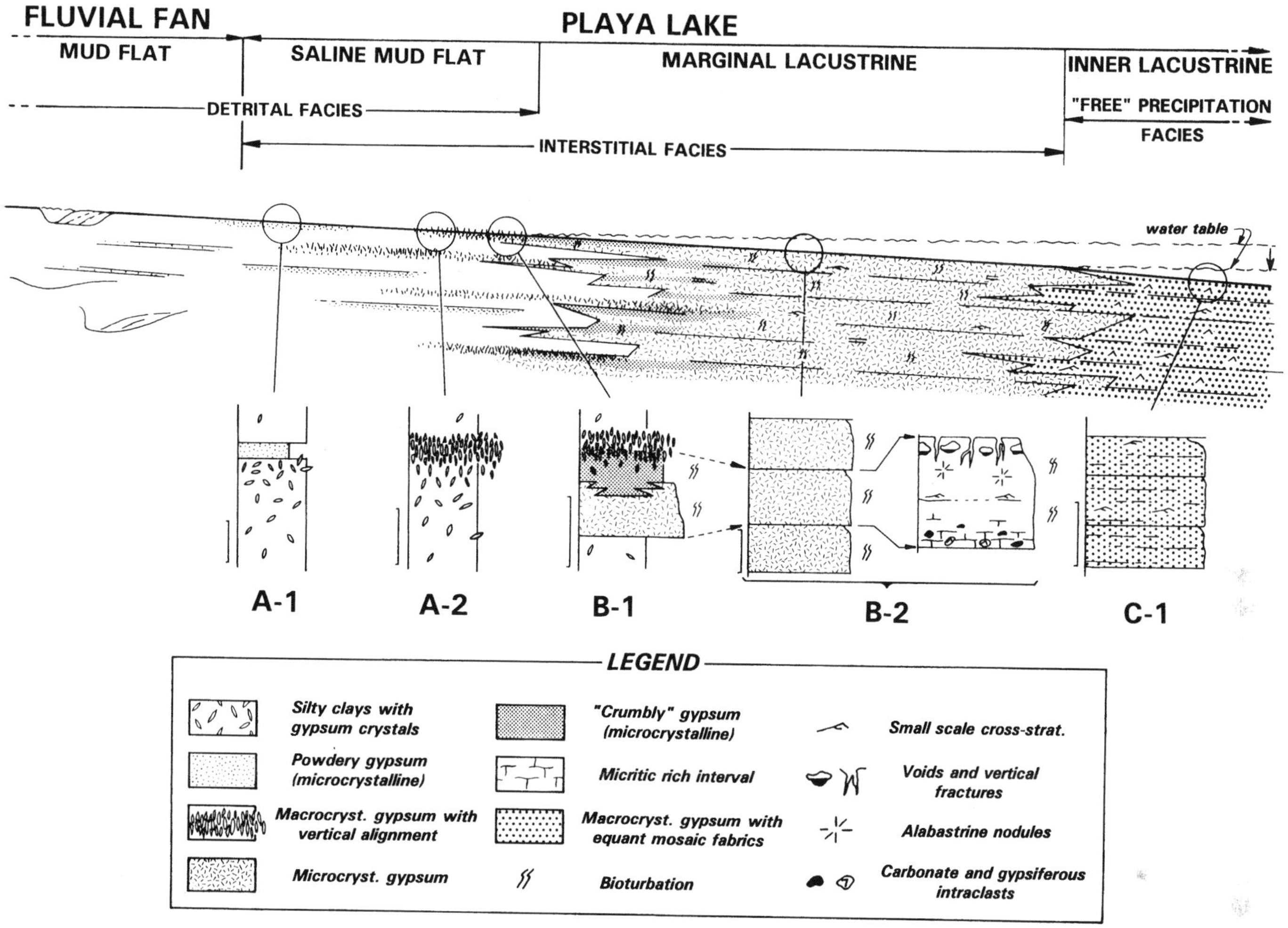

Fig. 3. Detrital and saline facies distribution of stage 2 on a playa lake sedimentation model, showing the principal sequences that characterize each subenvironment. See text for explanation of sequences. Vertical scale bar on sequence schemes: 0.5 m approximately.

tion, the saline units can be considered as evaporitic megasequences (4th order cycles), which were initiated with thick silty-clay deposits representing time intervals when the fans were more active, and consequently with the maximum saline dilution. On the basis of the palaeontological localities and faunas (see Chapter C6), the stratigraphic time duration of the four saline units together is estimated as 1 my, which gives an average length of 0.25 my for each saline unit.

The lateral activity of the fans influenced the saline episodes (Figs. 1 and 2), providing fine-grained detritus and diluting waters (surface and ground-waters) to the lacustrine environments. In turn, tectonics affected the differential activity of the fans and the environmental distribution (see Chapter C5); saline facies distribution is asymmetric in the basin. The activity of the Tórtola fan was confined to the first saline unit, while the Villalba de la Sierra fan was still active during deposition of saline units 3 and 4. Facies that need more stable or deeper water bodies (marls, macrocrystalline gypsum with equant mosaic) occur close to the Villalba de la Sierra fan deposits.

The evolution of the lacustrine environments shows an initial stage (first saline unit) of low salinity (carbonate lacustrine) restricted to the central part of the cross-section, showing asymmetric facies distribution (Fig. 2). An important expansion of lacustrine environments is exhibited during the second saline unit with an increase of water salinity. This saline unit is characterized by deposits generated in more-stable water bodies. Salinity remained almost constant during gypsum sedimentation of the third and fourth saline units. In addition, saline environments migrated to the north as the fluvial system of the Villalba de la Sierra fan died.

In spite of the relationship of the saline units to humid-area fans, the nature of the source area, the progressive aridity, and the endorheism of the basin are the main factors controlling the development of saline lakes during stage 2 of the Upper Unit in the Loranca Basin. The transitional character of this stage is clear from the establishment of more extensive saline environments during stage 3.

Acknowledgments

This research was supported by the CICYT PB85–0022 project (1987–1989).

References

Arakel, A.V. (1980). Genesis and diagenesis of Holocene evaporitic sediments in Hunt and Leeman lagoons, Western Australia. *J. Sediment. Petrol.*, **50**: 1305–1326.

Arribas, J., Bustillo, M.A. and Díaz-Molina, M. (1991). Chert in bioturbated sediments of sabkha paleoenvironment. *VI Flint International Symposium*: 29–33.

Arribas, M.E. (1986). Petrología y análisis secuencial de los carbonatos lacustres del Paleógeno del sector N de la Cuenca Terciaria del Tajo (Provincia de Guadalajara). *Cuad. Geol. Ibérica*, **10**: 295–334.

Bustillo-Revuelta, A. and Díaz-Molina, M. (1980). Silex tobáceos en el Mioceno inferior continental (provincia de Cuenca). Un ejemplo de silicificaciones de paleosuelos en ambiente de lago-playa. *Bol. R. Soc. Esp. Hist. Nat. (Geol.)*, **78**: 227–241.

Daams, R. and Meulen, A.J. van der (1984). Paleoenvironmental and paleoclimatic interpretation in the Upper Oligocene and Miocene of north central Spain. *Paléobiol. Cont.*, **14**: 241–257.

Díaz-Molina, M., Arribas-Mocoroa, J. and Bustillo-Revuelta, A. (1989). The Tórtola and Villalba de la Sierra fluvial fans: Late Oligocene–Early Miocene, Loranca Basin, Central Spain. *4th International Conference on Fluvial Sedimentology*, Sitges, Spain. Field trip **7**, 74 pp.

Haq, B.U., Hardenbol, J. and Vail, P.R. (1988). Mesozoic and Cenozoic chronostratigraphy and eustatic cycles. In C.K. Wilgus *et al.* (eds.), *Sea-level changes: an integrated approach, S.E.P.M. Special Publication*, **42**: 71–108.

Hardie, L.A., Smoot, J.P. and Eugster, H.P. (1978). Saline lakes and their deposits: a sedimentological approach. In A. Matter and M.E. Tucker (eds.), *Modern and ancient lakes sediments, I.A.S. Special Publication*, **2**: 7–41.

Lowenstein, T.K. and Hardie, L.A. (1985). Criteria for the recognition of salt-pan evaporites. *Sedimentology*, **32**: 627–644.

Ortí, F. (1979–82). Características deposicionales y petrológicas de las secuencias evaporíticas continentales en las cuencas terciarias peninsulares. *Temas Geológico-Mineros*, **6, II**: 485–606. IGME, Madrid, 1982.

Ortí, F. (1987). La zona de Villel–Cascante–Javalambre. Introducción a las formaciones evaporíticas y al volcanismo jurásico. *XXI Curso de Geología Práctica de Teruel*, Universidad de Verano de Teruel, 56–95.

Ortí, F., Salvani, J.M., Rosell, L. and Ingles, M. (1989). Sistemas lacustres evaporíticos del Terciario de la Cuenca del Ebro. *Geogaceta*, **6**: 103–104.

Reeves, C.C. (1978). Economic significance of playa lake deposits. In A. Matter and M.E. Tucker (eds.), *Modern and ancient lakes sediments, I.A.S. Special Publication*, **2**: 279–290.

Rosen, M.R. and Warren, J.K. (1990). The origin and significance of groundwater-seepage gypsum from Bristol Dry Lake, California, USA. *Sedimentology*, **37**: 983–996.

Schreiber, B. Ch. (1978). Environments of subaqueous gypsum deposits. In *Marine evaporites, S.E.P.M. Short Course* no. 4, 43–73.

Stoops, G. and Ilaiwi, M. (1981). Gypsum in arid soils: morphology and genesis. *Proceedings 3rd International Soil Classification Workshop*, ACSAD, 175–185.

Warren, J.K. (1982). The hydrological setting, occurrence and significance of gypsum in late Quaternary salt lakes in South Australia. *Sedimentology*, **29**: 609–637.

Watson, A. (1985). Structure, chemistry and origin of gypsum crusts in southern Tunisia and the central Namib Desert. *Sedimentology*, **32**: 855–875.

C9 Shallow carbonate lacustrine depositional controls during the Late Oligocene – Early Miocene in the Loranca Basin (Cuenca Province, central Spain)

M.E. ARRIBAS, R. MAS AND M. DÍAZ-MOLINA

Abstract

Shallow carbonate lakes with ramp-type margins were developed during the deposition of the Upper Unit of the Loranca Basin in two different sedimentary realms controlled by tectonics: restricted and extensive carbonate lacustrine environments. The restricted ones, located in the central part of the basin, were created by tectonic uplifting of compressive structures (Huete anticline) that provoked a general rise of base level in the Tórtola alluvial system. The extensive lacustrine environments were developed in isolated areas as a result of topographic threshold formation which separated these lakes from the Tórtola system and caused them to form in the southern part of the basin.

Introduction

The Upper Unit is part of the fill of the Loranca Basin (Díaz-Molina *et al.*, 1985, 1989). This Unit was deposited during the Late Oligocene and Early Miocene and is formed by two coalescing alluvial depositional systems, the Tórtola and the Villalba de la Sierra fluvial fans (Díaz-Molina *et al.*, 1989). Stratigraphically, the Upper Unit may be divided into three stages which reflect the rate of diastrophism and changes in the base level of the fluvial systems (Díaz-Molina *et al.*, 1989; see also Chapter C5).

In the Upper Unit, both stages 1 and 2 display some carbonate lithosomes interbedded with the siliciclastic facies. These carbonate deposits correspond to shallow lacustrine episodes developed in different sedimentary realms in the basin. Basically, two types of lacustrine environment can be differentiated: 1. shallow lakes which were directly related to the alluvial fan system; and 2. lakes which were not connected to the alluvial system and which represented a different depositional system. An example of each of these lacustrine environments is studied in two different areas in the Loranca Basin: Huete sector lakes for the first case and El Hito sector lakes for the second (Fig. 1).

Lakes associated with the alluvial fan system: Huete sector

Lacustrine carbonate deposits appear at the top and bottom of stages 1 and 2 respectively and they are separated by an unconformity in this area related to a break in the infilling of the basin (Fig. 2). These form two separated restricted lacustrine episodes: the first and the second lacustrine episodes (Arribas *et al.*, 1992). The carbonate deposits of the second episode change to saline sediments towards the north (see Chapter C8). Four facies characterize both lacustrine episodes: gastropodal limestones, oncolitic limestones, marls and lutites.

The gastropodal limestones are exclusively present in the first lacustrine unit and are composed of bioclastic and intraclastic wackestones with oncolites. Bioclasts include: gastropods, charophytes and cyanobacterial colonies. Besides these components it is possible to see many pedogenic micritic nodules and root traces which represent the desiccation of the lacustrine basin. At the top of this limestone a well-developed paleokarstic surface is present (Arribas *et al.*, 1992). These limestones are interpreted as lacustrine deposits with superimposed pedological features.

The oncolitic limestones are characteristic in the second lacustrine unit and display a quite homogeneous aspect with abundant oncolites as the most important features. The deposits of this facies are oncolitic mudstones and wackestones with bioclasts. At the top of these limestones a high concentration of stromatolites surrounding calcified stems of plants in life position occurs. Sometimes, in the upper part of these limestones, phytoturbation and early diagenetic structures are present due to subaerial exposure. These limestones are interpreted as shallow lacustrine deposits affected by pedogenesis.

The marls are interbedded with the gastropodal and oncolitic limestones. They are homogeneous, generally silty and they can contain gypsum crystals and oncolites. These facies are interpreted as lake-bottom sediments produced by intermixing of locally produced carbonate and fine siliciclastic sediments coming from the surrounding floodplains.

The lutites lie above the gastropodal limestones as a relict soil. Phytoturbation structures (root-traces) and large-scale mottling are common. The lutites have been interpreted as floodplain sediments which formed over a carbonate substrate of gastropodal limestone.

Very similar facies are typical in many Mesozoic and Tertiary shallow lacustrine carbonate sequences (e.g. Freytet & Plaziat, 1982; Cabrera *et al.*, 1985; Arribas, 1986; Alonso *et al.*, 1991), which

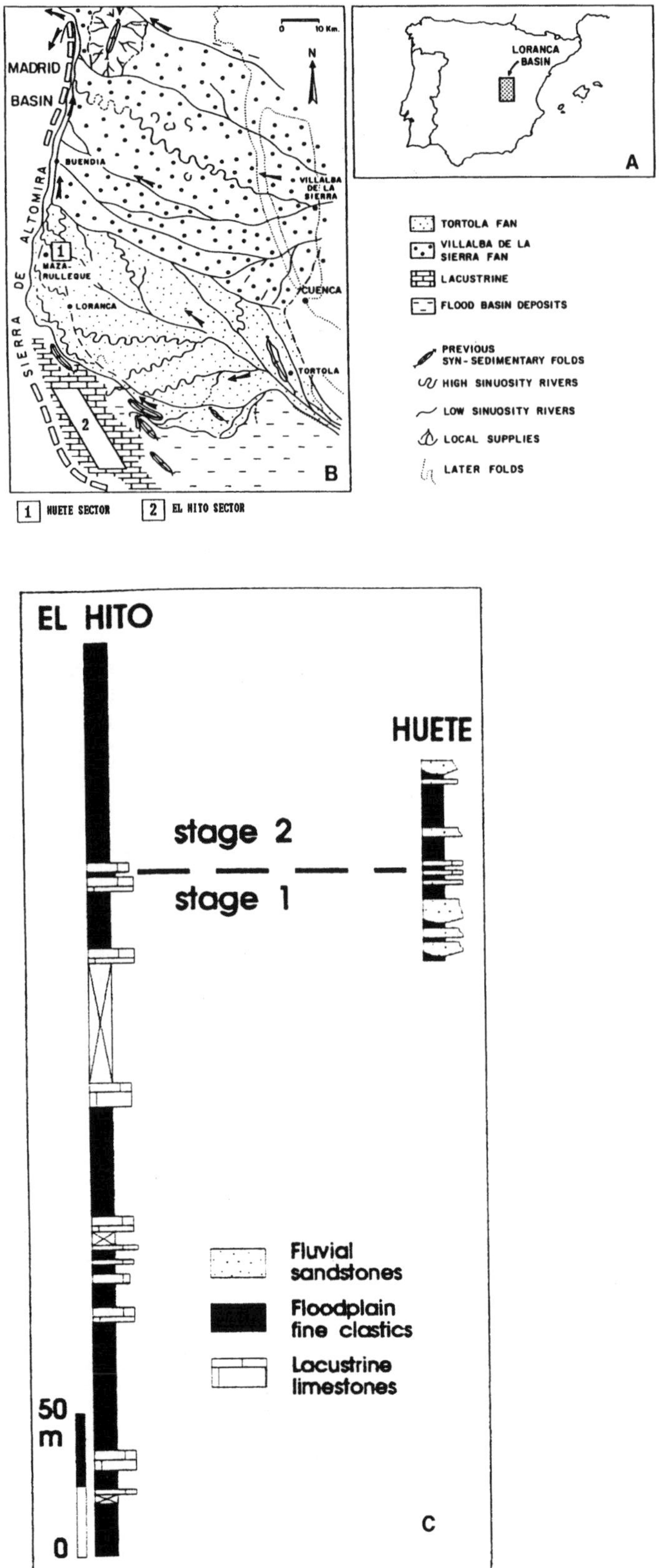

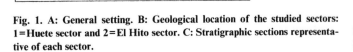

Fig. 1. A: General setting. B: Geological location of the studied sectors: 1=Huete sector and 2=El Hito sector. C: Stratigraphic sections representative of each sector.

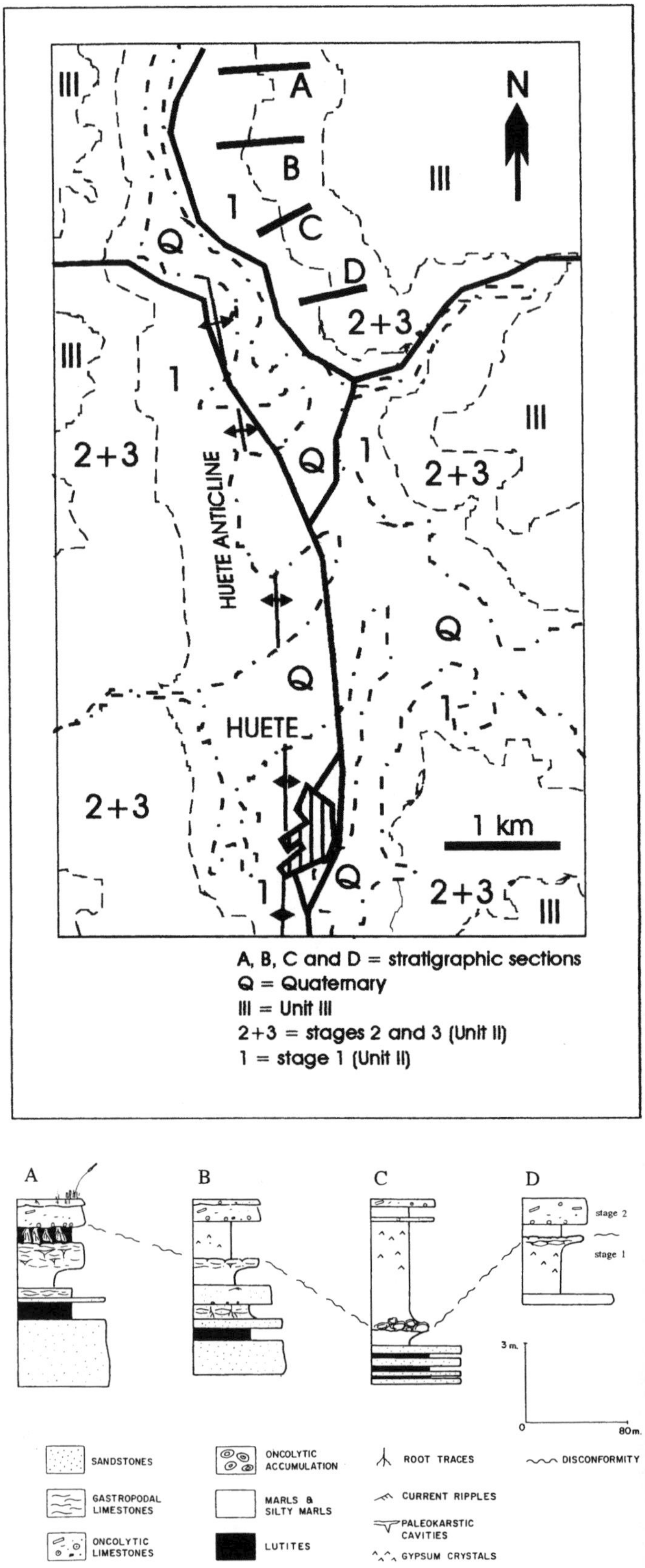

Fig. 2. Location map and stratigraphic correlation of the detailed sections (A–D) in the Huete sector.

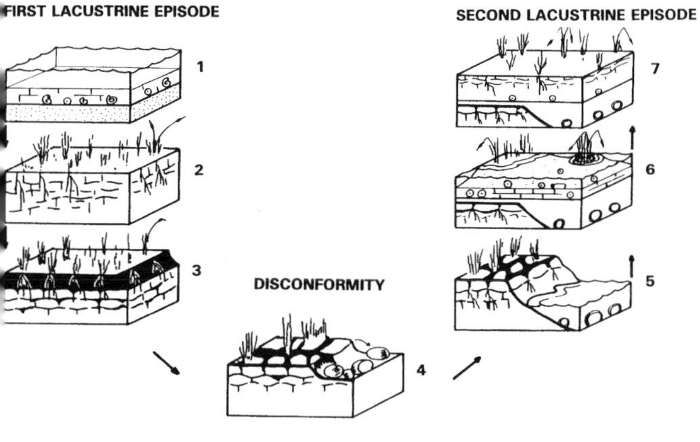

Fig. 3. Evolution of near-surface sediments during the two lacustrine episodes in the Huete sector.

would characterize dominantly low-energy lakes with low-gradient ramp-type margins (Platt & Wright, 1991).

Lacustrine evolution

All these facies form two separate episodes of shallow lacustrine sedimentation (Fig. 3). The sedimentary evolution of these lacustrine episodes was controlled by syn-sedimentary tectonic deformation, which led to the development of the Huete anticline. The sedimentary interval existing between the two lacustrine events is considered to be the local stratigraphic boundary between stages 1 and 2 of the Upper Unit.

The development of the first lacustrine episode was related to a lateral displacement of the active channels and an extensive flooding of the previous floodplain sediments (sandstones and lutites) (1 on Fig. 3), produced by the tectonic deformation of the anticline system of the Huete sector. Gradually, the lacustrine basin dried up, changing towards a palustrine basin where many pedological processes were developed (2 on Fig. 3). Later, with renewed movement of the anticline, a karstic surface was formed over the first lacustrine unit (3 and 4 on Fig. 3).

Following this tectonic reactivation, which produced the important topographic modification and erosion, the second lacustrine unit was created during a renewed ascent of the water table (5 and 6 on Fig. 3). Quick progressive replenishment and colonization led to the development of a cover of vegetation towards the centre of the lake that resulted in the desiccation of the basin (7 on Fig. 3). Sedimentation evolved quickly from a lacustrine–palustrine environment to an alluvial environment at the beginning of stage 2, and a new alluvial plain was created over the site of the former lake.

Lakes disconnected from the alluvial system: El Hito sector

These lakes occupied extensive areas during stage 1 in the most southern part of Loranca Basin (Fig. 1). These areas were isolated from the Tórtola system by a topographic threshold created by the tectonic activity with a series of anticlines located between this part of the basin and the alluvial system (Fig. 1). In these areas four well-developed carbonate lacustrine episodes took place without detrital input coming from the Tórtola fluvial system.

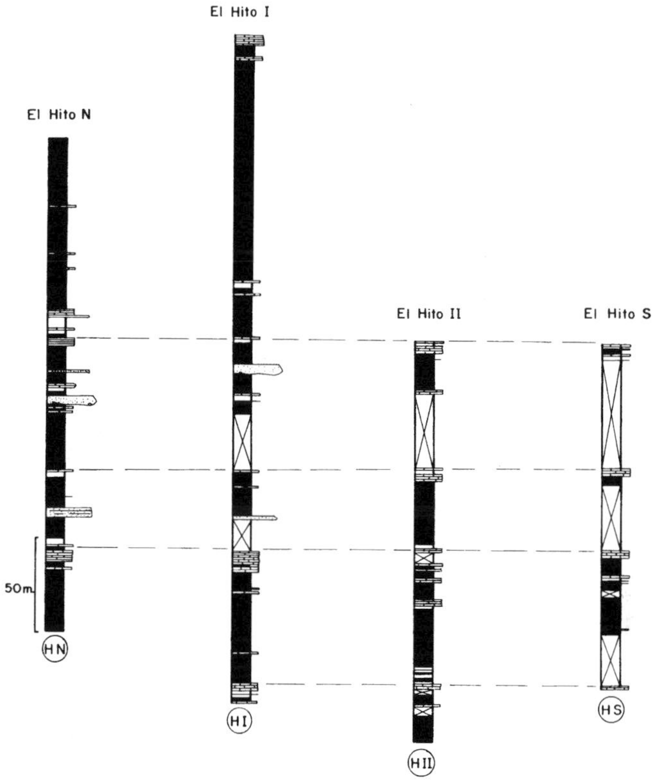

Fig. 4. Stratigraphic sections in the El Hito sector (see also Fig. 5).

In the El Hito sector, these lacustrine episodes are exposed in a N–S-trending outcrop (approximately 15 km long) of four E-dipping carbonate units (*cuestas*) which alternate with muddy–silty units. This outcrop is oriented parallel to the Altomira Range (Fig. 1). All these carbonate lithosomes display very similar facies assemblages in this part of the basin and their maximum thickness in the depocentre areas is up to 10–12 m for each one (Fig. 4). Four stratigraphic sections are presented for the sedimentological analysis of the carbonate episodes and three main facies assemblages can be distinguished: laminated limestones, massive early limestones and calcretes (Fig. 5).

The laminated limestones are ostracodal–charophytic wackestones–packstones displaying dark–light lamination as a consequence of the alternation of sparry laminae very rich in ostracods and charophytes and micritic–cryptalgal laminae exhibiting stromatolitic fabric. In the upper part of these facies, root traces are frequent and occasionally calcitic pseudomorphs of gypsum. Sometimes beds of massive intraclastic packstone are intercalated. These facies are interpreted as the result of carbonate deposition in shallow lakes characterized by quiet and clear waters. Root traces and scarce evaporites are related to desiccation episodes. The scarce massive intraclastic levels could correspond to storm events in these generally quiet shallow lakes.

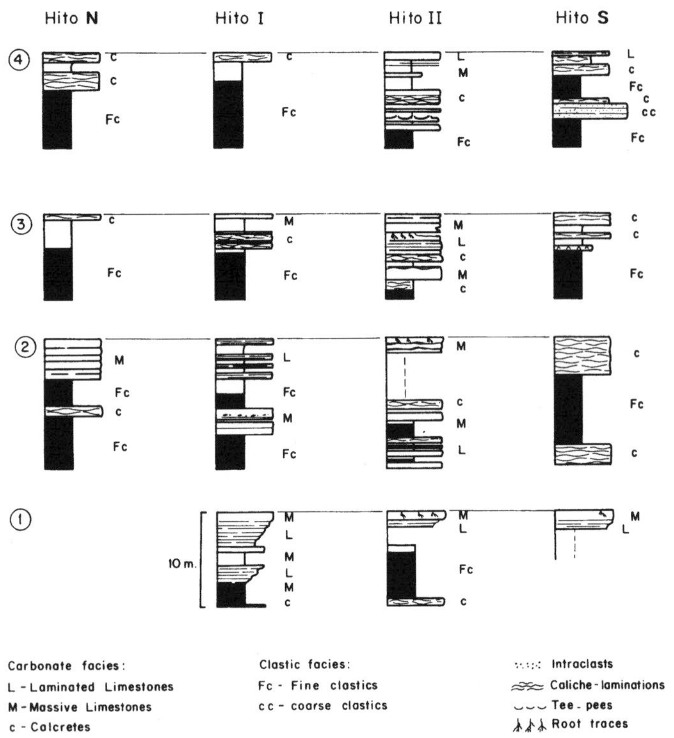

Fig. 5. Detailed correlation of the four lacustrine episodes in the El Hito sector.

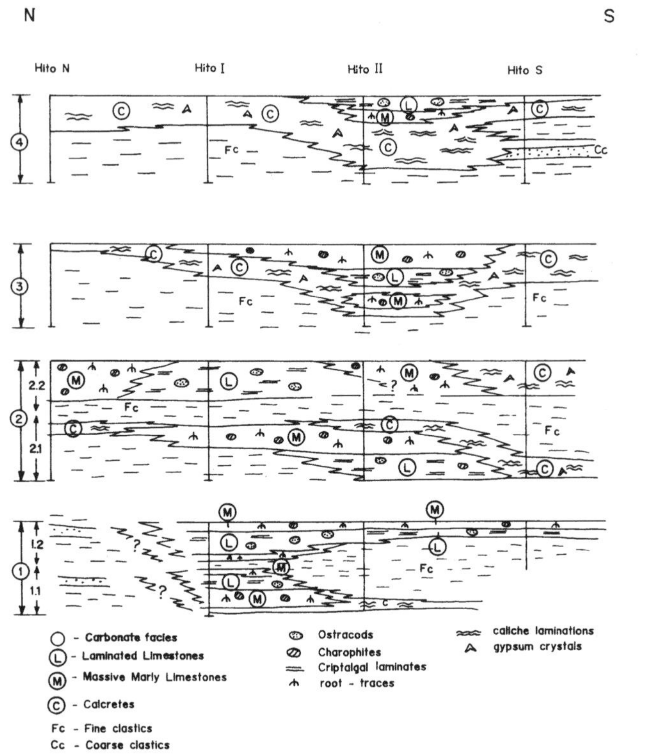

Fig. 6. Facies distribution in the four lacustrine episodes (El Hito sector).

In the first and second carbonate units laminated facies are well represented, exhibiting a better development towards the north-central part of the area (Hito I); however, in the third and fourth units, laminated carbonates are progressively displaced towards the south showing thin sequences there (Fig. 6).

The massive marly limestones are wackestones with charophytes frequently displaying fine clastic influx. Root traces, bioturbation and calcitic pseudomorphs of gypsum crystals are present. Well-bedded limestones with intercalated levels of marls are quite characteristic. Although this facies is better represented towards the central areas in the second and third carbonate units, it occurs in all units transitionally (vertically and laterally) related to the laminated limestones facies (Fig. 6). This facies would correspond to very shallow marginal areas of the lakes which would be colonized by charophytes and undergo frequent desiccation periods. Fine clastic sediments would be the consequence of flood episodes in the distributary fluvial systems.

Calcretes. These are principally represented by mudstones–wackestones with charophytes and ostracods which have lost most of their primary fabrics by early diagenetic processes. The primary fabrics corresponded to those of the aforementioned laminated and massive limestones. Diverse overprinted early diagenetic features are recognizable: nodulization, recrystallization, brecciation and tepees structures, root-traces, authigenic growth of gypsum crystals, crustose caliche-lamination, dolomitization and silicification. All these facies are interpreted as the result of early diagenetic processes modifying the original shallow lacustrine facies during subaerial exposure and desiccation. From an environmental point of view the calcretes formed mainly in palustrine areas that rimmed the carbonate shallow lakes during deposition of laminated and massive facies.

Calcretes are better represented in the third and fourth carbonate units in all sections. In the first and second units they are only significantly developed towards the southernmost areas (Hito II and Hito South) (Fig. 6).

In both this sector, and the Huete sector, the carbonate facies correspond to classical low-energy shallow carbonate lakes which displayed ramp-type margins as defined by Platt & Wright (1991).

Lacustrine evolution

During the development of the first lacustrine episode two separate stages can be differentiated (1 on Fig. 7). In the first one the carbonate lake was exclusively located towards the northern part of the El Hito area (1.1 on Fig. 7). During the second one the lake extended across the whole north-central part of the basin, where the more permanently subaqueous zone was developed (1.2 on Fig. 7).

The second lacustrine episode was also characterized by two separate stages during its development (2 on Fig. 7). In the first stage the carbonate lake changed its position dramatically towards the south-central part of the area, where a deeper and permanently subaqueous environment was dominantly developed (2.1 on Fig. 7). During the second stage the deeper part of the lake was located in the north-central zone, while palustrine areas with calcretes were

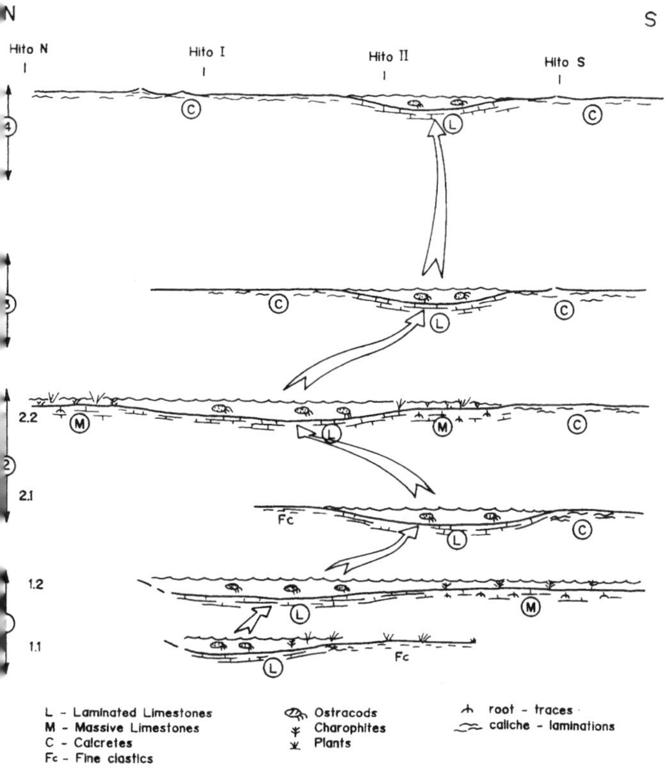

Fig. 7. Lacustrine evolution in the El Hito sector.

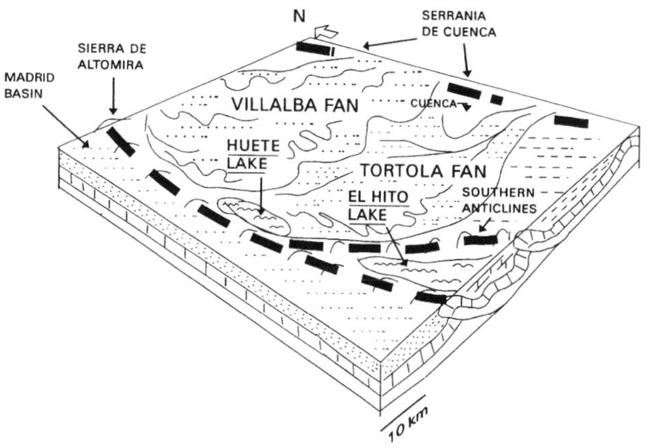

Fig. 8. Geological realm for the carbonate lacustrine episodes during deposition of the Upper Unit in the Loranca Basin.

ocated in the south. The shallow charophytic marginal areas of the ake developed between these subenvironments (2.2 on Fig. 7).

The third lacustrine episode was characterized by a new displacement of the more permanent part of the lake towards the southern one (3 on Fig. 7). This southward position of the deeper part of the ake was also repeated during the fourth lacustrine episode (4 on Fig. 7). An important development of palustrine areas where calcretes were formed was a characteristic feature of the third, and particularly the fourth, episodes.

This evolution shows a progressive displacement of the lakes towards the south. This is interpreted as the result of compressive movements in the southern anticlines of the Loranca Basin. Moreover, these tectonic movements caused the isolation of these acustrine areas from the Tórtola system.

Although at first sight the type of lake and lake evolution appear to correspond essentially to those of the classical shallow carbonate akes, there are some peculiarities which are important to note. As was mentioned above, the types of facies and their arrangement are quite similar for each episode but also their fossil content is remarkably homogeneous. This is characterized by an ostracodal–charophytic (oogonia exclusively) association, in the deeper subenvironment, and charophytes (oogonia and stems) with very scarce ostracods in the more shallow subaquatic areas. In the subaerial palustrine areas authigenic growth of gypsum crystals occurred. Moreover, gastropods were absent in the subaquatic environments and very thin stromatolitic mats grew over the bottom of the lakes in their more permanent and deep lacustrine zones.

All these features indicate relatively high saline conditions in the waters of the lakes, favouring a restricted biota for each lacustrine episode. These relatively high saline conditions would have been related to the endorheic environment created by the tectonic activity of the southern anticlines which isolated this area from the rest of the Loranca Basin. This area, located between these anticlines and the Altomira high, was a sub-basin separated from the clastic influx of the Tórtola fan.

Lakes distribution related to tectonic controls: discussion

The Loranca Basin is a foreland basin limited by two important compressive tectonic zones with southwestward vergency: the Serrania de Cuenca (Iberian Range) to the east and the Sierra de Altomira to the west. As well as controlling the alluvial systems, both compressive structures played a very important role in the lacustrine deposition coeval with Tórtola alluvial fan development during the Loranca Basin infilling. As in many other foreland basins, tectonically controlled shallow lakes with ramp-type margins (Platt & Wright, 1991) are a typical feature in this basin. This tectonic influence controlled the development of the two different types of lake: those directly related to the fluvial fan system and those which were separated from the system (Fig. 8).

Tectonics acted in two different ways. For the Huete sector lakes, which formed in the more distal areas of the Tórtola fluvial fan system, the tectonic uplift of the compressive structures parallel to the Sierra de Altomira (e.g. the Huete anticline) caused a general rise of local base level in the fluvial system. So, the volume of water retained in this area of the basin was increased and lacustrine environments were developed as a result (Fig. 8).

In the El Hito sector, where the lacustrine system was more extensive and continuous, tectonics was responsible for the isolation of these lakes from the Tórtola fluvial system. A system of compressive structures (the southern anticlines) was created between the El Hito area and the Tórtola fan. These structures caused a NW–SE topographic threshold which blocked detrital influx from the Tórtola System during each lacustrine episode (Fig. 8).

Conclusions

The development of the shallow carbonate lakes with ramp-type margins during the deposition of the Upper Unit of the Loranca Basin took place in two different sedimentary realms: restricted lacustrine environments, which were located in the distal part of the alluvial fan, and extensive lakes, which were located in isolated areas separated from the fluvial system.

The magnitude of these episodes varied conspicuously depending on the geological realm. The greatest development of lacustrine carbonate deposits is located in the southern Loranca Basin. This area constituted an isolated sub-basin created as a consequence of the existence of an uplifted area (the southern anticlines) between the main fluvial system (Tórtola fan) and the El Hito depression.

However the less extensive lakes directly associated with the Tórtola system (Huete area) are related to a general rise of the base level in the alluvial system as a consequence of the uplifting of compressive structures like the Huete anticline in the central part of the basin.

Acknowledgments

This research was supported by the CICYT PB85–0022 project (1987–1989) and by the commission of the European Communities, JOULE Programme, sub-programme: Hydrocarbons (1989–1992).

References

Alonso, A., Meléndez, N. and Mas, J.R. (1991). Sedimentación lacustre durante el Cretácico en la Cordillera Ibérica, España. *Acta Geol. Hisp.*, **26** (1): 35–54.

Arribas, M.E. (1986). Petrología y análisis secuencial de los carbonatos lacustres del Paleógeno del sector N de la Cuenca Terciaria del Tajo (prov. Guadalajara). *Cuad. Geol. Ibérica*, **10**: 295–334.

Arribas, M.E., Díaz-Molina, M. and Mas, R. (1992). Covered subsoil karst development in shallow lacustrine carbonate deposits: Late Oligocene – Early Miocene, Loranca Basin (Cuenca province, Central Spain). *III Congreso Geológico de España y VII Congreso Latinoamericano de Geología*, **1**: 185–196.

Cabrera, L., Colombo, F. and Robles, S. (1985). Sedimentation and tectonics interrelationships in the Paleogene marginal alluvial systems of the SE Ebro Basin: transition from alluvial to shallow lacustrine environments. In *Exc. guide-book I.A.S. 6th. European Regional Meeting, Lérida, Spain*, 393–492.

Díaz-Molina, M., Bustillo-Revuelta, M.A., Capote, R. and López-Martínez, N. (1985). Wet fluvial fans of the Loranca basin (central Spain). Channel models and distal bioturbated gypsum with chert. In *Exc. guide-book I.A.S. 6th. European Regional Meeting, Lérida, Spain*, 149–185.

Díaz-Molina, M., Arribas-Mocoroa, J. and Bustillo-Revuelta, M.A. (1989). The Tórtola and Villalba de la Sierra fluvial fans: Late Oligocene–Early Miocene, Loranca Basin, Central Spain. In *Exc. guide-book. 4th International Conference on Fluvial Sedimentology, Barcelona–Sitges, Spain*, 74 pp.

Freytet, P. and Plaziat, J. (1982). Continental carbonate sedimentation and pedogenesis. *Contributions to Sedimentology*, **11**: 216 pp.

Platt, N.H. and Wright, V.P. (1991). Lacustrine carbonates: facies models, facies distributions and hydrocarbon aspects. *Spec. Publ. Int. Ass. Sediment.*, **13**: 57–74.

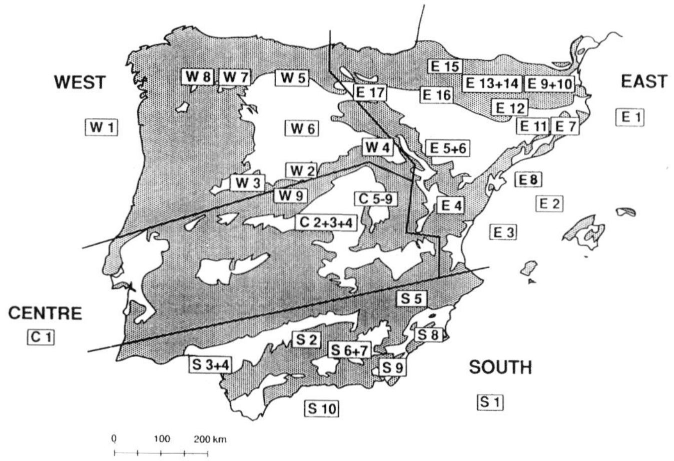

Part S
South

South of the Granada Basin, on the river Cacin. This unconformity of Lower Tortonian (Middle Miocene) sediments overstepping vertical Alpujarride marbles (probably Triassic in depositional age) of the Internal Zone of the Betic Cordillera, illustrates the major components of the structure of this part of southern Spain (photo C. Sanz de Galdeano).

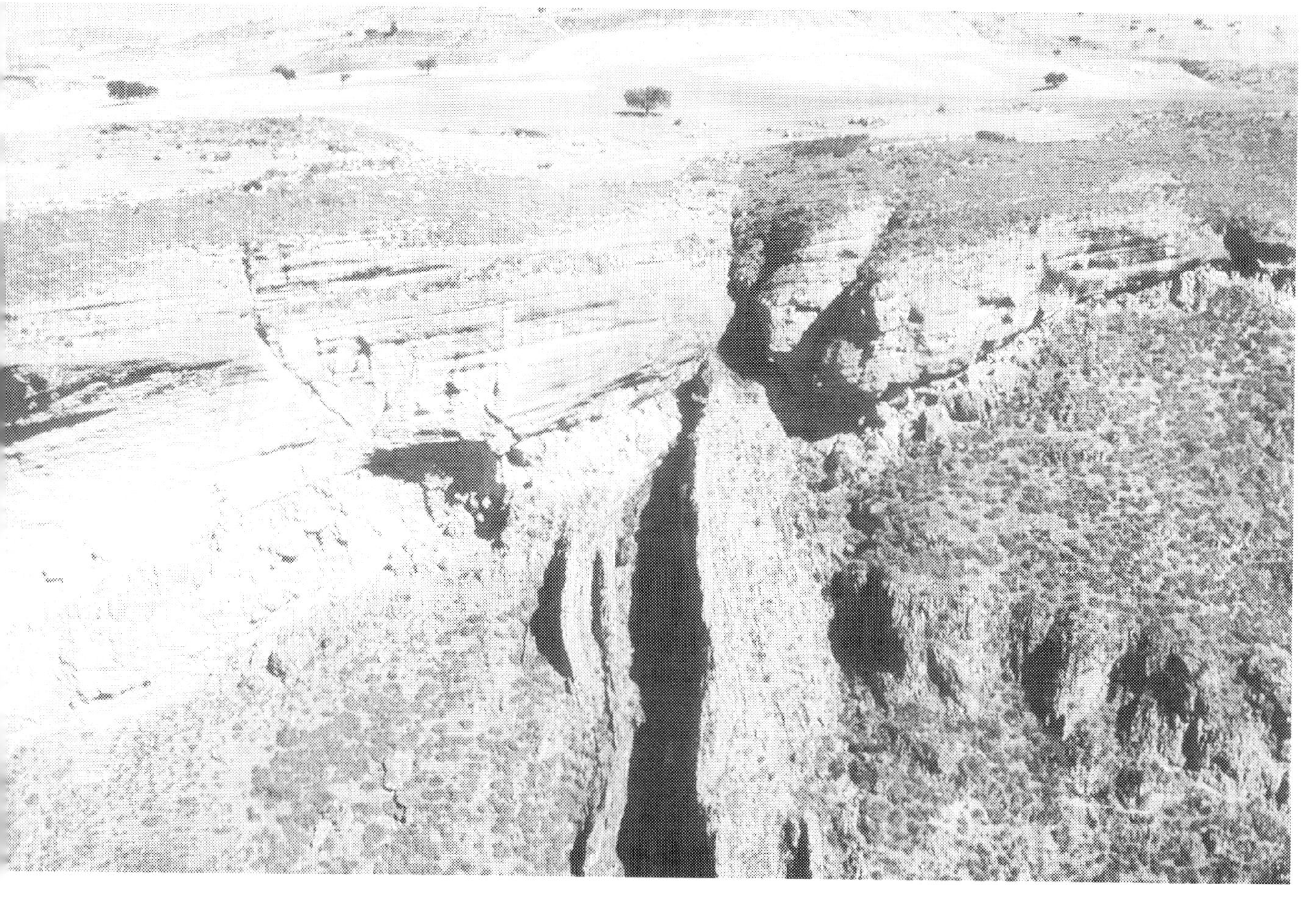

S1 The Betic Neogene basins: introduction

CH. MONTENAT

The Betic Cordilleras stretch across the southern part of Iberia for about 650 km, from Cabo de la Nao (south of Valencia), in the east, to Cadiz, in the west. These Cordilleras consist of a complex assemblage of different geological domains (Figs. 1 and 2).

The *External* and *Internal Zones* correspond to different continental blocks, the southern Iberian paleomargin and the Alboran block respectively, juxtaposed along the complex, ENE–WSW-trending wrench corridor, the North Betic Wrench fault (Fig. 2A).

The *External Zone* includes the following domains: 1. the Prebetic autochthonous folded domain, linked northwards to the Meseta; and 2. the Subbetic allochthonous units overthrusting the Prebetic northwards.

The Prebetic and Subbetic are composed of sedimentary series ranging from Triassic to Miocene in age. The Late Triassic evaporites acted as a major level of detachment and have played an important part in the tectono-sedimentary development.

The *Internal Zone* is composed of numerous allochthonous units: the Flysch, the Dorsal unit, and a pile of nappes of sedimentary and varied metamorphic rocks, mainly altered Palaeozoic and Triassic rocks (e.g. from the base to the top, the Nevado–Filabrides, Alpujarrides and Malaguides nappes).

Neogene successions are widely involved in the polyphase structures of the different domains; they display a great diversity of facies and deformation. It must be emphasised that the Betic orogen was deformed to a major extent during Miocene times, when large areas of various Betic domains were still drowned by the sea, often producing deep marine environments. Sedimentary and tectonic processes then acted jointly, and gave rise to a remarkable range of synsedimentary structures. Because of this, it is possible to establish an accurate chronology for the development of the orogen, the different Miocene stages being related to N–S to NW–SE compression, resulting from the collision of Africa and Iberia. The Betic Cordilleras also offer unusual opportunity for the study of Miocene basin dynamics, and this is illustrated in the following chapters.

Distinct tectono-sedimentary sequences correspond to a variety of basin patterns that evolved in the different Betic domains during Neogene times (Chapter S2), the most significant of which are picked out below.

The *eastern Prebetic* domain provides an example of synsedimentary folded basins developing alongside extensional areas (Chapter S5).

The vicinity of the *Subbetic front* is characterised by the occurrence of large-scale gravity-slide phenomena, particularly olistostromes of Triassic materials formed when the Subbetic nappes were emplaced in the southern Prebetic basin during late Seravallian times. These phenomena were particularly important in the large Guadalquivir depression (Fig. 1) which is a well-documented example of a foredeep basin (Chapters S3 and S4).

The *boundary between the External and Internal Zones* provides a record of particularly complex tectono-sedimentary evolution, characterised by intense deformation and large-scale transcurrent movements.

The *central and eastern parts of the Internal Zone* display a variety of Neogene sedimentary basin patterns, largely related to wrench tectonics (Chapters S7, S8 and S9). In the easternmost sector, important magmatic activity was linked to the basin development.

Comparisons with the adjacent offshore domains (Chapter S10) suggest that this Betic area offers great scope for future multidisciplinary work.

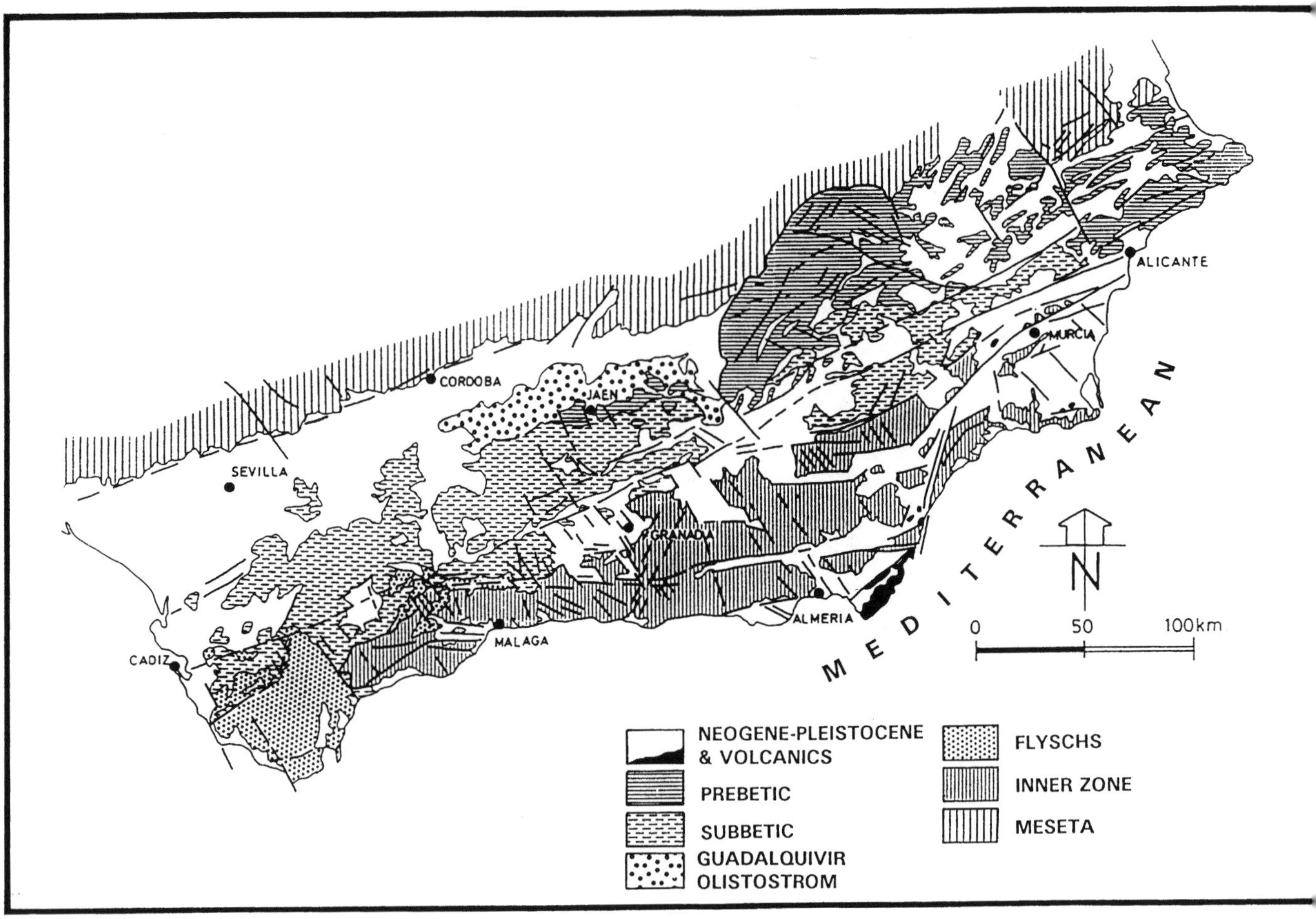

Fig. 1. Simplified geological map of the Betic area of southern Spain.

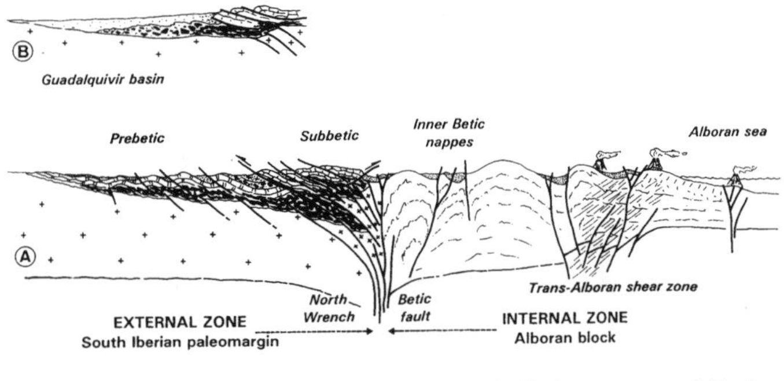

Fig. 2. Schematic cross-sections of A, the main Betic orogen, and B, the Guadalquivir foredeep basin. Neither cross-section is to scale, nor do they show individual structures precisely. The Trans-Alboran shear zone, marked by hatching, was characterised by various Miocene thermo-magmatic events and a significant thinning of the crust towards the Alboran Sea.

S2 Neogene palaeogeography of the Betic Cordillera: an attempt at reconstruction

C.M. SANZ DE GALDEANO AND J. RODRÍGUEZ-FERNÁNDEZ

Abstract

The Neogene palaeogeography of the Betic Cordillera is reviewed in the context of the following western Mediterranean events: the initial opening of the Algerian basin and the Alboran Sea, the progressive westward displacement and continuous stretching of the Betic–Rifian Internal Zone, the collision with the External Zones and formation of the North-Betic Strait, the destruction of the Flysch Basin and the beginning of the formation of the Gibraltar Arc. Depending on the particular event, this complex sequence began between the Early Aquitanian (or Late Oligocene?) and the end of the Early Burdigalian and continued with lesser intensity during the Middle Miocene, becoming practically paralysed in the Late Miocene. Then, the Prebetic zone was particularly deformed and emerged, the North-Betic Strait closed and considerable uplift of the entire Cordillera began. These last events, together with those taking place in the Rif and the eustatic lowering, brought about the Messinian salinity crisis in the Mediterranean. By this last time there was already a clear distinction between the Atlantic and Mediterranean basins in the Betic Cordillera. The situation in the Pliocene and more especially in the Quaternary was very similar to the present.

Introduction: regional setting of the Betic Cordillera

In the present chapter we present an interpretation of the Neogene palaeogeography of the Betic Cordillera in the context of the western Mediterranean.

Together with the Rif, the Betic Cordillera forms the westernmost part of the Alpine Mediterranean chains. It occupies southern and southeastern Spain (Fig. 1) and is made up of several important domains: the External Zones (divided into Subbetic and Prebetic), the Internal Zone or Betic *sensu stricto*, the Campo de Gibraltar Complex and the Neogene basins.

The 'Campo de Gibraltar Complex' (Flysch Units on Fig. 1) is part of the allochthonous units originally deposited in the North African Flysch Basin to the south of the South-Sardinian Domain (Sanz de Galdeano, 1990), between Calabria and the Rif and Tell (see Wildi, 1983).

The Neogene basins contain synorogenic and postorogenic sediments. The basins that formed during the Early and Middle Miocene have a more typically synorogenic (and, in some cases, in the External Zones, preorogenic) character. This does not mean that the later basins are not deformed; they are deformed, and locally to a high degree.

Numerous papers have been published on the Neogene evolution of the Betic Cordillera (Andrieux *et al.*, 1971; Andrieux & Mattauer, 1973; Weijermars, 1985a, b; Doblas & Oyarzun, 1989a, b; Platt & Vissers, 1989; Sanz de Galdeano, 1990; Sanz de Galdeano *et al.*, 1990; Frizon de Lamotte *et al.*, 1989, 1991, etc.). The Groupe de Recherche Néotectonique (1977) reconstructed the basic features of the Late Neogene palaeogeography. Also, Rodríguez-Fernández & Sanz de Galdeano (1990, 1992) and Sanz de Galdeano & Vera (1992) described the Betic palaeogeography during the Middle and Late Miocene.

Our reconstruction follows the model of Sanz de Galdeano (1990) and Sanz de Galdeano *et al.* (1990). It accepts the ideas of Van Bemmelen (1954, 1972, 1973) regarding the importance of the extensional processes which took place in the western Mediterranean, and also accepts the expulsion of the present Betic–Rifian Internal Zone towards the west proposed by Andrieux *et al.* (1971). It can also incorporate the subduction of Africa beneath the Mediterranean as proposed by Boillot *et al.* (1984) and Blanco & Spackman (1993).

Palaeogeography of the Betic Cordillera

The situation in the Aquitanian is represented by Fig. 2. In Fig. 2A the approximate position of the Internal Zone is shown. It was previously nappe structured. The Subbetic and Prebetic are shown to the west with their original palaeogeographic directions, at least from the Late Cretaceous on.

The scarcity and the type of Oligocene–Early Aquitanian sediments outcropping at present in the Betic–Rifian Internal Zone (Rodríguez-Fernández & Sanz de Galdeano, 1992), suggest that this Internal Zone was, to a large extent, emergent (Fig. 2B) and that the submerged parts were very shallow. The Betic External Zones (and those of the Rif and Tell) were, in general, submerged,

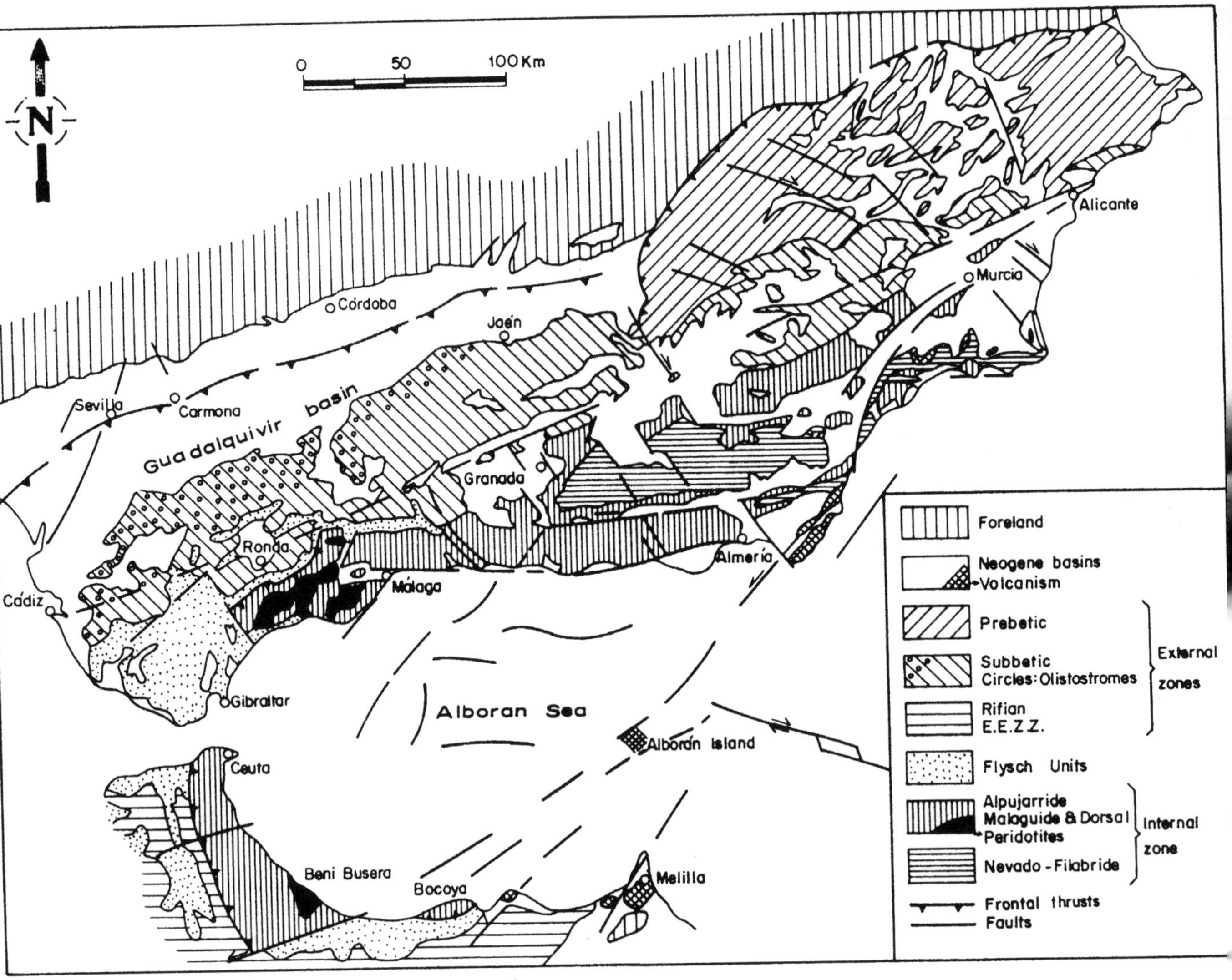

Fig. 1. General map of the Betic Cordillera.

although some islands were present. The unconformities present in the Subbetic and Prebetic indicate tectonic instability. The Flysch basin, possibly lying on oceanic crust, in turn received a large amount of mainly turbiditic sediments (Fig. 2B) (Durand-Delga, 1980; Wildi, 1983; Martín-Algarra, 1987).

Towards the end of the Aquitanian an important marine transgression began in the Betic–Rifian Internal Zone. This is clear from the present existence of Viñuela Group sediments (Latest Aquitanian–Early Burdigalian) both in sectors of the Internal Zone close to the External Zones, and within the Internal Zone itself (Sanz de Galdeano *et al.*, 1993) and even in the present Alboran Sea, where Comas *et al.* (1990a, b) detect Early Burdigalian deposits. We interpret this as due to the subsidence caused by the beginning of the extension and crustal thinning affecting the Alboran Sea and its vicinity, and which spread from the Algerian basin. As a result of this subsidence, a large marine area of several hundred metres deep, with marly sedimentation, became established (Sanz de Galdeano *et al.*, 1993). This process continued until the end of the Early Burdigalian.

At this time the Algerian basin was already opening strongly and the thinning of the continental crust in the Alboran Sea was considerable. Similarly, the westward displacement of the Betic–Rifian Internal Zone was very active. It was approximately at this time, at the end of the Early Burdigalian (Hermes, 1985), when the transpressive collision of the Internal Zone with the southern and southeastern margins of the Iberian Massif and with the margin situated to the northeast of the Atlas Mountains took place (so these margins then became the Betic and Rifian External Zones). This Early Burdigalian age seems to be indicated by the data of González Donoso *et al.* (1988), who determined the first appearance of detritus inherited from the External Zones in sediments of the Fuente–Espejos Formation (situated on the edge of the Internal Zone).

The Flysch basin was destroyed and its fragments were emplaced

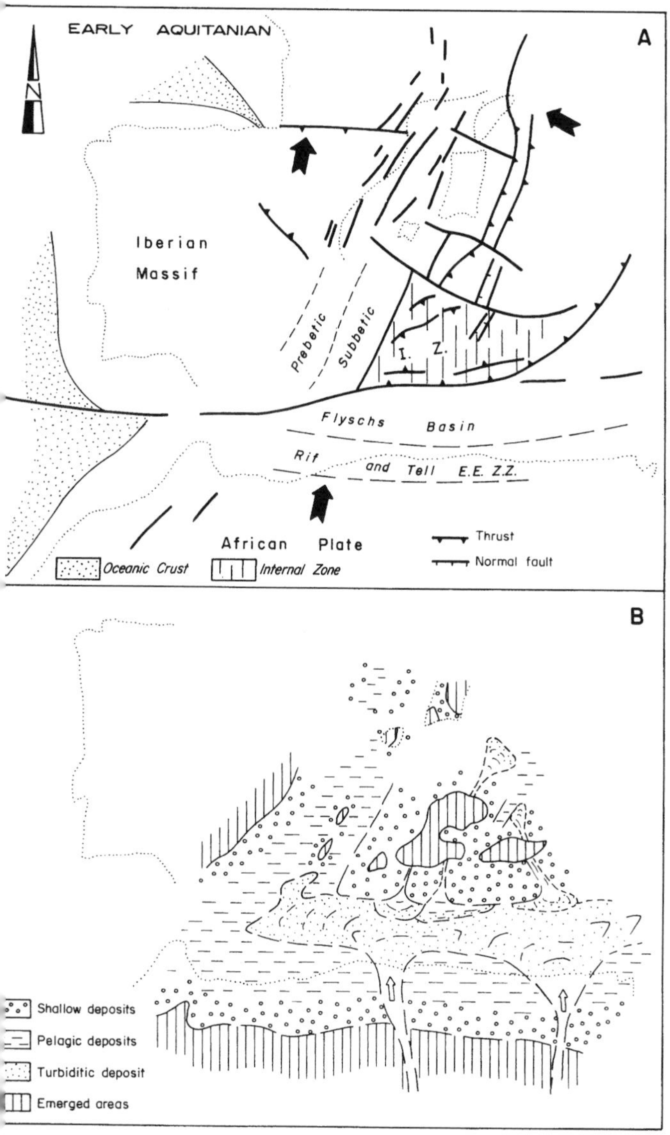

Fig. 2. Early Aquitanian reconstructions. A: Location of the principal domains of the western Mediterranean. I.Z.: Internal Zone. Black arrow: direction of the African and Iberian displacement. B: Principal sedimentary environments. The arrows indicate the main supply lines of detrital material.

on the External Zones, particularly in the sector situated to the west of the advancing Internal Zone (Campo de Gibraltar area and the Gulf of Cadiz). These fragments of the Flysch basin also underwent gravitational back-thrustings and then covered the abovementioned Viñuela Group sediments, as in the vicinity of Malaga and nearby areas (Bourgois, 1978; Sanz de Galdeano *et al.*, 1993), where deposition was abruptly curtailed, in the western sector at least.

During the Middle Burdigalian (Fig. 3A) the Subbetic was highly deformed (cf. Fig. 2A) as a result of the collision of the Internal Zone. The Subbetic was to a large extent displaced westwards and large olistostromic masses formed, particularly in the more western sectors. It was superposed on part of the Prebetic (which was likewise partially displaced) and overall its units rotated approximately 30 degrees clockwise, which agrees with the data obtained by Osete *et al.* (1988, 1989). The attempted reconstruction of the Middle Burdigalian palaeogeography can be seen in Fig. 3B. In this reconstruction the appearance of the North-Betic Strait is noteworthy (Colom, 1952; Calvo *et al.*, 1978), as it formed for the first time and continued to exist until the Late Miocene.

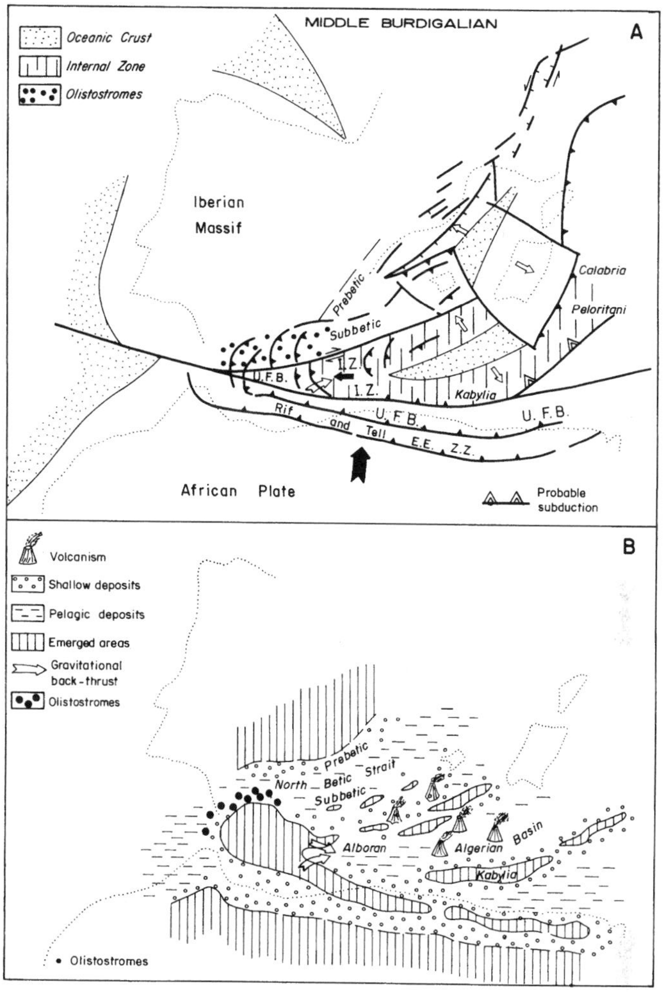

Fig. 3. Middle Burdigalian reconstructions. A: Location of the principal domains of the western Mediterranean. U.F.B.: Units of the Flysch basin. The opposing white arrows indicate extensions. The large black arrow indicates the main direction of tectonic displacement. B: Principal sedimentary environments.

The westward displacement of the Betic–Rifian Internal Zone continued, albeit less actively, during the Middle Miocene, as too did the opening of the Alboran Sea and the stretching and thinning of the complexes marking up the Internal Zone (García-Dueñas and Balanyá, 1991). Fig. 4A shows the reconstruction made for the beginning of the Langhian. It should be pointed out that fragments of the Flysch basin, together with some fragments of the External Zones, were still displaced towards the Gulf of Cadiz. Fig. 4B shows the hypothetical reconstruction of the palaeogeography of the Betic–Rifian domain at the beginning of the Langhian. It is noteworthy that a large part of the North-Betic Strait was very

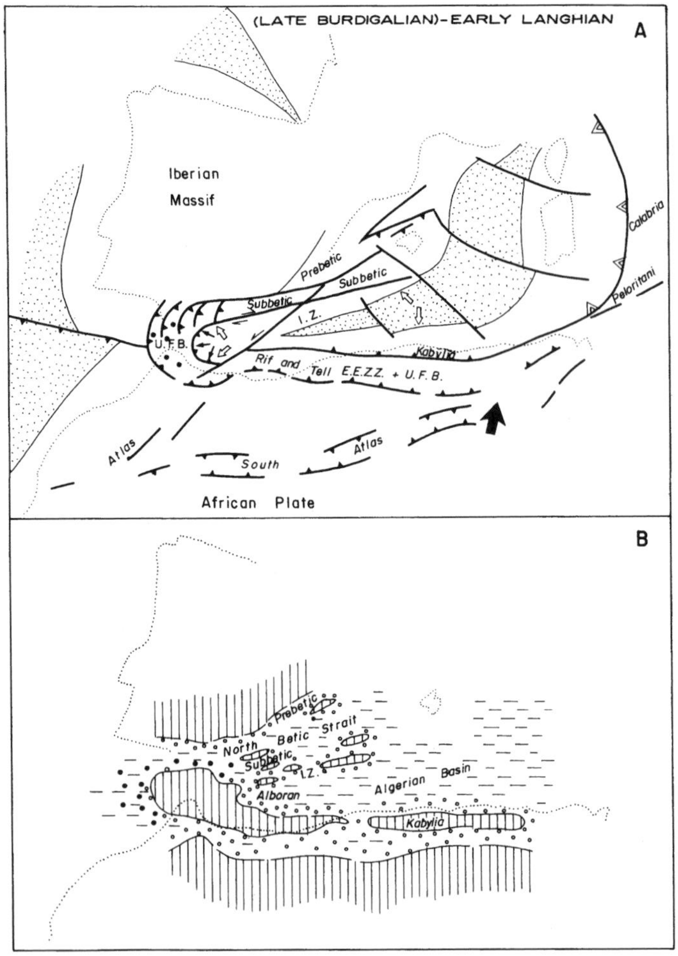

Fig. 4. (Late Burdigalian) – Early Langhian reconstructions. A: Location of the principal domains of the western Mediterranean. B: Principal sedimentary environments. Same key as Fig. 3.

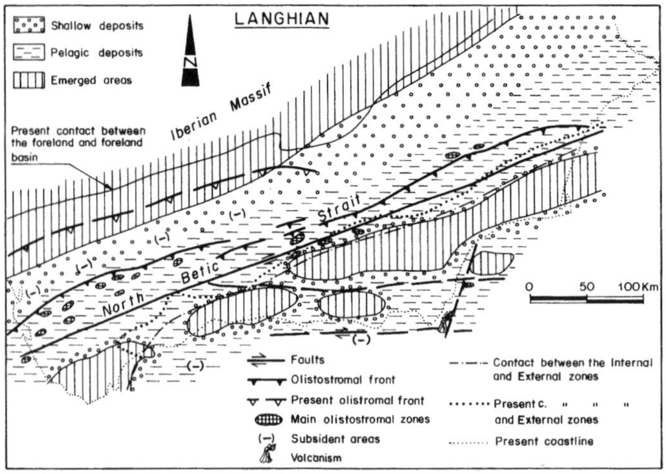

Fig. 5. Palaeogeographic reconstruction of the Betic Cordillera during the Langhian.

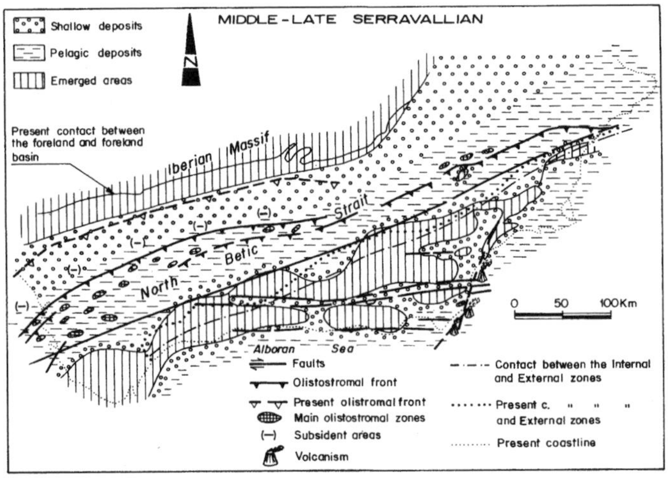

Fig. 6. Palaeogeographic reconstruction of the Betic Cordillera during the Middle–Late Serravallian.

mobile and continued to receive olistostromic materials (Sanz de Galdeano, 1983; Martín Algarra, 1987; Sanz de Galdeano & Vera, 1992).

In our interpretation the main displacement of the Internal Zone had already been completed in the Early Langhian. The arrangement of the principal domains has not changed substantially since that time, and for this reason Fig. 4A is the last of the general reconstructions presented here. Fig. 5 shows the palaeogeography of the Betic Cordillera during the Langhian in greater detail. The absence of significant relief features is noticeable in the Internal Zone, although there was considerable tectonic instability (Rodríguez-Fernández & Sanz de Galdeano, 1990).

Fig. 6 shows our reconstruction for the Middle–Late Serravallian. Displacements of olistostromic units were still taking place in the North-Betic Strait (Portero García & Alvaro López, 1984). The creation of relief features becomes more noticeable in the Internal Zone and movements took place on some important fault sets, such as those of the Alpujarran Corridor (Sanz de Galdeano *et al.*, 1985; Rodríguez-Fernández *et al.*, 1990). The westward displacements of the Internal Zone had almost ceased at this time, the Alboran Sea had practically stopped opening, and the Internal Zone complexes underwent hardly any stretching and extension towards the WSW. An important uplift was about to take place, coinciding with a significant eustatic lowering (Haq *et al.*, 1987 and 1988).

With the attenuation of the westward displacement of the Internal Zone and the end of the expansion of the Alboran Sea, the generally compressive situation existing between Africa and Iberia was restored in the area studied, with an approximately NNW–SSE direction.

Fig. 7 shows the palaeogeographic reconstruction for the Early Tortonian, when the important Serravallian–Tortonian marine regression had ceased (Rodríguez-Fernández & Sanz de Galdeano, 1992). The last emplacement of olistostromic units in the North-Betic Strait took place in the Early Tortonian (Portero García & Alvaro López, 1984).

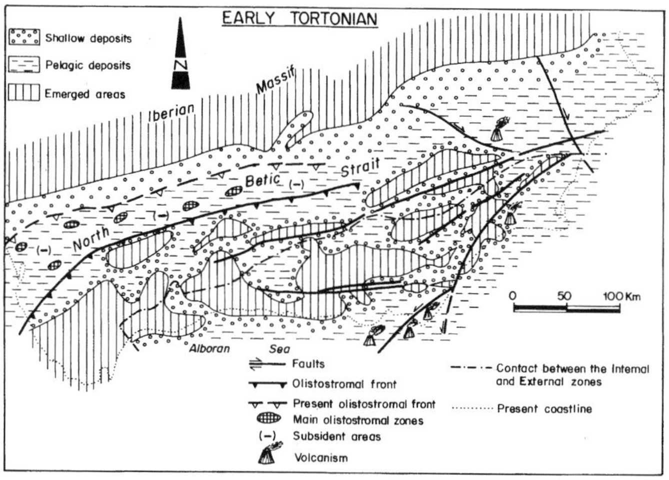

Fig. 7. Palaeogeographic reconstruction of the Betic Cordillera during the Early Tortonian.

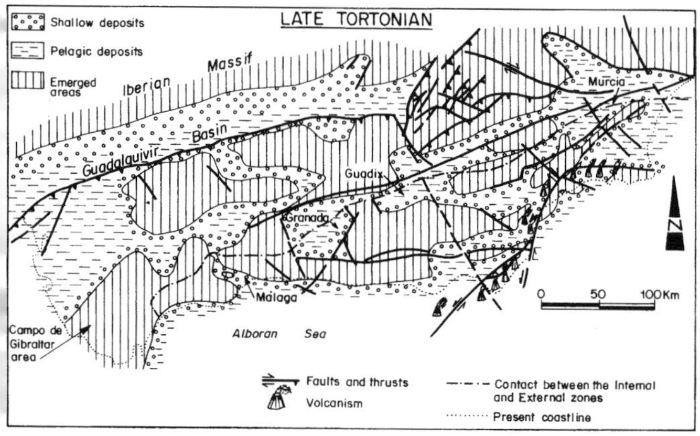

Fig. 8. Palaeogeographic reconstruction of the Betic Cordillera during the Late Tortonian.

During the Late Tortonian–Earliest Messinian in the outermost part of the Cordillera, thrusts formed towards the foreland in the Prebetic, which emerged almost completely, causing the disappearance of the North-Betic Strait (which was reduced to form the Guadalquivir Basin). Some sectors emerged at a slightly later date (Servant-Vildary *et al.*, 1990). In addition, in the sectors where the crust was still abnormally thickened, in particular in the area of contact between the Internal and External Zones, there was a clear trend to uplift. Many relief features rose, especially the Sierra Nevada and Sierra Filabres, and at the same time these developed a clear, almost E–W antiformal shape, partly as a result of the almost N–S compression, and partly due to differential uplift using previously existing faults. NW–SE and NE–SW to NNE–SWW faults in turn moved with dextral and sinistral strike–slip displacements respectively, in many cases with important throws, sometimes as purely normal faults.

Fig. 8 shows the reconstruction of the Late Tortonian palaeogeography, in which numerous intramontane basins appear. At this time in the Late Tortonian two events coincided: (a) an important marine regression (due to an important eustatic fall), which left exposed areas which had been covered by the sea during the Early Tortonian; and (b) the trend to uplift of the Cordillera. The previously mentioned intramontane basins thus became progressively defined. Moreover, at the end of the Tortonian the passage between the Atlantic and the Mediterranean in the Betic Cordillera was interrupted. A similar process must have taken place in the Rif (Aît Brahim, 1991). At the same time basins such as those of Granada and Guadix–Baza changed in character to continental.

During the Messinian the trend to uplift continued, so that the lack of contact between the Atlantic and the Mediterranean in the Betic Cordillera was constant. On the Atlantic side the large Guadalquivir Basin remained. On the Mediterranean side, the Alicante Basin and parts of the Murcia and Almeria basins still continued to be marine (Montenat *et al.*, 1987, 1990; Ott d'Estevou and Montenat, 1985), but large basins such as the Guadix-Baza and Granada basins presented lacustrine and fluvial sediments (Rodríguez-Fernández *et al.*, 1995).

The complete isolation of the Mediterranean throughout the whole of its extent cannot be easily explained only by tectonic causes. Study of the Antarctic ice cap suggests that it may well have been partially due to an important eustatic fall (already starting in the Late Tortonian) which coincided with the accretion of the Antarctic ice cap (McKenzie *et al.*, 1984).

The palaeogeographic situation in the Pliocene was very similar to the present (see Fig. 4 of Montenat, 1977). Some sectors on the Atlantic side in the provinces of Cadiz and Huelva that are now emergent were covered by a shallow sea. The appearance of the Strait in the Gibraltar Arc towards the end of the Messinian is noteworthy, because it definitively restored marine communication and brought to an end the formation of evaporitic deposits in the Mediterranean. On the Mediterranean border some subsiding coastal basins such as those of Fuengirola, Malaga and Almeria were marine.

In the Quaternary the differences with regard to the present position of coastline are very small. In comparison with the Pliocene, the presence of the sea decreased in the southwestern part of the Guadalquivir Basin and in the Fuengirola and Malaga Basins, among other sectors. Since then, movements of uplift and subsidence have continued and are particularly noticeable in some locations.

Acknowledgments

This paper is a contribution of the Projects PB88–0059 and PB91–0079 (DIGCYT) and the Working Group 4085, *Análisis y dinámica de cuencas (Junta de Andalucía)*.

References

Aît Brahim, L. (1991). Tectoniques cassantes et états de contraintes récents au Nord du Maroc. Thesis Univ. Rabat. 273 pp. (Unpubl.).

Andrieux, J. and Mattauer, M. (1973). Précisions sur un modèle explicatif de l'Arc de Gibraltar. *Bull. Soc. Géol. France*, **7**, **XV**, 115–118.

Andrieux, J., Fontboté, J.M. and Mattauer, M. (1971). Sur un modèle explicatif de l'Arc de Gibraltar. *Earth Planet. Sci. Lett.*, **12**, 191–198.

Blanco, M.J. and Spackman, W. (1993). The *P*-velocity structure of the mantle below the Iberian Peninsula: evidence for subducted lithosphere below southern Spain. *Tectonophysics*, **221**, 13–34.

Boillot, G., Montadert, L., Lemoine, M. and Biju-Duval, B. (1984). *Les margins continentales actuelles et fossiles autour de la France*. Masson, Paris, 342 pp.

Bourgois, J. (1978). La transversale de Ronda. Cordillères Bétiques, Espagne. Donnes géologiques pour un modèle d'èvolution de l'Arc de Gibraltar. Thesis Univ. Besançon, 445 pp.

Calvo, J.P., Elizaga, E., López-Martínez, N., Robles, F. and Usera, J. (1978). El Mioceno superior continental del Prebético externo: evolución de estrecho nord-bético. *Bol. Gol. Min.*, **89**, 9–28.

Colom, G. (1952). Aquitanian–Burdigalian diatom deposits of the North Betic Strait, Spain. *J. Paleontol.*, **26**, 867–885.

Comas, M.C. and Jurado, M.J. (1990). The sedimentary record of the Iberian Alboran margin. *IX R.C.M.N.S. Congress (Global events and Neogene evolution of the Mediterranean)*, Barcelona, abstract, 105.

Comas, M.C., García-Dueñas, V., Maldonado, A. and Megías, A.G. (1990). The Alboran basin: tectonic regime and evolution of the northern Alboran Sea. *IX R.C.M.N.S. Congress (Global events and Neogene evolution of the Mediterranean)*, Barcelona, 107–108.

Doblas, M. and Oyarzun, R. (1989a). Neogene extensional collapse in the western Mediterranean (Betic–Rif Alpine orogenic belt): implications for the genesis of the Gibraltar Arc and magmatic activity. *Geology*, **17**, 430–433.

Doblas, M. and Oyarzun, R. (1989b). 'Mantle core complexes' and Neogene extensional detachment tectonics in the western Betic Cordilleras, Spain: an alternative model for the emplacement of the Ronda peridotite. *Earth Planet. Sci. Lett.*, **93**, 76–84.

Durand-Delga, M. (1980). La Méditerranée occidentale: étape de sa genèse et problèmes structuraux liés à celle-ci. *Livre Jubilaire de la soc. Géol. de France*, 1830–1980.

Frizon de Lamotte, D., Guézou, J.C. and Albertini, M.A. (1989). Deformation related to Miocene westward translation in the core of the Betic zone. Implications on the tectonic interpretation of the Betic orogen (Spain). *Geodinam. Acta*, **3–4**, 267–281.

Frizon de Lamotte, D., Andrieux, J. and Guézou, J.C. (1991). Cinématique des chevauchements néogènes dans l'Arc Bético-Rifain: Discussion sur les modèles Géodynamiques. *Bull. Soc. Géol. Fr.*, **162**(4), 611–626.

García Dueñas, V. and Balanya, J.C. (1991). Fallas normales de bajo ángulo a gran escala en las Béticas occidentales. *Geogaceta*, **9**, 33–37.

Gonález Donoso, J.M., Linares, D., Molina, E. and Serrano, F. (1988). El Mioceno inferior de Chirivel (Almería): Bioestratigrafía, cronoestratigrafía y significado tectosedimentario de las formaciones Ciudad Granada y Fuente-Espejos. *Rev. Soc. Geol. Esp.*, **1**, 53–71.

Groupe de Recherche Néotectonique de l'Arc de Gibraltar. (1977). L'histoire tectonique récente (Tortonien à Quaternaire) de l'Arc de Gibraltar et des bordures de la mer d'Alboran. *Bull. Soc. Géol. Fr.*, **19**(3), 575–614.

Haq, B.U., Hardenbol, J. and Vail, P.R. (1987). Chronology of fluctuating sea levels since the Triassic. *Science*, **235**, 1156–1167.

Haq, B.U., Hardenbol, J. and Vail, P.R. (1988). Mesozoic and Cenozoic chronostratigraphy and eustatic cycles. In *Sea-level changes: an integrated approach* (C.K. Wilgus, B.S. Hastings, C.G.S.C. Kendall, H. Posamentier, C.A. Ross & J.C. Van Vagoner, eds.), *Soc. Econ. Paleontol. Mineral.*, Spec. Publ., **42**, 71–108.

Hermes, J.J. (1985). Algunos aspectos de la estructura de la Zona Subbética (Cordilleras Béticas, España Meridional). *Estud. Geol.*, **41**, 157–176.

Martín-Algarra, A. (1987). Evolución geológica alpina del contacto entre las Zonas Internas y las Zonas Externas de la Cordillera Bética. Thesis Univ. Granada, 1171 pp.

McKenzie, J.A., Mueller, D. and Oberhänsli, H. (1984). Paleoceanographic expressions of the Messinian salinity crisis: A correlation of isotopic and sedimentologic events from the Atlantic Ocean and the Mediterranean Region (Spain and Sicily). *5th European Regional Meeting of Sedimentology*. Abstract.

Montenat, Ch. (1977). Chronologie et principaux événements de l'histoire paléogéographique du Néogene récent. *Bull. Soc. Géol. Fr.*, **7**, **XIX**, **3**, 577–583.

Montenat, Ch., Ott d'Estevou, Ph. and Masse, P. (1987). Tectonic sedimentary characters of the Betic Neogene basins evolving in a crustal transcurrent shear zone (SE Spain). *Bull. Cent. Rech. Explor. Prod. Elf Aquitaine*, **11**, 1–22.

Montenat, Ch., Ott d'Estevou, Ph. and Coppier, G. (1990). Les Bassins néogènes entre Alicante et Cartagena. *Doc. et Trav. IGAL*, **12–13**, 313–368.

Osete, M.L., Freeman, R. and Vegas, R. (1988). Preliminary palaeomagnetic results from the Subbetic Zone (Betic Cordillera, southern Spain): kinematic and structural implications. *Phys. Earth Planet. Interiors*, **52**, 283–300.

Osete, M.L., Freeman, R. and Vegas, R. (1989). Palaeomagnetic evidence for block rotations and distributed deformation of the Iberian-African Plate boudary. In *Paleomagnetic Rotations and Continental Deformation* (C. Kissel & C. Laj, eds.), Kluwer Academic Publ., 381–391.

Ott D'Estevou, Ph. and Montenat, Ch. (1985). Evolution structurale de la zone bétique orientale (Espagne) du Tortonien a l'Holocène. *C.R. Acad. Sci. Paris*, **II**, **300**, **8**, 363–368.

Platt, J.P. and Vissers, R.L.M. (1989). Extensional collapse of thickened continental lithosphere: a working hypothesis for the Alboran Sea and Gibraltar Arc. *Geology*, **17**, 540–543.

Portero García, J.M. and Alvaro López, M. (1984). La depresión del Guadalquivir, cuenca de antepaís durante el Neógeno: génesis, evolución y relleno final. *I Congreso Español de Geología*, Segovia, **III**, 241–252.

Rodríguez-Fernández, J. and Sanz de Galdeano, C. (1990). The palaeogeography of the Betic Cordilleras during the Middle and Upper Miocene. *IXth R.C.M.N.S. Congress, Global events and Neogene evolution of the Mediterranean, Barcelona*, 287–288.

Rodríguez-Fernández, J. and Sanz de Galdeano, C. (1992). Onshore Neogene Stratigraphy in the north of the Alboran Sea (Betic Internal Zones). Paleogeographic implications. *Geo-Marine Lett.*, **12**, 123–128.

Rodríguez-Fernández, J., Sanz de Galdeano, C. and Serrano, F. (1990). Le Couloir des Alpujarras. *Doc. et Trav. IGAL*, **12–13**, 87–100.

Rodríguez-Fernández, J., Sanz de Galdeano, C. and Vera, J.A. (1995). The Granada Basin. *Doc. et Trav. IGAL*, 14.

Sanz de Galdeano, C. (1983). Los accidentes y fracturas principales de las Cordilleras Béticas. *Estud. Geol.*, **39**, 157–165.

Sanz de Galdeano, C. (1990). Geologic evolution of the Betic Cordilleras in the Western Mediterranean, Miocene to the present. *Tectonophysics*, **172**, 107–119.

Sanz de Galdeano, C. and Vera, J. (1992). Stratigraphic record and palaeogeographical context of the Neogene basins in the Betic Cordillera, Spain. *Basin Res.*, **4**, 21–36.

Sanz de Galdeano, C., Rodríguez-Fernández, J. and López-Garrido, A.C. (1985). A strike–slip fault corridor within the Alpujarra Mountains (Betic Cordilleras, Spain). *Geol. Rundschau*, **74**, 642–655.

Sanz de Galdeano, C., Rodríguez-Fernández, J. and López Garrido, A.C. (1990). Les Cordillères Bétiques dans le cadre géodynamique néoalpin de la Méditerranée occidentale. *Riv. Ital. Paleontol. Stratigr.*, **96**(2–3), 191–202.

Sanz de Galdeano, C., Serrano, F., López Garrido, A.C. and Martín Pérez, J.A. (1993). Palaeogeography of the Late Aquitanian–Early Burdigalian in the Western Betic Internal Zone. *Geobios*, **26**, 1.

Servant-Vildary, S., Rouchy, J.M., Pierre, C. and Foucault, A. (1990). Marine and continental water contributions to a hypersaline basin using diatom ecology, sedimentology and stable isotopes: an example in the Late Miocene of the Mediterranean (Hellin Basin, southern Spain). *Palaeogeogr., Palaeoclimatol., Palaeoecol.*, **79**, 189–204.

Van Bemmelen, R.W. (1954). *Mountain building*, Nijhoff, The Hague, 177 pp.

Van Bemmelen, R.W. (1972). Driving forces of Mediterranean orogeny (Tyrrhenian test-case). *Geol. Mijn.*, **51**(5), 548–573.

Van Bemmelen, R.W. (1973). Geodynamic models for the Alpine type of orogeny (Test-case II: The Alps in Central Europe). *Tectonophysics*, **18**, 33–79.

Weijermars, R. (1985a). In search for a relationship between harmonic resolutions of the geoid, convective stress patterns and tectonics in the lithosphere: a possible explanation for the Betic orocline. *Phys. Earth Planet. Interiors*, **37**, 135–148.

Weijermars, R. (1985b). Uplift and subsidence history of the Alboran Basin and a profile of the Alboran Diapir (Western Mediterranean). *Geol. Mijn.*, **64**, 349–356.

Wildi, W. (1983). Les chaînes tello-rifaines (Algérie, Maroc, Tunisie): structure, stratigraphie et évolution du Trias au Miocène. *Rev. Géogr. Phys Géol. Dyn.*, **24**, 201–297.

S3 Depositional model of the Guadalquivir – Gulf of Cádiz Tertiary basin

C. RIAZA AND W. MARTÍNEZ DEL OLMO

Abstract

The Gualdalquivir–Gulf of Cádiz Neogene Basin started to form at the end of the Early Miocene compressional phase in the Betic System. Seven stratigraphic sequences are identified in the Basin: 1–3. Preolistostromic sequences (Atlantida Group), A1, A2 and A3; 4. Synolistostromic sequence (Betica Sequence); 5. Andalucia sequence; and 6–7. Post-erosional phase sequences (Marismas and Odiel).

Introduction

After the Early Miocene Betic compressional phase, an extensional period began, affecting the Guadalquivir Valley and the Gulf of Cádiz area (Fig. 1) and forming a Neogene foreland basin with the following general characteristics:

- Langhian to Quaternary sediments infilling a WSW–ENE basin widening and deepening to the west.
- A passive margin to the north, resting on a Paleozoic basement and a residual Mesozoic platform.
- An active margin to the south, which is compressional and gravitational, with a large olistostromic mass infilling this side of the basin during Late Tortonian times, partly destroying and masking the previous sediments of this margin. Some olistostromic events have been reported before that time, but they are less important than the Late Tortonian gravitational advance.

The main tectonic phase affecting the basin shape occurred during the Late Miocene, when downwarping of the basement took place in the south followed by the extension of a system of faults. These processes created an important trough where a thick sedimentary sequence quickly accumulated. The incompetent southern Betic margin then collapsed into the basin, infilling a large volume of it. After that phase only halokinetic movements into the olistostrome and the consequent adjustments affected the basin.

The area has been intensively explored by oil companies, and this has provided us with an important quantity of seismic lines and well data. This information has been studied in order to define the stratigraphic sequences and their relationship with tectonic and eustatic movements (Vail, 1987; Vail & Sangree, 1988).

This study allows us to identify at least seven depositional sequences which can be related to the main olistostromic and erosional events. Only in one of these sequences are LST sediments well developed, showing the external position of this segment of the Betic foreland basin in Neogene times.

Pre-olistostromic sediments (Atlantida group)

According to the seismic, well and field data, the age of the main olistostromal emplacement has been established as Late Tortonian. Nevertheless, in the upper part of the Guadalquivir Valley, frequent outcrops of Lower Tortonian sediments appear. The facies include fluviodeltaic to turbiditic sediments, within a basin largely developed to the south of the valley.

Deltaic systems on the northern margin in the upper part of the valley have been described by Marín Señán (1988), and shallow marine sediments were also studied by Sierro *et al.* (1992).

In this part of the basin other outcrops of turbiditic facies, previously included in the Upper Tortonian sequences, are now considered to be older. Channelled turbiditic bodies in the Ubeda area, described by Suarez *et al.* (1989), could represent the relicts of the lower Tortonian basin, which was almost destroyed by the olistostrome advance. The turbiditic fans could represent the deep facies at the foot of the prograding prodelta complex, which are related to the Highstand System Tract (HST).

Transgressive calcarenitic facies from the Transgressive System Tract (TST), grading upwards into silts and marls from the HST, have been studied in outcrops in the Guadalquivir area. To the west, on the seismic lines, the plunge of these series can be observed, under the overlying Bética Sequence.

At the only well drilled in this part of the basin, the sediments of this age can be seen to consist of a detrital sequence, where only the TST and HST are recognised.

At that time the basin had a SW–NE orientation, oblique to the actual olistostrome front, which allowed these sediments to be preserved only in the NE sector of the basin. Only one well has been

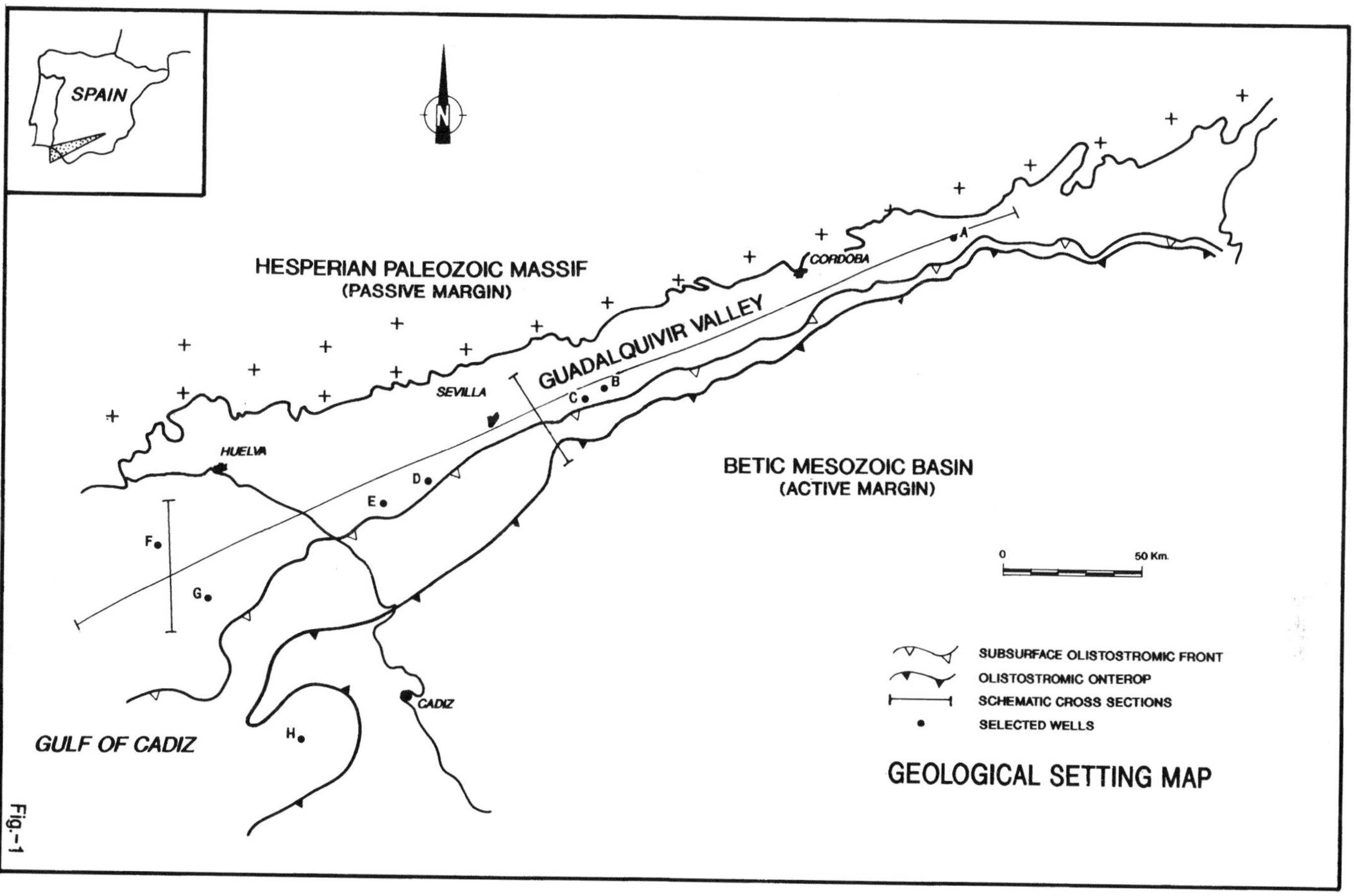

Fig. 1. Map of the geological setting. A–H refer to sites of selected wells, several of which are shown in the following figures.

drilled through these sediments and the seismic record is not of good enough quality to provide a better control below the olistostrome.

In the Gulf of Cádiz area, sediments older than late Tortonian have also been recognised. On the well logs at least three Miocene sequences, dated as Langhian to Early Tortonian, have been interpreted below the Late Tortonian Bética sequence. Only the TST and HST are represented in them. (Fig. 2).

These sequences of the external northern margin of the basin represent the old Atlántida Group as defined by Martínez del Olmo *et al.* (1984). Because of the difficulty in separating the sequences, the name is retained to cover the whole group of sequences.

Two margins have been recognised, the northern one, against the autochthonous unit, is mainly calcareous, and the southern one, mostly detrital. These sequences were also involved in the olistostromic overthrusts and were drilled either at the autochthonous unit or at the olistostromic overthrusted mass, at the present southern margin.

The syn-olistostromic Betica sequence

The Bética Group was defined by Martínez del Olmo *et al.* (1984) to include the Upper Tortonian and the Lower Messinian sediments. The group includes both the turbiditic fans and the related facies described by these authors in previous publications as the Guadalquivir and Guadiana Sands (Fig. 3).

Seismic profiles and well logs allow the Bética sediments to be studied in the Guadalquivir on-shore area. The group is now interpreted as a depositional sequence including only TST and HST sediments.

Because of new correlations between the sequences, the Gulf of Cádiz model has to be modified. The Guadiana Sands have now to be included in a younger sequence, and the Bética sequence is represented by hemipelagic sediments. (Fig. 4).

The Guadalquivir turbiditic sands are interpreted as deep sediments at the foot of the prograding HST complexes, coming from both margins. As the southern margin appears to have been the most active one, the progradation of the turbiditic channelled bodies has a south to north polarity with a westward axial direction. These bodies, linked to the turbiditic system, show multiple forms with channelled and lobulate facies prograding westward and northward. Using the seismic data these channels have been studied in detail, recognising the northern currents cutting the previous southern deposits with polarity indications of the distal sediments. Finally, the axial westward current erodes, transports and redeposits these sediments, creating multiple scars on the earlier deposits.

The basal detrital sediments, which appear at the base of the

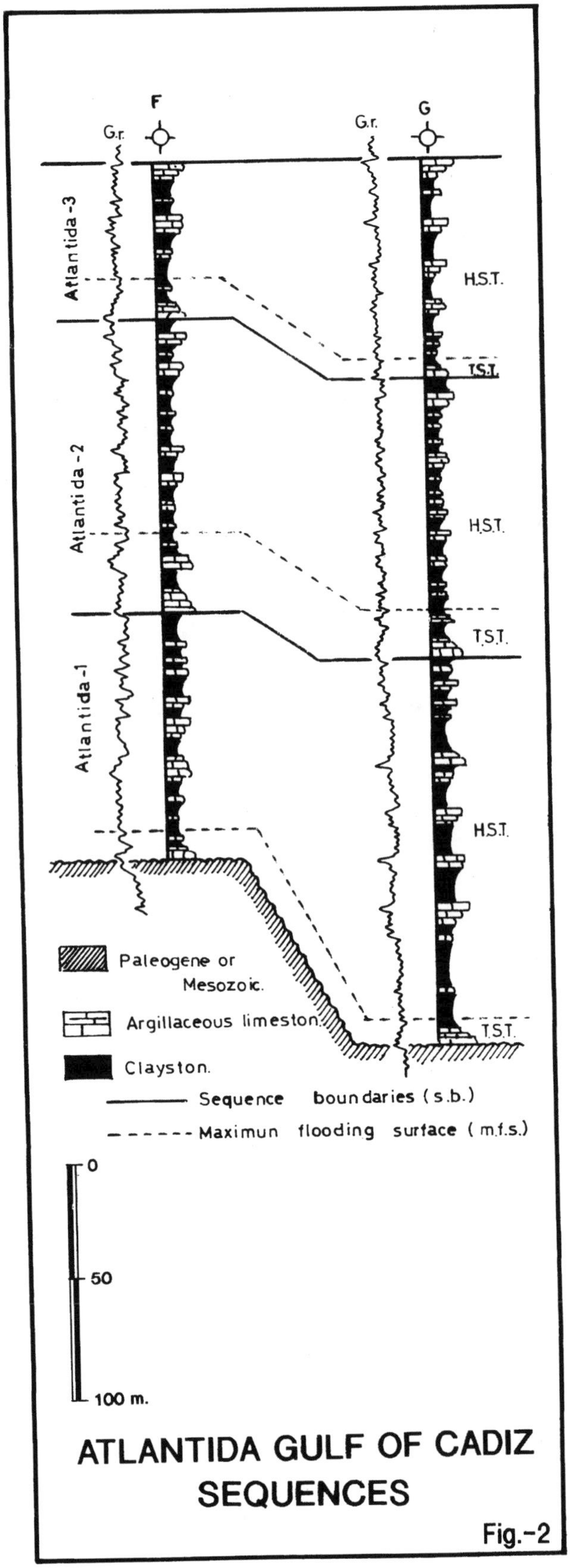

Fig. 2. Atlantida Gulf of Cádiz Sequences.

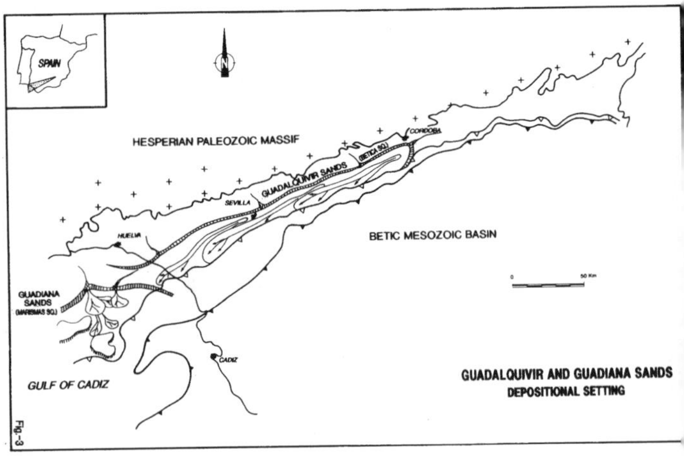

Fig. 3. Guadalquivir and Guadiana Sands. Depositional setting.

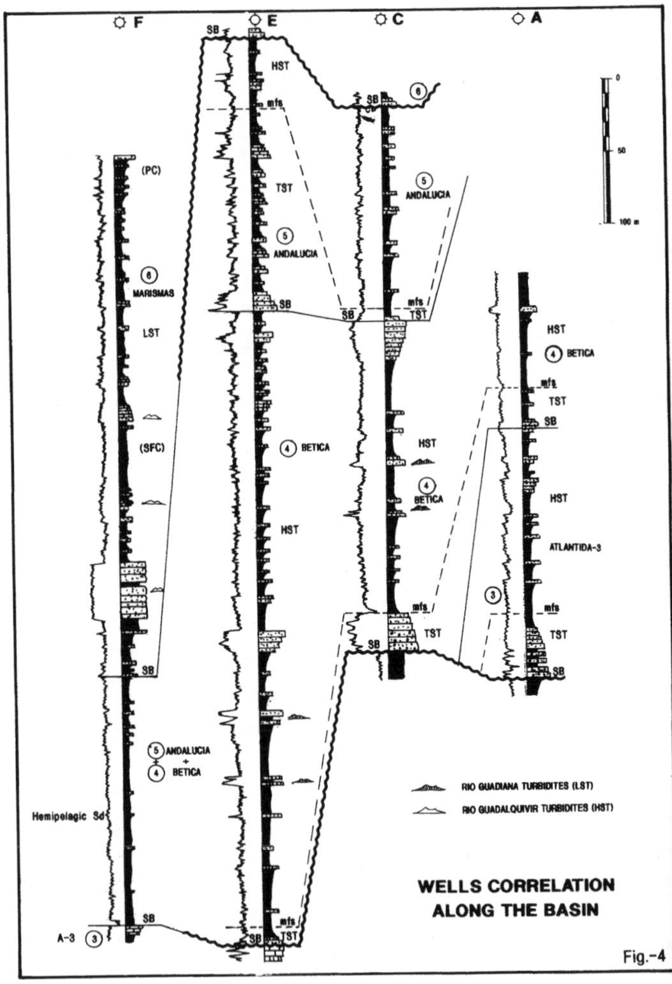

Fig. 4. Correlation of wells along the basin (see Fig. 1 for locations). Numbers refer to correlatable sequences.

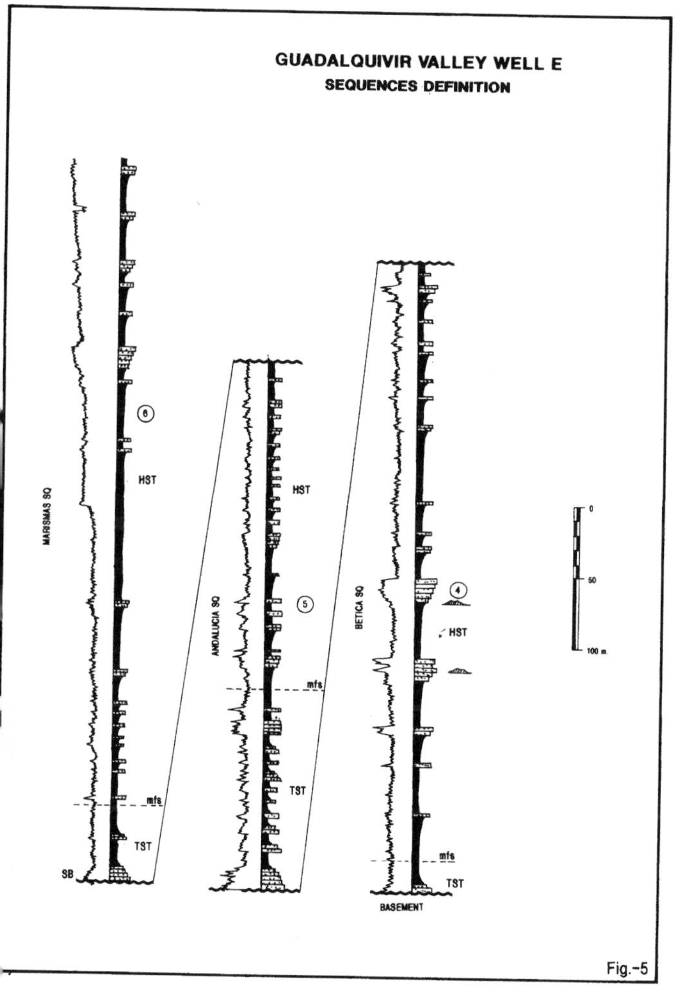

Fig. 5. Guadalquivir Valley. Well E. Sequences definition.

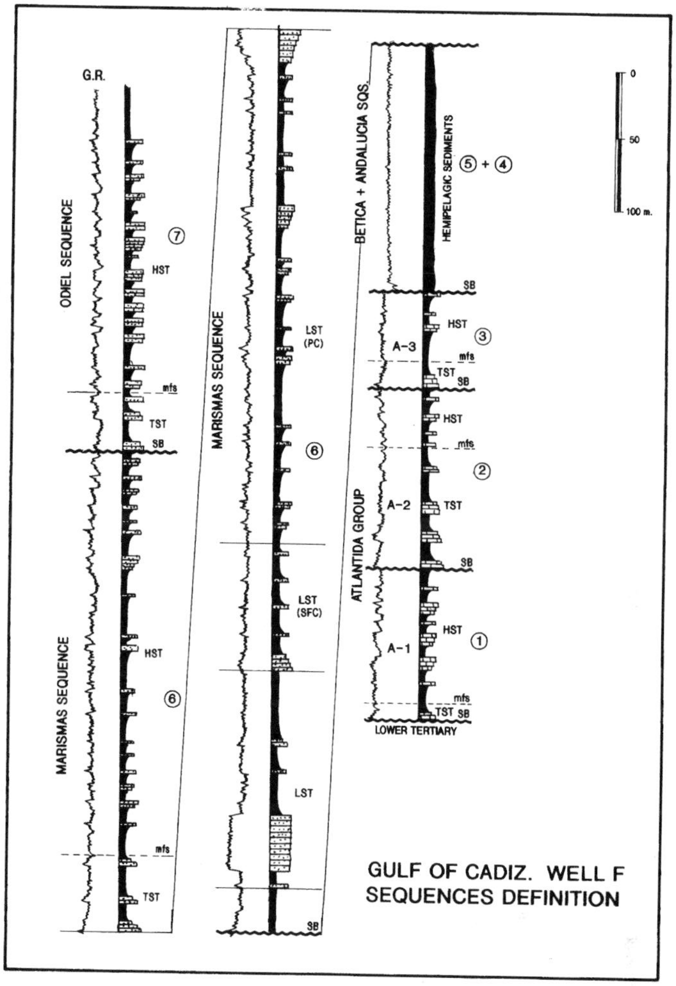

Fig. 6. Gulf of Cádiz. Well F. Sequences definition.

Miocene series in most of the Guadalquivir Valley area, may represent the coastal transgressive system of this Bética Sequence. (Fig. 5).

At this stage the olistostromic masses from the southern active margin arrived in the basin, partially destroying the previous sediments and forming a detrital mega-element in the basin.

The HST prograding system is well defined on both margins. The northern one is siliciclastic with regular southward accretion. The topset is represented by a siliciclastic platform that crops out at the edge of the basin. An important section of blue marls represents the foreset facies of this system. A by-pass is observed between the north prograding sediments and the turbiditic facies at the axis of the basin. The southern margin is more calcareous and the topset and foreset can also be identified, but it is partially destroyed by the olistostromic advance.

So, the turbiditic system is limited to the north by a sedimentary ramp, to the south by an active olistostrome and to the east by an important talus feature defined at the meridian of Córdoba (Fig. 3).

At the northern margin some incised valleys are observed. From the seismic profiles the Almonte and Moguer channels can be distinguished; they represent infills of some erosive valleys of this sequence. At the outcrops, there is an infill canyon at the Fuente de la Corregidora (Jaen) in the upper valley, described by Sierro *et al.* (1992).

The Andalucía depositional sequence

The Andalucía Group was defined by Martínez del Olmo *et al.* (1984). It includes most of the Messinian sediments. It has been recognised in the drilled wells and at the northern and southern margin outcrops.

A more detailed recent analysis of this group using the well data allows us to identify two different systems of a new depositional sequence.

- The TST is represented by a thin interval in the central area, where most of the wells were drilled. Nevertheless, this system thickens up towards the margins, cropping out extensively on the olistostromic mass.
- The HST, identified in the seismic profiles and logs, is partly eroded at the central Guadalquivir Valley. It is represented in the wells studied by thick marly shales,

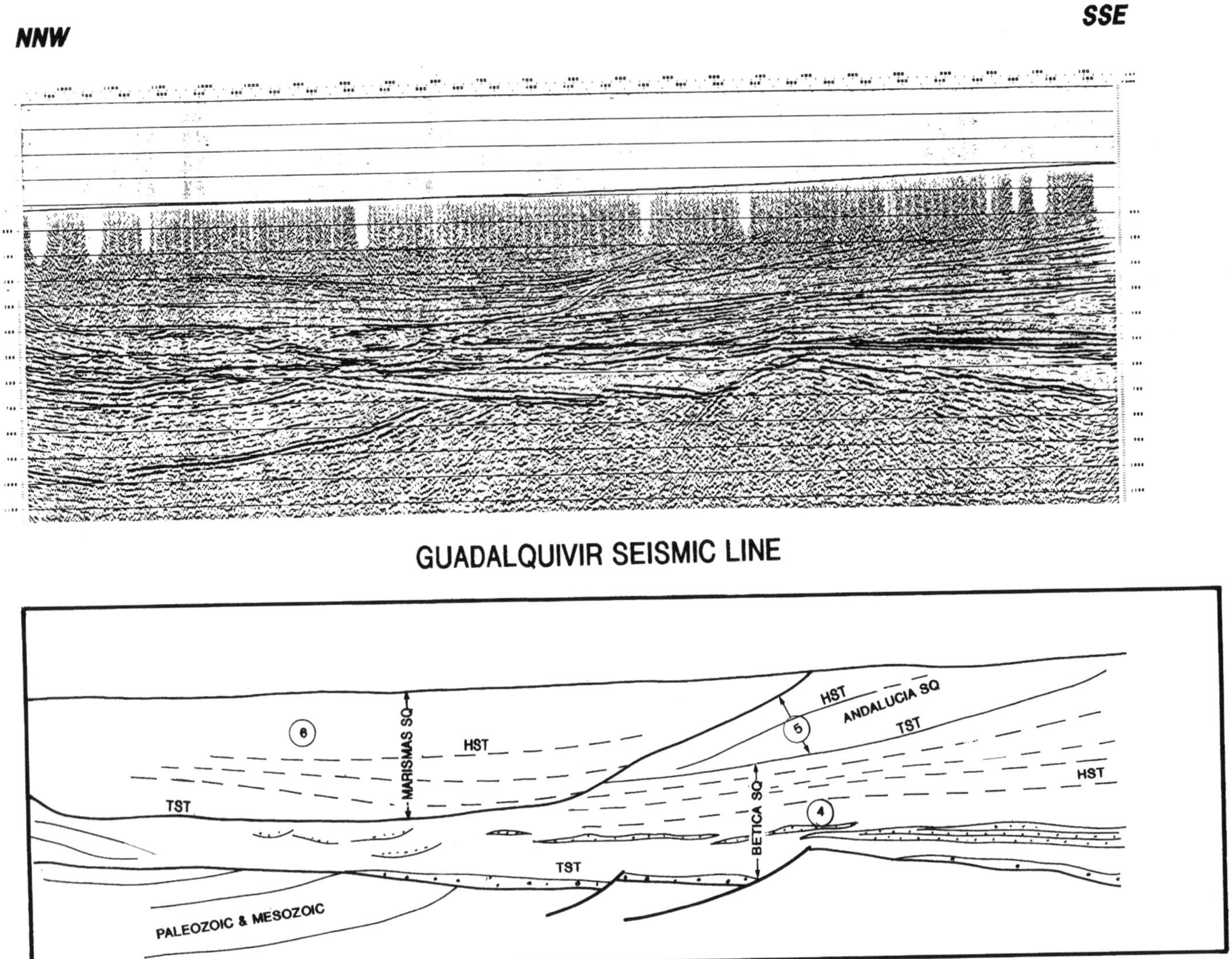

Fig. 7. Guadalquivir seismic line.

grading upwards into siltstones and very fine sandstones. At the southern margin these sediments thicken up, developing important calcarenitic bodies (Carmona calcarenites).

The southern margin is more developed and was unstable because of the olistostromic movements. The topsets are calcarenitic because of the more calcareous Betic foldbelt source rock. The southern margin prograded into the deep basin, overlapping the turbiditic complex of the Bética sequence.

In the Gulf of Cádiz area, below the turbiditic Guadiana complex, the sequence also contains hemipelagic facies and cannot be separated from the Bética hemipelagic sediments (Fig. 6).

The Post-olistostromic sediments (Marismas Group)

This group, defined by Martínez del Olmo *et al.* (1984), includes Pliocene and Quaternary sediments mainly. It is characterised by an important erosive base, very well represented in the seismic profiles in the upper part of the Guadalquivir Valley and in the Gulf of Cádiz northern margin (Fig. 7). This erosion even locally affects the turbiditic Guadalquivir Sands and cuts most of the Andalucía sediments in the northern side of the basin, showing an important sea-level fall.

Until now the Guadiana Sands in the Gulf of Cádiz area were considered equivalents to the Guadalquivir Sands in the Guadalquivir Valley, but recent studies lead us to consider them to be younger. The important sea-level fall that occurred at this time produced deep erosion in the upper part of the basin, generating an important quantity of detrital material. These detrital sediments were eroded during the Late Messinian(?)–Early Pliocene times and were deposited downstream in the Gulf of Cadiz area. So, the Guadiana turbiditic complex represents the Lowstand System Tract (LST) of this new sequence (the Marismas Sequence). (Fig. 8).

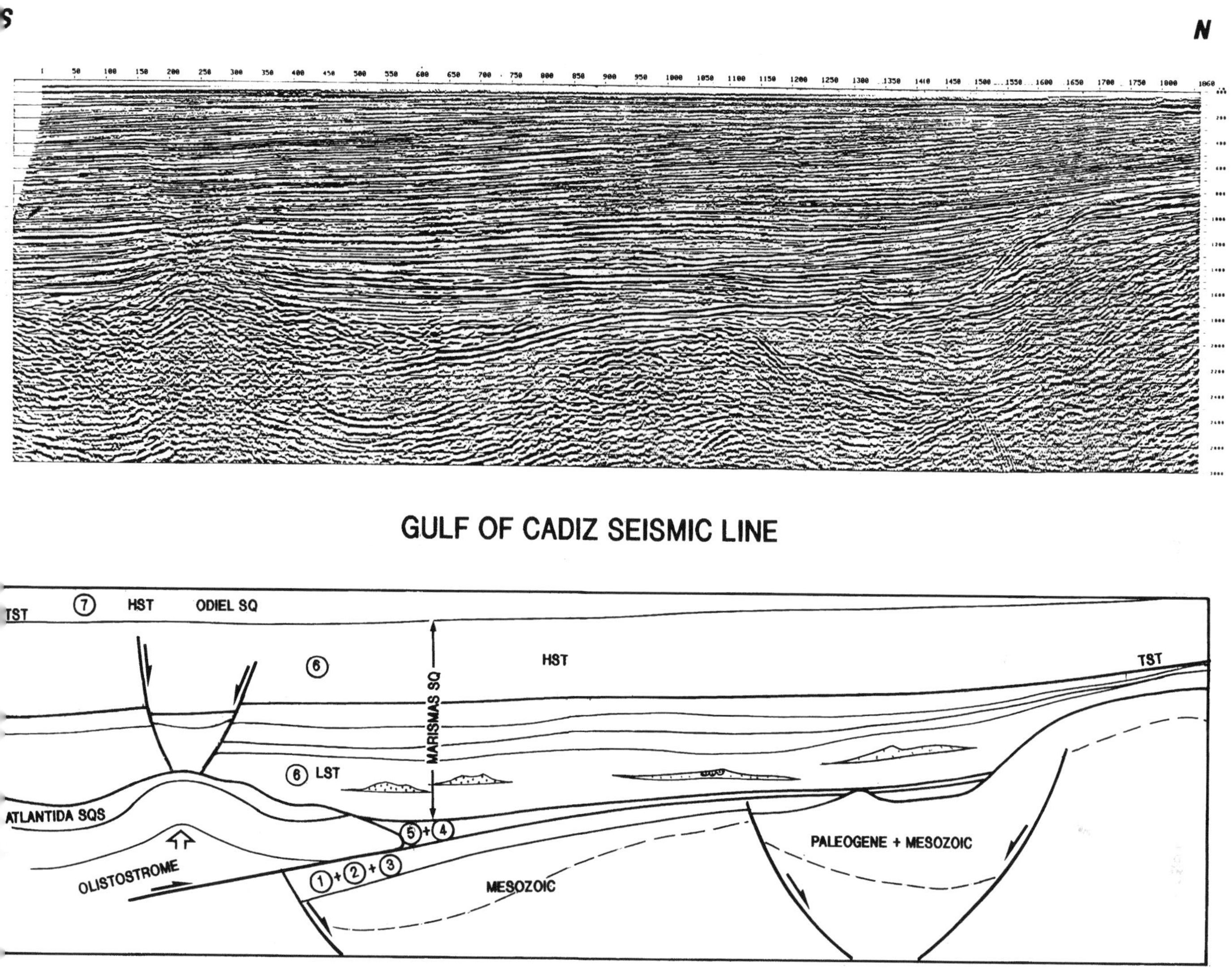

Fig. 8. Gulf of Cádiz seismic line.

Marismas Sequence

Very well characterised in the lower Guadalquivir area, his sequence only includes the transgressive and the high stand ystems overlapping the partly eroded previous sequences. The ithology is mostly shale with a few thin fine sandstones.

According to the new interpretation of the Gulf of Cádiz, the ST of the sequence is represented by a prograding complex and a lope fan complex (Guadiana Sands), with a very-well-developed urbiditic system, retrograding to the northern margin from where t was fed by several canyons coming from the north and northeast.

The prograding complex of this LST and the TST and HST eposits are well recognised on the well logs, characterised by very haly sediments with minor fine sandstones and siltstones. (Figs. 5 nd 6).

The older sediments of this sequence, belonging to the LST, may e correlative with the Latest Messinian sediments that occur in ther Spanish Tertiary basins.

Odiel Sequence

In the Gulf of Cádiz wells, above the Pliocene Marismas Sequence, a new sequence is recognised on the logs (Fig. 6). Only the transgressive and highstand systems are represented. The age of this new sequence can be considered as Late Pliocene – Pleistocene and it has been named the Odiel Sequence.

Due to its local extent and its reduced thickness, it is not easy to separate it from the Marismas Sequence on the seismic profiles.

The basin and the sea-level changes

In Middle Miocene times the Guadalquivir–Gulf of Cádiz basin represented a marginal setting of a larger basin that extended southwards. Three sea-level cycles are recognised from the Langhian to the Early Tortonian.

The actual shape of the basin only appeared after the down warp and extensional Late Tortonian phase, which affected the Betic

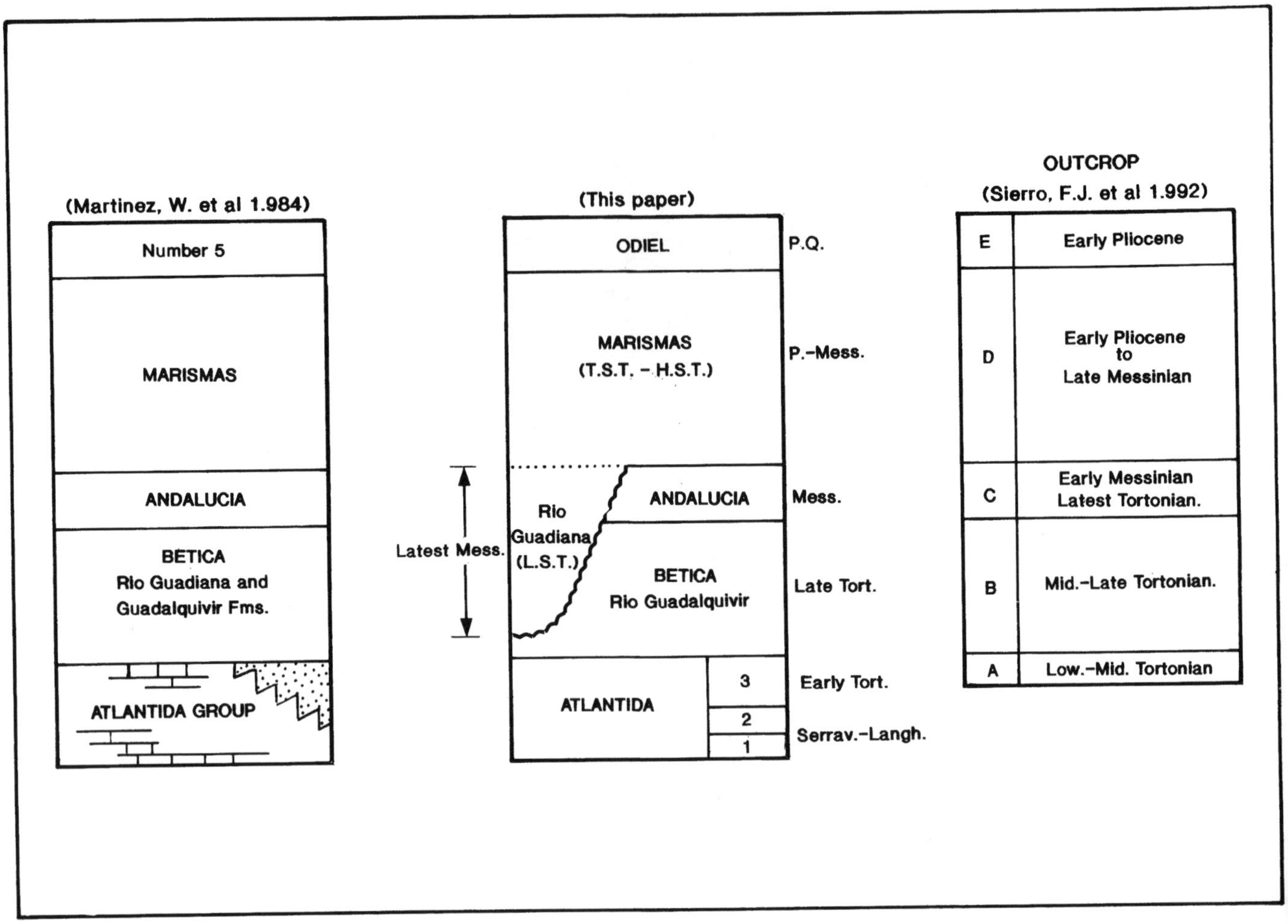

Fig. 9. Equivalences and correlations with previous works.

foldbelt and its margins. A long and narrow corridor, parallel to the northern edge, appears, open to the west, with lateral progradations from the northern and southern margins, infilling the central trough, during the regressive phase (Bética sequence). At the basin axis, and connected to the prograding system, turbiditic facies appear (Guadalquivir Sands) extending to the west.

The topographic slope produced gravitational movements of plastic Triassic and associated sediments, which accumulated at the Betic southern margin.

Further eustatic movements caused a new transgressive–regressive cycle, represented in the Andalucía Sequence.

But the most important sea-level fall occurred in Late Messinian times, when very active erosion has been recorded. That sea-level cycle produced the only complete sequence recognised in the basin (Marismas Sequence), with LST deposits (Guadiana Sands). In the rest of the sequences only the TST and HST sediments are well represented, with hemipelagic thin sediments in the distal zones.

The last sea-level cycle is represented by the Odiel Sequence, only present in the Gulf of Cádiz area. (Fig. 8).

Conclusions

From the seismic profiles and well data, seven stratigraphic sequences have been identified. In relation to the main basin events: the olistostromic phase movements in Late Tortonian times and the strongest sea-level fall in latest Messinian times, the following sequences can be distinguished (Fig. 9):

- Pre-olistostromic Sequences (Atlántida group), Langhian to Early Tortonian in age, with three identified sequences, named A-1, A-2 and A-3.
- Syn-olistostromic sequence (Bética Sequence), partly equivalent to the old Betica Group (Martínez del Olmo *et al.*), including the Guadalquivir sand complex, but not the Guadiana Sands, and belonging to Late Tortonian times.
- The Andalucía Sequence, Messinian in age, represents a partly eroded sequence, equivalent, only in the Guadalquivir Valley area, to the old Andalucía Group.
- Post-erosional phase sequences: Marismas Sequence (Latest Messinian–Pliocene), which represents the most complete and complex sequence, including the

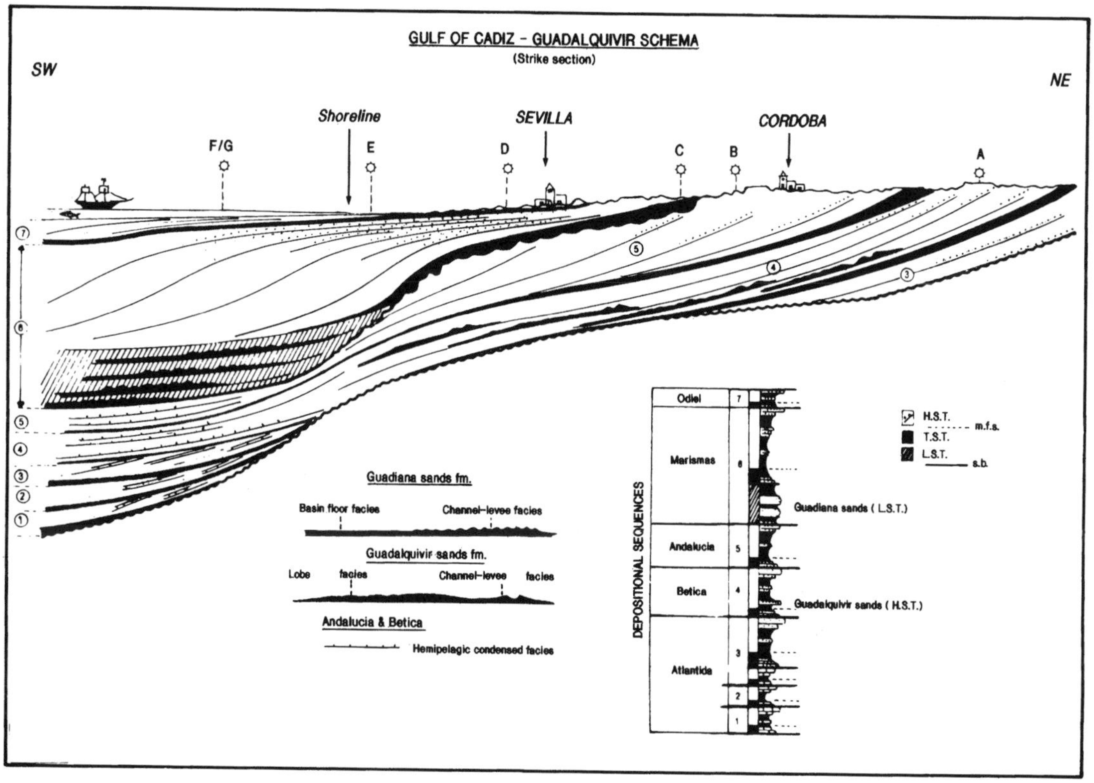

Fig. 10. Gulf of Cádiz – Guadalquivir Valley strike section.

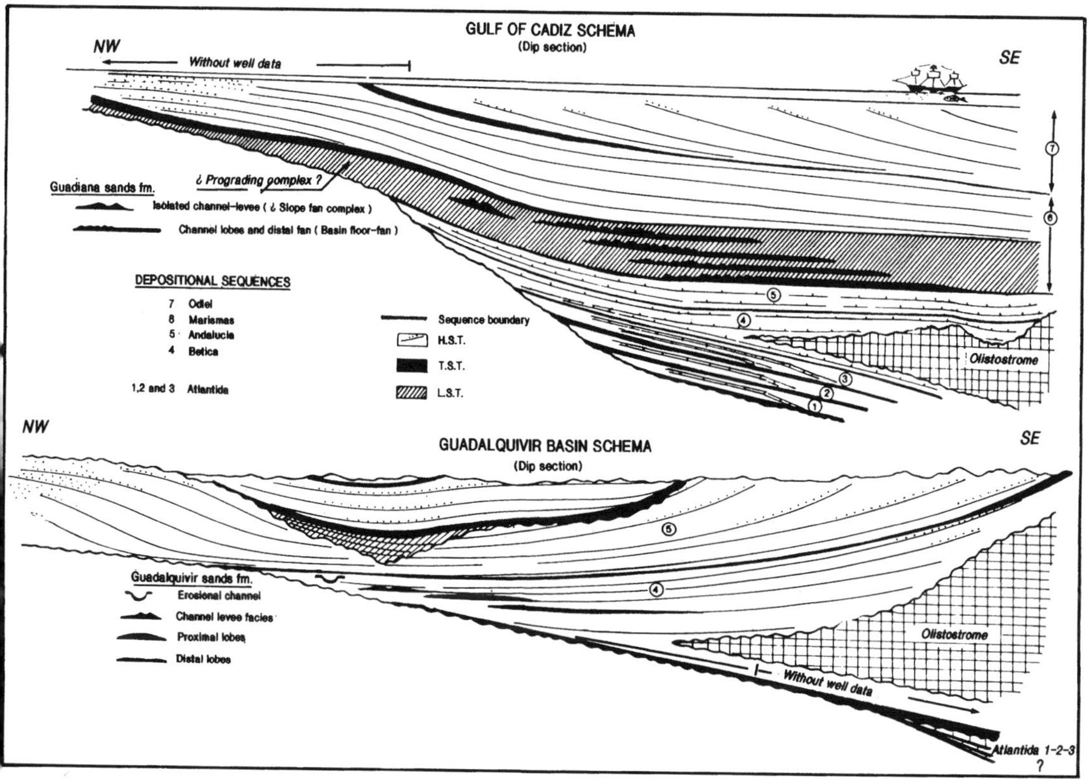

Fig. 11. Gulf of Cádiz and Guadalquivir Basin dip sections.

Guadiana turbiditic sands, and the Odiel Sequence, which includes sediments up to the present day (Late Pliocene–Pleistocene).

For all of these sequences a general model is presented in Figs. 9, 10 and 11.

References

Galloway, W.E. (1989). Genetic stratigraphic sequences in basin analysis. I: Architecture and genesis of flooding surface bounded depositional units. *A.A.P.G. Bull.*, **73**(2): 125–142.

Marín Señán, J.M. (1988). Sedimentación detrítica en el borde norte de la depresión del Guadalquivir (Sector de Villanueva de la Reina, prov. de Jaen). *II Congresso Geológico de España*, **I**: 123–126.

Martínez del Olmo, W. (1990). Secuencias deposicionales a través de diagrafías de pozo. Curso de Doctorado. Universidad Complutense. Madrid.

Martínez del Olmo, W., García Mallo, J., Leret, G., Serrano, A. and Suárez, J. (1984). Modelo tectosedimentario del Bajo Guadalquivir. *I Congresso Español de Geologia*, **I**: 199–213.

Martínez del Olmo, W., Suárez, J., Serrano, A. and Leret, G. (1986). Los sistemas turbidíticos del Guadalquivir–Golfo de Cádiz: Fms Rio Guadiana y Rio Guadalquivir. *Abstract X Congreso Español de Sedimentologia. Barcelona.*

Megías, A.G., Leret, G., Martínez del Olmo, W. and Soler, R. (1980). La sedimentación Neógena de las Béticas: Análisis tectosedimentario. *Congreso Grupo Español de Sedimentologia. Salamanca.*

Sierro, F.J., González Delgado, J.A., Dabrio, C., Flores, J.A. and Civis, J. (1992). The Neogene of the Guadalquivir Basin (SW Spain). *III Congreso Geológico de España. Guia de excursiones geologicas*: 180–236.

Suárez, J., Martínez del Olmo, W., Serrano, A. and Leret, G. (1989). Estructura del sistema turbidítico de la formación Arenas del Guadalquivir, Neógeno del Valle del Guadalquivir. *Libro homenaje a R. Soler, AGEEP*, 123–132.

Vail, P.R. (1987). Seismic stratigraphy interpretation procedure. In *A.A.P.G. Atlas of seismic stratigraph.* (Bally, A.W., ed.) **27 (II)**: II

Vail, P.R. and Sangree, J.B. (1988). Sequence stratigraphy workbook, fundamentals of sequence stratigraphy. *A.A.P.G. Annual Convention Short Course: Sequence stratigraphy interpretation of seismic, wells and outcrop data.* Houston, Texas.

S4 Late Neogene depositional sequences in the foreland basin of Guadalquivir (SW Spain)

F.J. SIERRO, J.A. GONZÁLEZ DELGADO, C.J. DABRIO, J.A. FLORES AND J. CIVIS

Abstract

This chapter presents a reconstruction of the geometry of the Guadalquivir Basin based upon the spatial position and morphology of isochronous surfaces defined by means of bio-events (i.e. discontinuities in the palaeontological record) apparent in the associations of calcareous plankton. It is possible to interpret the depositional history of the basin and to define five depositional sequences to correlate with cycles of global sea-level change.

Introduction

The Guadalquivir Basin is an ENE–WSW-elongated depression filled with Neogene sediments that crop out more than 800 m above sea level at the far eastern end. The topographic surface descends gradually to the west until it reaches the present coast of the Gulf of Cádiz. Sedimentation continues today below sea level in the Gulf of Cádiz.

The Guadalquivir basin separates an emergent foreland made up of Palaeozoic and Mesozoic rocks (the Variscian massif of the Spanish Meseta) to the North and the Betic Cordillera with Mesozoic and Cainozoic rocks to the South (Fig. 1). The Betic Cordillera (also referred to as the Betics) in the southern Iberian peninsula and the Rif in north Africa forms the westernmost major structure of Alpine age. The last Alpine deformation occurred during Serravallian and Early Tortonian times and several 'post-orogenic' basins have developed under tectonic control of the major structural patterns of the Betics.

The Internal Zone of the Betics exposes very mobile crust with rocks of Palaeozoic and Mesozoic ages. The External Zone (Mesozoic and Cainozoic rocks) corresponds to the margin of the European plate. Some of these areas, such as the Subbetic and Penibetic Zones, suffered intense deformation.

The Betic foredeep originated between the Betic orogen in the south and the Iberian Foreland to the north. Deposition in the External Zone during the Early and Middle Miocene took place in the so-called North Betic Strait that connected the Atlantic and the Mediterranean. Like other typical foreland basins, the passive northern margin is characterised by a gradual deepening of the basement towards the south due to flexure of the substratum (the formerly and wrongly called 'Guadalquivir fault'). The deepest part of this strait was located to the south, next to the Betic orogen. The southern margin is very steep because of its position at the front of the active subbetic thrust belt, and olistostromes of chaotic Mesozoic and Cainozoic rocks slid down towards the north to be incorporated in the deposits.

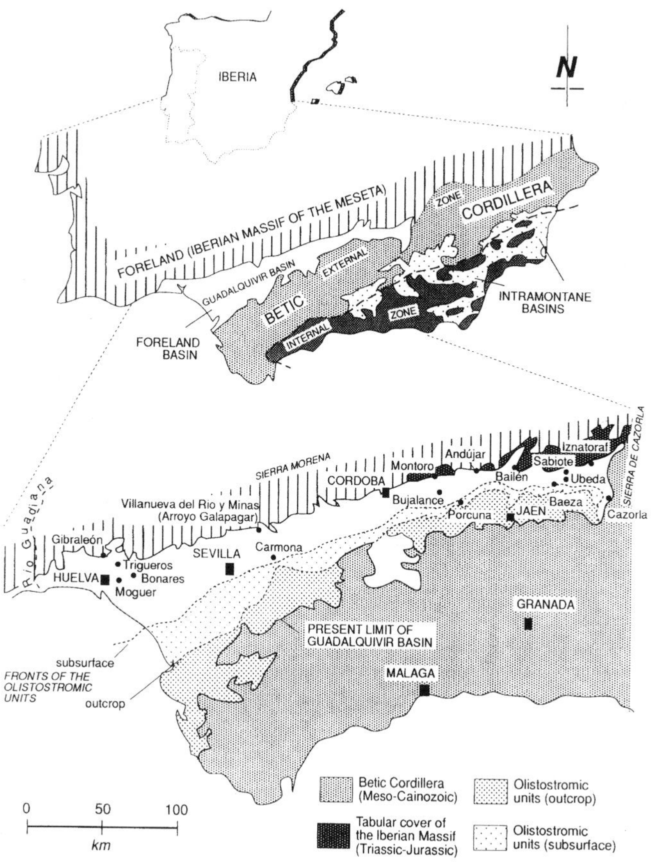

Fig. 1. Location map.

Throughout the major part of the Miocene a northward migration of the foreland depocentres took place because of the displacement of the Subbetic thrust belt in the same direction. A major palaeogeographic change took place in the Lower Tortonian when the North Betic Strait was closed due to folding and uplifting of the central and eastern part of the foredeep. The blocking of the former strait caused erosion of the sedimentary cover in this region. The western part of the ancient foredeep evolved into a triangular-shaped foreland basin that separated the folded chain (and its intramontane basins) from the emergent foreland, forming the present geography of the Guadalquivir basin.

Thus, the evolution of the Betic foredeep has a special significance because it was the key that controlled Atlantic–Mediterranean communication, and probably provided a deep passage between both realms during most of the Neogene.

At the eastern end of the basin are the Sierra de Cazorla mountains, which consist of Prebetic Mesozoic materials rising to more than 2000 m. In contrast, the southern margin of the basin is defined by the low hills of the Prebetic to the east and the Subbetic to the west. Autochthonous sediments crop out in the northern half of the basin, while the southern half consists of large olistostromes. The front of the olistostromic masses lies more to the north in subsurface surveys as compared with outcrop geology (Fig. 1).

Martínez del Olmo *et al.* (1984) divided the sedimentary fill of the western Guadalquivir Basin into five tectonosedimentary (TSU) units based on seismic data and drill cores. Sierro *et al.* (1990) distinguished several depositional sequences for the whole basin that partly coincide with the TSU limits of Martínez del Olmo *et al.* (1984). Such units have been established by means of detailed palaeontologic correlation based upon careful micropalaeontologic surveys that allow the definition of a succession of calcareous plankton events (Sierro, 1984, 1985; Flores, 1985; Flores & Sierro, 1987). These discontinuities of the palaeontological record permit the recognition of synchronous reference surfaces (large clinoforms prograding to the west) which limit sedimentary units. However, it should not be forgotten that southward and northward progradations exist as well. There follows a brief description of this methodology.

Calcareous plankton event stratigraphy

The physical properties of the sediments used in these studies are rather wider than those measured with conventional instruments (grain size, porosity, permeability, radioactivity, etc.) because they include the properties of the biogenic fraction. These are measured with macro- or micropalaeontologic techniques and depicted using a graph named *Eco-log* (a contraction from Ecological Log).

Pelagic ecosystems are very susceptible to small oceanographic variations. Small changes in the dynamics of the ocean–atmosphere system affect the temperature and salinity of sea water and the distribution of nutrients in large areas of the ocean. These changes usually affect plankton and benthos communities.

Exhaustive micropalaeontological study of samples from drilling in the Atlantic and the Guadalquivir and other basins of the Betic–Rifean system and western Mediterranean has allowed us to

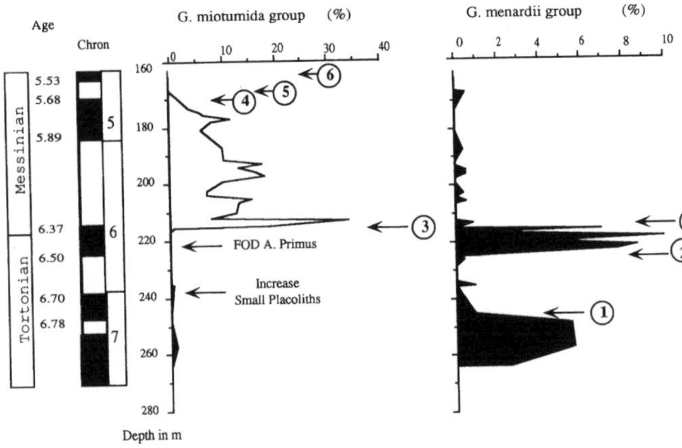

Fig. 2. Pattern of variations in the calcareous plankton assemblages in DSDP Site 410 (North Atlantic). Modified from Sierro *et al.* (1993). This pattern is considered in this work as a reference 'Eco-log', useful for stratigraphic correlation in the NE Atlantic and Mediterranean (Fig. 3). The main changes (isochronous discontinuities of the paleontological record) were identified in some Globorotaliid groups (events 1 to 6) or in the Calcareous Nannoplankton associations. The recognition of these events in the Guadalquivir basin was used to reconstruct the geometry of the basin as shown in Fig. 4. The definition of the events and its correlation with the Magnetostratigraphic Scale of Berggren *et al.* (1985) may be seen in Sierro *et al.* (1993). FOD is the first occurrence datum.

reconstruct the evolution of calcareous plankton during the Late Miocene and Pliocene in this region. A series of successive quantitative, and sometimes qualitative, changes of micropaleontological associations were recognised (Sierro *et al.*, 1993; Flores & Sierro, 1989) (Fig. 2).

These episodes in the calcareous plankton were related to palaeo-oceanographic changes in the Northern Atlantic and the Mediterranean but unrelated to the tectonic, local climatic or palaeoenvironmental evolution of the basin. Such episodes have been calibrated with the current magnetostratigraphic scales and correlated to isotopic and/or geochemical events, some of them of global extent, and relative changes of sea level using Haq *et al.* (1988) curves (Sierro *et al.*, 1990).

Accordingly, we consider these events as isochronous in the studied area, and their recognition in the various parts of the basin allows the recognition of successive surfaces defining isochronous rock bodies (Fig. 4).

Geometry of the basin fill

Eventostratigraphic studies suggest a sedimentary model for the Guadalquivir basin quite similar to the present one. The model consists of a shelf–talus system that received sediment supply from the east and prograded towards the WSW. Most of the filling of the basin takes place longitudinally with clinoforms dipping westwards (Fig. 4).

Four sedimentary environments have been recognised in the basin (Fig. 5): (a) **Basin facies**: hemipelagic deposits rich in calcareous microfossils with deep benthic foraminifera. In some cases

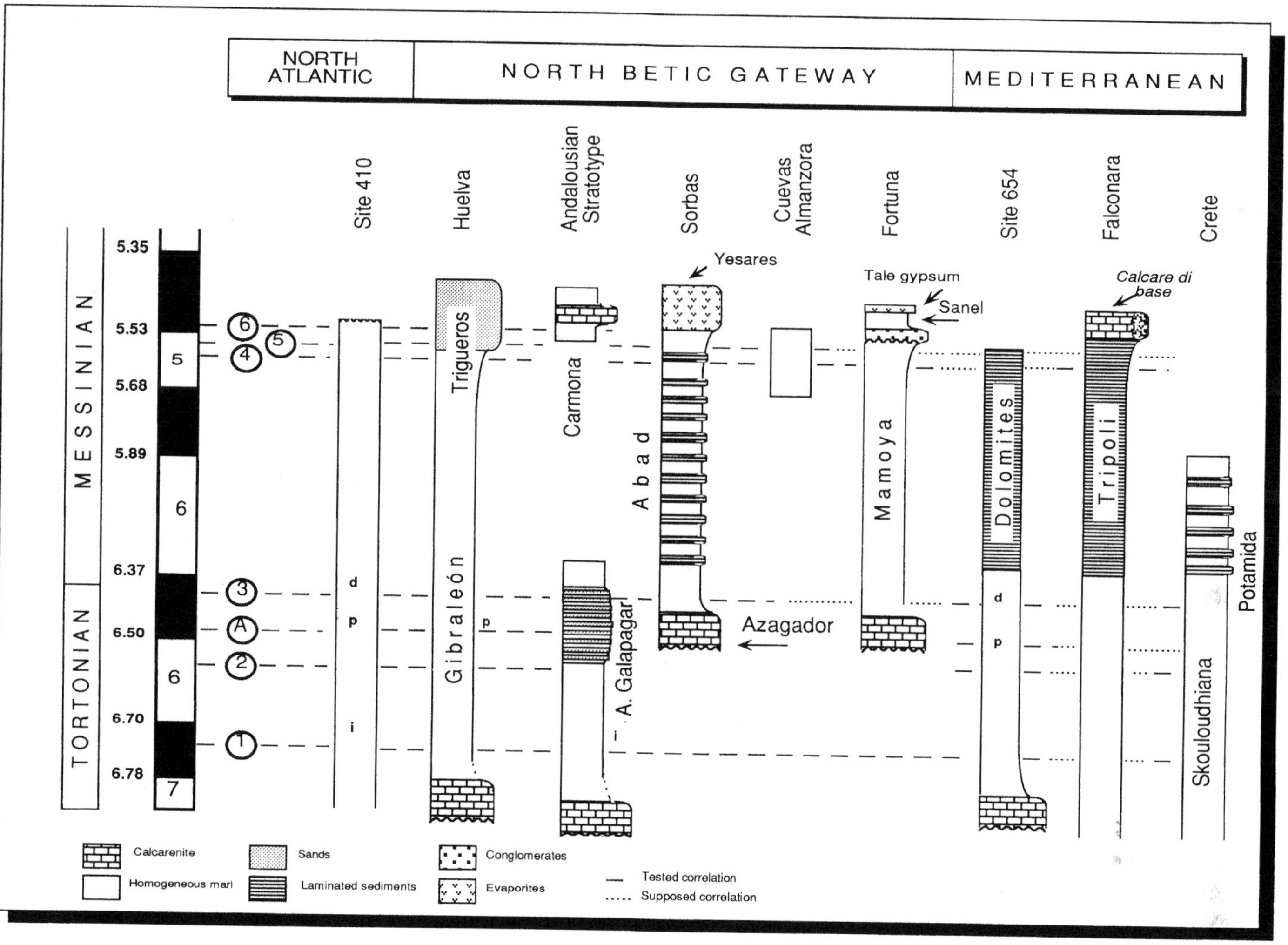

Fig. 3. Correlation of the calcareous plankton events in basins from the North Atlantic to the Eastern Mediterranean, through the North Betic Gateway. Nannofossil events: d – first occurrence datum (FOD) of *Amaurolithus delicatus*; p and A – FOD of *Amaurolithus primus*; i – increase of small placoliths; 1 to 6 – changes in Globorotaliid groups. Modified from Sierro *et al.* (1993).

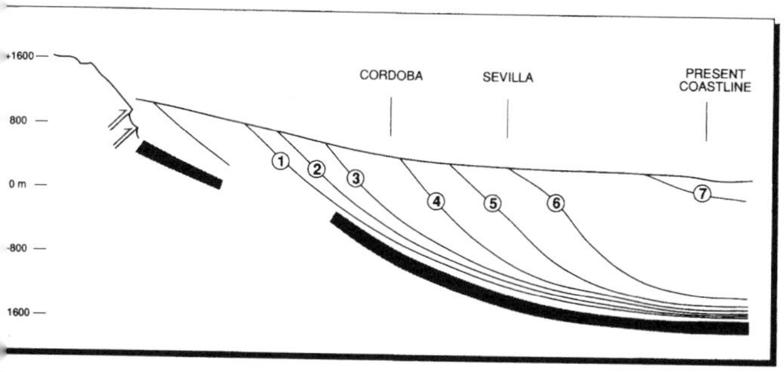

Fig. 4. Clinoforms in the foreland basin of Guadalquivir identified and dated by some of the main calcareous planktonic events. 1–7 are the clinoform boundaries.

bottom currents reworked these deposits, as in the present base of the slope of the Gulf of Cádiz, which is swept by the Mediterranean outflow (Sierro & Flores, 1989). (b) **Base of slope facies**: terrigenous turbidites related to the axial areas of the basin, including channel, channel levee and lobe deposits as seen in seismic profiles (Martínez del Olmo *et al.*, 1984; Suárez *et al.*, 1989) and outcrop in the eastern end of the basin (Sierro *et al.*, 1990b). (c) **Slope facies**: mostly clay and silty clay. Fine-grained terrigenous clastics predominate over biogenic deposits. Eventostratigraphic analysis reveals high rates of deposition and steep slope gradients, often exceeding 10%. (d) **Shelf facies**: interbedded clay and silts or fine sands passing upwards into sands, usually rich in benthic (and nektonic) macrofauna. As a whole there is a coarsening-upwards trend. These deposits are best observed at the western end of the basin due to intense erosion of the eastern end.

Progradation began near the Sierra de Segura area (Fig. 1) about 8 my ago and is still continuing at present in the Gulf of Cádiz.

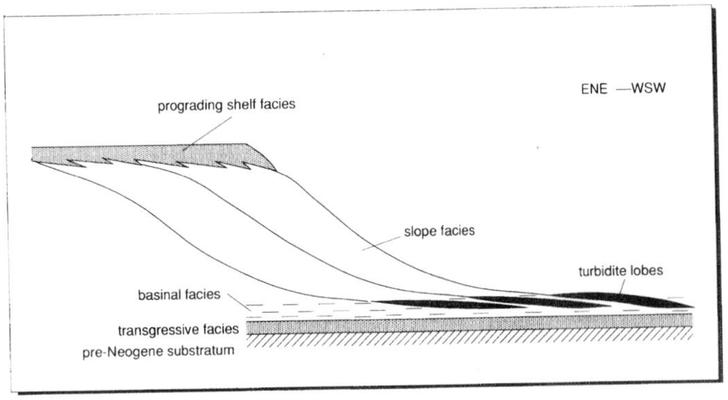

Fig. 5. A general model of deposition with the four sedimentary environments recognised in the foreland basin of Guadalquivir.

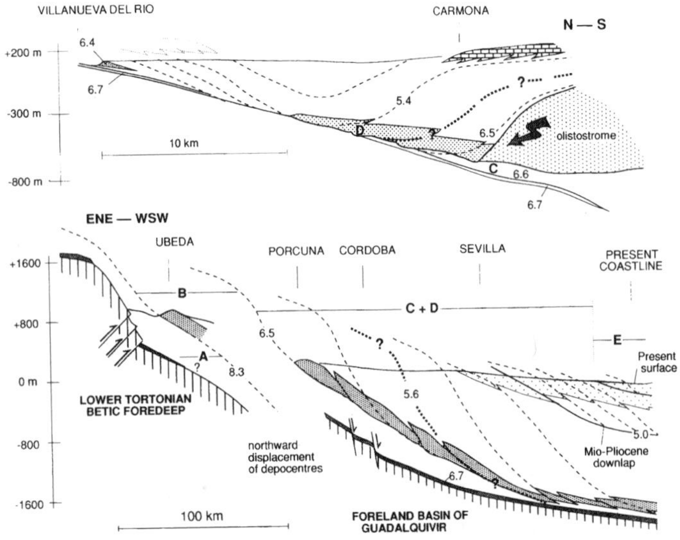

Fig. 6. North–South and ENE–WSW cross-sections of the Guadalquivir basin illustrating the five offlapping depositional sequences, A–E, which are progressively younger towards the west. Erosion has removed the tops of the successive sequences, and the resulting surface dips gently towards the west. The discontinuity between sequences C and D is inferred (see text).

During the Messinian, rapid progradation, probably of an episodic type, took place in the Guadalquivir basin (Fig. 5). It was characterised by turbiditic deposits at the slope base (Suárez *et al.*, 1989) which seem to have the cyclic character defined by channel incision and erosion during relative falls of sea level, channel and levee deposits and turbidite lobe development during lowstands, and abandonment of the turbidite system during transgressive and highstand stages. The mapped and described turbidite bodies suggest that this evolution was repeated up to six times during Messinian times (Fig. 7). These 'cycles' are comparable to those recognised using stratigraphic and palaeo-oceanographic criteria in the eastern North Betic Strait (Santisteban & Taberner, 1983; Muller & Hsü, 1987; Muller & Schrader, 1989). In particular, the deposition of turbidite bodies Gv-3 to Gv-6 (called Gv from 'Guadalquivir' in Suárez *et al.*, 1989) may be coeval with the deposition of the lower evaporites in the Mediterranean.

Four coexisting episodes influenced the rhythm of westward progradation: (a) the uplift of the Sierra del Segura; (b) the northwards displacement of depocentres during Middle Tortonian times; (c) the emplacement of huge olistostromic masses from the southeast (Roldán García & García Cortés, 1988) which rapidly filled up a large part of the basin; and (d) the global eustatic changes.

Depositional sequences

Five offlapping depositional sequences are observable in an east–west cross-section of the basin. These sequences are progressively younger towards the west (Figs. 6 and 8). Erosion denuded the tops of the successive sequences and the resulting erosion surface dips gently towards the west.

Sequence A (Lower to Middle Tortonian) only crops out in the eastern part of the basin, between Bailén and Iznatoraf. It begins at the northern margin with shallow marine deposits, transgressive over the Mesozoic basement, which rapidly pass upwards into rich planktonic foraminiferal silts (Sabiote, Fig. 8) deposited in a deep environment and overlain by grey marls and clays that are very difficult to study in the field because of the absence of good outcrops. Good sections occur only in quarries. The basal calcarenitic or terrigenous sediments and the marls represent the Transgressive and Highstand Systems Tract and the basal rich planktonic foraminiferal marls represent the condensed section (as summarised in Posamentier *et al.*, 1988). The deposition probably occurred during the global sea-level rise of Cycle 3.1 (Haq *et al.*, 1987).

Sequence B (Middle to Late Tortonian) is well exposed over the whole basin. A major palaeogeographical change took place between the deposition of Sequences A and B related to the closing and uplift of the eastern North Betic Strait when the Sierras de Cazorla and Segura were formed. In the western half of the Betic foredeep such a change produced a dextral rotation of depocentres (i.e. a displacement to the north-west) and the individualisation of the new foreland basin of Guadalquivir. The migration of depocentres to the north-west meant the drowning (transgression) of large areas of the emergent foreland where the new deposits rest directly upon Palaeozoic and Mesozoic rocks in the central and western regions of the basin. A similar, symmetrical migration appears to have happened in northern Africa (South Rif Foreland Basin, Morocco).

In the eastern part of the basin, Sequence B began with turbiditic sedimentation (Lowstand Systems Tract, probably correlated to Cycle 3.2 of Haq *et al.*, 1987) resting upon the deep-water marls of Sabiote (Sabiote, Baeza, Fig. 8). However, in the central and western parts, these deposits do not crop out. Here, the base of the sequence is a coastal transgressive deposit ('calcarenite') overlying pre-Neogene rocks in the northern margin.

The Transgressive Systems Tract is represented by a calcarenite unit, and the Highstand Systems Tract by greyish blue marls, whereas the glauconitic, pelagic silts correspond to the condensed section (Montoro and Arroyo Galapagar, Fig. 8). The age of this glauconite layer is very close to 7 my and strikingly coincides with the global condensed section of Cycle 3.2.

Sequence C (Latest Tortonian to Early Messinian) comprises the

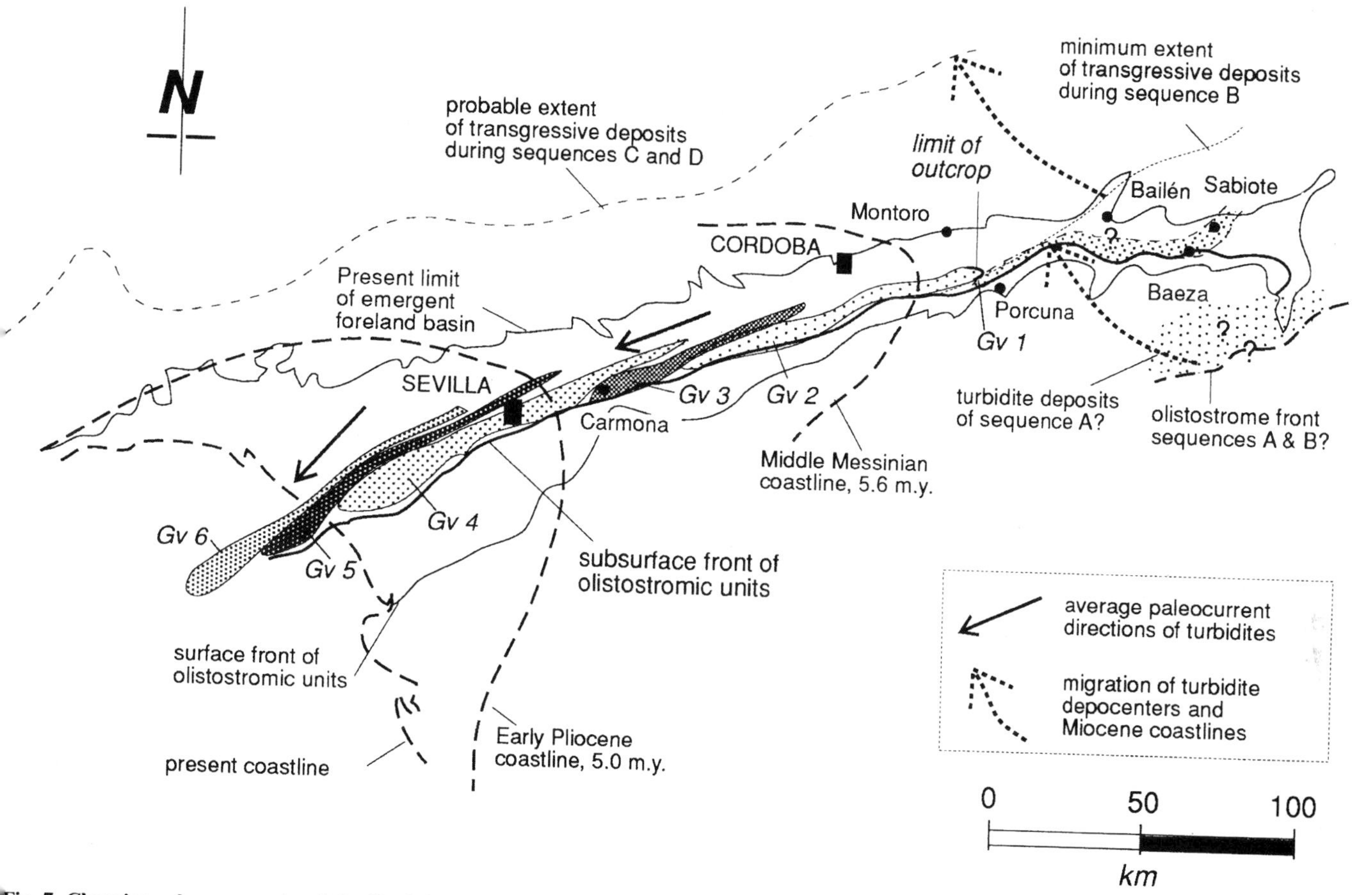

Fig. 7. Changing palaeogeography of the Guadalquivir basin and successive coastlines (the precise position of the older coastlines is only roughly inferred). Position of the successive turbidite bodies based on Suárez *et al.* (1989).

sandstone Unit of Porcuna that forms the base of the sequence constituting the Lowstand Systems Tract. The overlying Lower Messinian blue marls of Carmona are representative of the Transgressive and Highstand Systems Tracts. Materials of this sequence crop out in the northern edge (central and western parts) and in the axis (central part) of the basin.

We estimate that the age of the first turbidite sediments in the Bujalance–Porcuna area is 6.6 my; this is comparable with Cycle 3.3. Suárez *et al.* (1989) described turbidite units of the same age in other locations in the Guadalquivir Basin and Gulf of Cádiz (Gv-2).

Sequence D (Late Messinian to Early Pliocene) consists of the axial turbidites from Carmona to the Gulf of Cádiz (Suárez *et al.*, 1989), the younger upper part of the Gibraleón Clays, and the Huelva Sands.

The lower boundary coincides with a relative fall of sea level, which is responsible for the deposition of the rapid axial progradation of the turbidite bodies. The age of this event closely coincides with that proposed for the limit of global Cycles 3.3 and 3.4 by Haq *et al.* (1987). In the axial region, these turbidites are overlain by a thick series of clays which pass upwards to sands (Huelva Sands), progressively later towards the west.

The transition between the clays and the overlying sands is progressively younger towards the west. In the Province of Huelva there is a glauconite layer marking the boundary between these two units. This layer may be interpreted as a condensed section deposited during the maximum transgression of Latest Messinian or Earliest Pliocene age (Cycle 3.4). However, in this case, the relative sea-level rise must have been slower than the rate of sedimentation, causing a regressive sequence to be generated in the context of a global sea-level rise.

Distinction between depositional sequences C and D should be easy and clear if we blindly follow the ideas of Haq *et al.* (1987), because a relative fall of sea level occurred between 6.5 and 5.5 Ma (at approximately the base of sequence D separating Cycles 3.3 and 3.4). However, we still have some doubts because we cannot determine any direct relationship (or lack of relationship) between these deposits either from surface or drill samples or seismic surveys. The calcareous plankton analysis allows us to suggest that an important change in the rate of turbidite progradation occurred in this time, if we take into account that these turbidite bodies were probably deposited between 5.5 and 5.2 my, whereas the age of the base of turbidite body Gv-2 is 6.6 my. The only indications of stratigraphic discontinuity are the age of turbidite bodies and the fill of submarine channels near Córdoba (Cuesta del Espino section). It should be noted that sequence D (Cycle 3.4) includes at least four or five turbidite bodies that are stacked against each other (Fig. 6) with erosional surfaces. The precise relationships between turbidite bodies Gv-1 and Gv-2 with the others (Gv-3 to 6) remain obscure.

Sequence E (Early Pliocene) crops out in the western part of the

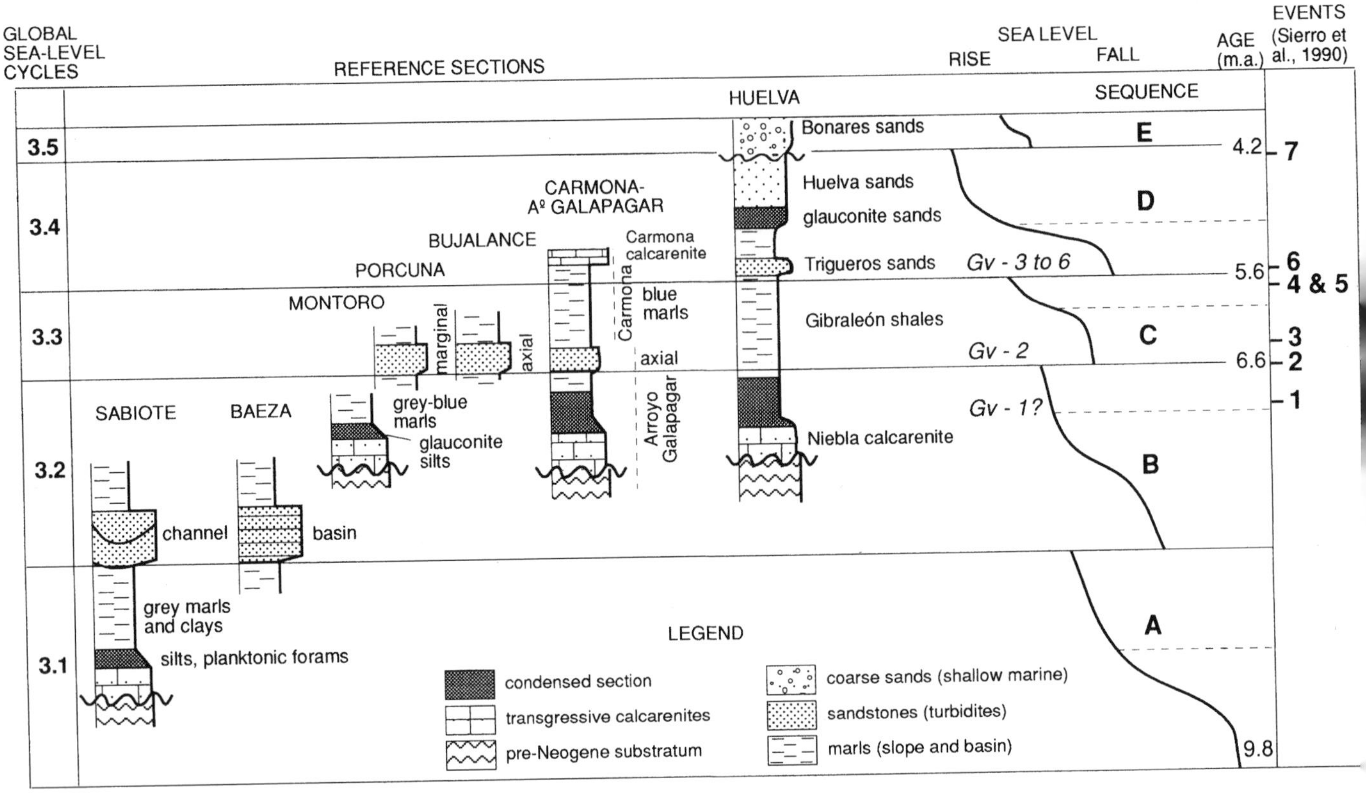

Fig. 8. Depositional sequences distinguished in the Guadalquivir Basin and correlation with the Global Sea Level Cycles of Haq *et al.* (1987). Gv 1 to Gv 6 are turbidite bodies of the Guadalquivir Sands Formation (Suárez *et al.*, 1989). Modified from Sierro *et al.* (1990).

basin as the Bonares Sands. The base of the sequence is a gentle unconformity detected by palaeontological studies, probably related to a fall of sea level that caused the erosion of the underlying sequence particularly in its eastern part. The first record of *Globorotalia puncticulata* (4.24 Ma), several metres below the unconformity, strongly suggests that this unconformity is of the same age as the global sea-level fall separating Cycles 3.4 and 3.5 (4.2 Ma). This is the last marine sequence in this region, though younger marine sediments crop out in some localities of the Cádiz Province.

The palaeobathymetric interpretation of some sequences leads us to discuss an interesting feature of the basin's depositional history. A noticeable increase in the estimated depth to the west occurs in sequences (such as the upper part of sequence B) traced for some distance in an east–west section. This pattern seems contradictory to the interpretation given above if we do not take into account the erosional nature of the line that presently defines the northern margin of the basin. This erosion has removed a large part of the sedimentary cover that once existed in the eastern part.

In contrast, shallow sediments still crop out far away from the depocentres in the western part. The northern margin of the olistostrome units, which defines the line of depocentres of the Late Tortonian immediately after the sliding, is exposed in the eastern part, whereas in the western region this line progressively sinks to reach more than 1000 m depth near Carmona (Sierro *et al.*, 1990). These data suggest a general uplift of the eastern part of the basin that took place after deposition of the oldest sequences.

Acknowledgements

Financial support for this work came from Spanish DGICYT Projects No. PB89–0398 and PB91–0097.

References

Berggren, W.A., Kent, D.V. and Van Couvering, J.A. (1985). The Neogene: Part II. Neogene geochronology and chronostratigraphy. In N.J. Snelling (ed.), *The chronology of the geological record. Geol. Soc. (London). Mem.* **10**: 211–250.

Flores, J.A. (1985). Nanoplancton calcáreo en el Neógeno del borde noroccidental de la Cuenca del Guadalquivir (SO de España). Doctoral thesis, Universidad de Salamanca: 1–714.

Flores, J.A. and Sierro, F.J. (1987). Calcareous nannoplankton in the Tortonian/Messinian transition series of the northwestern edge of the Guadalquivir basin. *Abh. Geol. B-A.* **39**: 67–84.

Flores, J.A. and Sierro, F.J. (1989). Calcareous nannoflora and planktonic foraminifera in the Tortonian/Messinian boundary interval of East Atlantic DSDP Sites and their relation to Spanish and Moroccan sections. In J. Crux and S.E. Van Heck (eds.) *Nannofossils and their applications. British Micropaleontological Society Series*. Ellis Horwood: 249–266.

Flores, J.A., Sierro, F.J. and Glaçon, G. (1992). Calcareous plankton analysis in the preevaporitic sediments of the ODP Site 65- (Tyrrhenian Sea, Western Mediterranean). *Micropaleontology* **38**(3): 279–288.

Haq, B.U., Hardenbol, J. and Vail, P.R. (1987). Chronology of fluctuating sea levels since the Triassic. *Science*, **235**: 1156–1167.

Martínez del Olmo, W., García Mallo, J., Leret, J., Serrano Oñate, A. and Suarez Alba, J. (1984). Modelo tectosedimentario del Bajo Guadalquivir. *I Congreso Español de Geología*. **I**: 199–213.

Müller, D.W. and Hsü, K.J. (1987). Event stratigraphy and paleoceanography in the Fortuna basin (Southeast Spain): a scenario for the Messinian salinity crisis. *Paleoceanography*, **2**(6): 679–696.

Posamentier, H.W., Jervey, M.T. and Vail, P.R. (1988). Eustatic controls on clastic deposition. I: Conceptual framework. In *Sea level changes: an integrated approach. SEPM Spec. Publ.*, **42**: 109–124.

Roldán García, F.J. and García Cortes, A. (1988). Implicaciones de materiales triásicos en la Depresión del Guadalquivir, Cordilleras Béticas (Provincias de Córdoba y Jaén). *Congreso Geológico de España 1988. Comunicaciones*. Vol. **I**: 189–192.

Santisteban, C. and Taberner, C. (1983). Shallow marine and continental conglomerates derived from coral reef complexes after dessication of a deep marine basin: the Tortonian–Messinian deposits of the Fortuna basin, SE Spain. *J. Geol. Soc. London*, **140**: 401–411.

Sierro, F.J. (1984). Foraminíferos planctónicos y bioestratigrafía del Mioceno superior-Plioceno del borde occidental de la Cuenca del Guadalquivir (SO de España). Doctoral thesis, Universidad de Salamanca: 391 pp.

Sierro, F.J. (1985). The replacement of the '*Globorotalia menardii*' group by the *Globorotalia miotumida* group: an aid to recognizing the Tortonian Messinian boundary in the Mediterranean and adjacent Atlantic. *Mar. Micropaleontol.*, **9**: 525–535.

Sierro, F.J. and Flores, J.A. (1989). Winnowed sediments in the Guadalquivir basin: evidence of an Atlantic/Mediterranean water flow exchange before the Mediterranean salinity crisis? *Third International Conference on Paleoceanography. Cambridge, Sept. 1989. Terra Abstracts*, **1**: 36 pp.

Sierro, F.J., González Delgado, J.A., Flores, J.A., Dabrio, C. and Civis, J. (1990a). Global sea level changes and deposition in the Atlantic–Mediterranean North Betic Strait (Guadalquivir Basin). *Abstracts IX R.C.M.N.S. Congress, Barcelona, 1990*. 321–322.

Sierro, F.J., González Delgado, J.A., Dabrio, C.J., Flores, J.A. and Civis, J. (1990b). The Neogene of the Guadalquivir basin (SW Spain). *Field Trip Guidebooks. IXth Congress R.C.M.N.S., Barcelona 1990*. 209–250.

Sierro, F.J., Flores, J.A., Civis, J., González Delgado, J.A. and Francés, G. (1993). Late Miocene globorotaliid event-stratigraphy and biogeography in the NE Atlantic and Mediterranean. *Marine Micropaleontology*, **21**: 143–168.

Suarez, J., Martínez del Olmo, W., Serrano, A. and Leret, G. (1989). Estructura del sistema turbidítico de la Formación Arenas del Guadalquivir, Neógeno del Valle del Guadalquivir. *Libro Homenaje R. Soler, AGGEP*: 123–132.

S5 Miocene basins of the eastern Prebetic Zone: some tectono-sedimentary aspects

CH. MONTENAT, P. OTT D'ESTEVOU AND L. PIERSON D'AUTREY

Abstract

The Miocene series of the eastern Prebetic domain include Aquitanian to early Tortonian marine deposits and a late Tortonian to Messinian–Early Pliocene continental sequence. These deposits are contemporaneous with different episodes of folding and faulting related to the variations of the compressional axis, between N–S and NW–SE. Numerous unconformities and synsedimentary or sealed structures illustrate the different stages of development of Miocene synorogenic basins and give an accurate chronology of the tectonic history of the eastern part of the External Betic Cordilleras.

Introduction

The eastern Betic Cordilleras include three major geological domains, from the North to the South (Fig. 1A): 1. the Prebetic autochthonous domain; 2. the Subbetic allochthonous units, overthrusting northwards the Prebetic domain; Prebetic and Subbetic are referred to as the external zone of the Betic Cordilleras; and 3. the internal allochthonous zone composed of a pile of alpine nappes, including sedimentary and various metamorphic rocks; this domain is separated from the external zone by a major ENE–WSW-trending wrench-fault corridor (the North Betic wrench fault).

This study is dedicated to the Miocene basins evolving in the eastern part of the Prebetic domain (Fig. 2).

The eastern segment of the Prebetic domain is a folded ENE–WSW-trending zone. The folds are elongated structures often faulted and thrusted northwards. Some minor fold axes are twisted with a NE–SW to N–S orientation. NE–SW faults have recorded sinistral or reverse-sinistral movements. NW–SE faults cut through the folds; they acted as normal faults, often reactivated with dextral movement. Triassic evaporites play an important part in the Prebetic tectonics as a major detachment level within the pile of pre-Miocene materials (García Rodrigo, 1971); they generate diapiric structures and extrusions along faults.

Widespread Miocene series, mainly of marine environments, were deposited during the structural development of the Prebetic and Subbetic areas. These sediments reveal the tectonic history of a synorogenic compressional basin. On the other hand, the great diversity of synsedimentary or sealed structures provides an accurate chronology of the structural evolution of the external zone (Ott d'Estevou *et al.*, 1988). These different aspects are mainly illustrated here from the studies of the Alcoy and Ibi-Elda areas (Pierson d'Autrey, 1988; Ladure 1992).

Stratigraphic framework

Miocene series of the eastern Prebetic area are analysed in terms of various ages and facies and partitioned by unconformities of a regional spread (Fig. 3).

Aquitanian to Early Tortonian series were deposited in a marine environment and the late Tortonian to Messinian–Early Pliocene deposits belong to a continental sequence.

Aquitanian to Early Burdigalian levels display scattered outcrops. Biogenic carbonates with Miogypsinids predominate to the north (Alcoy) whilst planktonic-rich marls and marly limestone (*Globigerinoides primordius*, *Globigerinita dissimilis* and *Globigerinoides trilobus* zones) have a better development to the south (between Tibi and Elda, for example).

Transgressive late Burdigalian limestones and marls (*Globigerinoides bisphaericus* = *G. sicamus* zone) mark the beginning of a thick marine succession (≤700 m) including Langhian (*Praeorbulina glomerosa*, *Orbulina suturalis* and *G. peripheroronda* zones) and Serravallian (*Globorotalia mayeri*, *G. praemenardii* and *G. menardii* zones) deposits, which represent the major part of the Miocene sedimentation. In most cases, the Serravallian beds appear widely transgressive and rest unconformably on Miocene or pre-Miocene rocks, (Figs. 4, 5 and 1B). Langhian and Serravallian basinal deposits display a quite similar facies, the so-called 'Tap' facies consisting of grey or whitish chalky pelagic marls very rich in planktonic foraminifera, coccoliths and siliceous nannofossils indicating an open marine deep environment. The 'Tap facies' may reach the base of the Tortonian, as evidenced by the presence of *Globorotalia acostaensis* in its uppermost part (Río Vinalopó area, Ladure, 1992).

Early Tortonian biocalcarenites and sandy limestones, with abundant *Amphistegina*, Pectinids, Bryozoa and Echinids (Montenat, 1975, 1977), conformably overly the preceding sediments, but

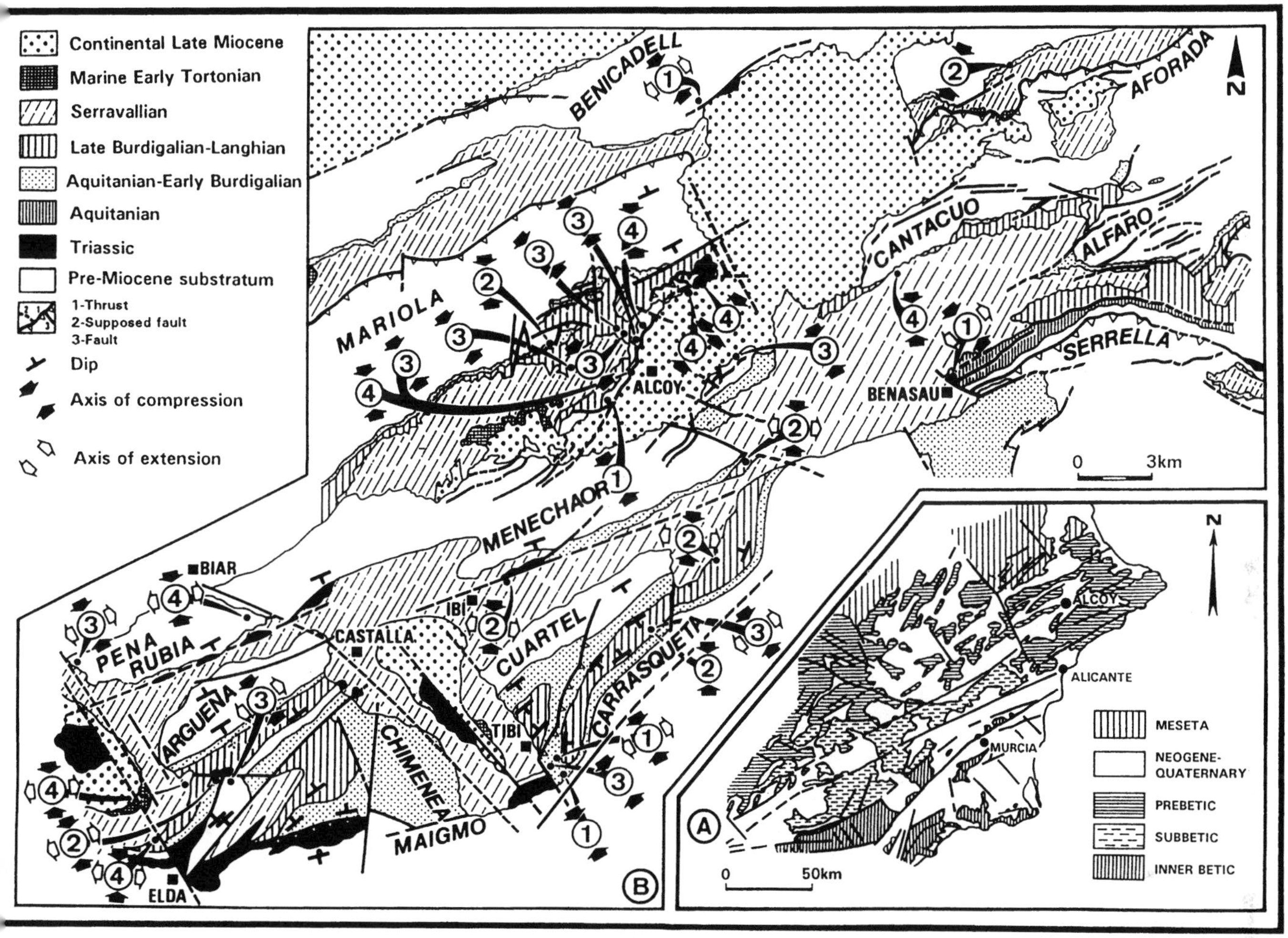

ig. 1. A. Simplified geological map of the eastern Betic Cordilleras. B. Geological sketch-map of the Miocene series (Alcoy–Ibi area) with indications of the tress tensor data. Numbers within circles refer to the stages of structural development: 1 – Aquitanian – Early Burdigalian; 2 – Late Burdigalian – Langhian *pro arte*; 3 – Late Langhian–Serravallian to Early Tortonian; 4 – Late Tortonian–Messinian to Early Pliocene (after Ott d'Estevou *et al.*, 1988).

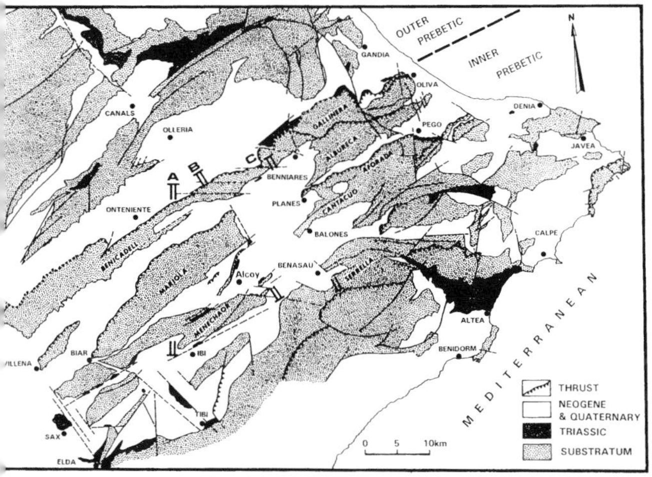

ig. 2. Geological sketch-map of the eastern Prebetic domain. Note the evelopment of faulted overfolds thrusting northwestwards in the Inner rebetic and the more tabular structure of the Outer Prebetic. A, B and C, ocation of sections, in Fig. 4. (after Pierson d'Autrey, 1987).

are separated by an erosional surface. A type-section of these different stratigraphic units is exposed near to Alcoy (Durand-Delga *et al.*, 1964; Montenat 1975; Pierson d'Autrey, 1987) (Fig. 4).

Late Miocene series include alluvial fans, flood plain and channel sediments, lacustrine deposits and local accumulations of lignite. They rest unconformably on Miocene or pre-Miocene rocks. A Messinian age is inferred from the presence of the Ostracods *Cyprideis* and *Ilyocypris* (Pierson d'Autrey, 1987). The lignites of Alcoy, located in the uppermost part of the series, yield an Early Ruscinian mammal fauna (see discussion in Mein & Agustí, 1990); on this evidence, the continental sequence may reach the Early Pliocene.

Evolution of the Prebetic Miocene basins

Tectonic–sedimentary relationship

Four stages of the tectonic–sedimentary evolution may be distinguished (Figs. 3 and 1B).

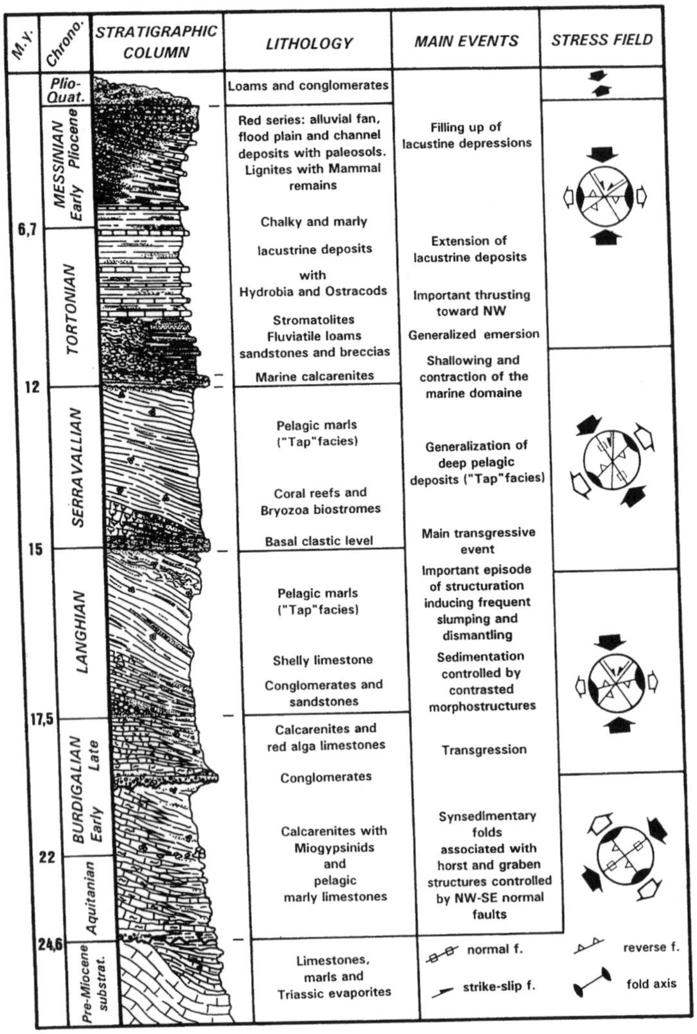

Fig. 3. Eastern Prebetic Miocene. Main sedimentary and tectonic events.

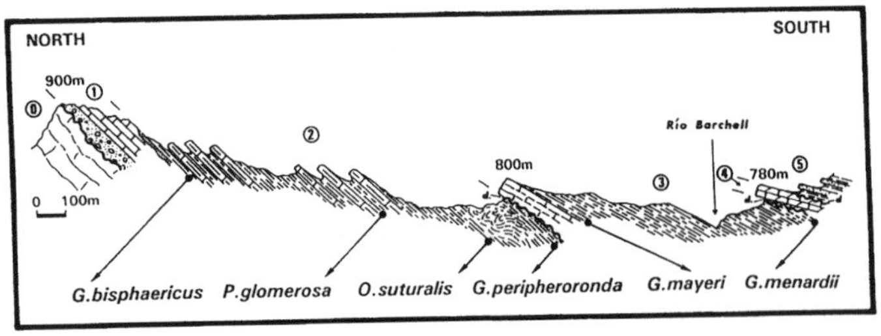

Fig. 4. A type-section of the Prebetic Miocene with indication of the first appearance of significant planktonic foraminifera (between Bocairente and Barchell, near Alcoy). Note the different unconformities (d); (after Pierson d'Autrey, 1987). 0–5 refer to informal unit numbers.

Aquitanian–Early Burdigalian

Within the area concerned, the pre-Miocene folded structures are poorly documented and are only slightly apparent. Nevertheless, there is evidence of some N070–N080-trending folds with gentle dips, of late Paleogene age (Azema, 1977). These pre-existing structures influenced the early Miocene depositional pattern: Aquitanian carbonates with Miogypsinids and red algae developed on the top of eroded anticlines (up to the Cretaceous) whilst pelagic marls with planktonic foraminifera were deposited in synclines, often conformably on the marine Oligocene deposits.

The Aquitanian succession is followed by the Early Burdigalian deposits without a notable break.

During that period, the synsedimentary structural development is characterized by the initiation of N050-trending folds which display progressive unconformities (Arguena syncline, Figs. 6A and 2B) and generate gravity flows and olistoliths flowing basinwards. Some folds evolved into reverse faults during the Early Burdigalian (Arguena, Castalla). At the same time, large NW–SE normal faults bounded horst and graben structures which directly controlled the sedimentation (for example, the uplifted block of Chimenea, the rifts of Tibi and the Río Vinalopó, which accumulated thick pelagic series; Fig. 6A). Some of these faults were sealed during the Early Burdigalian.

Late Burdigalian–Langhian

The Late Burdigalian limestones extend transgressively on the pre-Miocene formations. In the pre-existing depressions, an erosional surface and/or a low-angle unconformity mark the contact with the Early Burdigalian. The pelagic 'Tap' sedimentation began close to the Burdigalian–Langhian boundary. The deposits were strongly deformed prior to the deposition of the Serravallian sediments (Fig. 6B); they have recorded two stages of structural development. In the first stage, N070–N080 trending folds were formed (Fig. 7A). They often evolved toward faulted folds and reverse faults, generally overthrusted northwards. The NW–SE faults, previously activated as normal faults, display evidence of Langhian dextral movement; some NE–SW faults recorded left-reverse movement (Ladure, 1992).

In the second stage, during the Late Langhian, folding associated with the activation of reverse faults occurred along a NE–SW trend. The most spectacular expression of that stage of structural development is illustrated by the large synsedimentary fold of the 'Barranco del Sing' near Alcoy, shown in Fig. 8. By the end of that second stage, the NW–SE faults turn once more to normal faults and cut sharply through the folded structures (Tibi graben, for example, Fig. 6B). Extrusions of Triassic variegated marls and evaporites are located at the intersection of N080- and N140-trending faults. The extruded material, removed as mud-flows, was interbedded with the pelagic Langhian sediments.

Serravallian – Early Tortonian

Serravallian pelagic deposits ('Tap') buried many Miocene folded structures (see for example Fig. 8) and sealed many fault scarps. In some cases, the angular unconformity between Langhian and Serravallian rocks reaches 90°, for example along NE–SW faults, and it may be underlain by breccias and olistoliths produced from dismantled Langhian structures. Throughout, the Serravallian extends transgressively over the previously emergent or subemergent pre-Miocene basement (Figs. 5, 7B and 1B). This period corresponds to a maximum of submersion of the eastern Prebetic

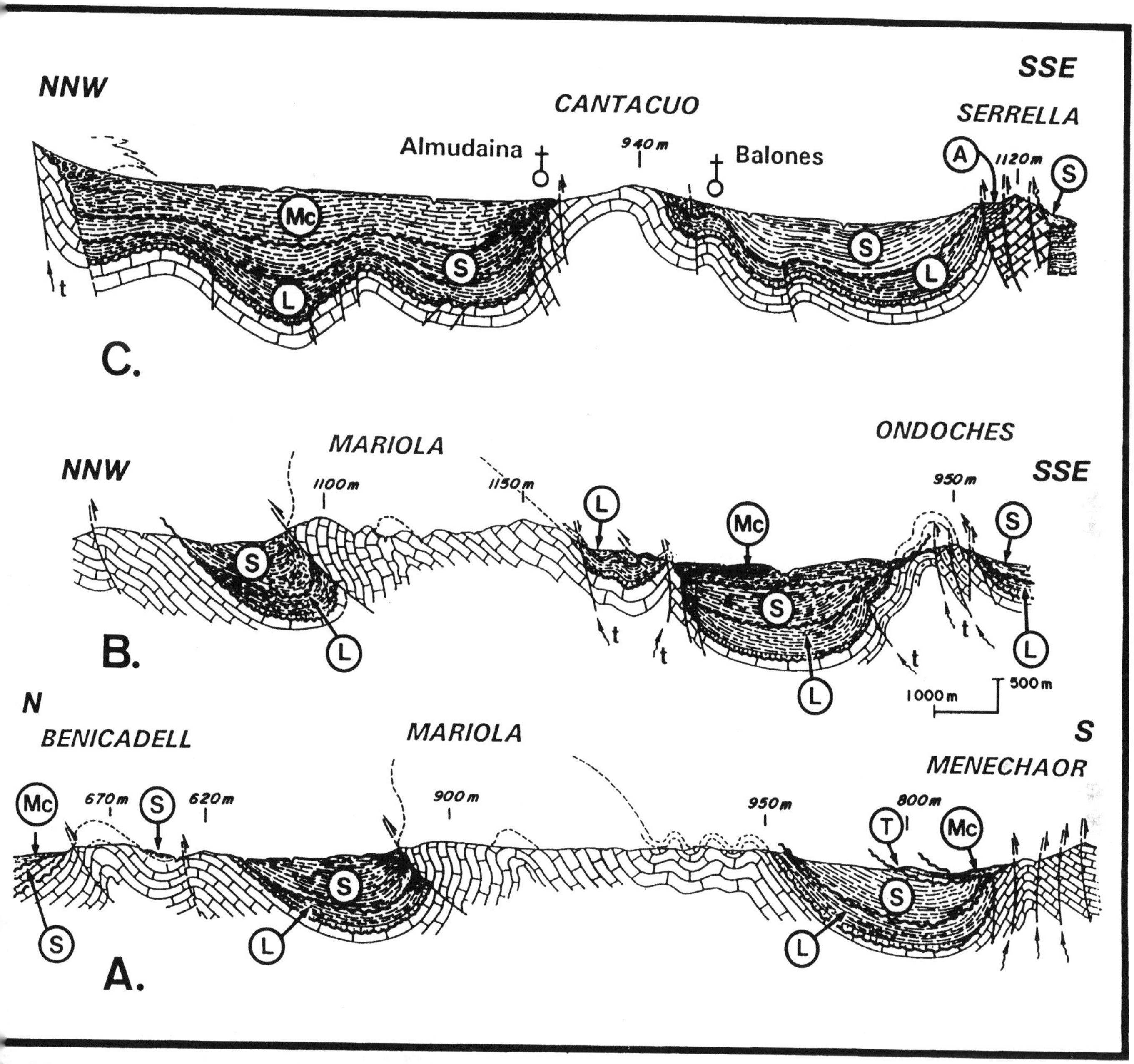

'ig. 5. Cross-sections of the Miocene basins. A, B and C, see location on Fig. 1. Indications of the Miocene series: A – Aquitanian; L – Burdigalian–Langhian; S – ;erravallian; T – marine Early Tortonian; Mc – Tortonian–Messinian; continental deposits. t – extrusion of Triassic material along fault planes. Note the vidences of synsedimentary folding (north of Ondoches) and the sealing of faults and folds by the late Miocene series (adapted from Pierson d'Autrey, 1987).

one under deep marine waters. It may be related to an important ustatic event, which, significantly, is recorded in different struc-ural zones of the Betic orogen (see Montenat, 1990; Soria, 1993).

The synsedimentary Serravallian structural development is more iscrete than during the preceding periods. The formation of some olds and flexures with a NE–SW axis indicates the persistence of ompressional strain. In many places, normal movements along IW–SE faults are recorded. Resedimentation of Triassic material ; still frequent, mainly along NNE–SSW and NW–SE faults (Ibi, or example). Highly deformed blocks of pre-Miocene rocks (metric to decametric in scale) are often enclosed within the Triassic marls. These almond-shaped tectonic blocks are crushed within the Miocene shear zones.

Tortonian and Messinian

The paleogeography was entirely transformed during Early Tortonian times. A drastic and sudden shallowing of the sea resulted in a notable contraction of the marine domain, prelude to a general emergence. Marine Late Tortonian, Messinian and locally

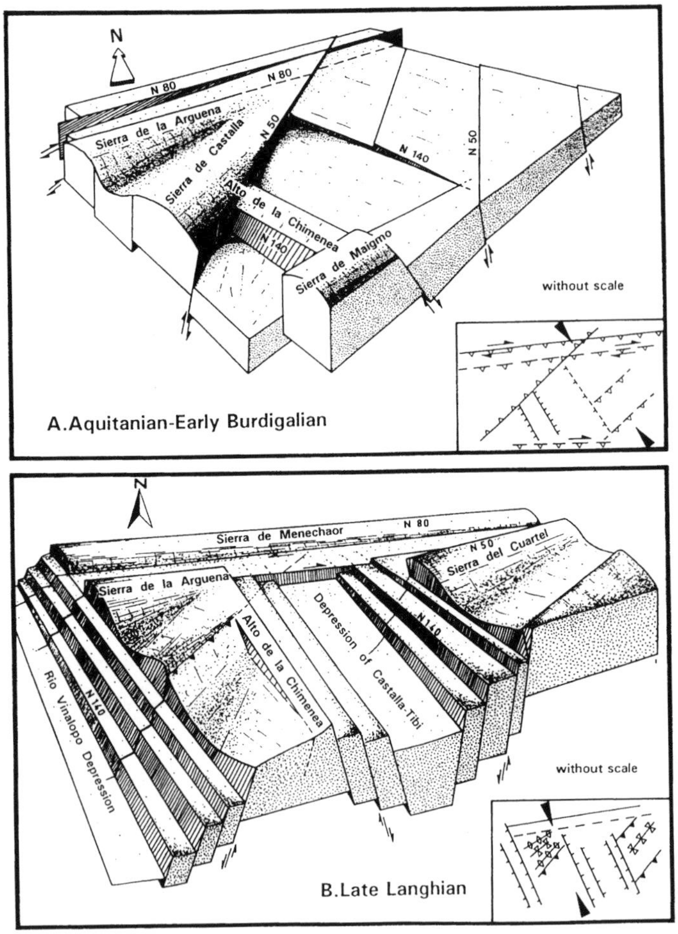

Fig. 6. **Diagrammatic representation of the structuration of the pre-Miocene substratum in the Ibi-Tibi area (without scale). A – Aquitanian – Early Burdigalian. B – Langhian–Serravallian boundary. Main structures and axis of compression are figured in the cartouches (after Ladure, 1992).**

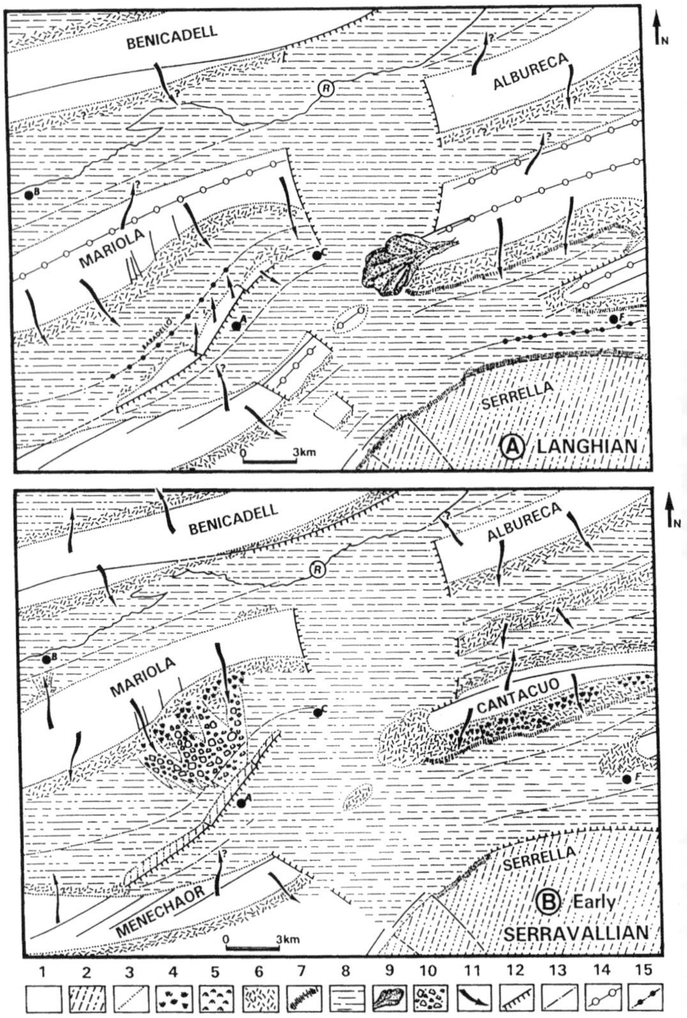

Fig. 7. Example of relationship between the structural framework and the sedimentary processes. A. Langhian; B. Early Serravallian. 1 – Emergent zone; 2 – shallows; 3 – shoreline; 4 – coral reef (mainly *Porites*); 5 – *Bryozoa* biostrome; 6 – platform calcarenitic deposits; 7 – talus; 8 – planktonic ooze ('Tap' facies); 9 – deltaic fan; 10 – debris-flow; 11 – supplying of clastic and calcarenitic sediments (turbidites or grain-flows); 12 – fault scarps (without indications of movement); 13 – axis of subsidence; 14 – synsedimentary anticline; 15 – synsed. syncline. R (within circle): present-day southern border of the Sierra Benicadell (for comparison). Localities: A – Alcoy; B – Bocairente; C – Cocentaina; F – Famorca (modified after Pierson d'Autrey 1987).

Pliocene deposits are then confined to the southern fringe of the eastern Prebetic (Alicante, Elche) where they are connected with the Neogene basins of the internal zone.

In the area concerned, Tortonian and Messinian sediments recorded a notable increase in clastic material; the reworking of Triassic rocks is frequent.

In some places, the Early Tortonian marine calcarenites were faulted and dismantled prior to the deposition of the continental series: N050 reverse sinistral faults and N080 reverse faults are observed (Pierson d'Autrey, 1987).

Late Miocene continental deposits are slightly deformed; the deformation is generally concentrated near to some major faults. For example, the NW–SE faults which bounded the Serravallian grabens (cf. above) are reactivated with a dextral movement (Tibi, Alcoy).

In many places faults and folds are sealed by the continental series (Fig. 6). In the southern region (Alicante, Elche), synsedimentary E–W-trending folds were developed during late Tortonian to Pliocene times; some of the anticline cores involved Triassic material (Montenat, 1977).

Evolution of the Miocene stress field

The analysis of synsedimentary or sealed megastructures and associated small fractures indicates that the domain considered was subjected to a *continuum* of compressive strain during the whole of the Miocene.

However, the succession of tectonic–sedimentary features clearly shows that the directions of the compressional axis (σ_1 horizontal

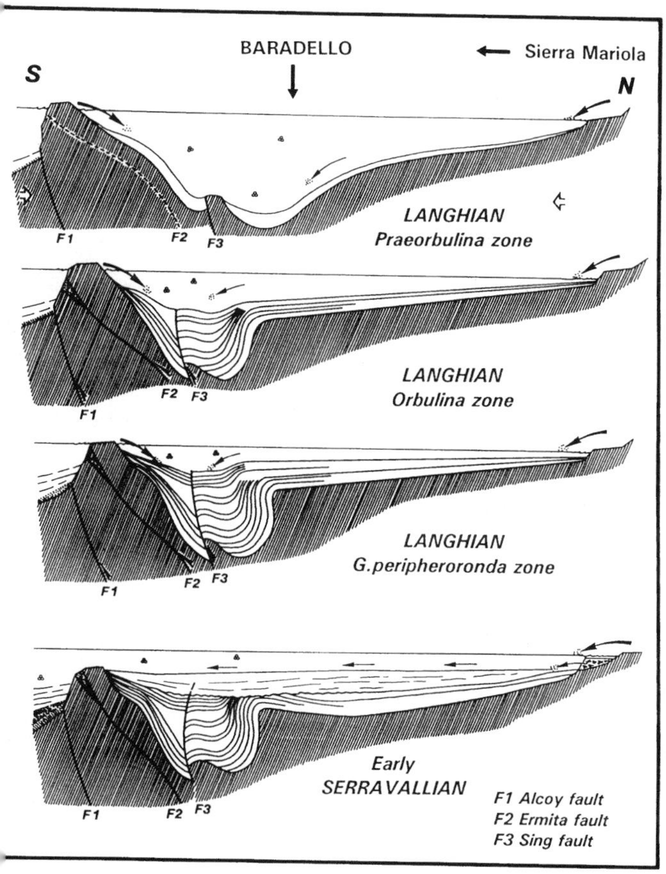

Fig. 8. Interpretative reconstruction of the Barranco del Sing folded structure (near Alcoy) during Langhian and Serravallian times (after Pierson d'Autrey, 1987).

was not stable but oscillated between NE–SE and N–S. The compression induced a slight perpendicular extension (σ_3 horizontal). On that account, folds and grabens, with appropriate orientations, may be associated in the region during the same tectonic–sedimentary episode. The stress field evolved as follows (Figs. 3 and 1B):

- Aquitanian–Early Burdigalian, compressional axis directed NW–SE;
- Late Burdigalian–Langhian (*pro-parte*), the compressional axis rotated to N–S;
- Late Langhian–Serravallian and Early Tortonian. The direction of compression turned to NW–SE; the associated NE–SW extension is well expressed by the development of grabens bounded by NW–SE normal faults. Note that the changes in the stress field do not exactly coincide with the unconformities bounding the Serravallian series;
- Late Tortonian–Messinian, the compressional axis directed N–S. The effects of the compression appear more pronounced to the south, close to the contact with the internal zone.

These data may be compared with those obtained for the eastern internal zone, which was also subjected to compressive strain during Neogene times (Coppier *et al.*, 1989; Montenat, 1990; Rodríguez-Fernández *et al.*, 1990). Note that there is the same NW–SE direction of compression during the Serravallian and Early Tortonian, which changes to N–S during late Tortonian to early Pliocene times (see Chapter S8). These variations in the stress field, for short periods (some millions of years), may be regarded as a characteristic feature of the Neogene tectonics in the Betic–Rif orogen.

Conclusion

The eastern Prebetic Miocene basins have recorded a polyphase structural evolution, marked by the permanency of compressive strain. Synsedimentary folds play an outstanding part in the tectonic–sedimentary development, basically associated with extensional structures (grabens). Triassic evaporites acted as an important level of detachment between the basement and the post-Triassic series, which greatly favoured the development of folding. Some of the large Miocene folds were induced by the reactivation of ENE–WSW faults, probably inherited from the basement and generating an important raising of the disharmonic Triassic material. It that way, the kinematics of the Miocene basins evolving in the external and internal Betic zones show notable differences. In the internal domain, the basins developed on a rigid, often metamorphic basement display much larger faulted structures, directly inherited from the basement tectonics; in the same domain, folds have a minor role and are essentially associated with strike–slip movements (see Chapter S8).

The Eastern Prebetic Miocene basins are a part of a much wider marine domain with a deep marine environment, stretching all along the Prebetic zone, the so-called 'North Betic Strait', which connected the Atlantic and the Mediterranean during the early and middle Miocene times (see Chapter S2).

To the north, the shores of this strait may be traced through the northern Prebetic regions where little-deformed, subhorizontal calcarenites, sands and conglomerates take the place of the pelagic 'Tap' facies (North of Nerpio, Almansa, Elche de la Sierra, etc.) (Jerez Mir, 1973; Usera, 1972).

The southern border, however, which is contiguous with the Subbetic domain, recorded a complex structural evolution. The Subbetic allochthonous units widely overthrusted the Miocene series of the Prebetic basin. For that reason, its southern margin is no longer exposed. In some places, it may be observed that the setting of the allochthonous Subbetic front (units made of Cretaceous and Jurassic rocks) began with the sliding of an important mass of Triassic material during the Late Serravallian. This olistostrome, made of successive mass- and mud-flows (see Hoedemaeker, 1973, for the Moratalla region), originated from a southern zone. It flowed into the basin and gave rise to very spectacular striations of the Miocene sea bottom, looking like glacial erosional surfaces (the most interesting examples are exposed in the El Sabinar – Nerpio area to the west of Moratalla (Montenat & Ott d'Estevou, 1984)).

The setting of the Subbetic units was established at the beginning of the Tortonian. At that time, shallow marine calcarenites and other clastic sediments, produced by the erosion of the newly uplifted Subbetic relief, were deposited at the front of the nappes.

The previously noted sharp contacts between the early Tortonian calcarenites and the Serravallian 'Tap' deposits may be regarded as an attenuated and distant recording of that major event across the whole Prebetic basin.

References

Azema, J. (1977). Etude géologique des zones externes des Cordillières Bétiques aux confins des provinces d'Alicante et de Murcia (Espagne). Thesis Sci., Univ. of Paris VI: 396 pp.

Coppier, G., Grievaud, P., Larouzière, F.D. de., Montenat, C. and Ott d'Estevou, Ph. (1989). Example of Neogene tectonic indentation in the Eastern Betic Cordilleras: the arc of Aguilas (Southeastern Spain). *Geodinam. Acta*, **3**(1): 37–51.

Durand-Delga, M., García Rodrigo, B., Magne, J. and Polveche, J. (1964). A propos du Miocène de la région d'Alcoy (province d'Alicante, Espagne). *2eme réunion Com. Neog. Medit., Sabadell. Curs. y Conf., Madrid*, **9**: 213–217.

García Rodrigo, B. (1971). Sur la structure du Prébétique au Nord d'Alicante. In *Livre à la Mémoire du Professeur Paul Fallot. Mem. h. Ser. Soc. Géol. France*, Paris, **I**: 137–141.

Hoedemaeker, Ph. J. (1973). Olisthostromes and other delapsional deposits and their occurrence in the region of Moratalla (province of Murcia, Spain). *Scripta Geol. Publ.* Amsterdam, **19**, 207 pp.

Jerez Mir, L. (1973). Geología de la Zona Prebética en la transversal de Elche de la Sierra y sectores adyacentes (provincias de Albacete y Murcia). Thesis, Univ. of Granada, 750 pp.

Ladure, F. (1992). Evolution tectono-sédimentaire de la région d'Ibi (Prébétique oriental, Espagne) au Néogène. Thesis, Univ. of Bordeaux, **I**, 281 pp.

Mein, P. and Agusti, J. (1990). Les gisements de Mammifères néogènes de la zone bétique. *Doc. Trav. IGAL*, Paris, **12–13**: 81–84.

Montenat, C. (1975). *Le Néogène des Cordillères bétiques: essai de synthèse stratigraphique et paléogéographique*. Rapport BEICIP, 187 pp.

Montenat, C. (1977). Les bassins néogènes du Levant d'Alicante et de Murcia: stratigraphie, paléogéographie et evolution dynamique. *Docum. Lab. Géol. Fac. Sci.* Lyon, **69**, 345 pp.

Montenat, C. (ed.) (1990). Les bassins néogènes du domaine bétique oriental (Espagne). *Doc. Trav. IGAL*, Paris, **12–13**, 392 pp.

Ott d'Estevou, Ph. and Montenat, C. (1984). Déformations d'un fond marin miocène, consécutives au glissement de nappes gravitaires (Front subbétique – Espagne). *5eme Congrès Europ. Sédiment., Marseille, Abstracts*: 337–338.

Ott d'Estevou, Ph., Montenat, C., Ladure, F. and Pierson d'Autrey, L. (1988). Evolution tectono-sédimentaire du domaine prébétique oriental (Espagne) au Miocène. *C.R. Acad. Sci., Paris*, **307**, **II**: 789–796.

Pierson d'Autrey, L. (1987). Sédimentation et structuration synsédimentaire dans le bassin Miocène d'Alcoy (Cordillères bétiques externes orientales – Espagne). Thesis, Univ. of Paris XI, 315 pp.

Rodríguez-Fernandez, J., Sanz de Galdeano, C. and Serrano, F. (1990). Le couloir des Alpujarras. *Doc. Trav. IGAL* Paris, **12–13**: 87–100.

Soria Mingorance, J.M. (1993). La sedimentacion neógena entre Sierra Araña y el Rio Guadiana Menor (Cordillera Bética Central). Thesis, Univ. of Granada, 292 pp.

Usera, J. (1972). Paleogeografia del Mioceno en la provincia de Valencia. *Bol. Real. Soc. Esp. Hist. Nat.*, **70**: 307–315.

S6 Stratigraphic architecture of the Neogene basins in the central sector of the Betic Cordillera (Spain): tectonic control and base-level changes

J. FERNÁNDEZ, J. SORIA AND C. VISERAS

Abstract

Stratigraphic and sedimentological analysis of the Granada and Guadix basins establishes the time when they became distinct (Upper Tortonian) and their different phases of evolution, up to the entrenchment of the present fluvial network during the Holocene. Six depositional sequences have been differentiated. These are delimited by major unconformities, representing tectonic and/or eustatic events. The two oldest sequences (Tortonian) correspond to infilling during the phase of marine sedimentation, the third (uppermost Tortonian) corresponds to the marine–continental transition, and the remaining three (post-Tortonian) correspond to the continental infilling.

Introduction

The Granada and Guadix basins are intramontane basins developed during and after the Upper Miocene, after westward displacement of the Alborán realm had ceased (Sanz de Galdeano & Vera, 1992). The presence of the remains of Lower and Middle Miocene marine materials on the basement suggests that these and other basins were superimposed on what was a sedimentation area which, from its inception and throughout its evolution, played a part in the geological history of the Alborán basin. The sedimentary cover recognized on the basement of the Alborán basin (Comas *et al.*, 1992), which characterizes the sedimentary infilling of the Betic and Rifian Neogene basins, allows us to visualize a more or less continuous sedimentary deposition during part of the Neogene in an 'ancient Alborán basin'.

These basins are located in the central sector of the Betic Cordillera, lying partially on the contact (N60–70E in direction) between the South Iberian palaeomargin (External Zones) and the Alborán realm (Internal Zones), so that the northwestern half of its basement is made up of Mesozoic and Tertiary materials from the External Zones, and the southeastern half consists of Palaeozoic and Triassic materials from the metamorphic complexes of the Internal Zones. Their boundaries present geometries controlled by the most recent fault systems (NW–SE, NNE–SSW and E–W) which intersect the main domains of the Cordillera (Fig. 1).

The application of new methods of basin analysis, concerning both the differentiation of depositional sequences and the three-dimensional description of their stratigraphic organization, represent an advance in our understanding of stratigraphic architecture, the recognition of the major unconformities related to global (tectonic, eustatic and/or climatic) changes, and the factors determining the sedimentary record.

We present here a comprehensive sedimentological and stratigraphic study of the infilling of these basins, as regards the nature and organization of the sequences and depositional systems. We also make a comparative analysis of both basins and propose an integral scheme for their tectono-sedimentary development in the context of global evolution.

Stratigraphy

The oldest materials of the sedimentary infill in these basins correspond to the Lower Tortonian. The most recent have been dated as Middle Pleistocene in the Granada basin (Ruíz-Bustos *et al.*, 1990) and as Middle–Upper Pleistocene boundary in the Guadix basin (Martín Penela, 1987). During the Holocene and probably also part of the Upper Pleistocene, the processes of erosion and sedimentation were related to the present fluvial system (Viseras & Fernández, 1992).

From the stratigraphic point of view, the data on surface geology and on the subsurface, as known from seismic profiles and boreholes, allow us to distinguish six depositional sequences (Figs. 2, 3 and 4). The two oldest sequences (Tortonian) correspond to infilling during the phase of marine sedimentation, the third (uppermost Tortonian) corresponds to the marine–continental transition, and the remaining three (post-Tortonian) correspond exclusively to the continental infilling.

Marine infilling

The marine infilling is represented by depositional sequences I and II (Fig. 5). Depositional sequence I (Lower Tortonian) represents the infilling of the very irregular palaeogeography that existed after the first phases of relief creation and

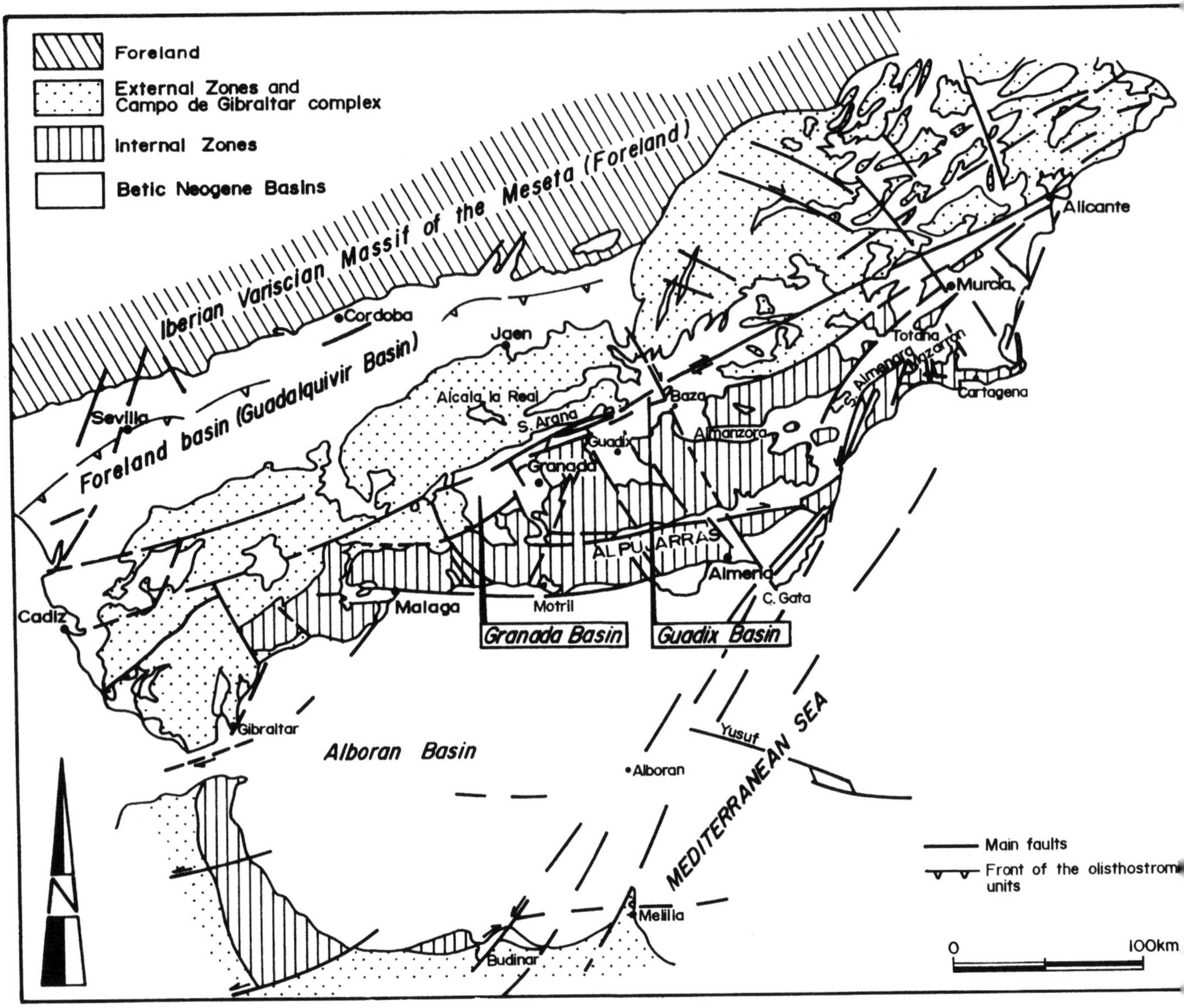

Fig. 1. Main faults of the Betic Cordillera (modified from Sanz de Galdeano, 1983) and positions of Granada and Guadix basins.

individualization. It therefore presents large variations in thickness, with onlap geometry on the margins and relative highs inside the basin.

The most representative sedimentary association formed on carbonate ramps with bioclastic calcarenite deposits in the shallowest areas and foraminiferous marls inside the basins. The deposits of coastal environments linked to the transgression have only been preserved locally and their characteristics vary according to the lithological nature of the basement: cliffs and conglomeratic beaches on Alpujarride dolostones from Sierra Arana (on the southwestern margin of the Guadix basin), or estuaries and sandy beaches on Alpujarride relief made up of schists and micaschists on the southern margin of the Granada basin (Fernández & Rodríguez-Fernández, 1989). The materials of this sequence constitute a transgressive system tract (Fig. 6).

Depositional sequence II (Upper Tortonian) is made up of marls in the centre of the basins, changing towards the southern margins to fan deltas and prograding platforms supplied from the Sierra Nevada (Dabrio *et al.*, 1978; Vera & Rodríguez-Fernández, 1988; Fernández & Guerra, 1993). The lower boundary is an unconformity which, from a tectonic point of view, coincides with an intra-Tortonian event of fracturing and folding detected at many sites in the Cordillera (Rodríguez-Fernández, 1982), which may coincide with the limit between third-order cycles TB 3.1 and TB 3.2 (Haq *et al.*, 1987, 1988) and, from a sedimentary point of view, is marked by an abrupt change in sedimentary polarity, involving a change from retrograding to prograding trends. This in turn coincides with a situation of maximum relative sea-level rise, and so the deposits of this sequence can be considered a highstand system tract (Fig. 7).

Towards the top, communication with the open sea became

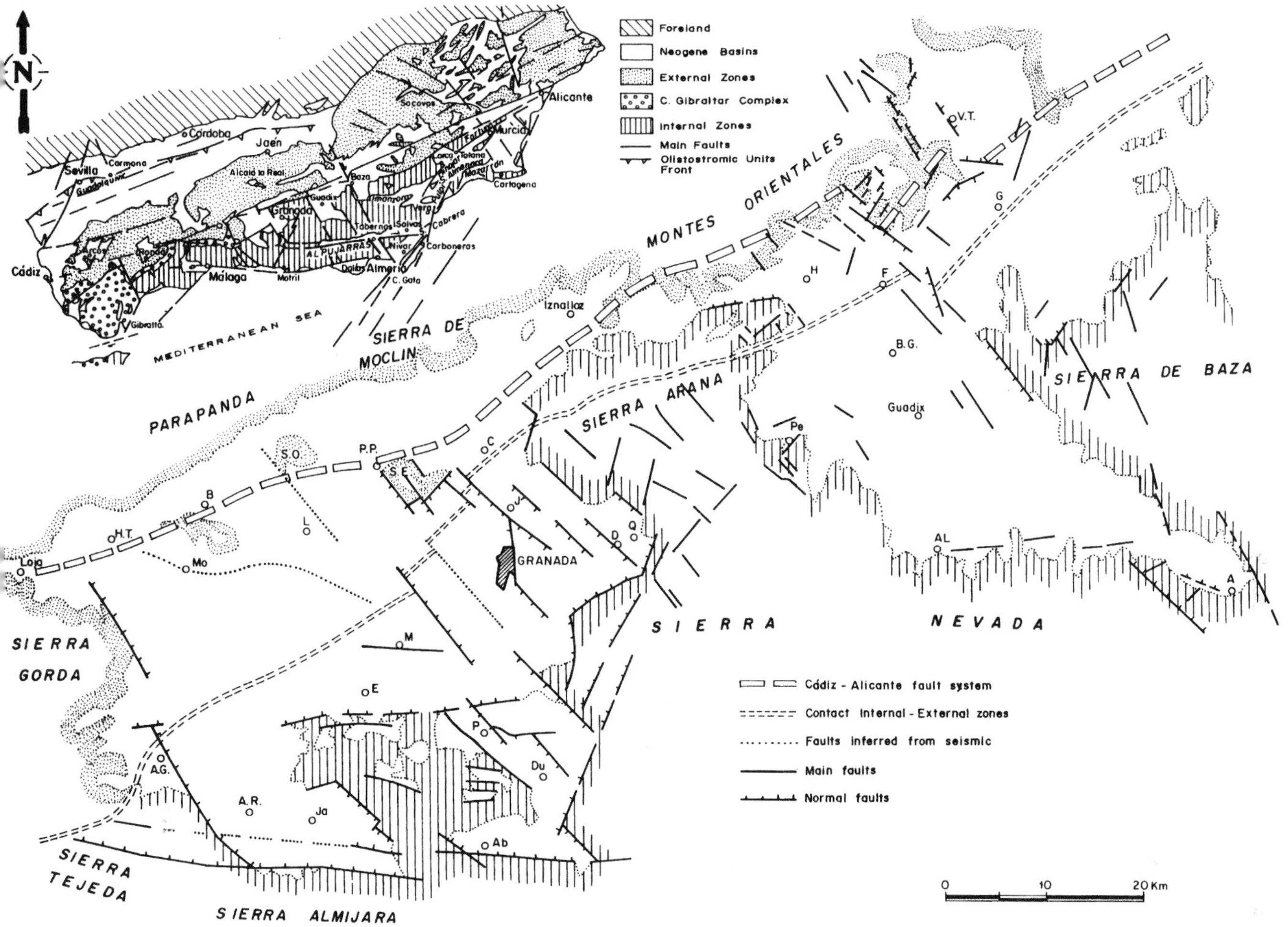

Fig. 2. Main tectonic features of the Granada and Guadix basins. HT: Huétor-Tájar; B: Brácana; SO: Sierra de Obeilar; SE: Sierra Elvira; PP: Pinos Puente; Mo: Moraleda; L: Láchar; M: La Malá; E: Escúzar; AB: Albuñuelas AG: Alhama de Granada; AR: Arenas del Rey; Ja: Jayena; Du: Durcal; P: Padul; D: Dúdar; Q: Quéntar; J: Jun; C: Calicasas; H: Huélago; Pe: La Peza; AL: Alquife; A: Abla; BG: Benalúa de Guadix; F: Fonelas; VT: Villanueva de las Torres; G: Gorafe.

progressively more restricted, thus causing a rise in temperature in some basins such as the Granada basin and the appearance of the first patch-reefs colonizing the topsets of the fan deltas and the shallower sectors of the deeper slopes (Dabrio *et al.*, 1978; Braga *et al.*, 1990).

Marine–continental transition

Towards the Tortonian–Messinian boundary a eustatic fall took place, coinciding approximately with the boundary between third order cycles TB 3.2 and TB 3.3. This fall caused the major withdrawal of the sea from many Betic intramontane basins and the complete disconnection of the Mediterranean and Atlantic. As a consequence of this eustatic fall a palaeogeography was formed characterized by emerged marginal areas and shallow marine basins restricted to the central zones. The sedimentary record of this situation is the third depositional sequence (Figs. 8 and 9), in which materials of different lithology are found deposited in different environments developed in conditions of sea-level fall (lowstand system tract).

In the Granada basin a facies association can be recognized characterizing the following distribution of sedimentary environments from the margin to the centre of the basin (Dabrio *et al.*, 1982):

- Alluvial fans related to drainage of Sierra Nevada and Sierra Arana and represented by conglomerates changing distally to sands.
- Lutite plain with development of stromatolites mineralized by celestine.
- Shallow marine basin with selenite deposits in marginal areas and halite in the centre.

In the Guadix basin the zonation of facies from margin to centre includes alluvial fans, marine deposits corresponding to small Gilbert-type deltas and shallow calcarenite platforms.

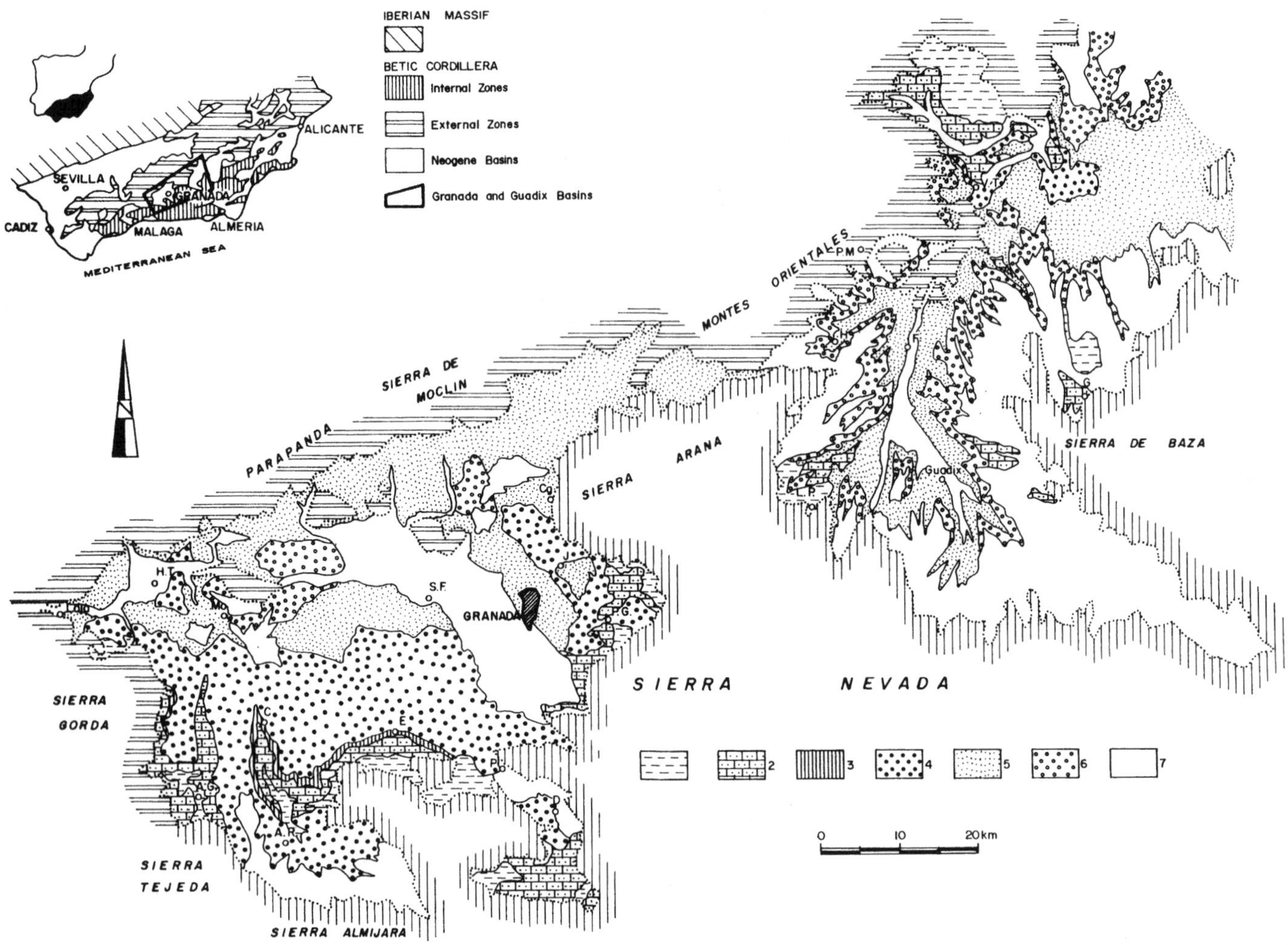

Fig. 3. Map of depositional sequences in the Granada and Guadix basins. 1: depositional sequence I (Lower Tortonian); 2: depositional sequence II (Upper Tortonian); 3: depositional sequence III (Uppermost Tortonian); 4: depositional sequence IV (Upper Turolian–Ventian); 5: depositional sequence V (Lower Pliocene); 6: depositional sequence VI (Upper Pliocene–Pleistocene); 7: recent Quaternary. From north to south: VT – Villanueva de las Torres; PM – Pedro Martínez; F – Fonelas; G – Gor; Cg – Cogollos; LP – La Peza; HT – Huétor Tájar; J – Jun; SF – Santa Fé; PG – Pinos Genil; C – Cacín; E – Escúzar; P – Padul; D – Dúrcal; AG – Alhama de Granada; AR – Arenas del Rey.

Continental infilling

The continentalization of these basins took place in a context basically defined by the eustatic fall at the end of the Upper Tortonian and tectonic structuring that brought about an important physiographic change, basically involving expansion of the fomerly marine basins and migration of the depocentres (Fig. 10). This stage of tectonic structuring was associated with the activity of the main fault systems and an isostatic adjustment involving the uplift of the Sierra Nevada together with its folding as a large antiform, all of which took place in a context of NNW–SSE compression (Sanz de Galdeano, 1983; Fernández *et al.*, 1991).

During continental infilling the basins were markedly asymmetrical, with their longitudinal axis close to the northern margin, coinciding with the areas of lacustrine sedimentation. Their stratigraphic architecture is a result of the stacking of depositional sequences IV, V and VI (Fig. 11). Depositional sequence IV is represented by Upper Turolian–Ventian materials (Rodríguez-Fernández *et al.*, 1990; Ruíz-Bustos *et al.*, 1990; Soria & Ruíz-Bustos, 1992). It is unconformable and onlapping over the deposits of the previous depositional sequences, forming onlaps towards the edges and relative highs. It presents large variations in thickness with wedging normally towards the centre.

Depositional sequence V involves extension relative to sequence IV and its age is approximately Lower Pliocene. The depocentre outlined in the preceding sequence became accentuated and migrated northwards in both basins, thus emphasizing their asymmetry. Finally, depositional sequence VI, which also involves expansion, has a considerably lower volume of materials, although it occupies a similar time interval (most of the Upper Pliocene and the Pleistocene).

The internal organization of these depositional sequences is the result of the interrelation of several depositional systems (Viseras 1991):

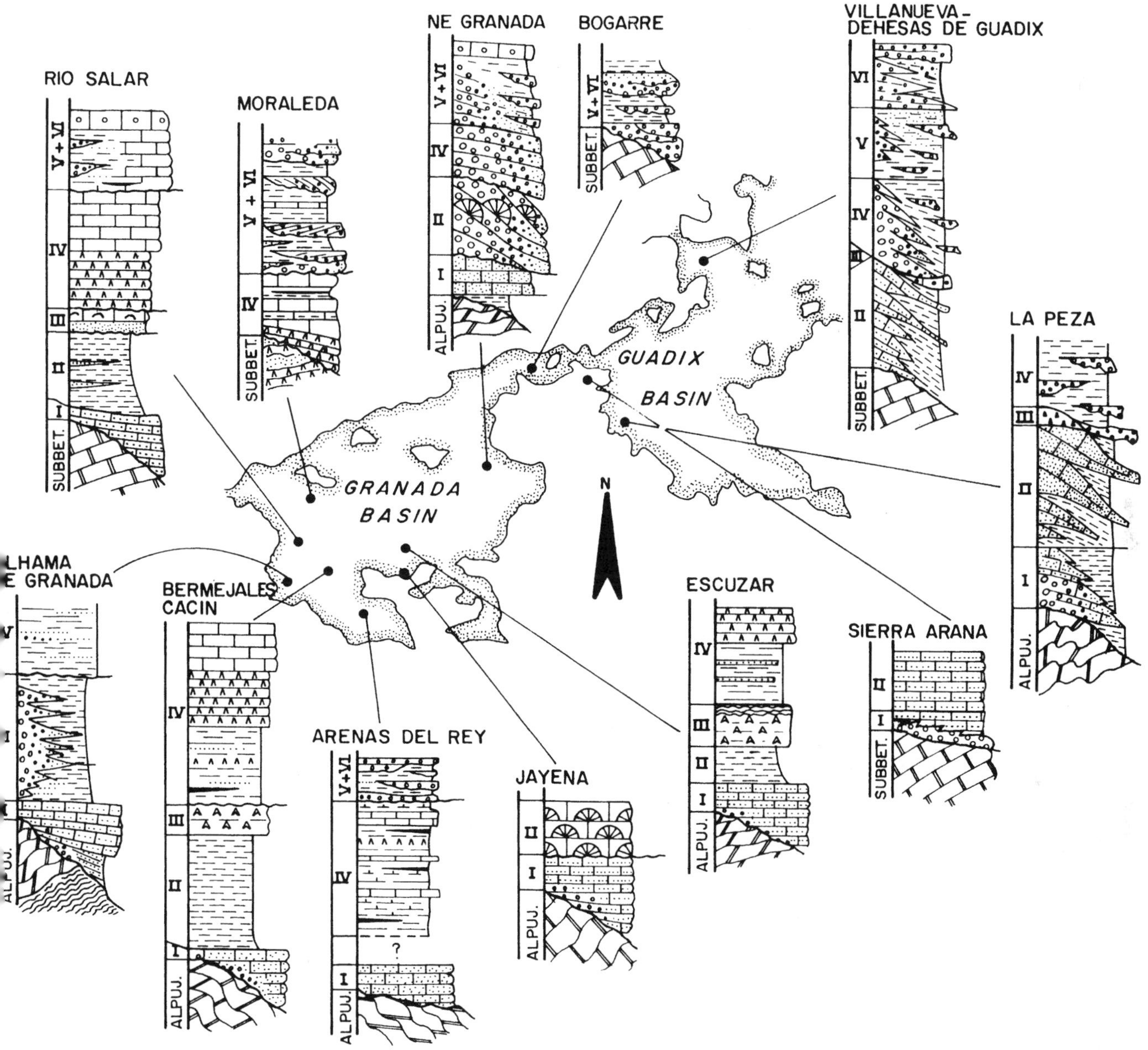

ig. 4. Representative stratigraphic profiles of the Granada and Guadix basins (depositional sequences are shown in each case).

- The longitudinal system originated in the Sierra Nevada. It is represented by alluvial fan deposits changing distally to fluvial deposits of different characteristics according to their situation (Fernández & Soria, 1986–87; Fernández *et al.*, 1989).
- Transverse drainage systems from the Internal and External Zone (referred to as the Internal and External transverse systems). The Internal transverse system, located on relatively low gradients, developed fluvial systems of low sinuosity with important conglomeratic channels and broad lutite flood plains. On the other hand, the External transverse system, located on steep gradients, developed lacustrine fan deltas and alluvial fans with considerably less longitudinal evolution (Fernández & Dabrio, 1983; Fernández *et al.*, 1991; Viseras *et al.*, in press).
- The lacustrine system, which is represented by different subenvironments of fluvial–lacustrine sedimentation (Arribas *et al.*, 1988), is coincident with the longitudinal system and with the palaeogeographic axis of the basin.

Although the detailed characteristics of these systems are different for each sequence and basin, the regional model of integration

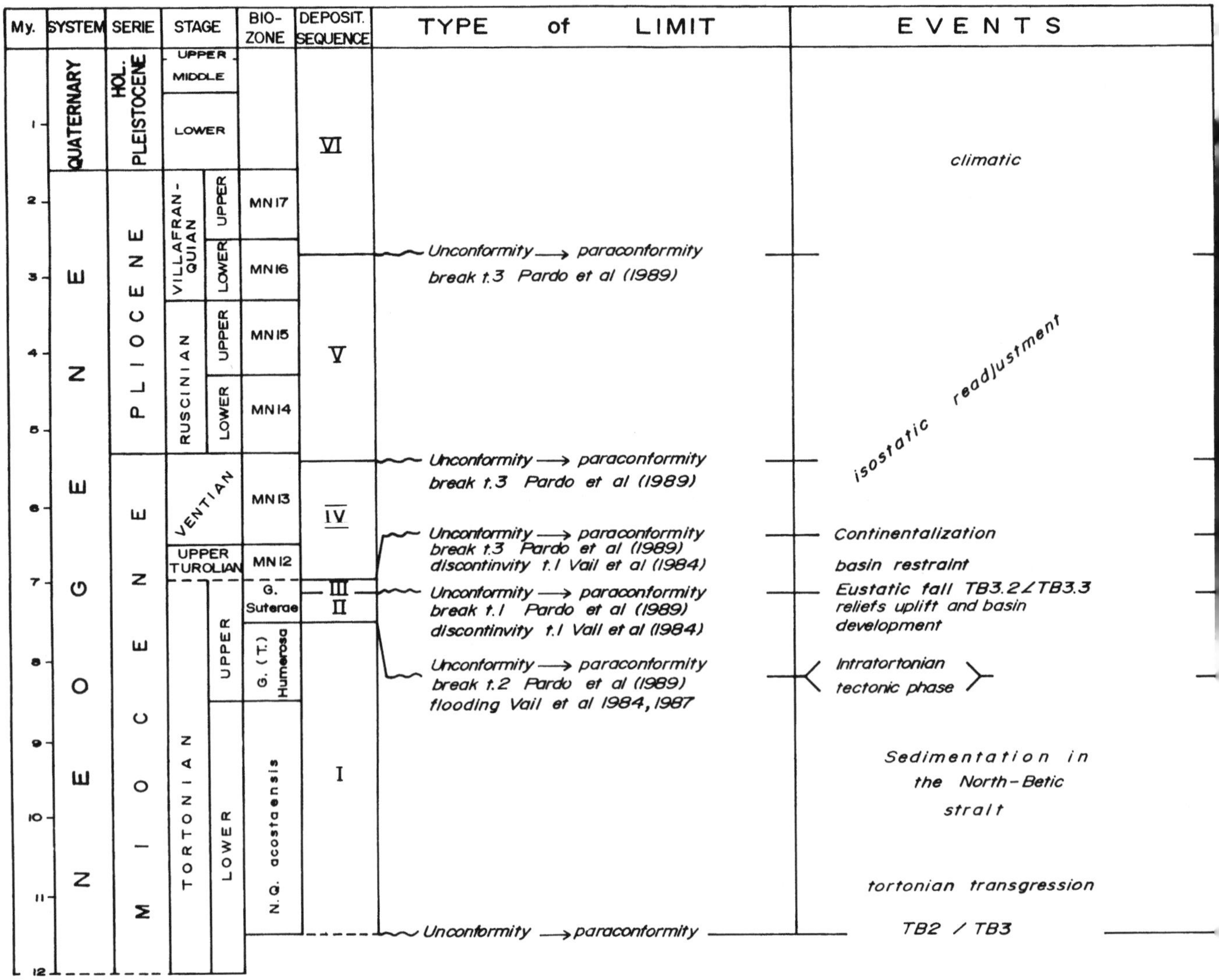

Fig. 5. Depositional sequences differentiated in the Granada and Guadix basins. Type of limits, biostratigraphic data and main tectonic, eustatic and/or climati events are shown.

of all the environments in one scheme is similar (Fig. 12). The datum by which this palaeogeography is established is the lake level, as its fluctuations (changes in base level) control both the sequences formed by the transverse systems and those formed by the longitudinal system and the lakes themselves (Fernández *et al.*, 1993).

Moreover, the differentiation between several transverse drainage systems and the longtudinal system, on which the areas of lacustrine sedimentation are superposed, leads to understanding of the spatial–temporal evolution of the basin axis. As regards supply, this evolution reflects the relative influence of each of the transverse systems, with the consequent implications concerning the supply/subsidence ratios on the basin margins and tectonic or climatic activity in the respective source areas (Viseras, 1991).

In general, throughout the evolution of the continental margins there was northward displacement of the areas of maximum sediment accumulation and of the palaeogeographic axis. This also involved withdrawal from Sierra Nevada, which may have bee related to the formation of large anticlinal structures in the Interna zones of Sierra Nevada and Sierra de Filabres (and its continuatio towards Sierra de Baza).

The more vertical geometry of the facies change surfaces i depositional sequences IV and V may be the result of both th higher subsidence rates during the sedimentation period and th rising sea-level, which was a generalized phenomenon in marin areas close to these basins during the Early Pliocene. Depositiona sequence VI is an accommodation unit in both basins. However, implies an accentuation of their asymmetry, as in places the axis completely displaced towards the External Zones margin. It shoul be remembered that the general eustatic rise had by this time cease and, in addition, the basins were subjected to lower sedimentatio rates. Moreover, extreme asymmetry developed between the tw source areas in this depositional sequence. The cold periods relate

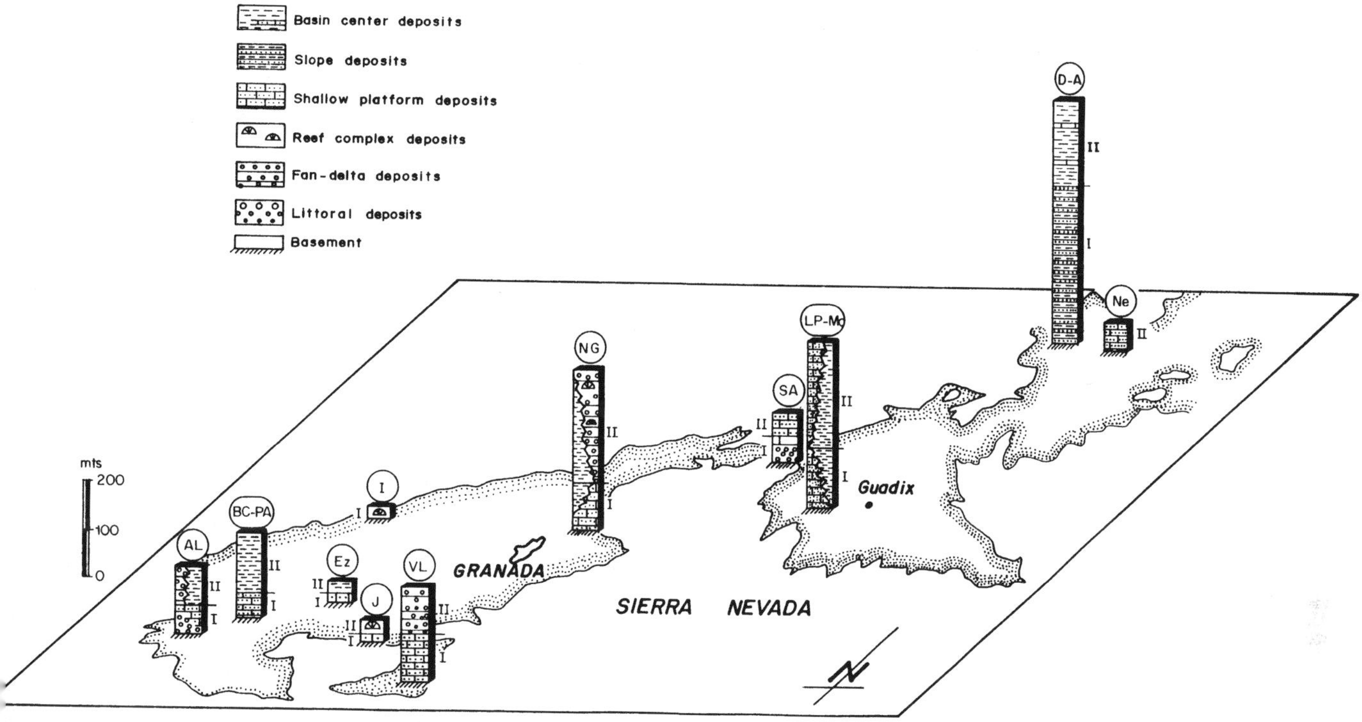

Fig. 6. Columns diagram of the marine infilling (depositional sequences I and II) of the Granada and Guadix basins. AL: Alhama de Granada; BC-PA: Brácana-Río Alhama; Ez: Escúzar; I: Illora; J: Jayena; VL: Valle de Lecrín; NG: NE Granada; SA: Sierra Arana; LP-Mo: La Peza-Molicias; D-A: Dehesas-Alicún; Ne: Negratín.

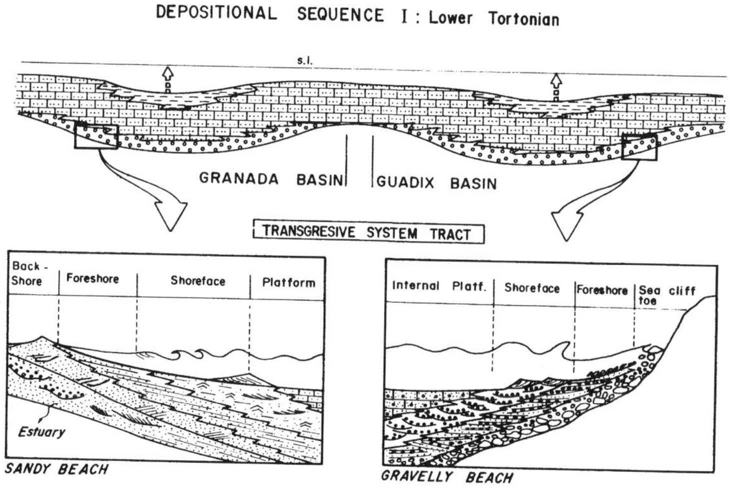

ig. 7. Palaeogeographic and sedimentary outline for depositional sequence I. At that time (Lower Tortonian) the Granada and Guadix basins were the only main epocentres of a big sedimentary basin referred to as the North Betic Strait. The coastal deposits related to the transgression correspond to gravelly or sandy eaches, depending on the characteristics of the basement.

glaciation were responsible for a much higher volume of precipi-tion in the high relief of the Internal Zones than in the lower xternal Zones margin. This produced a considerable dispropor-on in the volume of supply entering the basin on either margin, so at the alluvial systems from the Internal Zones caused displace-ent of the longitudinal axis towards the NNW, given the lower olume of the alluvial fans of the External Transverse System.

Conclusions

Surface and subsurface data show that during the Lower Miocene the region in which the Granada and Guadix basins later developed was part of the Alborán basin, whose geological history it therefore shares.

This stratigraphic and sedimentological analysis permits recogni-

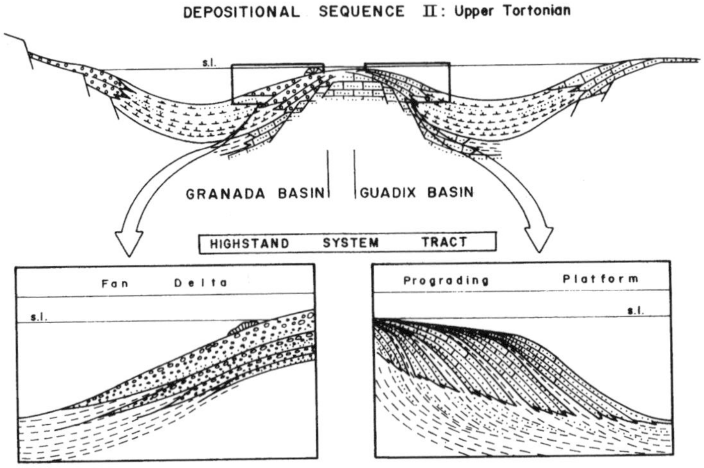

Fig. 8. Palaeogeographic and sedimentary setting for depositional sequence II (Upper Tortonian). Fan deltas and prograding platforms are the most characteristic systems.

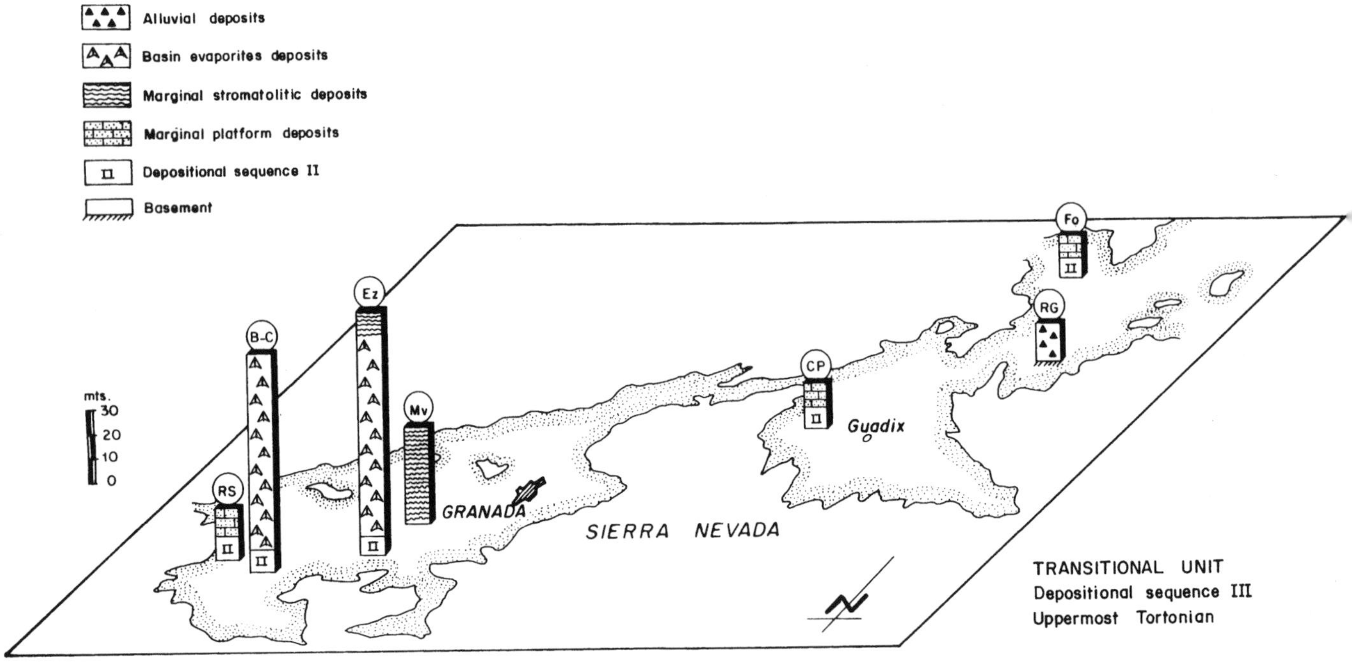

Fig. 9. Column diagram corresponding to the record of the eustatic fall, pointing out the marine–continental transition in the Granada and Guadix basins (depositional sequence III). RS: Río Salar; B-C: Bermejales-Cacín; Ez: Escúzar; Mv: Montevives; CP: Cuesta de Purullena; Fo: Forruchu; RG: Río Gor.

tion of the main unconformities on a regional scale and, therefore division of the sedimentary infilling into six depositional sequences, whose internal organization reflects the, at times, complex interrelation of different depositional environments (Fig. 14). The characteristics of these depositional sequences and their depositional systems reveal the main tectonic, eustatic and climatic events that took place during infilling of these basins. It is therefore a useful means of understanding the Neogene–Quaternary evolution of the Cordillera and of the other Neogene basins located in it (Fig. 15).

The first depositional sequence, developed in a transgressive context, coincided with a phase of relief formation related to a period of extensional tectonics coinciding with one of the last rifting episodes recognized in the Alborán basin. Sedimentation took place in an extensive marine basin, with broad platforms. The transition to the Upper Tortonian was marked by an intra-Tortonian tectonic stage coinciding with the last Miocene rifting episode. This stage resulted in important relief formation and the individualization of the peripheral basins of Alborán. Depositional sequence II was formed in this highstand context and is characterized in particular by the development of fan deltas. The evolution of extensional

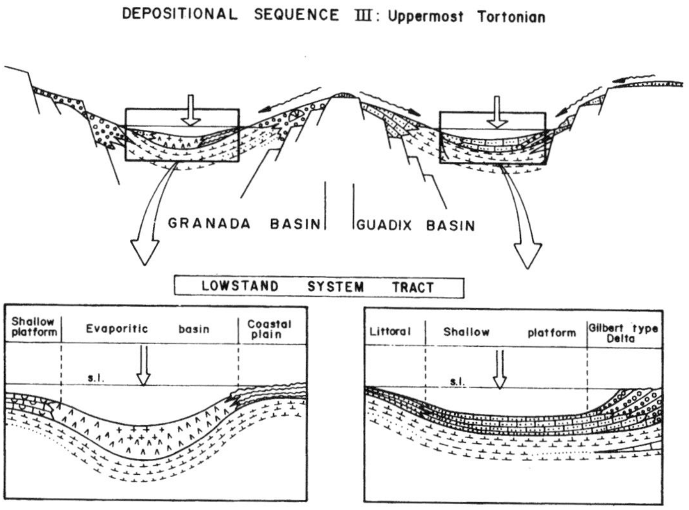

Fig. 10. Palaeogeographic and sedimentary setting for depositional sequence III. This corresponds to the sedimentary record of the eustatic fall at the Uppermost Tortonian (lowstand system tract).

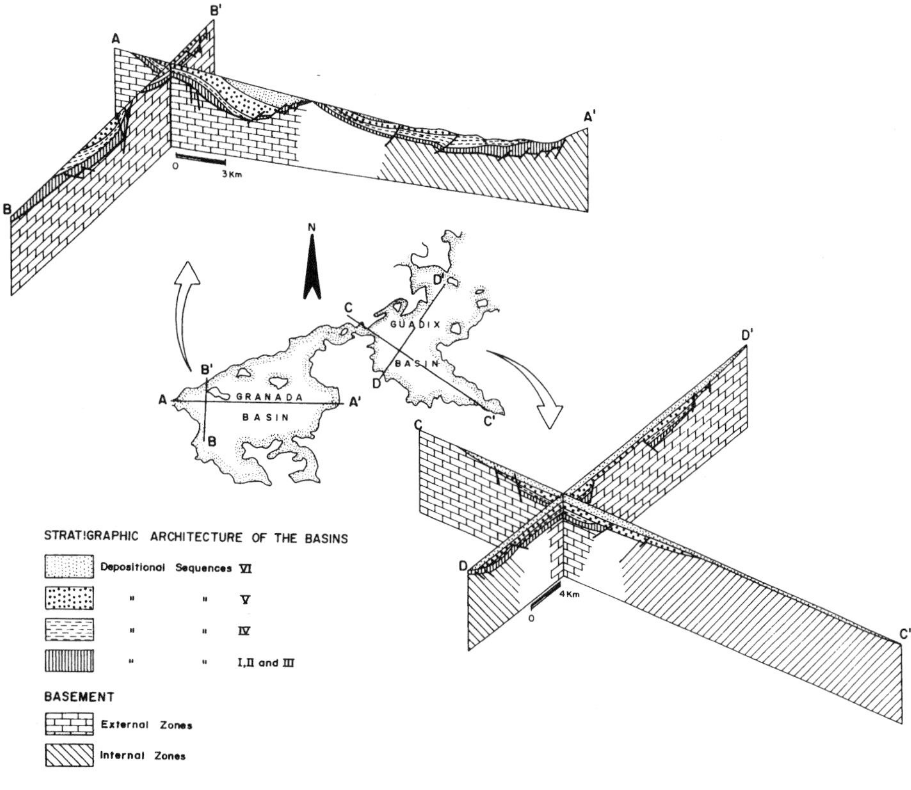

Fig. 11. Location of depocentres for every depositional sequence in two selected sections of the Granada and Guadix basins. These sections have been obtained using seismic profiles, boreholes and surface geology data.

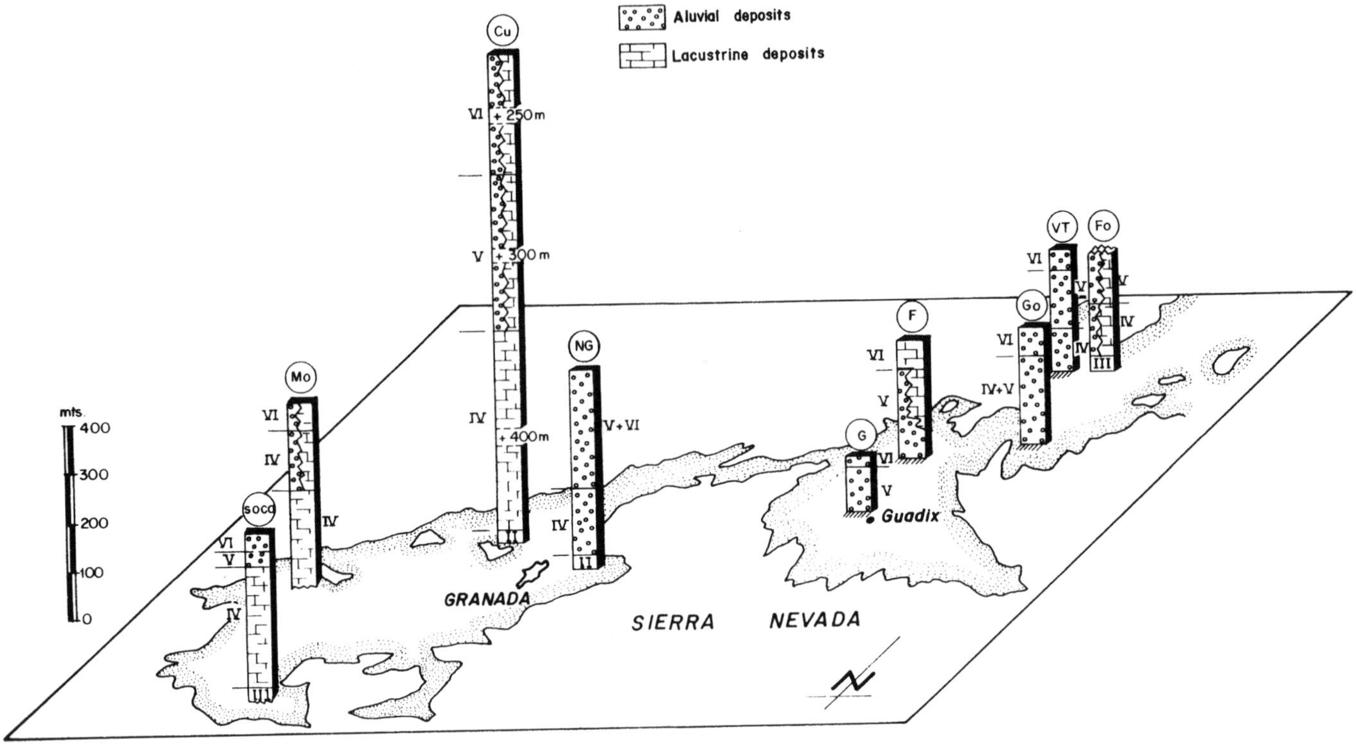

Fig. 12. Column diagram for the continental infilling of the Granada and Guadix basins. Spatial and subsurface distribution of every depositional sequence are shown. SWGB: SW of Granada basin; Mo: Moraleda; Cu: Cubillas; NG: NE Granada; G: Guadix; F: Fonelas; Go: Gorafe; VT: Villanueva de las Torres; Fo: Forruchu.

tectonics was interrupted towards the end of the Upper Tortonian and an important compressional event took place. A considerable eustatic fall occurred in this context, together with the uplift of most of the cordillera and the continentalization of the Granada and Guadix basins. Depositional sequence III represents the sedimentary record of the eustatic fall.

Three depositional sequences (IV, V, VI) can be distinguished in the continental infilling, the internal organization of which reflects the interrelation of various depositional systems (transverse system, longitudinal system and lacustrine system). The most common context during continental infill was north–south compression and activity of the NW–SE, NNE–SSW and N70E fault systems with largely vertical movements and isostatic uplift of the Sierra Nevada. In these circumstances the basins were markedly asymmetric and the depocentres migrated progressively northwards across the successive depositional sequences. The tectonic behaviour of the margins and the fluctuations in lake level (base level of the drainage systems) controlled the resulting stratigraphic architecture. From the Upper Pleistocene onwards, the processes of erosion and sedimentation were related to the activity of the present fluvial network.

Finally, comparison of the curve of the relative sea-level changes in the central sector of the Betic Cordilleras with the curve of global fluctuations (EXXON curve) reveals their lack of connection, as is to be expected in a context of high tectonic activity.

Acknowledgments

The text has been improved thanks to the suggestions of P.F. Friend. This paper is part of the results of the project PB91–0080–C02–01 granted by the Spanish DGICYT and research team JA4085 *Basin Analysis* granted by the *Junta de Andalucía*.

References

Arribas, M.E., Fernández, J. and García-Aguilar, J.M. (1988). Análisis sedimentológico de los materiales lacustres (Formación de Gorafe–Huélago) del sector central de la depresión de Guadix. *Estud. Geol.*, **44**, 61–73.

Braga, J.C., Martín, J.M. and Alcalá, B. (1990). Coral reef in coarse-terrigenous sedimentary environments (Upper Tortonian, Granada Basin, southern Spain). *Sediment. Geol.*, **66**, 135–150.

Comas, M.C., García-Dueñas, V. and Jurado, M.J. (1992). Neogene tectonic evolution of the Alborán Sea from MCS data. *Geo–Mar. Lett.*, **12**, 157–164.

Dabrio, C.J., Fernández, J., Peña, J.A., Ruíz-Bustos, A. and Sanz de Galdeano, C. (1978). Rasgos sedimentarios de los conglomerados miocenos del borde noreste de la depresión de Granada *Estud. Geol.*, **34**, 89–97.

Dabrio, C.J., Martín, J.M. and Megias, A. (1982). Signification sedimentaire des evaporites de la depression de Granada (Espagne). *Bull Soc. Geol. Fr.*, **24**, 705–710.

Fernández, J. and Dabrio, C.J. (1983). Los conglomerados de Moraleda Un modelo de sistema fluvial de tipo braided (depresión d Granada, España), *Estud. Geol.*, **39**, 53–69.

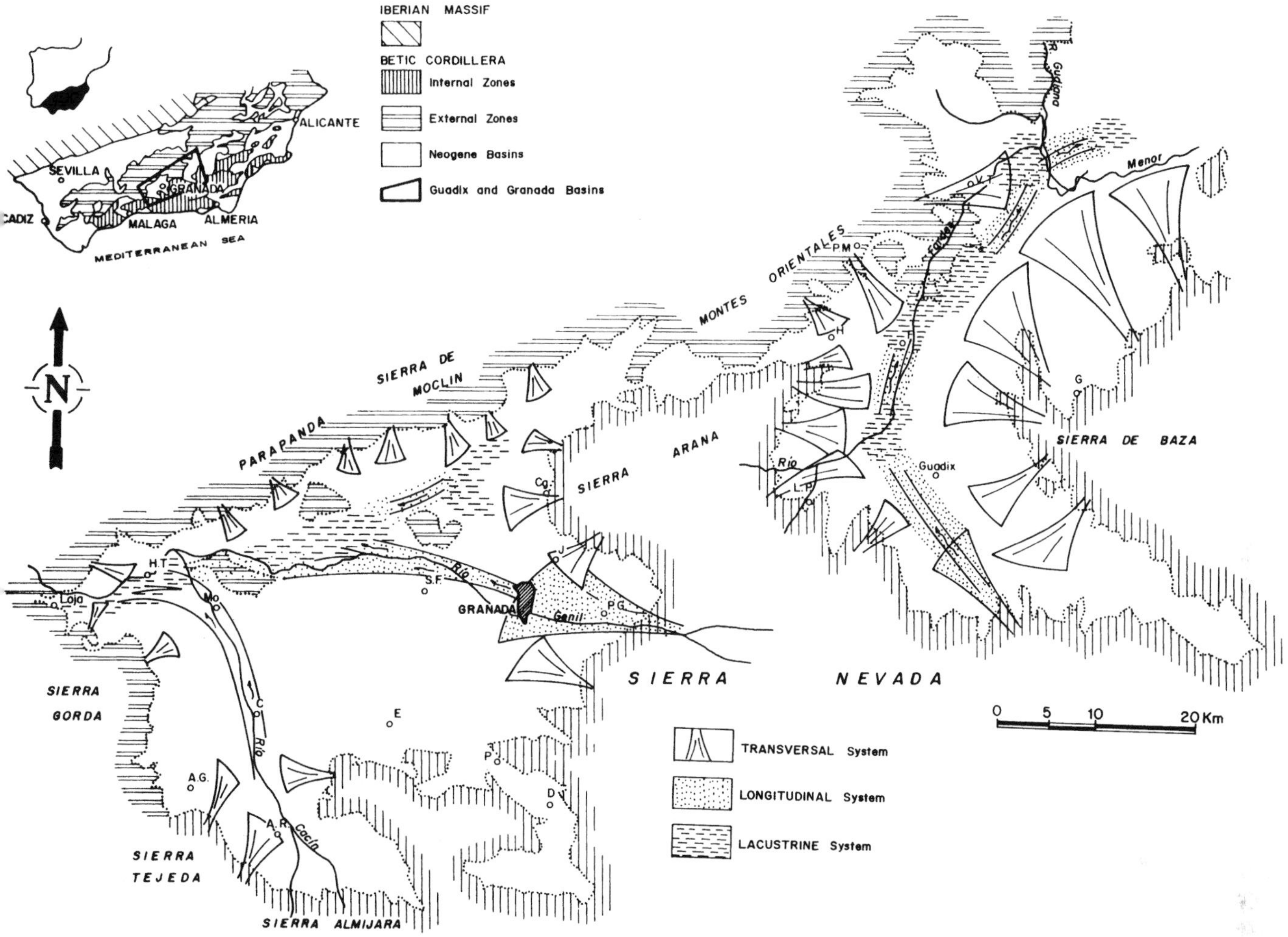

Fig. 13. Map showing the distribution of depositional systems for the continental infilling of the Granada and Guadix basins.

Fernández, J. and Guerra, A. (1993). Coarsening-upward megasequence generated by Gilbert-type fan–delta in a tectonically controlled context. Betic Cordillera (in press).

Fernández, J. and Rodríguez-Fernández, J. (1989). Facies evolution of nearshore marine clastic deposits during the Tortonian transgression–Granada Basin, Betic Cordilleras, Spain. *Sediment. Geol.*, **71**, 5–21.

Fernández, J. and Soria, J. (1986–87). Sedimentación en el borde norte de la depresión de Granada durante el Plio-Cuaternario. *Acta Geol. Hisp.*, **21–2**, 73–81.

Fernández, J., Viseras, C. and Bluck, B.J. (1989). Changes in evolution of Guadix basin as documented by alluvial architecture (Betic Ranges, Spain). *Publ. Ser. Geol. Catalunya*, **6**, 80 pp.

Fernández, J., Bluck, B.J. and Viseras, C. (1991). A lacustrine fan–delta system in the Pliocene deposits of the Guadix Basin (Betic Cordilleras, South Spain). *Cuad. Geol. Ibérica*, **15**, 299–317.

Fernández, J., Bluck, B.J. and Viseras, C. (1993). The effects of fluctuating base level on the structure of fan and associated fan–delta deposits: an example of the Tertiary of the Betic Cordillera (Spain). *Sedimentology*, **40**, 879–893.

Haq, B.U., Hardenbol, J. and Vail, P.R. (1987). Chronology of fluctuating sea levels since the Triassic. *Science*, **235**, 1156–1167.

Haq, B.U., Hardenbol, J. and Vail, P.R. (1988). Mesozoic and Cenozoic chronostratigraphy and Eustatic Cycles. In *Sea-level changes: an integrated approach* (ed. C.K. Wilgus, B.S. Hastings, C.G.S.C. Kendal, H. Posamentier, C.A. Ross & J.C. Van Vagoner). *Soc. Econ. Paleontol. Mineral., Spec. Publ.*, **42**, 71–108.

Martín Penela, A. (1987). Los grandes mamíferos del yacimiento achelense de la Solana del Zamborino (Fonelas, Granada). *Antropol. Paleoecol. Humana*, **5**, 29–188.

Pardo, G., Villena, J. and González, A. (1989). Contribución a los conceptos y a la aplicación del análisis tectosedimentario. Rupturas y unidades tectosedimentarias como fundamento de correlaciones estratigráficas. *Rev. Soc. Geol. Esp.*, **2**, 199–219.

Rodríguez-Fernández, J. (1982). *El Mioceno del sector central de las Cordilleras Béticas*. Thesis, Universidad Granada, 224 pp.

Rodríguez-Fernández, J., Sanz de Galdeano, C. and Fernández, J. (1989). Genesis and evolution of the Granada Basin (Betic Cordillera, Spain). In *Intermontane basins: geology and resources* (ed. T.P. Thauasuthipitak & P. Ounchanum), 294–305. Chiang-Mai, Thailand.

Ruíz-Bustos, A., Fernández, J., Morales, J., Rodríguez-Fernández, J. and Vera, J.A. (1990). Los materiales Plio-Pleistocenos del borde norte de la depresión de Granada. *Estud. Geol.*, **46**, 270–290.

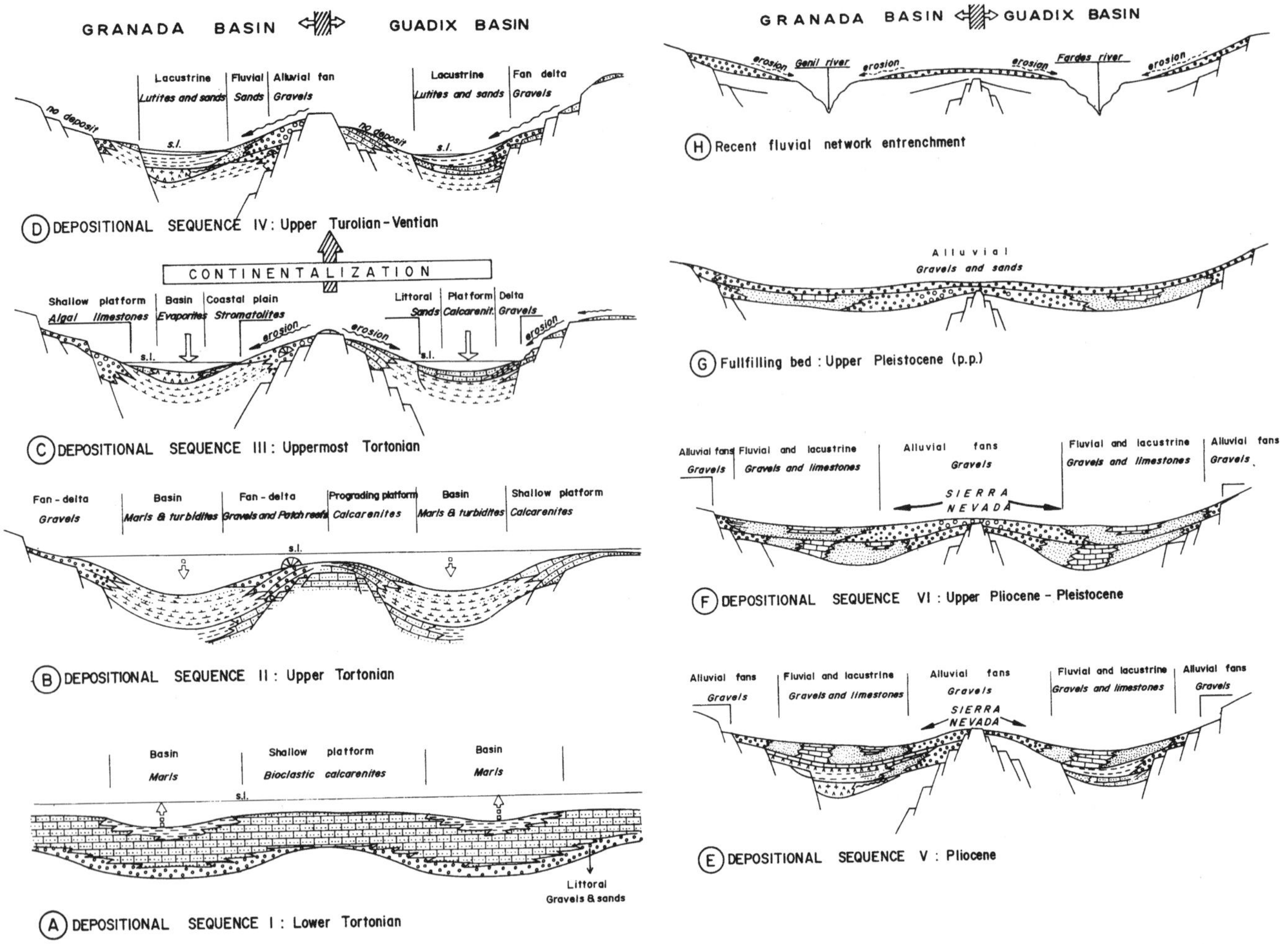

Fig. 14. Sketch showing the evolution of the Granada and Guadix basins from the Lower Tortonian (when they were the only depocenters in the North-Betic strait basin) until the Holocene entrenchment of the fluvial network. Special attention has been paid to the sedimentary environments at each stage.

Sanz de Galdeano, C. (1983). Los accidentes y fracturas principales de las Cordilleras Béticas. *Estud. Geol.*, **39**, 157–165.

Sanz de Galdeano, C. and Vera, J.A. (1992). Stratigraphic record and palaeogeographical context of the Neogene basins in the Betic Cordillera, Spain. *Basin Res.*, **4**, 21–36.

Serrano, F. (1979). Los foraminíferos planctonicos del Mioceno superior de la cuenca de Ronda y su comparación con los de otras áreas de las Cordilleras Béticas. Thesis, Universidad de Málaga, 272 pp.

Soria, J.M. and Ruíz-Bustos, A. (1992). Nuevos datos sobre la edad del inicio de la sedimentación continental en la cuenca de Guadix. Cordillera Bética, *Geogaceta*, **11**, 92–94.

Vail, P.R. (1987). Seismic stratigraphy interpretation procedure. In *Atlas of seismic stratigraphy* (ed. A.W. Bally). *Am. Assoc. Petrol. Geol., Stud. Geol.*, **27**, 1–10.

Vail, P.R., Hardenbol, J. and Todd, R.G. (1984). Jurassic unconformities, chronostratigraphy, and sea level changes from seismic stratigraphy and biostratigraphy. In *Interregional unconformities and hydrocarbon accumulation* (ed. I.S. Schlee), *Am. Assoc. Petrol. Geol., Mem.*, **36**, 129–137.

Vera, J.A. and Rodríguez-Fernández, J. (1988). Una modificación al modelo genético para la formación Molicias (Tortoniense superior, depresión de Guadix, S. de España). *Geogaceta*, **5**, 26–29.

Viseras, C. (1991). Estratigrafía y sedimentología del relleno aluvial de la Cuenca de Guadix (Cordilleras Béticas). Thesis, Universidad de Granada, 327 pp.

Viseras, C. and Fernández, J. (1992). Sedimentary basin destruction inferred from the evolution of drainage systems in the Betic Cordillera, southern Spain. *J. Geol. Soc. London*, **149**, 1021–1029.

Viseras, C., Fernández, J. and Bluck, B.J. (in press). Autocyclic processes of a Pliocene alluvial fan in the Guadix Basin, Spain. *Trans. R. Soc. Edinburgh.*

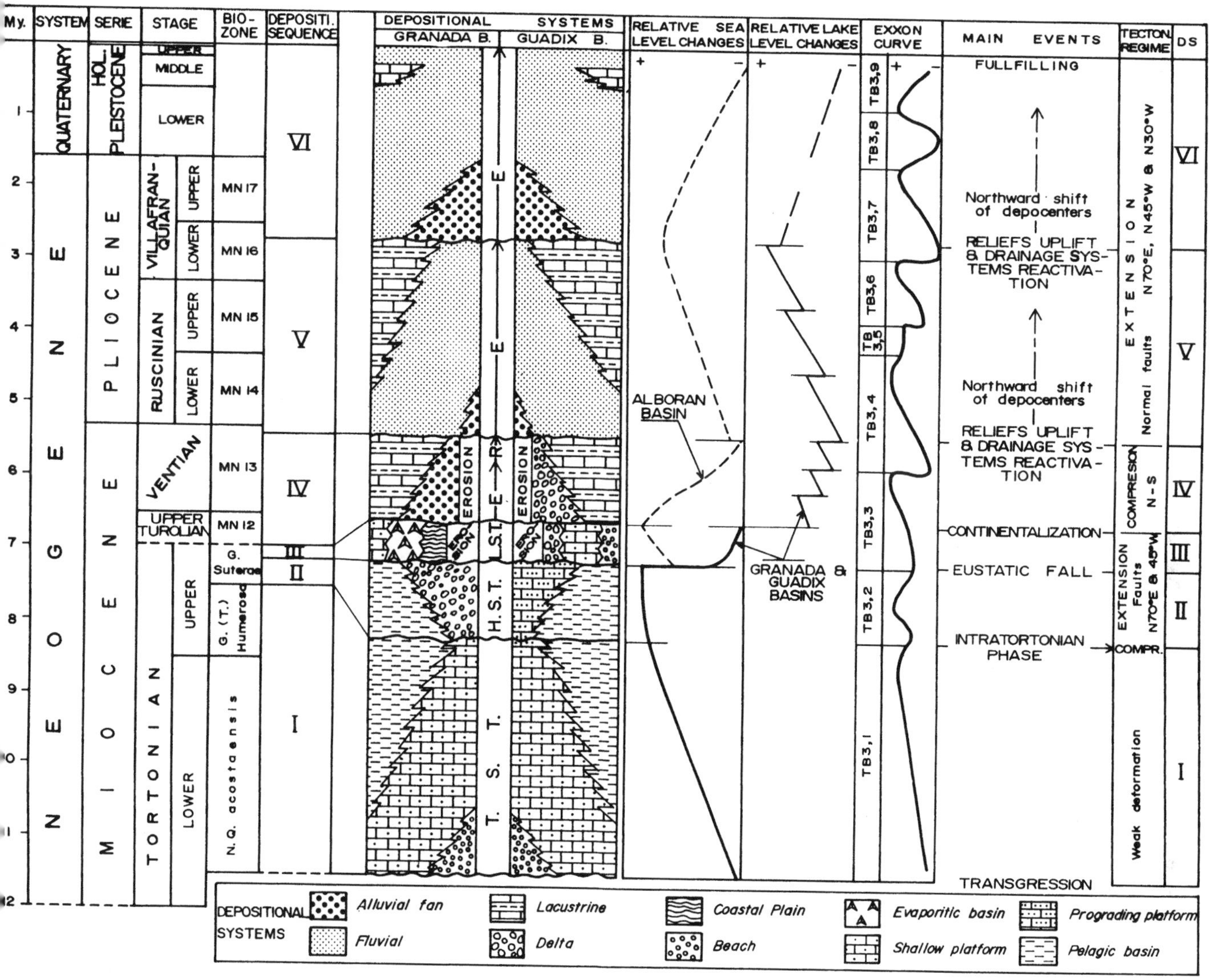

Fig. 15. Summary sketch of the sedimentary infilling of the Granada and Guadix basins. T.S.T: transgressive system tract; H.S.T: highstand system tract; L.S.T: lowstand system tract; E: expansion; R: retraction.

S7 Pliocene–Pleistocene continental infilling of the Granada and Guadix basins (Betic Cordillera, Spain): the influence of allocyclic and autocyclic processes on the resultant stratigraphic organization

J. FERNÁNDEZ, C. VISERAS AND J. SORIA

Abstract

The stratigraphic architecture of the continental infilling is the result of the stacking of three depositional sequences, controlled by tectonics and climate that affected the relief of the margins, and by subsidence and by variations of the general base level. Each depositional sequence arises out of a complex interrelation between several depositional systems (the longitudinal system, coinciding with the axis of the basin and the areas of lacustrine sedimentation, and the transverse systems), and its sequential organization depends on the dynamic behaviour of the different depositional systems and on the fluctuations in base level (lake level).

Introduction

The Granada and Guadix basins are intramontane basins that became distinct from each other towards the Late Miocene, after westward displacement of the Alborán realm had ceased (Sanz de Galdeano & Vera, 1992). They are located in the central sector of the Betic Cordillera, lying on the contact between the Internal and External Zones (see Fig. 1 in Chapter S6). The northwestern half of the basement is therefore made up of Mesozoic and Tertiary materials from the External Zones, while the southeastern half consists of Palaeozoic and Triassic materials from the metamorphic complexes of the Internal Zones.

Six depositional sequences (see Fig. 4 in Chapter S6) have been distinguished in the sedimentary infilling (see Chapter S6). The two oldest sequences (Tortonian) correspond to infilling during the phase of marine sedimentation, the third (uppermost Tortonian) corresponds to the marine–continental transition, and the three remaining sequences (post-Tortonian) correspond exclusively to the continental infilling.

These basins assumed their continental character towards the Tortonian–Messinian boundary, and this coincided with a period of important eustatic and tectonic events. An important fall in sea level at this time caused the major withdrawal of the sea from many Betic intramontane basins and the complete disconnection of the Mediterranean from the Atlantic. As regards tectonics, the isostatic uplift of the Sierra Nevada and the activity of broadly NW–SE and NNE–SSW faults brought about an important physiographic change consisting of expansion of the formerly marine basins and northward migration of depocentres.

This chapter presents a stratigraphic and sedimentological study of the continental infilling of the Granada and Guadix basins, and shows the complex interrelation between autocyclic and allocyclic processes that defined the resultant stratigraphic architecture.

Stratigraphic architecture

The stratigraphic architecture of the continental infilling is the result of the stacking of depositional sequences IV, V and VI (see Figs. 4 and 11 in Chapter S6). In all cases, the lower boundary is an unconformity at the edges, becoming a paraconformity towards the centre.

The continental basins are markedly asymmetrical with a tectonically active northern margin and a tectonically passive southern margin. Their axis is located near the northern margin, coinciding with the lacustrine areas and the longitudinal drainage (Viseras & Fernández, 1992). Transverse systems with a base level in the lacustrine areas developed from the drainage of the border relief. These transverse systems from the north, where the basin margin had high gradients, formed small alluvial fans and fan deltas, whereas the southern systems, on shallower gradients, developed bigger alluvial fans and fluvial systems. The internal organization of each sequence is, therefore, the result of the interrelation of the aforementioned depositional systems (longitudinal drainage system, transverse drainage systems and the lacustrine sedimentation system; see Fig. 12 in Chapter S6).

Although the detailed characteristics of these systems are different in each sequence and basin, the overall model of integration of all the environments in one scheme is similar. We therefore limit our description to the characteristics of these depositional systems in the Guadix basin and only for one of the depositional sequences (V), which is more highly developed and presents better outcrop conditions.

Internal transverse system

The Moraleda Conglomerate in the Granada basin (Fernández & Dabrio, 1983; Dabrio & Fernández, 1986) or the Arroyo de Gor section of the Guadix Fm. in the Guadix basin (Vera, 1970; Fernández *et al.*, 1986–87) are good examples to illustrate the characteristics of this system.

The Arroyo de Gor section in the Guadix Fm. presents a cross-section of a Lower Pliocene fan made of three stratigraphically superposed sets delimited by palaeosol horizons. Analyses of palaeocurrents and clast composition shows that there were three growth stages of the same fan, whose radius increased progressively from approximately 6 km in stage 1 (the lowest) to approximately 9 km in stage 3 (the uppermost), since its distal boundary can be established with relative precision where it connects with the aforementioned longitudinal axial system, whose compositional and sedimentological characteristics are clearly different (Viseras, 1991; Viseras & Fernández, 1989, 1994).

Each growth stage of the fan deposits is made of between three and five 15–20 m thick sequences, in which coarse (channel complexes) and fine (overbank complexes) components can be distinguished.

- *Channel complexes.* These are sheet bodies (Friend, 1983). Their base is erosional and stepped, following a constant pattern of westward displacement. Their internal structure consists of gravels with cross-bedding in which backsets and foresets can be identified, corresponding respectively to the heads and tails of bank-attached bars (Bluck, 1976, 1982; Haughton, 1989) in a channel migrating laterally westward.
- *Overbank complexes.* These are fine facies of lutites and sands with horizontal or cross lamination. Coarse lenses corresponding to crevasse splays, occasionally intercalate.

These channel and overbank facies complexes are organized in FTU (fining- and thinning-upward) sequences throughout the stratigraphic succession.

We propose a braided fluvial system as our interpretation of this section, characterized by a main channel (trunk channel) surrounded on both sides by secondary channels (entourage channels) and broad flood-plain areas. This system would have moved across the fan in a pendular way, as described for present and ancient examples (Thorrarinson, 1956; Gole & Chitale, 1966; Bluck, 1976, 1979, 1980; Wells & Dorr, 1987), as a result of the preferential development of bank-attached bars on one of the banks of the main channel (Amundson & Hendry, 1989; Todd & Went, 1989; Viseras *et al.*, in press).

According to this description, therefore, the passage of this channel system over any one point in the course of its lateral displacement would first have caused a coarsening- and thickening-upward sequence (CTU) because of the approach of the trunk channel preceded by the entourage channels on one of its sides, followed by an FTU sequence, due to withdrawal of the trunk channel, followed by the passage of the entourage channels on the other side, and ending with fine flood-plain deposits corresponding to overbank phases in the now distant channels (Viseras & Fernández, 1994). The degree of preservation of these sequences varies according to their position in the fan and the period of recurrence.

In a situation close to the margin of the cone described by the channel system, the FTU sequence caused by migration towards the nearby margin is eliminated by passage of the channels in the opposite direction, as only a short space of time intervened, which was insufficient for this sequence to become fossilized by a thick bed of overbank deposits. However, this new FTU sequence created by migration towards the centre of the cone was buried by a thick bed of fine deposits from the numerous floods occurring while the channel system moved towards the opposite margin and returned to this position. On returning, the channel system thus eliminated the flood deposits and possibly also the uppermost part of the FTU sequence on which they lie. In a marginal position, therefore, it is mainly the FTU sequences representing passage of the channel system towards the centre of the cone that is preserved.

Fig. 1 presents a graphic example of two hypothetical columns in central and marginal positions without taking into account the erosional processes, together with the result after application of the idea proposed above. The sequences are shown to the left and right of a central line depending on whether they represent movement of the channel system towards the left (sequences A and B, Fig. 1) or right (sequences C and D).

External transverse system

This system is represented by the Villanueva de las Torres complex (Fernández *et al.*, 1991) and is the result of a retrogressive trend of stacking of three alluvial fan – lacustrine fan delta units. Five transitional zones can be distinguished from proximal to distal positions, according to the internal structure of the alluvial fan – fan delta system.

Proximal alluvial fan. This is represented by tabular units built by debris flow deposits. They show no channelling and there are no fine overbank deposits.

Mid fan. In this part of the fan the debris flow deposits are replaced by tabular beds of clast-supported conglomerates, with horizontal or cross-bedding. They form lobes or interlobes deposited by turbulent flows. There are thin beds of silts and clays associated with these deposits. As in the proximal facies, there are few channels and no systematic vertical change in grain-size is observed.

Fan fringe. This area is very different from the two foregoing.The gravels are organized in tabular beds and deep channels. Thick beds of silts and clays, some of lacustrine origin, alternate with the gravels. From a sequential point of view, several CTU cycles are recognized in each phase of stacking of the alluvial fan – fan delta complex.

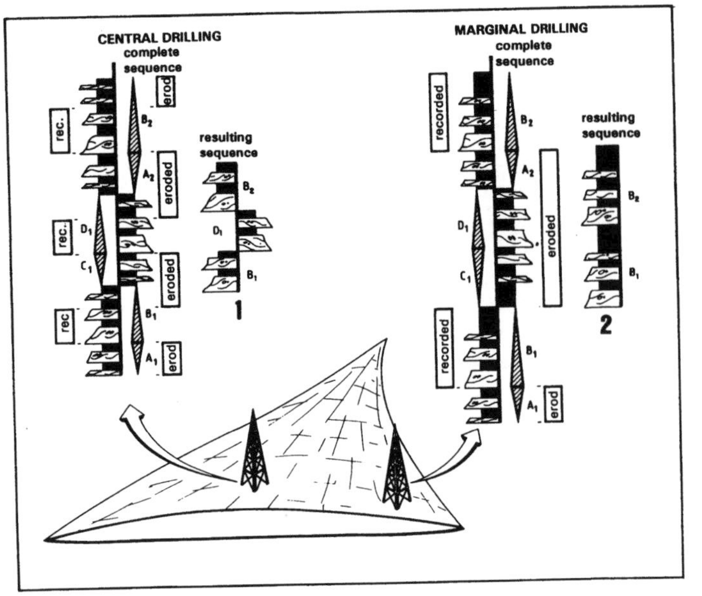

Fig. 1. Complete theoretical successions and real records considering erosion linked to channel network migration. 1 – A central position of the fan can be recognized by a high number of not very thick fining and thinning upward (FTU) cycles, scarcity of overbank sediments and backsets dipping alternately in opposite directions. 2 – A marginal position is characterized by fewer but thicker FTU cycles, a higher proportion of overbank fines and backsets dipping invariably towards the centre of the cone (modified from Viseras & Fernández, 1993). A1 to D1 represent units defined by stratigraphic trends.

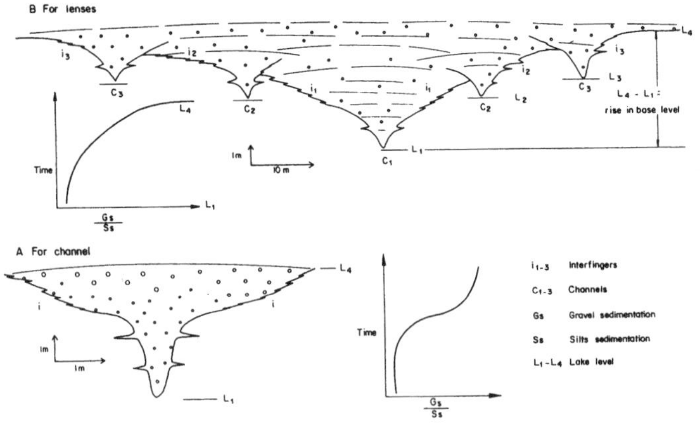

Fig. 2. Genetic model of isolated channels (A) and lenticular bodies (lenses) (B). The presence of wings and interfingered contacts indicates that the growth of these bodies was concomitant with that of the adjacent lacustrine areas and took place with a rising base level. Moreover, the form of the lithosome is a function of the G_s/S_s ratio throughout time. In the case of (B) lenses, the channel C_1, situated near the centre, was the first, and the others are secondary radial channels related to phases of overflow in the main channel. The internal structure (flat or slightly convex stratification) shows frequent internal unconformities caused by amalgamation of the levees of adjacent channels during their progradation.

Proximal fan delta. The stratigraphic succession is here characterized by the presence of lenticular beds of conglomerates with abundant channels at the base and channel-like bodies alternating with thick lacustrine beds. As in the fan fringe zone, these too are cyclical.

Distal (lacustrine) fan delta. The sections throughout this zone are dominated by lacustrine muds with thin beds of fine-grained conglomerates and sands. They show no cyclical character.

We analyse the fan fringe and fan delta zones where the succession shows cyclicity characterized by the alternation of silts and conglomerates. The silts correspond to stratified lacustrine deposits, which are either massive or have wave-ripple lamination and bioturbation. In distal areas they include evaporites, micritic limestones and organic matter. The conglomerates are present as isolated or amalgamated channel bodies, and lenticular or tabular bodies.

1. Isolated V-shaped channels with well-developed wings (Fig. 2).
2. Amalgamated channels. Two zones approximately 200 m wide, each with abundant channels, can be recognized between the fan fringe zone and the fan delta. The channels can be up to 6 m deep and are entrenched on one another, forming complex bodies of interconnected channels. Laterally and distally they change to isolated channels, forming the main distributary system on the fan delta.
3. Gravel lenticular bodies (Fig. 2). These have a characteristically lenticular shape with channels at the base and an approximately flat top. Their internal structure is flat or slightly convex stratification. The channels at the base maintain their identity, producing interdigitated contacts with the lacustrine sediments and internal unconformities as regards the growth of adjacent channels. They present a radial distribution converging some distance upstream, and are therefore interpreted as the result of flows radiating from a main channel zone near the centre of the lens. Alluviation in the main channel produced interdigitated contacts with the lacustrine sediments and encouraged the development of secondary radial channels upstream. The growth of levees in the main and secondary channels was caused by amalgamation in a lenticular body. The shape of these bodies and of the isolated channels in cross-section indicates that they developed in response to the relative growth of either the gravels or the lacustrine lutites with a rise in base level (Fig. 2).
4. Tabular bodies. These have flute and groove casts and abundant lineations on their bases. The fact that the base is flat indicates that they did not compete with the growth of lacustrine lutites during their deposition. On the contrary, their rapid emplacement prevented the development of interdigitated contacts and/or lenticular bodies.

There were fluctuations in lake level and structure of the fan delta. Sedimentation in this area formed CTU sequences tens of metres thick, in which a gradual change can be observed from isolate

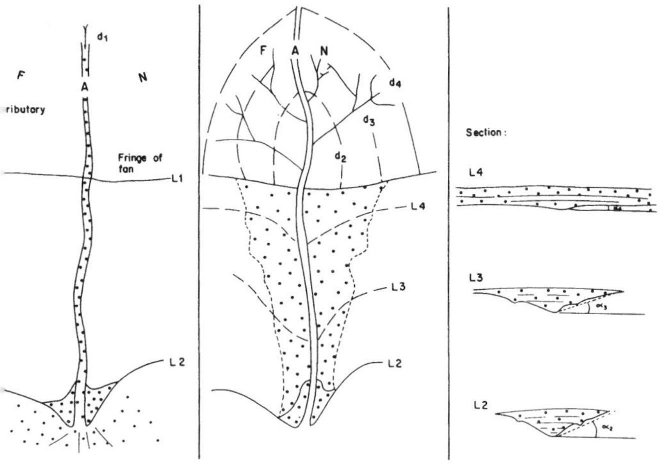

ig. 3. Explanatory model of the geometry of the gravel bodies from distal to roximal areas in the fan delta. This evolution occurs with a rising lake level nd a progressively more extensive drainage basin. Beginning with a situation f fall in lake level, L2, L3 and L4 represent successive positions of lake level, nd d2, d3 and d4 the limits of the respective drainage basins. The inclination of he gravel–silt interface (α2, α3, and α4) is a function of the G_s/S_s ratio of ravel and silt sedimentation.

hannels to lenticular bodies in distal areas of the fan delta and from malgamated channels to tabular bodies in proximal areas.

The thickness and genesis of these sequences was controlled by luctuations in lake level (Fernández *et al.*, 1993). An initial fall in his level caused lengthening and progressive entrenchment of the hannels from distal to proximal areas, together with a continuous ncrease in gravel supply from erosion in the drainage basin located n the subareal fan (Fig. 3). This increase in supply to the lacustrine rea brought about a progressive rise in lake level, which became lower as the lake occupied a wider area.

The rise in lake level encouraged the formation of lobes at the hannel endings and alluviation and overbanking of gravel sheets pstream, thus giving rise to the development of radial channels nd, finally, lenticular bodies. Meanwhile, in proximal areas (fan ringe), where there was a greater amount of sediment available, if he rise in lake level was equal to the sediment supply, tabular odies formed with flat stratification produced by vertical aggradaion. If the rise was higher than the supply, Gilbert-type deltas were ormed.

Another fall in lake level is marked by thinning of the sheet-like eds towards the lake. During this same episode these beds accumuated with their stratification dipping slightly in the same direction. f the fall continued, the gravel became concentrated in the chanels, which then extended to the new coastline, at which point the ycle began again.

Lacustrine system

The petrological and sedimentological study of the Gorfe–Huélago Fm. (Arribas *et al.*, 1988) allows us to distinguish ifferent types of facies characterizing different subenvironments of fluvial–lacustrine sedimentation: 1. mud flat (silts, carbonatic silts and sandstones); 2. swamp zone (carbonate marls); 3. carbonate marshy rim (root-bioturbated limestones, nodular limestones and palaeosols, calcareous crusts and nodular marls); 4. lacustrine zone (algal limestones, fossiliferous limestones, marly limestones and marls).

Longitudinal axial system

This system originated on the southern margin of the basin and the proximal facies correspond to the deposits of an alluvial fan dominated by debris flow processes. Due to the decrease in gradient occurring on connection with the fluvial valley, it rapidly acquired the characteristics of a coarse-grained meandering fluvial system, presenting lower and upper point bar structures in proximal areas. In more distal positions, where it coincides with lacustrine areas, simple point bar structures are found, consisting of lacustrine sediments that were eroded when the longitudinal system was active in conditions of lake-level fall.

Multistorey channels are also found in these areas of lacustrine sedimentation. Here, periods of entrenchment (developed during moments of lake-level fall) and infilling, with channel-fill structures (developed during moments of lake-level rise) can be distinguished.

Proposal of regional synthesis

The proposed model (Fig. 4) is based on the following premises: 1. A situation of stable lake level, with deposition of marls in the centre and development of mud flats in marginal areas. 2. Tectonic movements stimulated drainage of the edges and therefore also the progradation of the alluvial fans of the Internal and External transverse systems which, on coalescing, encouraged development of the lacustrine areas in the axial zone of the basin. 3. The progressive supply of sediments would have led to shallowing and expansion of these lacustrine areas, in a global context of rising lake level.

The CTU sequences observed in the fan deltas on the northern margin and the expanding sequences in the lacustrine areas would have developed in this context.

FTU sequences predominate in the alluvial fans of the southern margin, although at the base of these sequences and in distal areas V-shaped channels can be observed which are identical to those found at the base of the CTU sequences of the fan deltas on the northern margin. This means that both sequences began in a lowstand situation of lake level, which implies lengthening and entrenchment of the channels, after which their infilling would have taken place by vertical accretion with a rising lake level. On the northern margin, with a steep gradient, the nature of the resultant sequence at each point would have been determined by the lake-level trend. On the other hand, on the southern margin, with a lower gradient, the character of the sequences would have been determined by processes of erosion and sedimentation associated with the migration of the fluvial channels in the context of the model of pendular movement proposed above. For this reason the V-shaped

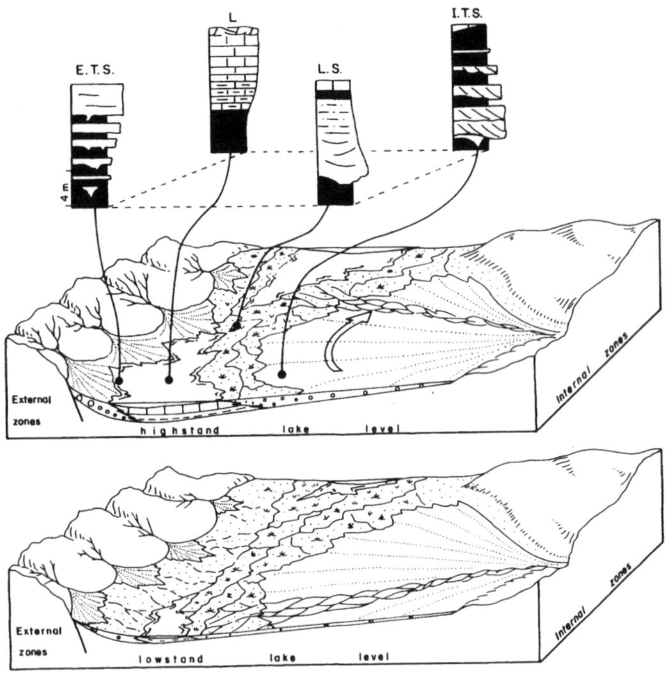

Fig. 4. Conceptual comprehensive model of component sequences in the several sedimentary systems considered in the Guadix basin. Every sequence starts under lowstand lake level conditions and finishes under highstand lake level conditions. Letters identify particular logs.

channels of the base of the sequences are immediately cut off by erosional scars representing the base of a laterally migrating channel, with a rising base level, as shown by its stepped base. Moreover, the top of the sequences, which shows palaeosol development in proximal parts, is a level of lacustrine expansion in distal areas.

The deposits associated with the longitudinal system are not cyclical. Here, the fluctuations of the lake level gave rise to differing degrees of erosion and infilling. Thus, the fall in lake level, caused by its overcoming the dam effect produced by the progradation of the fans associated with the transverse systems, caused erosion of the lacustrine sediments by activity of the longitudinal system; the lake-level rise produced different types of fluvial deposits and, in general, a decrease of energy towards the top, concluding with shallow lacustrine deposits of highstand lake level.

In addition, depositional sequence V is the result of the stacking of three units similar to that described above. This stacking, which was controlled by vertical movements of the margins, involved expansion and occurred together with northward displacement of the basin axis, which in all probability reflects the greater importance of supply from the Sierra Nevada and the Sierra de Baza than supply from the External Zone relief.

In conclusion, depositional sequence V is the result of the stacking, controlled by allocyclic processes, of three megasequences, which are in turn made up of several (3–5) component sequences basically controlled by autocyclic processes, and which reflect the characteristics and dynamic behaviour of each depositional system.

Conclusions

The stratigraphic architecture of the continental infilling is the result of the stacking of depositional sequences IV, V, and VI, each of which is characterized by the complex interrelation of several depositional systems. The stacking of the depositional sequences was the result of the interrelation between the factors controlling the volume of material entering each of the margins (tectonics and climate in both source areas) and those controlling the volume of material retained in the basin and not removed by the fluvial systems (subsidence and movements of the general base level). On a smaller scale, the sequences characterizing each depositional system reflect the dynamic behaviour of that system, are largely autocyclically controlled, and form part of a global model defined by the fluctuations of the base level (lake level).

Differentiation between the two transverse drainage systems and the longitudinal system, on which the areas of lacustrine sedimentation are superposed, is useful when establishing the spatial–temporal evolution of the palaeogeographic axis of the basin (longitudinal and lacustrine system), which reflects the relative asymmetry between the basin margins as regards the parameters mentioned above.

In general, throughout the period of development of the continental basins, we can detect northward displacement of the areas of maximum sediment accumulation and the paleogeographic axis and also movement away from Sierra Nevada, which could be related to the formation of large anticlinal structures in the Internal Zones (Sierra Nevada and Sierra de Filabres and its continuation towards Sierra de Baza).

Acknowledgments

The text has been improved by the suggestions of P.F. Friend. This study was made possible by funding approved for project PB91–0080–C02–01 granted by the Spanish DGICYT and research group JA4805 granted by the *Junta de Andalucía*.

References

Amundson, L. and Hendry, H. (1989). Lateral accretion in the braided South Saskatchewan river, Canada. *4th International Conference on Fluvial Sedimentology*, Abstracts, 62.

Arribas, M.E., Fernández, J. and García-Aguilar, J.M. (1988). Análisis sedimentológico de los materiales lacustres (Formación de Gorafe–Huélago) del sector central de la depresión de Guadix. *Estud. Geol.*, **44**, 61–73.

Bluck, B.J. (1976). Sedimentation in some Scottish Rivers of low sinuosity *Trans. R. Soc. Edinburgh*, **71**, 29–46.

Bluck, B.J. (1979). Structure of coarse grained stream alluvium. *Trans. R Soc. Edinburgh*, **70**, 181–221.

Bluck, B.J. (1980). Structure, generation and preservation of upward fining braided stream cycles in the Old Red Sandstone of Scotland *Trans. R. Soc. Edinburgh*, **71**, 29–46.

Bluck, B.J. (1982). Texture of gravel bars in braided streams. In *Gravel-bed rivers* (ed. R.D. Hey, J.C. Bathurst & C.D. Thorne), Wiley, 339–355.

Dabrio, C.J. and Fernández, J. (1986). Depósitos de rios trenzados conglomeráticos plio-pleistocénicos de la depresión de Granada. *Cuad. Geol. Iberica*, **10**, 31–53.

Fernández, J. and Dabrio, C.J. (1983). Los conglomerados de Moraleda. Un modelo de sistema fluvial de tipo braided (depresión de Granada, España), *Estud. Geol.*, **39**, 53–69.

Fernández, J., García-Aguilar, J.M. and Vera, J.A. (1986–87). Evolución de facies abanico aluvial-fluvial-lacustre en el Plioceno de la depresión de Guadix-Baza. *Acta Geol. Hispánica*, **21–22**, 83–90.

Fernández, J., Bluck, B.J. and Viseras, C. (1991). A lacustrine fan-delta system in the Pliocene deposits of the Guadix Basin (Betic Cordilleras, South Spain). *Cuad. Geol. Ibérica*, **15**, 299–317.

Fernández, J., Bluck, B.J. and Viseras, C. (1993). The effects of fluctuating base level on the structure of fan and associated fan-delta deposits; An example from the Tertiary of the Betic Cordillera, Spain. *Sedimentology*, **40**, 879–893.

Friend, P.F. (1983). Towards the field classification of alluvial architecture or sequence. In *Modern and ancient fluvial systems* (ed. J.D. Collinson & J. Lewin). Blackwell, Oxford. *Spec. Publ. Int. Assoc. Sediment.*, **6**, 345–354.

Gole, C.V. and Chitale, S.V. (1966). Inland delta building activity of the Kosi River. *Proc. Am. Ass. Civ. Eng., J. Hydraul. Div.*, **12**, 111–126.

Haughton, P.D.W. (1989). Structure of some lower Old Red Sandstone conglomerates, Kincardineshire, Scotland: deposition from late-orogenic antecedent streams? *J. Geol. Soc. London*, **146**, 509–525.

Pardo, G., Villena, J. and González, A. (1989). Contribución a los conceptos y a la aplicación del análisis tectosedimentario. Rupturas y unidades tectosedimentarias como fundamento de correlaciones estratigráficas. *Rev. Soc. Geol. España*, **2**, 199–219.

Ruíz-Bustos, A., Fernández, J., Morales, J., Rodríguez-Fernández, J. and Vera, J.A. (1990). Los materiales Plio-Pleistocenos del borde norte de la depresión de Granada. *Estud. Geol.*, **46**, 270–290.

Sanz de Galdeano, C. and Vera, J.A. (1992). Stratigraphic record and palaeogeographical context of the Neogene basins in the Betic Cordillera, Spain. *Basin Res.*, **4**, 21–36.

Serrano, F. (1979). *Los foraminíferos planctonicos del Mioceno superior de la cuenca de Ronda y su comparación con los de otras áreas de las Cordilleras Béticas.* Thesis, Universidad de Málaga, 272 pp.

Thorrarinson, S. (1956). The thousand years struggle against ice and fire. Reyjavik: Bokavge Menningarsjads.

Todd, S.P. and Went, D.J. (1989). Causes and effects of lateral migration of low-sinousity sand-bed rivers, with examples from the Slea Head Formation (Devonian) of SW Ireland and the Alderney Sandstone Formation (Cambrian) of the Channel Islands. *4th International Conference on Fluvial Sedimentology*, Abstracts, 232.

Vail, P.R. (1987). Seismic stratigraphy interpretation procedure. In *Atlas of seismic stratigraphy* (ed. A.W. Bally) *Am. Assoc. Petrol. Geol. Stud. Geol.*, **27**, 1–10.

Vail, P.R., Hardenbol, J. and Todd, R.G. (1984). Jurassic unconformities, chronostratigraphy, and sea level changes from seismic stratigraphy and biostratigraphy. In *Interregional unconformities and hydrocarbon accumulation* (ed. I.S. Schlee), *Am. Assoc. Petrol. Geol., Mem.*, **36**, 129–137.

Vera, J.A. (1970). Estudio estratigráfico de la depresión de Guadix-Baza. *Bol. Geol. Min.*, **81**, 429–462.

Viseras, C. (1991). *Estratigrafía y sedimentología del relleno aluvial de la Cuenca de Guadix (Cordilleras Béticas)*. Tesis, Universidad de Granada, 327 pp.

Viseras, C. and Fernández, J. (1989). Sistemas de drenaje transversales y longitudinales en el relleno aluvial de la Cuenca de Guadix (Cordilleras Béticas). *XII Congeso Español de Sedimentología*, Comunicaciones, 63–66.

Viseras, C. and Fernández, J. (1992). Sedimentary basin destruction inferred from the evolution of drainage systems in the Betic Cordillera, southern Spain. *J. Geol. Soc. London*, **149**, 1021–1029.

Viseras, C. and Fernández, J. (1994). Channel migration patterns in some alluvial fan systems. *Sediment. Geol.*, **88**, 201–217.

Viseras, C., Fernández, J. and Bluck, B.J. (in press). Autocyclic processes of a Pliocene alluvial fan in the Guadix Basin, Spain. *Trans. R. Soc. Edinburgh.*

Wells, N.A. and Dorr, J.A. (1987). A reconnaissance of sedimentation of the Kosi alluvial fan of India. In *Recent developments in fluvial sedimentology* (ed. F.G. Ethridge, R.M. Flores and M.D. Harvey), *Soc. Econ. Paleontol. Mineral. Spec. Publ.*, **39**, 51–61.

S8 Late Neogene basins evolving in the Eastern Betic transcurrent fault zone: an illustrated review

CH. MONTENAT AND P. OTT D'ESTEVOU

Abstract

Late Neogene sedimentary basins evolved within a wide transcurrent shear zone which passes through the Eastern Betic zone from the NE to the SW, and constitutes a segment of the major trans-Alboran shear zone. During the whole Neogene this structural corridor was subjected to a near-North–South compression resulting from the Iberia–Africa collision. As usual in wrench tectonics, compressional and extensional areas evolved at the same time within the structural corridor and controlled the development of two types of basin: groove-shaped faulted synclines and grabens. During the past ten million years, the direction of regional shortening varied between NW–SE and N–S, with important consequences for the kinematics of faulting and the sedimentary processes. Diversified magmatic activity, concentrated within the shear zone, interfered with the tectonic and sedimentary processes, especially during Tortonian times. The different magmatic products, originating from crustal (anatexis) and mantle lineages were telescoped within the corridor (e.g. Cabo de Gata), without any geochemical polarity indicative of a Miocene subduction zone. The major NE–SW trend of faults is deeply rooted in the lithosphere and marks the boundary between two different crustal slabs. The eastern slab is thinner and denser and has recorded an important late Miocene thermal anomaly. The basin evolution was greatly influenced by the properties of this hot and thin crust, which was particularly malleable under tectonic stress.

Introduction

During the past two decades, numerous studies have been dedicated to the Neogene geology of the Betic Cordilleras, covering a wide range of topics: stratigraphy, paleontology, sedimentology, tectonics, magmatism, etc. – most of the works are related to the central and eastern parts of the Chain.

This chapter is devoted to the Neogene basins of the eastern zone, located between Almería and Alicante (Fig. 1); it presents mainly the results of work by researchers of the Albert-de-Lapparent Geological Institut (IGAL) during the last decade. An overview of these studies was published by Montenat (1990). This gives a description of the different basins and includes a synthesis of the geological mapping of the area (scale 1/100 000). The present chapter aims to give an illustrated description of the most characteristic aspects of these basins.

Structural setting

The region concerned belongs to the Inner structural zone of the Betic orogen. It is composed of a pile of Alpine nappes, including sedimentary and various metamorphic rocks, which are, from the bottom to the top, the Nevado-Filabrides, Alpujarrides and Malaguides nappes. These units are mainly composed of Paleozoic and Triassic materials; the Malaguide nappes have a thin post-Triassic cover (Jurassic to early Miocene). The allochthonous series record a complex polyphase structural evolution, from Mesozoic to Miocene times (see a review in Díaz de Federico *et al.*, 1990).

These materials and their strongly developed deformational structures (thrust planes, wrench faults, foliation, etc.) form the basement of the Neogene basins, the first generations of which developed during the early and middle Miocene. Their deposits were severely deformed and eroded before Tortonian times (late Serravallian tectonic phase) and are now patchily preserved (Alpujarran Corridor, Rodríguez-Fernández *et al.*, 1990; Aguilas, Coppier *et al.*, 1989).

The basins discussed here correspond to the last generation, developed between the early Tortonian and the late Pliocene (about 10 my). The deposits fill intra-mountain, tectonically controlled depressions. The morphostructural pattern formed by the depressions filled with sediments and the 'sierras' exposing Betic materials roughly corresponds to the original paleogeographic disposition (Fig. 1).

The structural framework of the Neogene basins illustrated on Fig. 1A points out the major part played by NE–SW-trending faults. It corresponds to the general development of a large left lateral shear zone crossing the western end of the Mediterranean: the Trans-Alboran shear zone (Larouzière *et al.*, 1988).

The diverse groups of faults (NE–SW, NW–SE, sub-E–W and sub-N–S) have recorded different types of movement in response to

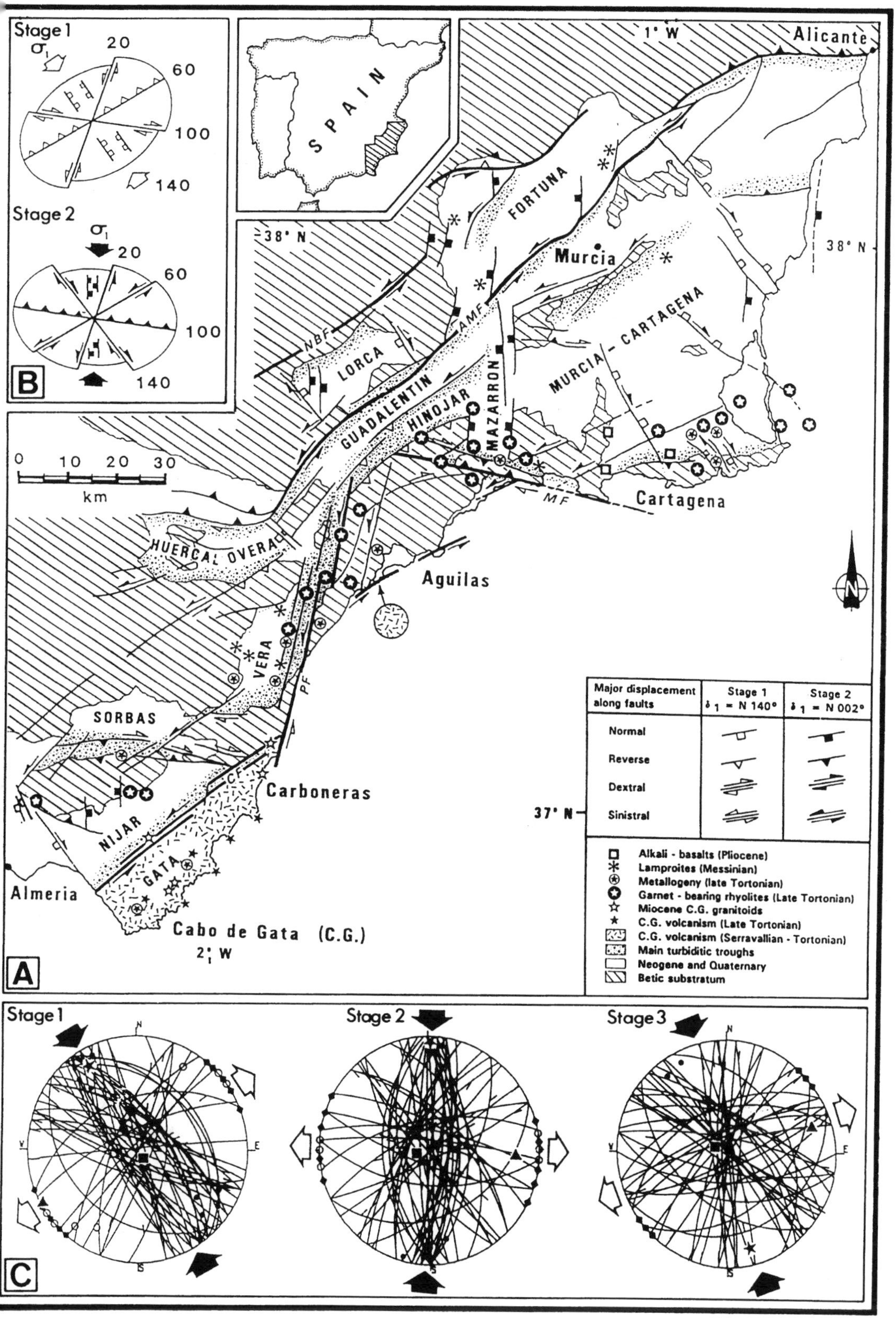

ig. 1. A. Structural framework of the late Neogene basins (Eastern Betic zone). Note the prevailing NE–SW-trending branch of faults formed by the Carboneras CF), Palomares (PF) and Alhama de Murcia (AMF) wrench faults. Most of the Neogene magmatic activity is located along this branch and within its eastern ompartment. Late Miocene turbiditic troughs are developed along the main wrench faults: sinistral faults (e.g. the previously quoted NE–SW branch); dextral ults, for example the Moreras fault (MF) to the South of Mazarrón; compressional tip of sinistral fault (North of Huercal Overa). B. Variations of fault ovements related to changes in stress field. Most of the fault pattern is inherited from pre-Tortonian tectonics. Note the formation of normal N–S faults in stage . Variations of fault movements are reported in A. C. Variations of stress fields: computation of the main stress axes from the processing of numerous nsedimentary or sealed faults. Stage 1: Tortonian; stage 2: uppermost Tortonian to early Pliocene; stage 3: late Pliocene to Holocene (after Montenat *et al.*, 990a). NBF=North Betic Fault.

variations in stress fields during Neogene times (Fig. 1B, C). Kinematic implications on basinal evolution are discussed below.

Stratigraphic aspect: chronology of the basinal development

In spite of frequent variations of facies, the different late Neogene series display the same general tendencies, illustrated on Fig. 2.

Widespread open marine deposits give an opportunity for accurate biostratigraphical dating, using planktonic foraminifera. Numerous levels abounding with micromammals provide correlations between marine and continental stratigraphic units (de Bruijn *et al.*, 1975; Mein & Agusti, 1990). Volcanic materials have yielded a large number of radiometric ages (see Fig. 11; Bellon *et al.*, 1983; Di Battistini *et al.*, 1987). The joint use of these data provides a confident chronological framework for the sedimentary, structural and magmatic events which characterized the basinal evolution (Fig. 3).

Sedimentological aspect

Late Miocene and Pliocene sedimentation occurred within mobile areas closely controlled by tectonic activity. The sedimentary dynamics are illustrated for the two main types of sediment, which are related to distinct structural controls:

- the clastics, predominant during Tortonian times in various environments, from alluvial fans to deep sea gravity flows (Bedu, 1990);
- the carbonates and evaporites, which prevailed during the Messinian (Garcin, 1987).

Clastic deposits

The Tortonian was a period of intense deformation (folding and wrenching): uplifted-relief features were deeply eroded and provided large amounts of clastic sediments that accumulated in the subsiding depressions (up to 3000 metres thick).

Thick alluvial fans are present in the lower Tortonian (Tortonian I), generally located along the downthrow side of active fault scarps. The deposits of alluvial or deltaic plains are poorly developed (except in the Huercal Overa basin; Briend *et al.*, 1990), as a result of the pre-eminence of steep, tectonically controlled morphologies.

Most of the widespread upper Tortonian marine clastic sediments were deposited in a basinal environment by various types of gravity flow (debris-flow to turbidites). Fig. 4 gives an illustration of the different depositional sequences constituting the Tortonian clastic series in the Sorbas basin (Almería). Fig. 5 is a reconstruction of the depositional process for the same series (Ott d'Estevou *et al.*, 1990). It points out the important combination of lateral and longitudinal transfers in deep clastic sedimentation.

The reappearance of coarse continental clastics by the end of the Messinian coincides with more humid weather (see also the development of lignites and lacustrine limestones in the uppermost Messinian). Most of these conglomerates are located to the west of the Alhama de Murcia fault (Fig. 1); they resulted from a general uplift of the inland domain (Meseta).

Carbonates and evaporitic sedimentation

The short interval of time related to the Messinian (about 1.2 my) recorded important changes compared with the Tortonian period:

- decrease in terrigenous input in most parts of the zone, probably due to climatic changes (aridity, Fig. 6) and tectonics;
- change in the basinal kinematics to predominant 'free-sliding' wrench movements with reduced development of folded and uplifted areas (see below and Fig. 3);
- gradual lowering of sea level resulting in progressive deterioration of the open marine environments, establishment of anoxic conditions and deposition of evaporitic sediments.

Pre-evaporitic and evaporitic Messinian sedimentations are summarized in Fig. 7. Widespread coral reefs prograded basinwards as a consequence of both tectonic uplift of the margins and lowering of sea level. In that general context, the tectonic behaviour of the margins gave rise to different types of architecture in the prograding reefs (Fig. 8). Evaporites accumulated in discontinuously subsiding depressions. Selenitic sulphate facies represent the basinal evaporitic deposits; they often interfinger with stromatolitic carbonates or calcarenites towards the margin (Garcin, 1987) (Fig. 9).

Magmatic aspects

An important and diverse Neogene magmatic activity is closely related to the 'Trans-Alboran' shear zone (Hernández *et al.*, 1987; Larouzière *et al.*, 1988). It interferes with Neogene sedimentation and tectonics. Four magmatic 'belts' are distinguished:

1. A calc-alkaline sequence (Cabo de Gata and Carboneras suite; Figs. 1 and 10) including various magmatic units (groups A to D; Bellon *et al.*, 1983; Bordet, 1985), from low-K and from low-Na series to those strongly enriched in K_2O and hydrothermalized rocks (Rodalquilar). The suite is late Serravallian–Tortonian in age (Bellon *et al.*, 1983; Di Battistini *et al.*, 1987) (Fig. 11). Drilling data (Cabo de Gata) indicate volcanics of more than 1100 m in thickness. These materials have a large extent offshore.

   Miocene plutonic equivalents (including Tortonian granodiorites and trondhjemites) are shallow batholiths which reached the surface by the way of extrusions such as pebble-pipes or pebble-dykes (Montenat *et al.*, 1984) (Fig. 11).
2. A calc-alkaline sequence with high-Al and high-K contents (cordierite-bearing dacites and rhyolites with abun-

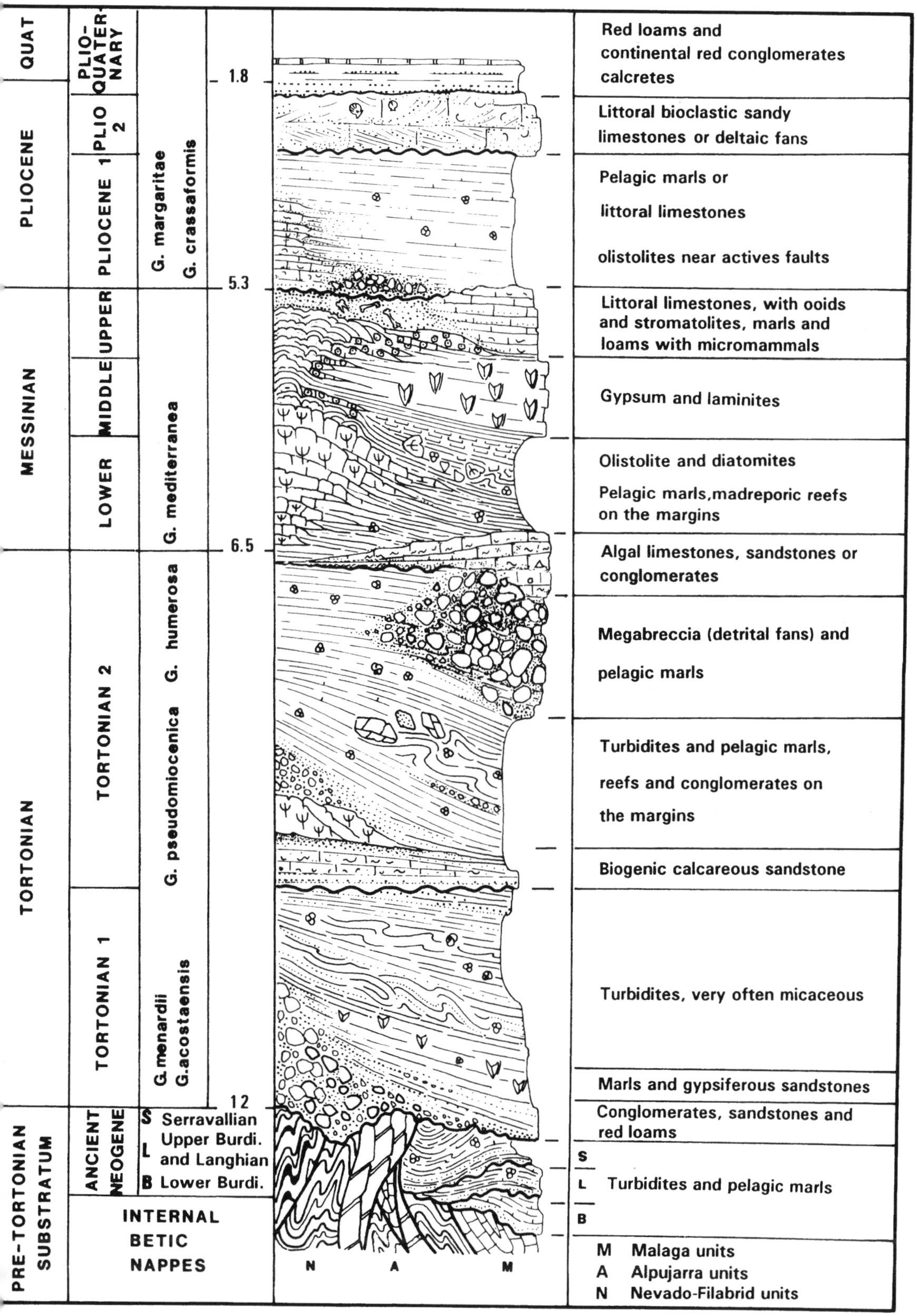

ig. 2. Synthetic stratigraphic column of the Eastern Betic late Neogene series. Note the thickness of Tortonian clastic deposits (usually more than 1000 m). hick red conglomerates and grits are often present in the lower Tortonian. Gravity phenomena (turbidites, slumping, olistoliths) are frequent in Tortonian deep arine environments. Open marine deposits are restricted to the lower half part of the Messinian column; Messinian evaporates have discontinuous distributions. Marine Pliocene sediments are localized in depressions close to the present Mediterranean coast. Pliocene epibathyal marls have little development in outcrop. lio-Pleistocene continental sedimentation is pellicular in most cases.

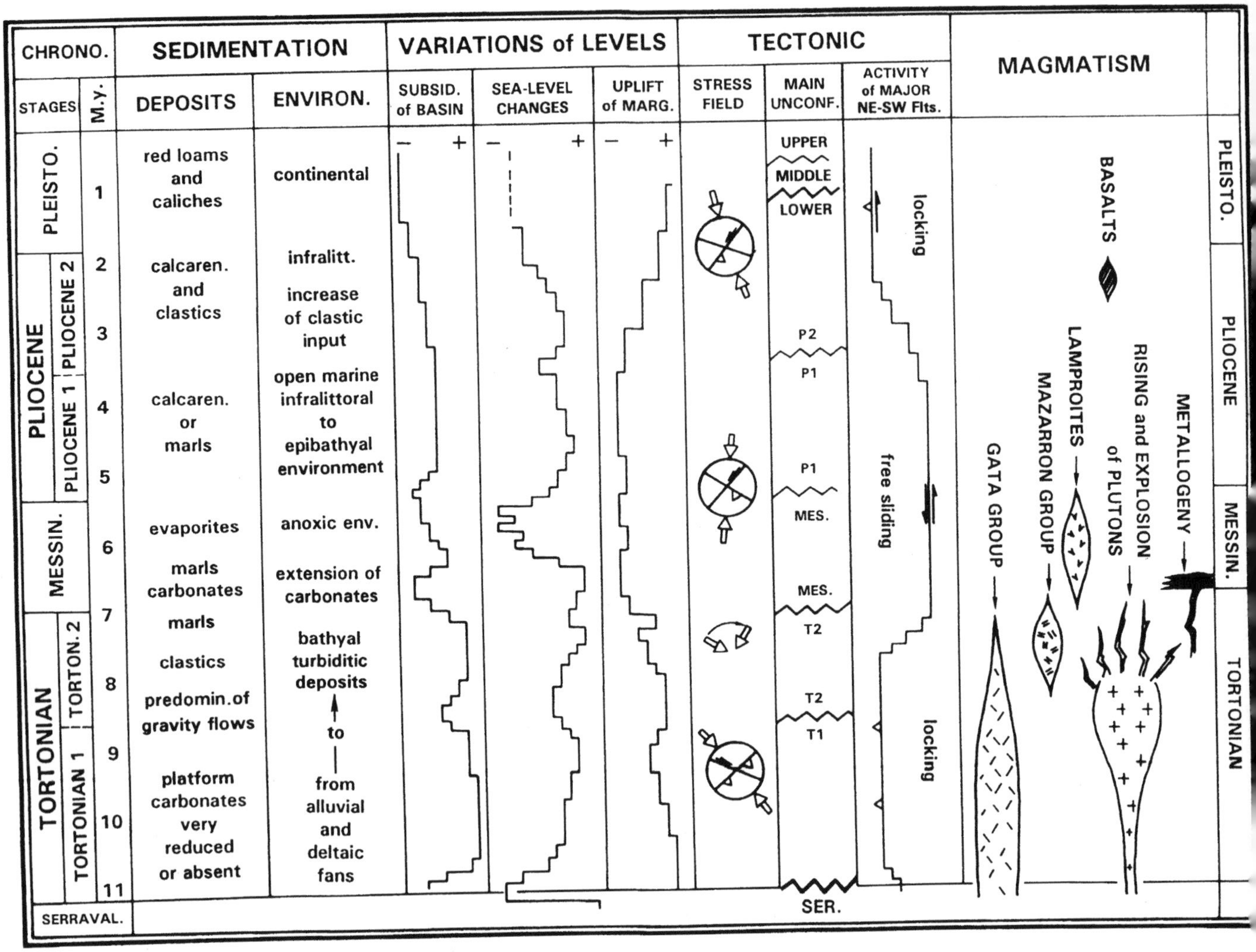

Fig. 3. Chronodiagram bringing together the different kinds of event which interacted during late Neogene basin evolution.

dant xenoliths and metamorphic rock inclusions) has formed numerous domes and associated breccias in the Níjar, Vera, Mazarrón and Cartagena areas (Fig. 10). These materials are late Tortonian in age (8.2 to 6.2 my) (Fig. 11). They were derived from a high-temperature shallow crustal anatexis from the gneissic Betic basement. (Zeck, 1968).

3. Lamproites display a variety of dark, glass-rich, high-K and high-Mg lavas, generally scattered as small groups of bodies (dykes, pipes, cones and flows). Their emplacement began by the end of the Tortonian and was mainly Messinian (7 to 5.7 Ma). They originated from a mantle source with crustal contamination indicated in some cases (Hernández *et al.*, 1987).
4. Alkali basalts with peridodite and granulite inclusions are located near Cartagena; they are late Pliocene in age (2.8 to 2.6 Ma).

An important metallogenic event gave rise to considerable or deposits (Rodalquilar, Vera-Garrucha, Aguilas, Mazarrón, La Unión; Fig. 1). The most frequent paragenesis developed siderite Fe–Mn oxides, barite, blende, galena with Ag–Pb, pyrite with gol (Rodalquilar) and, locally, casiterite and cinnabar–realgar. The or deposits are veins, stockwerks, or stratiform, enclosed withi Miocene volcanics or sedimentary rocks (limestones, turbidites) Ore bands are closely related to the structural pattern whic controlled basinal evolution. The metallogenic event occurred ove a short time, close to the Tortonian–Messinian boundary.

This large spectrum of thermo-magmatic phenomena may b related to a strong thermal anomaly, culminating during the lat Tortonian (Tortonian–Messinian boundary), with varied evidenc of shallow crustal anatexis, and spatially linked to the Easter branch of strike–slip faults (Carboneras, Palomares and Las Mo eras faults; see Fig. 1). The crustral implications of these data a discussed below.

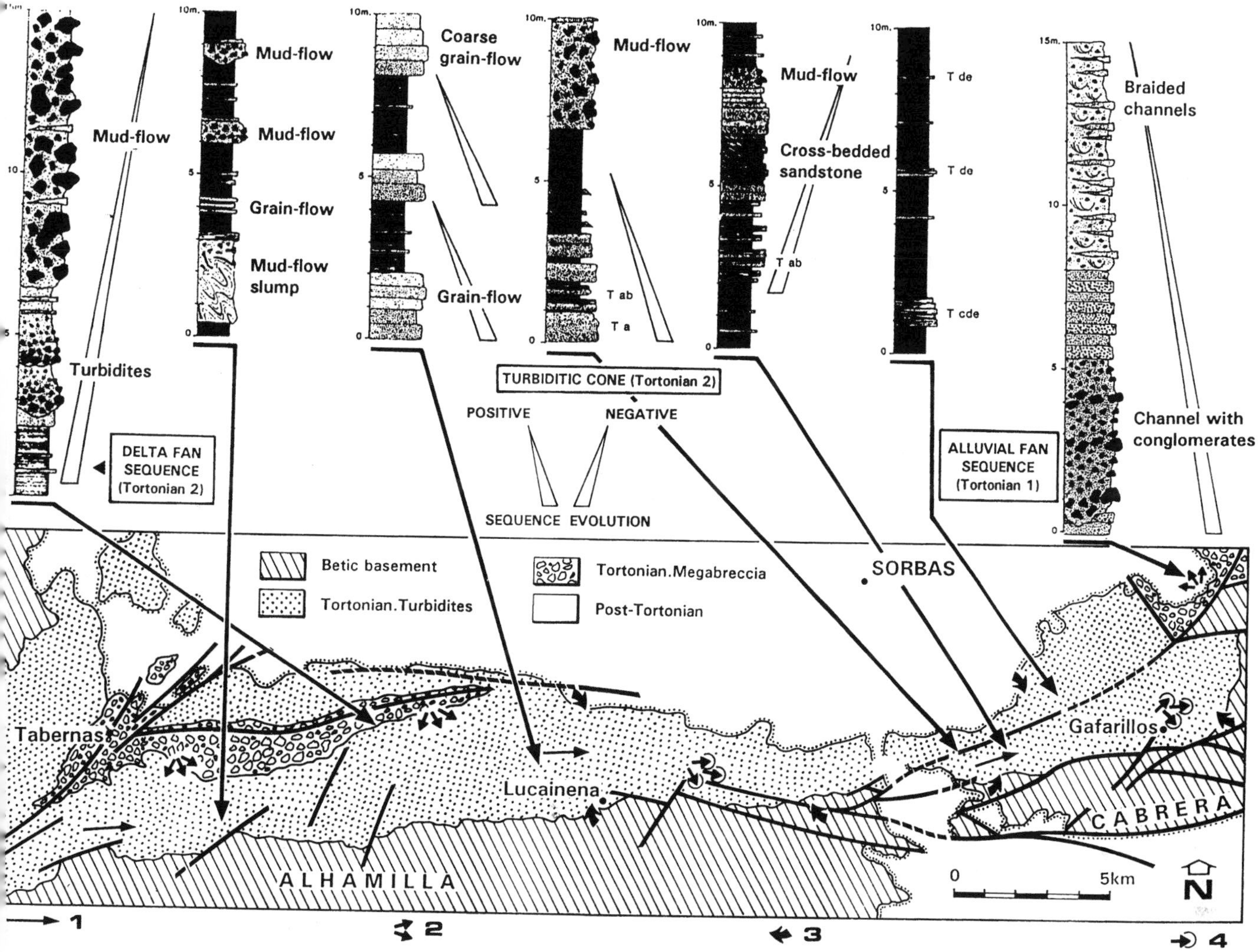

ig. 4. Example of Tortonian clastic sequences from the Sorbas basins (after Ott d'Estevou & Montenat, 1990, modified). 1. longitudinal transit; 2. main lateral ischarge; 3. lateral diffuse input of clastics; 4. lobe. Continental deposits were accumulated in discontinuous depressions during the early Tortonian (Tortonian I; ee alluvial fan sequence on the right). During the upper Tortonian (Tort. II) the basin was a narrow E–W trending subsiding trough, filled with deep sea clastics mud-flows, turbidites, grain-flows). Coarse discharges mainly originated from the northern faulted margin (debris- and mud-flows). Clastic material is then edistributed by longitudinal currents from West to East, with local lobe accumulations. Compare the geological sketch-map of the basin with the cross-section in ig. 16.

Basin kinematics

During the whole of the Neogene, the Eastern Betic zone 'as subjected, like the surrounding domains, to a compressional egime. The direction of regional shortening alternated between IW–SE and N–S (Fig. 1C). The deformation was mainly expressed y wrenching; folds have minor development and are often associated with strike–slip faults. As usually observed in a wrench ectonic regime, the direction of compression was associated with n orthogonal, relatively minor, horizontal extension. (Fig. 13).

Changes in the direction of shortening, applied to an inherited tructural pattern, obviously influenced fault movements (see Fig. B). For example, when the direction of the major stress was rthogonal to the main NE–SW shearing trend (Alhama de Murcia or Carboneras faults), the left lateral displacement was more or less prevented. These faults then had a strong reverse component: the greater part of the compressional deformation (including large-scale folding and uplift) took place during such a period (for example during most of Tortonian times).

This resulted from the combination of: 1. an inherited structural pattern; and 2. variations of the stress field. The shearing deformation appears to have been a polyphased process, with alternating periods of 'locking' and 'free-sliding' of the lateral movements (Montenat *et al.*, 1990a).

Two types of sedimentary basin evolved within the Eastern Betic shear zone, related to compressional and extensional areas which both coexisted during wrenching (Figs. 12 and 13):

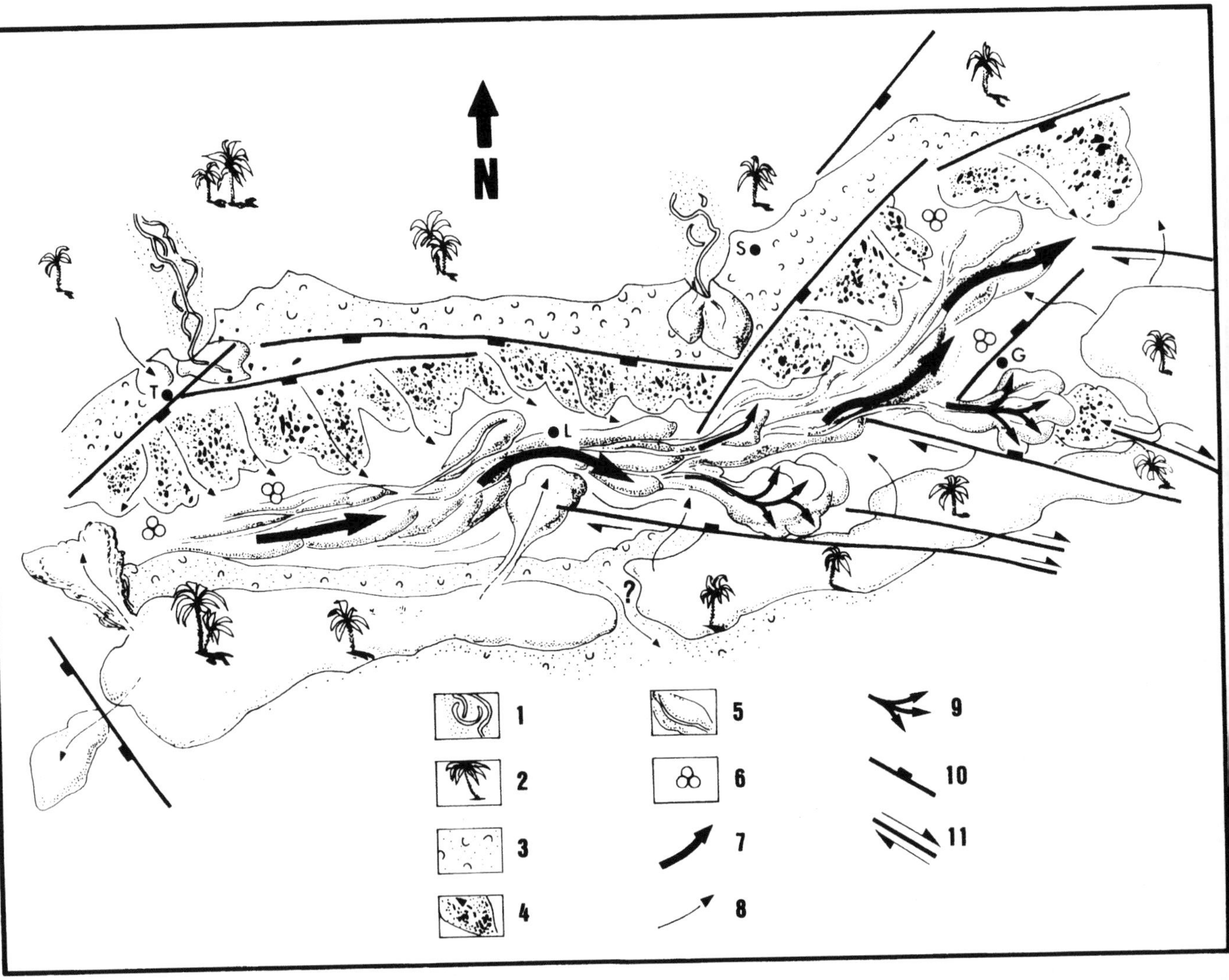

Fig. 5. Reconstruction of the Tortonian sedimentation in the Sorbas basins (after Ott d'Estevou & Montenat, 1990). 1. alluvial supplying; 2. emergent land; 3. littoral calcarenites; 4. megabreccias (debris- and mud-flows) close to active fault-scarp; 5. turbiditic channel; 6. epibathyal environment; 7. longitudinal current; 8. lateral input; 9. lobe; 10. fault-scarp; 11. dextral movement of fault. G. Gafarillos, L. Lucainena, S. Sorbas, T. Tabernas. The eastward (or northeastward) direction of the main longitudinal currents is recorded at a regional scale, from Almería to the Murcia area (Bedu, 1990) and may be related to the general circulation of sea water between the Atlantic and the Mediterranean basins. Note the location of lobe deposits in 'cul-de-sacs' controlled by dextral faults (for example, Gafarillos area; cf. Fig. 4).

– **Groove-shaped synclines** are narrow faulted synclines that developed along the main strike–slip faults, whatever their orientations and movement (NE–SW to sub-N–S left-lateral faults and sub-E–W right-lateral faults). The strongly subsiding furrows were filled with deep-water clastics (turbidites; cf. Sorbas basins, Figs. 4 and 5). They recorded strong deformation of the sediments, particularly seen on the margins (Figs. 14A, 1B, 15 and 16) and generating large 'flower-structure' type structural systems (Fig. 14B).
This spectacularly strong deformation induced a lateral migration of the basinal areas (Figs. 15, 16 and 17).
– **Flat-bottomed grabens** with a rectangular shape record a relatively low subsidence and appear to be only slightly deformed. They are situated between major wrench faults (Lorca basin for example); NW–SE or N–S normal faults controlled the margins (Mazarrón; see Figs. 13 and 14C).

Paleogeography

The late Neogene paleogeography of the Eastern Betic zone displays a mosaic of deep depressions linked by straits alternating with an archipelago of emergent lands. Both emergent and submarine morphologies are generally steep.

It is difficult to draw relevant paleogeographic maps because the part played by transcurrent faults appears to have been so important (up to several tens of kilometres for the last 10 My) and the

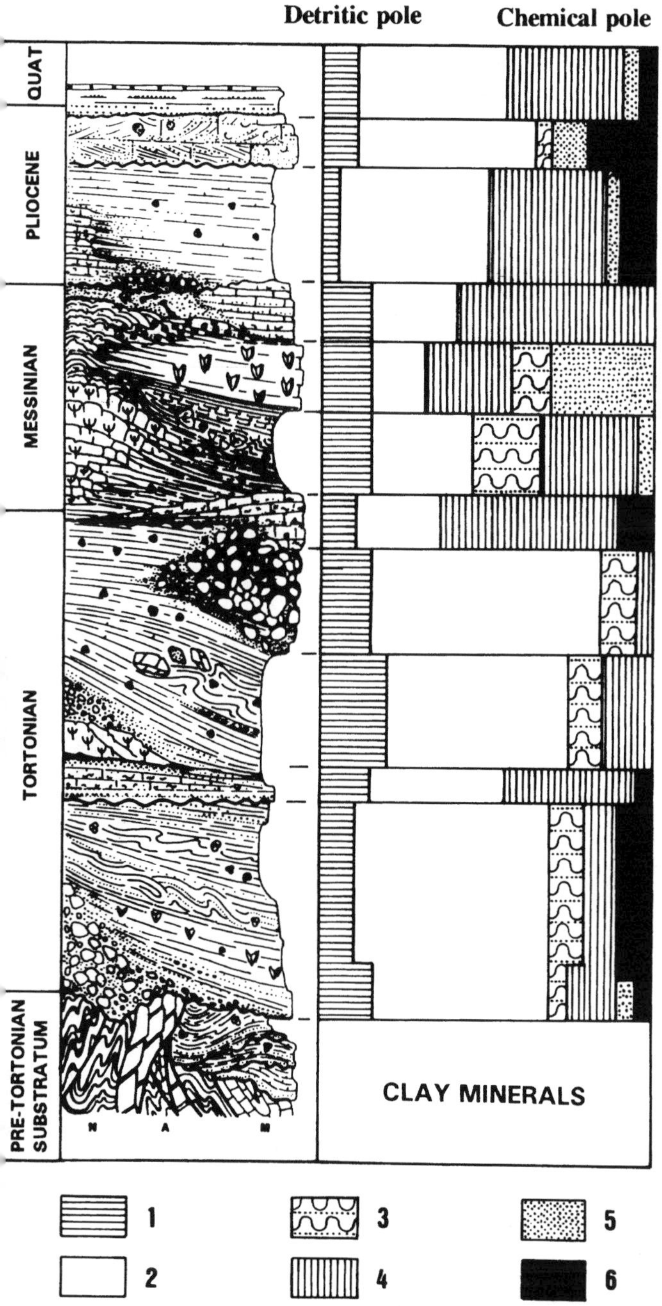

ig. 6. Composition of clay minerals in the late Neogene sediments; compiled om the analysis of various sections (after Griveaud-Marchal, 1992). Note the redominance of smectite in the Messinian interval. 1. chlorite; 2. illite; 3. iterlayered clays; 4. smectite; 5. fibrous clays; 6. kaolinite.

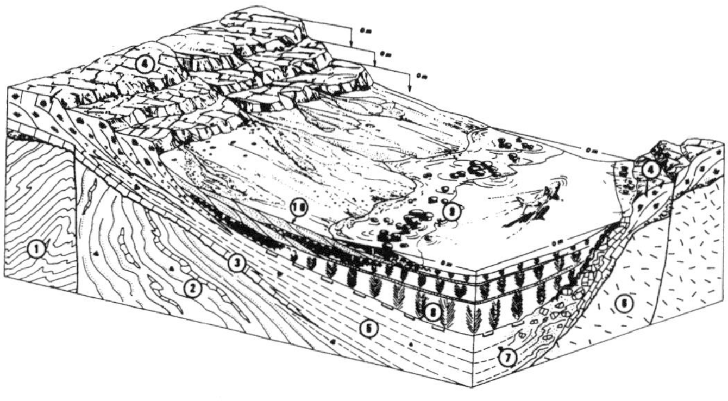

Fig. 7. Synthetic illustration of the Messinian sediments (after Montenat *et al.*, 1987). *Substratum*: 1. Betic basement; 2. highly structured Tortonian deposits; 3. Tortonian volcanics (Cabo de Gata); 4. 'Calcaire à Algues' (Tortonian/ Messinian boundary). *Pre-evaporitic deposits*: 5. planktonic-rich ooze; 6. different generations of coral reefs developing basinwards. The reef-flats recorded the gradual lowering of the Messinian sea-level; the higher reefs display precocious karstic weathering; 7. olistolites on steep unstable margins. *Evaporitic deposits*: 8. selenitic gypsum alternating with marls including marine fauna; diatomitic episodes are often present at the transition between preevaporitic marls and selenite; 9. stromatolitic carbonates contemporaneous with gypsum; 10. spreading of calcareous sands. 9 and 10 represent coastal sandy carbonate deposits facing 'basinal' evaporites (8). Note the important drop in sea-level (estimated to about 250 m in the Betic basins) which occurred between the first generation of reefs and the deposition of evaporites.

Fig. 8. Schematic illustration of different types of Messinian reef building, depending on morphostructures of the basin margins. A. Widely prograding reefs on a relatively stable, gentle slope. B. Narrow reef fringing a steep margin controlled by transcurrent faulting: note the steep reef front and a limited progradation basin-wards. 1. basement; 2. Tortonian deposits deformed prior to Messinian sedimentation; 3. Messinian marls; 4. Messinian reefs. (From the example of the Sorbas basin; after Ott d'Estevou & Montenat, 1990.)

ıovement is not precisely known for the different faults and the ifferent intervals of times.

An attempt to reconstruct the paleogeographic evolution and inematics of the Sorbas, Níjar–Carboneras and Vera basins is resented by the palinspastic sketchmaps in Fig. 18. The amounts f horizontal movement from one stage to another are general ıdications: 30 km for the Carboneras faults and about 35 km for the Palomares faults (Weijermars, 1987), both sinistral, since the beginning of late Miocene times. Moreover, the rate of shortening due to folding and crushing is not precisely evaluated (about 10 km for the Nijar – Cabo de Gata area, including the Sierra Alhamilla). Therefore, the inferred displacements may be regarded as minimum values. Fig. 18A illustrates the general movement of the Gata volcanic block during late Neogene times (see the successive positions, C1 to C6 of the locality of Carboneras). Variations in the direction of displacement may have been related to changes in stress-field orientations. An example of paleogeographic configuration is presented on Fig. 18B (see the different maps in Montenat *et al.*, 1990b).

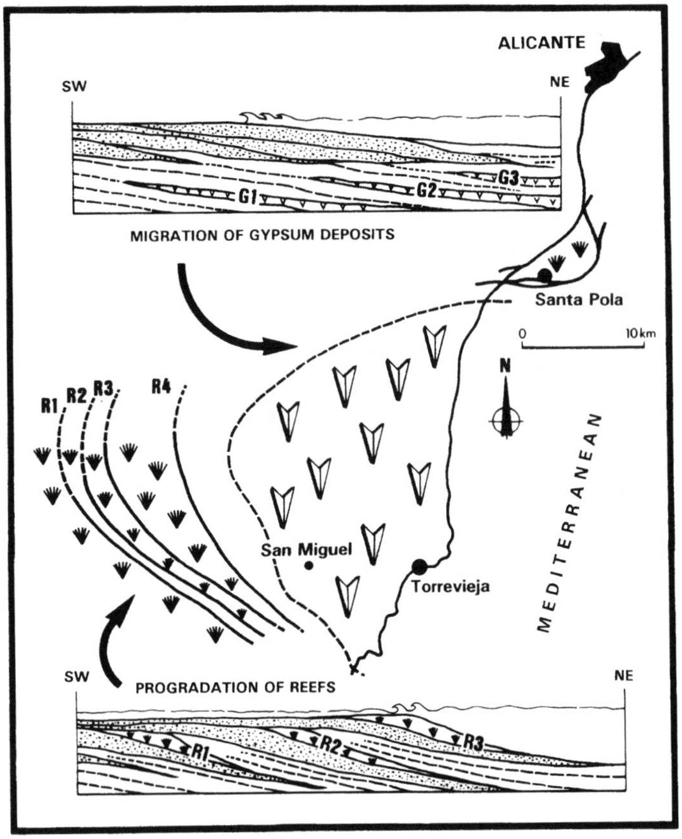

Fig. 9. Centripetal migration of Messinian reefs (R1 to R4), occurring before the evaporitic episode; NE part of the Murcia basin. Progradation of reefs (R1 to R3) on a gentle slope. Contraction of the marine realm accompanied the lowering of sea-level. Centripetal migration of selenitic bodies and progradation of calcareo-detritic coastal deposits (filling up sequence). (After Garcin, 1987, modified.) G1 to G3 represent stages in the migration of gypsum deposits through time.

Crustal aspects

The nature and structure of the crust lying beneath the Eastern Betic basins is documented in various sources (see a review of data in Larouzière *et al.*, 1988).

- The existence of distinct crustal slabs on each side of the Carboneras – Palomeras – Alhama de Murcia trend of faults is indicated by seismic refraction profiles (Figs. 10 and 19) and confirmed by gravimetric and magnetic data. The crust is thinner and denser within the eastern compartment, where it displays also a higher geothermal gradient as shown by oil industry drilling.
- Experimental petrologic studies on xenoliths raised by volcanism give information concerning a deep thermal anomaly and related anatexis processes.

These different data, in conjunction with the studies of the magmatism, illustrate lithospheric implications of the Eastern Betic wrench corridor; Fig. 19 synthesizes the results.

Conclusion

The Eastern Betic zone provides a detailed and comprehensive model of sedimentary basins evolving in a wrench regime within the Trans-Alboran corridor during late Neogene times. The following must be emphasized:

- the diversity of basinal depressions, displaying various geometries and kinematics;
- the high amplitude of vertical and horizontal movements (respectively several kilometres and several tens of kilometres) recorded during the basin evolution;
- the importance and diversity of the magmatic activity, indicative of a strong shallow thermal anomaly, obviously expressed during late Tortonian times.

These diverse tectono-thermal phenomena are the manifestation of a deep rooted transcurrent zone of lithospheric importance, which marked the boundary between two different crustal slabs.

References

Bedu, P. (1990). Evolution des environnements sédimentaires dans un couloir de décrochement: les bassins néogènes du domaine bétique oriental (Espagne). Thesis, University of Caen: 1–264.

Bellon, H., Bordet, P. and Montenat, C. (1983). Le magmatisme néogène des Cordillères bétiques (Espagne): Chronologie et principaux caractères géochimiques. *Bull. Soc. Géol. France*, **(7)**, **2**, 205–218.

Bordet, P. (1985). Le volcanisme miocène des Sierras de Gata et Carboneras. *Doc. Trav. IGAL*, Paris, **8**: 70, 1 colour map.

Briend, M., Montenat, C. and Ott d'Estevou, P. (1990). Le Bassin de Huercal Overa. *Doc. Trav. IGAL*, Paris, **12–13**: 239–259.

Coppier, G., Griveaud, P., Larouziere, F.D. de., Montenat, C. and Ott d'Estevou P. (1988). Example of Neogene tectonic indentation in the Eastern Betic Cordilleras: The arc of Aguilas (Southeastern Spain). *Geodinami. Acta*, **3**, **1**: 37–51.

De Bruijn, H., Mein, P., Montenat, C. and Van de Weer, D. (1975). Les gisements de Mammifères du Miocène supérieur d'Espagne méridionale (Provinces d'Alicante et de Murcia). Corrélations avec les formations marines du Miocène terminal. *Proc. Kon. Ned. Akad. Wetensch.*, **B**, **78**, **4**: 1–32.

Díaz de Federico, A., Torres-Roldán, R. and Puga, E. (1990). The rock-series of the Betic substratum. *Doc Trav. IGAL*, Paris, **12–13**: 19–29.

Di Battistini, G., Toscani, L., Iaccarino, S. and Villa, I. (1987). K/Ar Ages and Geological Setting of Calc-Alcaline Volcanic Rocks from Sierra de Gata, SE Spain. *Neues Jhrb. Mineral Mh.*, **H8**, 369–383

Garcin, M. (1987). Le bassin de San Miguel de Salinas (Alicante, Espagne) relations entre contexte structuro-sédimentaire et dépôts évaporitiques et carbonatés au Messinien. Thesis, University Paris-Sud 297 pp.

Grieveaud-Marchal, P. (1992). Evolution de la sédimentation argileuse des bassins néogènes des Cordillères bétiques orientales. Description minéralogique et interprétation géologique. *Mem. Géol. IGAL* Cergy, no. **45**: 1–254.

Hernández, J., Larouzière, F.D. de, Bolze, J. and Bordet, P. (1987). Le magmatisme néogène bético-rifain et le couloir de décrochement 'Trans-Alboran'. *Bull. Soc. Géol. France* **8**, **3**, **2**: 257–267.

Larouzière, F.D. de., Montenat, C., Ott d'Estevou, P. and Griveaud, P. (1987). Evolution simultanée de bassins néogènes en compression et en extension dans un couloir de décrochement: Hinojar e

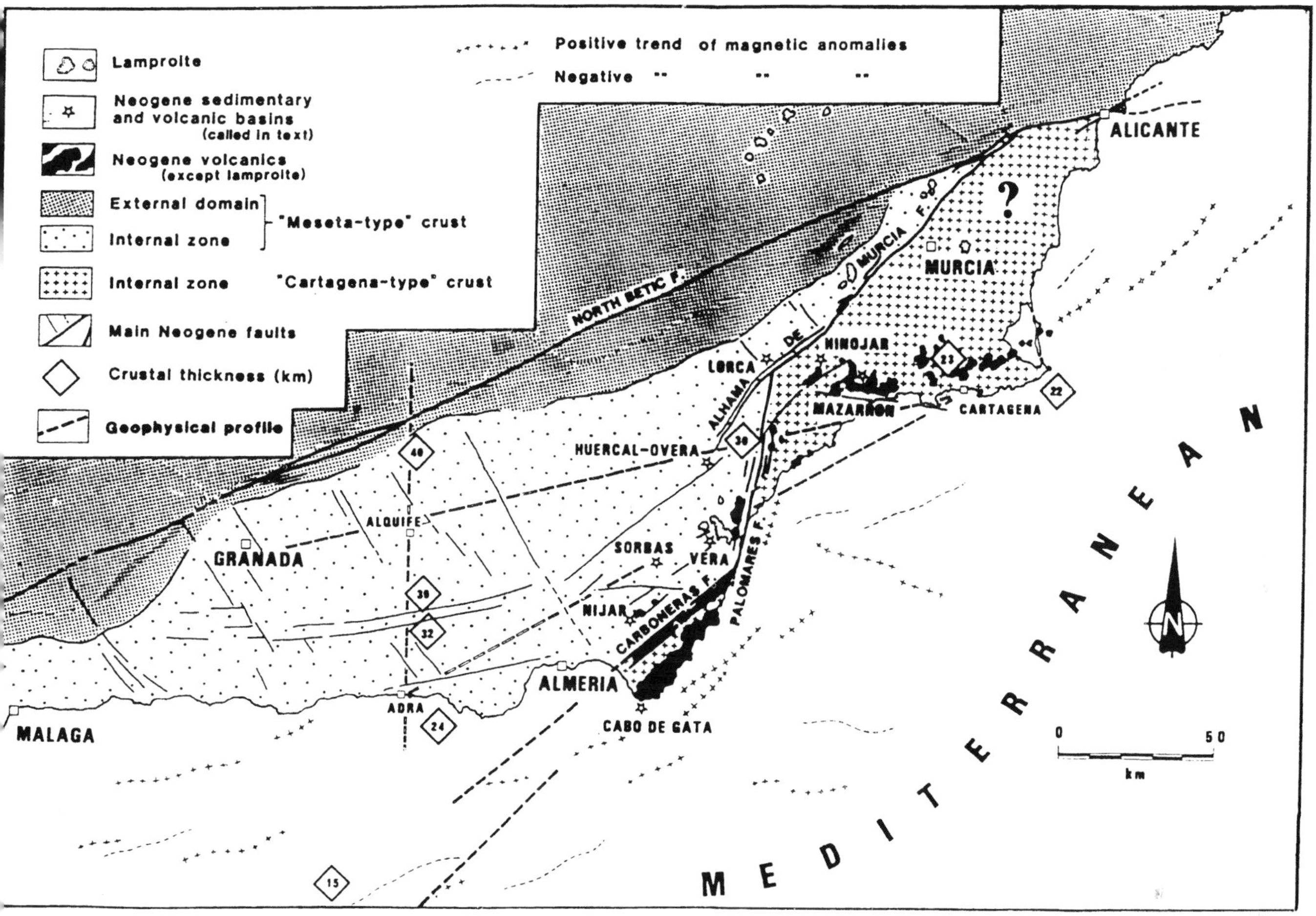

Fig. 10. The late Neogene volcanism of the Eastern Betic zone, with indication of the main structural and geophysical features. Note the existence of different continental crust on each side of the major NE–SW-trending wrench system (cf. Fig. 19).

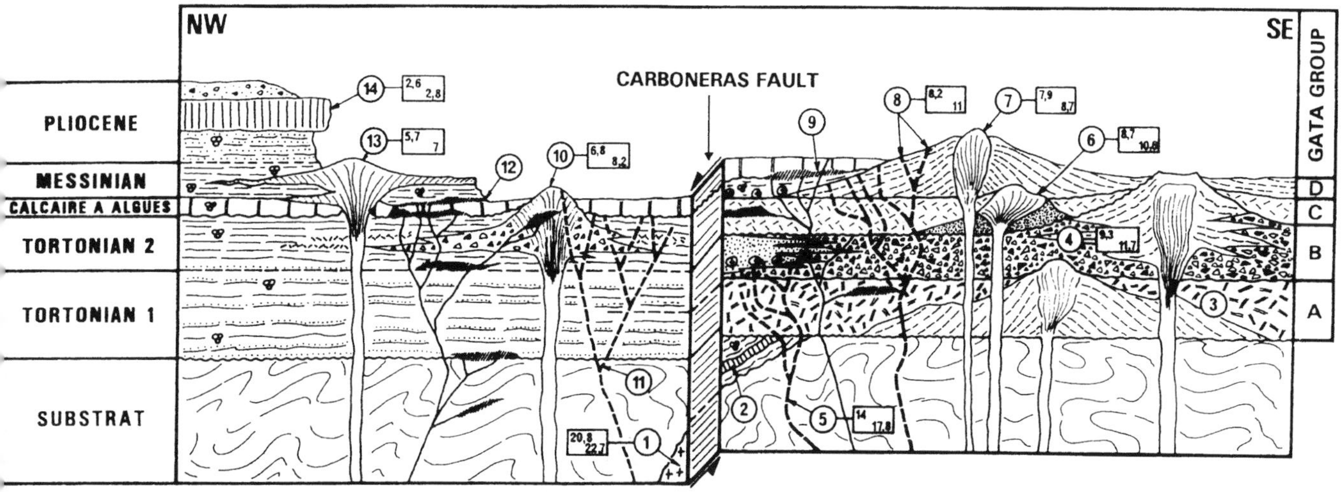

Fig. 11. Compilation of data concerning the late Neogene magmatism (after Montenat *et al.*, 1987). The southeastern compartment is an illustration of the Cabo de Gata magmatic complex; the northwestern one is a synthesis of the other magmatic occurences. Within rectangles: radiometric ages in my. 1. granite veins injected into the basement, close to the Carboneras fault; 2. basalt flow interbedded within late Burdigalian–Langhian pelagic marls; 3. Cabo de Gata, group A (Serravallian to early Tortonian); 4. Cabo de Gata, group B; 5. granophyres (blocks originating from pebble dykes) included within group B material; 6. Cabo de Gata, group C; 7. Cabo de Gata, group D; 8. granitoid blocks originating from pebble dykes (with indications of ages); 9 and 12. metallogenic event (Tortonian/Messinian boundary) indicated by dark hatched lines in both compartments; 10. anatectic rhyodacite (Mazarrón group); 11. metal-bearing (Sn) pebble-pipes of the Cartagena area; 13. Lamproitic volcanism; 14. basalts of Cartagena.

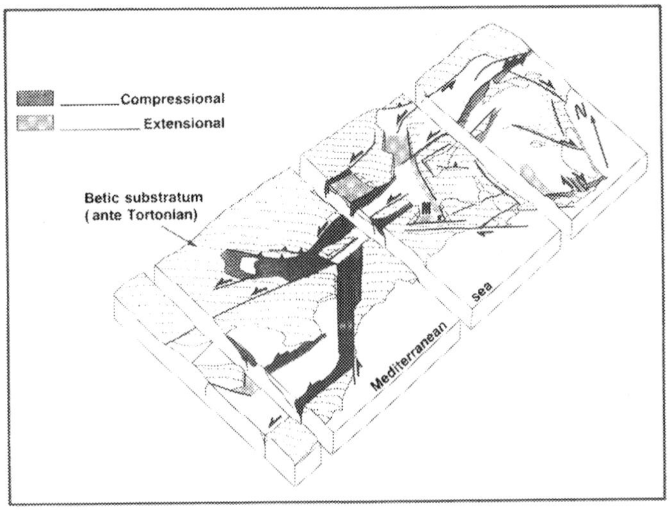

Fig. 12. Block-diagram of the Neogene basins with an indication of the main compressional and extensional areas (cf. Fig. 1; M: Mazarrón, see Fig. 13A) (after Montenat *et al.*, 1990a, modified).

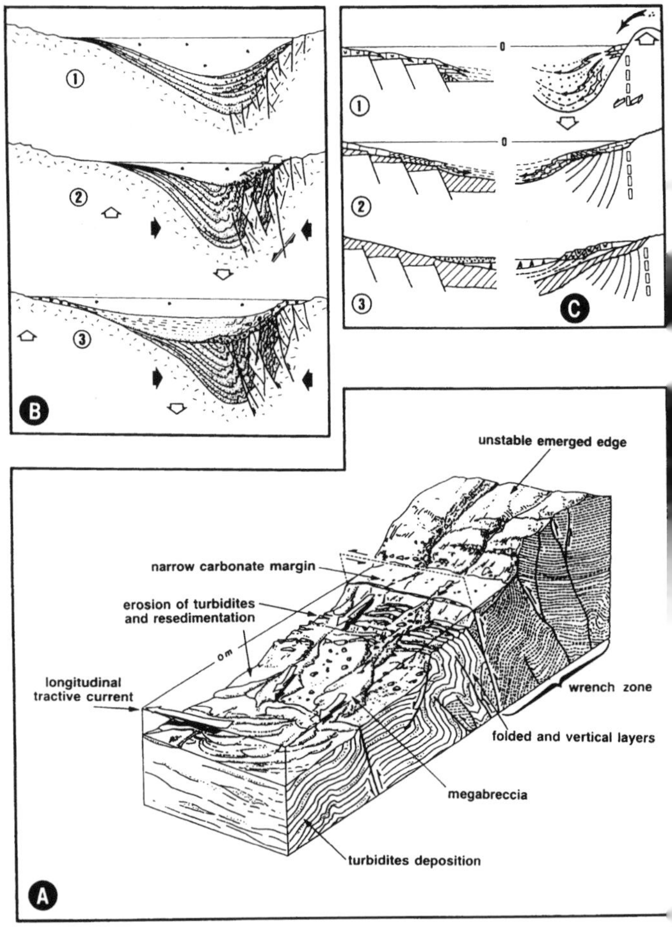

Fig. 14. Relationship between basinal structural evolution and sedimentation. A. Characteristic sedimentary processes on a wrench faulted margin (after Montenat *et al.*, 1990a). B. General evolution of a groove-shaped syncline (inspired by the late Miocene Vera basin). Note: the asymmetrical profile of the groove (1); the importance of resedimentation (2); the precocious structural evolution and the migration of the area of subsidence (3). C. Schematic illustration of the different behaviours of a margin controlled by normal faulting (left) or wrenching (right). 1. Tortonian; 2. Tortonian/Messinian boundary; 3. Messinian. The graben margins recorded a lower subsidence (3) and a best development of carbonate deposits (1 and 2) (after Montenat *et al.* 1987).

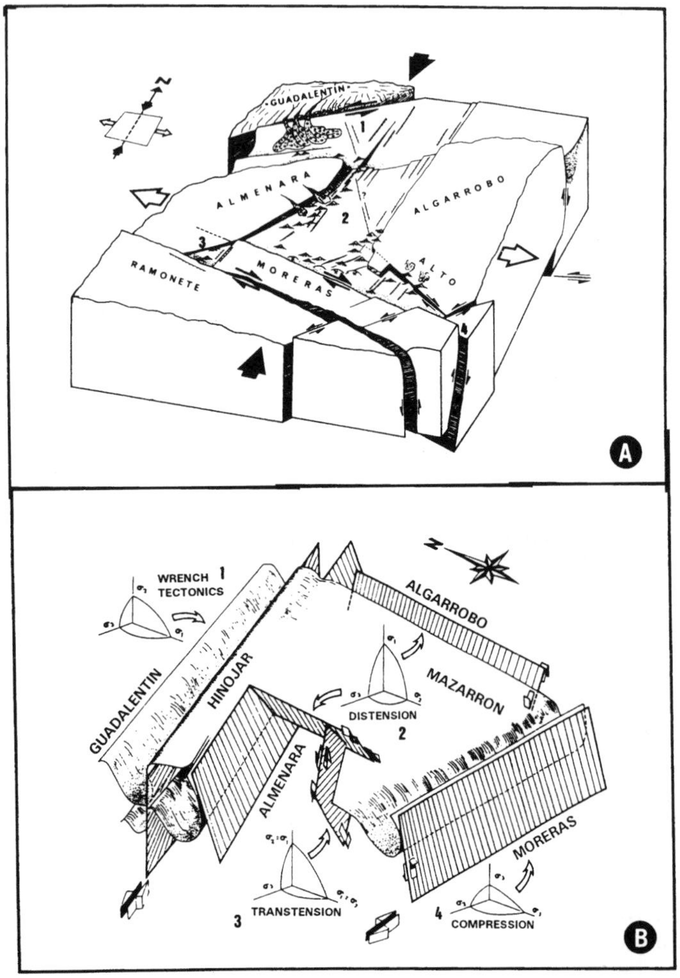

Fig. 13. Different types of basins evolving jointly in the Mazarrón area during late Tortonian times (cf. Fig. 12). A. Block-diagram illustrating the structural framework of the contemporaneous basins: Hinojar and Moreras groove-shaped synclines (1 and 4); Mazarrón graben (2) and a small transtensive depression (3). Note the development of volcanoes (late Tortonian rhyodacitic material) around the Mazarrón graben. B. Illustration of the states of stress for the different basins (1 to 4: as in 7A). The different types of contemporaneous deformations are related to a regional submeridian shortening (after Montenat *et al.*, 1987).

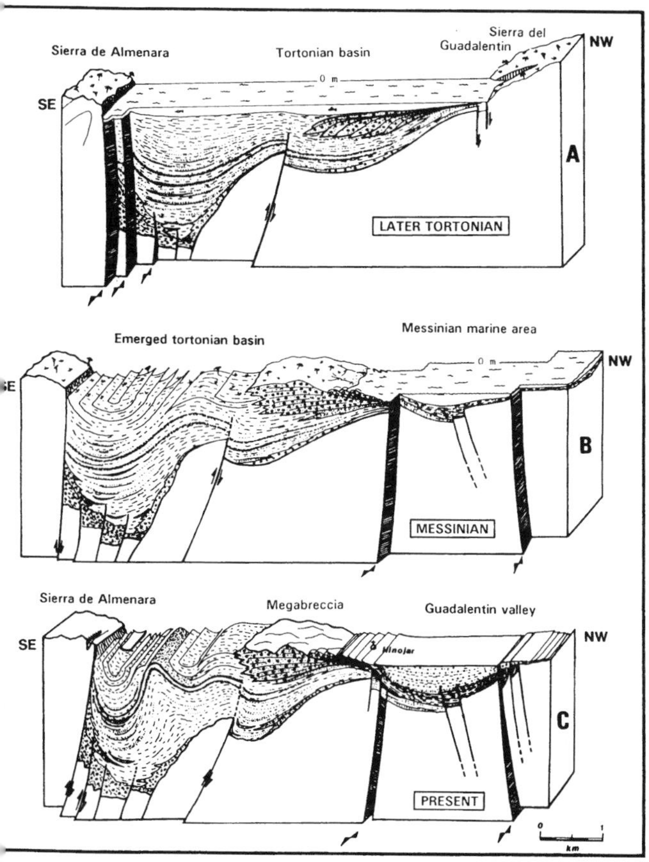

ig. 15. Three stages of evolution of the Hinojar basin controlled by the NE–W wrench faults (see location on Fig. 1). The Tortonian deposits have ecorded strong structural evolution before the Messinian sedimentation egan. Note the inversion of relief between the late Tortonian (A) and the 1essinian (B), due to the shifting of the previously emerged Sierra del uadalentin along sinistral faults. The reverse component of wrench faults is ssociated with folding of the Tortonian series, during Plio-Pleistocene times C) (after Larouzière *et al.*, 1987).

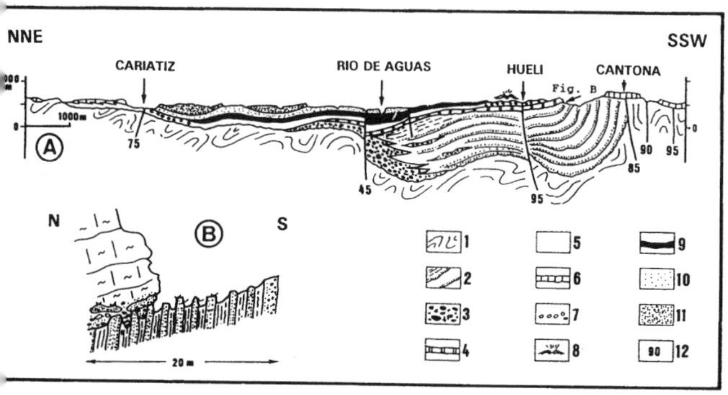

ig. 16. A. Cross-section of the Sorbas basin (for location see Figs. 1 and 4). asement – 1; Late Tortonian – 2; turbiditic series – 3; megabreccias (mud-ows) – 4; 'calcaire à Algues', Messinian – 5; marls – 6; reefs – 7; olistolites – 8; ibmarine hydrothermal spout – 9; gypsum – 10; loams, gravels and lacustrine mestones, Pliocene – 11; orientation of faults (in degrees) – 12. B. detail of the ructural unconformity between the Tortonian turbidites and the 'Calcaires à lgues' (Tortonian/Messinian boundary) (after Ott d'Estevou & Montenat, 990). Note the asymmetrical profile of the basin, the location of the egabreccias close to an active fault (still active during gypsum deposition), e strong pre-Messinian structuration and the northward migration of the sin during the late Neogene.

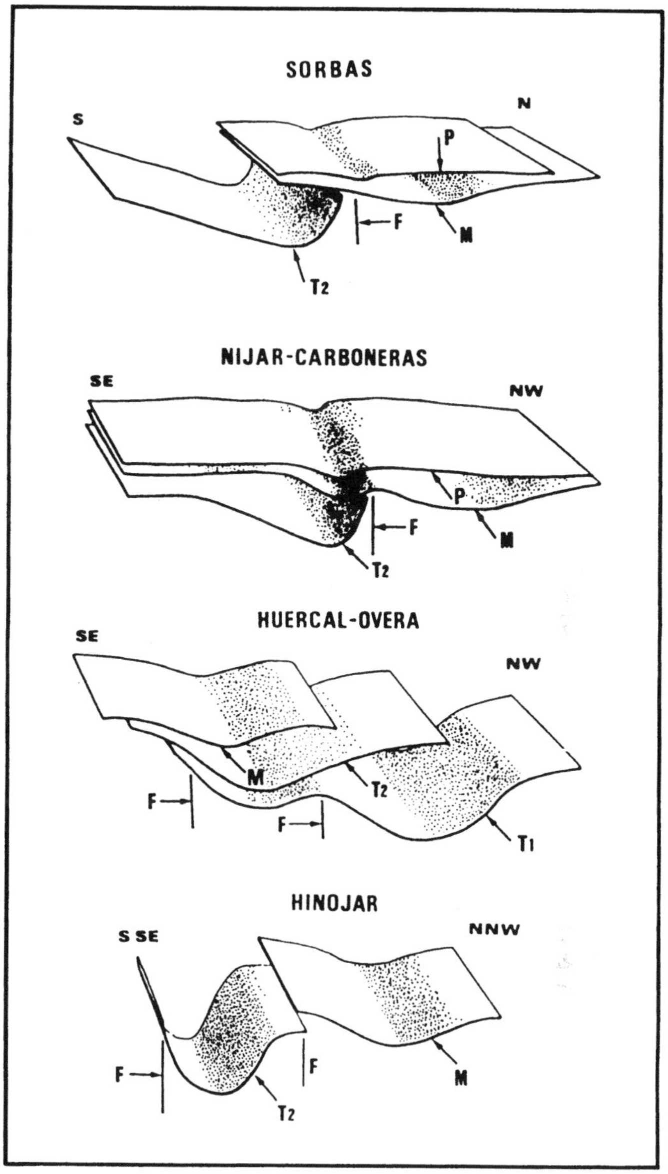

Fig. 17. Examples of migrations of basinal areas resulting from wrench tectonics. T_1 and T_2: Tortonian; M: Messinian; P: Pliocene; F: major strike–slip faults (after Montenat *et al.*, 1990a, modified). For location of the basins see Fig. 1.

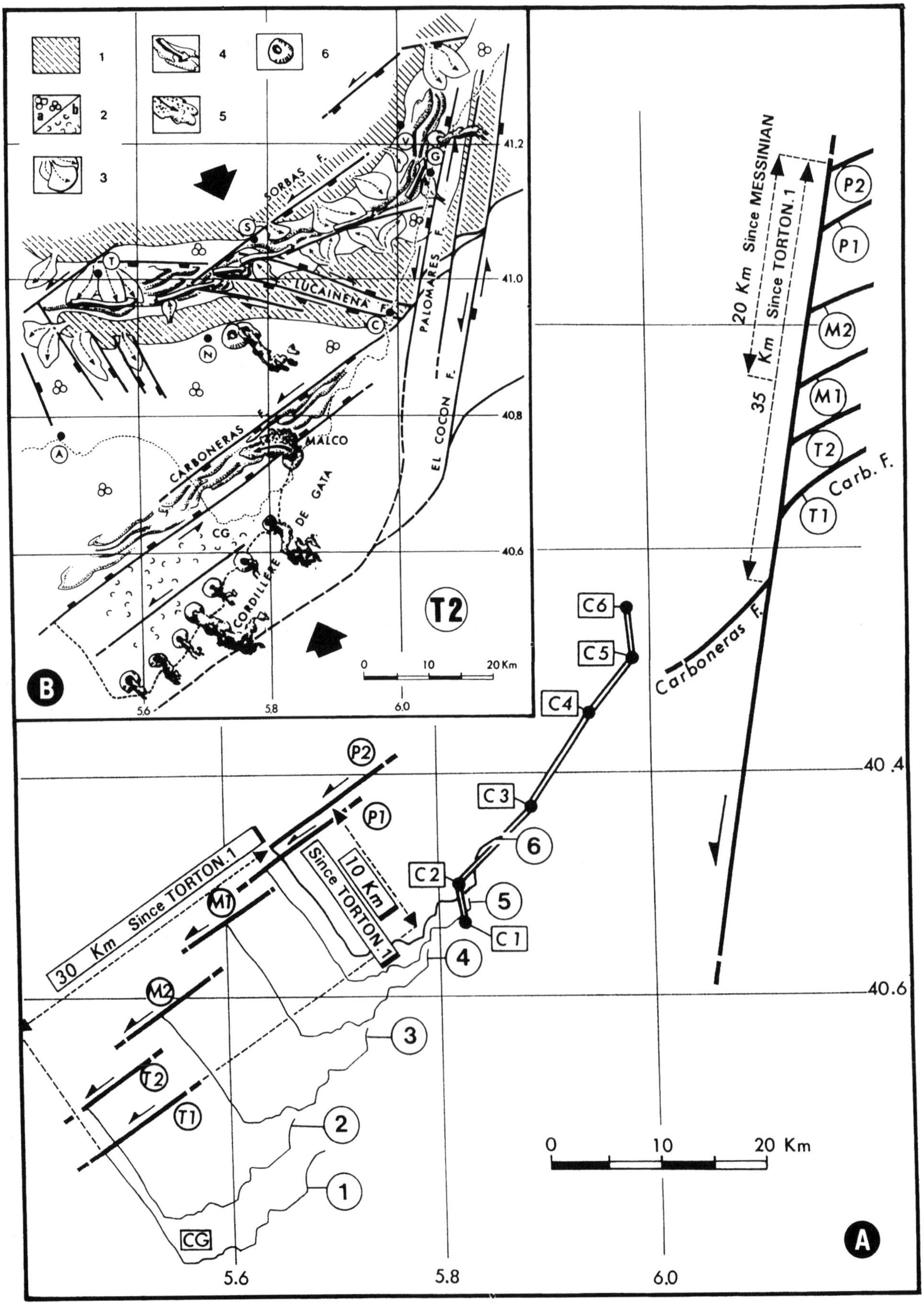

1
2
3
4
5
6
SORBAS F.
LUCAINENA F.
PALOMARES F.
EL COCON F.
CARBONERAS F.
MALCO
CORDILLERE DE GATA
CG
T2
B
41.2
41.0
40.8
40.6
5.6
5.8
6.0
0 10 20 Km
20 Km Since MESSINIAN
35 Km Since TORTON.1
P2
P1
M2
M1
T2
T1
Carb. F.
Carboneras F.
C6
C5
C4
C3
C2
C1
30 Km Since TORTON.1
10 Km Since TORTON.1
CG
40.4
40.6
6.0
A

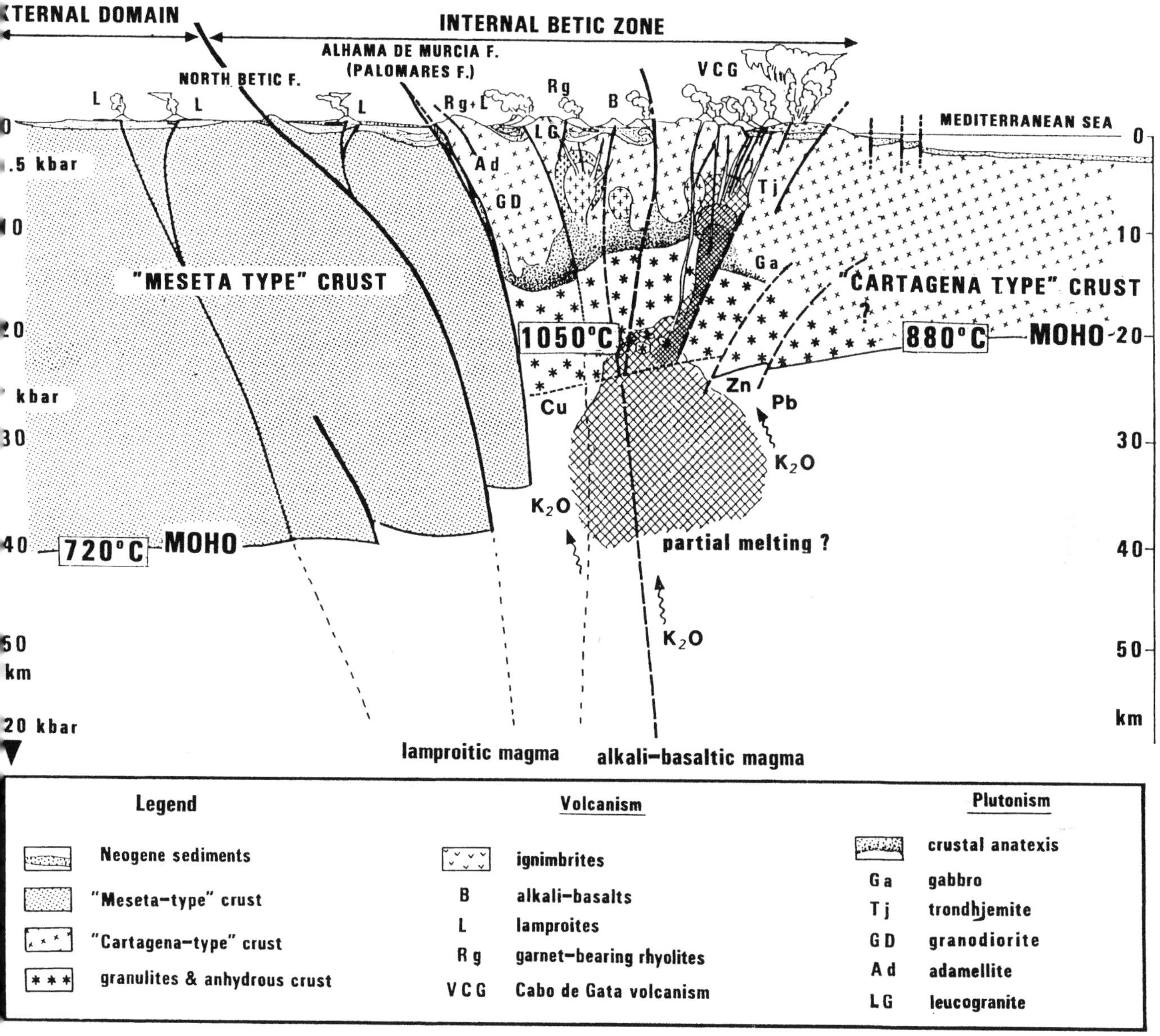

g. 19. Interpretative cross-section of the Eastern Betic crustal shear-zone. This is an attempt to synthesize the different data concerning crust properties and ickness, magmatism thermicity, structures, etc. (after Montenat *et al.*, 1987).

g. 18. An attempt to reconstruct the paleogeography and kinematics of the late Neogene basin. The example of the Almería zone (for location see Fig. 1) (after ontenat *et al.*, 1990b). A. Northeastward shifting of the Cabo de Gata block from early Tortonian to late Pliocene times and offsetting of the Carboneras fault the submeridian sinistral Palomares fault. T1 and T2, M1 and M2, P1 and P2 correspond to early and late Tortonian, Messinian and Pliocene respectively. ne six stages of migration are numbered 1 to 6 (within circles); C1 to C6 are the corresponding positions of the locality of Carboneras. B. Example of leogeographic configuration during late Tortonian (T2; stage 2) including the Sorbas, Nijar-Carboneras and Vera basins. 1. emergent basement; 2. pelagic (a) d littoral (b) deposits; 3. submarine fan; 4. turbiditic flow with indication of transit; 5. olistostrome of volcanic material ('Brèche rouge'); 6. active volcanoes. rge black arrows: direction of regional shortening.

Mazarrón (S.E. de l'Espagne). *Bull. Centres Rech. Explor. Prod. Elf-Aquitaine*, **11**, **1**: 23–38.

Larouzière, F.D. de, Bolze, J., Bordet, P., Hernández, J., Montenat, C. and Ott d'Estevou, P. (1988). The Betic segment of the lithospheric Trans-Alboran shear zone during upper Miocene. *Tectonophysics*, **152**, 41–52.

Mein, P. and Agusti, J. (1990). Les gisements de Mammifères néogènes de la zone bétique *Doc. Trav. IGAL*, Paris, **12–13**: 81–84.

Montenat, C. (ed.) (1990). Les Bassins néogènes du domaine bétique oriental (Espagne). Tectonique et sédimentation dans un couloir de décrochement. Première partie: étude régionale. *Doc. Trav. IGAL*, Paris, **12–13**: 1–392, 3 colour maps. (This volume gives a great number of references concerning the Eastern Betic zone and the late Neogene basins.)

Montenat, C., Bolze, J., Bordet, P. and Ott d'Estevou, P. (1984). Extrusion de type 'pebble dyke' à éléments plutoniques miocènes, dans le Tortonien des Cordillères bétiques orientales (Espagne). *C.R. Acad. Sci.*, Paris, **299**: 343–346.

Montenat, C., Masse, P., Coppier, G. and Ott d'Estevou, P. (1990a). The sequence of deformations in the Betic shear zone (SE Spain). *Ann. Tect.* Special Issue, **IV**, **2**, 96–103.

Montenat, C., Ott d'Estevou, P., Rodríguez-Fernández, J. and Sanz de Galdeano, C. (1990b). Geodynamic evolution of the Betic Neogene intramontane basins (S and SE Spain). Guide Book, IXe RCMNS Congr., Barcelona. In *Iberian Neogene basins, Paleontol. Evol., Mem.* sp. **2**: 5–59.

Ott d'Estevou, P. and Montenat, C. (1990). Le bassins de Sorbas–Tabernas *Doc. Trav. IGAL*, Paris, **12–13**: 101–128.

Rodríguez-Fernández, J., Sanz de Galdeano, C. and Serrano, F. (1990. Le couloir des Alpujarras. *Doc. Trav. IGAL*, Paris, **12–13**: 87–100.

Weijermas, R. (1987). The Palomares brittle–ductile Shear Zone of southern Spain. *J. Struct. Geol.*, **9**, **2**: 139–157.

Zeck, H.P. (1968). Anatectic origin and further petrogenesis of almandine-bearing biotite-cordierite-labradorite dacite with many inclusions of restites and basaltoid material, Cerro el Hoyazo, SE Spain. Thesis, University of Amsterdam, *Medit. geol. Inst.*, no **361**: 1–161.

S9 Tectonic signals in the Messinian stratigraphy of the Sorbas basin (Almería, SE Spain)

J.M. MARTÍN AND J.C. BRAGA

Abstract

The Messinian (Late Miocene) stratigraphic record of the Sorbas basin (SE Spain) comprises two major sedimentary cycles that can be correlated with the TB 3.3 and TB 3.4 cycles of Haq *et al.* (1987). The lower cycle consists of a temperate-carbonate unit assigned to the Low-Systems Tract and bioherm and fringing reef units representing the Transgressive and High-Stand parts respectively. The upper cycle consists of a selenite–gypsum unit (Low-Stand), a marine, stromatolite-bearing unit corresponding to the Transgressive Systems Tract, and an alluvial, continental unit representing the High-Stand.

Tectonic movements were involved in the formation of the unconformities at the base of both major, third-order cycles as well as at the boundaries and interruptions of some high-order ones. The unconformity at the base of the lower cycle reflects the combined effects of the sea-level fall at the end of the last Tortonian cycle (TB 3.2 of Haq *et al.*, 1987) and a pulse of tectonic uplifting affecting the southern margin of the basin (Sierra Alhamilla). As a consequence of this, temperate, shallow-water carbonates lie directly on top of deep-water, submarine-fan deposits around the Sierra Alhamilla.

The unconformity separating both major Messinian cycles may be correlated with a global sea-level fall and may perhaps reflect the main evaporite event of the Mediterranean region. Detailed analysis of the sedimentary evolution of the unit immediately underneath this unconformity (the 'fringing reef' unit) reveals that tectonics also played an important role in the progressive restriction and isolation of the Mediterranean leading to its final dessication.

The Sorbas basin is a narrow, elongated, east–west-trending basin within the Betic Cordillera of southeastern Spain (Fig. 1). The Nevado–Filábride and Alpujarride Complexes constitute the basement of the basin. These two complexes occur within the Internal zones of the Betic Cordillera and consist of metamorphic rocks, mainly micaschists, phyllites, quartzites, amphibolites and marbles. They comprise the Sierra de Filabres to the north and the Sierra Alhamilla/Sierra Cabrera to the south.

The fill of the basin is made up of a series of sedimentary units ranging in age from the Middle Miocene to the Quaternary. Its

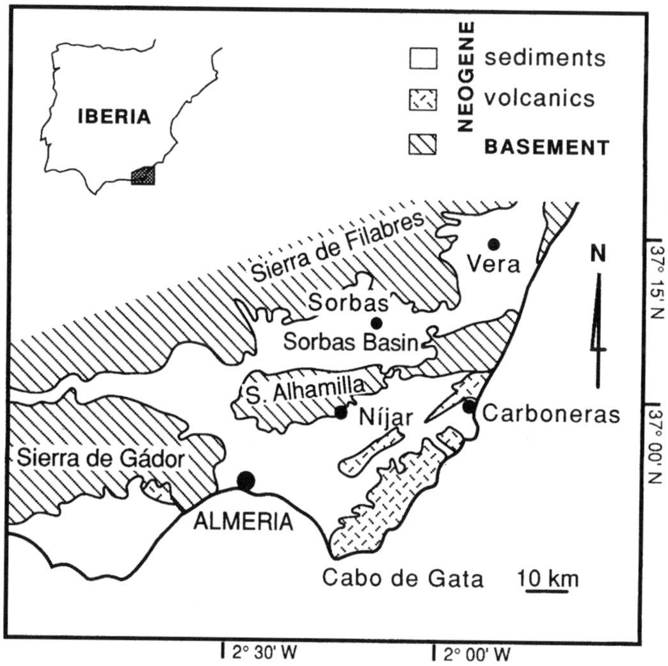

Fig. 1. Geographic and geologic setting of the Sorbas basin.

lower part consists of red, continental conglomerates, possibly Serravallian in age, which locally exceed 100 m in thickness. These conglomerates are unconformably overlain in the northern margin of the basin by up to 50 m of Upper Tortonian fan-delta siliciclastics and platform carbonates with local coral reefs. In the southern part, close to the Sierra Alhamilla, Upper Tortonian rocks consist of up to 700 m of debris-flow conglomerates, turbidites, and plankton-rich silty marls. These sediments have been studied in detail in the Tabernas area (western part of the basin) by Kleverlaan (1987, 1989), who interpreted them as submarine-fans and related channel deposits derived from the Sierra de Filabres to the north. The channels extended across platform carbonates of the northern margin, southward to the mouths of submarine canyons to the area of the present-day Sierra de Alhamilla, which, at that time, was submerged by several hundred metres of sea water. The asymmetry

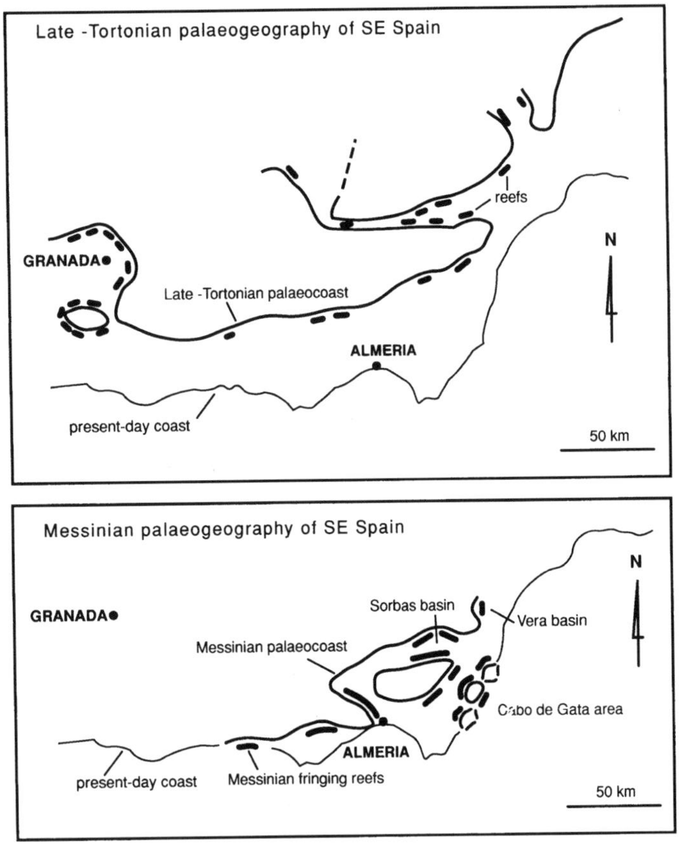

Fig. 2. Late-Tortonian and Messinian palaeogeography of SE Spain delineated by reef distribution (from Braga & Martín, 1992). Note that the Sorbas basin only existed as such since the Early Messinian, while in the Late Tortonian it was completely open to the south. A major change in palaeogeography, concomitant with the uplifting of Sierra Alhamilla, occurred during the Tortonian–Messinian transition within the Almería area.

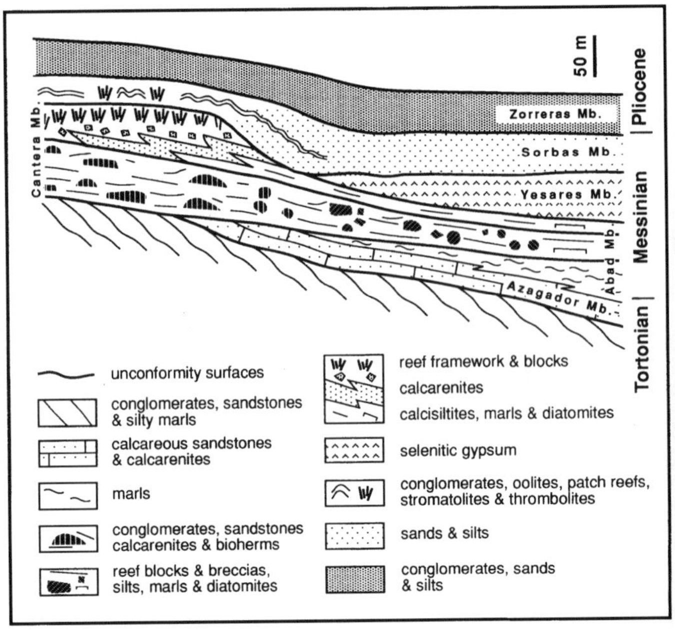

Fig. 3. Detailed Messinian stratigraphy of the Sorbas basin (from Martín & Braga, 1994; modified from Braga & Martín, 1992). Lithostratigraphic names are those of Ruegg (1964).

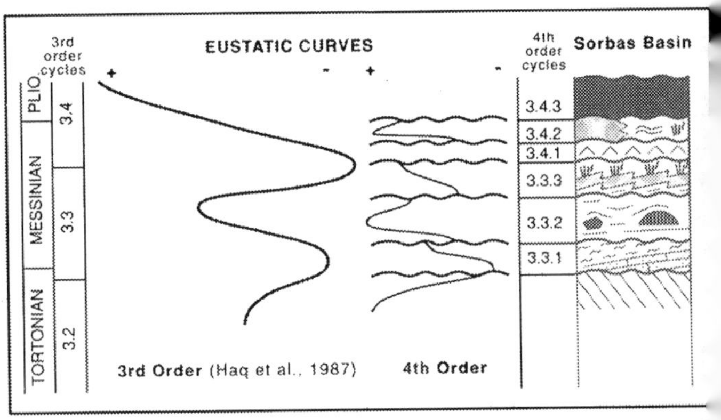

Fig. 4. Third- and fourth-order sedimentary cycles represented in the Messi nian of the Sorbas basin and the inferred eustatic curve. In some of the fourth order units, such as the temperate–carbonate unit (3.3.1) and the fringing-ree unit (3.3.3), tectonic pulses aborted their internal evolution and only their Low Stand and part of their Transgressive Systems Tracts are present. In othe units, however, such as the bioherm unit (3.3.2) and the Sorbas unit (3.4.2) whole fourth-order sedimentary cycles, with well-developed Low-Stand Transgressive and High-Stand Systems Tracts, occur. The internal evolutio in the case of the gypsum unit (3.4.1) is unknown.

of the basin deposits and their petrology indicate that the basin was open to the south (Fig. 2). Only the northern margin, which is marked by the Sierra de Filabres, existed during Late Tortonian as an emergent relief feature. The basin was partitioned at the end of the Tortonian by tectonic uplift of the Sierra Alhamilla (Weijermars *et al.*, 1985).

The Messinian stratigraphy (Fig. 3) is very similar in both margins of the basin. Several units separated by unconformities are believed to comprise two major sedimentary cycles that can be correlated with the TB 3.3 and TB 3.4 cycles of Haq *et al.* (1987) (Fig. 4).

The lower cycle begins at its base with temperate, platform calcarenites and calcareous sandstones that contain bivalves, bryozoans, and coralline algae – the Azagador Member of Ruegg (1964) – and still belong, at least partly, to the Upper Tortonian (Sierro *et al.*, 1993). These deposits pass upward and laterally into white marls and marly limestones – the lowermost part of the Abad Member of Ruegg (1964) – which contain abundant calcareous nanoplankton and planktonic foraminifera (e.g. *Globorotalia mediterranea* Catalano & Sprovieri) of unquestionable Messinian age (Iaccarino *et al.*, 1975; Serrano, 1979; Sierro *et al.*, 1993). Messinian reefs of th Cantera Member lie unconformably above. The reefs and relate deposits comprise two units. The lower one consists of platforr carbonates and isolated bioherms. Coral-block olistoliths an breccias occur seaward at the platform margin (Martín & Brag 1990, 1991). The upper unit consists of a coral–stromatolit prograding, fringing reef (Riding *et al.*, 1991). Basinward, bot ınits change to yellowish, silty marls – the tobacco-coloured marls 'orming the upper part of the Abad Member – with intercalated diatomites (Martín & Braga, 1990). Near the basin margins the ontact between the bioherm unit and the overlying fringing reef nit is marked by an unconformity.

Taking into account the possible tectonic origin of the ınconformity at the base of the bioherm unit (see below), the ٰzagador Member could be assigned to the Low-Systems Tract of he lower cycle (i.e. the one representing TB 3.3), with the bioherm ınd fringing reef units representing the Transgressive and High-Stand parts respectively (Fig. 4).

The upper cycle (representing TB 3.4) consists of the Yesares, iorbas, and Zorreras Members (terminology after Ruegg, 1964) Fig. 3). The Yesares Member, restricted to the centre of the basin, onsists of selenite gypsum that locally exceeds 100 m in thickness Dronkert, 1977). The variable Sorbas Member is relatively thin at he basin margins (30–40 m), where it consists of conglomerates, ands, oolitic limestones, small coral patch-reefs and stromatolites the 'Terminal Complex' of Riding *et al.*, 1991). In the basin centre he Sorbas Member is up to 70 m thick and consists of silts and ands. Locally, they represent prograding beach-sands (Roep *et al.*, 979; Dabrio *et al.*, 1985). The Zorreras Member, which is pre-umed to be Pliocene in age, is up to 60 m thick. It consists of red, ontinental, alluvial-fan conglomerates that change toward the entre of the basin to silts with intercalated conglomerates (flood-lain deposits) and ostracod-bearing lacustrine limestones.

In this upper cycle the gypsum deposits are assigned to the Low-tand, the Sorbas to the Transgressive, and the Zorreras Member to he High-Stand (Fig. 4). The intra-Messinian unconformity,which ccurs at the basin margins between the fringing reef and the 'erminal Complex, is perhaps a reflection of the main evaporite vent of the Mediterranean region.

The basin fill is capped by a few metres of Pliocene calcareous andstone containing marine bivalves (Civis *et al.*, 1977), along with n interval of Quaternary fluvial conglomerates, sandstones and aliches up to 20 m thick.

The tectonics are clearly superimposed on the eustatic sea-level hanges controlling these two major Messinian sedimentary cycles. . transpressive regime prevailed in the Sorbas basin (Ott d'Estevou : Montenat, 1990) in a general compressive tectonic setting for the outheastern part of the Betic Cordillera during the Messinian. A 1140E major axis of compression produced folds with an axis of 160E and faulting along N100E and N20E fractures (Ott d'Este-ou & Montenat, 1990) (Fig. 5). In this context tectonic movements nd uplift have modified the stratigraphy of the Sorbas basin at arious levels.

The unconformity at the top of the Upper Tortonian siliciclastics the southern margin of the basin reflects the combined effects of ıe sea-level fall at the end of the last Tortonian cycle (TB 3.2 of Haq *al.*, 1987) and the tectonic uplifting of the Sierra Alhamilla. As a onsequence of this, Uppermost Tortonian–Messinian shallow-ater carbonates lie directly on top of Upper Tortonian deep-ater, submarine-fan deposits around the Sierra Alhamilla.

The unconformity between the two major Messinian cycles may be correlated with the global sea-level fall separating the Messinian third-order cycles (Vail *et al.*, 1977). This global sea-level fall has been considered to be the main cause of the 'Salinity Crisis' and dessication of the Mediterranean, which would have been completely isolated from the Atlantic at low sea levels (see Esteban, 1995). Nevertheless, detailed analysis of the sedimentary evolution of the Sorbas basin indicates that it was tectonics that triggered the isolation of the Mediterranean (as suggested by Weijermars, 1988) and started the complex phenomena involved in the 'Salinity Crisis'. The fringing reef is the sedimentary unit just below the unconformity separating both major cycles. In the best outcrop of the reef along La Rambla de los Castaños, in the northern margin of the basin, reef geometries and facies distribution show three orders of cyclicity (Braga & Martín, 1992; Bosence *et al.*, 1992). As can be seen by following the facies distribution and geometries along the section (Fig. 6), the succession consists of one complete lowest (largest) order cycle and the beginning of a second one which is suddenly interrupted during the rising sea-level phase. This sudden

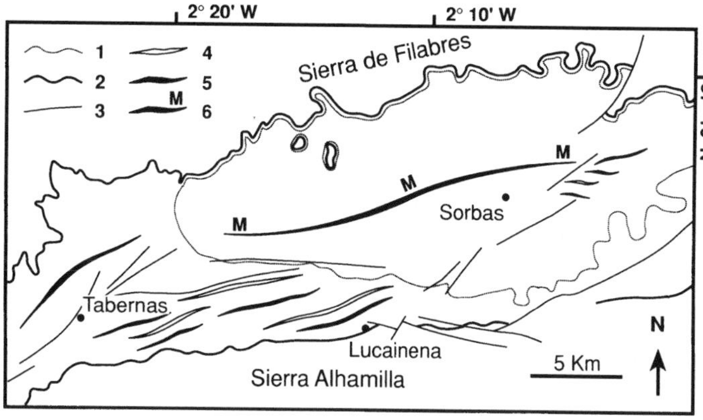

Fig. 5. Simplified structural sketch-map of the Sorbas basin (after Ott d'Estevou & Montenat, 1990). 1: base of the Messinian; 2: limit of the basement; 3: major faults; 4 and 5: anticlines and synclines affecting Tortonian beds; 6: axis of the Messinian basin.

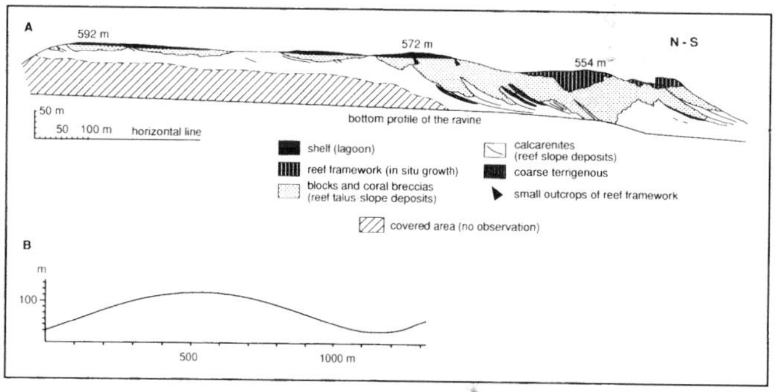

Fig. 6. A: Fringing-reef section along the Rambla de los Castaños, near Cariatiz at the northern margin of the Sorbas basin, showing reef progradation and internal cyclicity (after Braga & Martín, 1992). B: Inferred sea-level curve. The lowest order of cyclicity comprises a complete cycle and part of the following one. Reef evolution was clearly aborted in the aggradation, rising sea-level phase of the latter.

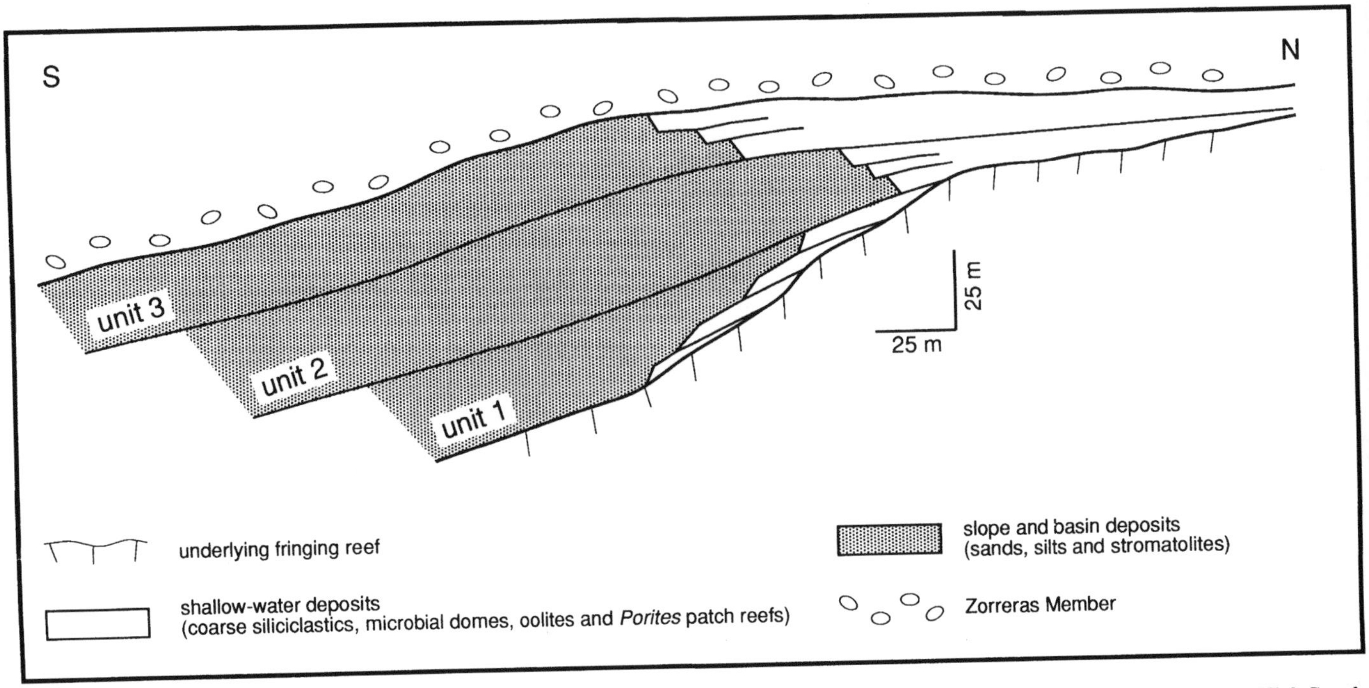

Fig. 7. Section of the basinal Sorbas Member and the equivalent marginal conglomeratic Terminal Complex near Gochar on the northern margin of the Sorbas basin showing well-defined Low-Stand, Transgressive and High-Stand Systems Tracts (units 1, 2 and 3 respectively) (after Martín *et al.*, 1993).

interruption in an aggradation phase of reef growth can only be explained by tectonic causes. The reef was afterwards eroded and this erosion surface is probably the temporal equivalent in the marginal Sorbas basin of evaporite deposition in the centre of the Mediterranean. In our view, this phase of tectonic uplift of relief features was also responsible for the restriction and final closing of the Rifian corridors in northern Morocco, thus disconnecting the Mediterranean from the Atlantic. The Betic corridors had very probably already been closed during the Late Tortonian (Esteban *et al.*, 1993). Later sea-level fall increased and prolonged the isolation of the Mediterranean. This tectonic event, which affected the westernmost end of the Mediterranean, is also reflected in the Sorbas area by tilting and faulting of pre-evaporite Messinian deposits.

Some carbonate sedimentary units inside these two major Messinian cycles represent an entire higher-order cycle (presumably 4th order) with relatively well-developed Low-Stand, Transgressive, and High-Stand Systems Tracts. This is well exemplified by the bioherm unit and the Sorbas Member (Fig. 7) (Martín *et al.*, 1993). In the first carbonate sedimentary unit, however, this high-order cyclicity is interrupted by another pulse of tectonic uplifting of the Sierra Alhamilla. An erosional unconformity separates the folded and faulted temperate carbonates and marls from the bioherm unit lying on top. In this interrupted cycle only its own Low-Stand Systems Tract (platform temperate carbonates – Azagador member) and part of the Transgressive Systems Tract (white marls) were deposited.

In summary, unconformities and interruptions of sedimentary cycles are reflective of tectonic activity during the Messinian in the Sorbas basin. Tectonic events affected the Messinian stratigraphy of the basin at least at two different hierarchy levels. Tectonic movements were involved in the formation of the unconformities at the base of both major, third-order cycles as well as at the boundaries and interruptions of some higher-order ones.

Related to these phases of tectonic uplifting there was a major change in palaeogeography within the Almería area, clearly exemplified by Late Tortonian and Early Messinian reef distribution (Fig. 2). However, the major effect of this tectonic activity, which affected not only the Sorbas and nearby basins in southeastern Spain but the entire western Mediterranean area, was the progressive restriction and isolation of the Mediterranean Sea which finally led to its dessication and the deposition of huge masses of evaporites in its deepest parts.

References

Bosence, D.W., Braga, J.C., Martín, J.M. and Hardy, S. (1992). Computer-modelling depositional sequences in Late Miocene platforms Sorbas Basin, Spain (Abstract). *SEPM/IAS Research Conference on Carbonate Stratigraphic Sequences: Sequence Boundaries and Associated Facies*, August 30–September 3, La Seu, Spain.

Braga, J.C. and Martín, J.M. (1992). Messinian carbonates of the Sorbas basin: sequence stratigraphy, cyclicity and facies. In *Late Miocene carbonate sequences of Southern Spain: a guidebook for the Las Negras and Sorbas area*, in conjunction with the SEPM/IAS Research Conference on Carbonate Stratigraphic Sequences: Sequence Boundaries and Associated Facies, August 30–September 3, La Seu, Spain, pp. 78–108.

Civis, J., Martinell, J. and de Porta, J. (1977). Precisiones sobre la edad del miembro Zorreras (Sorbas, Almería). In *Messinian Seminar no 3*, Abstracts, Málaga, 1977.

Dabrio, C.J., Martín, J.M. and Megías, A.G. (1985). The tectosedimentary evolution of Mio-Pliocene reefs in the Province of Almería. In M.D. Milá and J. Rosell (eds.), *6th European Regional Meeting of Sedimentologists, Excursion Guidebook*, Llerida, Spain, pp. 269–305.

Dronkert, H. (1977). The evaporites of the Sorbas basin. *Rev. Instit. Inv. Geol. Dip. Provincial Univ. Barcelona*, **32**, 55–76.

Esteban, M. (1995). An overview of Miocene reefs from the Mediterranean areas: General trends and facies models. In C. Jordan, M. Colgan and M. Esteban (eds.), *Miocene reefs: a global comparison*. Springer-Verlag, Heidelberg (in press).

Esteban, M., Braga, J.C., Martín, J.M. and Santisteban, C. (1993). An overview of Miocene reefs from the Mediterranean areas: Miocene reefs of the Western Mediterranean. In C. Jordan, M. Colgan and M. Esteban (eds.): *Miocene reefs: a global comparison*. Springer-Verlag, Heidelberg (in press).

Haq, B.U., Hardenbol, J. and Vail, P.R. (1987). Chronology of fluctuating sea levels since the Triassic. *Science*, **235**, 1156–1167.

Iaccarino, S., Morlotti, E., Papani, G., Pelosio, G. and Raffi, S. (1975). Litostratigrafia e biostratigrafia di alcune serie neogeniche della provincia di Almería (Andalusia orientale-Spagna). *Ateneo Parmense, Acta Nat.*, **11**, 237–313.

Kleverlaan, K. (1987). Gordo megabed: a possible seismite in a Tortonian submarine fan, Tabernas basin, Province Almeria, southeast Spain. *Sediment. Geol.*, **51**, 165–180.

Kleverlaan, K. (1989). Three distinctive feeder-lobe systems within one time slice of the Tortonian Tabernas fan, SE Spain. *Sedimentology*, **36**, 25–45.

Martín, J.M. and Braga, J.C. (1990). Arrecifes messinienses de Almería. Tipologías de crecimiento, posición estratigráfica y relación con las evaporitas. *Geogaceta*, **7**, 66–68.

Martín, J.M. and Braga, J.C. (1991). Lower Messinian patch reefs and associated sediments, southeastern Spain. In *Dolomieu Conference on Carbonate Platforms and Dolomitization. Val Gardena, The Dolomites (Italy). September 1991*. Abstracts book, p. 161.

Martín, J.M. and Braga, J.C. (1994). Messinian events in the Sorbas basin of Southeastern Spain and their implications in the recent history of the Mediterranean. *Sediment. Geol.* **90**: 257–268.

Martín, J.M., Braga, J.C. and Riding, R. (1993). Siliciclastic stromatolites and thrombolites, late Miocene, S.E. Spain: *J. Sediment. Petrol.*, **63**, 131–139.

Ott d'Estevou, P. and Montenat, C. (1990). Le bassin de Sorbas–Tabernas. *Doc. Trav. IGAL.*, **12–13**, 101–128.

Riding, R., Martín, J.M. and Braga, J.C. (1991). Coral-stromatolite reef framework, Upper Miocene, Almería, Spain. *Sedimentology*, **38**, 799–818.

Roep, Th.B., Beets, D.J., Dronkert, H. and Pagnier, H. (1979). A prograding coastal sequence of wave-built structures of Messinian age, Sorbas, Almería, Spain. *Sediment. Geol.*, **22**, 135–163.

Ruegg, G.J.H. (1964). Geologische onderzoe-kingen in het bekken van Sorbas, S Spanje. *Amsterdam Geological Institut, University of Amsterdam*, Holland, 64 pp.

Serrano, F. (1979). Los foraminíferos planctónicos del Mioceno superior de la cuenca de Ronda y su comparación con los de otras áreas de las Cordilleras Béticas. Thesis, Universidad de Málaga, 272 pp.

Sierro, F.J., Flores, J.A., Civis, J., González-Delgado, J.A. and Francés, G. (1993). Late Miocene globorotaliid event-stratigraphy and biogeography in the NE–Atlantic and Mediterranean. *Mar. Micropaleontol.*, **21**, 143–168.

Vail, R.R., Mitchum, Jr, R.M., Todd, R.G., Wildmier, J.M., Thompson, S. III., Sangree, J.B., Bubb, J.N. and Hatfield, W.G. (1977). Seismic stratigraphy and global changes of sea level. In C.E. Payton (ed.), *Seismic stratigraphy: applications to Hydrocarbon exploration*. Am. Assoc. Petrol. Geol. Mem. **26**, 49–212.

Weijermars, R. (1988). Neogene tectonics in the Western Mediterranean may have caused the Messinian Salinity Crisis and an associated glacial event. *Tectonophysics*, **148**, 211–219.

Weijermars, R., Roep, Th.B., Van Den Eeckhout, B., Postma, G. and Kleverlaan, K. (1985). Uplift history of a Betic fold nappe inferred from Neogene–Quaternary sedimentation and tectonics (in the Sierra Alhamilla and Almería, Sorbas and Tabernas Basins of the Betic Cordilleras, SE Spain). *Geol. Mijn.*, **64**, 379–411.

S10 Basinwide interpretation of seismic data in the Alborán Sea

C. DOCHERTY AND E. BANDA

Abstract

The Alborán Sea basin contains one of the largest Neogene sedimentary accumulations in the western Mediterranean, distributed in several sub-basins separated by structural highs. Depth conversion and isopach maps of interpreted reflection seismic sections attest to the complexity of the basement architecture and basin infill.

Introduction

The Alborán Sea basin is located at the westernmost extreme of the Mediterranean Sea, closed to the Atlantic Ocean in the west at the Gibraltar Strait, and open to the east, where it passes into the Algerian Sea basin (Fig. 1). Its sedimentary sequence, of marine origin, is of Neogene age, the oldest sediments penetrated at well sites on the basin's northern margin being of upper Aquitanian/lower Burdigalian age (Jurado and Comas, 1992). The sedimentary sequence is interrupted by several unconformities, of which the Messinian is the most prominent. The basin is underlain by thinned continental crust (Hatzfeld, 1976; Banda and Ansorge, 1980) which is the continuation of the metamorphosed Alboran domain units (Balanyá and García-Dueñas, 1987) that crop out on the mainlands of southern Spain and NW Morocco as the internal complexes of the Betic and Rif Cordilleras respectively. The thinned continental crust is host to volcanic centres that occur in basins and ridges (Galdeano *et al.*, 1974), but are present also in wells on the northern basin margin (Jurado and Comas, 1992). The basin is complex structurally, being segmented into sub-basins by prominent ridges that cross the basin. The deepest sub-basins lie to the west, where basement depths in excess of 7 km are reported (IGME, 1990). The western sub-basins are associated with diapirism deep in their sedimentary succession.

Major extension has been experienced at the site of the Alborán Sea basin, occurring while the Mediterranean was under a regionally compressive regime. This apparent paradox has led to many hypotheses on the basin's formation. Multichannel seismic reflection profiling images basin structure and its sedimentary succession, and as such provides a powerful tool in the study of basin evolution, which in turn contrains any model of formation.

More than 3400 km of seismic reflection profiles were collected for the Alboran Sea region. Of this total, some 2400 km is of commercial, publically released material. Non-commercial data consist of a seismic survey carried out by Lamont–Doherty Geological Observatory of the University of Columbia, New York, USA, in the Research Vessel *Robert Conrad* in November 1988 (Watts *et al.*, 1993). Well tie lines were used to extrapolate well horizons onto adjoining profiles, using downhole logs. In this chapter we outline some results of a detailed study of the seismic data (Docherty, 1993) and present maps produced from depth-converted seismic data relevant to the evolution of the basin.

Seismic interpretation

To provide a basinwide analysis, the reflectors should be interpretable throughout the basin. Thus, the selection of reflectors was limited by the basin architecture as well as by reflector continuity. In the proximity of the northern margin, determination of a number of reflectors was relatively straightforward, with a well-developed sedimentary series, good seismic quality, and distinguishable seismic reflections that correspond to angular unconformities. Jurado and Comas (1992) distinguish six units based on seismic facies analysis for the north Alborán Sea. Approaching the centre of the basin, however, the quality of seismic imaging diminishes as the basement topography becomes more irregular, and diapirism in the western basins strongly attenuates the seismic signal and introduces diffractions at the domed top surfaces of the diapirs. In addition to the seafloor and basement reflector (H4) only the most widespread reflector (reflector H1), which also corresponds to a reflector that can be traced regardless of topography due to its distinct character, was selected. It represents the top Miocene (Messinian) unconformity (reflection R1 of Jurado and Comas 1992). Where present, a third layer has been mapped – representing a diapiric layer (unit VI of Jurado and Comas, 1992) – but it was only interpreted for the purpose of depth conversion. The top of this unit is labelled H3. Locally, other reflectors are shown on the interpretations – H2 (mid-Serravallian), P1 (base upper Pliocene and P2 (top Pliocene).

The reflection that corresponds to the top Miocene (H1) represents a strong erosional unconformity and is an irregular surface seen in both dip and strike profiles that very often displays an onlap

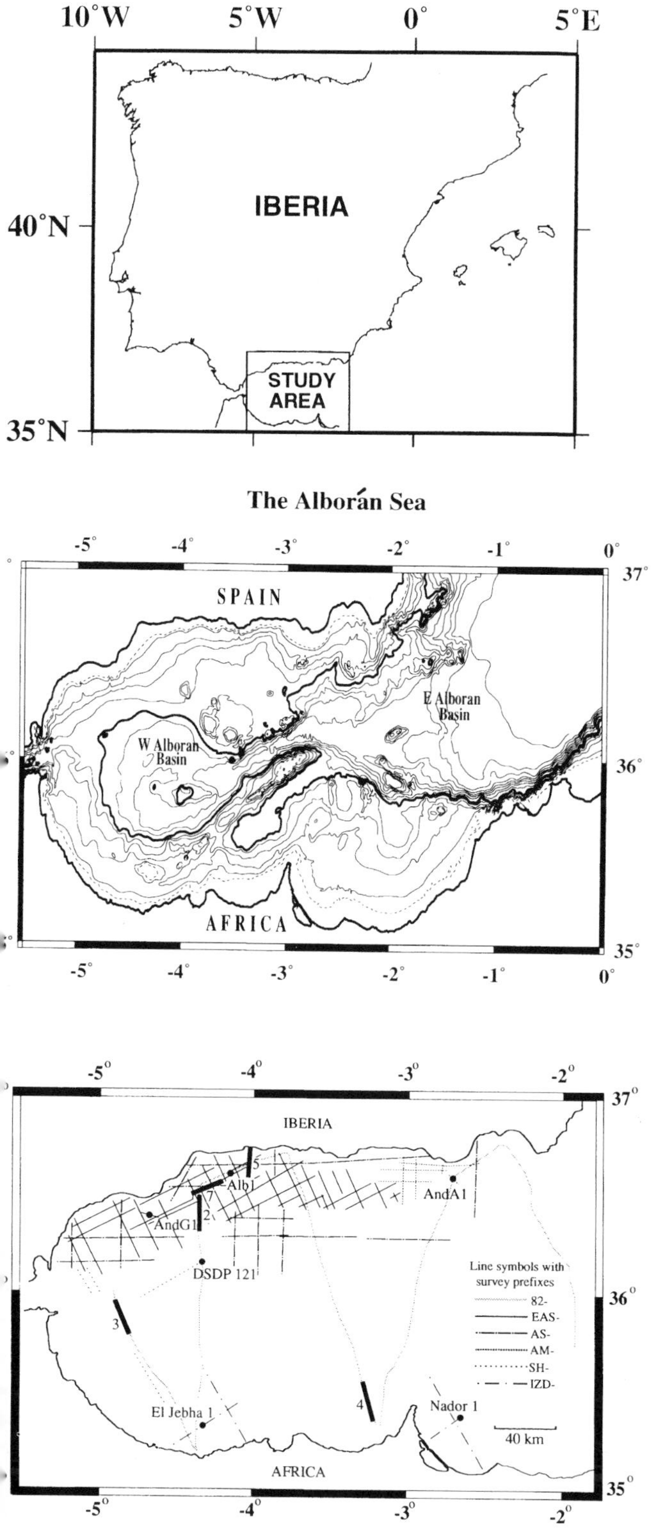

g. 1. (a) Bathymetry map of the Alborán Sea basin. Contour interval 200 m, cept 100 m contour (dashed line). 1000 m is marked as thicker contour. The E–SW-trending Alborán ridge forms the partition between the east and west borán basins. (b) Seismic data base of the Alborán Sea used in this study. ell abbreviations: AndA1 (Andalucía A1); AndG1 (Andalucía G1) Alb1 lborán A1). Other well locations marked. Locations of seismic data shown Figures 2–5 and 7 are given as broader lines with corresponding figure mbers.

surface with overlying reflections terminating against it. Reflection H1 corresponds to an erosional truncation in the northern margin area (Fig. 2). At the Moroccan margin, the Miocene sequence appears paraconformable with the lower Pliocene. This difference between margins is probably due to segmentation of the basin by basement blocks, allowing the sedimentary sequences to form independently within the individual sub-basins. The erosional character of the top Messinian is dramatically demonstrated in Fig. 3.

Away from coastal areas, and in areas unaffected by tectonism, the post-Miocene sequence is typified by a monotonous series of flat-lying, parallel to sub-parallel reflectors. They are characterised by their good continuity and varying amplitudes. Close to the basin margin, this facies may be represented by a more discontinuous series of reflections. In strike section, this discontinuity is clearly observed.

Figs. 4 (Moroccan margin) and 5 (Spanish margin) show examples of the Pliocene to Recent sequence. These figures show the coastal area infill to be characterised by similar reflection patterns: the top Pliocene (P2), with dipping, sub-parallel reflectors of Pliocene age overlain by oblique clinoforms that form a prograding wedge, passing basinward of the shelf break to a divergent reflection configuration. This type of coastal reflection pattern has earlier been recognised on the Moroccan margin by Gensous *et al.* (1986) where reflector P2 forms the top of a set of flat parallel reflectors that dip 5–7° northward. Fig. 6 is a sketch of the generalised growth forms as interpreted here. The observed configuration is interpreted as the response to tectonic tilting of parallel horizontal layers formed by aggradation (Fig. 6a) that occurred at the Pliocene–Quaternary boundary, resulting in the deepening of the margins basinward and the uplift of the margins onshore. This change produced the P2 unconformity recognised throughout the basin (Alonso and Maldonado, 1992; Ercilla, 1992). Upon tilting, the sedimentary sequence would image tectonism in the form of reflections offshore that diverge basinward, whilst uplifting the near onshore area would bring about the migration of the coastlines basinward (Fig. 6b). Finally, the present-day configuration reflects the new coastline with its shelf area characterised by prograding clinoforms, sourced by the erosion of the uplifted onshore area (Fig. 6c). This evolution has been suggested for the western Iberian margin where a similar configuration has been encountered (Mougenot *et al.*, 1983). Recognition of this type of feature implies that the basin as a whole has been subsiding throughout the Quaternary, and that the basin has reduced in width over the last few million years.

Reflection H4 is identified as the basement reflector separating an ordered, parallel series of reflections representing sedimentary strata, and an underlying, chaotic facies caused by crystalline basement rocks. It is not always readily identifiable, even though it must have a high reflection coefficient due to the velocity contrast between crystalline basement rocks and the overlying sediments. This occurs in some coastal areas and in basins where it is covered by diapiric clay, which causes dispersion of seismic energy, weakening the basement signal response below. Reflection H4 was therefore sometimes difficult to distinguish in these diapiric areas.

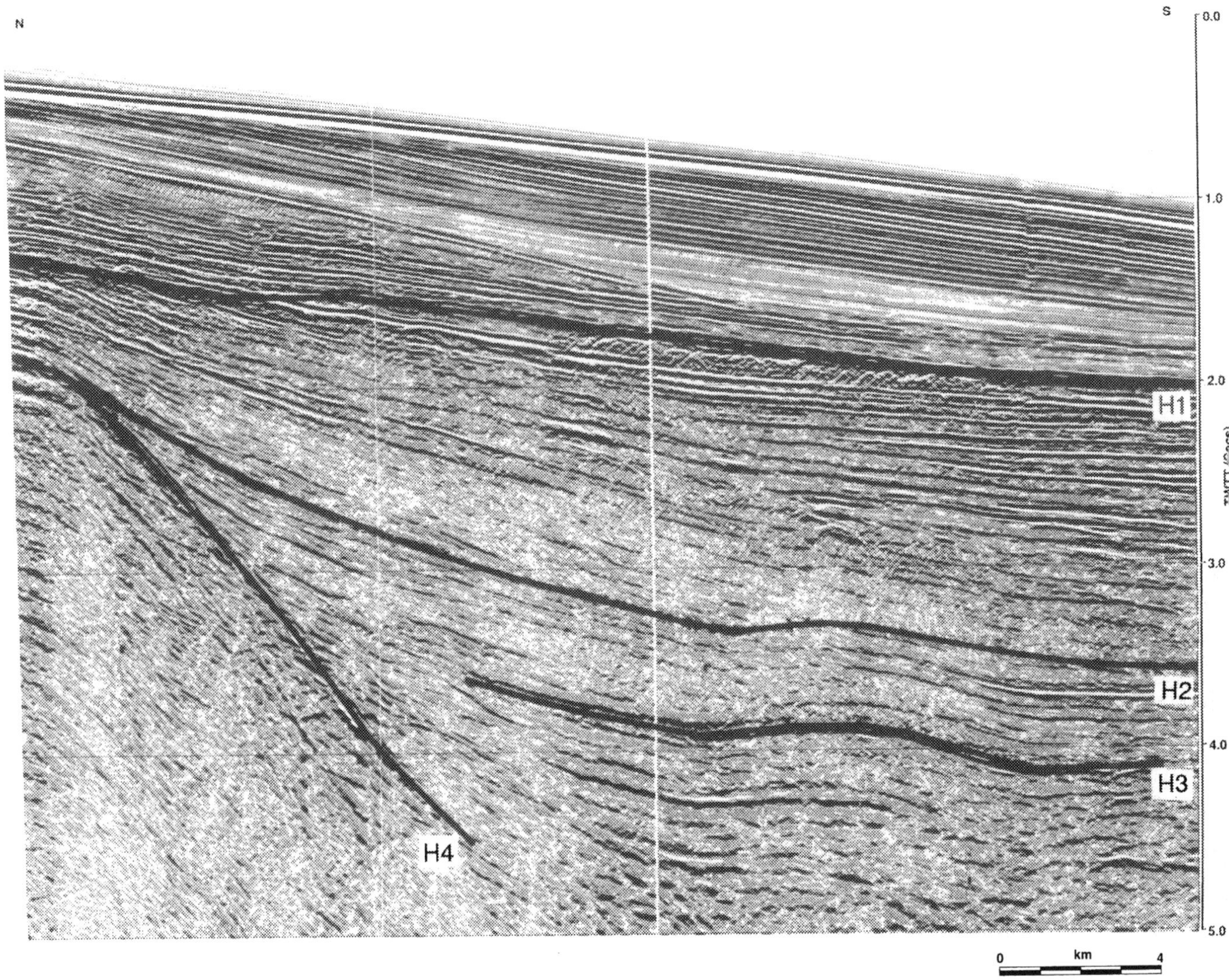

Fig. 2. Profile in the central western area (see Fig. 1 for location) of the Alborán Sea basin shows several features in its northern segment. Reflector H1 is seen a an erosional unconformity, characterised by small diffractions from its irregular surface. Reflector H2 (Serravallian) is also erosional. Further unconformitie can be interpreted between H1 and H2. Low in its sedimentary sequence (below 4 seconds TWTT), doming is observed (H3), caused by diapiric movement of clay

Reflection H4 on strike profiles is interpreted as a block-faulted basement surface close to the northern coast, as for example in Fig. 7. In dip profiles, reflection H4 is seen to dip from the respective coasts of Iberia (Fig. 2) and Morocco (Fig. 4) towards the basin centre, representing the basement of deep, relatively flat-lying, sub-basins compartmented by the basement forming fault-bounded blocks. In general, the basement dips much more sharply basinward at the Moroccan margin.

In contrast to the post-Messinian sedimentary sequence, the Miocene sequence is dominated by a series of reflectors that onlap the metamorphic Betic basement (H4) near the margins and diverge toward the depocentres, and contains intra-Miocene unconformities. The reflection pattern of the Miocene unit in the western area of the basin is chiefly governed by the presence of diapirs. The northern basin margin (Fig. 2) has a series of parallel, undulating reflections deep in the reflection sequence. The marked reflector above H3 in Fig. 2 corresponds to a reflector lying above what is interpreted as a relatively thin overpressured clay unit. This 'to diapir' reflector, H3, recognised as being the change in seismic facie between diapiric clay and surrounding argillaceous sediments, an equivalent to reflector R5 in Jurado and Comas (1992), is furthe shown in Fig. 5. Between H3 and H1, several unconformities ar distinguishable, one of which (H2) was mapped in the norther margin of the basin and corresponds to reflector R4 of Jurado an Comas (1992), and is given a mid-Serravallian age. The Miocen unconformities mark events in the basin history that correspond t rifting stages in the Alborán Sea.

In the eastern basin sector, the Miocene sequence commonl displays a disorganised seismic facies, especially in the centre of th basin, which is not evidenced in the western sector. This is possib due to the increase in volcanic material present within the sedimer tary sequence, as reflected in the Andalucía A1 well (Jurado an Comas, 1992).

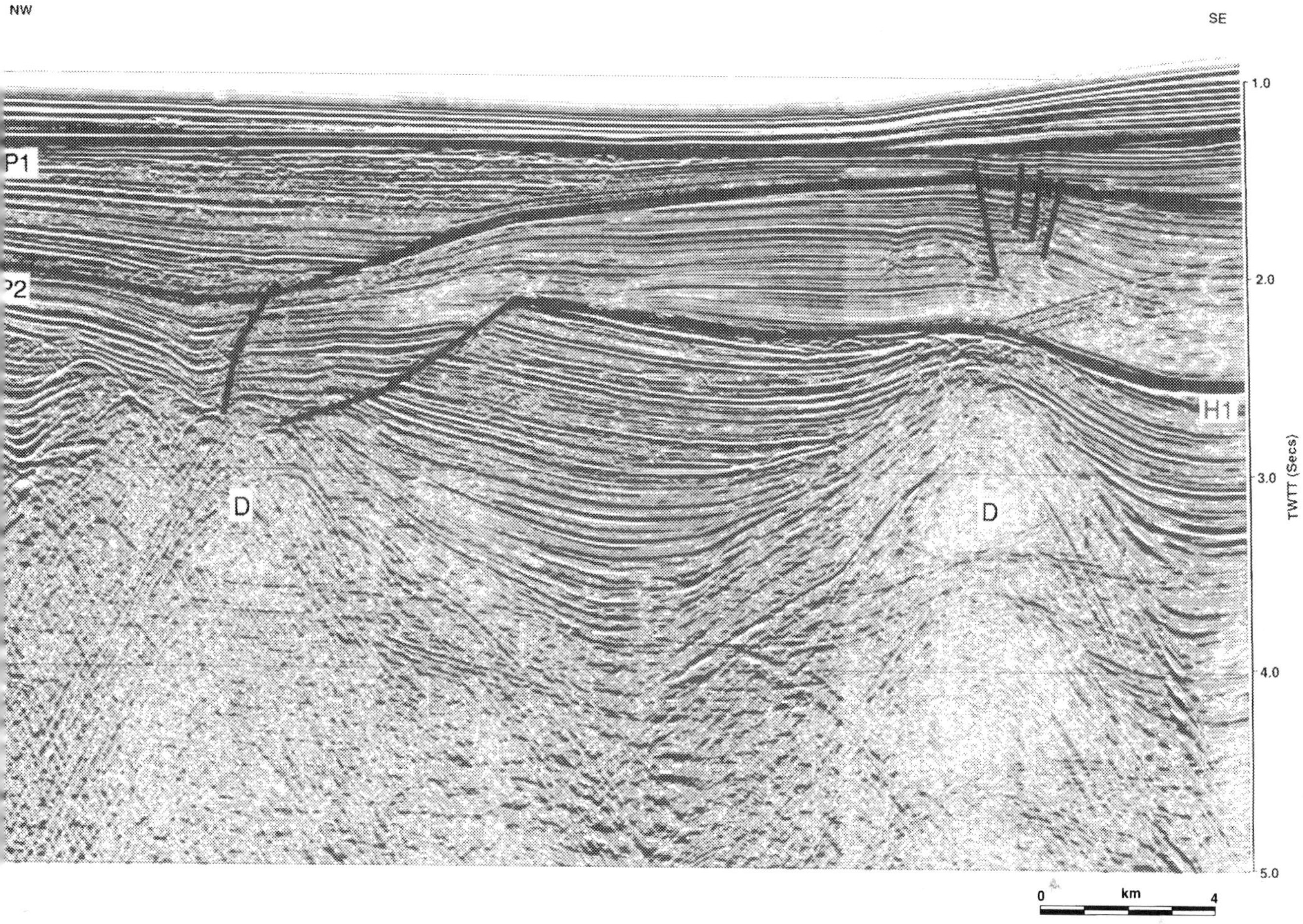

ig. 3. Profile in the western Alboran Sea basin (see Fig. 1 for location). Feature left of centre between 2.0 and 3.0 seconds TWTT probably represents a alaeogorge. Diapirs marked with a 'D'. Pliocene unconformities (P1 and P2) along the margin are apparent above.

Depth conversion

After interpretation of the selected seismic profiles, the icked reflectors of H1, H3 and H4 were digitised, using a series of rograms to arrive at a suitable format for further analysis. Average elocities were taken for the four layers: (a) sea-level–seafloor: 1500 /s; (b) seafloor–H1: 2100 m/s; (c) H1–H3: 2800 m/s; (d) H3–H4: 000 m/s. Velocities of seafloor–H3 horizons are rounded averages ased on the sonic well logs of Alborán A1, Andalucía A1, and ndalucía G1 (Fig. 1). The value used for overpressured clay (2000 /s) is below the range advocated by Musgrave and Hicks (1968) of 150–2800 m/s, which is a range that would overlap with the normal ressure sediments tested at the well sites, and is therefore not presentative.

The Dix equation was used to find the interval velocities, which ere used to depth convert the time sections. The low velocities of verpressured clay cause a velocity inversion in the sedimentary quence, so RMS velocities were used.

Isopach maps of the Miocene and Plio-Quaternary sequences are resented in Fig. 8, based on depth conversions of selected commer-al and academic surveys (shown in Fig. 1). The interpretation is computer aided, using the method devised by Slootweg (1978), whereby a digital filter is applied to the data. Owing to the spacing of the seismic profiles large data gaps exist, particularly for the southern margin. A filter with a cut-off wavelength of 40 km was considered optimum, although short wavelength detail is lost. In practice, this implies that the depth of basins and topography of highlands are underestimated in the contour maps.

Discussion

Due to incompleteness of the data set, there are several apparent incongruities in the isopach maps, namely in front of the Gibraltar Strait and the isopach configurations (for both maps) towards the Moroccan coast between 4° and 5° W. In front of the Gibraltar Strait, the Miocene isopachs should curve inward (basin-ward) as a strong erosional episode scoured deep into the Miocene sequence (see Mulder and Parry, 1977, Figure 3) as a result of the opening of the Gibraltar Strait at the end of the Messinian crisis. Nevertheless, conclusions can be drawn by comparing the two maps, of which the following are of note: 1. over broad areas, depocentres are in similar positions for the two periods, indicating

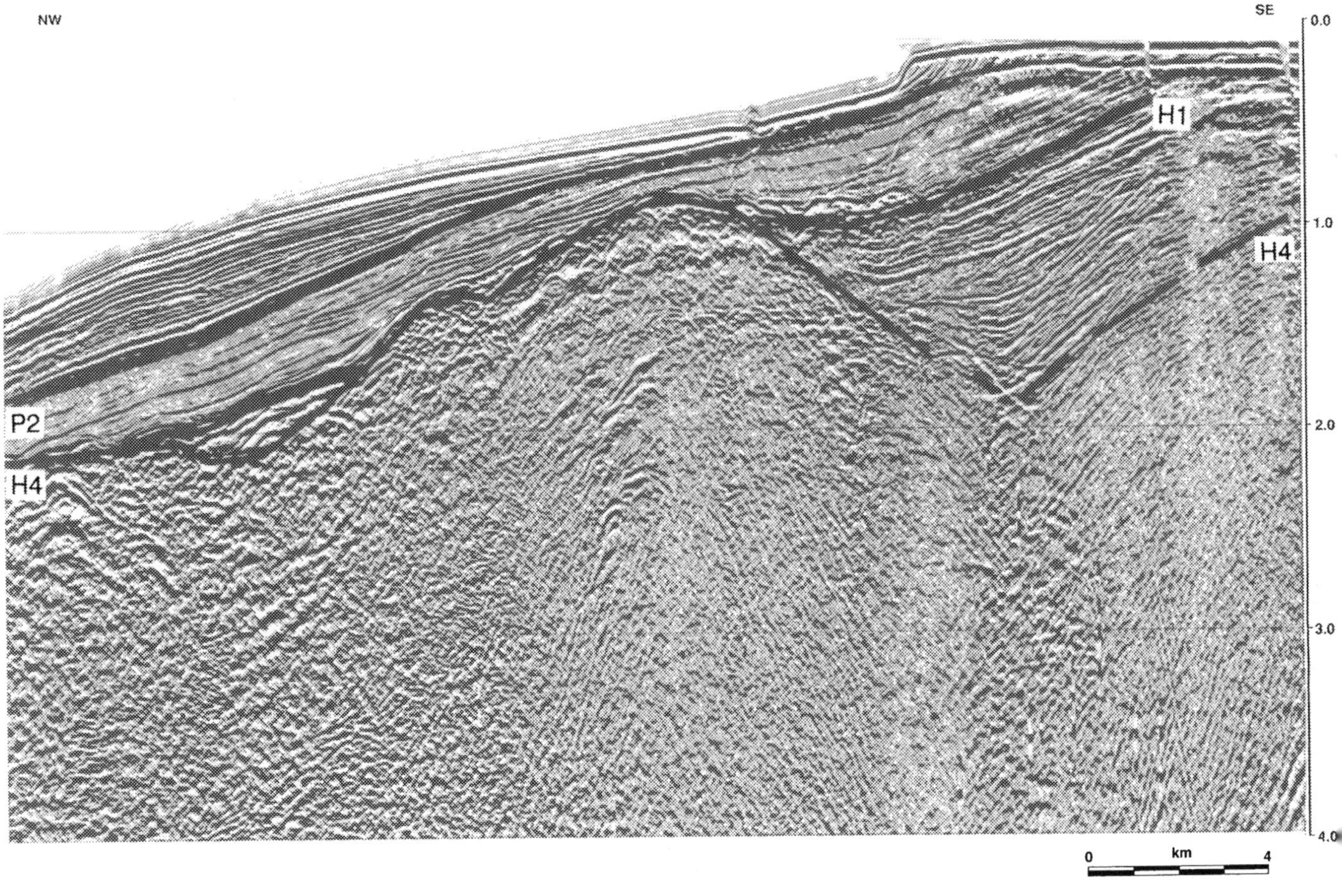

Fig. 4. Profile (see Fig. 1 for location) showing Reflector H1 traceable to the Moroccan margin. Above lie typical reflection configurations of the coastal Plio-Quaternary sequence. P2 indicates top Pliocene.

that they have not migrated significantly with respect to each other; 2. isopachs for the Miocene attain much greater thicknesses than for the Pliocene–Recent period, with the latter having a much smoother sediment distribution across the basin; 3. both Miocene and Plio-Quaternary sequences reach their greatest development in the western sub-basins, thinning toward the east; 4. thinning of the isopachs marks the positions of the structural highs. The isopachs in Fig. 8 reflect the positions of the fault-bounded structural highs. Faults in other areas covered by the seismic profiles, that may involve thickness contrasts across them, are lacking (Watts *et al.*, 1993) and as such do not show up on the isopach maps. Moreover, strike–slip faults, which have played an important role in structuring the basin during the oblique collision of Africa with respect to Europe, often do not have the major vertical displacement that would enable them to be distinguished on the isopach maps.

The complexity of the Miocene basins as opposed to the comparative smoothness of the Plio-Quaternary infill shown in the isopach maps accords with observations on seismic sections, where different reflection configuration patterns are distinguished for the sequences. It is interpreted here that the divergent reflection pattern for the Miocene sequence represents the extensional phase of the basin, and the flat-lying reflectors that correspond to the Pliocene sequence reflect a phase of basin sag, the thermal phase of subsidence related to the cooling of the lithosphere after major rifting ceased. Tectonism in the latest Pliocene has modified marginal areas.

As with the previous seismic studies, the interpretation presented here confirms the complex basement architecture and varied seismic reflection patterns of the Neogene Alboran basin, but it is the first to quantify interpretation of seismic data basinwide through depth conversion. Even with incomplete seismic coverage across the basin, important features are observed, the most salient of which include an overall thinning of the sedimentary units to the east, and the stability of the major depocentres through time, which may reflect the importance of vertical movements during basin evolution.

Acknowledgements

The authors thank A.B. Watts for making available Alboran Sea seismic data, and Peter Slootweg for his contouring program. Publically released seismic data were obtained from ERICO SA, Madrid. Partial financial support was received from DGICYT projects PB87–0363 and CE91–0013.

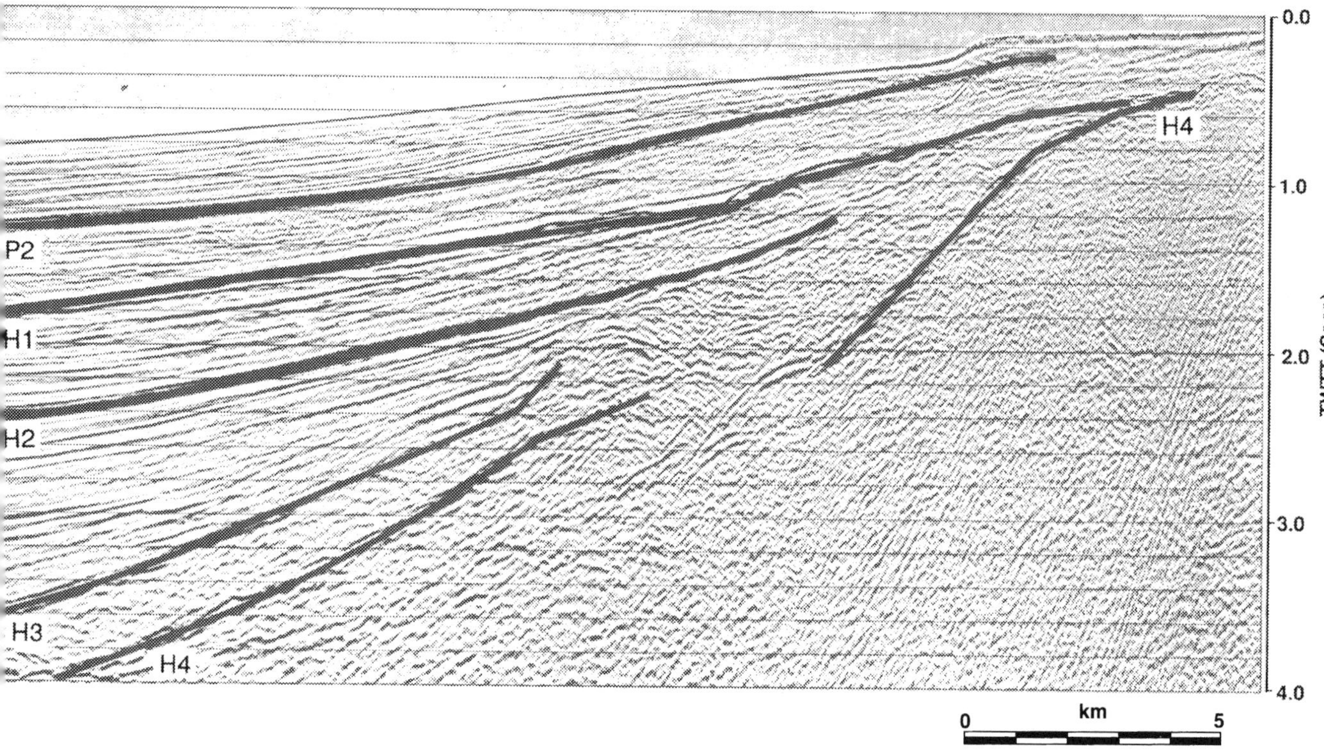

ig. 5. Profile near the centre of the Iberian margin (see Fig. 1 for location) showing Plio-Quartenary reflection configurations (compare with Figure 4). Top 'liocene (P2) along with H1 and H2 indicated.

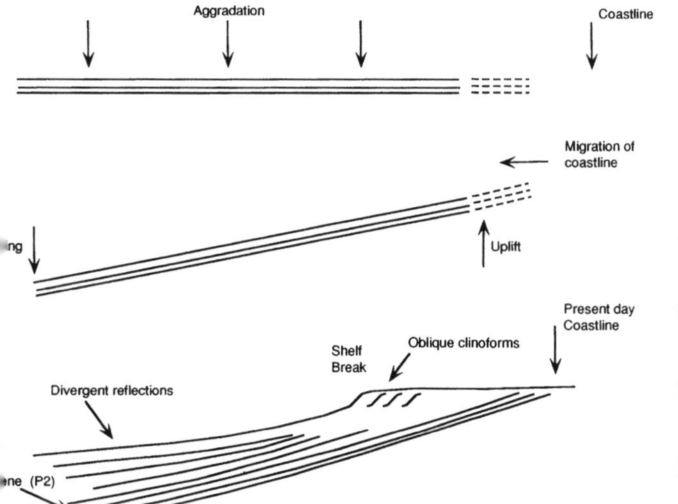

ig. 6. Interpretation of how reflection configurations evolved at the basin's argins: (a) formation of parallel, horizontal reflections by aggradation rocesses away from margins (late Pliocene); (b) after a tectonic event (latest 'liocene), layers become tilted, the nearshore margin uplifted, and the oastline migrated basinward; (c) present-day configuration, with reflections iverging basinward through the Quaternary, and a prograding platform equence.

References

Alonso, B. and Maldonado, A. (1992). Plio-Quaternary margin growth pattern in a complex tectonic setting: Northeastern Alboran Sea. *Geo-marine Letters* **12**: 137–143.

Balanyá, J.C. and García-Dueñas, V. (1987). Les directions structurales dans le Domaine d'Alborán de part et d'autre du Détroit de Gibraltar. *Comptes Rendues Académie de Sciences de Paris* **304**: 929–933.

Banda, E. and Ansorge, J. (1980). Crustal structure under the central and eastern part of the Betic Cordillera. *Geophysical Journal of the Royal Astronomical Society* **63**: 515–532.

Comas, M.C., García-Dueñas, V. and Jurado, M.J. (1992). Neogene tectonic evolution of the Alborán Basin from MCS data. *Geo-Marine Letters* **12**: 144–149.

Docherty, J.I.C. (1993). *Tectonic and subsidence history of the Alboran Sea basin, western Mediterranean*. Published PhD thesis, University of Barcelona, 211 pp.

Ercilla, G. (1992). *Sedimentación en márgenes continentales y cuencas del Mediterráneo occidental durante el Cuaternario (Peninsula Ibérica)*. Published PhD thesis, Universitat Politécnica de Catalunya, 567 pp.

Galdeano, A., Gourtillot, V., Leborgne, E., Le Mouel, J.L. and Rossignol, J.C. (1974). An aeromagnetic survey of the southwest of the western Mediterranean: description and tectonic consequences. *Earth and Planetary Science Letters* **23**: 323–336.

Gensous, B., Tesson, M. and Winnock, E. (1986). La marge meridionale de la Mer d'Alborán. Caractères structuro-sedimentaires et evolution recente. *Marine Geology* **72**: 341–370.

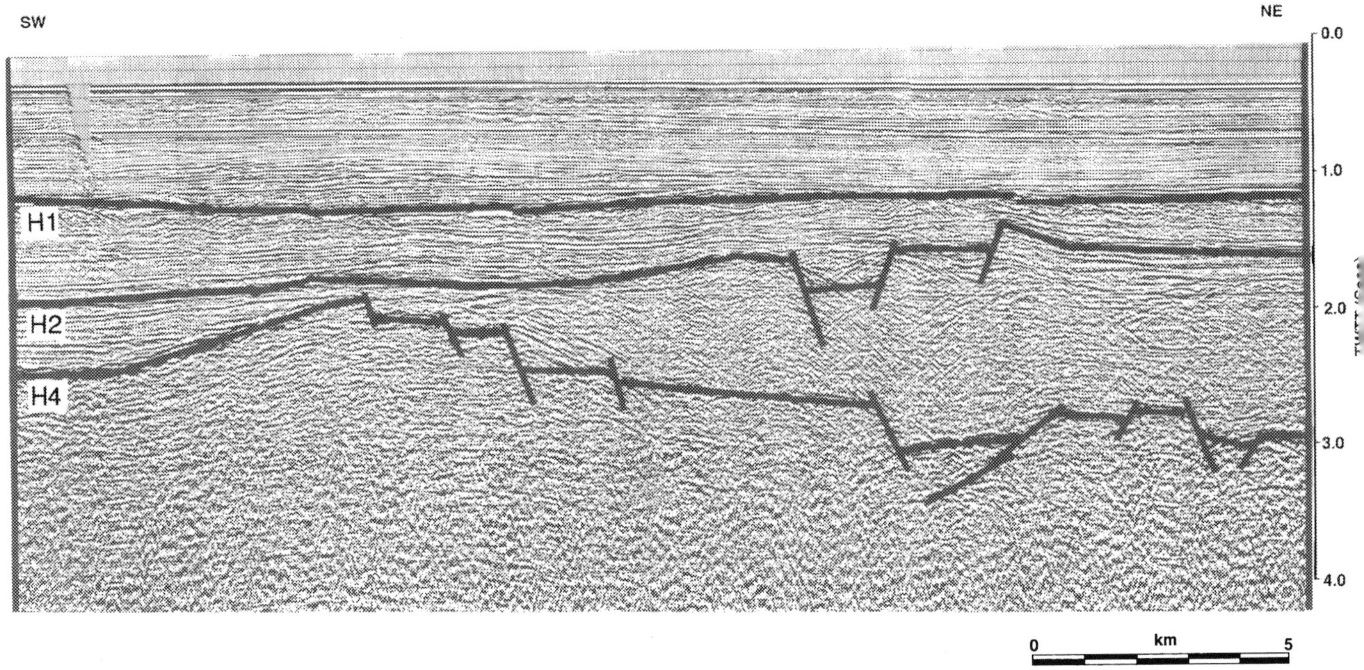

Fig. 7. Strike profile (see Fig. 1 for location) close to the coast showing structured basement and further dip–slip faulting affecting reflectors H1 and H2.

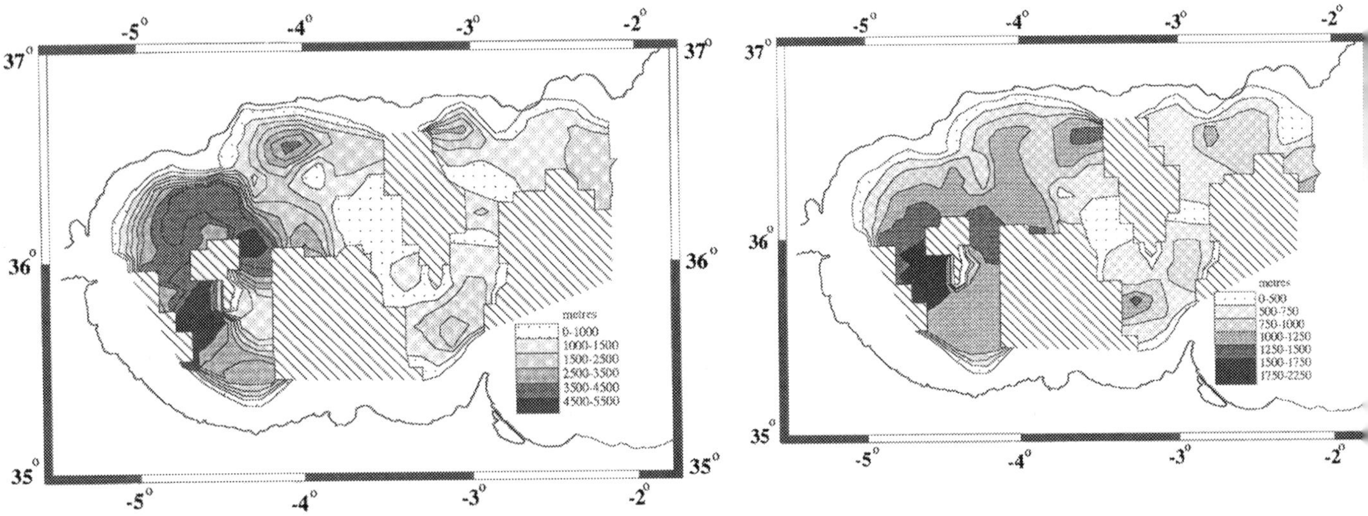

Fig. 8. Isopach maps of (a) Miocene, (b) Plio-Quaternary sequences. Hatched areas represent data gaps.

Hatzfeld, D. (1976). Etude seismologique et gravimetrique de la structure profunde de la mer d'Alborán: mise en évidence d'un manteau anormal. *Comptes Rendue Academie Science Paris* **283**: 1021–1024.

Instituto Tecnológico Geominero de España (TGE) (1990). Documentos sobre la geologia del subsuelo de España. Vol. II: plates 62–65.

Jurado, M.J. and Comas, M.C. (1992). Well log interpretation and seismic character of the Cenozoic sequence in the northern Alboran Sea. *Geo-Marine Letters* **12**: 129–136.

Mougenot, D., Boillot, G. and Rehault, J-P. (1983). Prograding shelfbreak types on passive continental margins: some European examples. *Special Publication of the Society of Economic Paleontologists and Mineralogists*, pp. 61–77.

Mulder, C.J. and Parry, G.R. (1977). Late Tertiary evolution of the Alborá Sea at the eastern entrance of the Straits of Gibraltar. *Inte national symposium on the structural history of the Mediterrane basins, Split*, 25–29 October, 1976. Biju-Duval, B. and Mont dert, L. (eds.), Editions Technip, Paris, pp. 401–410.

Musgrave, A.W. and Hicks, W.G. (1968). Outlining shale masses b Geophysical methods. In J. Braunstein, and G.D. O'Brien (eds *Diapirism and diapirs*, Memoir 8, American Petroleum Geol gists, pp. 122–136.

Slootweg, A.P. (1978). Computer contouring with a digital filter. *Mari Geophysical Researches* **3**: 401–405.

Watts, A.B., Platt, J. and Buhl, P. (1993). Tectonic evolution of the Albor Sea basin. *Basin Research*, **5**: 153–177.

Index